Molecular Mechanisms of Membrane Fusion

Molecular Mechanisms of Membrane Fusion

Edited by
Shinpei Ohki
Darrell Doyle and
Thomas D. Flanagan
State University of New York at Buffalo
Buffalo, New York

and

Sek Wen Hui and
Eric Mayhew
Roswell Park Memorial Institute
Buffalo, New York

Plenum Press • New York and London

Library of Congress Cataloging in Publication Data

International Symposium on Molecular Mechanisms of Membrane Fusion (1987: Buffalo, N.Y.)
Molecular mechanisms of membrane fusion.

"Proceedings of the International Symposium on Molecular Mechanisms of Membrane Fusion, held April 27–29, 1987, in Buffalo, New York"—Copr. p.
Includes bibliographical references and index.
1. Membrane fusion—Congresses. I. Ohki, Shinpei. II. Title.
QH601.I55 1987 574.87′5 87-29163
ISBN 0-306-42773-7

Proceedings of the International Symposium on Molecular Mechanisms of Membrane Fusion, held April 27–29, 1987, in Buffalo, New York

A Division of Plenum Publishing Corporation
233 Spring Street, New York, N.Y. 10013

Printed in the United States of America

PREFACE

Membrane fusion is an important molecular event which plays a pivotal role in many dynamic cellular processes, such as exocytosis, endocytosis, membrane biogenesis, fertilization, etc.

While many reports on the physico-chemical process involved in membrane fusion have appeared in the literature and much information has accumulated, there has been no comprehensive meeting of workers in the field. A recent symposium which took place at the Center for Tomorrow, State University of New York at Buffalo, New York, April 27-29, 1987, was the first meeting of its kind to specifically address the molecular mechanisms of membrane fusion.

The Symposium consisted of oral, workshop and poster presentation sessions. The papers presented in the oral and workshop sessions are collected here and arranged approximately in the order presented at the Symposium. These papers are the most up-to-date and representative work at the forfront of each aspect of membrane fusion.

Although the readers may find some differences in interpretations regarding the molecular mechanisms of membrane fusion, there is an over-all concensus that increased hydrophobicity and dehydration of the membrane surface are essential physico-chemical factors for membrane fusion to occur.

We trust that these papers will contribute to your further understanding of the mechanisms of membrane fusion.

June, 1987 Editors

CONTENTS

CALCIUM-INDUCED MEMBRANE FUSION: FROM LIPOSOMES TO CELLULAR MEMBRANES

Demetrios Papahadjopoulos[a,c], Paul R. Meers[a], Keelung Hong[a], Joel D. Ernst[b,e], Ira M. Goldstein[b,e] and Nejat Düzgüneş[a,d]

Cancer Research Institute[a], the Rosalind Russell Arthritis Research Laboratory[b], and Departments of Pharmacology[c], Pharmaceutical Chemistry[d] and Medicine[e], University of California, San Francisco, CA 94143-0128, U.S.A.

INTRODUCTION

Although Ca^{2+} is involved in many biological membrane fusion phenomena, the mechanism of its action is not well understood. It could participate directly in the fusion reaction by interacting with phospholipids; it could activate particular enzymes; it could induce a conformational change in a fusogenic protein; or it could merely initiate a cascade of reactions which eventually mediate fusion.

The observation that membrane vesicles composed of pure phospholipids can undergo fusion in the presence of Ca^{2+} raised the intriguing possibility that Ca^{2+} could mediate membrane fusion by interacting with negatively charged phospholipids in biological membranes (Papahadjopoulos et al., 1974, 1976, 1977). Since phosphatidylserine is the most abundant negatively charged lipid found in cell membranes, the fusion behavior of vesicles (liposomes) composed of pure phosphatidylserine has attracted much attention (Papahadjopoulos et al., 1979). The fusion characteristics of membranes composed of zwitterionic phospholipids in conjunction with phosphatidylserine, as well as other negatively charged phospholipids, such as phosphatidate (phosphatidic acid), phosphatidylglycerol, phosphatidylinositol and cardiolipin, have been studied by many investigators (Sundler, 1984; Wilschut and Hoekstra, 1984; Düzgüneş et al., 1985). The specificity of Ca^{2+} over Mg^{2+} in the fusion of pure phospholipid vesicles, and the modulation of fusion by low molecular weight molecules including phosphate (Fraley et al., 1980) and polyamines (Schuber et al., 1983) have pointed to the relevance of such simple systems for understanding the control of membrane fusion at the intracellular and intercellular level. Liposomes are very simple models of biological membranes. Yet they provide the only relevant system to study the role of individual membrane components in a very complex biological phenomenon, such as membrane fusion. Liposomes have been particularly useful in the development of sensitive assays to study the kinetics of membrane fusion, since they exhibit well-defined and reproducible fusion characteristics (Wilschut et al., 1980; Struck et al., 1981; Ellens et al., 1985; Düzgüneş and Bentz, 1987).

The complexity of the liposome fusion system can be increased to study the role of other molecules such as cytoplasmic and membrane proteins (Düzgüneş, 1985; Hong et al., 1987; Ohnishi, 1987). The discovery that a cytoplasmic protein, synexin, could mediate the aggregation of isolated secretory granules in the presence of low Ca^{2+} concentrations (Creutz et al., 1978), and could also facilitate the Ca^{2+}-induced fusion of phospholipid vesicles (Hong et al., 1981), suggested that such proteins could be molecular targets of Ca^{2+} in the cytoplasm. Other "synexin-like" molecules have since been identified. These include the chromobindins that bind to chromaffin granule membranes in the presence of Ca^{2+} (Creutz et al., 1983), the calelectrins which aggregate secretory vesicles (Südhof et al., 1982), and synexin-like proteins from human polymorphonuclear leukocytes (PMN) (Meers et al., 1987a).

FUSION LIPOSOMES INDUCED BY DIVALENT METAL IONS

Soon after the first demonstration that liposomes could be used as a model for membrane permeability studies (Bangham et al., 1965), it was observed that the addition of Ca^{2+} could induce aggregation and release of contents of phosphatidylserine liposomes (Papahadjopoulos and Bangham, 1966), although the reaction was not identified as fusion until several years later (Papahadjopoulos et al., 1974). Phosphatidylserine is the major negatively charged phospholipid in biological membranes, and is probably the most extensively studied lipid with respect to interactions with divalent cations and membrane fusion (Papahadjopoulos et al., 1979; Düzgüneş and Papahadjopoulos, 1983). Large unilamellar vesicles (LUV) composed of different acidic phospholipids aggregate and fuse in the presence of different threshold concentrations of divalent cations (Table 1).

Table 1. Threshold concentrations of Ca^{2+} and Mg^{2+} for inducing the aggregation and fusion of LUV composed of pure phospholipids at pH 7.4 and 25° C (Adapted from Düzgüneş et al., 1985).

Phospholipid	Aggregation		Fusion[a]	
	Ca^{2+}	Mg^{2+}	Ca^{2+}	Mg^{2+}
Phosphatidylinositol	3	6	-	-
Phosphatidylglycerol[b]	5	20	15	-
Phosphatidylserine	2	5	2	-
Phosphatidate	0.2	0.4	0.2	0.4

[a] The dashes indicate that no fusion is observed.
[b] Data from Sundler (1984), Rosenberg et al. (1983), and N. Düzgüneş, unpublished.

The amount of divalent cation bound to the membrane surface appears to be one of the critical determinants of fusion (Nir et al., 1980a; Düzgüneş et al., 1980, 1981a). Under ionic conditions where aggregation of small unilamellar vesicles composed of phosphatidylserine is rate limiting, i.e. in 100 mM Na^{+}, the bulk concentration of divalent cations which induce fusion at a threshold rate of 10 % maximal Tb fluorescence/ minute increases in the sequence Ba^{2+}, Ca^{2+}, Sr^{2+}, Mg^{2+} (Bentz et al., 1983b). This sequence is the same found for the aggregation of these vesicles (Ohki et al., 1982). Increasing the monovalent salt concentration renders the aggregation step very rapid compared to the fusion step, allowing for the direct determination of the rate of membrane fusion per se (Bentz et al., 1983b; Nir et al., 1983b). Under these conditions Ca^{2+} is more effective than Ba^{2+}; i.e. less Ca^{2+} is bound per phosphatidylserine molecule than Ba^{2+} at the fusion threshold.

Ca^{2+} has an absolute specificity over Mg^{2+} in inducing the fusion of phosphatidylserine LUV (Wilschut et al., 1981; Table 1). This specificity probably arises during interbilayer contact, since Ca^{2+} can form an anhydrous complex with phosphatidylserine only if the two interacting membranes are allowed to come into close contact (Portis et al., 1979). Ca^{2+} has a much higher affinity for phosphatidylserine in this "trans" (intermembrane) binding mode (Portis et al., 1979; Rehfeld et al., 1981; Ekerdt and Papahadjopoulos, 1982; Nir, 1984; Feigenson, 1986). The specificity of Ca^{2+} over Mg^{2+} is altered when phosphatidylethanolamine or cholesterol are present in the vesicle membrane; such vesicles undergo fusion in the presence of Mg^{2+}, although at a slower rate than with Ca^{2+} (Düzgüneş et al., 1981b; Düzgüneş, 1987). The biophysical basis for this Ca^{2+}/Mg^{2+} specificity has been discussed previously in detail (Düzgüneş et al., 1981b, 1985; Sundler et al., 1981; Wilschut et al., 1981).

Several phospholipids are capable of undergoing a transition from the lamellar to the inverted hexagonal phase (Cullis and de Kruijff, 1979; Verkleij, 1984; Gruner et al., 1985; Siegel, 1984). The formation of non-bilayer structures under conditions which also induce the fusion of membranes has led to the proposal that the bilayer-hexagonal transition is the driving force for membrane fusion (Cullis and Hope, 1978; Cullis and Verkleij, 1979). We have outlined elsewhere the evidence against the involvement of the bilayer-hexagonal phase transition in divalent cation-induced fusion of phospholipid vesicles (Düzgüneş et al., 1987a).

Lipidic particles have also been proposed to be intermediate structures in the fusion of liposomes of various compositions (Verkleij et al., 1979, 1980; Hope et al., 1983; Verkleij, 1984). Quick-freezing freeze-fracture electron microscopy of cardiolipin/phosphatidylcholine or phosphatidylserine/phosphatidylethanolamine LUV immediately after stimulation with Ca^{2+} reveals either smooth elongated bridges or a tight lip structure between fusing liposomes. Lipidic particles are not observed at this stage of fusion, and are visible only after prolonged incubation of the vesicles with Ca^{2+}, or if glycerol is present as a cryoprotectant (Bearer et al., 1982). Similar observations have been made by Verkleij et al. (1984). It appears, therefore, that lipidic particles as defined by their morphology in freeze fracture electron microscopy are not involved in membrane fusion even in membrane systems in which they are observed after long periods of incubation. However, non-bilayer structures may occur at the sites of fusion at rates too fast to be visualized by morphological studies, or may be confined to a small area in the contact zone between fusing membranes (Bearer et al., 1982; Düzgüneş et al., 1985; Siegel, 1984). This intermediate structure could be a local defect in membrane structure that allows the destabilization and fusion of the apposed membranes (Düzgüneş et al., 1984a; Düzgüneş, 1985). These non-bilayer intermediates could transform to more stable structures in time, such as lipidic particles, the hexagonal phase or the crystalline bilayer (Bearer et al., 1982).

Fluorescence assays for membrane fusion have indicated that for many of the liposome types studied, the rates of lipid mixing and aqueous contents mixing are faster than the leakage of the contents into the medium during fusion (Wilschut et al., 1980, 1983, 1985a,b; Düzgüneş et al., 1981a,b, 1987b). Some vesicles can fuse without any leakage (Wilschut et al., 1983), and some exhibit extremely fast rates of leakage when fusion is initiated (Sundler and Papahadjopoulos, 1981). Lipid mixing in the absence of contents mixing occurs under certain conditions (Rosenberg et al., 1983; Ellens et al., 1985). For phosphatidylserine LUV, lipid mixing proceeds at a faster rate than contents mixing, suggesting that the destabilization process leading to the mixing of the lipids in the outer monolayers of the vesicles is different than that leading to complete mixing of the bilayers and the aqueous contents (Düzgüneş et al., 1987b). Another possibility is that lipids exchange during the reversible aggregation of the vesicles before actual fusion occurs (Wilschut et al., 1985b).

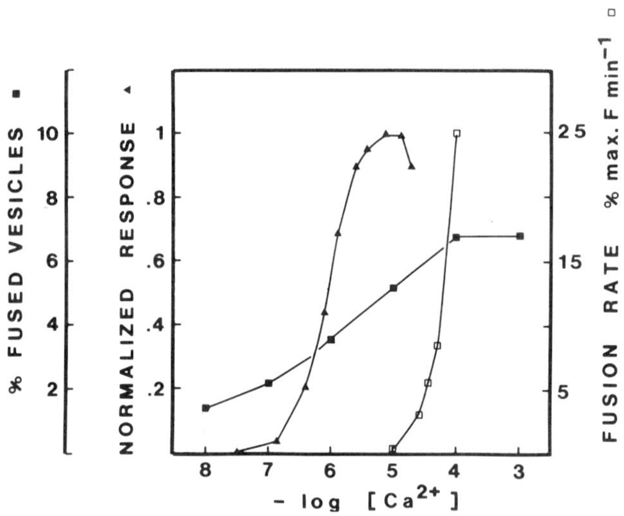

Figure 1. Ca^{2+} concentration dependence of exocytosis in leaky adrenal medullary cells (triangles, data from Baker et al., 1980), fusion of secretory vesicles isolated from adrenal medulla (solid squares, data from Dahl et al., 1979), and fusion of phospholipid vesicles in the presence of synexin and Mg^{2+} (open squares, data from Hong et al., 1982a). Reproduced from Düzgüneş (1985).

EFFECTS OF SYNEXIN AND OTHER CALCIUM-BINDING PROTEINS ON LIPOSOME FUSION

One major difference between liposome fusion and the fusion of certain biological membranes is the concentration of Ca^{2+} required (Gratzl et al., 1980; Ekerdt et al., 1981). The free cytoplasmic Ca^{2+} concentration at which exocytosis is induced in a variety of cellular systems is believed to be in the micromolar range (Figure 1; Baker, 1987; Baker and Knight, 1984; Dunn and Holz, 1983; Wilson and Kirshner, 1983). Isolated secretory granules can undergo fusion in the presence of micromolar concentrations of Ca^{2+}, and membrane glycoproteins are apparently involved in this process (Dahl et al., 1979; Grätzl et al., 1980). Liposomes, on the other hand, require millimolar Ca^{2+} concentrations for fusion (Table 1). The difference could be due to the necessity to overcome the electrostatic repulsion between liposomes, and the relatively low concentration of liposomes that must be used in most fusion experiments. If negatively charged liposome membranes containing glycolipids are brought into close contact by means of agglutinating lectins, the threshold Ca^{2+} concentration is reduced, particularly when the membrane also contains phosphatidate

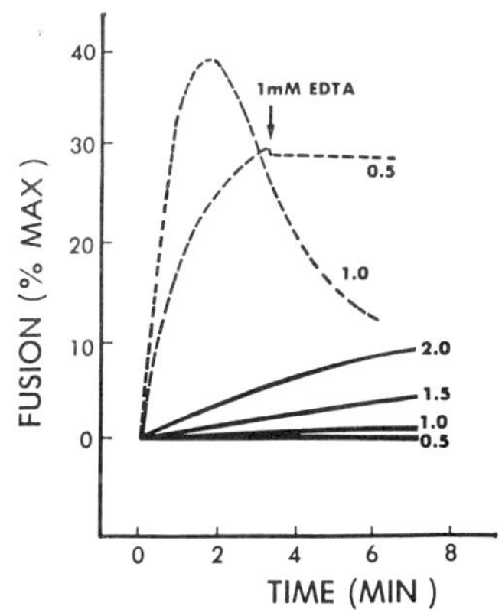

Figure 2. Time course of Ca^{2+}-induced fusion of phosphatidate/phosphatidylethanolamine (1:3) vesicles in the presence (dashed lines) and absence (solid lines) of synexin. Ca^{2+} was added at the indicated mM concentrations at time=0 (From Hong et al., 1982a).

(Sundler and Wijkander, 1983; Düzgüneş et al., 1984b; Hoekstra and Düzgüneş, 1986). Cytosolic components including Mg^{2+} and spermine can also lower the threshold Ca^{2+} concentration required for liposome fusion (Portis et al., 1979; Düzgüneş et al., 1981b; Schuber et al., 1983). Thus, many factors may be involved in mediating the fusion of cellular membranes at Ca^{2+} concentrations several orders of magnitude lower than that required for the fusion of liposomes composed of similar phospholipids.

Other cytoplasmic components such as Ca^{2+}-binding proteins may also mediate the response to Ca^{2+}. One such protein is synexin, which is a Ca^{2+}-dependent membrane-binding protein originally isolated from bovine adrenal medulla and subsequently found in several other tissues (Creutz et al., 1978, 1979). Synexin binds to chromaffin granule membranes in the presence of 4 µM Ca^{2+} (Creutz and Sterner, 1983), and aggregates the granules when the Ca^{2+} concentration exceeds 6 µM (Creutz et al., 1978).

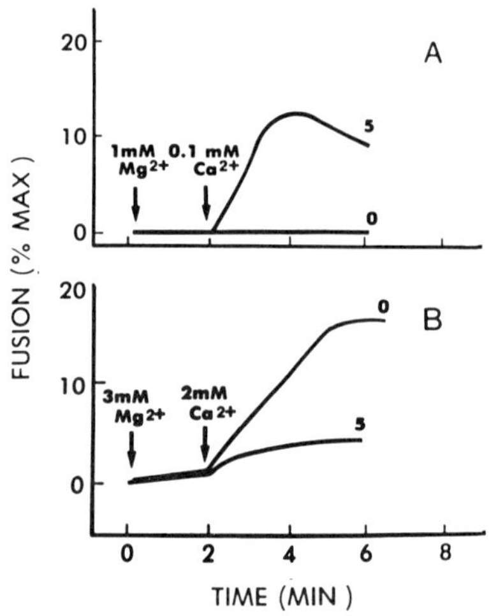

Figure 3. Fusion of (A) phosphatidate/phosphatidylethanolamine and (B) phosphatidylinositol/phosphatidylethanolamine (1:3) vesicles in the presence of 5 µg/mL synexin. The arrows indicate the order of addition of divalent cations. The free Ca^{2+} concentration in A was determined by a Ca^{2+}-selective electrode to be 30 µM (From Hong et al., 1982a).

Fusion of the granule membranes requires the addition of cis-unsaturated free fatty acids (Creutz, 1981). Synexin also facilitates the aggregation and fusion of negatively charged liposomes in the presence of Ca^{2+}, and reduces the threshold Ca^{2+} concentration for the fusion of certain types of liposomes to 10 µM (Figure 1; Hong et al., 1981, 1982a,b, 1987).

The action of synexin is phospholipid-specific. Synexin increases the rate of Ca^{2+}-induced fusion of liposomes containing phosphatidylserine and/or phosphatidate (Hong et al., 1981, 1982a,b). A particularly dramatic enhancement of the rate of fusion has been observed with phosphatidate/phosphatidylethanolamine liposomes, where the initial rate of fusion increases by three orders of magnitude (Figure 2; Hong et al., 1982a). On the other hand, synexin inhibits the Ca^{2+}-induced fusion of phosphatidylinositol-containing liposomes (Figure 3). The presence of phosphatidylcholine in the membrane reduces drastically any facilitation of fusion by synexin, although liposome aggregation is still enhanced (Hong et al., 1982b). Pure phosphatidylcholine liposomes do not aggregate in the presence of synexin and Ca^{2+}. The observation that synexin can mediate the aggregation of certain liposomes without fusion (Hong et al., 1981), suggests that it may be acting only as a Ca^{2+}-dependent membrane aggregating agent.

Synexin may facilitate membrane fusion either by enhancing the rate of aggregation of liposomes, or by acting as a fusogenic protein. Liposome fusion has been modeled kinetically as a two step process involving aggregation followed by the fusion reaction (Nir et al., 1980b, 1982, 1983a; Bentz et al., 1983a). As in any two step kinetic scheme, the slowest step is the rate-limiting step and controls the rate of the overall process. By altering the phospholipid composition and the ionic environment it is possible to convert liposome fusion systems that are largely rate-limited by aggregation into systems that are rate-limited by the fusion step (Bentz et al., 1983b; Bentz and Düzgüneş, 1985; Meers et al., 1987c). The effect of synexin on Ca^{2+}-induced fusion was observed only in aggregation rate-limited systems. In fusion rate-limited systems synexin exhibited either an inhibitory effect, or no effect at all. The synexin-mediated increase in the aggregation rate could also be modeled quantitatively using mass action kinetics (Hong et al., 1983; Meers et al., 1987c). Thus, synexin appears to catalyze the aggregation step by bridging two liposomes, but does not induce fusion by itself. Other factors, such as the membrane-bound products of phosphatidylinositol turnover accompanying exocytosis (Michell, 1975; Michell et al., 1981), may assist synexin in the fusion of intracellular membranes (Creutz, 1981).

One of these products, arachidonic acid, has a synergistic effect on the action of synexin. That is, it produces an enhancement of the overall rate of fusion in the presence of synexin under conditions where it has no effect by itself (P. Meers, D. Papahadjopoulos and K. Hong, in preparation). When tested under aggregation or fusion rate-limiting conditions, arachidonic acid, as well as other fatty acids, enhances the rate of fusion in the absence of synexin. Thus, fatty acids and synexin could exhibit apparently synergistic effects on the overall rate of fusion, synexin enhancing the rate of aggregation and fatty acids enhancing the rate of fusion. The effect of the fatty acids on the overall fusion rate becomes apparent only because synexin, by increasing the aggregation rate, renders the fusion reaction relatively slower and more rate-limiting, at which point the fatty acid effect appears. The observation that this effect is non-specific in terms of the fatty acid species also suggests that it is not a result of a direct interaction of fatty acids with synexin. Furthermore, the fatty acid synergism is observed with other promoters of the aggregation rate of liposomes besides synexin, such as spermine and Mg^{2+}. While the fatty acid effect is not specific for synexin, it is an example of one way in which phospholipid metabolites could play a role in fusion, along with synexin.

The properties of synexin are not shared by other Ca^{2+}-binding proteins, such as calmodulin, prothrombin or parvalbumin, which are either inhibitory or have no effect on the rate of fusion of liposomes (Hong et al., 1981, 1982a). Two other proteins that have synexin-like characteristics are the 67 kD and 32 kD calelectrins from bovine liver (Südhof et al., 1984). These proteins bind to phospholipid membranes at approximately 5 µM Ca^{2+}. The 32 kD calelectrin increases the rate of Ca^{2+}-induced fusion of liposomes composed of phosphatidate/phosphatidylethanolamine, but unlike synexin, inhibits the fusion of phosphatidylserine liposomes (P. Meers and K. Hong, unpublished data). The 67 kD calelectrin inhibits the Ca^{2+}-induced fusion of both types of liposomes, and may be identical with synhibin (Pollard and Scott, 1982; Creutz et al., 1987).

One important and unknown aspect of the function of synexin-like proteins is the mechanism by which they bind to membranes. While it is clear that the binding is Ca^{2+}-dependent and at least partially reversible (Creutz et al., 1983; Meers et al., 1987a), it is not known what the requirements for binding are and how they relate to the function of the proteins. Synexin most likely binds to the phospholipid segment of chromaffin granule membranes (Morris et al., 1982). Synexin and the calelectrins self-aggregate in the presence of Ca^{2+} (Creutz et al., 1979; Südhof et al., 1982), suggesting that they are bipolar molecules that can interact with and link two membranes. However, there is no clear evidence that the self-association relates to membrane binding. Self-association of synexin is actually inhibitory to its fusion rate enhancement (Hong et al., 1987; P. Meers, D. Papahadjopoulos, and K. Hong, in preparation). Calcium-dependent binding of synexin to phospholipid membranes has been measured by differential centrifugation of liposomes rendered dense by encapsulation of metrizamide, and assaying for synexin activity remaining in the supernatant (Meers et al., 1987b). Synexin binds preferentially to negatively charged phospholipids, and to phosphatidylethanolamine, but does not bind to phosphatidylcholine. Experiments to determine whether synexin penetrates into the phospholipid bilayer are in progress.

SYNEXIN IN HUMAN POLYMORPHONUCLEAR LEUKOCYTES

Polymorphonuclear leukocytes (PMN) are phagocytes involved in defending the host organism against invading microorganisms, which they ingest in phagocytic vesicles. The latter then fuse with granules containing digestive enzymes, in a process termed degranulation (Goldstein, 1984; Baggiolini and Dewald, 1984). Specific granules and azurophil granules fuse sequentially with the phagosome membrane (Bainton, 1973; Bentwood and Henson, 1980). Specific granules and gelatinase-containing secretory vesicles can also fuse with the plasma membrane and release their contents into the extracellular milieu (Bainton, 1973; Estensen et al., 1974; Goldstein et al., 1974; Dewald et al., 1982). The contents of azurophil granules may also be released extracellularly if phogocytosis is not completed before granule-phagosome fusion. A variety of stimuli, such as phorbol myristate acetate, chemotactic peptides and Ca^{2+} ionophores, induce degranulation, which is thought to be mediated by the release of membrane-bound or intracellularly sequestered Ca^{2+} into the cytoplasm (Goldstein, 1984; Baggiolini and Dewald, 1984). The Ca^{2+} concentration dependence of the degranulation of each of the granule types is different (Lew et al., 1986).

It was of interest, therefore, to investigate the possibility that synexin-like proteins may be involved in mediating degranulation in PMN. We have recently demonstrated the presence of Ca^{2+}-dependent phospholipid-binding proteins in PMN cytosol (Meers et al., 1987a). These synexin-like proteins were purified using an affinity column with covalently coupled liposomes composed of phosphatidylserine, phosphatidylcholine and phosphatidylethanolamine. Three major proteins were obtained with molecular masses of approximately 67, 47 and 28 kD, as determined by polyacrylamide gel electrophoresis. Rabbit antiserum to bovine liver synexin bound only the 47 kD band on Western blots of these proteins. In fact only a 47 kD band appeared in Western blots of whole PMN cytosol. Two dimensional electrophoresis showed that the 47 kD protein has an isoelectric point of approximately 7.0, as does bovine liver synexin (Ernst et al., 1986).

The identity of synexin-like proteins in human PMN was further verified by testing their activities in granule aggregation and liposome fusion. Both the PMN proteins and bovine liver synexin promoted the Ca^{2+}-induced fusion of liposomes composed of phosphatidate/ phosphatidylethanolamine as well as phosphatidylserine/phosphatidylethanolamine. Another measure of synexin activity was the Ca^{2+}-dependent aggregation of specific granules isolated from human PMN. In the absence of synexin, no aggregation of granules was observed up to 10 mM Ca^{2+}. By contrast, 0.4 mM free Ca^{2+} was sufficient to aggregate the granules in the presence of bovine synexin or the PMN proteins. The aggregation was also inhibited by 70% in the presence of 5 μM trifluoperazine (Ernst et al., 1986).

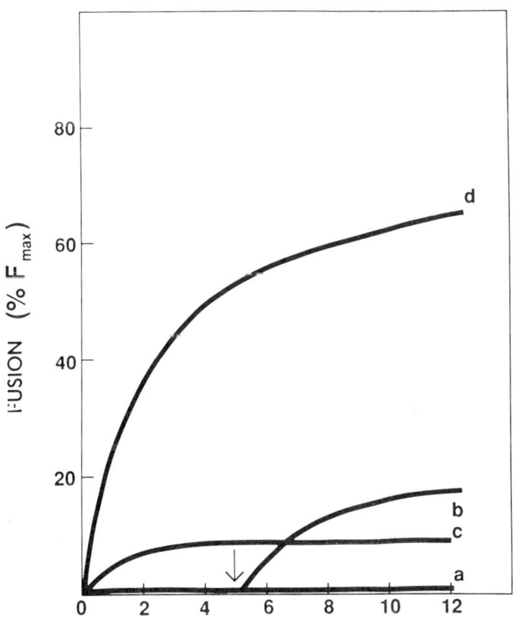

Figure 4. Fusion of PMN specific granules with liposomes composed of phosphatidate/phosphatidylethanolamine (1:3) in the presence of 1 mM Ca^{2+} added at time=0. The suspension was pre-incubated without any additions (a), or with 10 μg/mL bovine synexin (b), 10.6 μM arachidonic acid (c), or both (d) for 5 min. In (b) arachidonic acid was added at the time indicated by the arrow (from Meers et al., 1987a).

Finally, the ability of synexin to mediate the fusion of specific granules with liposomes was tested. In order to assay fusion, fluorescent lipid probes were incorporated into either the liposome membranes (Struck et al., 1981) or the granule membranes (Hoekstra et al., 1984). Dilution of these probes into unlabeled membranes upon fusion results in an increase in fluorescence intensity. In the presence of synexin and arachidonic acid, nitrobenzoxydiazol- and rhodamine-labeled phosphatidate/phosphatidylethanolamine liposomes fused with specific granules upon addition of Ca^{2+} (Figure 4). Similarly, octadecylrhodamine-labeled granules appeared to fuse with unlabeled liposomes. No fusion was apparent if one of the components of the mixture (Ca^{2+}, synexin or arachidonic acid) was omitted. The arachidonic acid requirement was particularly interesting in light of the fact that arachidonate is produced upon stimulation of PMN to degranulate (Stenson and Parker, 1979; Waite et al., 1979; Walsh et al., 1981). These results suggest that synexin and synexin-like proteins may be involved in mediating the fusion of PMN granules with phagosomes or the plasma membrane during degranulation.

ACKNOWLEDGEMENTS

This work was supported by NIH Grants GM28117 (DP, KH and ND) and AI23697 (JDE), a Grant-in-Aid from the American Heart Association with funds contributed by the California Affiliate (ND), an Arthritis Foundation Postdoctoral Fellowship (PRM), and a grant from the Treadwell Foundation (IMG). We thank Dr. Shlomo Nir for his comments on the manuscript, and Ms. Rose Antonucci for her assistance in the preparation of the manuscript.

REFERENCES

Baggiolini, M. and Dewald, B., 1984, Exocytosis by neutrophils, *in*: "Regulation of Leukocyte Function," R. Snyderman, ed., pp. 221-246, Plenum Press, New York.

Bainton, D.F., 1973, Sequential degranulation of the two types of polymorphonuclear leukocyte granules during phagocytosis of microorganisms, *J. Cell. Biol.*, 58:249.

Baker, P.F., 1987, Exocytosis in electropermeabilized cells: Clues to mechanism and physiological control, *in*: "Membrane Fusion in Fertilization, Cellular Transport and Viral Infection," N. Düzgüneş and F. Bronner, eds., Academic Press, New York (in press).

Baker, P.F., Knight, D.E, and Whitaker, M.G., 1980, Calcium and the control of exocytosis, *in*: "Calcium-Binding Proteins: Structure and Function," F.L. Siegel, E. Carafoli, R.H. Kretsinger, D.H. MacLennan and R.H. Wasserman, eds., pp. 47-55, Elsevier/North Holland, New York.

Baker, P.F., and Knight, D.E., 1984, Calcium control of exocytosis in bovine adrenal medullary cells, *Trends. Neurosci.* 7:120-126.

Bangham, A.D., Standish, M.M. and Watkins, J.C., 1965, Diffusion of univalent ions across lamellae of swollen phospholipids, *J. Mol. Biol.*, 13:238.

Bearer, E.L., Düzgüneş, D., Friend, D.S. and Papahadjopoulos, D., 1982, Fusion of phospholipid vesicles arrested by quick freezing. The question of lipidic particles as intermediates in membrane fusion, *Biochim. Biophys. Acta*, 693:93.

Bentwood, B.J. and Henson, P.M., 1980, The sequential release of granule constituents from human neutrophils, *J. Immunol.*, 124:855.

Bentz, J. and Düzgüneş, N., 1985, Fusogenic capacities of divalent cations and the effect of liposome size, Biochemistry, 24:5436.

Bentz, J., Nir, S. and Wilschut, J., 1983a, Mass action kinetics of vesicle aggregation and fusion, Colloids and Surfaces, 6:333.

Bentz, J., Düzgüneş, N. and Nir, S., 1983b, Kinetics of divalent cation induced fusion of phosphatidylserine vesicles: Correlation between fusogenic capacities and binding affinities, Biochemistry, 22:3320.

Bentz, J., Düzgüneş, N. and Nir, S., 1985, Temperature dependence of divalent cation induced fusion of phosphatidylserine liposomes: Evaluation of the kinetic rate constants, Biochemistry, 24:1064.

Creutz, C.E., 1981, cis- unsaturated fatty acids induce the fusion of chromaffin granules aggregated by synexin, J. Cell Biol., 91:247.

Creutz, C.E. and Sterner, D.C., 1983, Calcium dependence of the binding of synexin to isolated chromaffin granules, Biochem. Biophys. Res. Commun., 114:355.

Creutz, C.E., Pazoles, C.J., and Pollard, H.B., 1978, Identification and purification of an adrenal medullary protein (synexin) that causes calcium-dependent aggregation of isolated chromaffin granules, J. Biol. Chem., 253:2858.

Creutz, C.E., Pazoles, C.J., and Pollard, H.B., 1979, Self-association of synexin in the presence of calcium: Correlation with synexin-induced membrane fusion and examination of the structure of synexin aggregates. J. Biol. Chem., 254:553.

Creutz, C.E., Dowling, L.G., Sando, J.J., Villar-Palasi, C., Whipple, J.H., and Zaks, W.J., 1983, Characterization of the chromobindins: Soluble proteins that bind to the chromaffin granule membrane in the presence of Ca^{2+}, J. Biol. Chem., 258:14664.

Creutz, C.E., Zaks, W.J., Hamman, H.C. and Martin, W.H., 1987, The roles of Ca^{2+}-dependent membrane-binding proteins in the regulation and mechanism of exocytosis, in: "Cell Fusion," A.E. Sowers, ed., pp. 45-68, Plenum Press, New York.

Cullis, P.R. and Hope, M.J., 1978, Effects of fusogenic agent on membrane structure of erythrocyte ghosts and the mechanism of membrane fusion, Nature (Lond.), 271:672.

Cullis, P.R., and deKruijff, B., 1979, Lipid polymorphism and the functional roles of lipids in biological membranes, Biochim. Biophys. Acta, 559:399.

Cullis, P.R. and Verkleij, A.J., 1979, Modulation of membrane structure by Ca^{2+} and dibucaine as detected by ^{31}P NMR, Biochim. Biophys. Acta, 552:546.

Dahl, G., Ekerdt, R. and Gratzl, M., 1979, Models for exocytotic membrane fusion, Symp. Soc. Exp. Biol., 33:349.

Dewald, B., Bretz, U. and Baggiolini, M., 1982, Release of gelatinase from a novel secretory compartment of human neutrophils, J. Clin. Invest., 70:518.

Dunn, L.A. and Holz, R.W., 1983, Catecholamine secretion from digitonin-treated adrenal medullary chromaffin cells, J. Biol. Chem., 248:4989.

Düzgüneş,N., 1985, Membrane fusion, in: "Subcellular Biochemistry," Vol. 11, D.B. Roodyn, ed., pp. 195-286, Plenum Press, New York.

Düzgüneş, N., 1987, Cholesterol and membrane fusion, in: "Cholesterol in Biological Systems," P.L. Yeagle, ed., CRC Press, Boca Raton, Florida (in press).

Düzgüneş, N. and Papahadjopoulos, D., 1983, Ionotropic effects on phospholipid membranes: Calcium-magnesium specificity in binding, fluidity, and fusion, in: "Membrane Fluidity in Biology", Vol. 2, R.C. Aloia, ed., pp. 187-213, Academic Press, New York.

Düzgüneş N. and Bentz, J., 1987, Fluorescence assays for membrane fusion, in: "Spectroscopic Membrane Probes," L.M. Loew, ed., CRC Press, Boca Raton, Florida (in press).

Düzgüneş, N., Hong, K. and Papahadjopoulos, D., 1980, Membrane fusion: The involvement of phospholipids, proteins and calcium binding, in: "Calcium-Binding Proteins: Structure and Function," F.L. Siegel, E. Carafoli, R.H. Kretsinger, D.H. MacLennan and R.H. Wasserman, eds., pp. 17-22, Elsevier/North Holland, New York.

Düzgüneş, N., Nir, S., Wilschut, J., Bentz, J., Newton, C., Portis, A. and Papahadjopoulos, D., 1981a, Calcium- and magnesium-induced fusion of mixed phosphatidylserine /phosphatidylcholine vesicles: Effect of ion binding, J. Memb. Biol., 59:115.

Düzgüneş, N., Wilschut, J., Fraley, R., and Papahadjopoulos, D., 1981b, Studies on the mechanism of membrane fusion: Role of head-group composition in calcium- and magnesium-induced fusion of mixed phospholipid vesicles, Biochim. Biophys. Acta, 642:182.

Düzgüneş, N., Paiement, J., Freeman, K.B., Lopez, N.G., Wilschut, J. and Papahadjopoulos, D., 1984a, Modulation of membrane fusion by ionotropic and thermotropic phase transitions, Biochemistry, 23:3486.

Düzgüneş, N., Hoekstra, D., Hong, K. and Papahadjopoulos, D., 1984b, Lectins facilitate calcium-induced fusion of phospholipid vesiclces containing glycosphingolipids, FEBS Lett., 173:80.

Düzgüneş, N., Wilschut, J. and Papahadjopoulos, D, 1985, Control of membrane fusion by divalent cations, phospholipid head-groups and proteins, in: "Physical Methods on Biological Membranes and their Model Systems," F. Conti, W.E. Blumberg, J. DeGier, and F. Pocchiari eds., pp. 193-218, Plenum Press, New York.

Düzgüneş, N., Hong, K., Baldwin, P.A., Bentz, J., Nir, S., Papahadjopoulos, D., 1987a, Fusion of phospholipid vesicles induced by divalent cations and protons. Modulation by phase transitions, free fatty acids, monovalent cations, and polyamines, in: "Cell Fusion," A.E. Sowers, ed., pp. 241-267, Plenum Press, New York.

Düzgüneş, N., Allen, T.M., Fedor, J. and Papahadjopoulos, D., 1987b, Lipid mixing during membrane aggregation and fusion. Why fusion assays disagree, Biochemistry (submitted).

Ekerdt, R. and Papahadjopoulos, D., 1982, Intermembrane contact affects calcium binding to phospholipid vesicles, Proc. Natl. Acad. Sci. USA, 79:2273.

Ekerdt, R., Dahl, G., and Gratzl, M., 1981, Membrane fusion of secretory vesicles and liposomes. Two different types of fusion, Biochim. Biophys. Acta 646:10-22.

Ellens, H., Bentz, J. and Szoka, F. C., 1985, H^+- and Ca^{2+} -induced fusion and destabilization of liposomes, Biochemistry, 24:3099.

Ernst, J.D., Meers, P., Hong, K., Düzgüneş, N., Papahadjopoulos, D., and Goldstein, I.M., 1986, Human polymorphonuclear leukocytes contain synexin, a calcium-binding protein that mediates membrane fusion, Clin. Res., 34:722A.

Estensen, R.D., White, J.G. and Holmes, B., 1974, Specific degranulation of human polymorphonuclear leukocytes, Nature (Lond.), 248:347.

Feigenson, G.W., 1986, On the nature of calcium ion binding between phosphatidylserine lamellae, Biochemistry, 25:5819.

Fraley, R., Wilschut, J., Düzgüneş, N., Smith, C. and Papahadjopoulos, D., 1980, Studies on the mechanism of membrane fusion: The role of phosphate in promoting calcium-induced fusion of phospholipid vesicles, Biochemistry, 19:6021.

Goldstein, I.M., 1984, Neutrophil degranulation, in: "Regulation of Leukocyte Function," R. Snyderman, ed., pp.189-219, Plenum Press, New York.

Goldstein, I.M., Horn, J.K., Kaplan, H.B. and Weissmann, G., 1974, Calcium-induced lysozyme secretion from human polymorphonuclear leukocytes, Biochem. Biophys. Res. Commun., 60:807.

Gratzl, M., Schudt, C., Ekerdt, R., and Dahl, G., 1980, Fusion of isolated biological membranes: A tool to investigate basic processes of exocytosis and cell-cell fusion, in: Membrane Structure and Function, Vol. 3 (E.E. Bittar, ed.), pp. 59-92, Wiley, New York.

Gruner, S.M., Cullis, P.R., Hope, M.J. and Tilcock, C.P.S., Lipid polymorphism. The molecular basis of nonbilayer phases, Ann. Rev. Biophys. Biophys. Chem., 14:211.

Hoekstra, D. and Düzgüneş, N., 1986, Ricinus communis agglutinin-mediated agglutination and fusion of glycolipid-containing phospholipid vesicles. Effect of carbohydrate headgroup size, calcium ions and spermine, Biochemistry, 25:1321.

Hoekstra, D., de Boer, T., Klappe, K. and Wilschut, J., 1984, Fluorescence method for measuring the kinetics of fusion between biological membranes, Biochemistry, 23:5675.

Hong, K., Düzgüneş, N., and Papahadjopoulos, D., 1981, Role of synexin in membrane fusion, J. Biol. Chem., 256:3651.

Hong, K., Düzgüneş, N., and Papahadjopoulos, D., 1982a, Modulation of membrane fusion by calcium-binding proteins, Biophys. J., 37:296.

Hong, K., Düzgüneş, N., Ekerdt, R., and Papahadjopoulos, D., 1982b, Synexin facilitates fusion of specific phospholipid vesicles at divalent cation concentrations found intracellularly, Proc. Natl. Acad. Sci. USA, 70:4942.

Hong, K., Ekerdt, R., Bentz, J., Nir, S. and Papahadjopoulos, D., 1983, Kinetics of synexin-facilitated membrane fusion, Biophys. J., 41:31a.

Hong, K., Düzgüneş, N., Meers, P.R. and Papahadjopoulos, D., 1987, Protein modulation of liposome fusion, in: "Cell Fusion," A.E. Sowers, ed., pp. 269-284, Plenum Press, New York.

Hope, M.J., Walker, D.C. and Cullis, P.R., 1983, Calcium and pH-induced fusion of small unilamellar vesicles consisting of phosphatidylethanolamine and negatively charged phospholipids: A freeze-fracture study, Biochem. Biophys. Res. Commun., 110:15.

Lew, D.P., Monod, A., Waldvogel, F.A., Dewald, B., Baggiolini, M. and Pozzan, T., 1986, Quantitative analysis of the cytosolic free calcium dependency of exocytosis from three subcellular compartments in intact human neutrophils, J. Cell Biol., 102:2197.

Meers, P., Ernst, J.D., Düzgüneş, N., Hong, K., Fedor, J., Goldstein, I.M. and Papahadjopoulos, D., 1987a, Synexin-like proteins from human polymorphonuclear leukocytes. Identification and characterization of granule-aggregating and membrane-fusing activities, J. Biol. Chem., 262:7850.

Meers, P., Hong, K. and Papahadjopoulos, D., 1987b, Studies on the binding of synexin to phospholipid vesicles, in: "Proceedings of the Fifth International Symposium on Calcium Binding Proteins in Health and Disease," A.W. Norman, T.C. Vanaman and A.R. Means, eds., Academic Press, New York (in press).

Meers, P., Bentz, J., Alford, D., Nir, S., Papahadjopoulos, D. and Hong, K., 1987c, Kinetic mechanism of the modulation of liposome fusion by synexin (in preparation).

Michell, R.H., 1975, Inositol phospholipids and cell surface receptor function, Biochim. Biophys. Acta, 415:81.

Michell, R.H., Kirk, C.J., Jones, L.M., Downes, C.P. and Creba, J.A., 1981, The stimulation of inositol lipid metabolism that accompanies calcium mobilization in stimulated cells: defined characteristics and unanswered questions, Phil. Trans. R. Soc. Lond. B., 296:123.

Morris, S.J., Hughes, J.M.X., and Whittaker, V.P., 1982, Purification and mode of action of synexin: A protein enhancing calcium-induced membrane aggregation, J. Neurochem., 39:529.

Nir, S., 1984, A model for cation adsorption in closed systems: Application to calcium binding to phospholipid vesicles, J. Coll. Interface Sci., 102:313.

Nir, S., Bentz, J. and Portis, A.R., Jr., 1980a, Effect of cation concentrations and temperature on the rates of aggregation of acidic phospholipid vesicles. Application to fusion, Adv. Chem. Ser., 188:75.

Nir, S., Bentz, J., and Wilschut, J., 1980b, Mass action kinetics of phosphatidylserine vesicle fusion as monitored by coalescence of internal vesicle volumes, Biochemistry, 19:6030.

Nir, S., Wilschut, J. and Bentz, J., 1982, The rate of fusion of phospholipid vesicles and the role of bilayer curvature, Biochim. Biophys. Acta, 688:275.

Nir, S., Bentz, J., Wilschut, J. and Düzgüneş, N., 1983a, Aggregation and fusion of phospholipid vesicles, Prog. Surface Sci., 13:1.

Nir, S., Düzgüneş, N. and Bentz, J., 1983b, Binding of monovalent cations to phosphatidylserine and modulation of Ca^{2+}- and Mg^{2+}-induced vesicle fusion, Biochim. Biophys. Acta, 735:160.

Ohki, S., Düzgüneş, N. and Leonards, K., 1982, Phospholipid vesicle aggregation: Effect of monovalent and divalent ions, Biochemistry, 21:2127.

Ohnishi, S.-I., 1987, Fusion of viral envelopes with cellular membranes, in: "Membrane Fusion in Fertilization, Cellular Transport and Viral Infection," N. Düzgüneş and F. Bronner, eds., Academic Press, New York (in press).

Papahadjopoulos, D. and Bangham, A.D., 1966, Biophysical properties of phospholipids. II. Permeability of phosphatidylserine liquid crystals to univalent ions, Biochim. Biophys. Acta, 126:185.

Papahadjopoulos, D., Poste, G., Schaeffer, B.E. and Vail, W.J., 1974, Membrane fusion and molecular segregation in phospholipid vesicles, Biochim. Biophys. Acta, 352:10.

Papahadjopoulos, D., Vail, W.J., Pangborn, W.A. and Poste, G., 1976, Studies on membrane fusion. II. Induction of fusion in pure phospholipid membranes by calcium and other divalent metals, Biochim. Biophys. Acta, 448:265.

Papahadjopoulos, D., Vail, W.J., Newton, C., Nir, S., Jacobson, K., Poste, G. and Lazo, R., 1977, Studies on membrane fusion. III. The role of calcium-induced phase changes, Biochim. Biophys. Acta, 465:579.

Papahadjopoulos, D., Poste, G., and Vail, W.J., 1979, Studies on membrane fusion with natural and model membranes, Methods Memb. Biol., 10:1.

Pollard, H.B. and Scott, J.H., 1982, Synhibin: A new calcium-dependent membrane-binding protein that inhibits synexin-induced chromaffin granule aggregation and fusion, FEBS Lett., 150:201.

Portis, A., Newton, C., Pangborn, W. and Papahadjopoulos, D., 1979, Studies on the mechanism of membrane fusion: Evidence for an intermembrane Ca^{2+}-phospholipid complex, synergism with Mg^{2+}, and inhibition by spectrin, Biochemistry, 18:780.

Rehfeld, S.J., Düzgüneş, N., Newton, C., Papahadjopoulos, D. and Eatough, D.J., 1981, The exothermic reaction of calcium with unilamellar phosphatidylserine vesicles: Titration microcalorimetry, FEBS Lett., 123:249.

Rosenberg, J., Düzgüneş, N. and Kayalar, C., 1983, Comparison of two liposome fusion assays monitoring the intermixing of aqueous contents and of membrane components, Biochim. Biophys. Acta, 735:173.

Schuber, F., Hong, K., Düzgüneş, N. and Papahadjopoulos, D., 1983, Polyamines as modulators of membrane fusion: Aggregation and fusion of liposomes, Biochemistry, 22:6134.

Siegel, D.P., 1984, Inverted micellar structures in bilayer membranes: Formation rates and half-lives, Biophys. J., 45:399.

Stenson, W.F. and Parker, C.W., 1979, Metabolism of arachidonic acid in ionophore-stimulated neutrophils, J. Clin. Invest., 64:1457.

Struck, D. K., Hoekstra, D., and Pagano, R. E., 1981, Use of resonance energy transfer to monitor membrane fusion, Biochemistry, 20:4093.

Südhof, T.C., Walker, J.H., and Obrocki, J., 1982, Calelectrin self-aggregates and promotes membrane aggregation in the presence of calcium, EMBO J., 1:1167.

Südhof, T.C., Ebbecke, M., Walker, J.H., Fritsche, U., and Boustead, C., 1984, Isolation of mammalian calelectrins: A new class of ubiquitous Ca^{2+}-regulated proteins, Biochemistry, 23:1103.

Sundler, R., 1984, Role of phospholipid head group structure and polarity in the control of membrane fusion, Biomembranes, 12:563.

Sundler, R. and Papahadjopoulos, D., 1981, Control of membrane fusion by phospholipid head groups. I. Phosphatidate/phosphatidylinositol specificity, Biochim. Biophys. Acta, 649:743.

Sundler, R. and Wijkander, J., 1983, Protein-mediated intermembrane contact specifically enhances Ca^{2+}-induced fusion of phosphatidate-containing membranes, Biochim. Biophys. Acta, 730:391.

Sundler, R., Düzgüneş, N. and Papahadjopoulos, D., 1981, Control of membrane fusion by phospholipid head groups. II. The role of phosphatidylethanolamine in mixtures with phosphatidate and phosphatidylinositol, Biochim. Biophys. Acta, 649:751.

Verkleij, 1984, Lipidic intramembranous particles, Biochim. Biophys. Acta, 779:43.

Verkleij, A.J., Mombers, C., Gerritsen, W.J., Leunissen-Bijvelt, L. and Cullis, P.R., 1979, Fusion of phospholipid vesicles in association with the appearance of lipidic particles as visualized by freeze-fracturing, Biochim. Biophys. Acta, 555:358.

Verkleij, A.J., van Echteld, C.J.A., Gerritsen, W.J., Cullis, P.R. and de Kruijff, B., 1980, The lipidic particle as an intermediate structure in membrane fusion processes and bilayer to hexagonal H_{II} transitions, Biochim. Biophys. Acta, 600:620.

Verkleij, A.J., Leunissen-Bijvelt, J., de Kruijff, B., Hope, M. and Cullis, P.R., 1984, Non-bilayer structures in membrane fusion, in: "Cell Fusion, Ciba Foundation Symposium 103," pp. 45-59, Pitman Books, London.

Waite, M., DeChatelet, L.R., King, L. and Shirley, P.S., 1979, Phagocytosis-induced release of arachidonic acid from human neutrophils, Biochem. Biophys. Res. Commun., 90:984.

Walsh, C.E., Waite, B.M., Thomas, M.J. and DeChatelet, L.R., 1981, Release and metabolism of arachidonic acid in human neutrophils, J. Biol. Chem., 256:7228.

Wilschut, J. and Hoekstra, D., 1984, Membrane fusion: from liposomes to biological membranes, Trends Biochem. Sci., 9:479.

Wilschut, J., Düzgüneş, N. Fraley, R., and Papahadjopoulos, D., 1980, Studies on the mechanism of membrane fusion: Kinetics of Ca^{2+}-induced fusion of phosphatidylserine vesicles followed by a new assay for mixing of aqueous vesicle contents, Biochemistry, 19:6011.

Wilschut, J., Düzgüneş, N. and Papahadjopoulos, D., 1981, Calcium/magnesium specificity in membrane fusion: Kinetics of aggregation and fusion of phosphatidylserine vesicles and the role of bilayer curvature, Biochemistry, 20:3126.

Wilschut, J., Düzgüneş, N., Hong, K., Hoekstra, D., and Papahadjopoulos, D., 1983, Retention of aqueous contents during divalent cation-induced fusion of phospholipid vesicles, Biochim. Biophys. Acta, 734:309.

Wilschut, J., Nir, S., Scholma, J. and Hoekstra, D., 1985a, Kinetics of Ca^{2+}-induced fusion of cardiolipin-phosphatidylcholine vesicles: Correlation between vesicle aggregation, bilayer destabilization, and fusion, Biochemistry, 24:4630.

Wilschut, J., Scholma, J., Bental, M., Hoekstra, D. and Nir, S., 1985b, Ca^{2+}-induced fusion of phosphatidylserine vesicles: Mass action kinetic analysis of membrane lipid mixing and aqueous contents mixing, *Biochim. Biophys. Acta*, 821:45.

Wilson, S.P., and Kirshner, N., 1983, Calcium-evoked secretion from digitonin-permeabilized adrenal medullary chromaffin cells, *J. Biol. Chem.* 258:4994.

GEL PHASE FUSION OF DIPALMITOYL PHOSPHATIDYLCHOLINE SMALL UNILAMELLAR VESICLES

Thomas E. Thompson and Dov Lichtenberg

Department of Biochemistry, University of Virginia
Charlottesville, VA. 22908, U.S.A. and Department of
Physiology and Pharmacology, Tel Aviv University
Tel Aviv, 69978, Israel

SUMMARY

Small unilamellar vesicles produced by sonication of dipalmitoyl phosphatidylcholine and other saturated diacyl phosphatidylcholines above t_m, the gel to liquid crystal phase transition temperature, are stable indefinitely at temperatures above t_m in the absence of fusogens. However, gel phase vesicles when cooled below t_p, the pre-transition temperature, spontaneously fuse in the absence of fusogens to a remarkably homogeneous dispersion of large unilamellar vesicles about 700 Å in diameter. The rate of fusion increases with decreasing temperature and is strongly affected by the addition of polyols to the aqueous phase. The fusion process is preceded by vesicle aggregation that can be reversed by raising the temperature above t_m. The size of the aggregates is independent of vesicle concentration and temperature. However, the rate of conversion of aggregates to fused vesicles depends upon the concentration of aggregates in a manner consistent with a biaggregate collision process.

Below t_p, the small unilamellar vesicles are believed to take the form of irregular polyhedrons, each facet of which is a flat bilayer of gel phase phosphatidylcholine joined to neighboring facets by disorganized molecules in a glass-like gel phase. Aggregation occurs by apposition of facets of adjacent polyhedrons to give an aggregate of about 18 small unilamellar vesicles. It is possible that fusion of the members of an aggregate occurs when a collision takes place between two aggregates with sufficient energy to force a shape change in a vesicle pair within an aggregate. This shape change causes a transient bilayer rupture with a concommitant water flow which leads to the serial fusion of all members of the aggregate to give a large unilamellar vesicle.

INTRODUCTION

Huang in 1969 described the preparation and some of the physical properties of small unilamellar vesicles, about 200 Å in diameter, formed by the sonication above the gel to liquid crystalline phase transition temperature of multilamellar liposomes of egg yolk phosphatidyl-choline dispersed in aqueous systems. Aqueous dispersions of this type of vesicle prepared from a variety of phospholipids and their mixtures have been studied intensively in the years since as models for the bilayer component of biological membranes.

In the course of these studies it became apparent that, when vesicles were prepared from diacyl saturated phosphatidylcholines by sonication <u>above</u> the liquid crystalline-gel phase transition temperature, small unilamellar vesicles are formed which are stable when liquid crystalline, but in the gel phase fuse spontaneously in the absence of fusogens to form larger structures (Suurkuusk et al., 1976; Larrabee, 1979; Schullery et al., 1980; Schmidt et al., 1981; Lichtenberg and Schmidt, 1981; Wong and Thompson, 1982; Wong et al., 1982). Sonication of this type of phospholipid <u>below</u> the phase transition temperature results in unstable small vesicles which can, however, be annealed above t_m to form stable small vesicles (Lawaczeck et al., 1975). It was eventually shown that the fusion products that form from gel phase dipalmitoyl phosphatidylcholine small vesicles produced by sonication above t_m are large unilamellar vesicles with a surprisingly narrow size distribution (Schullery, 1980). Depending upon the conditions of formation, the mean size of the distribution ranges from about 700 to 950 Å (Wong et al., 1982; McConnell and Schullery, 1985). The unusual aspects of the fusion products are their unilamellar character, their large size and their narrow distribution in size. It has recently been shown that the large fusion vesicles form only below t_p, the transition midpoint of the so called pre-transition (McConnell and Schullery, 1985).

In the sections to follow, the phenomenological details of this fusion process and the mechanism by which fusion occurs will be discussed. Although the spontaneous fusion in the gel phase of other saturated diacyl phosphatidylcholines has been examined (Larrabee, 1979; Kantor et al., 1977; Parente and Lentz, 1984), attention will be focused on the dipalmitoyl phosphatidylcholine studies carried out at the University of Virginia.

THE PRODUCTS OF FUSION

In 1975 we undertook a differential scanning calorimetric study of the small unilamellar vesicles produced by the sonication of dipalmitoyl phosphatidylcholine (Suurkuusk et al., 1976). The small size of these vesicles, clearly established by molecular sieve chromatography and other physical techniques (Huang, 1969), suggested to us that their thermal properties should also be unusual. The scanning calorimeter available to us could only be used in a scan from low to high temperature and required a long equilibration at low temperature to establish initial thermal conditions. Thus before a run could be started, the small vesicle dispersion was incubated at 20°C for 3 to 6 h. When the temperature scan was completed about 2 h later, we were surprised to find not one but two thermal transitions, a rather sharp one centered at about 41°C and a broader one at 37°C. Subsequent studies showed that, as the time of incubation below the gel to liquid crystalline phase transition temperature increased, the enthalpy of the higher transition increased at the expense of that of the lower transition. These results indicated to us that the small unilamellar vesicles in the original scan were unstable in the gel phase and at these temperatures gradually transformed into a more stable form characterized by a t_m value close to that of multilamellar liposomes (41.2°C). Thus, we identified the lower temperature transition with the small vesicles and the higher transition with a new form resembling, but perhaps not identical to multilamellar liposomes. This two state model allowed us to assign a ΔH of 6.3 kcal mol^{-1} to the transition of the small vesicles and an enthalpic change of 8.4 kcal mol^{-1} to the second species. This interpretation was considerably strengthened by examination of the thermal properties of the system using fluorescence depolarization measurements of the probe diphenylhexatriene incorporated in the lipid bilayers. The fluorescence measurements could be made from high to low temperature and thus the low temperature incubation was avoided. Under these circumstances only a single low temperature transition was observed (Suurkuusk et al., 1976).

A more detailed investigation of the nature of the products that form spontaneously from small unilamellar vesicles in the gel phase was undertaken in my laboratory first by Steven Schullery and then by Martin Wong. A variety of techniques were used in these studies. It was clear from initial work using molecular sieve chromatography that the products consist of objects larger than the small unilamellar vesicles initially present.

Determination of the volume of the interior aqueous compartment, the so called trapped volume, over a period of 8 days at 21°C showed this parameter to increase substantially with time. Further characterization using molecular sieve chromatography and both freeze-fracture and negative stain electron microscopy strongly indicated that the products are not aggregates of small vesicles but are larger liposomes. The electron microscopy combined with NMR measurements of the ratio of ^{31}P nuclei accessible to Pr^{+3} to those inaccessible to this shift reagent clearly showed the products to be unilamellar vesicles. Thus these studies taken together lead to the conclusion that the products formed from small unilamellar vesicles composed of dipalmitoyl phosphatidylcholine at gel phase temperatures are large unilamellar vesicles with an average diameter of 700 Å and a surprisingly narrow size distribution. It seemed to us at that time, as it does now, that these large unilamellar vesicles are produced by the fusion of small unilamellar vesicles in the gel phase (Schullery et al., 1980).

These observations of the fusion product and the fusion process were considerably extended at Virginia by Martin Wong (Wong et al., 1982; Wong and Thompson, 1982). Wong observed that the fusion process is substantially faster at 4°C than at 21°C. This observation has recently been extended by McConnell and Schullery (1985). It is apparent that no detectable fusion occurs at 35.5°, the pre-transition temperature. At 30°C about 10% fusion has taken place after 7 days. In the range of 25° to 13°C about 2/3 of the lipid appears in large unilamellar vesicles after 2 days (McConnell and Schullery, 1985). It is clear that the fusion process has a negative temperature coefficient.

Wong also showed that the rate of formation of the large 700 Å vesicles increases as the concentration of small vesicles in the dispersion is increased and is approximately second order. Using the large increase in rate produced by a combination of low temperature and high vesicle concentration, it proved possible to observe the formation of new 950 Å diameter vesicles. These large vesicles were shown by molecular sieve chromatography, electron microscopy and trapped volume measurements to be unilamellar with a narrow size distribution. They were also shown to form not from small unilamellar vesicles, but from 700 Å vesicles. Their size suggested that they are the result of the fusion of two 700 Å vesicles. Subsequent studies by McConnell and Schullery (1985) have shown that the 950 Å vesicles begin to appear at a measurable rate only below 13°C.

The picture that has emerged from the studies outlined above is the following. Small unilamellar vesicles composed of dipalmitoyl phosphatidylcholine are stable above t_m. Below t_p in the gel phase they spontaneously fuse to form a population of unilamellar vesicles about

700 Å in diameter. At low temperature two 700 Å vesicles fuse to form a 950 Å unilamellar vesicle at a slower rate. There is no indication that intermediate-sized fusion species are present at any time. The sizes of the 700 Å and 950 Å fusion products suggest that about 18 small vesicles and two 700 Å vesicles, respectively, fuse to form them. Let us now turn to a more detailed examination of the process by which these structures form.

THE FUSION PROCESS

It has been suggested that the large vesicle species form not by fusion but by a spontaneous intervesicular transfer of phosphatidylcholine molecules (Lawaczeck, 1978). While intervesicular transfer is an established phenomenon in liquid crystalline systems (see Dawidowicz, 1987 for a review) it seems very unlikely that this process can play a significant role in the gel phase formation of 700 and 950 Å diameter vesicles. If spontaneous transfer were involved, a distribution of sizes would be expected in the product structures in the course of vesicle size-growth. That this is not the case is certain. Molecular sieve chromatography under all conditions of temperature and concentration gives bimodal size distributions of small and 700 Å vesicles or 700 and 950 Å vesicles. Electron microscopy also shows bimodal distributions (Schullery et al., 1980; Wong et al., 1982; McConnell and Schullery, 1985). Schmidt et al. (1981) analyzed the ^{31}P NMR line shape and found a conversion from an initial narrow line (14Hz) to a single broad line (100Hz) also suggesting a conversion from small to large vesicles with no intermediate-sized vesicles. In addition, the fact that the rate of generation of the large vesicles increases with increasing concentration of lipid strongly argues against involvement of the transfer process (Wong et al., 1982) since it is well established that the rate of spontaneous interbilayer transfer of lipids is concentration independent (Roseman and Thompson, 1980; Nichols and Pagano, 1981; McLean and Phillips, 1981).

As noted in the preceding section, the fusion of small unilamellar vesicles in the gel phase is approximately second order. This suggests a biaggregate collision mechanism, however the increased fusion rate with decreased temperature argues against simple vesicle-vesicle collision as the rate-limiting step. In addition, the small vesicle to 700 Å diameter vesicle fusion with no intermediate sized material clearly shows that about 18 small vesicles participate in the step which determines the size of the fusion product. The results discussed below will be used to establish a phenomenological model of the fusion process in this system which will reconcile these observations.

Quasi-elastic light scattering (QELS) measurements on small vesicle dispersions as a function of vesicle concentration and time over a 47 h period gave the following information (Wong and Thompson, 1982). At 50°C, above the phase transition temperature of dipalmitoyl phosphatidylcholine, the size is invariant and is that of small vesicles. At 4°C, in the gel phase, the size increases linearly with time for 4 h over the concentration range 0.25 to 112 mM inorganic phosphate. Beyond 4 h the QELS diameter, obtained at higher vesicle concentrations (20-112 mM P_i) gradually reaches a value of 950-1000 Å after 24-48 h. The vesicles formed under these conditions have been shown, as described above, to be the large unilamellar structures formed by the fusion of 18 small vesicles (Wong et al., 1982). For this to be the case it is necessary for aggregation to be faster than fusion. Comparison of the time course determined by QELS, which reflects both aggregation and fusion, with the time course determined by molecular sieve chromatography, which reflects only fusion, showed this to be the case.

At lower small vesicle concentrations (0.25-5 mM P_i) the QELS diameter continues to increase over the whole time interval of 47 h. That this linear rise in QELS diameter is due primarily to aggregation was shown by examining by 90° scattering the effect of raising the temperature to 50°C from 4°C and returning again to 4°C. The intensity of light scattered at 90° falls with increasing temperature to a value at 50°C equivalent to that given by authentic small vesicles at that temperature. Very little increase in the 90° scattering occurs when the temperature is quickly returned to 4°C. A similar result has been obtained using QELS measurements. These data make it clear that in the low concentration range (0.25 to 5 mM P_i) aggregation is much faster than fusion within the aggregates. This conclusion was further substantiated by freeze-fracture electron microscopy.

Additional experiments measuring 90° scattering as a function of time for various vesicle concentrations were also carried out (Wong and Thompson, 1982). Analysis of these data clearly showed that the aggregation process is not diffusion limited, but proceeds in a much slower time frame. Additional experiments also showed that the rate of aggregation of 700 Å diameter vesicles is much slower at 21° and 4°C than is the rate of small vesicle aggregation.

Recently it has been shown that in the presence of bilayer-impermeant solutes the rate of formation of 700 Å vesicles decreases, however membrane-permeant polyols greatly accelerate the rate. The inclusion of only 1% cholesterol stops the fusion process entirely (McConnell and Schullery, 1985). We have noted that the inclusion of only a few mole percent of a number of phospholipids or glycolipids frequently results in the formation of multilamellar liposomes or inhibits fusion.

AGGREGATION-FUSION MODEL

The observations discussed can be used to form a model for the process leading to the formation of large unilamellar vesicles by the fusion of small unilamellar vesicles in the gel phase. This model is shown in Figure 1 (Wong and Thompson, 1982). The time course of events is divided into three regimes.

Regime I. At low temperatures and low vesicle concentrations, aggregation proceeds with little or no fusion. Eventually the large aggregates of small vesicles settle out of the aqueous phase. Aggregation is essentially reversed by raising the temperature. This fact indicates that the disaggregation rate constant k_d' increases faster with increasing temperature than the aggregation rate constant k'.

Regime II. At low temperatures and high concentrations of small vesicles, aggregate size reaches about 18 small vesicles. Collision between aggregates promote fusion within aggregates to produce 700 Å diameter vesicles. This process is approximately second order in vesicle (aggregate) concentration (Schmidt et al., 1981; Wong et al., 1982). It is important to recall that the size of the fused vesicle is essentially independent of vesicle concentration and temperature, hence these variables do not affect the size of the aggregates. It is also important to note that the 700Å product is produced by fusion within each aggregate not between two aggregates. Thus when two aggregates of about 18 small vesicles collide, one or both aggregates fuse to form one or two 700 Å vesicles. This is the case, since the fused vesicle diameter and aggregate diameter as measured by QELS are about equal.

At 4°C k_d' is small but k_f is less than k', so that the overall rate of fusion is controlled by k_f. At 21°C, both k' and k_f have increased, but k_d' has increased by a relatively greater amount, thus decreasing the overall rate.

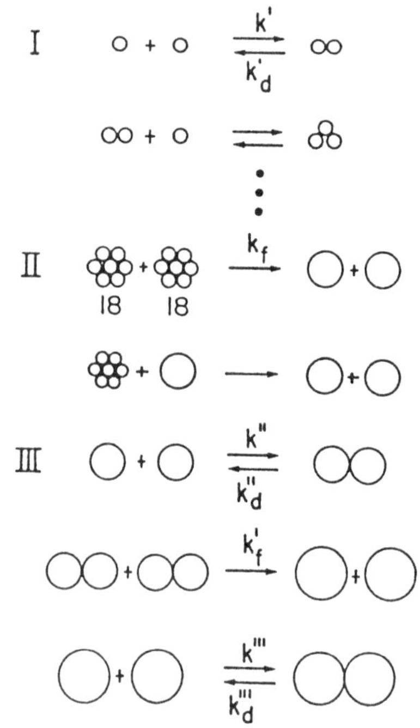

Figure 1. Model for dipalmitoyl phosphatidylcholine fusion of small unilamellar vesicles in the gel phase. k', k'' and k''' are the aggregation rate constants. The subscript d denotes disaggregation and f denotes fusion. Reprinted by permission of the American Chemical Society from Wong and Thompson, 1982.

Regime III. The conversion of 700 Å to 950 Å vesicles is portrayed as preceded by the formation of 700 Å dimers which in turn collide to produce 950 Å vesicles. In comparison with small vesicle fusion this is a very slow process. In this regime the rate limiting step is dimer-dimer collision resulting in fusion (Wong and Thompson, 1982).

The most interesting aspect of this overall process is the aggregation of small vesicles to form 18-mers that subsequently fuse to form 700 Å diameter unilamellar vesicle. A plausible, more detailed, but qualitative, description of this process may be as follows (Lichtenberg and Thompson, 1987). It seems quite possible, that below the gel to liquid crystalline phase transition, small unilamellar vesicles produced by sonication are not spherical but are irregular polyhedrons (Blaurock and Gamble, 1979). Each facet of a polyhedron is a flat bilayer of gel-phase phosphatidylcholine. The facets are joined to each other by disorganized phosphatidylcholine molecules in a glass-like gel phase. The energy minimum for any one polyhedron is obtained as a balance between the lower free energy of the gel phase facets and higher free energy of the joints between facets. In addition, the energy required to relocate the joint between facets by increasing the size of one facet and decreasing the size of an adjacent facet is small. Thus as far as the energetics of the bilayer itself is concerned, the shape of any one polyhedron can be changed by the application of distortion forces; it is compliant.

In aqueous dispersion these polyhedrons form aggregates of limited size which are stabilized by the energy minimum that exists between two bilayers at some critical distance from each other. This distance is likely to be less than 20 Å for gel phase phosphatidylcholines below t_p. The aggregation process can be imagined to begin when two vesicles find themselves with one facet of one polyhedron at a distance of about 20 Å or less from one facet of a second polyhedron. Additional vesicles then adhere to this initial dimer by the same process except that each additional vesicle can have two or more facets in apposition rather than just one facet per vesicle as in the initial dimer. Therefore the stabilization energy for the higher aggregates will be somewhat larger per vesicle than it was initially, in the starting vesicle dimer. This will be the case until the first shell of vesicles surrounding the initial dimer is filled. When the first shell is filled, the stabilization per vesicle for additional vesicles will be lower since, on the average, only one facet of each added vesicle can be apposed by a facet of a vesicle already in the aggregate. This, in conjunction with geometric considerations, suggests that the resultant aggregates will consist of about 18 small vesicles.

During the process of forming an aggregate, the compliance of the polyhedrons acts to maximize the facet to facet apposition, so that the final configuration of the aggregate may be quite regular.

Thus, the aggregate of small unilamellar vesicles from which a fused vesicle forms is comprised of about 18 rather regular polyhedral vesicles. The arrangement in the aggregate is such that the free energy of the aggregate is a minimum even if the distortions of some individual vesicles in the aggregate put them at a higher free energy than they would have if they were not part of the aggregate. The compliance of individual vesicles causes the aggregate as a whole to be rather easily deformed. The ease of deformation is, however, reduced by the fact that changes in shape of individual vesicles require concomitant changes in the aqueous volumes of the vesicle interiors which are resisted by the lower water permeability of gel phase phospholipid.

If, however, collisions occur between aggregates with sufficient energy to force a large shape change in a vesicle pair within the aggregate, the shape change must be accompanied by a transient bilayer rupture. This rupture permits bulk water to flow in order to accommodate the internal volume change coupled to the shape change. If the rupture occurs in two adjacent vesicles in the same place at the same time, fusion of these two vesicles becomes very likely. The fusion product will tend to round up while being leaky. That is, the whole bilayer of this new fusion vesicle will be transiently faulted in many places. If in adjacent vesicles transient spontaneous faults occur near the fusion product, a new fusion event is likely to take place, that is if the fault in the neighboring vesicle occurs near the original fusion product. In this way the fusion process is propagated throughout the whole aggregate. Since the process involves stepwise fusion of unilamellar vesicles only, the final product vesicle will be unilamellar also.

The fusion process as outlined depends upon the concentration of aggregates, because the initial nucleating event is a collision between two aggregates. The size of the product vesicle simply depends on the number of vesicles in the aggregate.

ACKNOWLEDGEMENTS

This work was supported by U.S. Public Health Service grants GM-14628, GM-23573 and a prize from the Alexander von Humboldt Foundation.

REFERENCES

Blaurock, A.E., and Gamble, R.C., 1979, Small phosphatidylcholine vesicles appear to be faceted below the thermal phase transition, J. Memb. Biol., 50:187.

Dawidowicz, E.A., 1987, Lipid exchange. Transmembrane movement, spontaneous movement and protein-mediated transfer of lipids, in: "Current topics in membrane transport" v.29, R.D. Klausner and J. van Renswoude, eds., Academic Press, N.Y. (in press).

Huang, C., 1969, Studies on phosphatidylcholine vesicles. Formation and physical characteristics, Biochemistry, 8:344.

Kantor, H.L., Mabrey, S., Prestegard, J.H., and Sturtevant, J.M., 1977, A calorimetric examination of stable and fusing lipid bilayer vesicles, Biochim. Biophys. Acta, 466:402.

Larrabee, A.L., 1979, Time-dependent changes in the size distribution of distearoyl phosphatidylcholine vesicles, Biochemistry, 18:3321.

Lawaczeck, R., 1978, Intervesicular lipid transfer and direct fusion of phospholipid vesicles: A comparison on a kinetic basis, J. Coll. Inter. Sci., 66:247.

Lawaczeck, R., Kainosho, M., Girardet, J.-L., and Chan, S.I., 1975, Effects of structural defects in sonicated phospholipid vesicles on fusion and ion permeability, Nature, 256:584.

Lichtenberg, D., and Schmidt, C.F., 1981, Molecular packing and stability in the gel phase of curved phosphatidylcholine vesicles, Lipids, 16:555.

Lichtenberg, D., and Thompson, T.E., 1987, Fusion of zwitterionic phospholipids, in: "Cellular membrane fusion," J. Wilschut and D. Hoekstra, eds., Marcel Dekker, New York (in press).

McConnell, D.S., and Schullery, S.D., 1985, Phospholipid vesicle fusion and drug loading: Temperature, solute and cholesterol effects, and, a rapid preparation for solute-loaded vesicles, Biochim. Biophys. Acta, 818:13.

McLean, L.R., and Phillips, M.C., 1981, Mechanism of cholesterol and phosphatidylcholine exchange or transfer between unilamellar vesicles, Biochemistry, 20:2893.

Nichols, J.W., and Pagano, R.E., 1981, Kinetics of soluble lipid monomer diffusion between vesicles, Biochemistry, 20:2783.

Parente, R.A., and Lentz, B.A., 1984, Phase behavior of large unilamellar vesicles composed of synthetic phospholipids, Biochemistry, 23:2353.

Roseman, M.A., and Thompson, T.E., 1980, Mechanism of spontaneous transfer of phospholipids between bilayers, Biochemistry, 19:439.

Schmidt, C.F., Lichtenberg, D., and Thompson, T.E., 1981, Vesicle-vesicle interactions in sonicated dispersions of dipalmitoyl phosphatidylcholine, Biochemistry, 20:4792.

Schullery, S.E., Schmidt, C.F., Felgner, P., Tillack, T.W., and Thompson, T.E., 1980, Fusion of dipalmitoyl phosphatidylcholine vesicles, Biochemistry, 19:3919.

Suurkuusk, J., Lentz, B.R., Barenholz, Y., Biltonen, R.L., and Thompson, T.E., 1976, Fusion of dipalmitoyl phosphatidylcholine vesicles, Biochemistry, 15:1393.

Wong, M., Anthony, F.H., Tillack, T.W., and Thompson, T.E., 1982, Fusion of dipalmitoyl phosphatidylcholine vesicles at 4°C, Biochemistry, 21:4126.

Wong, M., and Thompson, T.E., 1982, Aggregation of dipalmitoyl phosphatidylcholine vesicles, Biochemistry, 21:4133.

STRUCTURAL CHARACTERIZATION OF LAMELLAR Mg^{2+} COMPLEXES OF DILAUROYLPHOSPHATIDIC ACID USING ^{31}P and ^{13}NMR

M.P. Murari, M.P. O'Brien, and J.H. Presttegard

Department of Chemistry, Yale University
New Haven, CT 06511

INTRODUCTION

Among the numerous model membrane systems showing fusion phenomena, those containing anionic lipids and fusing in response to the addition of divalent cations have attracted the most attention[1]. In part, the reason for this is the observed correlation of fusion in natural systems with elevations in divalent ion concentrations, particularly Ca^{2+}. Most model systems display an enhanced activity of Ca^{2+} over ions such as Ba^{2+} and Mg^{2+} [2,3]. There are apparent exceptions to this rule[4], but these may involve cases in which aggregation rather than membrane rupture is the limiting step in fusion. It is a challenge to understand at a molecular level, cation-stimulated fusion of anionic lipid systems and the origin of ion selectivity. Structural studies of anionic lipid-cation complexes are a good place to begin this understanding.

The structural studies we present here deal with a phosphatidic acid-Mg^{2+} complex. Phosphatidic acid (PA) is not a major lipid of biological membranes, but it is an important intermediate in phosphorylation cycles and it is among the simplest of the anionic lipids in that it has a single negatively charged phosphate group attached to a diacylglycerol backbone. It is a logical target for an investigation in aqueous membrane dispersions because of the recent publication of an X-ray crystal structure of the sodium complex of dimyristoylphosphatidic acid (DMPA)[5]. The molecules in the crystal are packed in a membrane-like array, much as the molecules in previous crystal structures of dilauroylphosphatidylethanolamine (DLPE) and dimyristoylphosphatidylcholine (DMPC)[6,7]. However, the structure is distinctly different in having a headgroup region intricately connected by arrays of hydrogen bonds and coordinated ions, by having the glycerol backbone perpendicular rather than parallel to the bilayer normal, and by having the 2-acyl chain rather than the 1-acyl chain extended in an all-trans configuration into the depths of the bilayer (Figure 1). It is intriguing to ask whether this

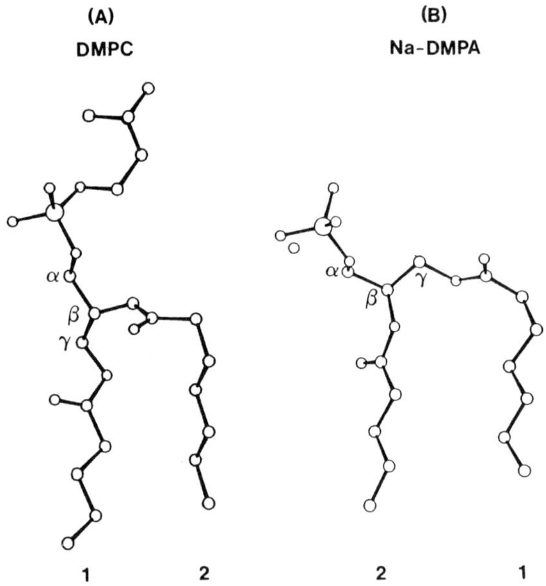

Figure 1. Crystal structure conformations for DMPC[7] (A) and Na-DMPA[5] (B). For DMPC only one of two conformations found is depicted. Hydrogens are not shown and acyl chains have been truncated at 5 or 7 carbons.

unusual structure persists in the less organized structures which exist in aqueous dispersions.

Studies of structure in partially ordered aqueous dispersions are often undertaken using solids NMR methods[8-10]. For phospholipids, ^{31}P NMR provides a directly applicable probe of headgroup orientation. The high natural abundance of the ^{31}P isotope makes ^{31}P NMR particularly convenient. We will also employ ^{13}C NMR as a probe of orientation of ester carbonyls. These experiments will require selective enrichment of carbonyls in the 1-acyl or 2-acyl positions. In both ^{31}P and ^{13}C NMR structural information comes from the partial averaging of chemical shift tensors. In the simplest models, that averaged tensor can be viewed as the result of axial averaging about the axis of the all trans acyl chain in structures such as those in Figure 1, and an additional order parameter describing wobble of this averaging axis. At this simple level of interpretation, proposed structural models can be screened for consistency with spectral data.

RESULTS AND DISCUSSION

Before delving into experiments directed at structure, it is necessary in ion-complex studies to determine stoichiometry. This is particularly important in phosphatidic acid systems since PA can exist in both singly and doubly charged states. Figure 2 presents high resolution ^{31}P NMR spectra of a sonicated DLPA dispersion, pH 6, to which successively higher amounts of Mg^{2+} have been added as the chloride salt. Initially a single sharp resonance is observed. On addition of small amounts of Mg^{2+}, this splits into three resonances; two of these could easily correspond to phosphates on inside and outside halves of vesicle bilayers. Dynamic light scattering results which suggest an average diameter of 250Å for particles in the initial dispersion are consistent with the existence of vesicle structures in the preparation. The third could represent a micelle fraction or a fraction of ruptured vesicles having poorly distinguished inside and outside resonances.

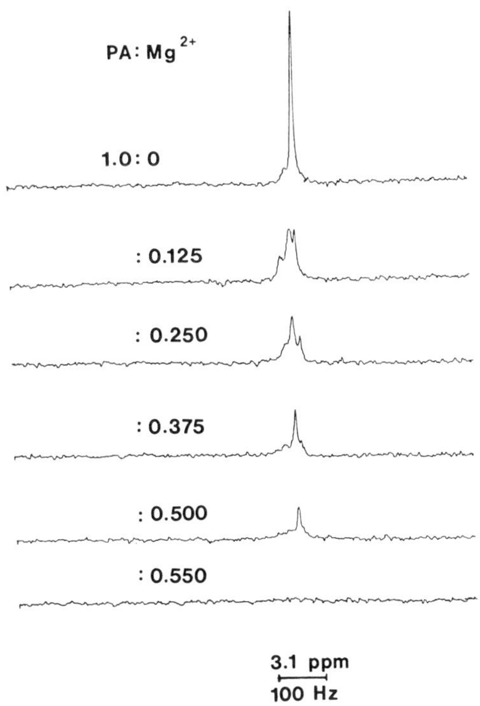

Figure 2. High resolution ^{31}P NMR spectra of dispersions of DLPA (32 mM, pH 6) to which increasing amounts of $MgCl_2$ have been added. Mole ratios of DLPA to Mg are indicated on the individual spectra.

On further addition of Mg^{2+}, the three resonances appear to lose area and eventually disappear. Samples become increasingly turbid as a new lipid phase forms. Except for a constant portion at low concentration, a plot of total resonance area versus $[Mg^{2+}]$ shows a nearly linear decrease in intensity (Figure 3). This is consistent with a model in which vesicles are kinetically or thermodynamically the more stable form up to a threshold of approximately 4 mM. Beyond this point, a 2:1 PA:Mg^{2+} complex forms which in a more highly aggregated state is not free to tumble isotropically on the timescale of NMR measurements (10^{-3} seconds). This 2:1 complex will be the object of our structural studies.

NMR resonances do not actually disappear at high Mg^{2+} concentrations. They simply become too broad to see in normal high resolution experiments. In Figure 4A, a ^{31}P chemical shift anisotropy powder pattern is shown which is typical of a system exhibiting rapid axial averaging. This pattern narrows slightly as temperatures are raised from 25 to 50°C. At 60°C the pattern changes abruptly to one half the width and opposite asymmetry. This change is typical of a transition from a bilayer to a hexagonal phase[11]. At lower temperatures or higher ion concentrations (as shown in Figure 4B), broader, less symmetric patterns are observed. The width of this pattern is, in fact, similar to that for crystalline Mg^{2+} complexes of glycerophosphate[12], suggesting a structure with little internal mobility.

The partially averaged lamellar phase signals seen between 25° and 60° can be interpreted on the basis of proposed structures in Figure 1 and assumed averaging with axial symmetry about the all trans acyl chain. One would begin with a molecular frame shielding tensor having components similar to those observed in crystalline glycerol phosphate complexes (-52, 18, 43)[12]. The most unique axis, that with shielding element -52, would be assumed to lie

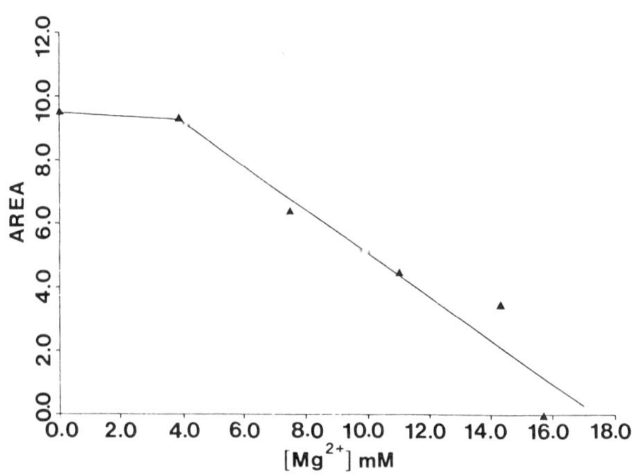

Figure 3. Plot of the area of resonances in Figure 2 versus Mg^{2+} concentration.

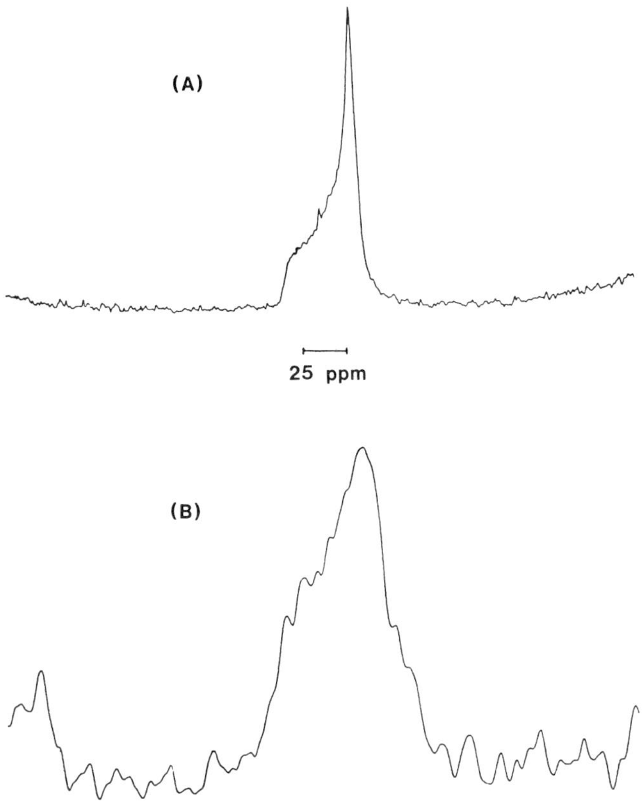

Figure 4. Solids ^{31}P NMR spectra of 2:1(A) and 1:1(B) DLPA:Mg^{2+} dispersions; pH 6, 25°C.

along the local symmetry axis of the phosphate group (the esterified P-O bond). Averaging then occurs first by axial rotation about the acyl chain axis. In structure B of Figure 1, that for the Na-dimyristoylglycerophosphate, the P-O bond makes an angle of 54° with this axis. This angle lies at what is called the magic angle. Because of the nature of the angular dependence of averaging, $(1-3\cos^2\theta)/2$, where θ is the angle described above, chemical shift anisotropy powder patterns in structures with θ near 54° are reduced to narrow lines. Any additional motions superimposed on axial rotation would simply reduce the pattern further. Structure B and the assumed motional model are clearly inconsistent with observation. Structure A, one of the structures found in the crystal structure of dimyristoyl PC[7], fares much better. The P-O bond sits at an angle of 34°. Axial averaging would produce a pattern of approximate width 50 ppm, having the proper symmetry. Some additional motion could reduce the width to that observed. The reduction observed is in fact similar to that seen for phosphatidylcholine systems in the Lα lamellar phase[9].

More definitive conclusions can be reached by examining ^{13}C spectra of molecules labelled at the carbonyl carbons. Figure 5 shows such a spectrum. There are two superimposed signals in the region expected for carbonyls, 180-120 ppm. One is a broad signal of near axial symmetry having a unique low field shielding element and a total width of 140 ppm. The other has a very narrow width of approximately 10 ppm. Single crystal studies of model compounds containing ester carbonyls lead us to expect static shielding tensor elements, referenced to an isotropic ester carbonyl to be -93, 39, 54 ppm[13]. The -93 ppm element is expected to be perpendicular to, and in the plane of the carbonyl, and the most shielded element is expected to be perpendicular to that plane. The observed broad signal would have to correspond to either a rigid component or a carbonyl oriented such that axial averaging had very little effect; ie., along a direction parallel to the -93 ppm element. In the crystal structure for DMPC, the carbonyl on chain 1 is oriented in a manner which would produce little reduction in a powder pattern on rotation about the average axis of the 1 chain. The same is true for the carbonyl on chain 2 for the DMPA structure. However, we hesitate to attribute the broad component to a uniquely oriented carbonyl on this basis alone because the cross-polarization methods used to acquire the spectra can lead to anomolously high intensities for small amounts of rigid components in the sample.

The narrow resonance is less prone to anomolous enhancement and is if anything under-represented in these experiments. This resonance is very reminiscent of narrow

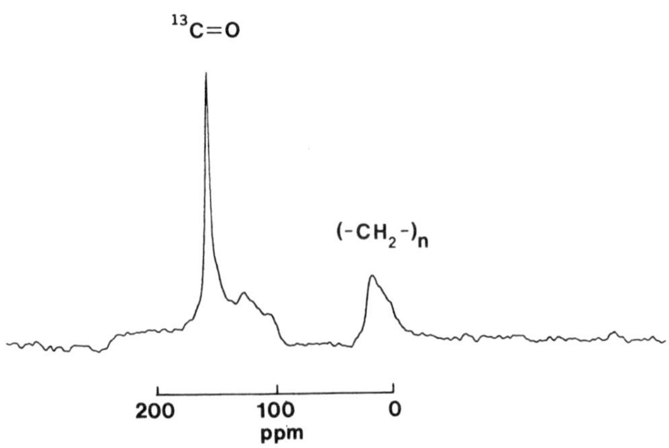

Figure 5. Solids ^{13}C NMR spectrum of a 2:1 DLPA:Mg^{2+} complex, pH 6, 40°C. The sample is enriched to 90% in carbonyls of both the 1- and 2-acyl chains.

resonances seen for phosphatidylcholine dispersions in lamellar liquid crystalline phases[14]. These resonances have been assigned to the 2-chain carbonyl and interpreted either as that carbonyl having a unique tensor element oriented at 54° relative to the averaging axis[15] or as the carbonyl having additional internal motions[14]. The conformation exhibited by the second molecule in the unit cell of the DMPC crystal structure does in fact show a 2-chain carbonyl with a 54° angle for the -93 ppm element[7]. The problem with our observation is that given a difference in chain geometries between DMPC and DMPA crystal structures, the orientation of the 1-chain carbonyl could just as easily give rise to the narrow resonance if the conformation seen in the DMPA structure dominated. Selective labelling of carbonyls solves this dilemma.

Figure 6 shows spectra taken under the same conditions for a sample selectively labelled at the 2-carbonyl of DLPA. The broad component is missing and the sharp component remains, clearly assigning this signal to the 2-chain. Lowering the temperature leads to broadening of the signal and eventual freezing out of motions leading to the reduced chemical shift powder pattern. The assignment of the narrow signal to the 2-position leads to the conclusion that in aqueous dispersions of 2:1 DLPA:Mg^{2+} complex, conformations similar to those seen in diacylglycerolphosphorylcholine crystals must be favored over conformations seen in the Na-DMPA crystal structure.

CONCLUSION

The implications of our observations for cation-induced membrane fusion require some speculation. However, it is clear that an ion which can associate with a bilayer surface without significantly perturbing the surface from its normal liquid crystalline configuration will do little to destabilize the structure and little to promote fusion. Mg^{2+} appears to be such an ion. This may be a rather fundamental property, connected with preferred coordination geometry. Even in simple crystalline complexes such as Mg-diethylphosphate, Mg^{2+} interacts with phosphates in a manner which places them in long rows 9.3Å apart. This spacing is long compared to spacing in Cd^{2+}, or presumably Ca^{2+}, complexes where the spacing is approximately 5.6Å[16]. It is interesting that phosphates in a liquid crystalline phosphatidylcholine bilayer can be arranged in rows with 9.2Å spacing. Bringing them closer would require a transition to a more condensed phase, a phase that would place more stress on the bilayer. It is quite possible that structures such as that found in the crystal structure of Na-DMPA are the basis of more condensed phases, and these structures may be found in complexes with more fusogenic cations such as Ca^{2+}.

MATERIALS AND METHODS

Dilauroylphosphatidic acid was purchased from Sigma Chemical Co., St. Louis, MO as the sodium salt and used without further purification. ^{13}C-labelled analogs were synthesized using lauric acid, ^{13}C-enriched to 90% at the C1

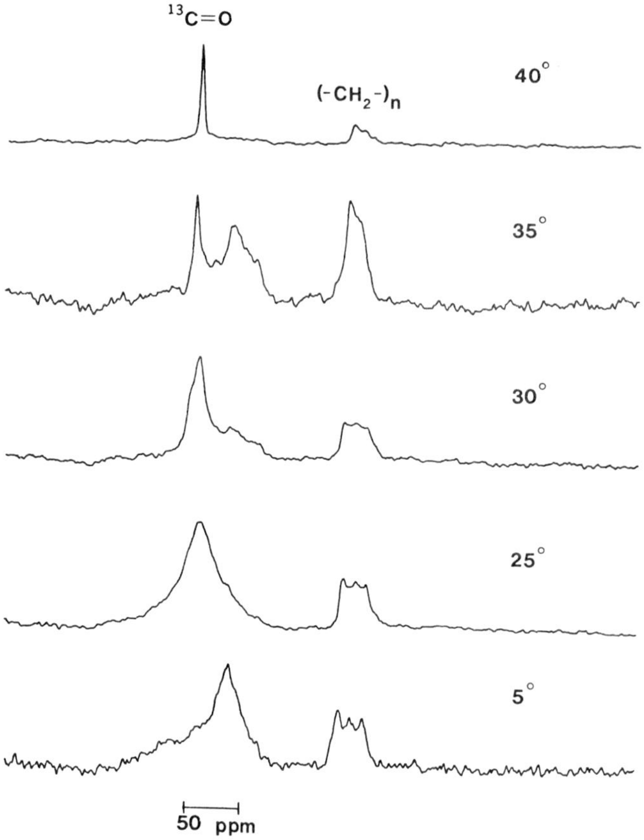

Figure 6. Solids ^{13}C NMR spectra of a 2:1 DLPA:Mg^{2+} complex, pH 6, as a function of temperature. The sample is enriched to 90% ^{13}C in the carbonyl of only the 2-acyl chain.

carbon, purchased from MSD Isotopes, St. Louis, MO and using the following procedures.

<u>1-(1'-^{13}C)-2-(1'-^{13}C)-L-α-DLPA, Monosodium Salt</u>

Acylation of the dimonocyclohexylammonium salt of <u>L</u>-<u>α</u>-glycerophosphate (Sigma Chemical Co., St. Louis, MO) to form the monosodium salt of 1-(1'-^{13}C)-2-(1'-^{13}C)-dilauroyl-3-<u>sn</u>-phosphatidic acid was carried out using a procedure similar to one previously reported[12].

2-(1'-^{13}C)-L-α-DLPC

1-Lauroyl-2-lyso-sn-glycero-3-phosphocholine, obtained by treatment of 1,2-dilauroyl-sn-glycero-3-phosphocholine (Avanti Polar Lipids, Inc., Birmingham, AL) with phospholipase A_2 (ophiophagus hannah snake venom; Miami Serpentarium Laboratories, Miami, FL), was acylated with the acylurea and/or anhydride[17] formed by reaction between 1-^{13}C-lauric acid and dicyclohexylcarbodiimide (DCC; Aldrich Chemical Co., Milwaukee, WI), utilizing the well-established acylation catalyst[18], N,N-dimethyl-4-aminopyridine (DMAP; Sigma Chemical Co., St. Louis, MO). The fatty acid (2.6 mmol) and DCC (1.3 mmol) were mixed in a small dry flask under nitrogen, melted (70°C), and maintained as a stirred melt for 30 min. After the reaction mixture resolidified at room temperature, the 2-lyso PC (0.2 mmol), DMAP (0.2 mmol) and dry chloroform (5 ml) were added. The flask was stoppered, sealed and protected from light and the reaction mixture stirred at room temperature until acylation was complete (1-3 hrs) as monitored by TLC.

The reaction mixture was dissolved in chloroform/methanol (1:1) (10 ml), washed with cold 0.9% KCL in 1 N HCl (10 ml) and then washed several times with water/methanol (1:1). The crude product was collected from the chloroform layer and was isolated by flash chromotography using a solvent gradient ($CHCl_3$; $CHCl_3$/MeOH, 1:1; $CHCl_3$/MeOH 1:2; $CHCl_3$/MeOH/H_2O, 65:35:8). Isolated yields were typically 80%. The purity was checked by analytical TLC (developing solvent: chloroform/methanol/water, 65:35:8). ^{13}C-NMR showed that greater than 90% of the ^{13}C had been incorporated into the 2-carbonyl with less than 10% incorporation into the 1-carbonyl due to acyl migration[19].

2-(1'-^{13}C)-L-α-DLPA, Monosodium Salt

The monosodium salt of 1-lauroyl-2-(1'-^{13}C)-lauroyl-3-sn-phosphatidic acid was made from 1-lauroyl-2-(1'-^{13}C)-lauroyl-sn-glycero-3-phosphocholine by enzymatic hydrolysis[20], using an extract from Savoy cabbage as the source of phospholipase D[21].

The DLPC (typically 0.2 mmol) was dispersed in an aqueous buffer (30 ml of 0.2 M acetate buffer, 67 mM in calcium chloride, pH 5.5)-diethyl ether mixture (3:1). Three parts of the enzyme extract were added and the mixture was sonicated at 30-35°C for 3-5 hrs. When the reaction had proceeded to completion as monitored by TLC, the Ca^{2+} salt of the PA was pelleted by centrifugation and the aqueous medium was removed by decanting. The remaining pellet was dispersed in chloroform and methanol (2:1) and converted to the sodium salt using cold (5°C) washes of 0.1 N HCl and 0.1 M NaCl. The crude product obtained by vacuum drying of the chloroform layer was recrystallized from cold (-20°C) dry chloroform/dry acetone (1:4). The purity was checked by analytical TLC (developing solvent, chloroform/acetone/methanol/acetic acid/water, 30:40:10:5:7). Retention of isotopic enrichment in the 2-carbonyl was confirmed to be greater than 90% by high-resolution ^{13}C NMR.

Complexation of the Monosodium Salt of L-α-DLPA with Mg^{2+}

A weighed amount of the monosodium salt of 1,2-dilauroyl-3-sn-phosphatidic acid (Sigma Chemical Co., St. Louis, MO; or as prepared above) was dispersed in the desired volume of buffer (pH 6.0, 20 mM cacodylate (^{13}C experiments) or 20 mM acetate (^{31}P experiments)) and sonicated below 45°C using a bath type sonicator until optical clarity was obtained. Several aliquots (usually 3-5) from a concentrated solution (400 mM) of magnesium chloride were added until the Mg^{2+} concentration in the sample was equivalent to 0.55 of the molar quantity of PA used. Between additions, the sample was sonicated for 30 min below 45_0C and the pH was adjusted back to 6.0 before the next Mg^{2+} aliquot was added. The precipitated complex was collected by centrifuging (2,500 rpm) at 20°C and removing the supernatant. Solids NMR spectra were acquired on the wet pellet or on a pellet that had been resuspended in 0.25-0.5 ml of the desired buffer. Stoichiometry of the complex in the pellet was checked by phosphate analysis[22,23] for lipid and atomic absorption for Mg^{2+}.

NMR Spectroscopy

Solids NMR spectra were obtained on a Bruker CXP-200 spectrometer operating at 80.98 MHz for ^{31}P and 50.30 MHz for ^{13}C. The ^{31}P spectra were acquired in the time domain after a $\pi/2$ pulse of 3.2 usec with proton decoupling during acquisition only. A 3 sec repetition time was used and spectra were acquired for periods of 10 to 13 hrs. Time domain spectra were weighted with a line broadening factor of 50 Hz for Figure 4A and 500 Hz for Figure 4B and Fourier transformed to a 1K point 30 kHz frequency domain.

^{13}C spectra were acquired using standard cross-polarization pulse sequences[24]. A $\pi/2$ proton pulse of 4.0 μsec began the sequence and acquisition followed a 5 msec contact time for which ^{13}C power was adjusted to satisfy the Hartmann-Hahn condition. Proton decoupling was maintained during acquisition of 18 msec and the entire sequence repeated after 3 sec. Time domain spectra were collected in quadrature over periods of 7 to 15 hrs, weighted by an exponential function yielding 100 Hz line broadening and transformed into a 1K point 30 kHz frequency domain.

High resolution spectra for sample analysis were acquired on a Bruker WM250 and Varian CFT20 spectrometer for ^{13}C and ^{31}P respectively. In all cases, 90° pulses were used and delays between acquisition were adequate to allow >90% recovery of magnetization. For ^{13}C spectra, proton irradiation occurred only during acquisition to eliminate distortion of intensity due to Nuclear Overhauser Effect.

ACKNOWLEDGEMENTS

This work was supported in part by a grant from the National Institute of General Medical Sciences, GM-19035.

REFERENCES

1. N. Duzgunes, J. Wilschut, and D. Papahadjopoulos,

Control of Membrane Fusion by Divalent Cations, Phospholipid Head-Groups and Proteins, in: "Physical Methods on Biological Membranes and their Model Systems," F. Conti, W. E. Blumberg, J. de Gier, and F. Pocchiari, ed., Plenum Publishing Corporation (1985).

2. J. Bentz, N. Duzgunes, and S. Nir, Kinetics of Divalent Cation Induced Fusion of Phosphatidylserine Vesicles: Correlation between Fusogenic Capacities and Binding Affinities, Biochem. 22:3320 (1983).
3. M-J. Liao and J. H. Prestegard, Ion Specificity in Fusion of Phosphatidic Acid-Phosphatidylcholine Mixed Lipid Vesicles, Biochim. Biophys. Acta. 601:453 (1980).
4. S. Ohki and H. Ohshima, Divalent Cation-Induced Phosphatidic Acid Membrane Fusion. Effect of Ion Binding and Membrane Surface Tension, Biochim. Biophys. Acta. 812:147 (1985).
5. K. Harlos, H. Eibl, I. Pascher, and S. Sundell, Conformation and Packing Properties of Phosphatidic Acid: The Crystal Structure of Monosodium Dimyristoylphosphatidate, Chem. Phys. Lipids. 34:115 (1985).
6. P. B. Hitchcock, R. Mason, K. M. Thomas, and G. G. Shipley, Structural Chemistry of 1,2 Dilauroyl-DL-phosphatidylethanolamine: Molecular Conformation and Intermolecular Packing of Phospholipids, Proc. Natl. Acad. Sci. USA 71:3036 (1974).
7. R. H. Pearson and I. Pascher, The Molecular Structure of Lecithin Dihydrate, Nature. 281:499 (1979).
8. R. G. Griffin, Solid State Nuclear Magnetic Resonance of Lipid Bilayers, Methods Enzymol. 72:108 (1981).
9. C. P. S. Tilcock, P. R. Cullis, and S. M. Gruner, On the Validity of ^{31}P-NMR Determinations of Phospholipid Polymorphic Phase Behaviour, Chem. Phys. Lipids. 40:47 (1986).
10. J. L. Browning and J. Seelig, Bilayers of Phosphatidylserine: A Deuterium and Phosphorus Nuclear Magnetic Resonance Study, Biochem. 19:1262 (1980).
11. P. R. Cullis and B. De Kruijff, The Polymorphic Phase Behaviour of Phosphatidylethanolamines of Natural and Synthetic Origin. A ^{31}P NMR Study, Biochim. Biophys. Acta. 513:31 (1978).
12. V. W. Miner and J. H. Prestegard, Structure of Divalent Cation-Phosphatidic Acid Complexes as Determined by ^{31}P-NMR, Biochim. Biophys. Acta. 774:227 (1984).
13. B. A. Cornell, Chemical shielding tensors of C^{13} in solid dimethyl oxalate, J. Chem. Phys. 85:4199 (1986).
14. B. A. Cornell, The Effect of the Bilayer Phase Transition on the Carbonyl Carbon-13 Chemical Shift Anisotropy, Chem. Phys. Lipids. 28:69 (1981).
15. R. J. Wittebort, C. F. Schmidt, and R. G. Griffin, Solid-State Carbon-13 Nuclear Magnetic Resonance of the Lecithin Gel to Liquid-Crystalline Phase Transition, Biochem. 20:4223 (1981).
16. V. W. Miner, J. H. Prestegard, and J. W. Faller, Cadmium Diethyl Phosphate: Structure Determination and Comparison to Cation Phospholipid Complexes, J. Inorg. Chem. 22:1862 (1983).

17. F. E. Ziegler and G. D. Berger, A Mild Method for the Esterification of Fatty Acids, Synth. Comm. 9:539 (1979).
18. C. M. Gupta, R. Radhakrishnan, and H. G. Khorana, Glycerophospholipid Synthesis: Improved General Method and New Analogs Containing Photoactivable Groups, Proc. Natl.Acad. Sci. 74:4315 (1977).
19. A. Pluckthun and E. A. Dennis, Acyl and Phosphoryl Migration in Lysophospholipids: Importance in Phospholipid Synthesis and Phospholipase Specificity, Biochem. 21:1743 (1982).
20. D. Papahadjopoulos and N. Miller, Phospholipid Model Membranes I. Structural Characteristics of Hydrated Liquid Crystals, Biochim. Biophys. Acta. 135:624 (1967).
21. H. Eibl and S. Kovatchev, Preparation of Phospholipids and Their Analogs by Phospholipase D, Methods Enzymol. 72:634 (1981).
22. G. V. Marinetti, Chromatographic Separation, Identification and Analysis of Phosphatides, J. Lipid Res. 3:1 (1962).
23. G. R. Bartlett, Phosphorous Assay in Column Chromatography, J. Biol. Chem. 234:466 (1959).
24. A. Pines, M. G. Gibby, and J. S. Waugh, Proton-enhanced NMR of dilute spins in solids, J. Chem. Phys. 59:569 (1973).

LIPID POLYMORPHISM, LIPID ASYMMETRY AND MEMBRANE FUSION

P.R. Cullis and M.J. Hope

Biochemistry Department
The University of British Columbia
Vancouver, B.C., Canada V6T 1W5

INTRODUCTION

In this review we describe the intimate relation between lipid polymorphism and membrane fusion, and briefly introduce recent work on factors leading to asymmetric transbilayer distributions of lipid. Such asymmetric distributions may be expected to play a major regulatory role in the ability of membranes to fuse.

LIPID POLYMORPHISM AND FUSION

Two distinct processes involved in membrane fusion are first those giving rise to close apposition of membranes, and second the interbilayer process giving rise to the completed fusion event. Here we focus on the second event, observing that it is topologically impossible for fusion to proceed without some local departure from lipid bilayer organization at the fusion interface. This has naturally led to an examination of the abilities of lipids to adopt non-bilayer structure (lipid polymorphism) to determine whether such structures may play an intermediate role in the fusion event.

In order for lipid polymorphism to be related to membrane fusion, four conditions must be satisfied. First, lipids which can adopt non-bilayer organization must be present in membranes. Second, factors which regulate or stimulate fusion in vivo must regulate the structural properties of appropriate lipid dispersions in a consistent manner. Third, the structural transitions demonstrated by lipids must provide logical intermediate structures. Finally, the structural intermediate available must be compatible with a non-leaky fusion event. In the next section we review studies supporting these conditions and subsequently relate these observations to fusion of selected model and biological membranes.

Lipid Polymorphism and Factors Regulating the Structural Preferences of Lipids

The ability of aqueous dispersions of lipids to assume a variety of structures on hydration is well-characterized and has been extensively reviewed elsewhere (Cullis & de Kruijff, 1979, Cullis et al., 1985). Four major points can be made. First, individual species of lipids found in

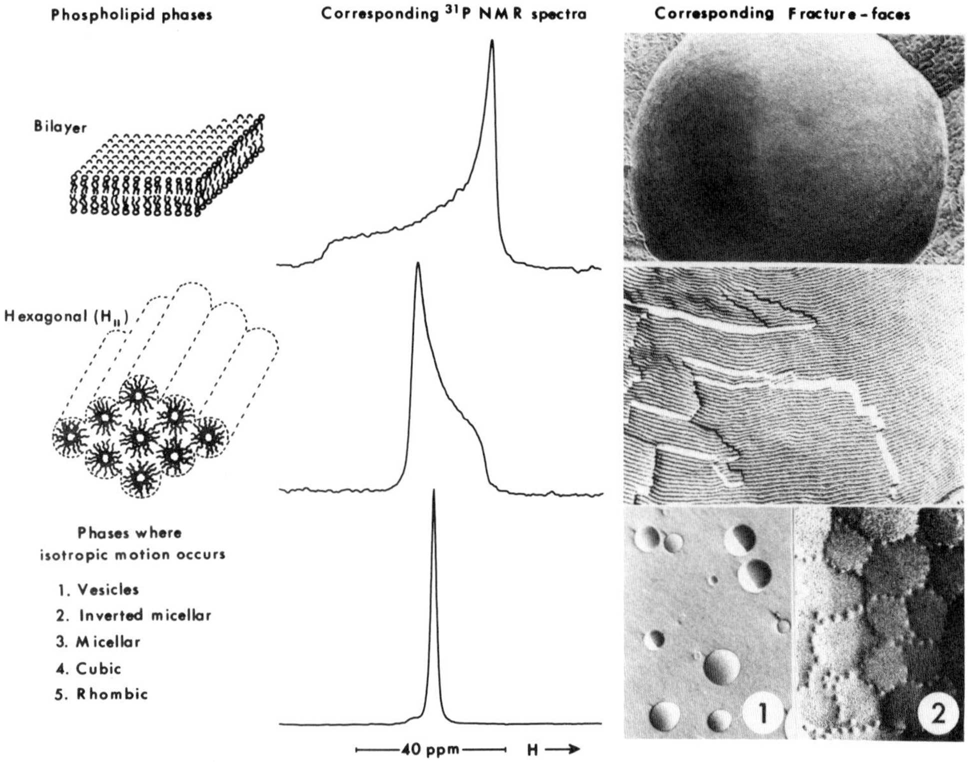

Figure 1. ^{31}P NMR and freeze-fracture characteristics of phospholipids in various phases. The bilayer ^{31}P NMR spectrum was obtained from aqueous dispersions of egg yolk phosphatidylcholine and the hexagonal (H_{II}) phase spectrum from phosphatidylethanolamine (prepared from soybean phosphatidylcholine). The ^{31}P NMR spectrum representing isotropic motion was obtained from a mixture of 70 mol% soya phosphatidylethanolamine and 30% egg yolk phosphatidylcholine after heating to 90°C for 15 min. All preparations were hydrated in 10 mM Tris-acetic acid (pH 7.0) containing 100 mM NaCl and the ^{31}P NMR spectra were recorded at 30°C in the presence of proton decoupling. The freeze-fracture micrographs represent typical fracture faces obtained from bilayer and H_{II} phase systems as well as structures giving rise to isotropic motional averaging. The bilayer configuration (total erythrocyte lipids) gives rise to a smooth fracture face, whereas the hexagonal (H_{II}) configuration is characterized by ridges displaying a periodicity of 6 to 15 mm. Common conformations that give rise to isotropic motion are represented in the bottom micrograph: (1) bilayer vesicles (~100 nm diameter) of egg phosphatidylcholine prepared by extrusion techniques and (2) large lipid structures containing lipidic particles. This latter system was generated by fusing SUVs composed of egg phosphatidylethanolamine and 20 mol% egg phosphatidylserine which were prepared at pH 7 and then incubated at pH 4 for 15 min to induce fusion.

membranes can adopt either the bilayer or hexagonal H_{II} phase indicated in Fig. 1. As indicated in Table 1, a large proportion of membrane lipids either adopt the hexagonal (H_{II}) phase on hydration at physiological temperatures or induce hexagonal structure in mixed lipid systems. The most notable examples are (eukaryotic) phosphatidylethanolamines (PE's) as well as cholesterol. Cholesterol has a remarkable ability to induce H_{II} phase organization in mixed lipid systems. Non-bilayer lipids such as PE can of course be stabilized in a bilayer structure by the presence of bilayer-preferring lipids such as phosphatidylcholine (PC), phosphatidylserine (PS) or sphingomyelin (SPM). It is usually found that between 20 and 50 mol% of the bilayer-preferring lipids is required to maintain a net bilayer organization when mixed with H_{II}-preferring lipids such as PE.

The second point is that the structural preferences of these pure and mixed lipid systems are very sensitive to a variety of factors known to regulate fusion events in vivo. Perhaps the clearest example is Ca^{2+}, which can trigger bilayer to H_{II} transitions in multilamellar vesicle (MLV) systems composed of pure cardiolipin (CL), and as shown in Fig. 2, in mixtures of PE with the negatively charged lipids PS, phosphatidylglycerol (PG), phosphatidic acid (PA), phosphatidylinositol (PI) or CL (Cullis et al., 1985) as well as PE-PS-cholesterol and PE-PS-cholesterol systems (Tilcock & Cullis, 1981; Bally et al., 1983; Tilcock et al., 1984; Cullis et al., 1985). This ability extends to MLV systems with lipid compositions corresponding to the erythrocyte inner monolayer (Hope & Cullis, 1979). Other examples of factors which can trigger bilayer to H_{II} transitions include low pH as observed for systems containing PS (Hope & Cullis, 1980), PA (Farren et al., 1983) and PS-PE (Tilcock & Cullis, 1981); high ionic strength as observed for PE-PS-cholesterol systems (Bally et al., 1983); and proteins such as cytochrome c (de Kruijff & Cullis, 1980) and gramicidin (van Echteld et al., 1982). A particularly illustrative example of the correlation between in vitro fusion processes and the ability of endogenous lipids to adopt H_{II} phase organization is given by the action of lipid soluble "fusogens" (Ahkong et al., 1973) on the structural preferences of MLV's composed of erythrocyte membrane lipids. As shown in Fig. 3, fusogens induce H_{II} phase structure in such dispersions, whereas structurally related non-fusogens do not. A strong correlation therefore exists between factors known to promote fusion in vivo and in vitro and their ability to promote H_{II} phase organization in various model MLV systems.

A third point concerns the relation between the bilayer to hexagonal H_{II} transition and membrane fusion processes. Unsaturated PE dispersions undergo a bilayer to H_{II} transition as the temperature is raised, which usually occurs in the region of 10°C for PE of eukaryotic origin. ^{31}P-NMR results (Cullis et al., 1980) indicate that the inverted cylinders characteristic of the H_{II} phase form from closely apposed bilayers which fuse together, trapping some of the interbilayer water into the inverted lipid cylinders. The analogy to fusion is clear.

The fourth point concerns the fact that the hexagonal phase itself is clearly not an attractive fusion intermediate, in that formation of such a macroscopic structure would be expected to lead to extensive membrane disruption and leakage. However, mixtures of bilayer and H_{II}-preferring lipids often give rise to 'lipidic particle' structure visualized by freeze-fracture electron microscopy (see Fig. 4). Available evidence (Verkleij, 1984) indicates that these particles correspond to inverted lipid micelles formed between intersecting bilayers, and that these particles are intermediates between bilayer and H_{II} phase structure. This is indicated in freeze-fracture studies (Verkleij et al., 1980) where the particles are observed to line up in rows as apparent precursors to the

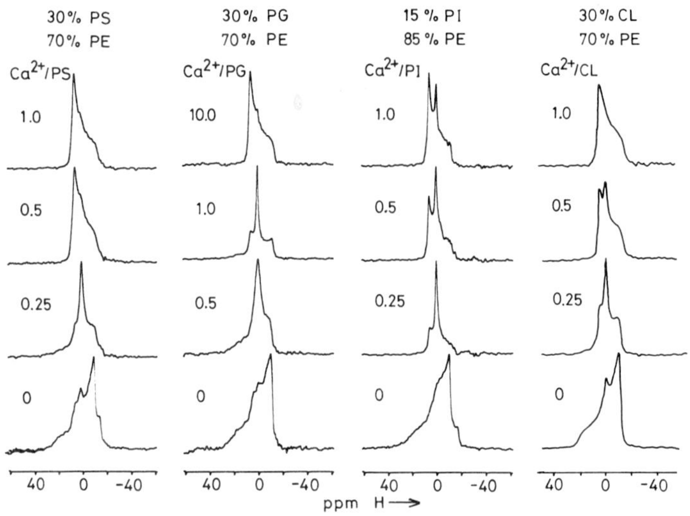

Figure 2. ^{31}P NMR spectra arising from mixtures of acidic phospholipids with soya phosphatidylethanolamine (a polyunsaturated PE derived from soya PC) in the presence of various molar ratios of Ca^{2+}. All samples were prepared from 50 μmol total phospholipid hydrated in MLV form by vortex mixing. The Ca^{2+} was added as aliquots from a 100 mM stock solution.

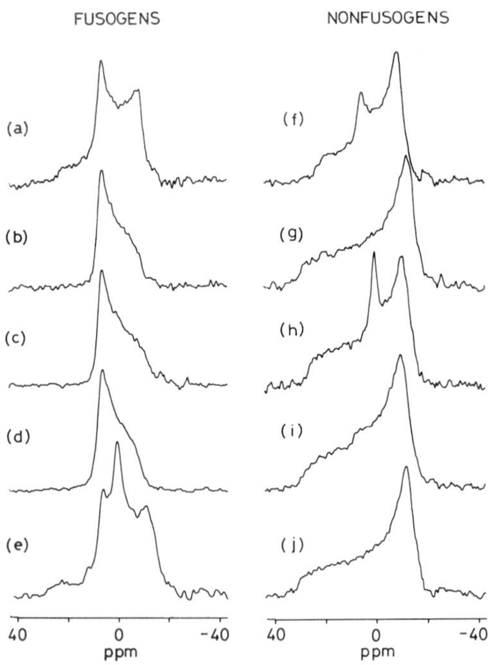

Figure 3. 81.0 MHz ^{31}P NMR spectra arising from reconstituted human erythrocyte lipids at 37°C in the presence of equimolar (with respect to phospholipid) fusogens (a–e) and non-fusogens (f–j). (a) Myristic acid, (b) palmitoleic acid, (c) retinol, (d) oleic acid, (e) glycerol monooleate, (f) palmitic acid, (g) triolein, (h) retinol palmitate, (i) stearic acid and (j) glycerol monostearate. Each sample consisted of 30 μmol of phospholipid, dispersed in 0.6 ml of buffer B (see Methods) and 10% ^2H_2O, 20 μmol of Ca^{2+} were added, following lipid dispersion. Accumulated free induction decays were obtained as described in the legend to Fig. 1.

Table 1. Polymorphic Phase Preferences of Liquid Crystalline, Unsaturated Lipids

Lipid	Phase Preferences: Physiological Conditions	Phase Preferences: Other conditions
Phosphatidylcholine	L	H_{II}, low hydration and high temp
Sphingomyelin	L	
Phosphatidylethanolamine	H_{II}	L, pH $\geq$8.5, low temp
Phosphatidylserine	L	H_{II}, pH$\leq$3.5
Phosphatidylglycerol	L	H_{II}, high temp, high salt conc.
Phosphatidylinositol	L	
Cardiolipin	L	H_{II}, divalent cations, pH$\leq$3 high salt
Phosphatidic acid	L	H_{II}, divalent cations, pH$\leq$3.5 high salt
Monoglucosyldiglyceride	H_{II}	
Diglucosyldiglyceride	L	
Monogalactosyldiglyceride	H_{II}	
Digalactosyldiglyceride	L	
Cerebroside	L	
Cerebroside sulfate	L	
Ganglioside	M	
Lysophosphatidylcholine	M	
Cholesterol		Induces H_{II} phase in mixed lipid system
Unsaturated fatty acids		Induce H_{II} phase

Note: L = Lamellar, H_{II} = Hexagonal, M = Micellar

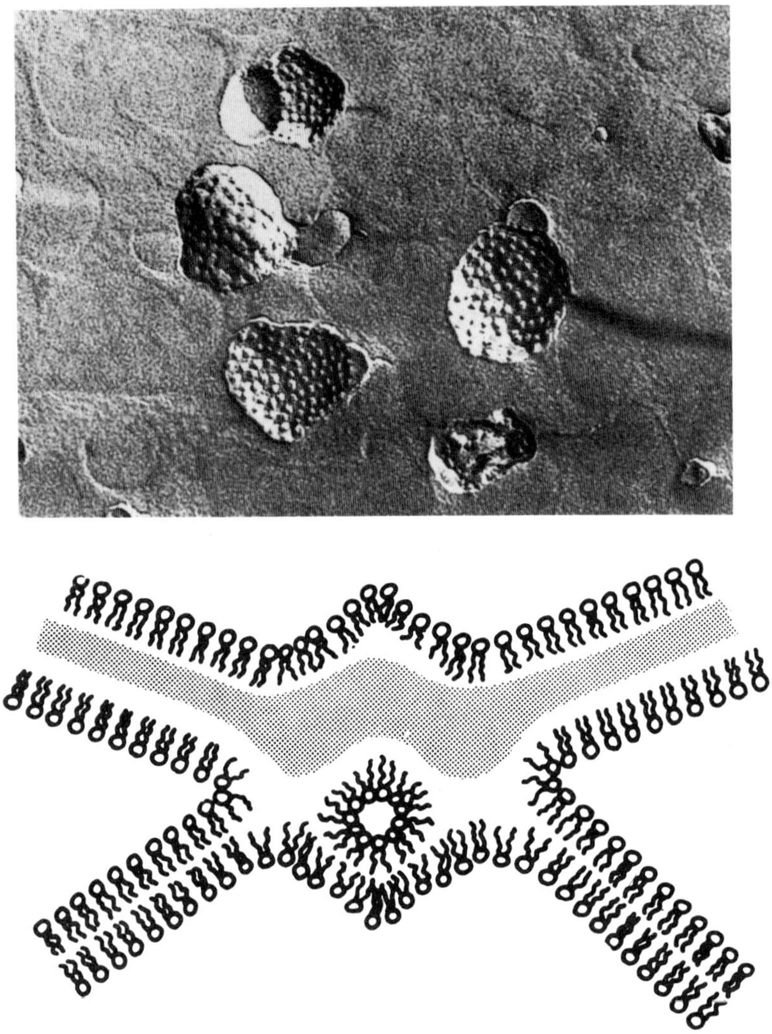

Figure 4. Freeze-fracture micrograph of lipidic particles induced by Ca^{2+} in a lipid system consisting of cardiolipin and soya phosphatidylethanolamine in the molar ratio of 1:4 (magnification X80,000). A model of the lipidic particle as an inverted micelle is depicted below the micrograph. The shaded area represents the fracture region.

inverted cylinders of the H_{II} phase. These inverted micelles are also attractive fusion intermediates, as they represent local, discrete departures from bilayer structure which are promoted by the same factors as promote bilayer to H_{II} transitions and which would not be expected to result in undue membrane disruption.

Fusion of Model and Biological Membranes

The results summarized above reveal a strong correlation between the ability of factors to promote membrane fusion and their ability to induce H_{II} phase or inverted micellar structure in MLV systems, as well as similarities between bilayer to non-bilayer transitions in model systems and membrane fusion events. This suggests that factors promoting H_{II} or inverted micellar lipid structure should promote fusion between model and biological membranes. Here we examine the validity of this prediction for unilamellar model vesicles, cell fusion induced by lipid-soluble fusogens and the fusion process involved in exocytosis.

As indicated in the previous section, a variety of MLV systems can be induced to undergo bilayer to H_{II} transitions by addition of Ca^{2+} or incubation at lower pH values. It would therefore be predicted that unilamellar vesicles with similar lipid compositions should first fuse to form larger vesicle systems before the large-scale aggregation which eventually gives rise to H_{II} phase formation. This prediction has been verified for small (sonicated) vesicle systems composed of PE-PS, PE-PA, PE-PG, PE-PI and PE-CL systems (Hope et al., 1983) as well as PC-CL systems (Verkleij et al., 1979) and an example for PE-PS systems is shown in Fig. 5. Similarly, it has been shown that incubation at pH <5 results in fusion of PE-PS and PE-PA vesicles (Hope et al., 1983). Fusion of large unilamellar vesicles (LUV's) composed of erythrocyte inner monolayer lipids (PC-PE-PS-cholesterol) in response to Ca^{2+} is also illustrated in Fig. 5.

An interesting feature of the Ca^{2+}-induced fusion of small unilamellar vesicles (SUV's) of PE-PS concerns the appearance of lipidic particles, often localized to regions corresponding to the fusion interface. Similar features are observed for other fusion systems (Verkleij, 1984; Hope et al., 1983) and it was originally proposed (Verkleij et al., 1979) that these particles correspond directly to inverted micellar fusion intermediates. However, further studies (Duzgunes et al., 1982) indicate that these lipidic particles are either formed subsequent to the actual fusion event or not observed at all, as in the case for the fusion of 'inner monolayer' LUV's (Fig. 5). It has been suggested (Siegel, 1984) that the actual fusion intermediates have a very short lifetime, which could preclude detection by freeze-fracture techniques.

With regard to lipid-soluble fusogens, the theory that fusion proceeds via a non-bilayer 'inverted' intermediate would predict that such agents should promote fusion at membrane concentrations sufficient to promote H_{II} phase formation. As previously indicated a variety of lipid-soluble fusogens do induce H_{II} phase formation in MLV systems composed of total erythrocyte lipids (as well as erythrocyte 'ghost' membranes) (Hope & Cullis, 1981), whereas chemically related non-fusogens do not. Further, the membrane concentrations of oleic acid (Cullis & Hope, 1978) and glycerol monooleate (Hope & Cullis, 1978) required to induce detectable H_{II} organization (approximately equimolar with respect to phospholipid) also correspond closely to the membrane concentrations required to induce erythrocyte fusion (Hope & Cullis, 1981).

The possible role of non-bilayer intermediates in fusion events involved in exocytosis has been investigated (Nayar et al., 1982) for the chromaffin granule system, which undergoes Ca^{2+}-dependent fusion with

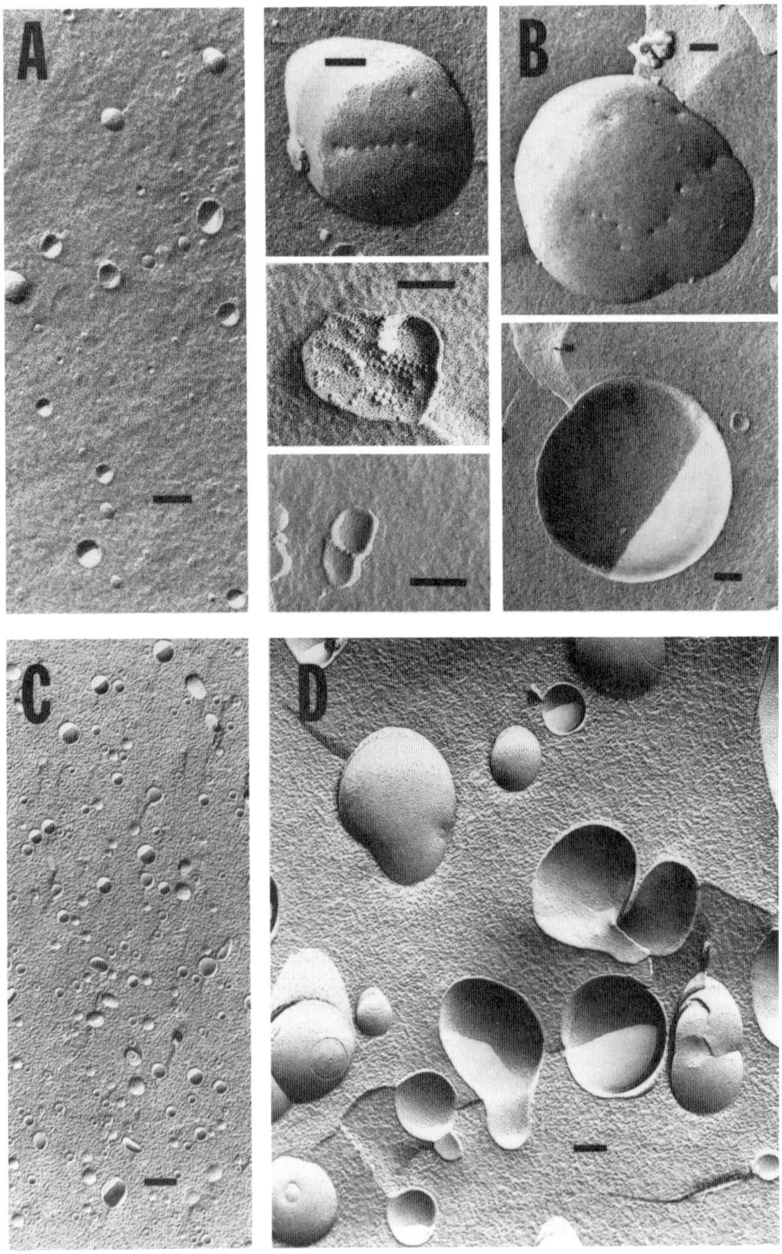

Figure 5. Freeze fracture micrographs of sonicated vesicle composed of PE-PS (A) were dialysed against Ca^{2+} to induce fusion (B). In (C) sonicated vesicles of extracted erythrocyte lipids, reconstituted in proportions corresponding to the inner monolayer lipid composition (PC/PE/PS/SPM cholesterol in the molar ratio 0.15:0.47:0.28:0.10:1), were dialysed against Ca^{2+} to induce fusion (D).

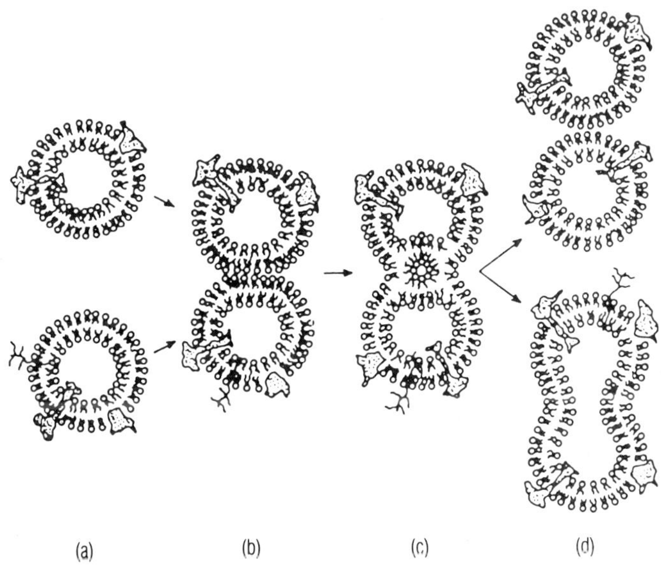

Figure 6. Membrane fusion proceeding via an inverted micellar intermediate.

the plasma membrane of chromaffin cells (Trifaro, 1977). This fusion event must initially involve granule fusion with the plasma membrane inner monolayer. By analogy with the erythrocyte membrane, it is likely that the inner monolayer is composed primarily of PE and PS. It may therefore be suggested that the presence of Ca^{2+} could trigger fusion by enhancing the proclivity of the inner monolayer for inverted micellar or H_{II} phase structure, resulting in fusion with closely apposed granule membranes. If this is the case, Ca^{2+}-stimulated fusion between granules and PE-PS liposomes should proceed readily. This has been observed for PE-PS SUV's incubated with chromaffin granules (Nayar et al., 1982), where levels of Ca^{2+} well below those required to induce granule-granule fusion induce massive PE-PS-granule fusion, with associated release of contents.

These and related (Cullis et al., 1985) observations provide strong evidence for a role of an inverted non-bilayer lipid organization in membrane fusion. The resulting fusion model we have proposed (Cullis & Hope, 1978) is illustrated in Fig. 6, where two membrane-bound systems are brought into close apposition (possibly a protein-mediated event), and an interbilayer inverted micelle (or short inverted cylinder) formed. Reversion of the intrabilayer inverted structure to bilayer structure could then either result in reformation of separate vesicles or a completed fusion event. It may be noted that the situation of Fig. 6(c), representing partially fused membranes, can apparently be a relatively stable situation, and may correspond to situations of arrested fusion noted for tight junctions (Kachar & Reese, 1982).

LIPID ASYMMETRY

Recently, our interest in fusion phenomena has been centred around the properties of lipid systems which exhibit asymmetric transbilayer distributions of lipids such as are observed in biological membranes. Such asymmetry poses interesting problems for fusion. For example, in the erythrocyte membrane and, apparently, other plasma membranes (Op den Kamp, 1979) the predominance of PC and sphingomyelin in the outer monolayer and PE and PS in the inner monolayer would suggest that the inner monolayer is more "fusogenic" than the outer. The question then arises as to whether the presence of bilayer lipids in one monolayer can influence the ability of the other to undergo fusion. While a system exhibiting asymmetric transbilayer distributions of phospholipids has not yet been achieved, LUV systems exhibiting asymmetric distributions of lipid which exhibit weak acid (e.g. fatty acids) and weak base (e.g. stearylamine) characteristics have been achieved (Hope and Cullis, 1987). These systems arose from initial speculations that lipid asymmetry may be related to ion gradients present across membranes, giving rise to membrane potential ($\Delta\psi$) and pH gradients (ΔpH). Here we summarize the methods involved in generating asymmetry due to transmembrane pH gradients and the characteristics of the resulting systems.

Influence of ΔpH (acidic inside) on the Transbilayer Distribution of Amino Lipids

Experiments were aimed at examining the influence of pH gradients on the transbilayer distributions of the amino-containing lipids, stearylamine and sphingosine. The reasoning behind these experiments is based on the abilities of water-soluble weak bases such as methylamine to permeate vesicle membranes (in the neutral form) to achieve transmembrane (aqueous) concentrations obeying the Henderson-Hasselbach relation.

$$\frac{[AH^+]_i}{[AH^+]_o} = \frac{[H^+]_i}{[H^+]_o} \qquad (1)$$

where AH^+ refers to the protonated amine and the subscripts i and o refer to the interior and exterior of the vesicle, respectively. In the case of amino lipids such as stearylamine and sphingosine, their localization to the membrane should result in similar redistributions where the transbilayer location of the protonated amine reflects the interior and exterior proton concentrations at the membrane interfaces.

These predictions were tested employing DOPC LUVs containing 5 mol% stearylamine. As shown in Fig. 7A when no transmembrane ΔpH is present (pH = 8.5 inside and out), the TNBS labeling of stearylamine proceeded to completion as assayed spectrophotometrically. This is consistent with a rapid redistribution of inner monolayer stearylamine to the outer monolayer as the outer monolayer stearylamine is depleted on reaction with TNBS. However, as shown in Fig. 7B, when a pH gradient is present (interior pH 5.0, exterior pH 8.5) no labeling of stearylamine can be detected. The subsequent addition of detergent to solubilize the vesicles resulted in complete labeling within 20 s. These results are clearly consistent with a localization of stearylamine to the inner monolayer in response to the pH gradient (interior acidic).

Influence of ΔpH (inside basic) on the Transbilayer Distribution of Fatty Acids

The ability of pH gradients to markedly influence the transmembrane distributions of these amine-containing lipids would imply that the transmembrane distributions of lipids which are weak acids, such as fatty acids, should also be strongly dependent on transmembrane pH gradients. Assuming that the neutral (protonated) form can permeate the membrane, this will result in transmembrane gradients according to

$$\frac{[RCOO^-]_i}{[RCOO^-]_o} = \frac{[H^+]_o}{[H^+]_i} \qquad (2)$$

where $RCOO^-$ refers to the unprotonated fatty acid. In vesicles exhibiting pH gradients where the interior is basic, fatty acids should therefore move to the interior monolayer. In order to determine whether this is the case, we employed an assay based on the well-documented ability of fatty acid-free BSA to deplete membranes of free fatty acid.

As shown in Fig. 8A for DOPC vesicles containing 10 mol% oleic acid, only a small fraction of the oleic acid can be removed by BSA when a transmembrane pH gradient (interior pH = 10.0, exterior pH = 7.0) is present. Alternatively, when the interior and exterior pH values are the same (Fig. 8B), 95% of the oleic acid elutes with BSA rather than the LUVs. This is consistent with a localization of oleic acid to the inner monolayer in the presence of a proton gradient (interior basic). Similar results were observed for stearic acid (results not shown).

These systems clearly offer attractive models for the influence of lipid asymmetry on fusion. For example, preliminary studies show that oleic acid can be localized to the outer monolayer when the vesicle interior is acidic. In vesicles with suitable lipid compositions the presence of this fusogen in the outer monolayer would be expected to promote fusion. Alternatively, systems with basic interiors would be

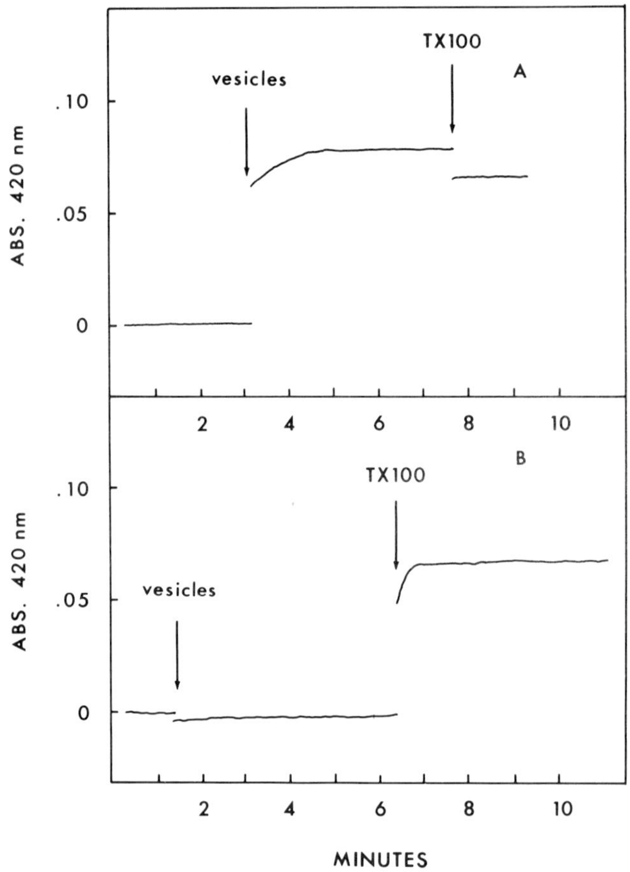

Figure 7. Influence of proton gradients (inside acidic) on the availability of stearylamine to TNBS in DOPC vesicles. (A) DOPE vesicles containing 5 mol% stearylamine. These LUVs do not exhibit a proton gradient (pH = 8.5) at the vesicle interior and exterior. Absorbance at 420 nm is monitored throughout the incubation. Following the addition of vesicles, Triton X-100 (TX100) is employed to solubilize the membranes and expose all available amines. A slight decrease in absorbance is observed due to dilution and solubilization effects. (B) DOPC LUVs exhibit a pH gradient (pH 5.0 inside and pH 8.5 outside).

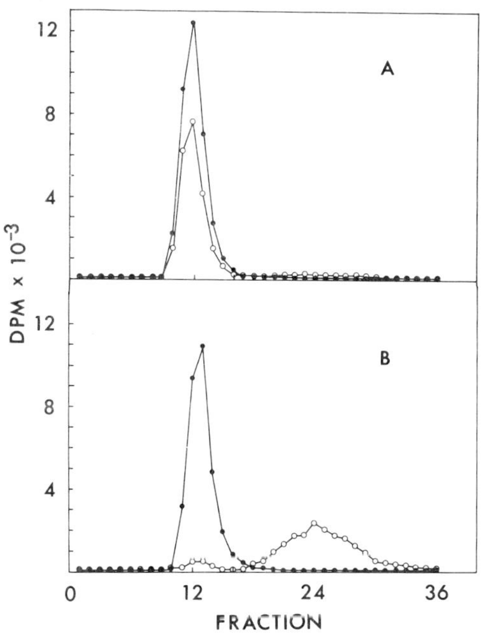

Figure 8. The effect of a membrane proton gradient on the ability of BSA to extract [^{14}C]oleic acid from a gel filtration column. (A) Vesicles contain 10 mol% oleic acid and exhibit an internal pH of 10.0. The external buffer is at pH 7.0. The peak represents vesicles (●) and oleic acid (o). BSA elutes between fractions 20 and 30. (B) Elution profile of vesicles with no proton gradient (pH 7.0 inside and outside). The majority of the oleic acid (o) now elutes with BSA, whereas the position of the vesicle peak (●) is unchanged.

rendered less susceptible to fusion events. These hypotheses are currently being investigated.

CONCLUDING REMARKS

In this article we have summarized evidence supporting a role of non-bilayer lipid structures in membrane fusion, and outlined initial studies leading to vesicle systems with asymmetric transbilayer distributions of lipid which more accurately reflect the biological situation. With regard to polymorphism, it may be concluded that lipid which can adopt non-bilayer structures are present in high proportions in membranes, that factors which initiate fusion in vivo induce non-bilayer (H_{II} phase) structure in appropriate model membrane systems, that the structures of H_{II} phase and inverted micellar organizations are consistent with an intermediary role in fusion and that the inverted micelle would provide an intermediate structure which may not give rise to induce membrane disruption. Some theoretical support for such a role of non-bilayer lipid structures is now available (see Siegel, D; this volume).

ACKNOWLEDGEMENTS

This research was supported by the Medical Research Council (MRC) of Canada. P.R.C. is an MRC Scientist.

REFERENCES

Ahkong, Q.F., Pampion, W. and Lucy, J.A., 1973, Biochem. J. 136: 147-155.
Bally, M.B., Tilcock, C.P.S., Hope, P.R. and Cullis, P.R., 1983, Can. J. Biochem., 61: 346-352.
Cullis, P.R. and Hope, M.J., 1978, Nature (London), 271: 672-675.
Cullis, P.R. and de Kruijff, B., 1979, Biochim. Biophys. Acta, 559: 399-420.
Cullis, P.R., de Kruijff, B., Hope, M.J., Nayar, R. and Schmid, S.L., 1980, Can. J. Biochem., 58: 1091-1100.
Cullis, P.R., Hope, M.J., de Kruijff, B., Verkleij, A.J. and Tilcock, C.P.S., 1985, in: "Phospholipids and Cellular Regulation," J.F. Kuo, ed., pp. 1-62, CRC Press, Boca Raton, FL.
Duzgunes, N., Bearer, E. and Papahadjopoulos, D., 1982, Biophys. J., 37: 25a.
Evered, D. and Whelan, J., eds., 1984, "Ciba Foundation Symposium on Cell Fusion," Pitman, London.
van Echteld, C.J.A., de Kruijff, B., Verkleij, A.J., Leunissen-Bijvelt, J. and de Gier, J., 1982, Biochim. Biophys. Acta, 692: 126-135.
Farren, S.B., Hope, M.J. and Cullis, P.R., 1983, Biochem. Biophys. Res. Commun., 111: 675-682.
Hope, M.J. and Cullis, P.R., 1979, FEBS Lett., 107: 323-326.
Hope, M.J. and Cullis, P.R., 1980, Biochem. Biophys. Res. Commun., 92: 846-852.
Hope, M.J. and Cullis, P.R., 1981, Biochim. Biophys. Acta, 640: 82-90.
Hope, M.J. and Cullis, P.R., 1987, J. Biol. Chem. 262: 4360-4366.
Hope, M.J., Walker, D.C. and Cullis, P.R., 1982, Biochem. Biophys. Res. Commun., 110: 15-22.
Kachar, B. and Reese, T.S., 1982, Nature (London), 296: 464-466.
de Kruijff, B. and Cullis, P.R., 1980, Biochim. Biophys. Acta, 602: 477-490.
Nayar, R., Hope, M.J. and Cullis, P.R., 1982, Biochemistry, 21: 4583-4590.
Siegel, D.P., 1984, Biophys. J., 45: 399-420.

Tilcock, C.P.S., Bally, M.B., Farren, S.B., Cullis, P.R. and Gruner, S.M., 1984, Biochemistry, 23: 2696-2703.

Trifaro, J.M., 1977, Annu. Rev. Pharmacol. Toxicol., 17: 27-54.

Verkleij, A.J., 1984, Biochim. Biophys. Acta, 779: 43-82.

Verkleij, A.J. Mombers, C., Gerritsen, W.J., Leunissen-Bijvelt, J. and Cullis, P.R., 1979, Biochim. Biophys. Acta, 555: 358-362.

MEMBRANE FUSION VIA INTERMEDIATES IN L_α/H_{II} PHASE TRANSITIONS

D.P. Siegel[+], H. Ellens[*], and J. Bentz[#]

[+]Procter & Gamble Co.
P.O. Box 39175
Cincinnati, Ohio 45247

Depts. of Pharmacology[*] and Pharmaceutical Chemistry[#]
University of California
San Francisco, CA 94143

INTRODUCTION

Many researchers have suggested that membrane fusion can occur via structures that form during bilayer-to-non-bilayer phase transitions (for reviews, see [1-3]). These structures, detected via electron microscopy and ^{31}P-NMR, were originally referred to as "lipidic particles" [4]. They seem to form between apposed lipid bilayers (generally in the "liquid crystalline", or L_α phase) when conditions are such that the lipid is undergoing a phase transition to the inverted hexagonal (H_{II}) phase, or is on the verge of doing so [1-3]. Such structures make the interfaces of apposed bilayers continuous, and this makes them attractive as postulated mediators of membrane fusion.

However, we must be cautious in making this hypothesis. These structures appear to be intermediates in the formation of the H_{II} phase [5-7], which can not encapsulate any significant volume of aqueous solution. It is not obvious that formation of such structures between two apposed liposomes can produce a single liposome without leakage of the encapsulated solution (which is the biologically-relevant definition of liposome fusion). Moreover, in some systems examined at temperatures near the equilibrium L_α/H_{II} transition temperature (T_H), Ellens, Bentz and co-workers [9-12] found that initial encounters of liposomes result in rapid leakage, not in fusion, although fusion is observed in other systems below T_H [8,10,12].

Here, we describe a model of liposome-liposome interactions near the L_α/H_{II} phase boundary based on recent experimental [8-15] and theoretical [7,16,17] work. This work indicates that fusion can occur via structures that form during L_α/H_{II} phase transitions, but that fusion occurs at observable rates in only a certain sub-set of systems with such phase transitions, and then only within a certain temperature range. Formation of L_α/H_{II} transition intermediates between aggregated liposomes can result in lipid exchange, contents leakage, or fusion. The relative rates of these processes are predicted as a function of temperature, liposome radius, and of a parameter related to the spontaneous radius of curvature of the lipid system [18,19]. This parameter can be measured via X-ray diffraction experiments [20,21].

The predictions of this model agree with observations of several lipid systems. It is based on a kinetic model of the L_α/H_{II} transition mechanism [7,16,17]. The proposed mechanism is compatible with the available morphological, NMR, DSC, and X-ray observations of systems undergoing this phase transition, and the estimates of transition times are close to observed values [22-24]. This theory predicts that a subset of L_α/H_{II} transitions should be particularly hysteretic and form so-called "isotropic" or inverted cubic phases of bicontinuous structure [17]. Such phases and the intermediates from which they form have characteristic ^{31}P-NMR resonances and morphologies in electron micrographs. The same intermediate structures that generate these phases appear to mediate fusion between apposed liposomes. Hence, evolution of such phases in bulk lipid samples should be associated with at least some fusion in unilamellar dispersions of the same lipid incubated under the same conditions. This seems to be true [12,13].

The striking features of our model are: first, that one can associate a specific defect structure with the fusion process; and second, that one can relate fusion kinetics to the phase behavior and phase transition hysteresis of lipid-water systems. In particular, it is the <u>hysteresis</u> in the L_α/H_{II} transition, and not the occurrence of the transition per se, that can lead to fusion via this mechanism. Although leakage after aggregation should generally be observed along with fusion via this mechanism, there are circumstances (moderate R_o systems, 0.1 μm-diameter vesicles; see below) in which the great majority of initial liposome-liposome encounters can result in fusion.

First we will outline the model of the L_α/H_{II} phase transition. We will then show how it can be used to predict the outcomes of initial liposome-liposome encounters when the lipids of the system are close to the L_α/H_{II} phase boundary, and detail the model's predictions. Then we will compare those predictions with some recent results. In conclusion we will describe some of the model's implications concerning biomembrane fusion and the design of membrane fusion experiments.

MECHANISM AND KINETICS OF L_α/H_{II} PHASE TRANSITIONS

The L_α/H_{II} phase transition is an interlamellar phenomenon: more than one bilayer must be apposed in order for the transition to occur [6,7,10]. The "lipidic particle" morphology, which is the first nonbilayer morphology observed as one approaches the L_α/H_{II} phase boundary along composition or temperature axes, is not observed in isolated unilamellar liposomes [25,26], while such morphology is ubiquitous in multilamellar preparations (references in [6,7]). There are a number of reasons to believe that these particles are the first structures to form during this phase transition [7], and that they are either inverted micelle-containing structures like those proposed previously by many authors (reviewed in [1-3,6,7]), or structures that form from them (ILAs; see below). We refer to the former structure as an inverted micellar intermediate, or IMI (Fig. 1, top).

The IMI is cylindrically symmetric about the dashed vertical axis in Fig. 1: it consists of a spherical inverted micelle enclosed in a semi-toroidal (hourglass-shaped) external monolayer. The chemical potential of lipid in an IMI is very nearly the same as that of lipid in the H_{II} phase under the same conditions. Because most L_α/H_{II} transitions have very low enthalpies, the chemical potential differences between lipids in IMIs and the L_α and H_{II} phases are very small (<< kT) even at temperatures as much as tens of degrees below the equilibrium transition temperature, T_H [7]. Moreover, the formation rate of IMIs in the L_α phase is estimated [7] to be rapid across this temperature range. Therefore, IMIs should begin forming in the L_α phase even when the H_{II} phase is not the equilibrium phase: the IMIs are metastable in this range, and might be regarded as a type of sub-critical fluctuation.

IMIs that form between apposed bilayers have three possible fates [7,16]. First, they are in dynamic equilibrium with the planar bilayers, and can revert to that geometry. The half-life with respect to this process is estimated to be on the order of 1 millisecond or less, depending on the temperature [7]. Second, when the H_{II} phase is the equilibrium phase (i.e., the temperature T is $\geq T_H$), IMIs existing between the same two apposed bilayer interfaces can aggregate into H_{II} phase precursors (Fig. 1, middle right). This process and its sequelae have been observed by many authors, and the details of the probable H_{II} formation mechanism are described in [7]. For our purposes, the most important attributes of this process are that it proceeds via aggregation of IMIs and that it can be very rapid (one second or faster; [7]) in systems with facile L_α/H_{II} transitions. The third fate is conversion of individual IMIs into structures that we call interlamellar attachments (ILAs, [7,16]; Fig. 1, middle left). ILAs can mediate membrane fusion [16] and, under some circumstances, assemble into isotropic and inverted cubic phases [17] (Fig. 1, bottom left).

The rate and outcome of the L_α/H_{II} transition depends on which of these latter two processes is faster; aggregation of pairs of IMIs into H_{II} precursors, or ILA formation from individual IMIs. In the vicinity of the phase transition, the number of IMIs per unit area of apposed bilayers rapidly reaches a steady-state value, n_I^o [7]. H_{II} precursor formation and ILA formation compete for this pool of IMIs. The kinetic model of this competition is represented in Fig. 2. The rate of H_{II} precursor formation is given by

$$k_I(n_I^o)^2, \qquad \text{(Eq. 1)}$$

where k_I is a second-order aggregation rate constant $\approx 10^{-11}$ cm^2/s [7]. The rate of ILA formation is

$$P_{ILA}n_I^o, \qquad \text{(Eq. 2)}$$

where P_{ILA} is a first-order rate constant [16]. Obviously, the relative rates of these two processes determine the predominant structures that form from IMIs:

$$\text{if } P_{ILA} < k_I n_I^o, \text{ then } H_{II} \text{ phase formation predominates;} \qquad \text{(Eq. 3)}$$

$$\text{if } P_{ILA} \gtrsim k_I n_I^o, \text{ then ILA formation predominates.} \qquad \text{(Eq. 4)}$$

In particular, if P_{ILA} is $<< k_I n_I^o$, hardly any ILAs will form.

P_{ILA} is extremely sensitive to an equilibrium structural property of the lipid known as Z [16,17]. Z is the ratio of the area per lipid head group in the L_α phase divided by the corresponding area in the H_{II} phase at equilibrium in excess water. P_{ILA} can change by two orders of magnitude when Z changes by 10% [17]. It is largely the value of Z that determines which of Eqs. 3 & 4 applies in a given system [16,17].

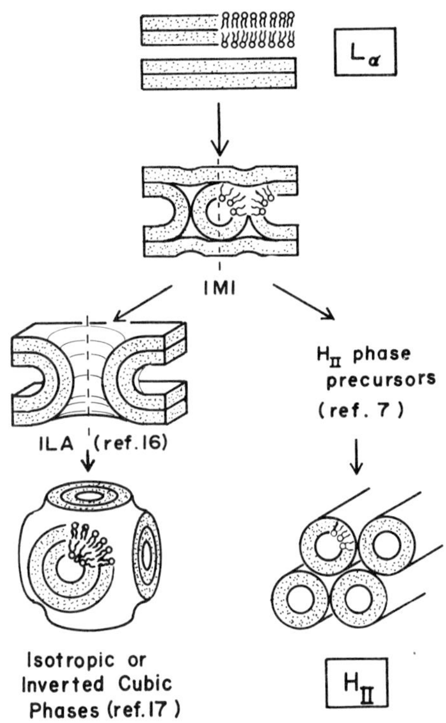

Figure 1. Formation and possible transformations of IMIs. The relative rates of H_{II} phase assembly (middle right) and ILA formation (middle left) are controlled by the spontaneous radius of curvature of the specific system (Z or R_o; see text) and the value of $k_I n_I^o$ (Eqs. 3,4).

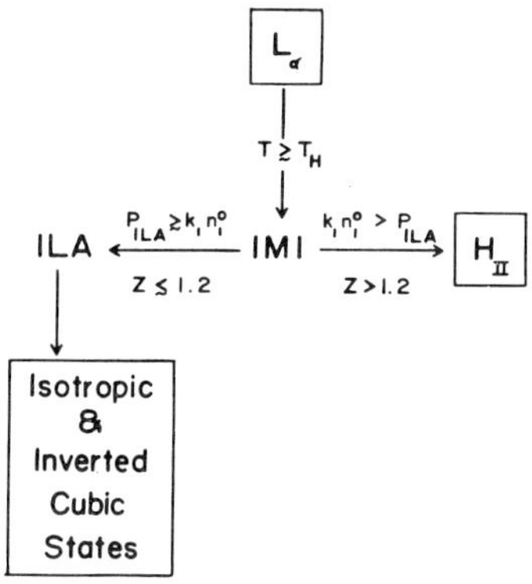

Figure 2. Summary of the kinetics of H_{II}, isotropic and inverted cubic phase formation. In some cases, what have been called isotropic or inverted cubic phases may actually be metastable states [17]; hence the usage "states" here. See text for explanation of symbols. Reproduced from [17], by permission.

The value of Z reflects the equilibrium curvature of the lipid-water interfaces. This parameter is important because a large lateral concentration fluctuation must occur in the lipid monolayers around the IMI in order for an ILA to form [16]. The frequency of the required fluctuation is strongly anticorrelated to its magnitude, and this is usually the rate-limiting step in ILA formation. The magnitude is determined by the number of lipids per unit area on the curved interfaces, which is determined, in turn, by Z. Z can be measured by X-ray diffraction studies of lipid-water systems as a function of increasing hydration, as in [27,28].

Large values of Z (in the range 1.4-1.8) correspond to systems with small equilibrium radii of curvature. In such systems, P_{ILA} is << $k_I n_I^o$ (Eq. 3), and hardly any ILAs form. In general, systems with this range of Z have smaller-diameter H_{II} tubes at equilibrium and comparatively rapid, non-hysteretic transitions. Egg PE is an example: using data from [28,29], Z for this system appears to be 1.8. As one would expect on this basis, no inverted cubic phases or pronounced transition hysteresis have been reported for this lipid-water system.

Smaller values of Z (ca. 1.2 or less) correspond to systems with larger radii. In such systems, P_{ILA} is $\approx k_I n_I^o$ (Eq. 4), and many ILAs form. In general, systems with small values of Z have larger-diameter H_{II} tubes at equilibrium and sluggish, hysteretic L_α/H_{II} phase transitions. The hysteresis is on account of the ILAs themselves. ILAs are kinetically stable, and represent kinetic traps for lipid: ILAs can neither revert to the L_α phase or form the H_{II} phase at significant rates, especially at $T < T_H$ [17]. An array of bilayers with many ILAs should have all the features of the "isotropic phases" often described in such systems (see [17] for references). In some cases, the ILAs can form inverted cubic phases of bicontinuous geometry [17]: it may be very hard to establish if such phases are true phases (i.e., chemical potentials very slightly less than that of the L_α & H_{II} phases in the temperature region just below T_H) or merely metastable structures. These issues are under investigation [22,27]. Almost all the systems in which isotropic or inverted cubic phases have been reported are in fact systems which one would expect to have small values of Z: mixtures of H_{II}-phase forming lipids with L_α phase lipids (reviewed in [1,2,17,30]), or series of systems with similar acyl chains but head groups of increasing size [27,31,32].

The dependence of P_{ILA} on the equilibrium curvature of the system can also be described using the spontaneous radius of curvature parameter of Gruner and co-workers [18-21]. They have developed a powerful description of the relative stability of bilayer and non-bilayer phases. A key parameter in this treatment is the spontaneous radius of curvature of the lipid-water interface in the lipid system, R_o. R_o is the radius the system would adopt when another lipid packing constraint is relaxed. It determines the zero point of the curvature elastic energy of the monolayer, and is a major contribution to the relative chemical potential difference between the bilayer and non-bilayer phases. Z [16,17] is a property closely-related to R_o. In general, systems with small values of R_o correspond to systems with large values of Z, and systems with "moderate" values of R_o (e.g., mono-methylated dioleoylphosphatidylethanolamine; [27]) have small values of Z ($\lesssim$ 1.2).

The kinetic model described here was derived for systems with thermotropic L_α/H_{II} transitions [7,16,17]. The same description should be qualitatively applicable to ionotropic systems, where counterion binding to ionized lipid triggers the L_α/H_{II} transition. However, in such cases we expect that there will be considerably more hysteresis in interconversion of the different types of intermediates. In particular, we would expect fusion to occur in a higher proportion of liposome-liposome encounters in such systems [3,16].

DESCRIPTION OF LIPOSOME-LIPOSOME INTERACTIONS

The model described in Figs. 1 & 2 can be used to predict the relative rates of different interactions of aggregated liposomes via L_α/H_{II} transition intermediates. The model applies only to systems under conditions such that the equilibrium phase of the lipids when the liposomes aggregate is the H_{II} phase, or such that the lipids are close to that phase boundary (e.g., temperature within 20° K or less of T_H). The predicted IMI/ILA-mediated interactions are diagrammed in Fig. 3 and summarized as a function of Z, T, and liposome radius in Table I.

IMI-induced Lipid Mixing

IMIs can form between the apposed bilayers of the two liposomes, just as they can form between adjacent bilayers in multilamellar aggregates. As discussed above and in [7,16], IMI should form in significant numbers as much as 20° K below T_H. Note that an IMI makes the outer leaflets of the two apposed bilayers continuous for the lifetime of the IMI. This permits lipid molecules on the surfaces of the two liposomes to mix (Fig. 3, top). It can be shown [16] that the average lifetime of an IMI is on the same order of magnitude as the random-walk diffusion time of a lipid molecule around the surface of a 0.1 μm-radius liposome. Hence, formation of a single IMI between two such unilamellar liposomes may result in extensive mixing of 50% of the total lipid.

The rate of IMI-mediated lipid mixing between apposed liposomes below T_H is controlled by the rate of IMI formation. The rate of IMI formation is a sharply increasing function of increasing temperature for $T \lesssim T_H$ [16]. The lipid mixing can occur within much less than 1 second for liposomes > 0.1 μm in radius when $T \approx T_H$, but this process may be much slower at lower temperatures. It is not clear if IMIs can mediate interactions between very small liposomes (ca. 0.01 μm radius) since IMI formation requires close apposition of bilayer patches about 0.005 μm in diameter, and it is not clear whether such small liposomes would be deformable enough to do this. The high curvature elastic strain in such systems might make them more susceptible to leakage upon aggregation due to adhesion energy-induced stress [33].

IMI- & H_{II} Phase-Induced Liposomes Leakage

If $T \geq T_H$, the IMIs that form between two apposed liposomes will aggregate and form H_{II} phase in the contact areas. This will induce stresses in the two apposed membranes for two reasons. First, the favorable chemical potential gradient towards H_{II} formation will drive close apposition of as much of the surface area of the two liposomes as possible. This will deform the membranes (Fig. 3, middle left). Since the membranes can stretch to only a very limited extent before rupture [e.g., 34], this develops a tension in the membranes that will eventually become large enough to rupture them [16]. This will leak their aqueous contents to the external medium (Fig. 3, lower left). Another stress is due to the difference in the number of lipids per unit area of apposed outer monolayers before and after formation of H_{II} tubes. This difference increases markedly when H_{II} tubes form, making the monolayers susceptible to rupture via osmotic or other stresses on the membranes [35], or by occurrence of lateral concentration fluctuations in these monolayers.

One or more IMIs by themselves may represent enough of a defect in the membranes to nucleate membrane rupture and leakage in response to infrequent lateral lipid concentration fluctuations or other stresses. The exact mechanism of rupture may be hard to determine, since the deformability of membranes can be a complicated function of liposome size, rheological coefficients, membrane-membrane adhesion energies, and transmembrane osmotic gradient [35].

Membrane Fusion via ILAs

The value of Z for a given system determines to what extent ILAs form between apposed liposomes. If Z is $\lesssim 1.2$, P_{ILA} is large enough for IMIs formed between apposed liposomes to transform into ILAs. ILA formation makes both the membranes of the two original liposomes continuous while also opening a channel between their two aqueous compartments (Fig. 3, middle right). This occurs without leakage of the contents at any point, and thus corresponds to membrane fusion. If Z is $\gtrsim 1.2$, the H_{II} precusor formation rate will become larger than the ILA formation rate, and leakage processes will occur faster than fusion.

Fusion will be fastest when H_{II} phase formation can not occur; i.e., at $T < T_H$. The rate of ILA-mediated fusion is determined by the ILA formation rate, which is proportional to the number of IMIs per unit area of apposed bilayers under the experimental conditions, $n_I(T)$. $n_I(T)$ is an exponentially increasing function of temperature near T_H [16]. Therefore, we expect that fusion should be most rapid at temperatures just below T_H, where IMIs cannot be consumed via H_{II} phase formation, yet the temperature is high enough for IMIs to be fairly numerous. We expect the fusion rate to drop off rapidly at $T \gtrsim T_H$, because H_{II}

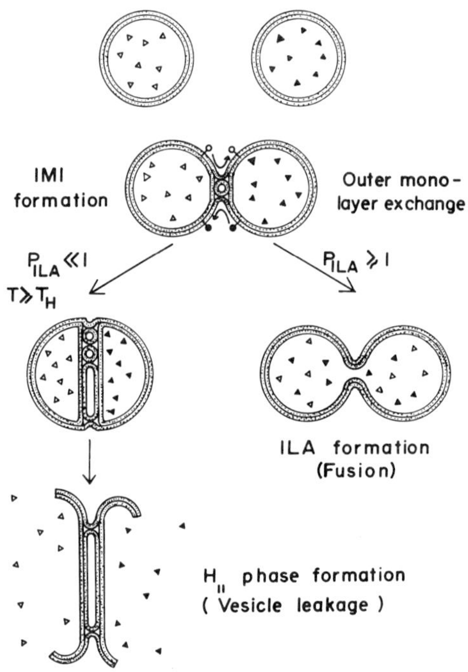

Figure 3. Liposome-liposome interactions mediated by L_α/H_{II} transition intermediates. Reproduced from [16], by copyright permission of the Biophysical Society.

TABLE I. PREDICTIONS OF L_{α}/H_{II} PHASE TRANSITION MODEL FOR LIPOSOME-LIPOSOME INTERACTIONS

Temperature	Z (R_o)	Interactions
$T_H-20 \leq T < T_H$	Any value of Z, ("small or moderate" R_o)	Mixing of lipids on the outer monolayers of apposed liposomes.
	$Z \lesssim 1.2$ ("moderate" R_o;[27])	Fusion. Rate increases with temperature to a maximum just below T_H.
$T \geq T_H$	$Z > 1.2$ ("small" R_o)	Lipid mixing and leakage after aggregation. Lipids mix completely due to destroyed sidedness of ruptured membranes. Rate of leakage increases rapidly with increasing temperature.
	$Z \lesssim 1.2$ ("moderate" R_o)	Fusion in some liposome-liposome encounters, leakage in others; ratio determined by specific values of Z, T, and liposome size.

Special Cases

(a) Liposome radius $\lesssim 0.1\ \mu m$

$T > T_H$	$Z > 1.2$ "small" R_o	Slow fusion in initial encounters until the av. radius is $> 0.1\ \mu m$, then leakage via H_{II} phase formation.

(b) Liposome radius $\approx 0.01\ \mu m$

Liposomes are probably too small for IMI formation: interactions probably occur via other mechanisms

---

NOTES

Fusion is defined as the simultaneous mixing of lipids and aqueous contents.

The critical value of Z for rapid fusion, 1.2, is only a theoretical estimate [16]. It corresponds to the value that makes $P_{ILA} > k_I n_I^o$ (Eq. 4), which should be on the order of 1 per IMI per second.

See Gruner et al. [27] for a discussion of the behavior of systems as a function of R_o.

phase formation will begin. This will rupture liposomes before they can fuse. Moreover, the rate of H_{II} phase precursor formation (Eq. 1) is proportional to $(n_I^o)^2$, whereas the ILA formation rate is only linear in n_I^o. Since n_I^o grows exponentially with increasing temperature, we expect the H_{II}-induced leakage rate to grow much faster than the fusion rate at $T > T_H$. There is also reason to expect P_{ILA} to decrease rapidly with increasing temperature [17].

Finally, we note a peculiar liposome size effect. Near T_H, n_I^o is estimated to be of the order of 10^{10}-10^{11}/cm^2 [7]. The contact area between two 0.1 μm-diameter liposomes will be of order 10^{-10} cm^2. For liposomes about this size or smaller, the steady-state number of IMIs existing in the contact area between two liposomes at any given time may be one or less. In such cases, according to the transition mechanism in [7,16], no H_{II} phase formation can occur. Therefore, we would expect to observe only lipid mixing and, if Z is in the appropriate range, fusion until the lipsomes grew large enough to appose more of their areas. At that point, rapid leakage via H_{II} phase formation would be possible.

Ellens et al. [12] proposed a very similar scheme to the one detailed above, based on their experimental study of the rates of the different interactions in mono-methylated dioleoylphosphatidylethanolamine systems. This scheme is reproduced in Fig. 4. T_I refers to the temperature range below T_H in which isotropic ^{31}P-NMR resonances are observed: we think this corresponds to the temperature range in which ILAs can form in significant numbers. At temperatures below T_I, whatever (slow) fusion there is probably occurs via non-H_{II} related mechanisms. At $T \geq T_I$, fusion via L_α/H_{II} transition intermediates begins. In [12], fusion was assumed to occur as a result of IMIs remaining isolated for sufficient lengths of time for them to rearrange into ILA-like structures, due to low values of n_I^o. Variation in n_I^o between systems was postulated to determine which systems exhibited fusion. This is similar to the present model, which shows that n_I^o, as well as P_{ILA}, controls the relative proportion of ILA and H_{II} phase production (Eqs. 3 & 4). n_I^o is sensitive in indirect ways to Z, and would be expected to decrease in systems with decreasing values of Z [7]. Hence, the model in [12] and the present model yield qualitatively similar predictions. However, theory [16] suggests that the extraordinary sensitivity of P_{ILA} to Z plays as large or a larger role in determining which lipid systems will exhibit fusion. At $T \geq T_H$, the model in Fig. 4 also predicts leakage of liposomes via H_{II} phase formation, as in the present model.

Liposome suspensions that exhibited fusion at $T \approx T_I$ were observed to undergo a curious "collapse" phenomenon (Fig. 4), which will be discussed below.

OBSERVED BEHAVIOR OF LIPOSOMES NEAR THE L_α/H_{II} PHASE BOUNDARY

We have recently studied liposome-liposome interactions near the L_α/H_{II} phase boundary [8-15]. Liposomal lipid mixing, contents leakage, and contents mixing were assayed by fluorescent resonant energy transfer and soluble fluorophore quenching assays. Results of most of these studies are reviewed in [3,12], and will be briefly summarized here. The systems studied included mixtures of phosphatidylethanolamines (PEs) with cholesterylhemisuccinate (CHEMS) [8,9,11,14], three different pure PEs [10], and mono-methylated PE and PE mixtures with phosphatidylcholine [12,13]. When the solution conditions of liposome suspensions are rapidly changed (e.g., pH jump) such that the lipid is either near or above T_H, lipid mixing, leakage, and, occasionally, fusion are observed. The lipid mixing and contents leakage processes were

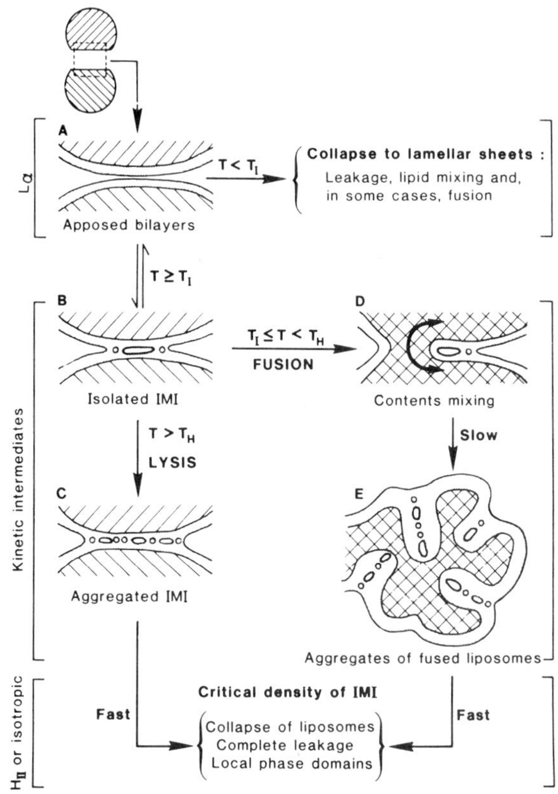

Figure 4. Model of liposome-liposome interactions via L_α/H_{II} transition intermediates of Ellens, Bentz & Szoka [12]. Reprinted with permission from [12], copyright (1986) American Chemical Society.

typically aggregation rate-limited due to the low lipid concentrations used [9,10,14]. Lipid mixing occurs across a wide temperature range [9,10], but the initial rate of this process increases rapidly with increasing temperature in the vicinity of T_H [9,10], and reaches a plateau a few degrees below T_H in CHEMS-containing systems [9]. The leakage rate increases dramatically in the immediate vicinity of T_H for all these systems [9,10,12]. However, no fusion was observed at any temperature in the CHEMS-PE systems where the transition was triggered by a drop in pH [8,9]. In three PE systems [10], fusion was observed at temperatures below T_H, but the rates and especially the extents of fusion fell off rapidly as temperatures approachedand exceeded T_H. The relative initial rates of fusion and leakage were compared carefully in the mono-methylated dioleoylphosphatidylethanolamine (DOPE-ME) and mixed PE/PC systems [12,13]. In those cases, fusion occurred at substantial rates in the temperature interval in which isotropic ^{31}P-NMR resonances were observed [13,31]; ca. 25-30°C below T_H to T_H. The initial rate was maximal just below T_H and fell rapidly thereafter, while the initial leakage rate increased rapidly in the immediate vicinity of T_H. The liposomes used in these studies were ca. 0.2-0.3 μm in diameter.

The behavior observed in these studies is very similar to the behavior predicted by the model described in the previous section. The lipid mixing rates become significant at temperatures below T_H and increase rapidly with increasing temperature in the region just below T_H, as one would expect if this process was mediated by structures forming as sub-critical fluctuations associated with the nearby L_α/H_{II} phase transition (i.e., IMIs). The leakage rates increased more sharply with temperature in the immediate vicinity of T_H, as one would expect for a cooperative leakage process involving more than one IMI per liposome-liposome contact and H_{II} phase production. Fusion was observed below T_H for the pure PE systems. In one case (egg PE produced by trans-esterification of egg PC) fusion was observed ca. 40°C below T_H: this is a temperature at which L_α/H_{II} transition intermediates are unlikely to form [16]. Markedly hysteretic transitions have not been reported for these PEs, nor do they exhibit isotropic ^{31}P-NMR resonances [36,37], which are necessary concomittants for ILA-mediated fusion [16]. They are unlikely to have small values of Z (e.g., Z may be as much as 1.8 for egg PE; [7]). We therefore interpret the data in these systems in terms of two interaction mechanisms: one which promotes fusion at T below T_H via a mechanism like that in Ca^{2+}-induced fusion of PS LUVs [38]; and an IMI-mediated mechanism that promotes leakage at $T \gtrsim T_H$.

Fusion with a temperature dependence consistent with the present model was observed in both the PE/PC and DOPE-ME systems. Fig. 5 is a plot of the initial rates of leakage and fusion in the DOPE/DOPC = 2/1 and DOPE-ME systems, reproduced from [12]. The temperature range of the calorimetrically-observed L_α/H_{II} phase transition is indicated. Both DOPE-ME and the PE/PC systems are expected to have small values of Z ("moderate" R_o) [27,31], because of the increased head group size of DOPE-ME relative to PE of the same acyl chain composition and the inclusion of L_α phase impurity, respectively [1,7,16-18,30]. Moreover, the DOPE-ME system is known to form inverted cubic phases upon prolonged incubations under these conditions [27], and both the DOPE-ME [13,31] and DOPE/DOPC = 2/1 systems [39] have been observed to yield isotropic ^{31}P-NMR resonances in the temperature range below T_H. These are the expected attributes of systems with Z values appropriate for ILA-mediated fusion [16,17]. As predicted by the present model, fusion and leakage are observed starting tens of degrees below T_H, and the maximum fusion rate is observed just below T_H. Fusion is observed in the same temperature ranges in which the isotropic ^{31}P-NMR resonances have been reported in both the systems in Fig. 5. Above T_H, the leakage rate dominates. This is as expected because of the increase in n_T^o with temperature [7,16] and the decrease in Z with temperature [16,17]. Z decreases with temperature because the radii of curvature (and thus R_o) of H_{II} phases decrease with temperature, corresponding to a decrease in the area per head group in the H_{II} phase, while the corresponding area in the L_α phase is probably much less temperature-sensitive.

The collapse phenomenon observed in [12] is not well-understood. After a varying incubation time that decreased with increasing temperature or lipid concentration, the aggregates in suspensions undergoing fusion and leakage would suddenly and completely leak their remaining contents [12]. This phenomenon may be related to the behavior predicted for 0.1 μm liposomes discussed in the previous section. Presumably, leakage is slower in initial aggregates because the 0.2-0.3 μm liposomes used in [12] were not large enough to form many IMIs in their contacts with neighbors, and this became more feasible the larger the liposomes became via the slow fusion reaction. The critical liposome size for rapid H_{II} phase formation may be fairly well-defined, leading to a rapid and suspension-wide wave of leakage.

However, this explanation is not totally satisfactory because the collapse process occurred even at temperatures below the calorimetrically-observed T_H. There is profound hysteresis in the L_α/H_{II} transition in DOPE-ME systems that might make it hard to determine the exact position of of this phase boundry [27], and it is possible that the circumstances in

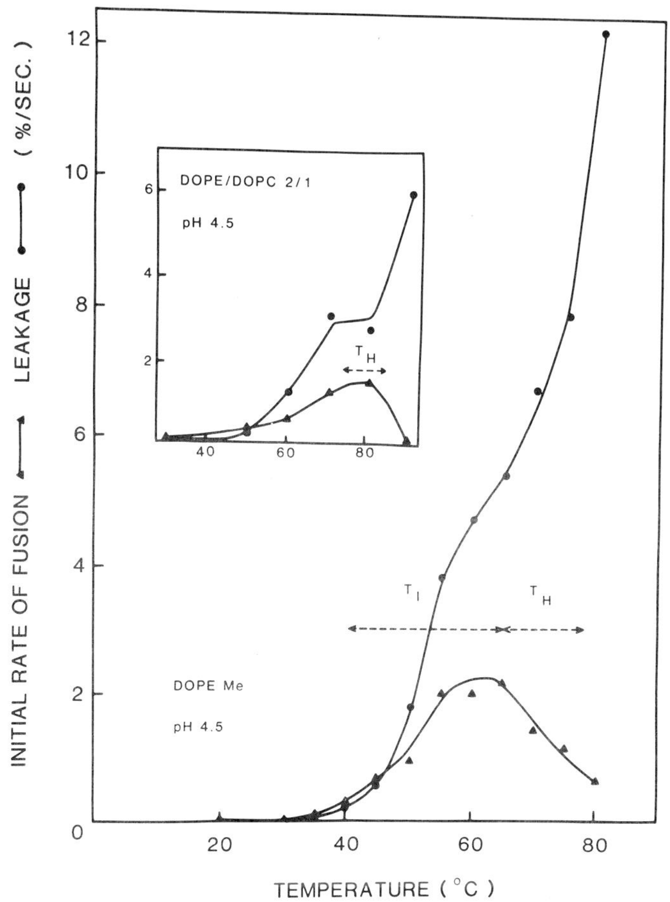

Figure 5. Initial rates of fusion and leakage (ANTS/DPX contents mixing and leakage assays) in suspensions of DOPE-ME liposomes and DOPE/DOPC= 2/1 (mol/mol) liposomes (inset), both in pH 4.5 acetate buffer 100 mM in NaCl and 0.1 mM in EDTA. Reprinted with permission from [12]; copyright American Chemical Society.

which H_{II} phase formation can occur are slightly different between bulk systems and unilamellar liposome dispersions [17]. Understanding of this process awaits physical characterization of the structures that form during this collapse, which is in progress. Rupert et al. [40] have recently observed a similar phenomenon in liposomes in the Calcium-didodecylphosphate system that were in the same 0.1 μm size range: upon exposure to Calcium, the liposomes fuse until they reach a larger size and then collapse into H_{II} phase aggregates.

IMPLICATIONS FOR MEMBRANE FUSION EXPERIMENTS AND BIOMEMBRANE FUSION

The model described in Figs. 3 & 4 and Table I is qualitatively consistent with the observed behavior of liposomal dispersions near the L_α/H_{II} phase boundary. It is particularly interesting that the dynamics in these suspensions can be related to the lipid-water phase diagram and bulk liquid crystalline properties of the system. In particular, marked hysteresis in the L_α/H_{II} phase transition (isotropic ^{31}P-NMR resonances at temperatures starting several tens of degrees below T_H, isotropic or inverted cubic phase formation in the same region) should be accompanied by at least some membrane fusion in liposomal dispersions under the same conditions. These principles should be of use in designing membrane preparations to fuse under defined circumstances.

Systems where the L_α/H_{II} transition is induced by ion binding to ionic lipids (e.g., Ca^{2+}-cardiolipin) should behave in a fashion similar to that predicted in Table I, but as a function of cation concentration with respect to the critical concentration for inducing the transition instead of temperature. However, there is reason to expect much more hysteresis [41] and more fusion [3,16] in these ionotropic systems.

It is important to note that IMIs and ILAs appear to be capable of mediating interactions between liposomes even when the equilibrium phase of the system is the L_α phase (e.g., temperatures ca. 20°C below T_H). Mechanistic studies of membrane fusion should be done with knowledge of the relevant phase diagram under the experimental conditions. It is also important to note that IMIs can mediate lipid mixing between liposomes even though no fusion with retention of aqueous contents occurs (e.g., CHEMS-PE systems; [8,9]): clearly, in many cases the observation of lipid mixing alone (via fluorescent resonant energy transfer probes) is insufficient evidence for fusion with retention of contents.

Relevance to Biomembrane Fusion

Several biomembrane preparations or biomembrane total lipid extracts have been reported to form either the H_{II} phase or exhibit lipidic particles and isotropic ^{31}P-NMR resonances when either dehydrated, heated, or exposed to divalent cation concentrations in excess of several millimolar. Examples are lobster muscle microsome and rat liver microsomes [42,43], rye leaf protoplasts [44], chloroplast or thylakoid membranes [45,46], retinal rod outer segment membranes [47-49], rat liver mitochondrial lipids [50] and native membranes [51], and membrane lipids of various bacteria [52-54]. Many have speculated that biomembrane lipids may be on the verge of undergoing bilayer/non-bilayer phase transitions under physiological conditions, and that one reason is that this makes them more susceptible to membrane fusion, as is required by many biological processes (e.g., [1-3,5,6,16]).

Here, we have noted that mixtures of bilayer and non-bilayer phase-forming lipids that roughly mimic biomembrane lipid compositions should be susceptible to fusion via L_α/H_{II} transition intermediates. Further, they are most susceptible to fusion when the equilibrium phase is still the L_α phase but the system is close to the L_α/H_{II} phase boundary, either in the sense of temperature near T_H, or in the sense of divalent cation concentration or hydration near the respective critical values for H_{II} formation. We note that close apposition of two membranes under outside influence (e.g., the action of proteins like actin or synexin) may correspond to local dehydration of the membrane interfaces. Hence, model system data [8-15] are consistent with the involvement of these phase transitions in some types of biological fusion.

T_H or the position of the L_α/H_{II} phase boundary along other axes is sensitive to small changes in membrane composition (see [1-3] for reviews). For example, it has recently been found that T_H is quite sensitive to low levels of diglycerides [55-57]. This seems to be particularly true for diglycerides with acyl chain substitution similar to those produced by phosphatidylinositiol hydrolysis in vivo [57]. When 2 mole % (lipid basis) of 1-oleoyl, 2-arachidonyl-glycerol is added to

DOPE-ME, for example, the onset temperature of the L_α/H_{II} phase coexistence region is lowered by 25° C and fusion is observed at correspondingly lower temperatures relative to pure DOPE-ME [57]. It is conceivable that the cell controls the susceptibility of relevant membranes to fusion (e.g., as in exocytosis) via this mechanism by making subtle and transient changes in the lipid composition, and that fusion via L_α/H_{II} transition intermediates is possible even when the the biomembrane lipid composition under steady-state physiological conditions is fairly far from the L_α/H_{II} phase boundary.

Biomembrane fusion via L_α/H_{II} intermediates is still a speculative hypothesis. The data in [8-15] are roughly consistent with IMI- and ILA-mediated processes, but these and other model systems have to be more thoroughly investigated. To date, we have always observed leakage with fusion in model systems. Fusion in vivo must occur without concomitant leakage or membranes would lose their integrity, resulting in cell death. However, preliminary data indicate that the ratio of contents mixing to leakage is a sensitive function of lipid composition [13]. In principle, lipid systems could exhibit ILA-mediated fusion with almost no leakage. Also, the phase transition mechanism on which we base our model remains to be unambiguously demonstrated. Studies of L_α/H_{II} transition kinetics via time-resolved X-ray diffraction (e.g., [22,24,27]) may allow us to infer the actual phase transition mechanism, particularly when done in parallel with appropriate NMR and morphological measurements. This would test the present model of membrane-membrane interactions. It has also been pointed out that a measurement of the activation volume for lipid mixing in unilamellar liposome dispersions near T_H might also permit a test of our model [3].

It is also possible that the state of biomembrane lipids is maintained near the L_α/H_{II} phase boundary for other reasons than modulating membrane fusion. It has been suggested that non-bilayer lipids are present in thylakoid membranes to accomodate protein incorporation [58]. It is also possible that membrane-bound enzymes function best when the lipids have the particular order parameter profile observed in this temperature region [59], or that the membrane permeability properties are most advantageous when the lipids are close to this phase boundary. Finally, it is possible that proximity of the lipids to the L_α/H_{II} phase boundary is required for optimal biological fusion, but for reasons related to the functioning of as yet-uncharacterized fusion proteins. One might expect such proteins, or the proteins controlling them, to behave differently depending upon the nature of the lipid environment.

The model described here should be of particular use in designing liposomal drug delivery systems. As reviewed in [3], the liposome compositions used in systems for endosomal drug delivery are such that they probably adopt the H_{II} phase after endocytosis upon exposure to the low pH of the endosomal lumen. This may be related to their ability to fuse with the endosomal membrane and release encapsulated drug to the cytoplasm or to leak their contents into the endosome (i.e., leakage after uptake into the target cell, as opposed to non-site-specific leakage). It may be objected that liposomes can be taken up into endosomes one at a time, and that interactions via L_α/H_{II} intermediates with the endosomal membrane are unlikely since it may be quite stable in the L_α phase. However, Bentz et al. [11] recently studied systems composed of two populations of liposomes at a temperature such that only one population was above T_H. Contact of the population that was above T_H with liposomes of the population stable in the L_α phase resulted in leakage from the unstable population. It is thus possible that contact-mediated leakage and perhaps fusion of such a liposome with an endosomal membrane occurs via L_α/H_{II} intermediates even if the endosomal membrane itself is far from the phase boundary.

By exploring the mechanism and hysteresis of the phase transitions between L_α, inverted cubic, and H_{II} phases we may be able to understand and predict many aspects of liposome-liposome interactions near these phase boundaries.

REFERENCES

1) A.J. Verkleij, 1984, Lipidic intramembranous particles, Biochim. Biophys. Acta, 779:43.
2) P.R. Cullis, M.J. Hope, B. de Kruijff, A.J. Verkleij, and C.P.S. Tilcock, 1985, Structural properties and functional roles of phospholipids in biological membranes, in: "Phospholipids and Cellular Regulation," J.F. Kuo, ed., CRC Press, Boca Raton, Florida.
3) D.P. Siegel, 1987, Membrane-membrane interactions via intermediates in lamellar-to-inverted hexagonal phase transitions, in: "Cell Fusion," A.E. Sowers, ed., Plenum Publ. Co., New York, N.Y..
4) A.J. Verkleij, C. Mombers, J. Leunissen-Bijvelt, P.H.J.Th. Ververgaert, 1979, Lipidic intramembranous particles, Nature, 279:162.
5) A.J. Verkleij, C.J.A. van Echteld, W.J. Gerritsen, P.R. Cullis, and B. de Kruijff, 1980, The lipidic particle as an intermediate structure in membrane fusion processes and bilayer to hexagonal H_{II} transitions, Biochim. Biophys. Acta, 600:620.
6) D.P. Siegel, 1984, Inverted micellar structures in bilayer membranes: formation rates and half-lives, Biophysical J., 45:399.
7) D.P. Siegel, 1986, Inverted micellar intermediates and the transitions between lamellar, cubic, and inverted hexagonal lipid phases. I. Mechanism of the L_α-H_{II} phase transitions, Biophysical J., 49:1155.
8) H. Ellens, J. Bentz, and F.C. Szoka, 1985, H^+-and Ca^{2+}-induced fusion and destabilization of liposomes, Biochemistry, 24:3099.
9) J. Bentz, H. Ellens, M.-Z. Lai, and F.C. Szoka, 1985, On the correlation between H_{II} phase and the contact-induced destabilization of phosphatidylethanolamine-containing membranes, Proc. Nat'l. Acad. Sci. (U.S.A.), 82:5742.
10) H. Ellens, J. Bentz, and F.C. Szoka, 1986, Destabilization of phosphatidylethanolamine liposomes at the hexagonal phase transition temperature, Biochemistry, 25:285.
11) J. Bentz, H. Ellens, and F.C. Szoka, 1987, Destabilization of phosphatidylethanolamine-containing liposomes: Hexagonal phase and asymmetric membranes, Biochemistry, (in press).
12) H. Ellens, J. Bentz, and F.C. Szoka, 1986, Fusion of phosphatidylethanolamine-containing liposomes and mechanism of the L_α-H_{II} phase transition, Biochemistry, 25:4141.
13) H. Ellens, D.P. Siegel, D.R. Alford, P.L. Yeagle, and J. Bentz, 1987, Fusion of liposomes mediated by L_α-H_{II} intermediates, Biophysical J., 51:355a (abstr.).
14) H. Ellens, J. Bentz, and F.C. Szoka, 1984, pH-induced destabilization of phosphatidylethanolamine-containing liposomes: Role of inter-bilayer contact, Biochemistry, 23:1532.
15) M.Z. Lai, W.J. Vail, and F.C. Szoka, 1985, Acid- and Calcium-induced structural changes in phosphatidylethanolamine membranes stabilized by cholesterylhemisuccinate, Biochemistry, 24:1654.
16) D. Siegel, 1986, Inverted micellar intermediates and the transitions between lamellar, cubic, and inverted hexagonal lipid phases. II. Implications for membrane-membrane interactions and membrane fusion, Biophysical J., 49:1171.
17) D. Siegel, 1987, Inverted micellar intermediates and the transitions between lamellar, cubic, and inverted hexagonal amphiphile phases.

III. Isotropic and inverted cubic state formation via intermediates in transitions between L_{α} and H_{II} phases, Chem. Phys. lipids, 42:279.
18) G.L. Kirk, S. M. Gruner, and D.L. Stein, 1984, A thermodynamic model of the lamellar to inverted hexagonal phase transition of lipid-water systems, Biochemistry, 23:1093.
19) S.M. Gruner, 1985, Intrinsic curvature hypothesis for biomembrane lipid composition: A role for non-bilayer lipids, Proc. Nat'l. Acad. Sci. (U.S.A.), 82:3665.
20) S.M. Gruner, V.A. Parsegian, and R. P. Rand, 1987, Directly measured deformation energy of phospholipid H_{II} hexagonal phases, Faraday Discuss. Chem. Soc., vol. 81 (in press).
21) M.W. Tate and S. M. Gruner, 1987, Lipid polymorphism of mixtures of dioleoylphosphatidylethanolamine and saturated and monounsaturated phosphatidylcholines of various chain lengths, Biochemistry, 26:231.
22) M. Caffrey, 1987, Kinetics and mechanism of transitions involving the lamellar, cubic, inverted hexagonal, and fluid isotropic phases of hydrated monoacylglycerides monitored by time-resolved X-ray diffraction, Biochemistry, (in press).
23) J.L. Ranck, L. Letellier, E. Shechter, B. Korp, P. Pernot, and A. Tardieu, 1984, X-ray analysis of the kinetics of E. coli lipid and membrane structural transitions, Biochemistry, 23:4955.
24) M.C. Caffrey, 1985, Kinetics and mechanism of the lamellar gel/lamellar liquid-crystal and lamellar/inverted hexagonal phase transition in phosphatidylethanolamine: A real-time X-ray diffraction study using synchrotron radiation, Biochemistry, 24:4826.
25) S.W. Hui, and T.P. Stewart, 1981, 'Lipidic particles' are intermembrane attachment sites, Nature, 290:427.
26) E. Bearer, N. Düzgünes, D.S. Friend, and D. Papahadjopoulos, 1982, Fusion of phospholipid vesicles arrested by quick-freezing: The question of lipidic particles as intermediates in membrane fusion, Biochim. Biophys. Acta, 693:93.
27) S.M. Gruner, M.W. Tate, G.L. Kirk, P.T.C. So, D.C. Turner, D.T. Keane, C.P.S. Tilcock, and P.R. Cullis, 1987, X-ray diffraction study of the polymorphic behavior of N-methylated phosphatidylethanolamine, Biochemistry, (in press).
28) F. Reiss-Husson, 1967, Stucture des phases liquide-crystalline de differentes phospholipides , monoglycerides, sphingolipides, anhydres ou en presence d'eau, J. Mol. Biol., 25:363.
29) L. Lis, M. McAlister, N. Fuller, R.P. Rand, and V.A. Parsegian, 1982, Interactions between neutral phospholipid membranes, Biophysical J., 37:657.
30) P.J. Quinn and W.P. Williams, 1983, The structural role of lipids in photosynthetic membranes, Biochim. Biophys. Acta, 737:223.
31) J. Gagné, L. Stamatatos, T. Diacovo, S.W. Hui, P.L. Yeagle, and J.R. Silvius, 1985, Physical properties and surface interactions of bilayer membranes containing N-methylated phosphatidylethanolamines, Biochemistry, 24:4400.
32) P.M. Brown, J. Steers, S.W. Hui, P.L. Yeagle, and J.R. Silvius, 1986, Role of head group structure in the phase behavior of amino phospholipids. 2. Lamellar and non-lamellar phases of unsaturated phosphatidylethanolamine analogues, Biochemistry, 25:4259.
33) V.A. Parsegian and R.P. Rand, 1983, Membrane interactions and deformation, Ann. N.Y. Acad. Sci., 416:1.
34) R. Kwok and E. A. Evans, 1981, Thermoelasticity of large lecithin bilayer vesicles, Biophys. J., 35:637.
35) E.A. Evans and V.A. Parsegian, 1983, Energetics of membrane deformation and adhesion in cell and vesicle aggregation, Ann. N.Y. Acad. Sci., 416:13.
36) P.R. Cullis and B. de Kruijff, 1978, The polymorphic phase behavior

of phosphatidylethanolamines of natural and synthetic origin, Biochim. Biophys. Acta, 513:31.

37) C.P.S. Tilcock and P.R. Cullis, 1982, The polymorphic phase behavior and miscibility properties of synthetic phosphatidylethanolamines, Biocehim. Biophys. Acta, 684:212.

38) J. Bentz and N. Düzgünes, 1985, Fusogenic capacities of divalent cations and effect of liposome size, Biochemistry, 24:5436.

39) G. van Duijn, C. Valtersson, T. Chojnacki, A.J. Verkleij, G. Dallner and B. de Kruijff, 1986, Dolichol phosphate induces non-bilayer structures, vesicle fusion and transbilayer movement of lipids: a model membrane study, Biochim. Biophys. Acta, 861:211 (1986).

40) L.A.M. Rupert, J.F.L. van Breeman, E.F.J. van Bruggen, J.B.F.N. Engberts, and D. Hoekstra, 1987, Calcium-induced fusion of didodecylphosphate vesicles. The lamellar to hexagonal II (H_{II}) phase transition, (submitted).

41) B. de Kruijff, A.J. Verkleij, J. Leunissen-Bijvelt, C.J.A. van Echteld, J. Hille, and H. Rijnbout, 1982, Further aspects of the Ca^{+}-dependent polymorphism of bovine heart cardiolipin, Biochim. Biophys. Acta, 693:1.

42) L.M. Crowe and J.H. Crowe, 1982, Hydration-dependent hexagonal phase lipid in a biological membrane, Arch. Biochem. Biophys., 217:582.

43) G. van Duijn, J. Luiken, A.J. Verkleij, and B. de Kruijff, 1986, Relation between lipid polymorphism and transbilayer movement of lipids in rat liver microsomes, Biochim. Biophys. Acta, 863:193.

44) W.J. Gordon-Kamm and P.L. Steponkus, 1984, Lamellar-to-hexagonal phase transitions in the plasma membrane of isolated protoplasts after freeze-induced dehydration, Proc. Nat'l. Acad. Sci. (U.S.A.), 81:6373.

45) K. Gounaris, A. Sen, A.P.R. Brain, P.J. Quinn, and W.P. Williams, 1983, The formation of non-bilayer structures in total polar lipid extracts of chloroplast membranes, Biochim. Biophys. Acta, 728:129.

46) K. Gounaris, A.P.R. Brain, P.J. Quinn, and W.P. Williams, 1983, Structural and functional changes associated with heat-induced phase separations of non-bilayer lipids in chloroplast thylakoid membranes, FEBS Letters, 153:47.

47) W.J. de Grip, E.H.S. Drenthe, C.J.A. van Echteld, B. de Kruijff, and A.J. Verkleij, 1979, A possible role of rhodopsin in maintaining bilayer structure in the photoreceptor membrane, Biochim. Biophys. Acta, 558:330.

48) S.M. Gruner, K.J. Rothschild, and N.A. Clark, 1982, X-ray diffraction and electron microscope study of phase separation in rod outer segment photoreceptor membrane multilayers, Biophys. J., 39:241.

49) A.D. Albert, A. Sen, and P.L. Yeagle, 1984, The effect of calcium on the bilayer stability of lipids from bovine rod outer segment disk membranes, Biochem. Biophys. Acta, 771:28.

50) P.R. Cullis, B. de Kruijff, M.J. Hope, R. Nayar, A. Rietveld, and A.J. Verkleij, 1980, Structural properties of phospholipids in the rat liver inner mitochondrial membrane, Biochim. Biophys. Acta, 600:625.

51) R. van Venetië and A.J. Verkleij, 1982, Possible role of non-bilayer lipids in the structure of mitochondria: A freeze-fracture electron microscope study, Biochim. Biophys. Acta, 692:397.

52) E. Burnell, L. van Alphen, A. Verkleij, and B. de Kruijff, 1980, ^{31}P-NMR and freeze-fracture electron microscopy studies on E. coli, Biochim. Biophys. Acta, 597:492.

53) P.J. Quinn, A.P.R. Brain, L.C. Stewart, and M. Kates, 1986, The structure of membrane lipids of the extreme halophile, H. cutirubrum, in aqueous systems studied by freeze-fracture, Biochim. Biophys. Acta, 863:213.

54) G. Lindblom, I. Brentel, M. Sjölund, G. Wikander, and A. Wieslander, 1986, Phase equilibria of membrane lipids from A. laidlawii: Impor-

tance of a single lipid forming nonlamellar phases, Biochemistry, 25:7502.

55) R.M. Epand, 1985, Diacylglycerols, lysolecithins, or hydrocarbons markedly alter the bilayer to hexagonal phase transition temperature of phosphatidylethanolamines, Biochemistry, 24:7092.

56) S. Das and R.P. Rand, 1986, Modification by diacylglycerol of the structure and interaction of various phospholipid bilayer membranes, Biochemistry, 25:2882.

57) D.P. Siegel, D. Alford, H. Ellens, L. Lis, P.J. Quinn, P.L. Yeagle, and J. Bentz, 1987, A 2-arachidonyl diglyceride is particularly effective in inducing L /H_{II} phase transitions and membrane fusion, Biophys. J., 51:355a (abstr.).

58) P.J. Quinn and W.P. Williams, 1983, The structural role of lipids in photosynthetic membranes, Biochim. Biophys. Acta, 737:223.

59) P.R. Cullis, M.J. Hope, and C.P.S. Tilcock, 1986, Lipid polymorphism and the roles of lipids in membranes, Chem. & Phys. Lipids, 40:127.

THE INFLUENCE OF POLAR GROUP IDENTITY ON THE INTERACTIONS BETWEEN PHOSPHOLIPID BILAYERS

R. P. Rand[1] and V.A. Parsegian[2]

[1]Biological Sciences, Brock University
St. Catharines, Ont., Canada
[2]DCRT, National Institutes of Health
Bethesda MD., U. S. A.

INTRODUCTION

Close apposition is a prerequisite for fusion. That apposition is determined by the interaction energies of the polar species populating the surfaces. Those energies are largely influenced by the hydration properties of the polar groups and hydration repulsion often precludes close apposition (1,2). We describe here large differences in the hydration properties of bilayers made of different phospholipid species and their mixtures; differences that greatly affect bilayer separation and may influence the facility of fusion. We have attributed some of the differences to varying contributions of an attractive hydration force (3,4). That force requires correlations between complementary polar groups on the apposing surfaces; correlations that are determined not only by the structure of the polar groups but also by the constraints imposed on them by the packing of their attached hydrocarbon chains. In another sense, hydration attraction reflects the tendency of the assembled polar groups to precipitate.

Under conditions that trigger close polar group contact through dehydration, such as the Ca-induced aggregation of phosphatidylserine (PS) vesicles, adhesion energies are so large they destroy the vesicles (2,5,6,7). When that adhesion energy is reduced by the addition of the strong hydration repulsion of phosphatidylcholine (PC), stable double-bilayer contacts between vesicles are formed (7,8). We describe preliminary evidence suggesting that in such cases even slight dehydration triggers vesicle destruction, demixing of the lipids and formation of the dehydrated Ca-PS lamellar phase. Little if any of the adhesion energy appears to go into dehydrating PC in the contact regions. Whether one of the less hydrated lipids such as phosphatidylethanolamine (PE), or digalactosyldiacylglyceride (DGDG) or diglyceride allows significant interaction without vesicle destruction remains to be seen, but it seems unlikely (2). Otherwise fusion seen within these disintegrating systems must be fundamentally different from the finely controlled cellular process.

For a selection of lipids we have summarized their hydration properties in terms of (i) the maximum amount of water imbibed by the lipid in the lamellar phase and (ii) the relation between the net interbilayer repulsive pressure and the water content of the lamellar phase. The former has been measured from the water-lipid phase diagram using X-ray diffracton (9); the latter by combining X-ray diffraction with controlled osmotic stress of the multilamellar lipid phase (9,10).

MAXIMUM WATER IMBIBITION BY LAMELLAR PHASES

The Table provides the following structural data on the lamellar phases formed by the lipids when equilibrated with excess water.

The Parameters[1]

- d : maximum lamellar X-ray repeat spacing.
- d_1 : bilayer thickness calculated according to the Luzzati tradition on a mass average basis assuming all and only the lipid is in this layer (11).
- $d_w = d-d_1$: water layer thickness or interbilayer distance assuming all and only water is in this space.
- A : area available per phospholipid molecule projected on a plane normal to the direction of the lamellar repeat.
- V_w : volume of water per lipid molecule.
- V_{wn} : volume of water normalized to a polar group mass equal to that of the PE molecule.
- M: molality of head groups in the interbilayer space.

The Lipids[2]

- SOPC : 1-stearyl-2-oleoylphosphatidylcholine
- POPE : 1-palymityl-2-oloeylphosphatidylethanolamine
- egg PC : PC extracted from hen eggs, or N-trimethylated eggPE-T.
- egg PE : PE extracted from hen eggs
- eggPE-T : PE produced by transphosphatidylation onto egg PC chains
- MMPE : N-monomethylated eggPE-T
- DMPE : N-dimethylated eggPE-T
- DGDG : digalactosyldiacylglyceride
- DOPS : dioleolylphosphatidylserine in 0.8-1M NaCl.

[1]	SOPC[2]	POPE	egg PE-T	MMPE	DMPE	egg PC	egg PE	DOPS	DGDG
d	64.6	53.3	52.4	61.8	63.1	61.9	52.8	53.7	52
d_1	37.8	41.6	36.7	37.1	36.3	35.1	32.3	39.0	39
d_w	26.8	11.7	15.7	24.7	26.8	26.8	20.5	14.7	13
V_w	900	334	521	826	933	982	770	492	520
V_{wn}	751	334	521	775	825	820	770	409	266
A	69	57	66	67	70	74	75	67	80
M	1.83	4.94	3.17	2.00	1.76	1.68	2.14	3.35	3.17
λ	2.29	0.84(1.0)	2.08	2.35	2.38	2.6(2.4)	2.03		2.16
$\log P_o$	10.1	12.4 (12)	9.6(10)	9.5	9.7	9.7(10)	10.6		9.87
$A_h \cdot 10^{14}$	1.2	6	6	1.0	1.2	1.2	1.2		
d_v	3	3	3	2	2	3	3		

Some Comparisons

(1) As has been well established in the past PE's hydrate less than PC's. These data show that both V_w and bilayer separation d_w are greater by more than a factor of two for SOPC compared with POPE. The difference is not in the molecular weights of the polar groups since V_{wn} shows that for equal polar mass PC is associated with more than twice the amount of water as POPE.

(2) PE's themselves vary in hydration by a factor of two. POPE/ eggPE/eggPE-T show a systematic increase in d_w and V_w that correlates with an increasing % polyunsaturation of their hydrocarbon chains, 0/17/34. This variation is greater than for PC's and suggests to us that PE hydration is remarkably sensitive to factors that would not be expected to affect either hydration repulsion or van der Waals attraction.

(3) A single methylation of eggPE-T results in enhanced hydration that then increases little with successive methylations : eggPE-T/ MMPE/DMPE/egg PC. This is consistent with a number of previous studies (18,19).

(4) DOPS bilayers with strongly reduced electrostatic repulsion achieve close apposition, hydrating only to levels comparable to the less hydrated PE's and very much less than the PC's.

(5) DGDG bilayers hydrate very little, on a polar mass basis even less than POPE. This was seen in the close apposition of galacto-lipid-coated mica sheets (28). The large polar group and molecular area confounds simple attempts at relating area to hydration.

(6) We think it is important that, where it has been measured, the molalities of the polar groups in the interbilayer space are between that of the maximum solubility of the free head groups in water and their concentration in crystalline form (12).

LAMELLAR PHASE DEHYDRATION or WATER IMBIBITION AGAINST OSMOTIC PRESSURE

Traditionally the net pressure between neutral bilayers, P, has been accounted for by the sum of van der Waals attraction, hydration repulsion and, more recently, fluctuation-enhanced repulsion F_{fl} (13).

Specifically

$$P(d_w) = P_o \exp(-d_w/\lambda) - A_h(1/6\pi(d_w - d_v)^3 + F_{fl} \qquad (1)$$

where

d_v and A_h are the distance the van der Waals plane projects from the lipid-water plane defined by d_w, and the Hamaker coefficient:

P_o and λ are the coefficient and decay distance of hydration repulsion.

As the bilayers are pushed closer together by withdrawing water, the relation between bilayer separation and net interbilayer pressure can be described by the exponential for much of the separation distance where hydration repulsion dominates (1,9). The exponential

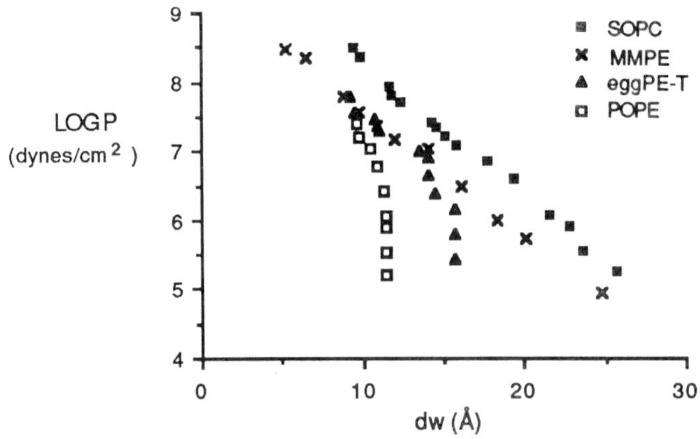

Figure 1. Plots of net interbilayer repulsion as it varies with interbilayer distance.

parameters estimated this way are shown in the Table. Examples of data that provide this description for POPE, EGGPE-T and MMPE and EGGPC are shown in Figure 1. The different degrees of hydration in excess water discussed above can be seen by the data near the d_w axis where net bilayer repulsion approaches zero and deviation from the exponential relation results as van der Waals attraction and fluctuation-driven repulsion become comparable to hydration repulsion. Equation (1) was applied to match the experimental data using the parameters shown in the Table. Any adjustments to the λ or P_o derived using the exponential portion only appear in brackets.

These curves illustrate some of the major differences among these lipids. Although these data can be fitted by equation (1) the parameters that are required to do this suggest the inadequacy of this theory. Consider the following comparisons.

(1) The PE's can only be fitted with (i) λ's that vary by a factor of 2, (ii) P_o's that are different by two orders of magnitude, and (iii) Hamaker coefficients that differ by a factor of about 5. One would not expect these variations from lipids that have identical head groups and differ only in the composition of their hydrocarbon chains. We do recognize however that POPE accounts for these differences and that its estimated λ and P_o are poor because of the small range of compression allowed before its chains crystallize. Nevertheless the data is difficult to match without these parameters.

(2) Methylation of eggPE-T does not result in a detectable increase in measured P_o that would account for its sudden increase in hydration. Further, it would be surprising if the Hamaker coefficient increased by a factor of 5 with the simple addition to the polar head of a methyl group. Successive methylations do not result in similar increases.

HYDRATION ATTRACTION

We suspect the action of an attractive hydration force. It could originate from complementary groups, ethanolamine and phosphate in the case of PE, to form hydrogen bonds. Such bonds have been directly observed in crystals of PE (14), and have been suggested as forming intermolecularly within bilayers and between them at low hydration (15,16,17). Under conditions where spatial correlations of such complementary groups exist between bilayers we suggest that water can

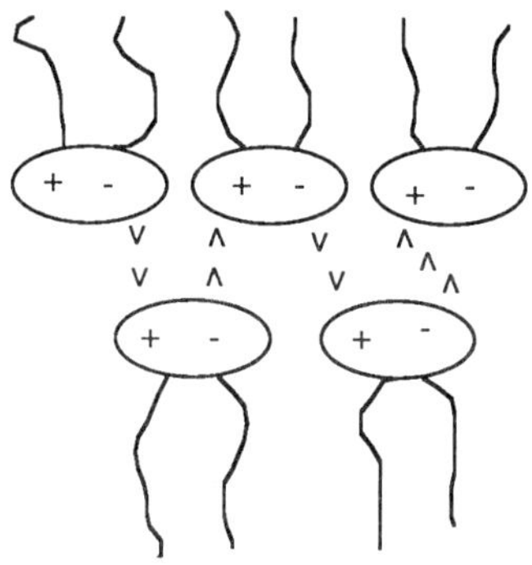

Figure 2. Schematic diagram of water molecules forming hydrogen-bonded bridges between spatially correlated complementary groups across the interbilayer space.

form interbilayer hydrogen-bonded bridges even at higher levels of hydration, figure 2. Such water suffers the same perturbation as that which accounts for hydration repulsion except the direction of polarization is the same across the whole interbilayer space. The same formalism (20) applies to both and this adds to equation (1) an attractive term

$$P_a \cdot \exp(-2d_w/\lambda).$$

λ is now a universal decay distance, characteristic of water, independent of lipid, and common with the repulsive term. P_a is an attraction coefficient sensitive to the spatial correlations that the complementary groups can form across the interbilayer space. In that view the PE's vary in hydration among themselves because chain heterogenity and degree of unsaturation sensitively affect those correlations through their packing. Similarly PC's and methylated PE's cannot form strong correlations because of the diminished ability of the methylated groups to hydrogen bond.

The added attraction term allows the data to be fitted with parameters more consonant with expectation. In particular the PE's can be fitted with combinations of parameters wherein λ, A_h, d_v are constant and P_0 and P_a adjusted within about an order of magnitude. The data do not allow distinction among combinations of parameters. The change in hydration with methylation can be seen as the removal of the hydration attraction term.

These data do not really allow clear distinction between or rejection of one of the alternative theoretical models. But we think hydration attraction is likely. Beside the arguments given above we question the treatment of water merely as a continuous dielectric at such low water contents. Importantly the concentrations of the maximally hydrated polar groups when assembled on bilayers (see Table) are higher than can be achieved in solution when unattached to their hydrocarbon chains and are considerably lower than when free polar groups are crystallized (12). It is as if, constrained by their attached chains, the polar groups are concentrated to supersaturating levels but are unable to precipitate and collapse across the interbilayer space. But that tendency to collapse would appear as an attractive force. We suggest that all solubility factors that govern polar group precipitation will contribute to a hydration attraction.

We suggest as well that the constraints on polar group packing by hydrocarbon chain packing will modulate that tendency.

In sum we expect both van der Waals and hydration attraction to act between bilayers, with PC's representing domination of the former and POPE representing domination by hydration attraction. The low hydration of shielded DOPS bilayers and of DGDG bilayers suggests that hydrogen bonding (21) or other solubility factors leads to hydration attraction for these lipids as well.

INTERACTING VESICLES

The complementary measurements of adhesion energy between phospholipid bilayer vesicles by Evans and his collaborators (21,22,23) indicate that they experience the same interactions as bilayers within multilayers. Although an order of magnitude difference in adhesion energy of PE and PC bilayers is measured (23), vesicles adhere very weakly without special conditions to overcome hydration repulsion. Enhanced adhesion can be effected by the addition of divalent cations to vesicles that contain acidic phospholipids. The response of such strongly adhered and subsequently deformed vesicles is sensitively dependent on the adhesion energy and the fragile mechanical properties of the bilayers (23,25). The destructive Ca-induced adhesion of PS vesicles (6) can be reduced by the addition of PC with its higher hydration repulsion or of PE with its more moderate hydration repulsion (7). In observing large unilamellar vesicles by electron microscopy and giant vesicles by differential interference contrast light microscopy we have defined three kinds of response after the

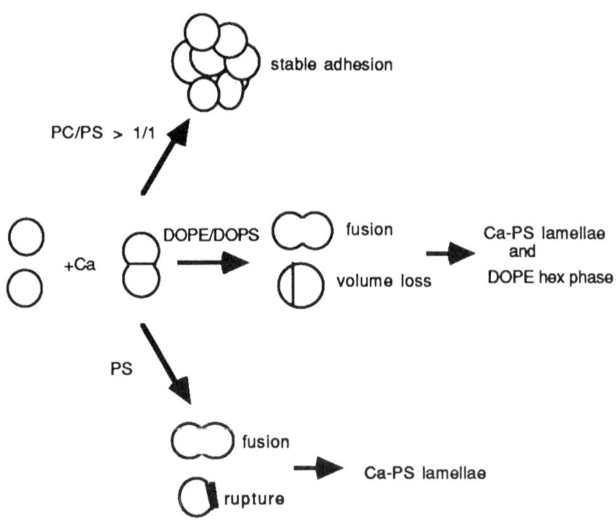

Figure 3. Schematic diagram of the changes that three kinds of bilayer vesicles undergo when induced to interact by the addition of 10 mM $CaCl_2$ (7). Their different reactions have been attributed to a competition between the rate at which bilayer tension builds as the vesicles deform and the release of that tension. Enough PC, with its strong hydration repulsion, prevents mechanical response beyond adhesion. Attempts to determine the properties of such stable bilayer contacts failed because even mild dehydration of them triggers lipid demixing and the formation of the Ca-PS dehydrated lamellae.

initial Ca-induced adhesion of such vesicles (6,7) (figure 3).

PS vesicles either fuse or burst, with about equal probability, and quickly form Ca-PS dehydrated multilamellae. DOPE-PS vesicles either fuse or deflate without bursting, i.e. leak their contents, with about equal probability; eventually dehydrated lamellar and DOPE hexagonal phases form. DOPC-DOPS vesicles at PC/PS ratios greater than 1/1 simply adhere in stable apposition. At higher PS contents a dehydrated lamellar phase similar to the Ca-PS complex forms on vesicle interaction.

Important in the understanding of such interactions are the composition, structure and interaction energy of the contact region of the adhered vesicles. We have recently attempted to measure the repulsive forces between the apposed bilayers evident in the stable apposition of Ca-PS-PC vesicles. The vesicles, aggregated in 10 mM $CaCl_2$, were osmotically dehydrated to different extents and observed by X-ray diffraction and freeze-fracture electron microscopy. For DOPC/DOPS ratios of 1/1 and 5/1 very stable vesicle aggregates formed. However even the mildest dehydration of these aggregates resulted in demixing of lipids with the formation of a lamellar phase with dimension similar to the dehydrated Ca-PS complex. We conclude that for DOPC/DOPS ratios above 1/1 although PC allows stable bilayer apposition, presumably through its strong hydration repulsion, slight dehydration triggers the formation of the Ca-PS complex and lipid demixing. The energy derived from the formation of a Ca-PS complex, which we previously estimated at about 27 ergs/cm^2, is revised upward to about 50 ergs/cm^2 in the light of the recent excellent measurements by Feigenson (27) of the very high affinity of interacting PS bilayers for Ca. This far exceeds the energy cost of demixing the lipids. Preliminary evidence from X-ray diffraction of PC/PS samples prepared with $CaCl_2$ suggests that the dehydrated phase contains little PC. If so, little energy from the Ca-induced interaction dehydrates PC and accommodates it in the contact region. We are attempting to establish whether the less hydrated lipids shown in the Table can be so accommodated. If not, and it seems unlikely (2), one must conclude that such phospholipid vesicle systems are so qualitatively different from, and so fail to mimic, the cellular process of fusion that fundamentally different mechanisms are involved.

CONCLUSIONS

Phosphatidylcholine appears to be the most hydrated of bilayer lipids tested. PS, with reduced electrostatic repulsion, and DGDG bilayers both have very low degrees of hydration and can achieve close apposition. PE has variable hydration sensitivity to its hydrocarbon chain composition. In all cases with negligible electrostatic repulsion, polar groups assembled into the interbilayer space of lamellar phases appear to be held at concentrations intermediate between the saturated solutions and the crystals of the free polar groups. We suggest that the large variation in polar group hydration may reflect a competition between the tendency of the polar groups to precipitate and the restraints put on precipitation by packing constraints of the attached hydrocarbon chains. The strength of the tendency to precipitate may depend on the ability to form hydrogen-bonded water bridges across the interbilayer space between complementary groups on apposing surfaces. But other factors in the chemistry of interacting polar groups that affect their solubility would also be expected to contribute to that tendency. Serious mismatch of the hydration properties of lipids in the contact area between bilayer vesicles can lead to lipid segregation and vesicle breakdown. Any fusion seen as such systems disintegrate must be fundamentally different from the cellular process.

ACKNOWLEDGEMENTS: We are grateful for the expert assistance of Mrs. Nola Fuller and for reference to the initial results of Jens Coorssen on lipid segregation. R. P. R. is supported by a grant from the Natural Sciences and Engineering Research Council of Canada.

REFERENCES

(1) Rand, R. P. 1981. Interacting Phospholipid Bilayers: Measured Forces and Induced Structural Changes. Ann. Rev.Biophys. Bioeng. 10:277-314.

(2) Rand, R. P. and V. A. Parsegian 1986. Mimicry and Mechanism in Phospholipid Models of Membrane Fusion. Ann. Rev. Physiol. 48:201-212.

(3) Rau, D. C., Lee, B. K., Parsegian, V. A. 1984. Measurement of the Repulsive Forces between Parallel DNA Double Helices. Proc. Nat Acad. Sci. USA. 81:2621-2625.

(4) Parsegian, V. A. Rand, R. P. and D. C. Rau 1985. Hydration Forces: What Next. Chemica Scripta 25:28-31.

(5) Parsegian, V. A., Rand, R. P. 1983. Membrane Interaction and Deformation. Ann. N. Y. Acad. Sci. 416:1-12.

(6) Rand, R. P., Kachar, B., Reese, T. S. 1985. Dynamic Morphology of Interacting Phosphatidylserine Vesicles. Biophys J.47:483-89.

(7) Kachar, B., Fuller, N., and R. P. Rand. 1986. Morphological Responses to Calcium-induced Interaction of Phosphatidylserine-containing Vesicles. Biophys J. 50:779-788.

(8) Duzgunes, N., Nir, S., Wilschut, J., Bentz, J., Newton, C., Portis, A., Papahadjopoulos, D. 1981. Calcium- and Magnesium-induced Fusion of Mixed Phosphatidylserine Vesicles: Effect of Ion Binding. J. Memb. Biol. 59:115-125.

(9) Parsegian, V. A., Fuller, N. L., Rand, R. P. 1979. Measured Work of Deformation and Repulsion of Lecithin Bilayers. Proc. Nat. Acad. Sci. USA 76:2750-2754.

(10) Parsegian V. A., Rand R. P., Fuller, N. L. and D. C. Rau (1986) Osmotic Stress for the Direct Measurement of Intermolecular Forces. in Methods in Enzymology 127:400-416.

(11) Luzzati, V., Husson, F. 1962. The Structure of the Liquid Crystalline Phases of Lipid Water Systems. J. Cell. Biol. 12:207-219.

(12) Schumacher, G. and H. Sandermann 1976. Solubility of Phospholipid Polar Group Model Compounds in Water. Biochim. Biophys. Acta. 448:642-644.

(13) Evans E. and V. A. Parsegian 1986. Thermal-mechanical Fluctuations Enhance Repulsion Between Bimolecular Layers. Proc. Nat. Acad. Sci. 83:7132-7136.

(14) Hauser, H., Pasher, I., Pearson, R. H. and Sundell, S. 1981. Preferred Conformation and Molecular Packing of Phosphatidylethanolamine and Phosphatidylcholine. Biochim. Biophys. Acta. 650:21-51.

(15) Kolber M.A. and D. H. Haynes, Evidence for a Role of Phosphatidyl Ethanolamine as a Modulator of Membrane-membrane Contact, 1979. J. Memb. Biol. 48:95-114.

(16) Seddon, J. M., Cevc, G., Kaye, R. D. and D. Marsh 1984. X-ray Diffraction Study of the Polymorphism of Hydrated Diacyl- and Diakylphosphatidylethanolamines. Biochemistry 23: 2634-2644.

(17) McIntosh, T. J. and S. A. Simon, 1986. Area per Molecule and Distribution of Water in Fully hydrated Dilauroylphosphatidylethanolamine Bilayers. Biochemistry 25:4948-4952

(18) Jendrasiak, G. L. and J. C. Mendible, 1976. The Phospholipid Head-Group Orientation: Effect on Hydration and Electrical Conductivity. Biochim. Biophys. Acta 424:149-158.

(19) Mulukutla, S. and G. G. Shipley 1984. Structure and Thermotropic Properties of Phosphatidylethanolamine and its N-Methyl Derivatives. Biochemistry 23:2514-2519.

(20) Marcelja, S., Radic, N., 1976. Repulsion of Interfaces due to Boundary Water. Chem. Phys. Lett. 42:129-130.

(21) Boggs, J., 1984. Intermolecular Hydrogen Bonding Between Membrane Lipids. Biomembranes 12:3-43.

(22) Evans, E., Metcalfe, M. 1984. Free Energy Potential for Aggregation of Giant, Neutral Lipid Bilayer Vesicles by van der Waals Attraction. Biophys. J. 46:423-425.

(23) Evans, E. A., Parsegian, V. A. 1983. Energetics of Membrane Deformation and Adhesion in Cell and Vesicle Aggregation. Ann N.Y. Acad. Sci. 416:13-33.

(24) Evans, E., and D. Needham, 1986. Giant Vesicle Bilayers Composed of Mixtures of Lipids, Cholesterol and Polypeptides. Farad. Soc. Discussions 81: (in press).

(25) Feigensen, G. W., 1986. On the Nature of Calcium Ion Binding between Phosphatidylserine Lamellae. Biochemistry 25:5819-582.

(26) Marra, J., 1985. Controlled Deposition of Lipid Monolayers and Bilayers onto Mica and Direct Force Measurements Between Galactolipid Bilayers in Aqueous Solutions. J. Coll. Interf. Sci. 107:446-458.

INTRINSIC COLLOIDAL ATTRACTION/REPULSION BETWEEN LIPID BILAYERS AND STRONG ATTRACTION INDUCED BY NON-ADSORBING POLYMERS

Evan Evans* and David Needham+

*Pathology and Physics, University of British Columbia, Vancouver, B.C. Canada; +Mechanical Engineering and Material Science, Duke University, Durham, North Carolina

INTRODUCTION

Surfactant bilayers in aqueous media interact non-specifically via long range electrostatic, electrodynamic, and solvation forces.[1-5] Even though the magnitude of each interaction can be very large, the free energy minimum at stable contact is usually small with a progressive increase in depth as bilayer separation decreases. Addition of large, non-adsorbing polymers (e.g. dextran or polyethylene oxide) can greatly augment the weak natural attraction between bilayers to force them into closer proximity (perhaps sufficiently close to facilitate fusion). With micromechanical experiments on giant bilayer vesicles, direct measurements are made of free energy potentials for assembly of two bilayers to adhesive contact in salt solutions and in concentrated polymer solutions.[6-14] Results for neutral and charged lipid bilayers in salt buffer correlate well with classical prescriptions for van der Waals attraction and electric double-layer repulsion based on structural data derived from published X-ray diffraction studies. Adhesion in concentrated solutions of non-adsorbing polymers is promoted by interaction of depletion layers due to polymer exclusion from the bilayer surfaces. For equilibrium exchange of polymer between the gap and bulk regions, the added attraction is shown to be simply the osmotic pressure difference between the bulk polymer concentration and the depreciated value at the mid-point of bilayer separation.[14] With only the virial coefficients derived from osmotic pressure measurements of the polymer solutions, a self-consistent mean field (SCMF) theory predicts the concentration-dependence of the adhesion energies very well for all molecular weights tested and <u>also</u> predicts the observed attenuation of adhesion energy for electric double-layer repulsion between charged bilayers in polymer solutions. For neutral lipids, the SCMF theory predicts that no polymer remains in the gap at equilibrium separation. A key finding is that the attraction in polyethylene oxide (PEO) solutions is three-fold greater than in dextran which leads to smaller bilayer separations at common polymer volume fractions.

DIRECT MEASUREMENT OF BILAYER-BILAYER ATTRACTION

The experimental approach is to create large bilayer capsules that can be manipulated to test surface-surface adhesivity. Lipids from two classes have been used: synthetic phospholipids (1-stearoyl-2-oleoyl-phosphatidylcholine SOPC, 1-palmitoyl-2-oleoyl-phosphatidylserine POPS, and dioleoyl phosphatidyl glycerol DOPG)) and a natural "sugar" lipid (digalactosyl diacylglycerol DGDG). These lipids were chosen to represent two types of neutral head group structures and two electrically charged species. Diacyl lipids spontaneously form closed vesicular capsules when hydrated from an anhydrous state in non-electrolyte buffers. A few of these capsules are of sufficient size ($10-20 \times 10^{-4}$cm) that they can be subjected to direct micromechanical adhesion tests. When resuspended in iso-osmotic salt solutions, the small difference in index of refraction between the interior and exterior of the vesicle greatly enhances the optical image as shown in Fig. 1. Because of the extremely low solubility of lipid molecules in aqueous media, single-walled vesicles form cohesive surfaces with limited area compressibility.[8–10] Further, because of the osmotic strength of solutes trapped inside the vesicles and the small pressure differentials applied to vesicles, vesicle volumes remain essentially constant during mechanical tests. Hence, vesicles

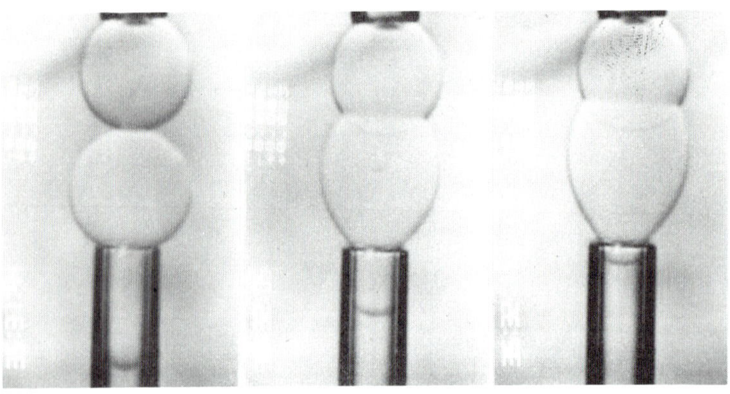

Fig. 1. Video micrographs of a vesicle adhesion test (Pipet calibre ~5 x 10^{-4}cm; vesicle diameters ~2 x 10^{-3}cm.).

deform as liquid-filled bags with nearly constant surface area and volume; spherical vesicles appear rigid and undeformable. Provided the surface is not forced to dilate, bilayers only oppose deformation with a small bending rigidity.[8–10] For big vesicles, the bending stiffness is so small that when the vesicle is slightly deflated by osmotic increases in the exterior solution, the vesicle becomes completely flaccid and deformable.

We have taken advantage of the general deformability properties of bilayers vesicles to establish a sensitive method for measurement of adhesion energy between vesicle surfaces.[6–14] First, two spherical vesicles are selected and transferred from the initial suspension in a chamber on a microscope stage to an adjacent chamber which contains a slightly more concentrated buffer (0.1 M pH 7.0 phosphate buffered NaCl–PBS) or PBS plus polymer at the desired concentration. There, the vesicles rapidly deflate to new equilibrium volumes. One vesicle is aspirated by a small micropipet and held with sufficient suction to form a rigid spherical segment outside of the pipet; this forms the "test" surface for adhesion. The second vesicle is aspirated by another pipet with low suction controlled to regulate the adhesion process. The second vesicle is then maneuvered into close proximity of the test vesicle surface (Fig. 1a) and the adhesion process is allowed to proceed in discrete (equilibrium) steps established by pipet suction (Fig. 1b,c). This experimental procedure yields the tension in the adherent vesicle bilayer as a function of the extent of coverage of the test vesicle. The extent of coverage x_c of the test vesicle is measured by the polar height z_c of the adhesion "cap" divided by the diameter $2R_s$ of the spherical test surface ($x_c = z_c/2R_s$). As illustrated in Fig. 2, adhesion in both salt buffer and buffer plus polymer is thermodynamically reversible.

The distance scale for forces that act between bilayer surfaces is much smaller than the macroscopic dimensions observable in these experiments; hence, only cumulation of the forces into an integral over distance is

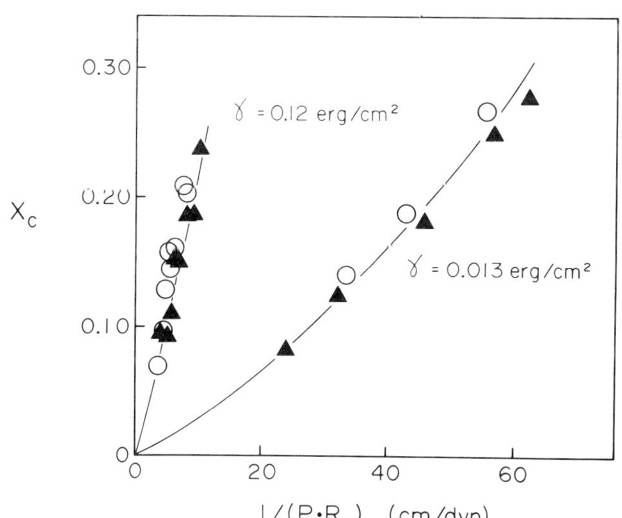

Fig. 2. Fractional area x_c of the rigid vesicle covered by the adherent vesicle *versus* the reciprocal of suction pressure P multiplied by pipet radius R_p. Triangles (▲) represent contact formation and the open circles (O) represent separation of the contact. Curves are predictions from mechanical analysis for uniform values of adhesion energy. The lower curve is for neutral SOPC vesicles in salt buffer. The upper curve is for neutral SOPC vesicles but in salt buffer *plus* 9.3 g/100cm^3 dextran – 36500 MW polymer.

measurable. This integral is the negative work or free energy reduction per unit area for assembly of the bilayers from infinite separation to stable contact where the force between the surfaces is zero,

$$\gamma \equiv - \int_{\infty}^{z_g} \sigma_n \cdot dz \tag{1}$$

The stress σ_n normal to the surfaces is the total action of both colloidal forces and the force induced by polymer in solution. Analysis of mechanical equilibrium for the adhesion process leads to a direct relation between the free energy potential for adhesion and the suction applied to the adherent vesicle,[11-12]

$$\frac{\gamma}{P \cdot R_p} = f(\text{geom}) \tag{2}$$

When the product of pipet suction P and radius R_p is converted to tension for the adherent vesicle bilayer, this equation takes the form of the classic Young equation where the geometric factor is $(1-\cos \Theta_c)$; Θ_c is the included angle between the bilayer surfaces. Because of the special deformation properties of vesicles outlined previously, the contact angle can be derived from measurements of the extent of encapsulation of the spherical test surface with a computational algorithm.[11] If the free energy potential for adhesion is uniform over the adhesion zone, then a single curve is predicted for the relationship between pipet suction and the fractional extent of coverage of the test vesicle (as shown in Fig. 2).

COLLOIDAL ATTRACTION AND REPULSION BETWEEN BILAYERS IN SALT SOLUTIONS

Separated by a medium with different polarizability properties, bilayers are drawn together by long-range van der Waals forces described by Lifshitz theory for continuous media.[15] Electrically-charged lipids in salt solutions oppose this attraction with double-layer forces at large distances; but at short distances, solvation and structure forces associated with the intervening liquid and surfactant head groups dominate repulsion for all lipids.[1-5] Reversible adhesion will occur if the attraction is sufficient to overcome long-range repulsion but not sufficient to overwhelm the short-range forces that prevent collapse and phase instability. X-ray diffraction studies of forced dehydration of multilamellar lipid-water phases show that the short-range repulsion follows a steep exponential-like decay.[2,16] Hence, the energetics of adhesion for electrically neutral bilayers can be conceptually viewed as approach along a soft van der Waals attraction to a limit determined by the magnitude and decay of the strong repulsion. The free energy potential for assembly of neutral bilayers to stable contact essentially represents the attractive van der Waals potential at the final separation distance. If the bilayers are unsupported, repulsion may be enhanced by secondary effects due to thermo-mechanical excitation of bilayer undulations.[17-18] For lipids with electrically-charged head groups, long-range repulsion follows an exponential decay determined by the characteristic screening distance of the ions in solution as described by Guoy–Chapman theory.[1] At a specific separation z_g, the total free energy per unit area of contact formation is given approximately by,[2,18]

$$\tilde{F} = \lambda_{sr} \cdot P_{sr} \cdot e^{-z_g/\lambda_{sr}} - A_H \cdot f(z_\ell/z_g)/(12 \cdot \pi \cdot z_g^{\,2}) + \lambda_{es} \cdot P_{es} \cdot e^{-z_g/\lambda_{es}} + \tilde{F}_{f\ell} \tag{3}$$

P_{sr} is the coefficient for the strong short-range repulsion and λ_{sr} is the decay length; A_H is the Hamaker coefficient for the van der Waals attraction and $f(z_l/z_g)$ is a weak retardation function of the ratio of the bilayer thickness z_l to the distance between bilayers; P_{es} is the coefficient for the electric double-layer repulsion and λ_{es} is the decay length. The last term in the free energy expression $\tilde{F}_{fl}$ is the free energy excess due to thermally-excited mechanical undulations of the bilayers. The stress applied normal to the bilayer surfaces is given by the derivative of Eq. (3) as,

$$\sigma_n^c = -P_{sr}\cdot e^{-z_g/\lambda_{sr}} + \frac{A_H\cdot f(z_g/z_\ell)}{6\pi\cdot z_g^3} - P_{es}\cdot e^{-z_g/\lambda_{es}} - \sigma_{f\ell} \qquad (4)$$

Equilibrium contact is specified by the condition at the secondary minimum where the stress σ_n is equal to zero.

Even though the interactions are given as functions of the same separation distance, the actual origin for each interaction is difficult to define because of molecular motion and architecture. The "mass-average" water gap between bilayers seems to be the best functional definition of distance because compositional and colligative properties are based on water content and work to remove water from lipid dispersions.[2,19] Compositional and x-ray diffraction studies of lamellar arrays naturally combine to yield the mean water gap with only the assumption of a layered structure.[20] This choice of common distance creates difficulty in the critical evaluation of van der Waals attraction and electric double-layer repulsion when the bilayers are in close proximity ($z_g < 3 \times 10^{-7}$ cm). To circumvent this difficulty, the usual approach is to introduce arbitrary displacements for the origins of these interactions which represent the location of head groups and charges relative to the mass-average water interface.

Free Energy Potentials for Neutral Bilayer Adhesion

Adhesion tests for bilayers composed of neutral PC's (egg PC, DMPC, SOPC) in the liquid state yield comparable values (0.01 – 0.015 erg/cm^2) of free energy reduction per unit area of contact.[10,13,21] Vesicles made from less-hydrated phosphatidylethanolamine POPE exhibit much stronger adhesion with energies that are an order of magnitude greater (0.12–0.15 erg/cm^2). Even larger free energy potentials (0.22 erg/cm^2) are measured for adhesion with DGDG (sugar-lipid) bilayers.[10,21] X-ray diffraction studies of forced dehydration (e.g. by osmotic or mechanical stress) of multilamellar arrays provide the essential measures of separation distance and the interlamellar stress at close range dominated by $P_{sr}EXP(-z_g/\lambda_{sr})$.[2,15,22] Properties for particular neutral lipids in the L_α (liquid) phase are given in Table 1. Obviously, larger values of adhesion energy are characteristic of smaller separations at full hydration. Note, however, DGDG and POPE form adhesive contact at similar separations but the adhesion energy for DGDG bilayers is nearly twice as strong as that for POPE bilayers. Since the phospholipids and the "sugar"-lipid possess similar acyl chain cores, the conclusion is that the augmented attraction is due to attraction between sugar head groups of DGDG across the water gap. Thus, a layered model must be introduced to represent the differential attraction properties of head groups and acyl chains across the water gap between bilayers.

Theoretical prescriptions for attraction between layered composites are well established and given approximately by,[23-24]

$$\tilde{F}_A \cong -A_1\cdot f_1/\left[12\pi\cdot(z_g-z_p)^2\right] - A_2\cdot f_2/\left[12\pi\cdot z_g^2\right] - A_3\cdot f_3/\left[12\pi(z_g+z_p)^2\right]$$

where again z_g is the "mass-average" water separation; $z_p/2$ is the head group projection beyond the mass-average water interface. The retardation functions (f_1, f_2, f_3) depend on the ratio of acyl chain thickness to water gap separation and polar head group thickness to water gap separation and are on the order of unity at close range.[23] The Hamaker coefficients (A_1, A_2, A_3) are related to correlations of differences in polarizability between adjacent layers integrated over the electromagnetic spectrum: $A_1 \sim$ (head group-water)^2; $A_2 \sim$ (head group-water)$\cdot$(acyl chains-head group); $A_3 \sim$ (acyl chains-head group)^2. Simple models of head group geometry indicate projections beyond the mass-average water interface of about 3 Å for the phospholipids and 4-5 Å for DGDG.[25-27] For bilayers in close proximity (e.g. POPE and DGDG), head group attraction across the water gap will be the major effect even for small Hamaker coefficients (A_1). At large separations away from equilibrium contact, attraction is characterized by the cumulative coefficient ($A_1 + A_2 + A_3$). Three coefficients for attraction plus geometric properties of the head group allow great freedom in selection of (A_1, A_2, A_3) consistent with measurements of adhesion energy. However, theoretical calculations indicate that the cross-coefficient A_2 is very small ($\lesssim 10^{-15}$ erg) and that the coefficient A_3 for hydrocarbon attraction across the head group region remains essentially constant ($\sim 3 \times 10^{-14}$ erg) for a wide range of head group properties.[23,28] Thus, only the coefficient A_1 is used to correlate the experimental results. For the phospholipids SOPC and POPE, A_1 is estimated to be about 1.2×10^{-14} erg and about 3×10^{-14} erg for the sugar-lipid DGDG. The significant observation is that head groups appear to play an important role in attraction at close range.

Table 1. Mass-Average Water Separation and Parameters for Short-Range Repulsion Derived from X-Ray Diffraction Studies[22]

Neutral Lipid	z_g ($\times 10^{-8}$cm)	P_{sr} (dyn/cm^2)	λ_{sr} ($\times 10^{-8}$cm)
SOPC	27	1.0×10^{10}	2.45
POPE	11.5	1.1×10^{12}	0.95
DGDG	13	1.3×10^{11}	1.50

These estimates for attraction coefficients must be refined because of the free energy contribution due to the short-range repulsion. If we accept an exponential form for the short-range repulsion, attenuation of the adhesion energy by repulsion is about 10-40% based on the parameters (P_{sr}, λ_{sr}) in Table 1; coefficients for attraction must be increased by about 50% to match the measured adhesion energies. If soft repulsive effects exist further out, the energy contribution could be much greater than deduced from $\lambda_{sr}P_{sr}\mathrm{EXP}(-z_g/\lambda_{sr})$. For example, softer repulsion is predicted from analysis of thermally excited bending undulations in unsupported bilayers.[17-18] Repulsion is enhanced by the work required to drive heat out of the system as the bilayers are forced together. For separations less than equilibrium contact, the free energy excess due to bilayer undulations is determined by the short-range repulsion,[18]

$$\tilde{F}_{f\ell} \sim (\pi \cdot k \cdot T/16)\sqrt{P_{sr}/B \cdot \lambda_{sr}} \cdot e^{-z_g/2 \cdot \lambda_{sr}}$$

where k is Boltzmann's constant and B is the bilayer curvature or bending elastic modulus. In this range, the primary effect of the added repulsion is to increase the observed decay length λ_{sr} and expand the separation at full hydration. For unsupported SOPC multilayers, analysis shows that the measured 2.45 Å decay would reduce to 2.1 Å if the layers were rigidly fixed to solid substrates. An auxiliary prediction is that the adhesion energy for <u>immobilized</u> PC bilayers should be larger than the value measured for unsupported PC vesicle membranes. This may account for the stronger adhesion energies measured for PC bilayers adsorbed to rigid mica sheets.[29] At distances well beyond equilibrium separation, the free energy excess from thermo-mechanical excitations is predicted to follow a weak steric repulsion given by,

$$\tilde{F}_{f\ell} \sim c \cdot (\pi \cdot k \cdot T/16)^2 / (B \cdot z_g^{\ 2})$$

where c ≅ 5.[17–18] For neutral bilayers, this weak steric repulsion should overcome the retarded van der Waals attraction at large separations ($>10^{-6}$ cm). However, the repulsive stress is predicted to be so small (<30 dyn/cm^2) that the barrier is insignificant.

<u>Double-Layer Repulsion Between Charged Bilayers in Salt Solutions</u>

Adhesion energies for mixtures of electrically charged (POPS, DOPG) and neutral (SOPC, DGDG) lipids in 0.1 M NaCl are shown in Fig. 3. A few percent of charged lipid overcomes the weak van der Waals attraction.[10] Based on

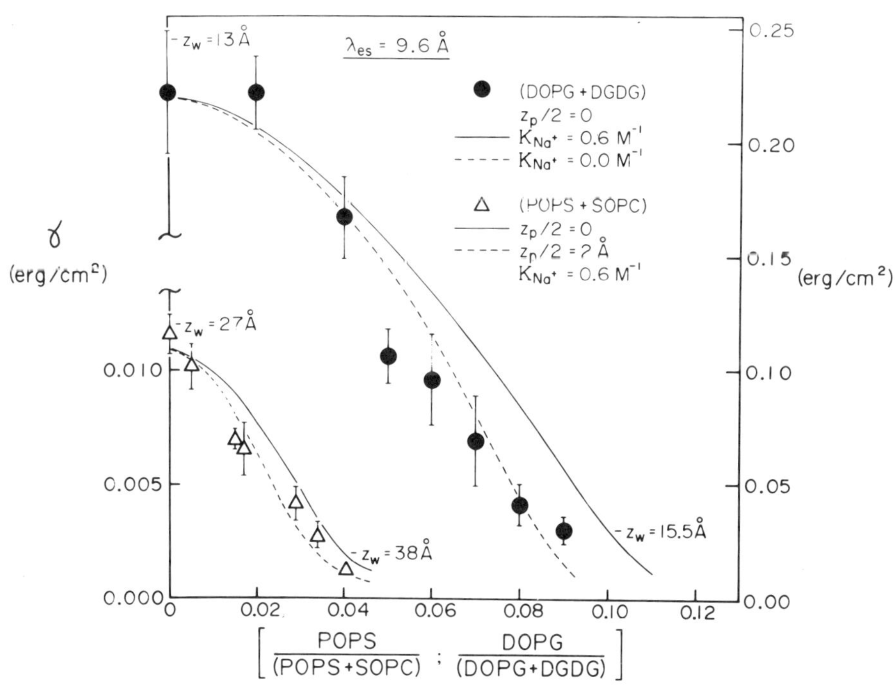

Fig. 3. Attenuation of adhesion energy by double-layer repulsion in 0.1M NaCl for electrically charged vesicles <u>versus</u> composition. Curves are predictions of electric double-layer theory.

area per lipid molecule derived from x-ray diffraction studies (e.g. $A_{SOPC} = 66–68\ Å^2$, $A_{DGDG} = 72\ Å^2$)[22] and one electrical charge per PS or PG, adhesion energies are calculated by inclusion of double-layer repulsion in the analysis for neutral bilayers.[30] The curves shown in Fig. 3 illustrate the effect of charge position and ion binding. Conceptually, the results can be viewed as reduction of the adhesion energy at essentially fixed separation – as the charge density is increased – followed by a "disjoining" transition where the bilayers separate to large distances. Ideal double-layer theory works well at this separation of 1–3 Debye lengths ($\lambda_{es} = 9.6$ A in 0.1 M NaCl). Furthermore, the outward displacement of the bilayers (implicit in the "disjoining" transition) is consistent with the combined distance dependence for attraction and double-layer repulsion.

POLYMER-INDUCED ATTRACTION BETWEEN BILAYERS IN SALT SOLUTIONS

The conceptual view of an adhesion process promoted by non-adsorbing polymers is that there is exclusion or depletion of polymer in the vicinity of the surface.[14,32–33] Theory shows that attraction is caused by interaction of the depleted concentration profiles associated with each surface which leads to a depreciated polymer concentration at the center of the gap. The concentration reduction in the gap relative to the exterior bulk solution gives rise to an osmotic effect that acts to draw the surfaces together.[14] When equilibrium exists between the gap and bulk, stabilization or approach to a plateau level is not anticipated in good solvents unless other interactions are present. The free energy potential for adhesion is expected to increase progressively with concentration even at large volume fractions. This is opposite to the behaviour predicted in the case of polymer adsorption.[34–39] Hence, (non-adsorption) depletion-based processes depend primarily on the colligative properties of the polymer in the solution. If polymer does not interact with the electric field between bilayers nor contribute significantly to the polarizability of the aqueous region between the surfaces, then these colloidal forces can be treated simply as external stresses that superpose on the action of the polymer, i.e.

$$\sigma_n = \sigma_n^c + \sigma_n^p \tag{5}$$

Adhesion tests for lipid bilayer vesicles in dextran and polyethylene oxide solutions are consistent with the (non-adsorption) depletion type of interaction.[14] This conclusion is based on: (i) the null observation that fluorescently labelled polymer cannot be detected in the gap between bilayers; (ii) the progressive increase in adhesion energy for neutral bilayers as the polymer concentration is increased to large volume fractions; and (iii) the transfer of adherent vesicle pairs with subsequent separation which shows that adhesion energy depends only on composition of the medium exterior to the contact zone. Hence, we have carefully developed a depletion-based theory for analysis of these experiments.[14] Here, we outline the mean-field approach and the attraction predicted for non-adsorbant polymers in good solvents.

Method of Analysis

Theoretical development over the past decade or so has been designed to predict the configurations and distribution of polymer segments in the vicinity of a solid surface or between surfaces and thereby to predict the deviation of the free energy of mixing from that in the adjacent bulk region.[40–46] The simplest approach is to introduce a "constitutive" relation for the free energy as a sum of the free energy for uniform concentration plus a term that represents the free energy excess due to configurational entropy gradients.[32,38,42,47-48] For linear flexible polymers, the relation can be deduced from a statistical equation for pair-correlation functions of segment distribution to give a self-consistent mean-field (SCMF) approximation.[42,45] Here, the free energy of the polymer confined to a gap thickness of z_g and area A is expressed by,

$$F_g = A \cdot \int_0^{z_g} \left(\tilde{F} + \frac{a_m^2}{6} \cdot \left| \frac{d\psi}{dz} \right|^2 \right) \cdot dz \tag{6}$$

where $\psi^2 = v/\nu_m$ is the expectation value for local segment concentration; v is the local volume fraction of polymer; ν_m is the molecular volume of a monomer segment; a_m is the effective length of a rigid segment of the flexible polymer. $\tilde{F}$ is the free energy density evaluated at the local concentration in the absence of gradients and given in terms of chemical potentials as,

$$\tilde{F} = \frac{v \cdot \mu_p}{N_p \cdot \upsilon_m} + \frac{(1 - v) \cdot \mu_s}{\upsilon_s}$$

or,

$$\tilde{F} = \frac{v \cdot \mu_p}{N_p \cdot \upsilon_m} - (1 - v) \cdot \Pi \tag{7}$$

where μ_p and μ_s are chemical potentials for the polymer and solvent; ν_s is the molecular volume of the solvent; N_p is the polymer index or number of segments per chain; Π is the osmotic pressure which is proportional to the chemical potential of the solvent.[49] A more useful expression for the free energy density is in terms of the chemical potential $\bar{\mu}_p$ for the polymer adjusted by the osmotic pressure work for displacement of solvent,

$$\frac{\bar{\mu}_p}{N_p \cdot \upsilon_m} \equiv \frac{\mu_p}{N_p \cdot \upsilon_m} + \Pi = \frac{\partial \tilde{F}}{\partial v} .$$

$$\tilde{F} = \frac{v \cdot \bar{\mu}_p}{N_p \cdot \upsilon_m} - \Pi \tag{8}$$

For bulk regions of large extent, the free energy F_B for the polymer solution is simply,

$$F_B = \left(\frac{\bar{v}_B \cdot \bar{\mu}_p^B}{N_p \cdot \upsilon_m} - \Pi_B \right) \cdot V_B$$

V_B is the total volume of the bulk region; $\bar{v}_B$ is the volume fraction; and Π_B is the osmotic pressure at the volume fraction $\bar{v}_B$.

If we assume "complete" equilibrium, the polymer is free to exchange between the gap and bulk regions and the free energy of the polymer in the gap is minimum. In this case, the segment distribution in the gap is specified by the variational requirement,[14]

$$\delta F_g \Big|_{(\bar{v}_g, z_g, A)} = \delta \left\{ \int_0^{z_g} \left[\frac{v \cdot \bar{\mu}_p}{N_p \cdot \upsilon_m} - \Pi + \frac{a_m^2}{6} \left| \frac{d\psi}{dz} \right|^2 - \lambda_c \cdot v \right] dz \right\} \Big|_{(z_g, A)} \equiv 0 \tag{9}$$

Further, the polymer-induced stress is explicitly given by the osmotic pressure difference between the bulk solution and the local composition at the mid-point ($z = z_g/2$) of the gap,[14]

$$\sigma_n^p = \Pi_B - \Pi_{z_g/2} \tag{10}$$

Hence, if the gap concentration is depleted with respect to the bulk, there will be attraction. On the other hand, if non-specific adsorption augments the gap concentration with respect to the bulk, there will be repulsion. In order to determine the volume fraction at the mid-point of the gap, Eq. (9) is solved to yield the optimum monomer distribution in the gap,

$$\frac{a_m^2}{6} \cdot \left| \frac{d\psi}{dz} \right|^2 = \Delta\tilde{F} - \Delta\tilde{F}_{z_g/2}$$

where,

$$\Delta\tilde{F} = \frac{v \cdot (\bar{\mu}_p - \bar{\mu}_p^B)}{N_p \cdot \upsilon_m} + (\Pi_B - \Pi)$$

which is integrated to give the mid-point relation,

$$z_g/2 = \frac{a_m}{\sqrt{6}} \int_0^{\psi_{z_g/2}} \frac{d\psi}{\left[\Delta\tilde{F} - \Delta\tilde{F}_{z_g/2} \right]^{1/2}} \tag{11}$$

We will particularize the rest of the analysis to the range of polymer concentrations dominated by the second and third virial coefficients, i.e. the "semi-dilute" range in good solvents.[45,50] The approach is clearly reasonable for solutions of the large polymers used in our adhesion tests over volume fractions of 0.01 to 0.1. For this range, the chemical potential differences (evaluated at a local volume fraction in the gap minus the reference value) are approximated by,[45,50]

$$\frac{v \cdot (\bar{\mu}_p - \bar{\mu}_p^B)}{N_p \cdot \upsilon_m} \cong \left(\frac{C_\chi}{\upsilon_s}\right) \cdot v \cdot (v - \bar{v}_B) + \frac{\beta}{2 \cdot \upsilon_s} \cdot v(v^2 - \bar{v}_B^2)$$

$$\tag{12}$$

$$(\Pi_B - \Pi) \cong \left(\frac{C_\chi}{2 \cdot \upsilon_s}\right) \cdot (\bar{v}_B^2 - v^2) + \frac{\beta}{3 \cdot \upsilon_s} \cdot (\bar{v}_B^3 - v^3)$$

where $C_\chi/2\cdot\upsilon_s$ and $\beta/3\cdot\upsilon_s$ are second and third virial coefficients in units of energy per volume.

An additional free energy contribution (from overall elastic extension – compression – of a polymer chain to fit into a gap less than twice the radius of gyration) must be taken into account because the free energy excess due to monomer concentration gradients – $a_m^2|d\psi/dz|^2$ – does not represent long-range changes in chain conformation. Hence, we add the elastic free energy $\tilde{F}_{el}$ of deformation[50] to the free energy of uniform mixing,

$$\tilde{F}_{e\ell} = \frac{\bar{v}_{B'}}{N_p\cdot\upsilon_m}\left[\left(\frac{2\cdot R_g}{z_g}\right) + \frac{1}{2}\cdot\left(\frac{z_g}{2\cdot R_g}\right)^2 - \frac{3}{2}\right] \quad ; \quad z_g \leq 2\cdot R_g$$

where $R_g = a_m N_p^{1/2}$. The effect of this ad-hoc inclusion of polymer deformation energy is to slightly change the reference state for the polymer in the gap (defined by the polymer configuration without monomer gradients). Hence, the adjusted chemical potential $\bar{\mu}_p^B$ for the reference state becomes $\bar{\mu}_p^{B'}$ given by,

$$\bar{\mu}_p^{B'} = \bar{\mu}_p^{B} - \tilde{F}_{e\ell}\cdot\upsilon_s/\bar{v}_{B'} \tag{13}$$

which defines a reference volume fraction $\bar{v}_{B'}$ in the absence of monomer gradients. In other words, $\bar{v}_{B'}$ would be the volume fraction of polymer in the gap for equilibrium exchange with the bulk region but without perturbation of the monomer distribution by the proximity of solid surfaces.

For "complete" equilibrium, Eq. (11) with the use of Eq. (12) and $\psi^2 = v/\nu_m$ determines the concentration at the midpoint of the gap,

$$z_g/2 = \frac{\zeta_m}{\sqrt{A_v}}\int_0^1 \frac{d\tilde{\psi}}{\left[(1-\tilde{\psi}^2)(1 - B_v\cdot\tilde{\psi}^2 - C_v\cdot\tilde{\psi}^4)\right]^{1/2}} \tag{14}$$

This equation is given in dimensionless form where the parameters are,

$$\zeta_m \equiv a_m\left[\frac{\upsilon_s}{6\cdot\upsilon_m\cdot\bar{v}_{B'}\cdot C_\chi}\right]^{1/2} \qquad B_v = \frac{v_{z_g/2}}{2\cdot\bar{v}_{B'}\cdot A_v}\cdot\left[1 + \frac{\beta\cdot v_{z_g/2}}{3\cdot C_\chi}\right]$$

$$A_v \equiv 1 - \frac{v_{z_g/2}}{2\cdot\bar{v}_{B'}} + \frac{\beta\cdot\bar{v}_{B'}}{2\cdot C_\chi}\cdot\left[1 - \frac{v^2_{z_g/2}}{3\cdot\bar{v}^2_{B'}}\right] \qquad C_v \equiv \frac{\beta\cdot v^2_{z_g/2}}{6\cdot C_\chi\cdot\bar{v}_{B'}\cdot A_v}$$

The distance scale ζ_m represents the depletion zone near the surface and the decay of depletion with distance. Hence, the volume fraction at the center of the gap is specified along with the normal stress induced by the polymer.

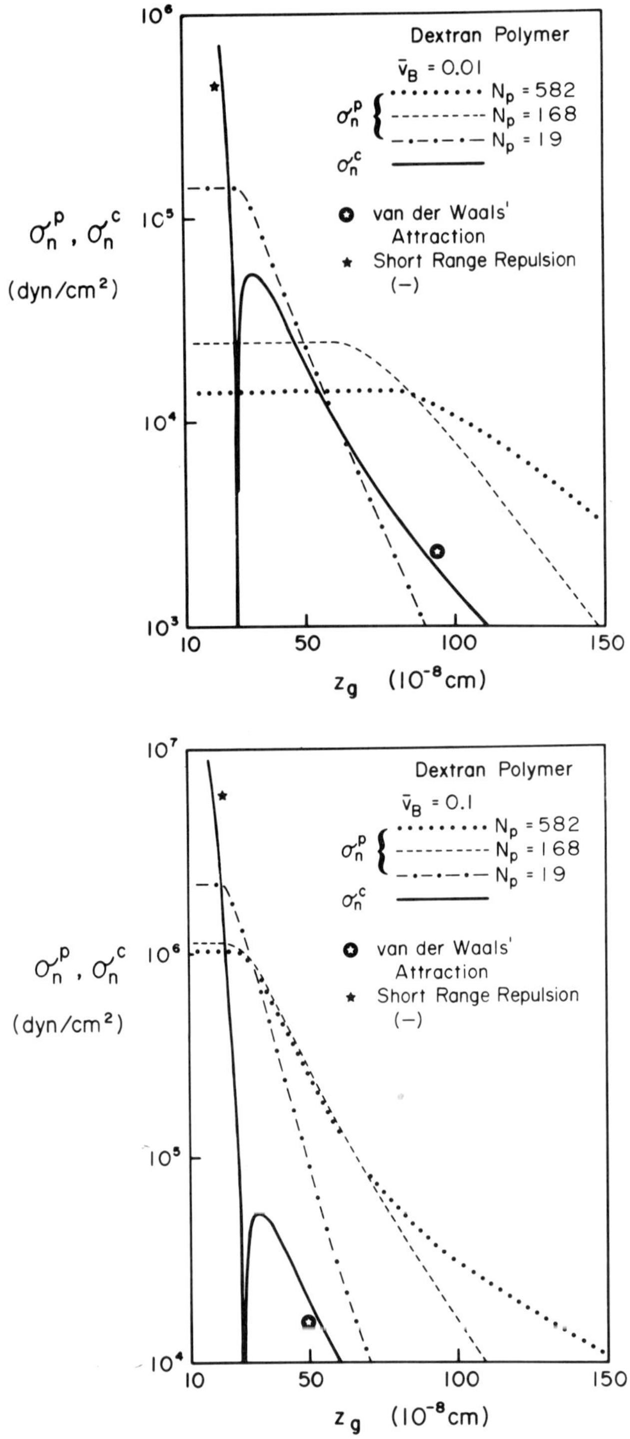

Fig. 4. Predictions of attractive stresses induced by non-adsorbant dextran polymers in water and the intrinsic stress for colloidal attraction/repulsion between bilayers *versus* separation.

Figure 4 shows the stress versus separation predicted for SOPC bilayers in dextran solutions of 0.01 and 0.1 volume fractions; plotted separately are the polymer-induced stresses and the natural colloidal interaction.[14,51] Since the polymer-induced attraction scales with osmotic pressure, the attractive stress in PEO solutions will be three-fold larger than those shown in Fig. 4 for dextran. Equilibrium separation is defined by zero total stress; hence it is apparent that the gap is reduced slightly from 27 to 24 Å at the higher polymer concentration. Integration of the total stress from infinity to equilibrium separation yields the adhesion energy.

Bilayer-Bilayer Adhesion in Polymer Solutions

Sharp-cut dextran fractions were provided by a gracious gift from Dr. K. Granath (Pharmacia, Sweden) through our colleague, Dr. D.E. Brooks (Univ. of British Columbia). Three fractions were used, labelled as 11200, 36500, and 147500 weight-average molecular weights. Sharp-cut PEO fractions were purchased from Scientific Polymer Products, Ontario, N.Y. Two fractions were used, labelled as 4948 and 19683 weight-average molecular weights. For solutions of these polymers up to 10% by volume, osmotic pressures were measured by freezing point depression and checked by vapour pressure osmometry at high concentration. Appropriate virial coefficients (listed in Table 2) were derived from the osmotic pressure data.[52] Also, the Flory[50] interaction parameter x which characterizes the quality of the solvent was derived from the second virial coefficient and found to be in the "good" solvent range, i.e. 0.43 for the dextran polymers and 0.37 for the PEO fractions.

Table 2. First, Second, Third Virial Coefficients for Dextran and PEO Fractions (from Osmometry)

Polymer Fraction (MW)	First Virial Coefficient[52] (atm-cm^3/g)	Second Virial Coefficient (atm-cm^6/g^2)	Third Virial Coefficient (atm-cm^9/g^3)	Number Average Molecular Weight[52] (M_n)
Dextran				
11,200	7.739	31.9	–	3,139
36,500	0.893	36.3	–	27,200
147,500	0.258	36.3	–	94,300
PEO				
4,948	5.39	126.2	647	4,506
19,683	1.655	117.7	647	14,677

Adhesion energies for neutral (SOPC) vesicle pairs at specific concentrations of these polymers in the range of 0–15 g/100cm^3 (wt/vol) are plotted in Fig. 5 as a function of volume fraction.[14] Even for these concentrated solutions, adhesion energies increased progressively (from 0.01 to 0.2 erg/cm^2 for dextran solutions and up to 0.6 erg/cm^2 for PEO solutions) with no tendency to plateau or saturate and with essentially no dependence on polymer size. To test the effect of molecular composition of the neutral surface, the free energy potential for adhesion of sugar-lipid (DGDG) vesicles was measured in solutions of 9.3 g/100cm^3 dextran (both

147500 and 36500 fractions) for comparison with the measurements at the same concentration with the neutral SOPC vesicles. Adhesion energies for the DGDG vesicles in both dextrans were 0.4 erg/cm^2, clearly much stronger than for the SOPC vesicles; but the differential-adhesion energies (in excess of the baseline van der Waals attraction potential) were comparable for both neutral lipids. Hence, there appeared to be no recognizable dependence on surface composition. Predictions for adhesion energies derived from the SCMF theory are plotted in Fig. 5 for each dextran polymer.[14] Similarly, adhesion

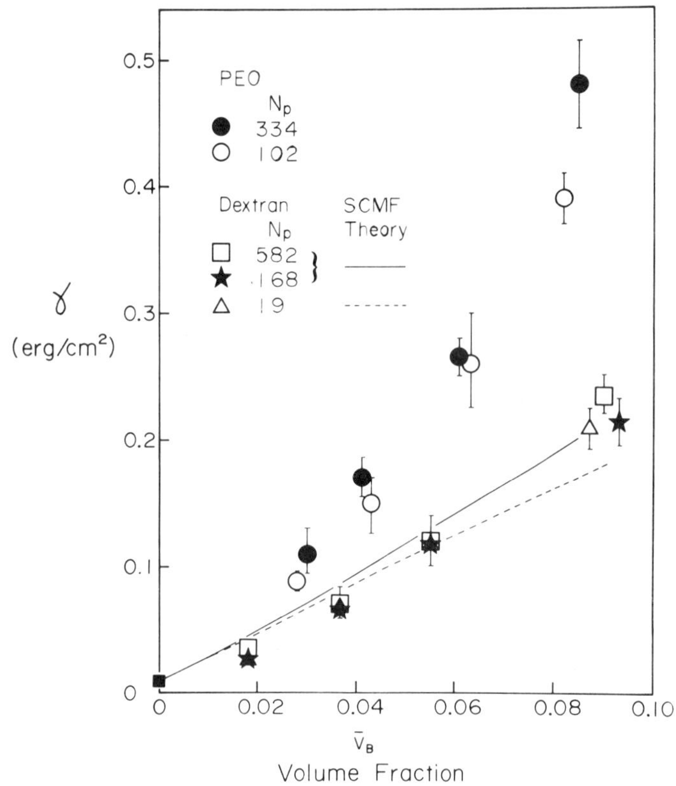

Fig. 5. Adhesion energies for neutral SOPC bilayers in 0.1M salt (PBS) plus dextran or PEO polymer at volume fractions up to 0.1. Curves are predictions for the dextran polymers obtained from Self Consistent Mean-Field theory.

between DGDG bilayers in dextran solutions have been calculated from the superposition of the colloidal attraction (i.e. 0.22 erg/cm^2 in the absence of polymer) plus the polymer-induced attraction. The polymer-induced attraction is predicted to include an added reduction (relative to the SOPC case) given by the osmotic pressure multiplied by the gap displacement from 24 Å down to 13 Å, i.e. $\gamma = 0.22 + 0.12 + (4 \times 10^5)(1.1 \times 10^{-7}) = 0.4$ erg/cm^2.

Attenuation of adhesion energy as a function of surface charge (shown in Fig. 6) was measured by assembly of mixed (SOPC + POPS) bilayer vesicles in dextran solutions fixed at 9.3 g/100cm^3 ($\bar{v}_B$ = 0.057). Unlike the behaviour of the adhesion energy for neutral lipid bilayers, there was an obvious difference between the small and higher molecular weight polymers. Adhesion could be prevented by surface charge contents greater than 15 mol% POPS in solutions of the small molecular weight polymer whereas even for POPS concentrations in excess of 30 mol%, there was a measurable level of adhesion between vesicles with the higher molecular weight polymers.[14] Predictions from SCMF theory for the attenuation of adhesion energy by the electric double-layer stresses are shown in Fig. 6.[14,30]

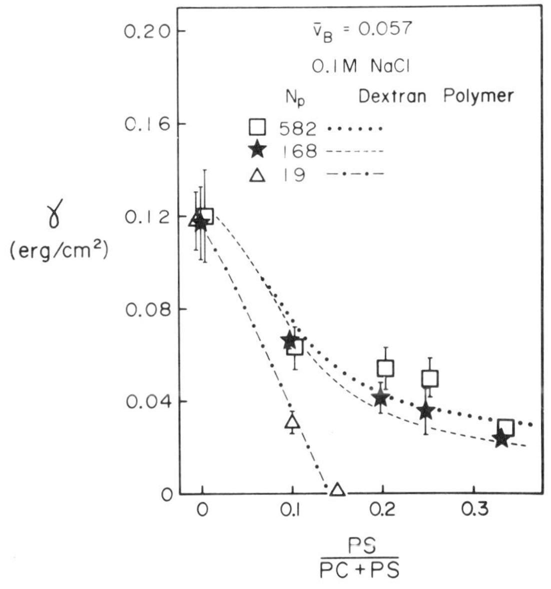

Fig. 6. Adhesion energies for charged bilayers in 0.1M salt (PBS) plus dextran polymer fixed at a volume fraction of 0.057 versus charge content given by composition. Curves are predictions from Self Consistent Mean-Field theory with electric double-layer repulsion.

The self-consistent mean-field theory for non-adsorbent polymers in good solvents agrees very well with our direct measurements of adhesion energy for neutral lipid bilayers in dextran solutions. Values for adhesion energy can be predicted directly from colligative (e.g. osmotic) properties of the polymer in solution with inclusion of the intrinsic attraction between bilayers in the absence of polymer. Also, the (SCMF) theory correlates adhesion energy attenuation measured for charged surfaces which implies that it models the distance-dependence of the polymer-induced field. The theory also works well even for the small polymer whose characteristics do not satisfy the approximations used in the analysis. This is most likely because the theoretical prediction for stress depends primarily on the scale of the depletion zone and the solution osmotic pressure. Decay of the depletion zone for each polymer size is illustrated in Fig. 4 and the effect is apparent in the separation between curves for adhesion energy with charged surfaces – Fig. 6. Although we have not yet calculated the adhesion energies predicted by SCMF theory for PEO solutions, we anticipate equally good correlation since the measured adhesion energies scale with the solution osmotic pressures in relation to dextran solutions. Clearly, the attractive stresses and energies will be three-fold larger for PEO than for dextran solutions. This may account for the observed enhancement of vesicle fusion in PEO solutions over that in dextran.

ACKNOWLEDGEMENT

This work was supported by the Medical Research Council of Canada through Grant MT 7477. EAE is grateful to the Alexander von Humboldt Foundation, Federal Republic of Germany, for a Senior Scientist Award.

REFERENCES

1. E.J.W. Verwey and J.Th.G. Overbeek: Theory of the Stability of Lyophobic Colloids (Elsevier, Amsterdam, 1948).
2. V.A. Parsegian, N. Fuller and R.P. Rand: Proc. Natl. Acad. Sci. USA, 76, 2750 (1979).
3. V.A. Parsegian: Ann. Rev. Biophys. Bioeng., 2, 221 (1973).
4. J.N. Israelachvili and R.M. Pashley: In Biophysics of Water, F. Franks Ed. (John Wiley and Sons, 1982) p. 183.
5. R.P. Rand: Ann. Rev. Biophys. Bioeng., 10, 277 (1981).
6. E. Evans: Colloids and Surfaces 1984, 10, 133.
7. E. Evans and M. Metcalfe: Biophys. J., 45, 715 (1984).
8. R. Kwok and E. Evans: Biophys. J., 35, 637 (1981).
9. E. Evans and D. Needham: Faraday Discuss. Chem. Soc., No. 81 (1987).
10. E. Evans and D. Needham: J. Phys. Chem. (in press).
11. E. Evans: Biophys. J. 1980, 31, 425.
12. E. Evans and V.A. Parsegian: In Surface Phenomena in Hemorheology: Theoretical, Experimental, and Clinical Aspects; Copley, A.L., Seaman, G.V.F., Eds. (N.Y. Acad. Sci., 1983) p. 13.
13. E. Evans and M. Metcalfe: Biophys. J., 46, 423 (1984).
14. E. Evans and D. Needham: Macromolecules (submitted).
15. E.M. Lifshitz: J. Exp. Theor. Phys. USSR, 29, 94 (1955); Sov. Phys. JETP, 2, 73 (1956).
16. L.J. Lis, M. McAlister, N. Fuller and R.P. Rand: Biophys. J., 37, 657 (1982).
17. W. Helfrich: Z. Naturforsch, 33a, 305 (1978).
18. E. Evans and V.A. Parsegian: Proc. Natl. Acad. Sci. USA, 83, 7132 (1986).
19. E. Evans and R. Skalak: Mechanics and Thermodynamics of Biomembranes (CRC Press, Boca Raton, Fla., 1980).
20. V. Luzzati: In Biological Membranes, D. Chapman Ed. (Academic Press, New York, 1968) p. 71.
21. E. Evans, D. Needham and R.P. Rand: Colloids and Surfaces (to be submitted).
22. R.P. Rand and V.A. Parsegian: (to be submitted).
23. D. Le Neveu, R.P. Rand, D. Gingell and V.A. Parsegian: Biophys. J., 18, 209 (1977).
24. B.W. Ninham and V.A. Parsegian: J. Chem. Phys., 53, 3398 (1970).
25. J. Marra: J. Colloid Interface Sci., 107, 446 (1985).
26. M. Hauser, I. Pascher, R.M. Pearson and S. Sundell: Biochim et Biophys. Acta, 650, 21 (1981).
27. T.J. McIntosh and S.A. Simon: Biochem., 25, 4058 (1986).
28. V.A. Parsegian: Private Communication; and V.A. Parsegian: Ann. Rev. Biophys. Bioeng. 1973, 2, 221.

29. J. Marra and J. Israelachvili: Biochem., 24, 4608 (1985).
30. Double-layer repulsion has been modelled by the full-nonlinear Poisson-Boltzmann equations and Guoy-Chapman theory.[1] For these small charge contents, linearized equations are completely acceptable. Also, Na^+ binding to the (PS + PC) bilayers is well-established and is characterized by an equilibrium constant of 0.6 M^{-1}.[31]
31. M. Eisenberg, T. Gresalfi, T. Riccio and S. McLaughlin: Biochem., 18, 3213 (1979).
32. J.F. Joanny, L. Leibler and P-G. de Gennes: J. Polym. Sci. Polym. Phys. Ed. 1979, 17, 1073.
33. G.J. Fleer, J.M.H.M. Scheutjens and B. Vincent: In Polymer Adsorption and Dispersion Stability, Goddard, E.D., Vincent, B., Eds. (Am. Chem. Soc., Washington, 1984) p. 245.
34. Th.F. Tadros: In Polymer Colloids, Buscall, R., Corner, T., Stageman, J.F., Eds. (Elsevier Applied Sci., London, 1985) p. 105.
35. B. Vincent: In Polymer Adsorption and Dispersion Stability, Goddard, E.D., Vincent, B., Eds. (Am. Chem. Soc., Washington, 1984) p. 1.
36. D.H. Napper: Polymeric Stabilization of Colloidal Dispersions, Academic Press, London, 1983.
37. P-G. de Gennes: J. Phys. (Paris) 1976, 37, 1445.
38. P-G. de Gennes: Macromolecules 1982, 15, 492.
39. J.M.H.M. Scheutjens and G.J. Fleer: Macromolecules 1985, 18, 182.
40. E. Mackor, and J. van der Waals: J. Colloid Sci. 1952, 7, 535.
41. S. Ash and G. Findenegg: Trans. Faraday Soc. 1971, 67, 2122.
42. S.F. Edwards: Proc. Phys. Soc. (London) 1965, 85, 613.
43. A. Silberberg: J. Chem. Phys. 1967, 46, 1105; 1968, 48, 2835.
44. R.J. Roe: J. Chem. Phys. 1974, 60, 4192.
45. P-G. de Gennes: Scaling Concepts in Polymer Physics (Cornell University Press, Ithaca, 1979).
46. J.M.H.M. Scheutjens and G.F. Fleer: Adv. Colloid Interface Sci. 1982, 16, 361; 1983, 18, 309.
47. J.W. Cahn and J.E. Hilliard: J. Chem. Phys. 1958, 28, 258.
48. G. Widom: Physica 1979, 95A, 1.
49. Note: all chemical potentials and free energies are assumed to be in units of thermal energy, i.e. normalized by kT.
50. P.J. Flory: Principles of Polymer Chemistry (Cornell University Press, Ithaca, 1953).
51. Note: The data required for these predictions are $\nu_s = 29.9 \times 10^{-24}$ cm^3; $\nu_m = 164 \times 10^{-24}$ cm^3; $a_m = \nu_m^{1/3} = 5.47 \times 10^{-8}$cm; $N_p = M_n/162$; and the virial coefficients for each polymer fraction taken from Table 2.
52. The number-average molecular weights for the larger fractions could not be accurately measured by osmotic pressure methods; hence the values given in Table 2 for the first virial coefficient are those measured by the producers.

MECHANISMS OF MEMBRANE FUSION

IN ACIDIC LIPID-CATION SYSTEMS

Robert C. MacDonald

Department of Biochemistry
Molecular Biology and Cell Biology
Northwestern University
Evanston, IL 60201

INTRODUCTION

Since the original discovery that Ca^{+2} converts small vesicles of phosphatidylserine into large sheets of spirally-wrapped membrane (1), the use of acidic lipid-divalent cation systems as models of membrane fusion has become widespread. There is now a wealth of experimental data available on the response of many different acidic lipids to a number of different cations under a wide variety of conditions (2). Many of these experimental investigations have utilized either a terbium-based fluorescence assay for mixing of vesicles contents (3) or a fluorescence energy transfer assay of membrane mixing (4). It is clear from results of both of these assays, as from early electron micrographs (1) that the membranes of many types of small vesicles merge with each other following divalent cation treatment, but the details of the pathway from initial vesicle to end product remain somewhat obscure. There is little doubt that fusion is involved at some stage, although the absence of precise terminology often leads to considerable ambiguity with regard to what kind of entity is fusing. The distinction is of some consequence, since the morphology of the species in which membranes become reconstructed places limits on the mechanical and chemical forces that may operate and hence influence the validity of the analogy that may be drawn to fusion of biological membranes. Thus, sentiment has been expressed for cautious extrapolation from these model systems to cells on the basis that extensive aggregation and leakage of contents, common features of model systems, cannot occur in cells (5-8). In spite of the differences in the overall pathway between the model and the biological systems, it remains possible that there are intermediate steps which share common mechanisms.

The elucidation of the mechanism of membrane merging in the model system is significant both for its relationship to biological systems and as a biophysical system of intrinsic interest. Several theoretical analyses have provided critically important information. In particular, the recognition of the relationship between membrane adhesion energies and the lysis tension of bilayers has provided a rational framework for understanding vesicle lysis (9-11). Similarly, finding a threshold surface tension that correlates with fusion points toward an important role of surface forces (12). In addition to the theoretical analyses, electron microscopic studies of interacting vesicles have revealed several features, particularly non-lamellar phases (13,14), which must be incorporated into a comprehensive mechanism.

A considerable amount of information has thus been amassed on a number of aspects of cation-induced fusion of acidic lipids. However, information has been lacking on π (surface pressure) and A (partial molar area) of the vesicle external surface, particularly as a function of vesicle radius. Such data are now available and they allow the formulation of mechanisms that account well for the observed differences between large and small phosphatidylserine vesicles as well as for different effects of Ca^{+2}, Mg^{+2} and other cations. Because the analysis involves the concept of surface pressure, the relevant aspects of surface chemistry will first be briefly reviewed. Subsequently, some recent measurements of the dependence of vesicle surface properties on vesicle diameter will be considered. Finally we will consider mechanisms for membrane fusion, based on an integration of experimental and theoretical results of others with the analysis of the behavior of lipid molecules in bilayers developed here.

THERMODYNAMICS OF SURFACES RELEVANT TO VESICLES AND MONOLAYERS

The free energy of a monolayer (G^{mono}) may be taken as the free energy of the surface of interest covered by a monolayer (G^s) minus the free energy of that surface lacking the monolayer, i.e.,

$$G^{mono} = G^s - G_o^s, \qquad (1)$$

so that

$$dG^{mono} = dG^s - dG_o^s \qquad (2)$$

From the First and Second Laws and the definition of Gibb's free energy, at equilibrium we have

$$dG = -SdT + VdP + \gamma da + \sum \mu_i dn_i, \qquad (3)$$

which, at constant temperature (T), pressure (P) and numbers of moles (n) of components (i) becomes,

$$\left.\frac{\partial G}{\partial a}\right)_{T,P,n_i} = \gamma, \qquad (4)$$

where a is the surface area under consideration and γ is surface tension. When equation 3 is applied to a clean surface and a surface covered by the monolayer, we have

$$\left.\frac{\partial G^{mono}}{\partial a}\right)_{T,P,n_i} = \frac{\partial G^s}{\partial a} - \frac{\partial G_o^s}{\partial a} = \gamma - \gamma_o \qquad (5)$$

Since surface pressure (π) is defined as

$$\pi = \gamma_o - \gamma, \qquad (6)$$

we have, from eqns 5 and 6,

$$\pi = -\frac{\partial G^{mono}}{\partial a}. \qquad (7)$$

Unless there are unusually large electrical or chemical interactions among head groups of the monolayer, a situation which does not appear to obtain with membrane phospholipids, the enthalpic portion of eqn. 7 is small and

$$\pi \cong \frac{T\partial S^{mono}}{\partial a}. \qquad (8)$$

This equation emphasizes the fact that the entropy of the monolayer molecules is reduced when the area available to each molecule is reduced. Because the number of configurations of alkyl chains are drastically reduced when molecules in a monolayer are compressed to the areas per molecule typically found in bilayers, the entropy term is a very steep function of area. Indeed, we have suggested elsewhere (15) that

$$\pi \cong \frac{\nu kT}{A}, \qquad (9)$$

where ν is the number of methylene groups in the alkyl chain and A is the partial molar area of the monolayer-forming molecule. The implication of this equation is that, for the relatively small area changes possible with coherent monolayers, each CH_2 group behaves as an independent entity. This relationship was derived from an analysis of the relationship between π and 1/A for dimyristoylphosphatidylchoine (DMPL) at the air/water interface, above the chain-melting temperature, as shown in diagrammatic form in Fig. 1.

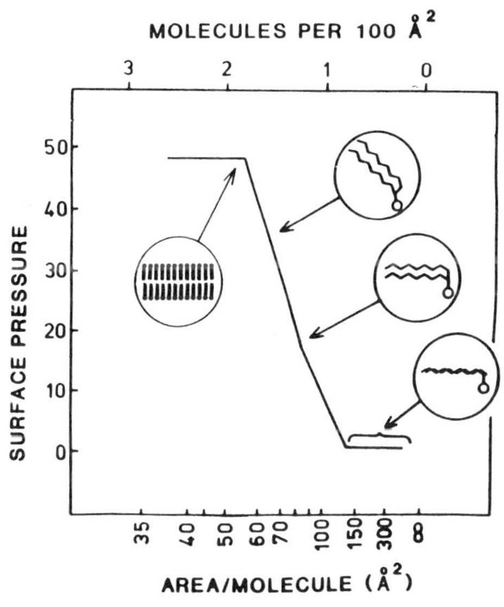

Fig. 1. Relationship between surface pressure and surface concentration for DMPC at the air/water interface.

RELATIONSHIP BETWEEN BILAYERS AND MONOLAYERS AT THE A/W INTERFACE

The π - 1/A relationship (as opposed to the π-A relationship) is dominated by straight lines (Fig. 1). At collapse, (the plateau of maximum pressure) which, for DMPC occurs at 49 dynes/cm, the monolayer is in equilibrium with bilayers in the subphase. For DMPC, the linear relationship between π and 1/A or Γ (moles per unit area; surface concentration) extends from 49 dynes/cm to 10-15 dynes/cm. The average inclination of the chains increases throughout this region. In the region below 10-15 dynes/cm the chains appear to be essentially horizontal to the water surface (15).

Our concern is with the monolayers that constitute bilayers, not monolayers at the air water interface, so we must inquire which regions of the π-1/A curve correspond to bilayers. Earlier theoretical (16,17) and experimental investigations (18) indicated that about 50 dynes/cm, corresponding to the collapse or equilibrium pressure, is the surface pressure of a monolayer of a bilayer. In addition, recently it has been found that the equilibrium pressure is the only pressure where the monolayers and bilayer vesicles of DMPC have the same transition temperature and area change at the transition (15). Finally, a simple analysis of the surface pressure of one monolayer of a bilayer relative to the other monolayer, reveals that for DMPC bilayers, the π per monolayer must be within a dyne or two of the numerical value of the oil/water tension between water and the alkyl chains that are exposed when one monolayer is peeled away from the other (19) Since oil/water interfacial tensions are typically about 50 dynes/cm, this analysis is consistent with the evidence of references 15-18.

FORCES THAT CONTROL MOLECULAR AREA IN BILAYERS

Before considering vesicle interactions *per se*, let us pause to identify the primary forces that determine the density of molecules in a bilayer in the absence of external forces.

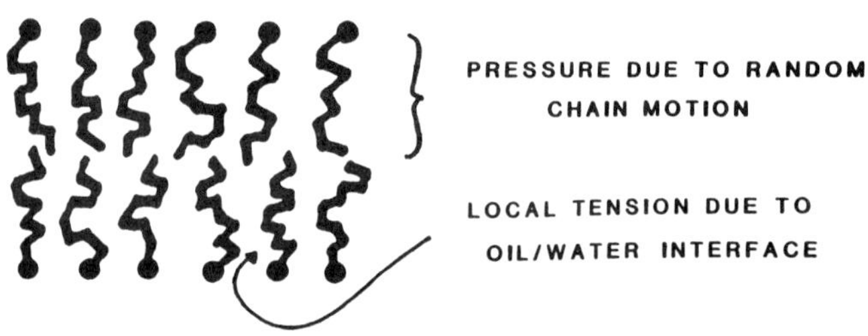

Fig. 2. Local pressure-tension balance in bilayers

As shown in Fig. 2, molecular spacing is controlled mainly by two opposing forces, the surface pressure due to the thermal motion of the molecule, primarily the alkyl chains, and the hydrophobic effect that acts to minimize the exposure of the hydrocarbon interior to water (17,26). The latter should produce a tension typical of oil/water interfaces, namely about 50 dynes/cm. This value is independent of area, whereas surface pressure is a steep function of area. The equilibrium area per molecule is maintained when pressure and local tension are equal and opposite; fluctuations to either higher or lower areas gives rise to a decrease or increase, respectively, in pressure, which brings the system back to equilibrium. It is thus obvious that surface pressure change will be the primary cause of area changes in bilayers. Hence, we will see below that Ca^{+2} reduces area per molecule by a mechanism that is basically an influence on surface pressure. When Ca^{+2} binds to 2 PS molecules, the latter no longer behave independently, so for a very brief period after binding, the pressure falls. During this short time, the tension due to the hydrophobic effect exceeds the magnitude of the pressure and the membrane contracts until the pressure rises to equal the hydrophobic effect tension again. It must be recognized that in SUV, this balance between surface pressure and local tension breaks down; pressure and local tension are not in balance, the result being that the inner monolayer is under net compression and the outer monolayer is under net tension.

PROPERTIES OF SUVs

Although the surface pressure of a monolayer in a bilayer (of DMPC) is near 50 dynes/cm, this is only true for bilayers that are flat, i.e., for large liposomes. Vesicles as small as SUVs have considerably different surface pressures. Because of the small radius of curvature in small vesicles, the polar groups on the external surface are necessarily more widely spaced than in a flat bilayer. One can estimate the additional area per molecule and the corresponding effect on monolayer surface pressure as well as vesicle surface tension. The procedure is clearly approximate, but provides a useful estimate of the relative change of bilayer organization as a function of vesicle radius. We begin by recognizing that the methyl termini of the alkyl chains of the external monolayer lipids cannot be packed more closely than the cross-sectional area of an alkyl chain, or about 20Å^2. We thus take the area per molecule as 40Å^2 at the methyl group plane and calculate the corresponding area at the vesicle surface. Using 20Å as the length of the molecule, we obtain the area data (A) shown in Table I. The value given for infinite diameter is measured on monolayers at the collapse pressure. There is no obvious reason why the $\pi A = 13kT$ relationship for DMPC monolayers at the air/water interface should not apply to the bilayer phase as well, so the A values given in the table can easily be converted to π as well as to γ. These values are shown in the table along with γA and πA, the surface energy of the vesicle on a per mole lipid basis and the energy of a mole of lipid in the surface due to surface pressure.

Table I. Surface energy parameters of DMPC SUVs

Diameter Å	A (Å^2/molecule)	π (dyne/cm)	γ (dyne/cm)	πA (kT)	γA (kT)
175	91	10	39	2.2	8.6
200	84	15	34	3.1	6.9
225	80	18	31	3.5	6.0
250	77	21	28	3.9	5.2
300	72	25	24	4.4	4.2
400	67	30	19	4.9	3.1
500	64	34	15	5.3	2.3
750	60	39	10	5.7	1.4
1000	59	41	8	5.9	1.1
2000	56	46	3	6.3	0.04
infinite	54	49	0	6.4	0

Before considering the data in this table further, it is prudent to inquire whether the accuracy of the calculated surface pressure can be verified by reference to experiment. Surface pressure of air/water monolayers of dioleoylphosphatidylcholine in equilibrium with vesicles have been determined as a function of vesicle diameter (20). Although the calculations of Table I were done for DMPC, they should be approximately in agreement with the experimental DOPC values if the procedure is valid. A comparison of experimental and calculated data is shown in Fig. 3.

While the agreement between measured and calculated pressure is only fair, perhaps partly because of the difference in fatty acyl groups, both sets of data clearly illustrate the significant effect of curvature on surface pressure. Although it is perhaps surprising that surface pressures for large and small vesicles can be so different, we have verified that the equilibrium surface pressure of DMPC SUVs can be as low as 15 dynes/cm, which corresponds to the usual lower size limit of SUVs. The surface pressure of small vesicles is sufficiently lower than that of large vesicles to rationalize the observed migration of lipid from large to small vesicles (21,19).

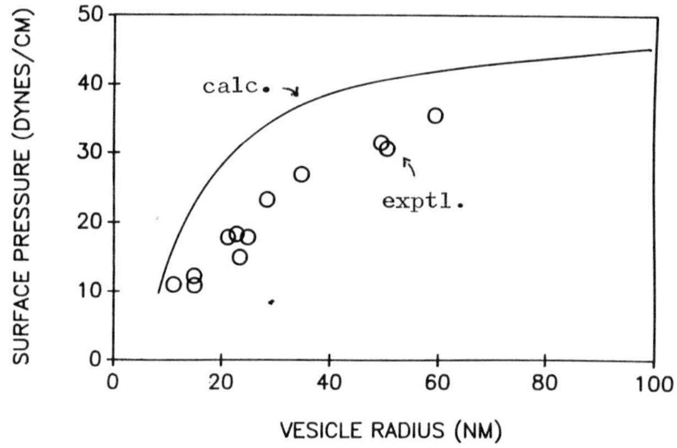

Fig. 3. Comparison of measured and calculated equilibrium surface pressure of phosphatidylcholine vesicles

Ohki has called attention to the importance of tension in fusion (11,21), and indeed, it is evident from the γA column that there is a considerable driving force for growth of small vesicles. There can be little doubt that their high surface energy accounts for the common observation that SUVs fuse much more readily than LUVs (2). The πA column reveals that although small vesicles have a higher surface energy than large vesicles, the surface pressure and hence, the chemical potential of the surface molecules, is lower in small than in large vesicles. Recognition of this distinction between γ of the vesicle and π of the molecules is essential to understanding the behavior of SUVs as well as other vesicles. Among numerous other phenomena (19), the low surface pressure of SUVs accounts for the fact that when molecular transport through the aqueous phase is the primary mode of vesicle size change (22), the SUVs grow by acquiring molecules from large vesicles which have higher surface pressures (21,19).

FUSION OF SUVs

The fusion of PS SUVs has most often been induced by Mg^{+2} or Ca^{+2}, but H^+, La^{+3} and polycations are also effective (2,23,24). Although PS SUVs will not have the same surface parameters as the PC SUVs that are the subject of Table I, it is evident that the same basic pattern must be followed; limit sonicated vesicles of all lipid types will be reduced to that size where further curvature is incompatible with a bilayer structure. All SUVs composed of a single lipid will hence have similar high surface energies and low surface pressures to those shown in Table I.

Fig. 4 depicts a pair of SUVs held in contact by bridging cations. Prior to contact, the two vesicles have uniform surfaces, the external monolayer under tension, the internal monolayer under compression. After contact, the external surface becomes anisotropic and the area of adhesion becomes a focal point for surface forces. The molecules outside the contact zone are at unusually low surface pressure and the entire external surface is under tension. By definition, the area of contact is dehydrated relative to the rest of the surface (whether or not *some* water remains). The consequence is that molecules of the external monolayer can move out of the contact area with little resistance, for as the external monolayer is pulled out of this region, the inner monolayers fill the vacated volume. The result is a partially fused pair of vesicles. From the point of view of surface pressure, the driving force for the hemifusion process is the reduction in the hydrophobic effect in the contact region. In the absence of bulk water, the local tension no longer balances surface pressure and the molecules separate, migrating, in part, to that portion of the external monolayer that is in contact with water, and where the surface pressure is low. The next step, rupture of the septum, completes the fusion process. It is driven by the high stress remaining on the still highly curved membranes. As shown in Table I, 250Å diameter vesicles are under about 4 dyne/cm more tension than are 300Å diameter vesicles, which would be the product of complete fusion of the former. These 4 dynes/cm will actually be approximately evenly distributed between the inner and outer monolayer, so that the former is under about 2 dynes compression. Even if withdrawal of molecules from the contact zone into the rest of the external monolayer reduces the tension to zero, the compressional forces remain. These combine and are sufficient (even given non-zero contact angle) to provide the 3 dyne/cm lysis force (9) to rupture the septal bilayer. The vesicle tension falls rapidly with increasing radius, so with curvature-dependent surface energy as the driving force for fusion, after a few rounds of fusion this will be dissipated. It is thus expected that SUVs can be induced to fuse by virtually any agent that causes aggregation but does not interfere with the displacement of the outer monolayer relative to the inner monolayer. Mg^{+2} seems to be the best example of such a cation; it does not induce the fusion of LUV (25) and hence must rely upon the surface energy of SUV to drive membrane merging.

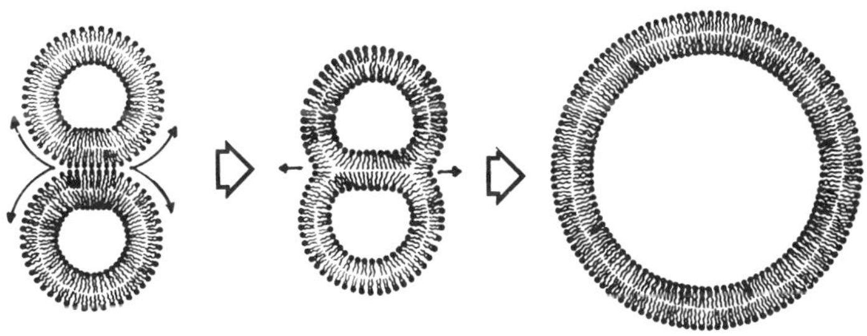

Fig. 4. Steps in fusion of SUVs.

LUV FUSION

As is evident from Table I, vesicles that are as large as about 750Å diameter have surface energies that are nearly as low as those of a flat bilayer. There is thus no energy to drive the membrane rearrangements that register as fusion by the common assays other than the energy of interaction of the fusing agent with the vesicle. We must therefore inquire as to the nature of that interaction. As in the case of analysis of SUV fusion, we make use of the equivalent monolayer at the air/water interface to assess the consequences of interaction of acidic lipids with cations. Because of the intense study that has been devoted to the Ca-PS system, we shall consider it first.

Figure 5 shows the π-Γ (π-1/A) relationship for PS in the presence and absence of 1 mM Ca^{+2}. The procedure was as described elsewhere (15).

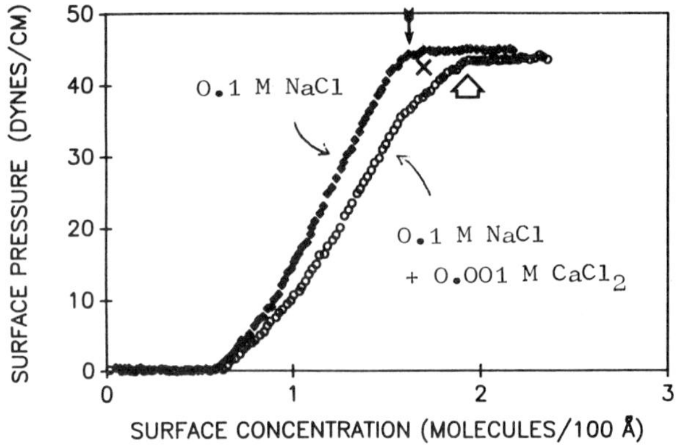

Fig. 5 Effect of Ca^{+2} on the π-Γ relationship of PS monolayers at the air/water interface

As is evident from the figure, Ca^{+2} causes a decrease of more than 15% in A at the equilibrium or collapse pressure. Since, as discussed above, at the equilibrium pressure, the air/water monolayer organization corresponds closely to that of the bilayer, the area per molecule in PS liposomes is at least 15% smaller in the presence than in the absence of 1 mM Ca^{+2}. The reason for this area reduction is that Ca^{+2} binds to two PS molecules. The result is a reduction in configurational entropy, which means that the π-Γ relationship (for monolayers at the a/w interface and in the bilayers) is displaced such that the balance between surface pressure and the local tension (the hydrophobic effect operating between lipid molecules) is restored. This is not possible when calcium ion is added after the vesicles have formed, since binding of the Ca^{+2} to the extent corresponding to an area/molecule marked by the open arrow in the figure would result in compression of the inner monolayer of the vesicle. Thus, the actual area per molecule on the external surface of intact vesicles after treatment with Ca^{+2} is given by the balance between binding energy and compression of the bilayer, i.e.,

$$2\ \pi(A)\Delta A = kT\ln K \qquad (10)$$

where the 2 is included to account for both monolayers of the bilayer, $\pi(A)$ is obtained from the π-Γ slope of Fig. 5 and K is the 2:1 (PS:Ca) binding constant. The value of the latter is approximately 0.6 M^{-1} (27), in which case equation 10 predicts the system would equilibrate at about the position marked x. It is thus evident that the effect of 1 mM Ca^{+2} on large PS vesicles is to reduce the average area per molecule by a few $Å^2$ and increase the surface energy by a few erg/cm^2, the latter being divided between compression of the inner monolayer and tension of the outer monolayer.

Were there a source of additional PS molecules available, the area of the external monolayer would fall to about 48 $Å^2$ per molecule (open arrow) and the inner monolayer should expand to about 60 $Å^2$ (filled arrow) i.e., that of the force-free state. In fact, a source is created when adhesion occurs. As shown in Fig. 6, in the adhesion zone, the hydrophobic effect no longer prevents the outer monolayer from shrinking to the normal Ca^{+2}-PS area per molecule of 48 $Å^2$, for the gaps thus created are exposed to oil, not water. The inner monolayer may, in contracting, leave sections stranded. These may reform into inverted micelles and migrate to the edge of the septum. Such a sequence of events would account for the lipidic particles sometimes observed in freeze-fracture electron micrographs of fused lipids (13,14).

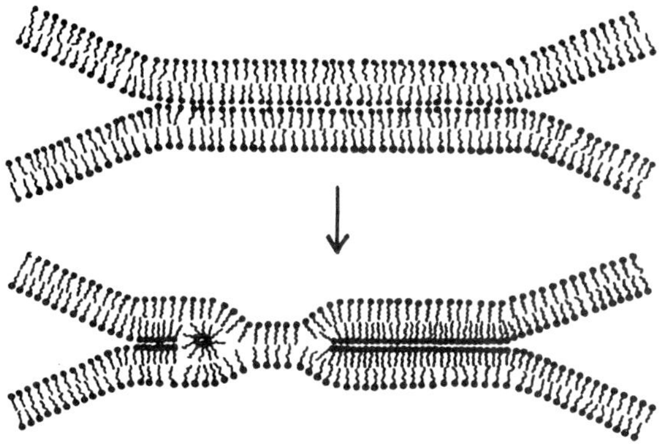

Fig. 6. Diagram of fusion of LUV induced by Ca^{+2}

The tendency for PS monolayers to contract in the presence of calcium accounts for the partial fusion of bilayers and is in agreement with the observations that membrane mixing precedes contents mixing (7,28,29). We next inquire as to forces which can account for the next step, namely, mixing of aqueous contents. This force has, in fact, been already identified as the tension placed on the entire vesicle bilayer by extension of the contact zone (10,11). As we have seen above, the energy of interaction of Ca^{+2} with PS membranes is sufficient to induce several dynes/cm stress. Thus, as the vesicles flatten against each other, the pressure rises enough that the tension in the septum reaches the lysis tension of about 3 dynes/cm (9). The only essential difference between the final stage of fusion as described here and the mechanism previously proposed (9,10) is that here the tension in both vesicles is concentrated on the single bilayer part of the septum. This considerably increases the possibility that the septum will rend before the vesicles lyse.

RELATIONSHIP BETWEEN MEMBRANE FUSION AND LIPID PHASE TRANSITION

It was clear as early as 1976 that, membranes must be in the fluid state to be fused by divalent cations (30). Nevertheless, most acidic lipids, when treated with Ca^{+2}, are converted to a crystalline Ca^{+2} salt and it was reasonable to ask whether the phase transition plays a role in fusion. Although the question is still debated (31), it is, as should be clear from the foregoing considerations, a question that is difficult to answer because it has not been properly formulated. The binding of Ca^{+2} to PS vesicles induces a reduction in area of a few Å^2 per molecule but condensation to the area per molecule of the crystalline phase is prevented by the resistance to compression of the inner monolayer. Only after the single bilayer septum forms and relieves that tension is tighter binding to the external monolayer possible, which, in turn, leads to further area reduction. Thus, fusion of LUV, at least, is driven by the area condensation that represents the beginning of the transition or phase separation, but it need not be completed until after membrane fusion is over.

COMPARISON OF SUV AND LUV FUSION

In both cases of membrane fusion, the initial step is postulated to be removal from the contact area of portions of the outer monolayers of both vesicles. The result is that a portion of the double bilayer septum is replaced by a single bilayer. In the case of SUV, the energy necessary to separate the outer monolayers is available from the high surface energy of the highly curved bilayer. This is also adequate to induce rupture of the single bilayer septum. In the case of LUV, binding of Ca^{+2} induces a tension in the outer monolayer and a compression in the inner monolayer. These stresses are relieved when the single monolayer region forms in the contact zone. An additional force must come into play to induce rupture of the septum. This is the isotropic tension placed upon the entire vesicle by the increase in contact area driven by the adhesion energy.

MORPHOLOGY OF THE FUSING SPECIES

Although in the discussion of both SUV and LUV, the mechanism was presented in the context of a pair of fusion vesicles, evidence for such a situation is rare. Instead, it appears likely that under the conditions of most published experiments, the fluorescence signals upon which the assessment of fusion is based, originate from within rather sizable aggregates. These are visible to the eye within 30 seconds of making a 50 μM PS dispersion 1 or 2 μM in Ca^{+2}, (6) and a careful study employing electron microscopy revealed a high proportion of vesicles 5-10 times larger than the original SUV's within 7 sec (25). Since a 10x increase in diameter represents 10^3 vesicles, pairwise fusion events may be rare. Although there remains the question of the proper description of breakage and reforming of membranes within an aggregate of numerous vesicles (16), "fusion" is not an unreasonable description provided there is general understanding of its meaning in a given context.

Given that fusion of a pair of isolated vesicles is not the means by which most fusion of membranes induced by cations occurs, what changes in the above hypothesis are required to cover the usual situation? The forces that operate to induce fusion of isolated vesicles are essentially unaltered when the vesicles are surrounded by other vesicles. The major difference is that the ratio of contact area to free area in an aggregate is smaller than in isolated dimers, so that the sink into which external monolayer that leaves the contact region can flow is reduced. The only consequence would be to reduce the number of rounds of fusion. Thus vesicle-vesicle fusion and membrane fusion within an aggregate can both be accounted for by the analysis presented here.

LYSIS VS. FUSION

It is evident that vesicles lyse under conditions of fusion (5-8). The distribution of lytic events relative to fusion events is poorly understood with respect to both place and time. What is clear is that even though the analysis given here predicts rupture of the septum more frequently than rupture of the membranes outside the contact region, the forces required to induce fusion are necessarily of a magnitude sufficient to cause lysis. It is thus likely that a membrane fusion process based on merging of purely lipid membranes cannot be modified so as to recreate the kind of efficient and leak-proof mechanisms that evidently operate in biological systems.

SUMMARY

Mechanisms for fusion of SUVs and LUVs are proposed, both of which involve a partially fused, intermediate state. In both cases, but for different reasons, the outer monolayer of the contacting vesicles is under more tension before than after fusion. The change in tension drives the first stage of the fusion process and is accomplished by withdrawal of molecules of the external monolayer from the zone of contact. Contraction of the outer monolayer relative to the inner monolayer produces gaps within the former, which are filled by transverse movement of the inner monolayers.

The inner layers come into contact at their methyl termini and so generate a portion of single monolayer in the septum between the two juxtaposed vesicles. In the case of SUV, the driving force for withdrawal of the external monolayer from the contact zone is the low surface pressure of the external monolayer, which is a consequence of the small radius of curvature of these vesicles. In the case of LUV, the formation of gaps in the external monolayer in the contact region is driven by Ca^{+2}-dependent contraction of portions of that region of the external monolayer. As with SUV, the internal monolayer of LUVs fills the gaps in the external monolayer to generate a septum that is, in part, a single monolayer. The septum of SUVs breaks primarily because of the force exerted on it by the curvature of the remainder of the vesicle, although some assistance may be provided by additional stress due to the binding of cations to the outside of the juxtaposed vesicles. The septum of LUVs breaks only when the area of contact increases to the extent that the internal hydrostatic pressure generates a tension in the vesicle membrane that exceeds the lysis tension. The partial and complete fusion of SUV and LUV can be fully rationalized on the basis of 1) the surface energies of the vesicles before and after treatment with the fusion-inducing cation and 2) the effect that vesicle-vesicle contact has on those energies. It is pointed out that the usual environment under which fusion occurs is not that of a pair of vesicles in contact but of a large number of vesicles in aggregates. This does not, however, fundamentally change the mechanism of membrane merging. In principle, all of the relevant energies can be obtained from measurement of the corresponding lipid in monolayers at the air/water interface.

REFERENCES

1. D. Papahadjopoulos, W. Vail, K. Jacobson and G. Poste, Biochim. Biophys. Acta 448:245-264 (1975).
2. N. Düzgüneş and J. Bentz, in *Spectroscopic Membrane Probes*, L. M. Loew, ed. CRC Press, in press.
3. J. Wilschut and D. Papahadjopoulos, Nature 281:690-692 (1979).
4. D. Hoekstra, Biochim. Biophys. Acta 692:171-175 (1982).
5. L. Ginsberg, Nature 275:758-760 (1978).
6. D.A. Kendall and R.C. MacDonald, J. Biol. Chem. 257:13892-13895 (1982).
7. S.J. Morris, C.C. Gibson, P.D. Smith, P.C. Greif, C.W. Stirk, D. Bradley, D.H. Haynes and R. Bumenthal, J. Biol. Chem. 260:4122-4127 ee(1985).
8. B. Kachar, N. Fuller and R.P. Rand, Biophys. J. 50:779-788 (1986).
9. R. Kwok and E. Evans, Biophys. J. 35:637-652 (1981).
10. E. Evans and V.A. Parsegian Ann. N.Y. Acad. Sci. 416:13-33 (1983).
11. V.A. Parsegian and R.P. Rand, Ann N.Y. Acad. Sci. 416:1-12 (1983).
12. S. Ohki, Biochim. Biophys. Acta 689:1-11 (1982).
13. A.J. Verkleij, C.J.A. Eckfeld, W.J. Gerritsen, P.R. Cullis and B. DeKruiff Biochim. Biophys. Acta 600:620-624 (1980).
14. S.W. Hu, T.P. Stewart, L.T. Boni & P.L. Yeagle, Science 212:921-923 (1981).
15. R.C. MacDonald and S.A. Simon, Proc. Natl. Acad. Sci. USA, in press.
16. J.F. Nagle, Ann. Rev. Phys. Chem 31:29-45 (1980).
17. J. Israelachvili, Marcelja and Horn. Quart. Rev. Biophys (1980).
18. S.W. Hui, M. Cowden, D. Papahadjopoulos and D. F. Parsons, Biochim. biophys. Acta 382:265-272 (1975).
19. R.C. MacDonald, to be published.
20. H. Schindler, FEBS Lett. 122:77-79 (1980).
21. S. Ohki and H. Ohshima, Biochim. Biophys. Acta 812:147-154 (1985).
22. F.J. Martin, and R.C. MacDonald, Biochemistry 15:321-327 (1976).
23. S. Ohki and J. Duax, Biochim. Biophys. Acta 861:177-186 (1986).
24. A. Walter, C.J. Steer and R. Blumenthal, Biochim. Biophys. Acta 861:319-330 (1986).

25. J. Wilschut, N. Düzgüneş, and D. Papahadjopoulos, Biochemistry 20:3126-3133 (1981).
26. E. A. Evans and R. Skalak, *Mechanics and Thermodynamics of Membranes*, CRC Press, 1980, p. 87-91.
27. S. McLaughlin, N. Mulrine, T. Gresalfi, G. Vaio and A. McLaughlin, J. Gen. Physiol. 77:445-473 (1981).
28. J. Wilchut, J. Scholma, M. Bental, D. Hoekstra and S. Nir, Biochim. Biophys Acta, 821:45-55 (1985).
29. L. Ababu and K. Hildenbrand, Chem. Phys. Lipids, 35:39-48 (1984).
30. D. Papahadjopoulos, W.J. Vail, W.A. Pangborn and G. Poste Biochim. Biophys. Acta 448:265-283 (1976).
31. J.R. Silvius and J. Gagné, Biochemistry 23:3241-3247 (1984).

ACKNOWLEDGEMENT

Supported by NIH grants NS20831 and AM36634.

POLAR INTERFACIAL INTERACTIONS, HYDRATION PRESSURE AND MEMBRANE FUSION

C.J. van Oss[1,2], M.K. Chaudhury[3], and R.J. Good[2]

Departments of [1]Microbiology and [2]Chemical Engineering
State University of New York at Buffalo, Buffalo, NY 14214
and [3]Dow Corning Corporation, Midland, MI 48686

SUMMARY

By rigorously distinguishing between the apolar (Lifshitz-van der Waals, or LW) and the polar (Lewis acid-base, or AB) components of the surface tension, it emerges that the AB component of the interfacial tension can become negative to such a degree that the total interfacial tension can also become negative. Net negative interfacial tensions with water occur most readily among certain polymers, biopolymers, phospholipids, etc., which have one predominant polar parameter (usually the electron-donor parameter) and virtually lack the other. The AB component of the surface tension of such materials then is close to zero, but these <u>monopolar</u> substances nevertheless strongly interact with <u>bipolar</u> liquids such as water.

Two monopolar materials of the same sign repel each other in water. This newly uncovered monopolar repulsive force causes, inter alia, phase separation of polymers in water, as well as the repulsion perceived as "hydration pressure". Monopoles become strongly hydrated in water, but that hydration actually causes a decrease in the net repulsion between hydrated monopoles, compared to the repulsion (ideally) possible between nonhydrated monopoles.

The role Ca^{++} plays in membrane fusion (apart from decreasing the negative ζ-potential, if any) lies mainly in its capacity for neutralizing the electron-donor monopolar energy of, e.g., phospholipid layers which, in the process, it appears to render more "hydrophobic". Thus, the admixture of Ca^{++} causes the monopolar repulsion to change into an apolar (LW) attraction, which favors fusion.

The role of polyethylene glycol in facilitating membrane fusion lies in its capacity to force two or more cells or vesicles together by phase separation caused by electron-donor monopolar repulsion, aided by its strong dehydrating power.

INTRODUCTION

Whilst the DLVO theory of stability of suspensions in liquid media, named after Dejaguin, Landau, Verwey and Overbeek[1,2], rather accurately described that stability in strictly apolar media, in polar *) liquids such as water, in addition to the classical van der Waals attractions and electrostatic repulsions treated by the DLVO theory, interfacial hydrogen-bonding interactions play a preponderant role. Such hydrogen-bonding (or more generally speaking, electron-donor/electron-acceptor) interactions, being inherently asymmetrical, can be attractive or repulsive, and they can be easily one or two orders of magnitude stronger than either van der Waals attractions or electrostatic repulsions. The quantitative determination of these interfacial hydrogen-bonding interactions has become possible, following the recent simultaneous development of their theoretical basis[3-7] and their experimental measurement methodology[4-7]. Attractive interfacial interactions in polar media are identical to the attractive forces that traditionally have been described as "hydrophobic" interactions, which could not however be treated quantitatively prior to our first attempts, of 1985[8,9]. The effects of repulsive interfacial interactions in polar media, which have been perceived as hydration pressure[10,11], actually are due to a strong monopolar repulsion (usually caused by electron-donor moieties that repel each other in water), attenuated by hydration[12].

Vesicles and cells repel each other by interfacial repulsion of a polar (i.e., hydrogen-bonding) nature, sometimes (but not necessarily always) enhanced by electrostatic repulsion. To effect fusion of such vesicles or cells, that repulsion must be overcome. Some of the mechanisms by which this may be achieved are discussed below.

THEORY

The derivation and/or proof of all the following equations has been given elsewhere (see e.g., Ref. 6) and will not be repeated here; only the equations necessary for the various applications, and those involved in

*) The expression "polar" is used here exclusively to designate either materials capable of engaging in electron-donor/electron-acceptor (or Lewis acid-base) interactions, or such interactions themselves.

the arguments leading up to them, are given below. Here follow first the equations that are valid for apolar, Lifshitz-van der Waals (LW), as well as for polar, Lewis acid-base (AB) interactions.

The energy of cohension of a (condensed) material i is:

$$\Delta G_{ii}^{coh} = -2\gamma_i \qquad [1]$$

where γ_i is the surface tension of material i. The energy of adhesion between materials 1 and 2 is:

$$\Delta G_{12}^{adh} = \gamma_{12} - \gamma_1 - \gamma_2 \qquad [2]$$

where $_1$ and $_2$ are the surface tensions of materials 1 and 2, and γ_{12} is their interfacial tension. (We will, below, omit the superscript, "adh" from the symbol for the free energy of adhesion.) The energy of interaction between materials 1 and 2 immersed in liquid 3 is:

$$\Delta G_{132} = \gamma_{12} - \gamma_{13} - \gamma_{23} \qquad [3]$$

and the energy of interaction between (macro) molecules or particles of material 1, immersed in liquid 3 is:

$$\Delta G_{131} = -2\gamma_{13} \qquad [4]$$

Apolar or Lifshitz-van der Waals Interactions

It was shown earlier, via the Lifshitz approach that all three electrodynamic interactions, i.e., the London-van der Waals (dispersion) as well as the "polar" (in the electrodynamic sense) Keesom-van der Waals (orientation) and Debye-van der Waals (induction) forces, decay with distance at the same rate, and can generally be treated with the same equations[3]. Together they represent the apolar, or Lifshitz-van der Waals (LW) interactions[3,4,6,8].

Apolar, or LW interfacial tensions between substances 1 and 2 are expressed as:

$$\gamma_{12}^{LW} = (\sqrt{\gamma_1^{LW}} - \sqrt{\gamma_2^{LW}})^2 \qquad [5]$$

Via eqns. [1-5] the various values for ΔG^{LW} can be obtained from the γ_i^{LW} and γ_{ij}^{LW} values. The decay with distance of LW interfacial interactions for semi-infinite parallel flat slabs may be expressed as:

$$\Delta G_{(d)}^{LW} = \Delta G_{(d_o)}^{LW} \left(\frac{d_o}{d}\right)^2 \qquad [6]$$

where d_o is the minimum equilibrium distance, taken to be ≈1.56 Å[13], and d may be any distance from 1.56 Å to ≈100 Å, beyond which distance retardation effects prevail[1,2] and eq. [6] becomes:

$$\Delta G_{(d)}^{LW} \quad (d>100\ \text{Å}) = \Delta G_{(d_o)}^{LW} \left(\frac{d_o}{d}\right)^3 \qquad [6A]$$

Polar or (Lewis) Acid-Base Interactions

Hydrogen-bonding or, more generally speaking, electron-donor/ electron-acceptor, and thus (Lewis) acid-base (AB) interactions decay with distance at a very different rate from LW interactions, and also in general obey an entirely different set of rules and equations, due in part to the intrinsic asymmetry of these polar interactions. We first must define γ_i^{AB} of a material i (in the condensed state), in terms of its electron-acceptor ($\gamma_i^{\oplus}$) and electron-donor ($\gamma_i^{\ominus}$) parameters:

$$\gamma_i^{AB} = 2\sqrt{\gamma_i^{\oplus}\gamma_i^{\ominus}} \quad [7]$$

where, in general, $\gamma_1^{\oplus} \neq \gamma_1^{\ominus}$. The polar component of the energy of adhesion between materials 1 and 2 is:

$$\Delta G_{12}^{AB} = -2\,(\sqrt{\gamma_1^{\oplus}\gamma_2^{\ominus}} + \sqrt{\gamma_1^{\ominus}\gamma_2^{\oplus}}) \quad [8]$$

and the polar component of the interfacial tension between materials 1 and 2 is:

$$\gamma_{12}^{AB} = 2\,(\sqrt{\gamma_1^{\oplus}\gamma_1^{\ominus}} + \sqrt{\gamma_2^{\oplus}\gamma_2^{\ominus}} - \sqrt{\gamma_1^{\oplus}\gamma_2^{\ominus}} - \sqrt{\gamma_1^{\ominus}\gamma_2^{\oplus}}) \quad [9]$$

The polar component of the energy of interaction between materials 1 and 2 immersed in liquid 3 is:

$$\Delta G_{132}^{AB} = 2\,[\sqrt{\gamma_3^{\oplus}}\,(\sqrt{\gamma_1^{\ominus}} + \sqrt{\gamma_2^{\ominus}} - \sqrt{\gamma_3^{\ominus}}) + \sqrt{\gamma_3^{\ominus}}\,(\sqrt{\gamma_1^{\oplus}} + \sqrt{\gamma_2^{\oplus}} - \sqrt{\gamma_3^{\oplus}}) - \sqrt{\gamma_1^{\oplus}\gamma_2^{\ominus}} - \sqrt{\gamma_1^{\ominus}\gamma_2^{\oplus}}] \quad [10]$$

and the polar component of the energy of interaction between (macro) molecules or particles of material 1, immersed in liquid 3 is:

$$\Delta G_{131}^{AB} = -4\,(\sqrt{\gamma_1^{\oplus}\gamma_1^{\ominus}} + \sqrt{\gamma_3^{\oplus}\gamma_3^{\ominus}} - \sqrt{\gamma_1^{\oplus}\gamma_3^{\ominus}} - \sqrt{\gamma_1^{\ominus}\gamma_3^{\oplus}}) \quad [11]$$

The decay with distance d of AB interfacial interactions for semi-infinite parallel flat slabs may be approximated for d>λ) as[14,15]:

$$\Delta G_{(d)}^{AB} = \Delta G_{(d_o)}^{AB} \exp \frac{(d_o - d)}{\lambda} \quad [12]$$

where λ is the correlation length typical for the liquid, and a measure of the dimension of the molecules of the liquid. For water λ ≈ 2 Å, but at high ionic strengths that value can be much higher (e.g.,≈ 12 Å) as the molecules of the liquid are then substantially represented by hydrated ions[16], and for water in which polymers have been dissolved at a considerable concentration, the value of λ can attain 100 Å, or more, which would extend the measurable range of ΔG^{AB} to distances well beyond 1,000 Å.

The Total Interfacial Interaction

The total free energy of interaction *) may be expressed as:

$$\Delta G^{TOT} = \Delta G^{LW} + \Delta G^{AB} \quad [13]$$

However, when both LW and AB components have been measured at the minimum

---

*) only LW and AB interactions are considered here; metal bonds are irrelevant to membrane fusion and electrostatic interactions, while relevant in certain cases, are determined by entirely different procedures[17] and will be mentioned here only qualitatively.

equilibrium distance d_o ($d_o \approx 1.56$ Å for LW and ≈ 2 Å for AB interactions), which is the case when that measurement has been done by contact angle determination (see below), the values for ΔG^{LW} and ΔG^{AB} at distances $d > d_o$ must be separately computed via equations [6 and 12], before arriving at $\Delta G^{TOT}_{(d)}$ by means of eq. [13].

From eq. [13] it also follows that:

$$\gamma^{TOT} = \gamma^{LW} + \gamma^{AB} \quad [14]$$

which is valid for surface tensions γ_i^{TOT}, as well as for interfacial tensions γ_{12}^{TOT}.

When $\Delta G<0$, an interfacial ("hydrophobic") attraction is favored; when $\Delta G>0$, a repulsion tends to occur.

Young's Equation

Both LW and AB interactions are incorporated in Young's equation[4-7,12] as follows:

$$(1 + \cos\Theta)\gamma_L^{TOT} = 2\left(\sqrt{\gamma_S^{LW}\gamma_L^{LW}} + \sqrt{\gamma_S^{\oplus}\gamma_L^{\ominus}} + \sqrt{\gamma_S^{\ominus}\gamma_L^{\oplus}}\right) \quad [15]$$

where the subscripts L and S stand for liquid and solid. It thus is possible to characterize a solid completely with respect to γ_S^{LW}, $\gamma_S^{\oplus}$ and $\gamma_S^{\ominus}$ when (advancing) contact angles are measured with three different liquids with known γ_L^{LW}, $\gamma_L^{\oplus}$ and $\gamma_L^{\ominus}$ values *)[4-7,12].

CONSEQUENCES OF POLAR INTERACTIONS

Repulsion Effects

Rewriting eq. [9]:

$$\gamma_{13}^{AB} = 2\left(\sqrt{\gamma_1^{\oplus}} - \sqrt{\gamma_3^{\oplus}}\right)\left(\sqrt{\gamma_1^{\ominus}} - \sqrt{\gamma_3^{\ominus}}\right) \quad [9A]$$

6

it becomes clear that $\gamma_{12}^{AB} < 0$ when :

$\gamma_1^{\oplus} > \gamma_3^{\oplus}$ and $\gamma_1^{\ominus} < \gamma_3^{\ominus}$, and when: $\gamma_1^{\oplus} < \gamma_3^{\oplus}$ and $\gamma_1^{\ominus} > \gamma_3^{\ominus}$

In all such cases $\Delta G_{131}^{AB} > 0$ (see eq. [10]), and if also $|\Delta G_{131}^{AB}| > |\Delta G_{131}^{LW}|$, particles or molecules 1, immersed or dissolved in liquid 3, will repel each other. This occurs especially readily when either $\gamma_1^{\oplus}$ or $\gamma_1^{\ominus} \approx 0$. In

---

*) In actual practice $\gamma_L^{\oplus}$ and $\gamma_L^{\ominus}$ are not yet known to any degree of accuracy for any liquid, but the $\gamma_{L1}^{\oplus}/\gamma_{L2}^{\oplus}$ and the $\gamma_{L1}^{\ominus}/\gamma_{L2}^{\ominus}$ ratios have been determined for various pairs of liquids, which allows the precise determination of ΔG_{12}^{AB}, ΔG_{132}^{AB}, ΔG_{131}^{AB} and γ_{12}^{AB} values[6,7]. For practical purposes, we postulate that for water $\gamma_W^{\oplus}/\gamma_W^{\ominus} = 2$ (although estimates as high as 3 for that ratio have been reported[20]), to enable us to give a quantitative approximation of various $\gamma_i^{\oplus}$ and $\gamma_i^{\ominus}$ values. It should be kept in mind however that these quoted values all are relative to the estimated $\gamma_W^{\oplus}/\gamma_W^{\ominus} = 2$, and are subject to correction as and when the exact value of that ratio becomes more precisely established.

such cases substance 1 may be called a monopole and its γ_1^{AB} will be zero, or close to zero, so that its $\gamma^{TOT} \approx \gamma_1^{LW}$. However its single polar moiety will stronger interact with a bipolar liquid, such as water.

In Nature, polar asymmetry, bordering on virtual monopolarity, with a sizeable $\gamma_1^{\ominus}$, and $\gamma_1^{\oplus} << 1$, appears to be rather common among poly- and oligosaccharides[4,6,12], proteins[4-6], RNA, cellulose esters[18], and phospholipids[19].

Hydration Pressure

Because the unequal distribution of the polar parameters of water, i.e.: $\gamma_W^{\oplus} > \gamma_W^{\ominus}$[6,7,19,20], $\gamma^{\ominus}$ monopoles become strongly hydrated when immersed in water, in a unidirectional, asymmetrical fashion, much like the model proposed by Parsegian *et al.*[21].

However, in the case of largely monopolar proteins with a $\gamma_1^{\ominus}$ of insufficient magnitude to allow γ_{13}^{AB} to overcome γ_{13}^{LW} (i.e., $|\Delta G_{131}^{LW}| > |\Delta G_{131}^{AB}|$; see eqs. [3] and [4]), this hydration substantially attenuates the force of the residual intermolecular attraction, thus favoring their solubility[5,22]. However, in the case of monopolar polysaccharides and polyethylene glycol with high $\gamma^{\ominus}$ values (i.e., $|\Delta G_{131}^{AB}| > |\Delta G_{131}^{LW}|$), the same type of hydration also mitigates the force of the intermolecular repulsion. Thus the repulsion observed as "hydration repulsion"[10,11] actually is the remnant of the much stronger underlying monopolar repulsion, attenuated by hydration, but also aided by hydration in exerting its influence to a greater distance from the particles' or molecules' aqueous interface than would have been possible in the unhydrated state. Figure 1 schematically illustrates the influence of various degrees of hydration (or dehydration) on the total interfacial attraction or repulsion, in the general case of a biopolymer with a $\gamma_1^{LW} \approx$ 40 mJ/m² (an often-encountered value), dissolved or suspended in water (γ_3^{LW} = 21.8 mJ/m²). In all such fairly typical cases, $\Delta G_{131}^{LW} \approx -5.5$ mJ/m². To overcome this amount of Lifshitz-van der Waals attraction, a (dry) monopolar electron-donor substance must have a $\gamma_1^{\ominus}$ parameter of at least 20 mJ/m² to overcome that attraction and begin to exert a net repulsion. Figure 1 shows the decrease in repulsive energy, as well as the decrease in attractive energy, as a function of hydration, for monopoles with $\gamma^{\ominus}$ 20 mJ/m² (leading to repulsion) as well as with $\gamma^{\ominus} > 20$ mJ/m² (resulting in a net attraction).

Solubility of Biopolymers

Polarity (or monopolarity) strong enough to give rise to a negative interfacial tension with water tends to lead to high solubility in, or miscibility with, water of the polar compound[6,7]. However even if the

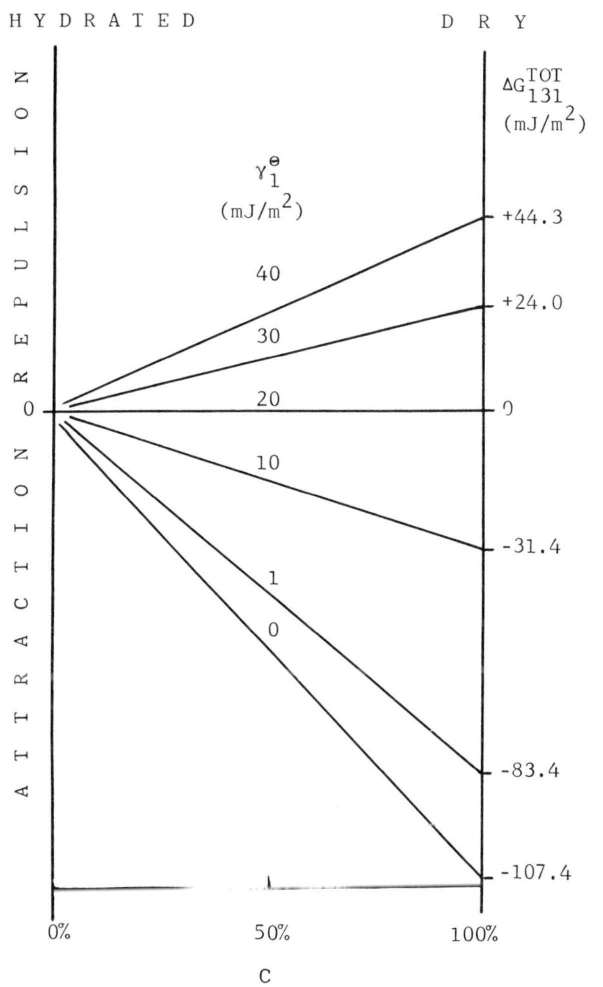

Fig. 1. Repulsion (top) and attraction (bottom) energies of monopolar electron-donor biopolymers (1) immersed in water (3), with a γ^{LW} of 40 mJ/m^2, and $\gamma_1^{\ominus}$ values of 0, 1, 10, 20, 30 and 40 mJ/m^2, as a function of the percentage (C) of dehydration (or of the biopolymer concentration C), in %.

polarity (or monopolarity) of a biopolymer does not suffice to cause a negative interfacial tension with water, and if the ΔG_{131}^{TOT} value for such a biopolymer immersed in water is sufficiently negative to make solubility of that biopolymer rather dubious, a relatively modest $\gamma_1^{\ominus}$ value usually will reduce the negative ΔG_{131}^{TOT} value of the hydrated biopolymer sufficiently to favor its solubility in water[5]; see also Figure 1.

Stability of Particles, Cells and Vesicles

A polarity (or monopolarity) situated at the aqueous interface of a particle, cell or vesicle, sufficient to give rise to a positive ΔG_{131}^{TOT} value, strongly favors the stability of such a particulate suspension, even if the repulsion it engenders is attenuated by hydration, see Figure 1. To achieve particle destabilization or membrane fusion of cells or vesicles, that repulsion must be overcome in some way.

MEMBRANE FUSION

There are a number of mechanisms by which the intercellular or intervesicular repulsion (due to a positive ΔG_{131}^{TOT}) may be surmounted. Two of these are discussed here.

Ca^{++}-Mediated Fusion

As the admixture of Ca^{++} is rather effective in reducing the negative ζ-potential of cells and vesicles, this method has gained some popularity for membrane fusion. However, as Gallez et al. pointed out[23], whilst Ca^{++} will indeed reduce the cellular ζ-potentials to negligible levels, that strongly reduced ζ-potential would not appear to allow quantitatively for a counterbalancing of the hydration pressure[10,11] to a degree necessary to achieve membrane fusion. However it was not realized until recently that underlying hydration pressure is an interfacial repulsion, usually of electron-donor monopolar origin[4-7]. Now Ca^{++} is an electron-acceptor and as such can sufficiently depress the electron-donor energy of phospholipids to reverse the sign of the initially positive ΔG_{131}^{TOT}, as evidenced, inter alia, by its capacity to render phosholipids hydrophobic[24]. The neutralization of a phospholipid's electron-donor capacity simultaneously decreases its hydration.

Thus Ca^{++} lowers the negative ζ-potential, diminishes the electron-donor capacity and attenuates the hydration of phospholipids to a sufficient degree to favor membrane fusion.

Polyethylene Glycol Facilitated Fusion

Polyethylene glycol (PEG), usually used as a polymer of a molecular weight of 6,000-8,000, does not materially influence the ζ-potential, nor the electron-donor capacity of phospholipids, as it has a negligible electrostatic potential, and as it is itself a strongly monopolar electron-donor[4,12]. However, PEG does two other things. First, PEG has an exceptionally strong capacity for causing phase separation by electron-donor monopolar repulsion of other electron-donors, in aqueous solution[12] or suspension[25], which is also exemplified by the negative adsorption of PEG with respect to vesicle or cell surfaces[26]. Second, PEG, due to its pronounced propensity for becoming hydrated, is a strong dehydrating agent[26] for most other solutes, which makes it unusually effective in precipitating, e.g., immunoglobulins, at relatively low PEG concentrations, by a combination of dehydration and exclusion[27].

Thus PEG can force two (or a few) cells[28] or vesicles together by monopolar electron-donor repulsion (phase separation). Further aided by its dehydrating capacity and by inducing the intercalation of individual polar phospholipid moieties in opposing phospholipid layers, PEG can materially facilitate membrane fusion[29].

REFERENCES

1. E.J.W. Verwey and J.Th.G. Overbeek, "Theory of the Stability of Lyophobic Colloids", Elsevier, Amsterdam (1948).
2. J.Th.G. Overbeek, The interaction between colloidal particles, in: "Colloid Science", vol. 1., pp. 245-277, H.R. Kruyt, ed., Elsevier, Amsterdam (1952).
3. M.K. Chaudhury, "Short-Range and Long-Range Forces in Colloidal and Macroscopic Systems", Ph.D. Dissertation, SUNY/Buffalo (1984).
4. C.J. van Oss, R.J. Good and M.K. Chaudhury, Determination of the hydrophobic interaction energy - Application to separation processes, Separ. Sci. Technol., 22: 1 (1987).
5. C.J. van Oss, R.J. Good and M.K. Chaudhury, Solubility of proteins, J. Protein Chem. 5: 385 (1986).
6. M.K. Chaudhury, C.J. van Oss and R.J. Good, Negative interfacial tension between polar liquids and some polar materials. I. Solid surfaces (submitted).
7. C.J. van Oss, L. Ju, M.K. Chaudhury and R.J. Good, Negative interfacial tension between polar liquids and some polar materials. II. Liquid surfaces: Estimation of the polar parameters of the surface tension of liquids by contact angle measurements on gels (submitted).
8. C.J. van Oss, R.J. Good and M.K. Chaudhury, The role of van der Waals forces and hydrogen bonds in hydrophobic interactions between biopolymers and low energy surfaces, Abstr. 5th Intl. Conf. Surface Colloid Sci., Nr. 65, Clarkson University, Potsdam (1985).
9. C.J. van Oss, R.J. Good and M.K. Chaudhury, The role of van der Waals forces and hydrogen bonds in "hydrophobic interactions" between biopolymers and low energy surfaces, J. Colloid Interface Sci., 111: 378 (1986).
10. D.M. LeNeveu, R.P. Rand, V.A. Parsegian and D. Gingell, Measurement and modification of forces between lecithin bilayers, Biophys. J. 18: 209 (1977).

11. A.C. Cowley, N.L. Fuller, R.P. Rand and V.A. Parsegian, Measurement of repulsive forces between charged phospholipid bilayers, Biochem. 17: 3163 (1978).
12. C.J. van Oss, M.K. Chaudhury and R.J. Good, The mechanism of partition in aqueous media, Separ. Sci. Technol. 22 (in press) (1987).
13. M.K. Chaudhury, R.J. Good and C.J. van Oss, Correlation between Lifshitz-Hamaker constants and surface tensions of liquids and solids, in preparation.
14. S. Marčelja and N. Radić, Repulsion of interfaces due to boundary water, Chem. Phys. Letters 42: 129 (1976).
15. D.W.R. Gruen, S. Marčelja and V.A. Parsegian, Water structure near the membrane surface, in: "Cell Surface Dynamics", A.S. Perelson, C. DeLisi and F.W. Wiegel, eds., pp. 59-91, Marcel Decker, New York (1984).
16. E. Ruckenstein and D. Schiby, On the origin of repulsive hydration forces between two mica plates. Chem. Phys. Letters 95: 439 (1983).
17. R.J. Hunter, "Zeta Potential in Colloid Science", Academic Press, New York (1981).
18. C.J. van Oss, R.J. Good and M.K. Chaudhury, Mechanism of DNA (Southern) and protein (Western) blotting on cellulose nitrate and other membranes. J. Chromatog. 391: 53 (1987).
19. C.J. van Oss and S. Ohki, unpublished observations.
20. V. Gutman, "The Donor-Acceptor Approach to Molecular Interactions", Plenum Press, New York (1978).
21. V.A. Parsegian, R.P. Rand and D.C. Rau, Hydration forces: What next?, Chemica Scripta 25: 28 (1985).
22. C.J. van Oss, L.L. Moore, R.J. Good and M.K. Chaudhury, Surface thermodynamic properties and chromatographic and salting-out behavior of IgA and other serum proteins, J. Prot. Chem. 4: 245 (1985).
23. D. Gallez, M. Prevost and A. Sanfeld, Repulsive hydration forces between charged lipidic bilayers. A linear stability analysis, Colloids Surfaces 10: 123 (1984).
24. S. Ohki, A mechanism of divalent ion-induced phosphatidylserine membrane fusion, Biochim. Biophys. Acta 689: 1 (1982).
25. W.T. Coakley and C.J. van Oss, Mechanism of different modes of erythrocyte stability and instability in the presence of various polymers (in preparation).
26. K. Arnold, A. Hermann, K. Gawrisch and L. Pratsch, Water-mediated effects of PEG on membrane properties, this volume, pp. 255.
27. A. Polson, A theory for the displacement of proteins and viruses with polyethylene glycol, Prep. Biochem. 7: 129 (1977).
28. T.D. Tilley, W. Coakley, R.K. Gould, S.E. Payne and L.A. Hewison, Real time observations of the agglutination of erythrocytes by polylysine, dextran and by polyethylene glycol, Europ. Biophys. J. (in press).
29. D. Hoekstra, Role of lipid phase separation and membrane hydration in phospholipid vesicle fusion, Biochem. 21: 2833 (1982).

SURFACE TENSION, HYDRATION ENERGY AND MEMBRANE FUSION

Shinpei Ohki

Department of Biophysical Sciences
State University of New York at Buffalo
Buffalo, N.Y. 14214

INTRODUCTION

Membrane fusion is one of the essential events involved in many biological cellular processes (Poste and Allison, 1973; Rasmussen, 1970). In spite of extensive research on membrane fusion, its molecular mechanism is not well understood. This is probably due to the fact that membrane fusion events undergo various possible molecular pathways and also biological membranes have complex molecular structures (Poste and Nicolson, 1978).

In order to gain insight into the molecular mechanisms of such biological membrane fusion events, model lipid membranes with defined molecular components and simpler molecular structures have been utilized for membrane fusion studies. Proposed mechanisms for membrane fusion observed in such membrane systems so far include the following:

1. Membrane micellization by incorporating short-chain fatty acids into membranes. Short-chain fatty acids may produce micelle type structures within the membrane that are thought to be the initiation sites for membrane fusion of the two closely apposed membranes (Lucy, 1970).
2. Membrane instability occurring possibly at the phase boundary between gel-liquid crystalline phases induced by divalent cations (Papahadjopoulos et al., 1977). These phase boundaries serve as the initiation sites of membrane fusion.
3. Membrane expansion due to increased temperature (Breisblatt and Ohki, 1975) or osmotic pressure gradients across membranes (Breisblatt and Ohki, 1975; Miller et al., 1976; Cohen, et al., 1980, Ohki, 1984).
4. Polygon formation of lipid vesicles by lowering temperature (Schullery et al., 1980) or membrane phase defects caused by mixed molecular species in membranes (Hui et al., 1981). These inhomogeneous molecular packing areas are thought to be the sites for membrane fusion.
5. The inverted micelle structures within membranes as an intermediate structure for membrane fusion (Cullis and Hope 1978; Verkleij et al., 1980).
6. Surface hydrophobicity (surface tension) increase in the two apposed membrane surfaces as a result of divalent or polyvalent cation binding (Ohki, 1982, 1984).

7. Enhancement of lipid membrane fusion by Ca^{2+} binding proteins (e.g. synexin, etc.) (Hong et al., 1983; Creutz, 1981) and poly-amino acids (Gad et al., 1982).
8. pH-dependent membrane fusion activity of amino acid peptides (Maeda and Ohnishi, 1980; Wang and Huang, 1984).
9. Some membrane proteins (such as F-protein in viral envelopes) mediate lipid membrane fusion (Haywood and Boyer, 1981; Klappe et al., 1986).
10. Dehydration of water from membrane surfaces by larger organic molecules e.g. polyethyleneglycol) (Hermann et al., 1983; Knutton, 1979).
11. Electric field-induced membrane fusion, where membrane fusion is induced by destabilizing membrane molecules in close contact by applying an electric breakdown potential across the two membranes (Zimmermann, 1982) Sowers, 1986).

Although the physical basis of molecular mechanisms for these fusion events has not been established, this author recently proposed a unified theory of membrane fusion. In this theory, the sufficiently increased surface free energy or increased hydrophobicity of the membrane surface is the main driving force of membrane fusion (Ohki, 1982, 1985).

This paper summarizes the experiments on cation-induced phosphatidylserine vesicle fusion and its associated membrane properties (interfacial tension, vesicle electrophoretic mobility) and discusses a physical basis fc a general theory of membrane fusion in terms of molecular interaction energy It is found experimentally that the degree of membrane fusion correlated well with the degree of increase in the interfacial tension of the membrane. For a theoretical study, the interaction energy for two closely apposed membranes was calculated in terms of molecular interaction energy as well as the surface hydration energy. It is found that the dehydration energy of the membrane, the magnitude of which depends on the nature of surface hydrophilic layers, plays a decisive role in membrane fusion. On the other hand, the van der Waals interaction energy between the two membranes, which depends on the nature of the surface hydrophilic layer as well as its thickness, contributes to the membrane fusion forces as a secondary factor. When the surface hydrophilic layer becomes sufficiently similar in nature to the bulk hydrocarbon phase of the membrane (i.e., increase in the interfacial tension of the membrane surface), the dehydration energy decreases and the attractive interaction energy increases. If the dehydration energy becomes smaller than a certain value so that the magnitude of the attractive interaction energy overcomes the dehydration energy, the membranes in close contact (having a few layers of water in between them) will be in a situation to be able to come further together and could fuse.

OBSERVATIONS

CATION-INDUCED VESICLE FUSION

Vesicle fusion induced by cation can be monitored by several methods. We have used these methods in our experiments: (a) the fluorescent Tb/DPA assay monitoring the fluorescent intensity changes due to mixing of the internal aqueous contents of lipid vesicles. The details are described in an earlier paper (Ohki, 1982), (b) vesicle-monolayer fusion assay where fusion between a lipid monolayer and vesicle is monitored by measuring the surface tension of the monolayer. The detailed experimental procedures are described in earlier papers (Ohki and Duzgunes, 1979; Ohki and Duax, 1986). (c) the turbidity measurements, where vesicles in a suspension solution are aggregated and/or fused by the addition of cations in the vesicle suspension solution and then the aggregation is redispersed by reducing fusogenic cation concentration. The residual amount of turbidity (non-reversal turbidity) is considered due to vesicle fusion (Ohki, 1984; Ohki and Leonard, 1982).

Threshold concentration of cation to induce membrane fusion is defined for each fusion assay as follows: (a) for the fluorescent fusion assay, the fluorescence intensity measured 15 sec after each change of cation concentration was plotted as a function of cation concentration. The concentration at which the fluorescence intensity-ion concentration curve gave the sharpest increase was defined as the threshold concentration of cation to induce vesicle fusion. (b) for the turbidity measurement fusion assay, the residual amount of turbidity of the suspension after the aggregated vesicles were redispersed, was plotted with respect to cation concentration. The sharpest increase in the residual turbidity occurring at a cation concentration was defined as the threshold concentration of cation that induced vesicle fusion. (c) for the vesicle-monolayer fusion assay, the threshold concentration of cation was determined by the concentration at which a sudden, large and progressive decrease in the monolayer surface tension occurred.

For most ion-induced fusion experiments, the fluorescent fusion assay was employed. Figure 1 shows typical changes in fluorescence intensity of phosphatidylserine vesicle suspensions in 0.1 M NaCl with respect to various cation concentrations in which two kinds of small unilamellar phosphatidylserine vesicles encapsulated with $TbCl_3$ or dipicolinic acid were suspended. All the experimental NaCl solutions contained 2 mM histidine, 2 mM TES, 0.01 mM EDTA (0 mM EDTA in the La^{3+} case), pH 7.0. The change in fluorescence intensity represents the degree of fusion of the two kinds of vesicles (Wilschut and Papahadjopoulos, 1979; Ohki, 1982). The threshold concentrations of cation that induced vesicle membrane fusion were 0.7 mM for Ba^{2+}, 1.0 mM for Ca^{2+}, 1.4 mM for Sr^{2+}, 6 mM for Mg^{2+}, and 8 μM for La^{3+}. Data for Ba^{2+} and Sr^{2+} are not shown in Figure 1. The threshold

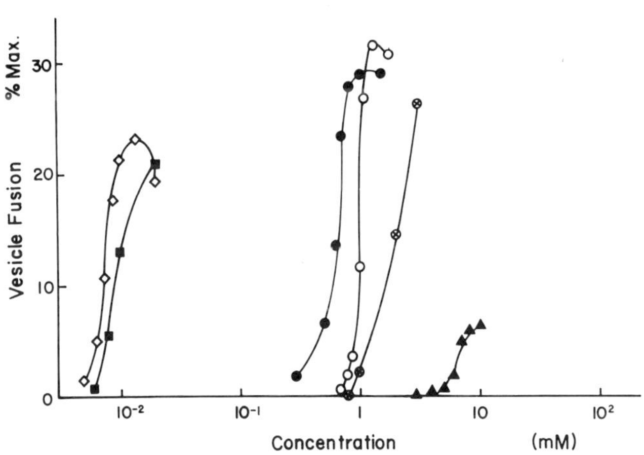

Figure 1. Fusion of phosphatidylserine vesicles in 0.1 M NaCl with various cation concentrations. The left ordinate refers to the fluorescence intensity at 492 nm and the abscissa refers to concentration of various cations. The fluorescence intensity was corrected for the background scattered light by subtracting the scattered light intensity from the total light intensity at 492 nm. The intensity expressed in percentage of the maximum fusion. The wavelength of excitation light was 272 nm. H^+-and Tb^{3+}-induced fusion were determined by measuring the non-reversal residual turbidity (measured with a spectrophotometer, Hitachi 100-60) of the vesicle suspension. The right ordinate refers to the turbidity. (◇) La^{3+}; (■) Tb^{3+}; (●) Mn^{2+}; (O) Ca^{2+}; (▲) Mg^{2+}; (⊗) H^+.

values obtained for Ca^{2+} and Mg^{2+} in this experiment correspond to those reported earlier (Wilschut et al., 1980, 1981). The threshold concentration for La^{3+} is much smaller than that (0.1 mM) observed by Hammoudah et al. (1979); however, they used other techniques to determine membrane fusion. For the La^{3+} case, the concentration of the vesicles in the suspension solution was one-twentieth of that for other divalent cation cases.

For H^+-induced vesicle fusion, the threshold concentration obtained from the turbidity measurement was about pH 2.8. The concentration corresponding to the threshold was not as narrow a range as those observed for polyvalent cation cases. The threshold concentration of H^+ determined from the vesicle-monolayer fusion assay was about pH 3.0 ± 0.2. For Tb^{3+}, the fusion assays used were those of the turbidity measurements and the monolayer-vesicle fusion assay. The turbidity measurements gave threshold concentration of Tb^{3+} to be approximately 10 μM. The monolayer-vesicle fusion assay gave about 8 μM. Other monovalent cations (Na^+, Li^+, and K^+) did not induce fusion of the small unilamellar phosphatidylserine vesicles up to concentration 2 M. It was also found that polyamines such as spermine, spermidine, and putrescine did not induce fusion of the phosphatidylserine vesicles.

EFFECT OF MEMBRANE CURVATURE ON VESICLE FUSION

The membrane curvature effect on membrane fusion has been studied by Ohki, 1984. The experimental results of further such studies are summarized in Figure 2 which shows the threshold concentrations of various cations with respect to various average sizes of phosphatidylserine vesicles. The experimental procedures were similar to those of small unilamellar vesicles except for the use of lower vesicle concentrations (about one quarter of that for the small vesicle). Large unilamellar lipid vesicles were prepared by a method modified (Ohki, 1984) from the Reverse Phase Evaporation method (Szoka and Papahadjopoulos, 1978). The mean size of vesicles was determined by quasielastic light scattering with a commercial particle sizer (Coulter submicron particle analyzer, N4-S). It was found that, as the size of vesicles increased, its threshold concentration became greater for all ions that induced vesicle fusion. The experimental results for Mn^{2+}, Ca^{2+} and Mg^{2+} are shown in Figure 2.

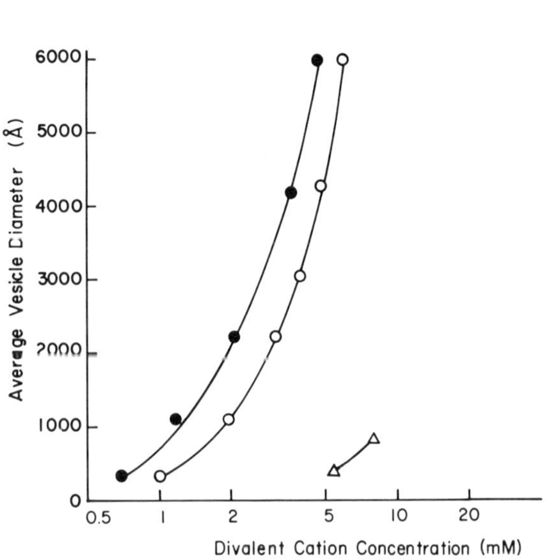

Figure 2. Fusion threshold concentrations of Mn^{2+} (●), Ca^{2+} (0) and Mg^{2+} (Δ) versus the sizes of phosphatidylserine vesicles.

MEMBRANE INTERFACIAL TENSION CHANGE DUE TO ION BINDING

It was observed that the interfacial tension of phosphatidylserine membranes increased when certain cations were adsorped (or bound) to the membrane surface. For measurements of the interfacial tension, lipid monolayers were prepared by placing an aliquot of the lipid spreading solution (approximately 1 mM lipid in hexane) by means of a microsyringe on an aqueous surface or at the oil (hexadecan)/aqueous interface of a constant area (64 cm^2 in a glass dish). Aqueous subphase solutions were 0.1 M NaCl/1 mM histidine/1 mM TES/0.01 mM EDTA, pH 7.4. The interfacial tension of a monolayer was measured with an electronic balance (Beckman, LM 500), using either a microscopic cover glass (9 x 9 x 0.2 mm) or a thin Teflon plate (11 x 11 x 1 mm) as a Whilhelmy plate (accuracy of 0.2 dynes/cm). The detailed procedures for measuring the surface tension were described previously, (Ohki, 1976; Ohki, 1982). The experiments were designed to measure the change in the surface tension of lipid monolayers as a function of various metal cation or polyamine concentrations in the subphase solution.

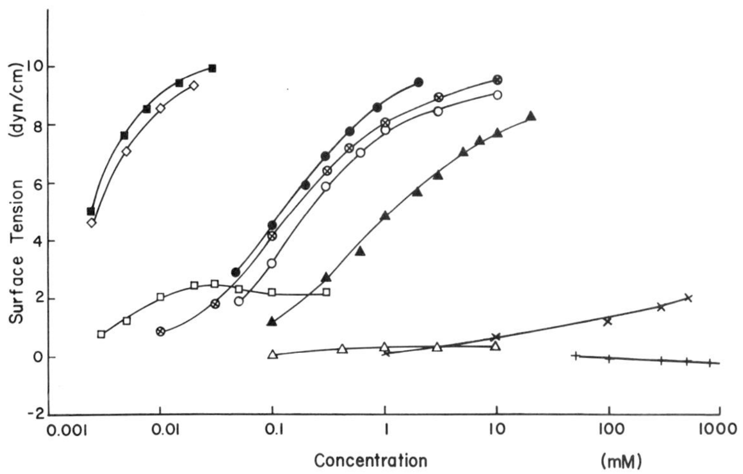

Figure 3. The increase in interfacial tension of a phosphatidylserine monolayer (70 A^2 per molecule) formed at the air-water (0.1 M NaCl/1 mM histidine/1 mM TES/ 0.01 mM EDTA (0 mM EDTA for La^{3+} and Tb^{3+})/pH 7.4) interface as a function of various metal cation and polyamine concentrations. (◊) La^{3+}; (■) Tb^{3+}; (●) Mn^{2+}; (O) Ca^{2+}; (▲) Mg^{2+}; (⊗) H^+; (x) Li^+; (+) Na^+; (□) $Spermine^{4+}$; (△) $Spermidine^{3+}$.

Figure 3 shows the surface tension increase in a phosphatidylserine monolayer (70 A^2 per molecule) as a function of various cation and polyamine concentrations. Trivalent cations (La^{3+} and Tb^{3+}) and divalent cations (Ca^{2+}, Mn^{2+}, Ba^{2+}, Sr^{2+}, and Mg^{2+}) caused large changes in the surface tension of the phosphatidylserine monolayer in the 5-20 μM and the 0.5-10 mM ranges, respectively. Data for Ba^{2+} and Sr^{2+} are not shown in Figure 3. Also, H^+ resulted in gradual but large changes in the surface tension of the monolayer in the range of pH 5-2. The surface tension increases showed nearly saturation at about 10 dynes for all cases. On the other hand, Na^+, Li^+, K^+ and other monovalent cations did not cause such increases in surface tension of the phosphatidylserine monolayer up to 1.0 M, although Li^+ resulted in a few dynes/cm increase in surface tension at 1 M. Spermine resulted in some increase in the surface tension of the phosphatidylserine monolayer but not as much as divalent cations do. However, other polyamines (spermidine and putrescine) did not cause any significant change in surface tension of the phosphatidylserine monolayer at any concentration (Fig. 3).

The surface pressure-area curves for a phosphatidylserine monolayer formed at the air/water and at the oil (hexadecan)/water interfaces are shown, respectively in Figure 4 where the water refers to 0.1 M NaCl/1 mM histidine/1 mM TES/0.01 mM EDTA, pH 7.4 with and without a certain concentration of divalent cation. In both experiments, the presence of divalent cations caused the increase in surface tension, about 8 dynes/cm for both 1 mM Ca^{2+} and 6 mM Mg^{2+} cases at a constant surface area (70 Å^2) per molecule. The similar surface tension change was observed in the range of area per molecule from 55 Å^2 to 80 Å^2 for both types of monolayers (the air/water and oil/water interfaces) (see Fig. 4).

VESICLE ELECTROPHORETIC MOBILITY - SURFACE CHARGES

The electrophoretic mobility of multilamellar phosphatidylserine vesicles was measured with a microelectrophoretic apparatus with respect to various cation and polyamine concentrations. The details are in the earlier paper (Ohki and Duax, 1986). Vesicle suspension solutions were all 0.1 M NaCl/2 mM TES/0.01 mM EDTA, pH 7.4. The relationship of the change in vesicle mobility with respect to the logarithm of ion concentration was almost linear for a considerable concentration range for most salts used (Fig. 5). The larger the valency of ions was, the less concentration of ion was needed to reduce the same mobility change of the vesicle among metal cations or polyamines, respectively. This is not necessarily the case between metal cation and polyamine ions.

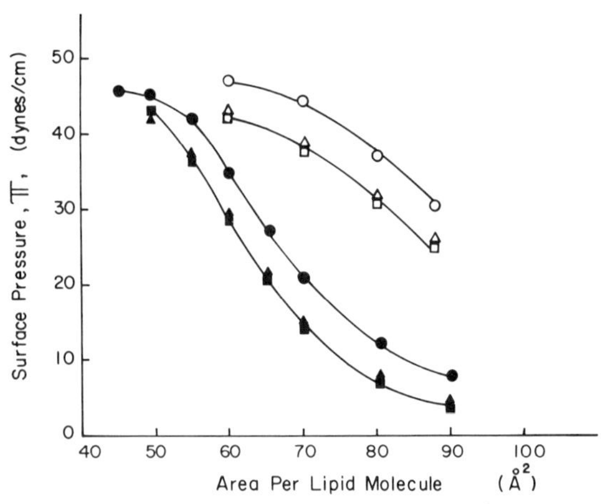

Figure 4. Surface pressure (Π)-area curves of phosphatidylserine monolayers formed at the air/water and the oil/water interfaces, respectively. The water was 0.1 M NaCl/1 mM histidine/1 mM TES/0.01 mM EDTA/pH 7.4 with or without divalent cations. The divalent cation concentrations were the same as those of the threshold concentration to induce membrane fusion. (●) for the air/water interface monolayer without divalent cation; (■) for the air/water monolayer with 1 mM Ca^{2+}; (▲) for the air/water monolayer with 6 mM Mg^{2+}; (○) for the oil/water monolayer without divalent cation; (□) for the oil/water monolayer with 1 mM Ca^{2+}; (△) for the oil/water monolayer with 6 mM Mg^{2+} in the subphase solution.

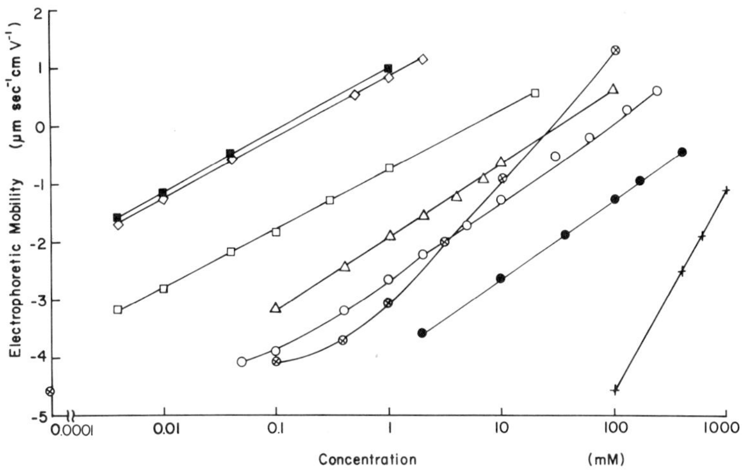

Figure 5. Electrophoretic mobility of multilamellar phosphatidylserine vesicles suspended in 0.1 M NaCl/2 mM histidine/ 2 mM TES/0.01 mM EDTA (0 mM EDTA for La^{3+} and Tb^{3+})/pH 7.4 as a function of various metal cation and polyamine concentrations; (◇) La^{3+}; (■) Tb^{3+}; (0) Ca^{2+}; (+) Na^{+} ; (□) Spermine^{4+}; (△) Spermidine^{3+}; (●) putrescine^{2+}; (⊗) H^{+}.

For La^{3+} and Tb^{3+}, the slopes of the vesicle mobility with respect to the logarithm of ion concentrations were 1.2 μm sec^{-1} cm V^{-1} per 10 fold change in salt concentration. For Ca^{2+} and Mg^{2+}, they were 1.4 μm sec^{-1} cm V^{-1} for the same change in salt concentration and 3.5 for both Na^{+} and Li^{+}. For H^{+}, it was rather gradual and somewhat between divalent and other monosalt cases. For polyamines, the slopes for spermine^{4+}, spermidine^{3+} and putrescine^{2+} were about 1.0, 1.2 and 1.4 μm sec^{-1} cm V^{-1}, respectively. For all trivalent and divalent cations examined and some polyamines (spermine), there were reversal points in the vesicle mobilities as the ion concentration reached a certain value; they were 120 μM for La^{3+},100 μM for Tb^{3+} and 90 mM for Ca^{2+}, etc.

From the above experimental results, it is clear that there is a good correlation between the degree of membrane fusion and that of the increased surface tension of the membrane, but not necessarily that of the reduction of surface charges.

THEORETICAL CONSIDERATION

In this section, we will consider the interaction forces involved in two closely apposed membranes in order to examine which interaction forces are responsible for the above mentioned lipid membrane fusion and how the increased surface tension is related to one of those forces. As a simple model representing a system of two interacting membranes, we consider two closely apposed, large flat hydrocarbon plates of infinite thickness which possess a thin hydrophilic layer of thickness, h, on their surfaces and are separated by an aqueous medium of distance R. These surface layers may be different from those of the interior of the flat hydrocarbon plates as well as the aqueous phase in molecular nature and its density (Fig. 6).

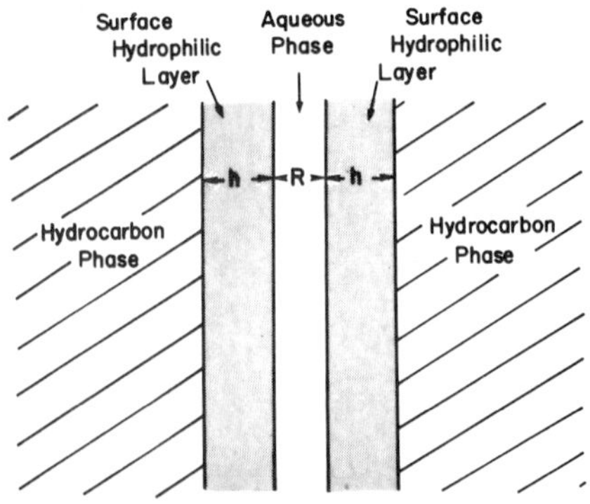

Figure 6. Schematic diagram of two closely apposed membranes. A membrane is represented as an infinite thickness of hydrocarbon phase having a hydrophilic surface layer of thickness h. The two membranes are separated by an aqueous phase at distance R.

The interaction energy, W, of two such flat plates can be expressed by the DLVO (Derjaguin, Landau, Verwey, and Overbeek) theory (Hiemenz, 1977; Verewy and Overbeek, 1948) as the sum of electrostatic repulsive interaction energy, V, the van der Waals attractive interaction energy, U, and the recent modifying terms (hydration energy V^h and short range attractive energy U^{SR}):

$$W = V + U + V^h + U^{SR}. \tag{1}$$

The interaction energy profile expressed by Eq. (1) usually has an energy barrier (the primary maximum) as the distance between the two plates becomes small (see Fig. 7). In addition to the four terms in Eq. (1), a term giving a short-range repulsive force due to the interaction of electron clouds surrounding molecules should be included in the total interaction energy. However, because this term is not important for membrane fusion, it is not expressed in the equation here. In some cases, the repulsive interaction energy term due to the structured water (LeNeveu et al., 1976; Rand, 1981), the term V^h in Eq. (1), which gives rise to the so-called "hydration pressure", is important. However, in the present paper, we consider the case in which there is no structured water extended to a long distance from the membrane surface but only a few layers of hydrated water at the membrane surfaces. If the two interfaces can overcome the primary maximum barrier by reducing their surface charges and taking off any existing structured water in the interspace between the two membranes, they will come close to the point of the primary minimum, which we may call the position of close adhesion. However, this close adhesion is not necessarily sufficient for the two interacting plates to fuse and become one.

When the separation distance is as small as in close adhesion, the electric field due to surface charges between the two layers is constant because of the Gauss theorem (Blum and Jancovici, 1984). Therefore, the electrostatic free energy changes very little with respect to the separation distance, compared with the van der Waals interaction energy. From this we may assume the electrostatic energy term to be a constant contribution to the total interaction energy at the close adhesion separation distance. Also, since the magnitude of surface charge density is reduced enough, the electrostatic energy contribution is small compared to the

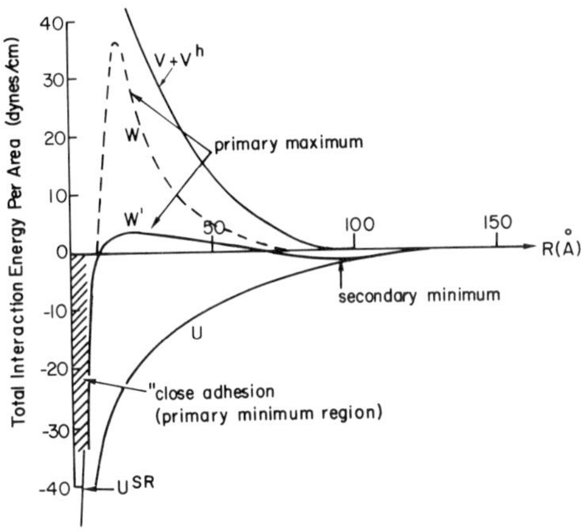

Figure 7. Schematic diagram of the interaction energy between the two membranes. V, Electrostatic repulsive energy; V^h, Hydration repulsive energy; U, van der Waals attractive energy; U^{SR}, short-range attractive energy. W = $V + V^h + U + U^{SR}$, total interaction energy having a large primary maximum; W, total interaction energy having a small primary maximum (e.g. due to ion binding). R, interlayer separation distance. The scales of the ordinate and the abcissa are rather arbitrary.

other attractive energies. Therefore, we consider only the van der Waals and the short-range attractive terms to describe the possible state for membrane fusion, or molecular cohesion. The van der Waals interaction energy may consist of the following parts: the London-dispersion, Debye, and Keesom terms. However, the latter two terms are, in general, small compared with the first term. Therefore, we consider only the London-dispersion term for van der Waals energy.

The van der Waals interaction energy between the two plates will be expressed as a function of the separation distance R, the thickness of the surface layer h, and the Hamaker constant A of two interacting phases (an interaction constant of the two interacting matters given in Eq. (3)) (Vincent, 1973; Vold, 1961):

$$U = -\Sigma_i \Sigma_j A_{ij} H(X_{ij}) \tag{2}$$

where A_{ij} is the Hamaker constant of interaction between the i-th and j-th phases and $H(X_{ij}) = 1/X_{ij}^2$, where X_{ij} is the distance between the two interfaces of either the surface hydrophilic layer or the bulk hydrocarbon phase-altogether four interfaces. The Hamaker constant (Hamaker, 1937) is approximately expressed in terms of the molecular polarization α, ionization potential I, and the molecular density ρ, of each phase as follows (London, 1937):

$$A_{ij} = \frac{3\pi^2}{2} \frac{I_i I_j}{I_i + I_j} \alpha_i \alpha_j \rho_i \rho_j . \tag{3}$$

Since the ionization potentials (Handbook of Chemistry and Physics, 1965) of the hydrocarbon unit C_2H_6 (12.8 eV), and some polar groups such as SO_2 (13.1 eV), and H_2O (12.6 eV) are similar quantities, eq. (2) can be expressed approximately by

$$U = -\frac{\pi I}{16} \{(\lambda_P - \lambda_W)^2 \frac{1}{R^2} + (\lambda_P - \lambda_W)(\lambda_H - \lambda_P) \frac{2}{(R+h)^2} + (\lambda_H - \lambda_P)^2 \frac{1}{(R+2h)^2} \} . \tag{4}$$

Here, $\lambda_i = \alpha_i \rho_i$, which we call the interaction coefficient, and suffixes W, P, and H refer to the aqueous, the polar hydrophilic layer and the hydrocarbon phases, respectively.

The interaction coefficients of water and hydrocarbon phases are 4.94×10^{-2} ($\alpha_{H_2O} = 1.48$ Å^3, $\rho_{H_2O} = 3.35 \times 10^{22}/cm^3$) and 7.72×10^{-2} ($\alpha_{C_2H_4} = 3.88$ Å^3, $\rho_{C_2H_4} = 1.99 \times 10^{22}/cm^3$), repsectively, (Handbook of Chemistry and Physics, 1965; Ohki and Fukuda, 1968) and the interaction coefficient of the surface hydrophilic layer is assumed to be somewhere between these two values. We can calculate approximately the van der Waals attractive energy between such two interacting membranes as a function of the interaction coefficient (λ_P) of the polar surface layer and its thickness (h) at a given separation distance (R) using the energy expression (Eq. 4) for the above mentioned model. We consider the separation distance to be the thickness of one or two layers of water between the two membranes. The short-range interaction energy U^{SR} arising from terms other than hydration energy may not be effective at such a separation distance (>2.5 Å) (C.J. van Oss, private communication). The results are shown as curves a, b, and c, respectively, in Figure 8.

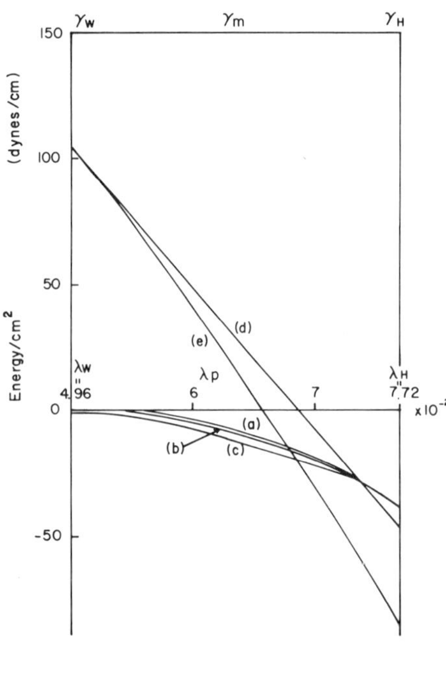

Figure 8. Van der Waals interaction energy between the two membranes and the dehydration energy per unit membrane area. The van der Waals energies for the two membranes (R = 2.5 Å and (a) h = 10 Å, (b) h = 6 Å, (c) h = 3 Å) are shown in Fig. 8 with respect to the interaction coefficient (λ_P) of the surface layer. λ_W (4.96×10^{-2}) and λ_H (7.22×10^{-2}) correspond to those of the aqueous (water) and hydrocarbon phases, respectively. Curve (d) is the effective dehydration energy E_{dh}^{eff} to remove the last layer of water from the membrane surfaces, with respect to hydrophobicity parameter p ($\gamma_m = p\gamma_H + q\gamma_W$). Curve (e) is the total energy of the van der Waals and dehydration energy, $U + E_{dh}^{eff}$.

It is reasonable to think that the attractive interaction energy must exceed a certain threshold value for the two membranes to fuse. This energy is assumed to be equal to the energy needed to dehydrate the last layer of water between the two interacting membranes. We may express such an energy E_{dh} in terms of the interfacial tension of the membrane surface. In order to dehydrate water from the membrane surface, first, it is necessary to separate the water from the membrane surface. This energy ΔG^{sep} is equal to the cohesion energy of the negative sign:

$$\Delta G^{sep} = \gamma_W + \gamma_m - \gamma_{m/W}, \tag{5}$$

where γ is the surface tension and $\gamma_{m/W}$ is the interfacial tension of the membrane/water interface. Then, the separated water surface should be removed from the occupied space. Therefore, the dehydration energy ΔG^{dh}:

$$\Delta G^{dh} = \gamma_{m/A} - \gamma_{m/W} \tag{6}$$

Since there are two such interfaces for the interacting membranes, the total dehydration energy is given by:

$$E_{dh} = 2(\gamma_{m/A} - \gamma_{m/W}). \tag{7}$$

Although $\gamma_{m/W}$ is experimentally obtainable to a certain extent since we do not have sufficient information about $\gamma_{m/A}$ and $\gamma_{m/W}$, we assume that the nature of the membrane surface layer is expressed by the sum of water and hydrocarbon portions:

$$\gamma_m = p\gamma_H + q\gamma_W,$$

$$\lambda_m = p\lambda_H + q\lambda_W \tag{8}$$

where p and q are the fractions of each portion ($p + q = 1$).

In such a case, we can calculate the dehydration energy of the membrane using the interfacial tensions of the water/air (≃72 dynes/cm), water/hydrocarbon (≃50 dynes/cm), and hydrocarbon/air (≃20 dynes/cm), respectively (Davies and Rideal, 1961). The results are shown as curve d in Figure 8. The effective dehydration energy $E^{eff}_{dh} = 0.75\ E_{dh}$.

When the sum of the attractive energy and the dehydration energy ($U + E^{eff}_{dh}$) shown in curve e becomes negative, the two interfaces (membranes) could fuse provided that no molecular structural constraint exists at the membrane surfaces. From our calculation, the interfacial tension of the membrane at this point is about 23 dynes/cm, which corresponds to the point of $\lambda_p = 6.5 \times 10^{-2}$, or $p \simeq 0.5$.

From the graphic results of the energy calculation (Fig. 8) it is seen that for membranes having similar molecular components and structures, the two factors (interaction coefficient and thickness h of the surface hydrophilic layers) are involved in membrane fusion. The membrane fusion should be facilitated by either increasing the magnitude of the interaction coefficient or reducing the thickness of the surface layers. However, the latter factor is less effective judging from Figure 8 (curves a-c). The important factor is the magnitude of dehydration energy, which is reduced as the hydrophilic layer becomes more hydrophobic (see Fig. 8, curve d).

Membrane Interfacial Tension and Hydrophobicity When the interfacial tension of a membrane is nearly zero, $\gamma_{m/W} = 0$, we may call the membrane interface completely hydrophilic or $p = 0$. When the interfacial tension increases and reaches about 50 dynes/cm which is that of the oil (hydrocarbon)/water interface, we may call the membrane interface as becoming hydrocarbon in nature, $p = 1$. The most lipid as well as biological membrane surfaces in monovalent salt environments have nearly zero interfacial tension. The lipid monolayer at the oil/water (0.1 M NaCl) interface has a low (a few dynes/cm) interfacial tension ($\gamma_{m/W} = \gamma_{o/W} - \Pi \cong 50$ dynes/cm $-\Pi$) at area for molecule of 60 ~ 65 Å^2 seen in Figure 4, which corresponds to the area for the lipid molecule for a flat lipid bilayer. However, the interfacial tension of such a monolayer increases as the area per lipid molecule increases or in the presence of Ca^{2+} or Mg^{2+} shown in Figure 4. For example, the interfacial tension at area per molecule 80 $\AA^2$ is about 15 dynes/cm for 0.1 M NaCl, pH 7.0 and 23 dynes/cm for 0.1 M NaCl + 1 mM $CaCl_2$, pH 7.0, or p = 0.46. If this lipid monolayer formed at the oil/water interface represents well the state of the interfacial tension of lipid bilayer membranes, we may use the results obtained from such monolayer studies to interpret the interfacial properties of the lipid bilayer membranes.

As we defined earlier, the membrane is composed of hydrophilic and hydrophobic components:

$$\lambda_m = p\lambda_H + q\lambda_w \quad \text{or} \quad \gamma_m = p\gamma_H + q\gamma_w$$
$$= p\lambda_o + q\lambda_w \ . \tag{9}$$

where at p = 0, the membrane is completely hydrophilic and at p = 1, the membrane is complete hydrocarbon or hydrophobic in nature. Therefore, when the value of p changes from 0 to 1, the membrane surface changes from hydrophilic to more hydrophobic in nature. The value of p may indicate the degree of hydrophobicity.

DISCUSSION

Experimental data on the cation-induced fusion of phospholipid membranes and the related physico-chemical properties of the membrane indicate that the degree of membrane fusion correlated well with the degree of increased surface tension or increased hydrophobicity of the membrane caused by cations (Ohki, 1977, 1982, 1984). A massive fusion of small unilamellar phosphatidylserine vesicles occurs at 1 mM Ca^{2+} in 100 mM NaCl, where the increase in interfacial tension of the membrane of about ≃8 dynes/cm from the reference state (0 mM Ca^{2+} and 0.1 mM NaCl). This is consistent with other cation cases. The increase in interfacial tension of the small unilamellar membrane is about ≃8 dynes/cm at the threshold concentration of each cation that induces the phosphatidylserine vesicle fusion. We may consider that this increase in interfacial tension is solely attributable to the change in nature of the surface hydrophilic layer. Our recent studies suggest that this energy increase arises mainly from a conformational change in the interfacial polar groups of the membrane molecules upon divalent cation chelation binding, rendering the hydrocarbon portion of the membrane molecule open to the water phase, but not due to the change in the electrostatic energy of the surface due to the reduction of surface charges (Ohki and Ohshima, 1984; Ohshima and Ohki, 1985).

The interfacial tension at the reference state (0.1 M NaCl) for the monolayer membrane of the plane phosphatidylserine bilayer is approximately a few dynes (2-4 dynes/cm) (Ohki, 1977 and also see Fig. 4). Compared with the flat bilayer the small lipid vesicle (300 Å in diameter) has approximately 12 dynes/cm, further, higher in surface tension due to the expanded area per molecule (80 $\AA^2$/molecule) of the outer monolayer of the vesicle because of its vesicle curvature and molecular geometry (Ohki, 1982). The total surface tension of this small phosphatidylserine vesicle in 0.1 M NaCl is about 15 dynes/cm from Figure 5. Therefore, the surface tension at the point corresponding to the threshold (where fusion occurs) for membrane fusion is about 23 dynes/cm. This value is approximately the same value as predicted in the present theory. It is not clear whether the short-range attractive force U^{SR} might contribute to the interaction energy or not (van Oss et al., 1986; Pashley et al., 1985).

Similar correlation between the interfacial tension increase and fusion of membranes is obtained for various monovalent and polyvalent cations including polyamines (see Table 1). Obviously, the monovalent metal cations (except for H^+) do not alter the hydrophilic nature of the surface layer by their bindings, as they neither bind in the chelating manner with lipid polar groups as divalent cations are known to do, nor dehydrate the hydrophilic polar surface layer as much as many other polyvalent cations and H^+ do (Ohki and Ohshima, 1984). Polyamines used in the experiments also do not cause an increase in interfacial tension of acidic phospholipid membranes. Although a considerable part of a polyamine is hydrophobic, the adsorption of many polyamines may effectively increase the thickness of the hydrophilic layer or the distance between the hydrophilic polar layers and therefore may not be effective in causing membrane fusion.

Table I. Effects of cations on fusion, aggregation and surface tension increase of phosphatidylserine membranes in 0.1 M NaCl

Ions or molecules	Concn.	SUV massive fusion	SUV massive aggregation	Increase in $\gamma_{m/W}$ of monolayer membranes (dynes/cm)
Na^+	<1 M	No	Yes	No change
K^+	<1 M	No	Slight	No change
Li^+	0.8 M	No	Yes	A few dynes
H^+	0.01 M	Yes	Yes	7-8
Ca^{2+}	1.0 mM	Yes	Yes	7-8
Mg^{2+}	6 mM	Yes	Yes	6-7
Ba^{2+}	0.8 mM	Yes	Yes	7-8
Mn^{2+}	0.7 mM	Yes	Yes	7-8
La^{3+}	6 μM	Yes	Yes	7-8
Tb^{3+}	8 μM	Yes	Yes	7-8
$Spermine^{4+}$	< 1 M	No	Yes	A few dynes
$Spermidine^{3+}$	<200 mM	No	Yes	No appreciable chg.
$Putrescine^{2+}$	<0.1 M	No	Yes	No appreciable chg.

Another example is seen in the case of membrane fusion induced by increasing temperature or an osmotic pressure difference applied across the membrane. It has been shown (Breisblatt and Ohki, 1975; Chaudhury and Ohki, 1981, 1984) that when the two phospholipid membranes are in close contact and are expanded by temperature or osmotic pressure change, the two membranes fuse after an increase of about 10 dynes/cm in interfacial tension of the membrane (Ohki, 1984). In this case, the lateral expansion of the membrane keeping its volume constant may result in thinning of the membrane, including the hydrophilic surface layer. This effectively increases the hydrophobicity in the hydrophilic surface layer region because the expansion of the hydrophilic surface layer in area would break up hydrogen bonds formed between the surface hydrophilic polar groups and water molecules. On an average, this would be expected to reveal more of the internal hydrophobic phase to the aqueous phase at the interface. This would result in the increase in interfacial tension of the membrane (Chaudhury and Ohki, 1982). The decrease in only the surface charge density of membranes is not a sufficient condition for two membranes to fuse (see Fig. 4).

The curvature of the membrane surface has another factor for surface hydrophobicity. It is seen from Figure 2, the larger the radium of curvature of phosphatidylserine vesicles is, the greater the threshold concentrations of divalent cations are. This can be explained by the surface tension values for the oil/water interface monolayer shown in Figure 5. As the radius of curvature of the membrane becomes larger, the area per lipid molecule will be smaller and therefore the surface tension will be smaller than that of the vesicle having a smaller radius of curvature at 0.1 M NaCl (see Fig. 5). A vesicle of larger size may need to have more adsorption of divalent cations on the surface than that of smaller size, in order to attain a certain value of surface tension necessary for membrane fusion.

The above argument was based on a physical quantity (i.e., the observed surface tension of the membrane) of the membrane averaged in time as well as space. However, some of the fusion processes could be due to a dynamical process or a local process: The separation distance between the two surfaces of the hydrocarbon phases at some portions of the membrane at a certain time could become smaller than the average separation distance of the two membrane hydrocarbon phases by either or both thermal fluctuations and mechanical disturbances applied on membranes. Hence, the strong interaction may occur through small but closely interacting portions of the two membranes. In such a case, the interaction energy per unit area in highly local portions could be greater than the average interaction energy per unit area and exceed a threshold energy required for membrane fusion, even though other parts of the membrane surface remain more hydrophilic. Membrane fusion may be caused by the phase defect through a local surface energy increase, occurring within the membrane (1) upon mixing of different membrane molecular components (Hui et al., 1981) or (2) by lowering temperature below a certain phase-transition temperature (Schullery et al., 1980), or (3) because these situations create non-uniform distribution of membrane molecules. In such a situation, the present proposed theory should be modified as a local event, and the physical quantities, which correlated to the membrane fusion, should be those of the local portions. However, the principal base for the theory presented here is unchanged. As previously proposed (Ohki, 1982, 1984), the membrane fusion events all have the following features in common: The increase in surface free energy or hydrophobicity and consequently the membrane becomes a dehydrated state at the site of membrane fusion, which is caused by ion binding to the membrane charged sites for ion-induced membrane fusion, or by the membrane expansion followed with the breakage of hydrogen binding for the increased temperature or osmotic pressure-induced membrane fusion, or by the formation of phase defect points or higher curvature in the membrane phase for ion or macromolecule-induced fusion or the lowering temperature-induced fusion. Thus, most of the observed membrane fusion can be explained in similar manner by the theoretical interpretation presented here.

Acknowledgments — This work was supported by a grant, GM24840, from the National Institutes of Health.

REFERENCES

Blum, L., and Jancovici, B. Exactly soluble model for the interaction of two parallel charged plates in an ionic medium. J. Phys.Chem. 88: 2294-2297 (1984).

Breisblatt, W., and Ohki, S. Fusion in phospholipid spherical membrane. I. Effect of temperature and lysolecithin, J. Membrane Biol. 23: 385-401 (1975).

Chaudhury, M., and Ohki, S. Correlation between membrane expansion and temperature-induced membrane fusion, Biochim. Biophys. Acta 642: 365-374 (1981).

Cohen, F.S., Zimmerberg, J., and Finkelstein, A. Fusion of phospholipid vesicles with planar phospholipid bilayer membranes. II. Incorporation of a vesicular membrane marker into the planar membrane, J. Gen. Physiol. 75: 251-270 (1980).

Creutz, C.E. Cis-unsaturated fatty acids induce the fusion of chromaffin granules aggregated by synexin, J. Cell Biol. 91: 247-256 (1981).

Cullis, P.R., and Hope, M.J. Effect of fusogenic agent on membrane structure of erythrocyte ghosts and the mechanism of membrane fusion, Nature (London) 271: 672-674 (1978).

Davies, J.T., and Rideal, E.K. Interfacial Phenomena, Academic Press, New York (1961).

Duzgunes, N., and Ohki, S. Calcium-induced interaction of phospholipid vesicles and bilayer lipid membranes, Biochim. Biophys. Acta 467: 301-308 (1977).

Gad, A.E., Silver, B.L., and Eytass, G.D. Polycation-induced fusion of negatively-charged vesicles, Biochim. Biophys. Acta 690: 124-132 (1982).

Hammoudah, M.M., Nir, S., Isac, T., Kornhauser, R., Stewart, T.P., Hui, S.W., and Vaz, W.L.C. Interaction of La^{3+} with phosphatidylserine vesicle: Binding phase transition, leakage and fusion, Biochim. Biophys Acta 558: 338-343 (1979).

Haywood, A.M., and Boyer, B.P. Initiation of fusion and disassembly of Sendai virus membranes into liposomes, Biochim. Biophys. Acta 646: 31-35 (1981).

Herrman, A., Pratsch, L., Arnold, K., and Lassman, G. Effect of poly (ethylene glycol) on the polarity of aqueous solutions and on the structure of vesicle membranes, Biochim. Biophys. Acta 733: 87-94 (1983).

Hiemenz, P.C. Principles of Colloid and Surface Chemistry, Marcel Dekker, New York (1977).

Hong, K., Schuber, F., and Papahadjopoulos, D. Biological modulators of membrane fusion, Biochim. Biophys. Acta 732: 469-472 (1983).

Hui, S.W., Stewart, T.P., Boni, L.T., and Yeagle, P.L. Membrane fusion through point defects in bilayers, Science 212: 921-923 (1981).

Klappe, K., Wilschut, J., Nir, S., and Hoekstra, D. Parameters affecting fusion between Sendai virus and liposomes. Role of viral proteins, liposome composition,and pH, Biochemistry 25: 8252-8260 (1986).

Knutton, S. Studies of membrane fusion. III. Fusion of erythrocytes with polyethylene glycol, J. Cell Sci. 36: 61-72 (1979).

LeNeveu, D.M., Rand, R.P., and Parsegian, V.A. Measurement of forces between lecithin bilayers, Nature (Lond.) 259: 601-603 (1976).

London, F. The general theory of molecular forces, Trans. Faraday Soc. 33: 8 (1937).

Lucy, J.A. The fusion of biological membranes, Nature (Lond.) 227: 814-817 (1970).

Maeda, T., and Ohnishi, S. Activation of influenza virus by acidic media causes hemolysis and fusion of erythrocytes, FEBA Lett. 122: 283-287 (1980).

Miller, C., Arvan, P., Telford, J.N., and Racker, E. Ca^{2+}-induced fusion of proteoliposomes: Dependence on transmembrane osmotic gradient, J. Membrane Biol. 30: 271-282 (1976).

Ohki, S., and Fukuda, N. Interaction energy between water and hydrocarbon phases J. Colloid Interface Sci. 27: 208-215 (1968).

Ohki, S., and Ohki, C.B. Monolayer at the oil/water interface as a proper model for bilayer membranes, J. Theor. Biol. 62: 389- (1975).

Ohki, S. A mechanism of divalent ion-induced phosphatidylserine membrane fusion, Biochim. Biophys. Acta 689: 1-11 (1982).

Ohki, S. Effects of divalent cations, temperature, osmotic pressure gradient, and vesicle curvature on phosphatidylserine vesicle fusion, J. Membrane Biol. 77: 265-275 (1984).

Ohki, S. Membrane fusion: Theory and Experiments, Studia Biophysica. 10: 95-104 (1985).

Ohki, S., and Leonards, K. Effects of proteins on phospholipid vesicle aggregation and lipid vesicle-monolayer interactions, Chemistry & Physics of Lipids 31: 307-318 (1982).

Ohki, S., and Ohshima, H. Divalent cation-induced surface tension increase in acidic phospholipid membranes: ion binding and membrane fusion, Biochim. Biophys. Acta 776: 177-182 (1984).

Ohki, S., and Duax, J. Interaction of polyamines with phosphatidylserine membranes, Biochim. Biophys. Acta 861: 177-186 (1986).

Ohshima, H., and Ohki, S. Effects of divalent cations on the surface tension of a lipid monolayer-coated air/water interface, J. Colloid Interface Sci. 102: 85-94 (1985).

Pashley, R.M., McGuigganm, P.M., Ninham, B.W., and Evans, D.F. Attractive forces between uncharged hydrophobic surfaces: Direct measurements in aqueous solution, Science 299: 1085-1089 (1985).
Papahadjopoulos, D., Vail, W.J., Newton, C., Nir, S., Jacobson, K., Poste, G., and Lazo, R. Studies on membrane fusion. III. The role of calcium-induced phase changes, Biochim. Biophys. Acta 465: 579-598 (1977).
Poste, G., and Allison, A.C. Membrane fusion, Biochim. Biophys. Acta 300: 421-465 (1973).
Poste, G., and Nicolson, G.L. Eds., Membrane Fusion, Elsevier/North-Holland, Amsterdam (1978).
Rand, P. Interacting phospholipid bilayers: Measured forces and induced structural charges, Rev. Biophys. Bioeng. 10: 277-314 (1981).
Rasmussen, H. Cell communication, calcium ion and cyclic adenosine monophosphate, Science 170: 404-412 (1970).
Rouser, G., Bauman, A.J., Kritchevasky, G., Heller, D., and O'Brien, J.S. Quantitative chromatographic fractionation of complex lipid mixtures: Brain lipids, J. Am. Oil Chem. Soc. 38: 544-555 (1961).
Schuber, F., Hong, K., Duzgunes, N., and Papahadjopoulos, D. Polyamines as modulators of membrane fusion: Aggregation and fusion of liposomes, Biochemistry 22: 6134-6140 (1983).
Schullery, S. E. Schmidt, C.F., Felgner, P., Tillack, T.W., and Thompson, T.E. Fusion of dipalmitoyl-phosphatidylcholine vesicles, Biochemistry 19: 3919-3923 (1980).
Sowers, A.E. A long-lived fusogenic state is induced in erythrocyte ghosts by electric pulses, J. Cell Biol. 102: 1358-1362 (1986).
van Oss, C.J., Good, R.J., and Chaudhury, M.K. The role of van der Waals forces and hydrogen bonds in "hydrophobic interactions" between biopolymers and low energy surfaces, J. Colloid Interface Sci. 111: 378 (1986).
Verkleij, A.J., Van Echteld, C.J.A., Gerritsen, W.J., Cullis, P.R., and DeKruijff, B. The lipidic particle as an intermediate structure in membrane fusion processes and bilayer to hexagonal H_{II} transitions, Biochim. Biophys. Acta 600: 620-624 (1980).
Verwey, E.J.W., and Overbeek, J.Th. G. Theory of the Stability of Lyophobic Colloids, Elsevier, Amsterdam (1948).
Vincent, B. The van der Waals attraction between colloid particles having adsorbed layers. II. Calculation of interaction curves, J. Colloid Interface Sci. 42: 270-285 (1973).
Vold, M.J. The effect of adsorption on the van der Waals interaction of spherical colloidal particles, J. Colloid Sci. 16: 1-12 (1961).
Wang, C., and Huang, L. Polyhistidine mediates an acid enhanced fusion of negatively charged liposomes, Biophys. J. 45: 700 (1984).
Weast, R.C., Ed., Handbook of Chemistry and Physics, The Chemical Rubber Co., Cleveland, Ohio (1965).
Wilschut, J., and Papahadjopoulos, D. Ca^{2+}-induced fusion of phospholipid vesicles monitored by mixing of aqueous contents, Nature (Lond.) 281: 690-692 (1979).
Wilschut, J., Duzgunes, N., and Papahadjopoulos, D. Studies on the mechanism of membrane fusion: Kinetics of Ca ion induced fusion of phosphatidylserine vesicles followed by a new assay for mixing of aqueous vesicle contents, Biochemistry 19: 6011 6021 (1980).
Wilschut, J., Duzgunes, N., and Papahadjopoulos, D. Calcium/magnesium specificity in membrane fusion: Kinetics of aggregation and fusion of phosphatidylserine vesicles and the role of bilayer curvature, Biochemistry 20: 3126-3133 (1981).
Zimmermann, U. Electric field-mediated fusion and related electrical phenomena, Biochim. Biophys. Acta 694: 227-277 (1982).

INTERACTIONS IN LIPOSOMAL-DRUG DELIVERY IN VIVO AND IN VITRO

Eric Mayhew

Departments of Experimental Pathology and
Experimental Therapeutics
Roswell Park Memorial Institute, Buffalo, NY 14263

Summary

In order for any drug carrier system to be successfully developed it needs to be shown that the overall therapeutic efficacy of the drug is improved after carrier associated administration compared to administration in the free form. Successful use of such carrier systems may involve more selective delivery of the drug to its site of action and/or more controlled release of the drug systemically or at the site of action. Liposomes have been investigated extensively as a delivery system in experimental systems and initial clinical trials of some liposome-drug combinations are underway. Liposomes can interact with the *in vivo* biological environment, which includes plasma components and cells, by a number of ways including exchange of liposome components with lipoproteins, direct interactions with lipoproteins or proteins, adsorption to cells followed by liposome degradation, or interiorization. Interiorization probably usually occurs by endocytosis although fusion may occur to a minor extent. The relative importance of these interactions depends both on liposome type and composition and the cells with which liposomes interact. It may be possible to alter liposomal composition to facilitate or inhibit different types of interaction. The consequences of these interactions, particularly with respect to interactions with plasma components and cells of the reticuloendothelial system for the use of liposomes as a drug-delivery system are discussed. In addition, the use of liposomes as an organ selective delivery system or a cell targeted delivery system for the delivery of drugs, such as cancer chemotherapeutic agents and biological response modifiers, for the treatment of metastatic cancer are discussed.

Introduction

Although liposomes were originally developed as a model membrane system their potential as a drug delivery system was realized shortly thereafter as they are able to sequester molecules including drugs within the aqueous space(s) or in the lipid bilayer(s). There are two main ways liposomes can deliver drugs directly and efficiently into cells; (a) into the cytoplasm or membranes after non-leaky fusion of the lipid bilayers with the cell membrane or (b) after endocytosis involving release into the cytoplasm via a lysosomal pathway (1,2). In vitro, where studies of mechanisms can be made under appropriate controlled conditions it has been shown that both of the above mechanisms of liposomal uptake can occur, usually after initial adsorbtion to the cell surface. In vivo however, most studies indicate that the preferred route of intracellular uptake of liposomes, where it occurs, is by endocytosis (3). The conditions needed to induce liposome-cell membrane fusion are not easy to duplicate in vivo. Thus, although membrane fusion is of basic biological interest in a number of areas, the in vivo application of liposomes and other model membrane systems for drug delivery is of practical importance and is the subject of this report. In the drug delivery area it is important that in vivo and in vitro studies proceed concurrently as results obtained in vitro may not be applicable in vivo or vice versa. Liposomes most likely to fuse are often liposomes which are the most unstable in vivo. The introduction of stabilizing molecules such as cholesterol to improve biological half-life (4) often reduces the potential for liposome-cell fusion to play a significant role in vivo. It would seem theoretically possible to design fusogenic liposomes for in vivo use such as using specific fusogenic proteins incorporated into cell-specific stable targeted drug or toxin carrying liposomes. However, the present available methodology for drug delivery using model membrane systems is not at such a level and this discussion will indicate some of the problems and possibilities involved in extending in vitro studies to the in vivo situation.

The following will be discussed:

1) Problems that need to be overcome in successful cancer therapy,
2) Areas where drug delivery systems may be of practical use,
3) Adriamycin-liposomes, an example of experimental studies which indicate that liposomal drug delivery may be of clinical use,
4) Results of some investigations into possible mechanisms of action of Adriamycin-liposomes,
5) Results of some investigations obtained on immunospecific targeting.

1) Problems associated with cancer chemotherapy

Table 1 summarizes a number of major problems associated with cancer chemotherapy which results in most of the individuals succumbing to the disease if they are not successfully treated by surgery or radiation. Taken together it can be seen why current therapy is relatively unsuccessful and it is also very unlikely that a single "magic bullet" is going to provide a solution to all the problems. Improvements in selectivity in drug delivery may, however, be expected to provide a quantitative improvement in therapeutic outcome combined with reduced side-effects.

Table 1

Major Problems Associated with Chemotherapy of Common Solid Cancers

1) Agents are not specifically selective for cancers,
2) Low therapeutic index occurs due to significant non-specific toxicity,
3) Cancers may be innately resistant to agent or aquire resistance during growth and during therapy,
4) Cancers are composed of inherently heterogenous cell populations, including cells with a range of drug sensitivities,
5) Cancers "progress", becoming more malignant, and sensitivity to therapeutic agent may change,
6) Metastases and primary tumors may differ in their drug sensitivities,
7) The agent may not be able to reach all the tumor cells, even though they may all be sensitive to it, due to anatomic considerations.

2) Areas where Drug Delivery may be of practical use

Improvements in drug delivery can include favorably altered pharmacokinetics and improved delivery to the specific target cells or the general site (e.g. a specific organ or tissue) where the target cells are located.

Although organ-selective delivery is not as specific a delivery method as cell-selective delivery it may have an advantage in that the delivery vehicle does not have to reach every target cell (5,6) although in order to eradicate cancer, the therapeutic agent itself would have to reach every cell. One hundred percent cell selective delivery may be difficult to achieve for macromolecular carriers due to anatomic considerations, but effective delivery of carriers to some organs sites can be achieved by a combination of delivery route and particle size.

Table 2 gives some examples of potential sites for organ selective delivery and as three of the most important sites involved in cancer spread are the liver, lungs and lymph nodes organ successful selective delivery of therapeutic agents to these sites may inhibit cancer spread. In particular, the spread of colon cancer has been shown to occur primarily via a stepwise process from the colon to the liver, then to the lungs and then to other sites. We have calculated the relative enhancement of uptake in organs such as the liver after different degrees of selective delivery to the target organ (6). For example, if 50% of the injected drug is delivered to the liver, which is attainable using stable liposomes, this results in an approximate 10 fold increase in delivery to that site. At the same time the amount of drug delivery to sites of non-specific toxicity may be decreased 50% (6). Therefore, the relative enhancement of delivery to the target site is approximately 20 times, overall, which should theoretically approximate the net therapeutic gain.

Table 2

Some Targets for Organ/Tissue Selective Delivery

Organ/Tissue	Delivery Route
Liver	vascular
Lungs	vascular (large particles), aerosol
Lymph Nodes	sub-cutaneous
Brain	intrathecal
Bladder	intravesicular
Skin	topical
Joints	intra-articular

3) Adriamycin - Liposomes

A number of laboratories have shown that entrapment of Adriamycin in liposomes reduces its toxicity in experimental animals (7,8,9). For example, we have shown that Adriamycin entrapped in 0.15μm diameter phosphatidylglycerol: hosphatidylcholine:cholesterol:α-tocopherol (1:4:4:0:1 molar ratio) liposomes is approximately 3 times less acutely or chronically toxic to mice than free drug under either a single dose or multiple (5 daily doses or 5 fortnightly doses) i.v. injection schedule (10). In addition, we have shown that Adriamycin-liposomes are more efficacious than free Adriamycin at equal doses in inhibiting the growth of several *in vivo* models of liver metastatic cancer (11,12). Therefore, Adriamycin-liposomes have a higher therapeutic index than the free drug. The question therefore arises is the gain in therapeutic index due to selective delivery as suggested?

4) Mechanisms of Adriamycin-Liposome Action

A necessary to preliminary to determining if the effects of Adriamycin-liposomes *in vivo* are due to selective delivery was to determine if there was any selectivity in uptake and effects on target cells dependant on dosage form. Table 3 shows that free and liposomal Adriamycin were equi-toxic *in vitro* to a panel of cultured cell types including tumor cells such as CT38 where Adriamycin liposomes show superior efficacy to free drug *in vivo* (liver metastases) (11). This and much other data obtained in culture systems suggests strongly that the effects of free or liposomal Adriamycin after intracellular uptake by tumor cells do not significantly differ. Other data shows that under similar incubation conditions less Adriamycin was taken up by cells after liposomal administration compared with free, including both endocytotic and non-endocytotic cell types (13), although DNA synthesis was inhibited similarly at similar initial free or liposomal Adriamycin concentrations. Fluorescence microscopy of cells incubated with free or liposomal Adriamycin showed that whereas free Adriamycin penetrated rapidly, liposomal Adriamycin was seen initially at the cell surface followed by intracellular uptake. At the fluorescence microscopy level there was no obvious difference in intracellular localization of Adriamycin; i.e. strong nuclear and perinuclear fluorescence. These results therefore suggest that, independent of dosage form, Adriamycin has similar effects when internalized although it is possible that some effects may be mediated at the cell surface for Adriamycin-liposomes.

Table 3

Toxicity of Free or Liposome Entrapped Adriamycin to Cultured Cells (ID_{50}, (M)).

Cell Type	Free Adriamycin	Adriamycin Liposomes
L1210	2×10^{-8}	1×10^{-8}
M5076	5×10^{-7}	3×10^{-7}
CT26	3×10^{-7}	3×10^{-7}
CT38	3×10^{-8}	3×10^{-8}
P388/0	1×10^{-8}	1×10^{-8}
P388/AM	$> 10^{-6}$	$> 10^{-6}$

(mean of 2-4 experiments, 2-4 wells/concentration/expt.)

When liposomes are injected into animals, a series of sequential interactions are possible as summarized in table 4. If the goal in administering liposomes is for selective delivery, intact liposomes must survive to steps 3-5, requiring liposomes which are not easily degraded in the circulation. Liposomes which are relatively stable in the circulation can be made (4,14) so that intact liposomes can reach cells at target organs or the reticulo-endothelial system. With Adriamycin-liposomes, several factors related to the overall alterations in therapeutic index need to be considered. Firstly, there is the question of distribution to the heart, the site of the dose-limiting cardiomyopathy.

Table 4

In Vivo Interactions of Liposomes with the Biological Environment after i.v. Injection

(1) Non-cellular plasma components
- (a) exchange
- (b) disruption
- (c) degradation/metabolism
- (d) salts/pH, etc.

(2) Cell components of plasma
- (a) erythrocytes
- (b) white blood cells, etc.

(3) Blood vessel linings/Fenestrations/Junctions

(4) Reticulo-endothelial cells
- (a) macrophages
- (b) lymphocytes, plasma cells, etc.

(5) Target and non-target cells
- (a) direct vascular access
- (b) extra vascular

Table 5 shows that 1 hour after administration, less Adriamycin was detectable in the heart after liposomal administration. At the same time, there was significantly greater total Adriamycin levels in the blood of animals receiving the liposomal drug (9). However, it can also be seen that there were differences in the ratio of free/liposome uptake when comparing total Adriamycin plus metabolites to Adriamycin only. There were more extractable metabolites in the hearts of mice treated with the free drug. If non-specific toxicity of Adriamycin was related to metabolite concentrations (toxic metabolites of Adriamycin are known) as well as, or instead of Adriamycin itself, it can be suggested that there could be qualitative as well as quantitative differences in Adriamycin uptake in tissues and could account for altered toxic effects.

Table 5

Adriamycin Content of Mouse Heart*

	ug/mg Protein Free Adriamycin	Adriamycin Liposomes	Free/ Liposomes
Adriamycin plus metabolites[a]	1.10±0.32	0.24±0.06	4.6
Adriamycin only[b]	0.47±0.14	0.21±0.09	2.2

*1 hour after i.v. injection of Free Adriamycin or Adriamycin Liposomes at 10mg/kg, into C57B16 mice (normal).

[a]Total fluorescence (acid extracted (15))

[b]Extracted sample run on HPLC (Adriamycin peak) (16)

Not only is the liver the major site of Adriamycin metabolism, it is the organ (along with the spleen) where liposomes are most effectively cleared and it is the site of metastasis where improved therapeutic effects have been shown in mouse model systems. Table 6 summarizes some data showing the disposition of Adriamycin in normal liver and liver metastases in the same animals at different times after administration in the two dosage forms. Firstly, it can be seen that at 1 hour there is only a marginal increase in the Adriamycin content of normal liver tissue after administration in the liposomal form whereas by 24 hours there is approximately 10 times as much after liposomal administration. Free Adriamycin is cleared from normal liver much more rapidly than after Adriamycin-liposome administration. Adriamycin uptake in liver metastases is less than in normal liver after both free and liposomal administration. The amounts of Adriamycin in liver tumors decreases between one and 24 hours after free drug administration whereas the levels after liposomal administration remain constant or rise slightly over the same time interval. It can be suggested that Adriamycin retained in normal liver, probably at least in part in RES cells, after liposomal administration acts as a reservoir releasing Adriamycin which sustains the levels in the tumor tissue, whereas the levels in the tumor after free Adriamycin administration decline rapidly. If maintenance of sustained levels of Adriamycin in tumors are related to the therapeutic effects then these results would suggest that the organ-selective delivery hypotheses may have some validity and explain the experimental therapeutic results.

Table 6

Amount of Adriamycin[a] in livers of mice bearing CT38LD liver metastases*

Tissue	Time (h)	ug/mg Protein Free Adriamycin	Adriamycin Liposomes	Lip/Free
Normal liver	1	1.2±0.5	1.4±0.4	1.2
" "	24	0.04±.02	0.39±.01	9.8
Liver metastases	1	0.05±.02	0.09±.03	1.8
" " "	24	0.015±.01	0.11±.005	7.3

*1 or 24 hours after i.v. injection of free Adriamycin or Adriamycin liposomes at 10mg/kg, into C57Bl mice bearing CT38LD liver metastases.

Tissues acid extracted and Adriamycin determined using HPLC (16).

However, there is an alternative or additional explanation of the improved therapeutic effects; that is that Adriamycin-liposomes can act as a more effective host defense modulator than the free drug resulting in enhanced tumoricidal activity by effector cells. A series of studies have shown that Adriamycin can modulate host defense systems (17,18), and the entrapment of other immunopotentiators in liposomes has been shown to augment their activity (19). Comparative determinations have been made of the magnitude and kinetics of the effects of free or liposomal Adriamycin on the tumoricidal activity of mouse peritoneal exudate cells used as a model system.

The results show that <u>in vivo</u> administration of liposomal Adriamycin did not enhance the magnitude of the <u>in vitro</u> tumoricidal response of peritoneal macrophages from the treated mice, when stimulated with lipopolysaccharide, against sensitive (P815) tumor cells but the duration of the response was extended from 14 to 18 days after administration. In addition, liposomal-Adriamycin augmented the magnitude of the NK tumoricidal response (YAC-1 cells) at 7, 11 or 14 days after administration compared to the free drug effects. It is possible that the results could be due to greater retention of liposomal-Adriamycin compared to free in the peritoneal exudate cells but evidence from other studies suggests that direct drug transfer was not responsible for the effects (19). In the liver, host defense systems similar to those in the peritoneal cavity may occur so it is possible that anti-tumor effects in the liver may be mediated by augmentation of host-defense systems in that organ.

It can be seen therefore, that there is suggestive evidence that alterations in therapeutic activity may result from the interplay of several mechanisms. It is probable therefore, that alterations in distribution, circulation time or mechanisms of interaction of liposomes due to variations in their components could be expected to exert different effects on the various possible effector mechanims.

5) Immunospecific targeting

It is pertinent to mention here some studies, which are in progress, where the aim is to attempt to immunospecifically cell target drug containing liposomes *in vivo* (20). The system investigated has been a B-cell lymphoma line, 2C3, which expresses at the cell surface idiotypic determinants on the immunoglobulin chains. Affinity purified Fab' anti-idiotype fragments which recognize a private antibody of the 2C3 cells were coupled to Cytosine Arabinoside containing liposomes (20). Cytosine Arabinoside is a drug used extensively in the clinic for treatment of leukemias and in addition the B cell lymphoma, 2C3, is sensitive to it.

In addition to the availability of a useful immunospecific system, the surface immunoglobulin is endocytosed after interaction with antibody, which may be expected to enhance cytotoxic effects of some liposome entrapped drugs. *In vivo* the B cell lymphoma, after i.v. injection, is mainly confined to the spleen and the vascular space so that macromolecular carriers should be able to reach the target sites. *In vitro* a number of experiments have shown, with appropriate controls, that highly selective killing of the specific target cell can occur. Under appropriate conditions all target 2C3 cells can be killed *in vitro* whereas a subline derived from the 2C3 but lacking the reference idiotype at its surface is spared. However, *in vivo*, the effects are not as convincing. Although in some instances specific inhibition of DNA synthesis of tumor cells in the spleen can be detected it has proven difficult to reproduce the results routinely. Prolongation of the life span of tumor bearing mice after treatment with immunospecific cytotoxic liposomes has not proven possible so far.

These studies indicate therefore, that although specific cytotoxicity after specific cell targeting can be demonstrated *in vitro* transfer of these results to the *in vivo* situation can be difficult. As discussed previously, the complexicity of the *in vivo* system reduces the chances of success. It is clear that even if elimination of tumors via immunospecific cell killing using macromolecular carriers is possible much optimization and refinement of the targeting systems used need to be done.

Although more selective anti-tumor therapy using macromolecular carriers is still at an early stage, the problems involved have become more recognizable. As even relatively unsophisticated systems such as Adriamycin-liposomes may have therapeutic advantages it would seem possible that with further development, utilizing results from studies at both the cellular and *in vivo* levels, that significant improvements in therapy for cancer and other diseases will occur.

Acknowledgements

The work described was supported in part by USPHS grants CA28494, 37304 and 33462. The author would like to acknowledge the collaboration of Y. Rustum, M.J. Ehrke, S. Radel and R. Bankert, and the technical assistance of R. Lazo, C. Bastian, D. Oleszek and D. Graham, in different aspects of the studies described.

References

1. R.E. Pagano and L. Huang, Interactions of Phospholipid Vesicles with Cultured Mammalin Cells. J. Cell Biol. 67:49-60, 1975.
2. E. Mayhew and D. Papahadjopoulos, Therapeutic Applications of Liposomes. In: "Liposomes". M. Ostro (ed.), M. Dekker, New York, pp. 289-341, 1983.
3. E. Wisse, G. Gregoriadis and W.Th. Daems, Electron Microscopic Cytochemical Localization of Intravenously Injected Liposome-Encapsulated Horse Radish Peroxidase in Rat Liver Cells. Adv. Exp. Med. Biol. 73:237-245, 1975.
4. E. Mayhew, Y. Rustum, F. Szoka and D. Papahadjopoulos, Role of Cholesterol in Enhancing the Anti-Tumor Activity of Cytosine Arabinoside Entrapped in Reverse Phase Evaporation Vesicles. Cancer Treat. Rep. 63:1923-1928, 1979.
5. E. Mayhew, Liposomes in Cancer In: "Liposome Letters", A.D. Bangham (ed.), Academic Press, London, pp. 357-367, 1983.
6. L. Weiss and E. Mayhew, An Approach to Therapy of Metastases of the Upper Rectum: A Working Hypothesis. Cancer Drug Delivery 2:19-33, 1985.
7. E.A. Forssen and Z.A. Tokes, In vitro and In vivo Studies with Adriamycin - Liposomes. Biochem. Biophys. Res. Comm. 91:1295-1301, 1979.
8. A. Rahman, A. Kessler, N. More, B. Sikic, G. Rowden, P. Woolley and P.S. Schein, Liposomal Protection of Adriamycin-Induced Cardiotoxicity in Mice. Cancer Res. 40:1532-1537, 1980.
9. F. Olson, E. Mayhew, D. Maslow, Y. Rustum and F. Szoka, Characterization, Toxicity and Therapeutic Efficacy of Adriamycin. Eur. J. Cancer 18:167-176, 1982.
10. E. Mayhew and Y.M. Rustum, The use of Liposomes as Carriers of Therapeutic Agents. In: "Molecular Basis of Cancer" Part, B, R. Rein (ed.), A. Liss, New York, pp. 301-310, 1985.
11. E. Mayhew, Y. Rustum and W.J. Vail, Inhibition of Liver Metastases of M5076 Tumor by Liposome Entrapped Adriamycin. Cancer Drug Delivery 1:43-58, 1983.
12. E. Mayhew, M.H. Goldrosen, J. Vaage and Y.M. Rustum, Effects of Liposome-Entrapped Doxorubicin on Liver Metastases of Mouse Colon Carcinomas 26 and 38. J. Natl. Cancer Inst. 78:707-713, 1987.
13. S. Radel and E. Mayhew, Uptake and Toxicity of Liposomal-Adriamycin in Cultured Cells (in preparation).
14. G. Gregoriadis and C. Davis, Stability of Liposomes *in vivo* and *in vitro* is Mediated by Their Cholesterol Content and the Presence of Blood Cells. Biochem. Biophys. Res. Comm. 89:1287-1291, 1979.

15. H.S. Schwartz, A Fluorometric Assay for Daunomycin and Adriamycin in Animal Tissues. Biochem. Med. 1:396-404, 1973.
16. J. Robert, Extraction of Anthracylines from Biological Fluids for HPLC Evaluation. J. Liquid Chromatog. 3:1561-1572, 1980.
17. M.J. Ehrke, V. Tomazic and V. Ryoyma, Adriamycin Induced Immunomodulation: Dependence upon Time of Administration. Int. J. Immunopharmacol. 5:43-48, 1983.
18. V. Mace, E. Mayhew, E. Mihich and M.J. Erke, Alteration in Murine Host Defense Functions by Adriamycin or Liposome-Encapsulated Adriamycin. Cancer Res. (in press).
19. I.J. Fidler, Z. Barnes, W.E. Fogler, R. Kirsh, P. Bugelski and G. Poste, Involvement of Macrophages in the Eradication of Established Metastases Following Intravenous Injection of Liposomes Containing Macrophage Activators. Cancer Res. 42:496-501, 1982.
20. R.B. Bankert, E. Mayhew, S. Jarlinski and Y.-H. Jou, Immunospecific Targeting of Drug-Containing Liposomes. 9th Int. Conv. on Immunology, Karger Press, NY, pp. 416-423, 1985.

DRUG DELIVERY BY IMMUNOLIPOSOMES

David Collins and Leaf Huang

Department of Biochemistry
University of Tennessee
Knoxville, TN 37996-0840

I. INTRODUCTION

Work in our laboratory has focused on immunoliposomes, vesicles bearing antibody specific for tumor or viral antigens on target cells. We have developed a system in which mouse IgG is reacted with the N-hydroxy-succinimide ester of palmitic acid (Fig. 1) (Huang et al., 1980) resulting in the attachment of 4-5 palmitoyl chains per IgG. Derivatization occurs primarily in the F_c region of the antibody (Huang et al., 1981; Ho and Huang, 1986a). Incorporation of the palmitoyl IgG into the liposome bilayer leads to target-specific uptake of the immunoliposomes by antigen-expressing cells. Here we describe our recent efforts at immunoliposome design for the targeted delivery of various drugs and macromolecules.

II. pH-SENSITIVE IMMUNOLIPOSOMES

We have shown that immunoliposomes are taken up through a receptor-mediated endocytosis following binding with the target cell (Huang et al, 1983). The eventual fate of the immunoliposome and its contents is delivery to the lysosome and degradation. Accordingly, we were able to deliver methotrexate but not cytosine arabinoside (araC) with immunoliposomes, since araC is rapidly degraded in the lysosomal compartment (Rustram et al., 1981).

In order to overcome the problems associated with lysosomal degradation, we have developed a pH-sensitive immunoliposome system capable of releasing encapsulated drug at a pre-lysosomal site. Liposomes composed of phosphatidylethanolamine and fatty acids (Huang et al., 1985; Connor and Huang, 1986; Düzgunes et al., 1985) acylated amino acids (Connor et al., 1984) or cholesterol hemisuccinate (Ellens et al., 1984) are stable at neutral or basic pH but unstable and fusion competent at pHs below 6.5. Immunoliposomes prepared in this fashion should release their contents at the level of the endosome since the pH of that organelle is in the range of 5-6.5 (Oukuma and Poole, 1978; Tycko and Maxfield, 1982).

These investigations were funded by NIH grant CA24553
L.H. was a Research Career Development Awardee (CA00718).

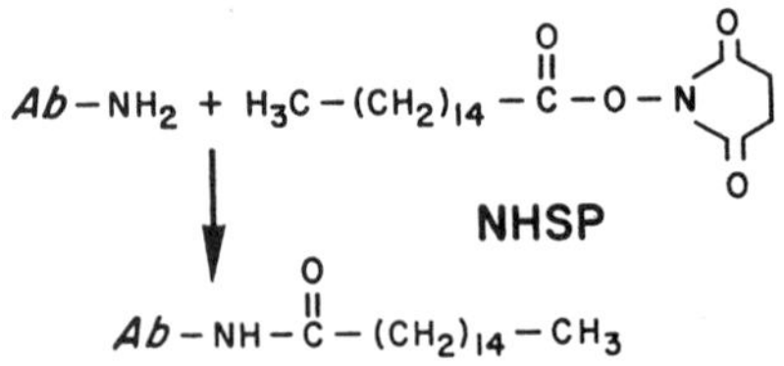

Fig. 1. Antibody derivatization using NHSP.

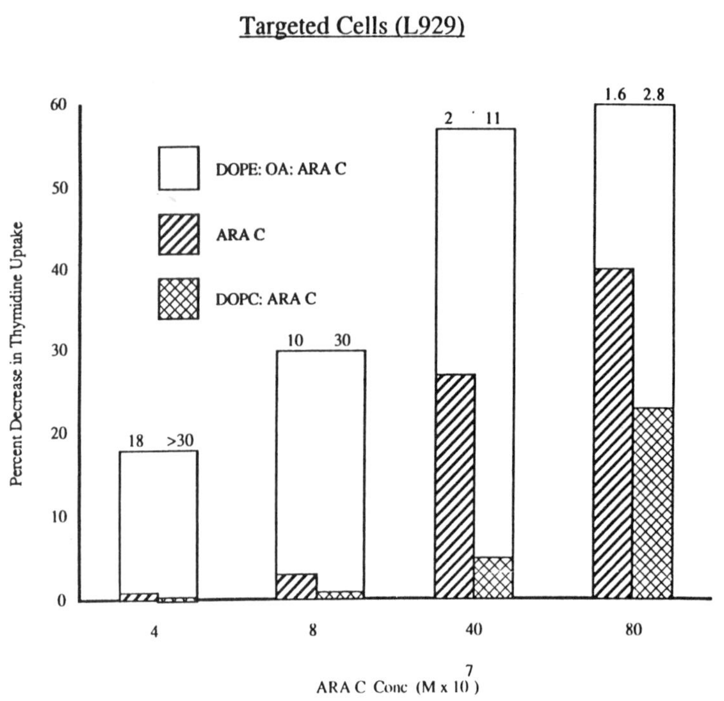

Fig. 2. Cytoxicity of immunoliposome-encapsulated araC. L929 cells were incubated with pH-sensitive immunoliposomes (DOPE:OA:ARA C), pH-insensitive immunoliposomes (DOPE:ARA C) both containing araC or free drug (ARA C). The cells were then pulsed with [^{3}H]-Thy and the inhibition of incorporation of label into DNA was evaluated. Numbers shown on the top of the bar represent the ratios of the fold enhancement of toxicity mediated by DOPE:OA:ARA C over DOPC:ARA C or ARA C (Connor and Huang, unpublished observation).

A. Delivery of Fluorescent Dye (Connor and Huang, 1985)

We prepared pH-sensitive immunoliposomes composed of dioleoylphosphatidylethanolamine (DOPE) and palmitoyl homocysteine to contain the self-quenching dye calcein. The immunoliposomes were targeted with a monoclonal antibody against the major histocompatability antigen H-$2K^k$ expressed on mouse L929 cells. When L929 cells were incubated with pH-sensitive immunoliposomes containing calcein, diffuse fluorescence throughout the cytoplasm was observed . This result indicated that pH-sensitive immunoliposomes were endocytosed by the target cells and released calcein into the cytoplasm. pH-insensitive immunoliposomes also bound and were endocytosed by the target cells but only punctate fluorescence was observed, indicating that release of dye from the endosomes had not occurred. Release of dye by pH-sensitive immunoliposomes was dependent upon endosome acidification since the lysosomotrophic amine chloroquine was able to block cytoplasmic delivery. Non target A31 cells, which express the H-$2K^d$ antigen, did not take up pH-sensitive immunoliposomes. Thus, specific binding of immunoliposomes to the target cells is required for cytoplasmic delivery of calcein.

B. Delivery of Cytosine Arabinoside (Connor and Huang 1986)

The ability to target calcein to L929 cells has obvious potential for delivery of cytotoxic agents to selected cells. Therefore, we prepared pH-sensitive immunoliposomes (DOPE:oleic acid=8:2) to contain araC. As mentioned above, araC is lysosome sensitive and cannot be delivered by conventional (pH-insensitive) immunoliposomes. When encapsulated araC is delivered with pH sensitive immunoliposomes, the drug is more toxic to target cells than free drug (Fig. 2). pH-sensitive immunoliposomes mediated cytotoxicity could be blocked by free antibody in excess or by immunoliposomes containing no araC. In addition, pH-sensitive immunoliposomes containing araC were nontoxic to A31 cells (H-$2K^d$ expressing). Liposomal toxicity could also be blocked by treatment of target L929 cells with chloroquine or NH_4Cl, weak bases which raise endosomal pH. Neither liposomes without antibody nor pH-insensitive immunoliposomes could mediate the cytoplasmic delivery of araC. Taken together, these results indicate that pH-sensitive immunoliposomes are target-specific and are able to release their contents at an acidified prelysosomal compartment (Connor and Huang, 1986).

C. Delivery of Diphtheria Toxin A Fragment (Collins and Huang, 1987)

Diphtheria toxin (DT) is a cytotoxic protein consisting of a binding subunit (fragment B) and an active subunit (fragment A or DTA). DTA inactivates eukaryotic protein synthesis by ADP ribosylating elongation factor 2 (Gill and Pappenheimer, 1971). Mouse cell lines have been shown to be resistant to DT, though they bind and internalize DT normally and possess a DTA sensitive elongation factor 2 (Keen et al., 1982; Uchida et al., 1979). Murine DT-resistance seems to result from a block in the translocation of DTA from the endosome to the cytoplasm. DTA alone has also been shown to be unable to cross lipid membranes in the absence of added fragment B (Donovan et al., 1985). Since pH-sensitive immunoliposomes are capable of delivering calcein and araC to the cytoplasms of target cells, they should be able to overcome the DTA translocation block in murine cells if delivery involves liposome-endosome fusion. We chose the mouse L929 cell line (H-$2K^k$ expressing) and the monoclonal anti H-$2K^k$ antibody targeted immunoliposome system to test this hypothesis.

We prepared pH-sensitive immunoliposomes (DOPE:oleic acid=8:2) to contain DTA. When incubated with target L929 cells, these immunoliposomes were able to deliver DTA to the cell cytoplasm as indicated by

Table 1. Effect of liposome composition on thymidine kinase activity* of cells+.

DOPC:cholesterol:OA (4:4:2)	DOPE:cholesterol:OA (4:4:2)
9.7 ± 3.6 (30)	76.8 ± 28.7 (30)**

*thymidine kinase activity is expressed as cpm/mg protein, mean ± S.D. (# of obs.)
**P<0.001.
+transformed cells were grown in MEM supplemented with 0.5 mM dibutyryl cAMP and 1 mM theophylline.

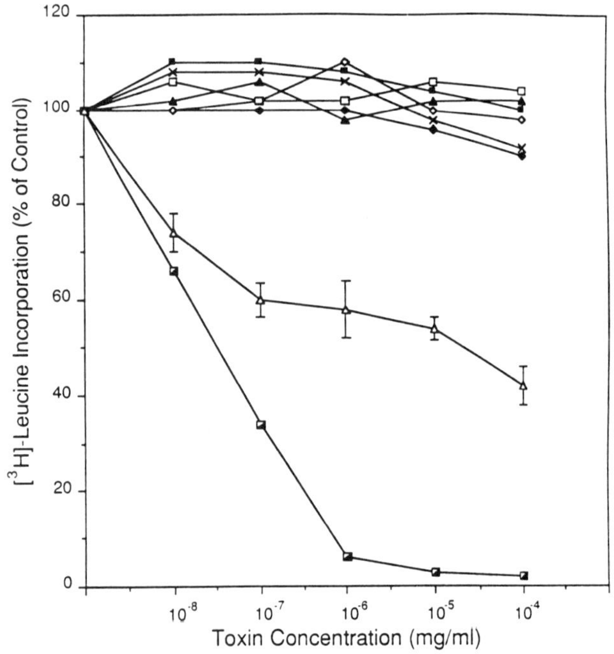

Fig. 3. Toxicity of free or encapsulated toxins. Target cells were incubated with DOPE:OA (8:2) immunoliposomes (△) DOPC immunoliposomes (□) DOPE:OA (8:2) nontargeted liposomes (▲) DOPC nontargeted liposomes (■) containing DTA. DTA (◆) nicked DT (X) and ricin toxin (◪) were incubated with the cells as were empty DOPE:OA (8:2) immunoliposomes (◇). Cytotoxicity was assayed as the inhibition of [^{3}H]-leu incorporation (Collins and Huang, 1987).

the inhibition of protein synthesis (Fig. 3). DTA in pH-insensitive immunoliposomes or pH-sensitive liposomes without antibody were not able to deliver DTA. The toxicity of encapsulated DTA could be blocked by excess free antibody or empty immunoliposomes as well as by chloroquine or NH_4Cl. DTA in pH-sensitive immunoliposome was nontoxic to either A31 cells or DT-sensitive Vero cells indicating the target specificity of the system.

D. Delivery of Plasmid DNA in vitro (Wang and Huang, 1987a)

The successful correction of genetic defects by gene therapy is currently one of the most sought after goals in biotechnology. Although gene therapy has great potential in both medicine and biotechnology, current methods for introducing foreign genes into cells are less than adequate. The main problems associated with current techniques are poor efficiency of DNA delivery and poor survival of DNA treated cells. As described above, pH-sensitive immunoliposomes are able to mediate efficient delivery of encapsulated drugs and macromolecules to the cytoplasm of target cells. Recently, we have applied pH-sensitive immunoliposome technology to improve the ability to deliver DNA to target cells.

We encapsulated the pPCTK-6A plasmid, containing the herpes simplex virus thymidine kinase gene, in pH-sensitive immunoliposomes composed of DOPE, cholesterol and oleic acid (4:4:2). The endogenous promoter of the HSVTK gene was replaced by the promoter of the rat phosphoenolpyruvate carboxykinase gene (PEPCK) which contains a cAMP regulatory region allowing cAMP control of HSVTK expression (Wynshaw-Boris et al, 1984). Our model system for this work made use of Ltk^- cells which were deficient in thymidine kinase activity and thus unable to grow in HAT (hypoxanthine, aminopterin, thymidine) media. We targeted the immunoliposomes to the $H\text{-}2K^k$ antigen expressed on the Ltk^- cells by using fatty acid derivatized monoclonal anti $H\text{-}2K^k$ antibody incorporated into the liposome membrane.

When pH-sensitive immunoliposomes containing the plasmid were incubated with Ltk^- cells the cells were transformed to a tk^+ phenotype. pH-insensitive immunoliposomes containing the plasmid were also able to induce the tk^+ phenotype in the cells, however the amount of tk activity was much lower than that for Ltk^- cells incubated with pH-sensitive immunoliposomes (Table 1). Antibody on the liposomes was also necessary for optimal expression of the activity; short-term expression mediated by antibody containing liposomes was 6-fold more effective than that mediated by liposomes without antibody. Cells transformed by plasmid in pH-sensitive immunoliposomes showed 2.3 fold higher in short term expression and about 10 fold higher in long term expression than for corresponding cells transformed by the $CaCl_2$ method. The viability of immunoliposome-transformed cells was also much higher than that of $CaCl_2$ -transformed cells indicating that liposomal delivery of DNA is much less toxic than $CaCl_2$ treatment.

E. Delivery of Plasmid DNA in vivo (Wang and Huang, 1987b)

As described above, pH-sensitive immunoliposomes can mediate the delivery and expression of encapsulated plasmid DNA in target cells. We have recently designed a model system to test the DNA delivery potential of pH-sensitive immunoliposomes *in vivo*. Our model uses mouse RDM4 cells expressing the $H\text{-}2K^k$ antigen grown as an ascites tumor in immunodeficient nude mice of $H\text{-}2K^k$ background. A plasmid, pBBO.6CAT, containing the *E. coli* chloramphenicol acetyltransferase (CAT) gene under control of a cAMP regulatory sequence in the rat PEPCK promotor (Wynshaw-Boris et al, 1984) was encapsulated in pH-sensitive immunoliposomes

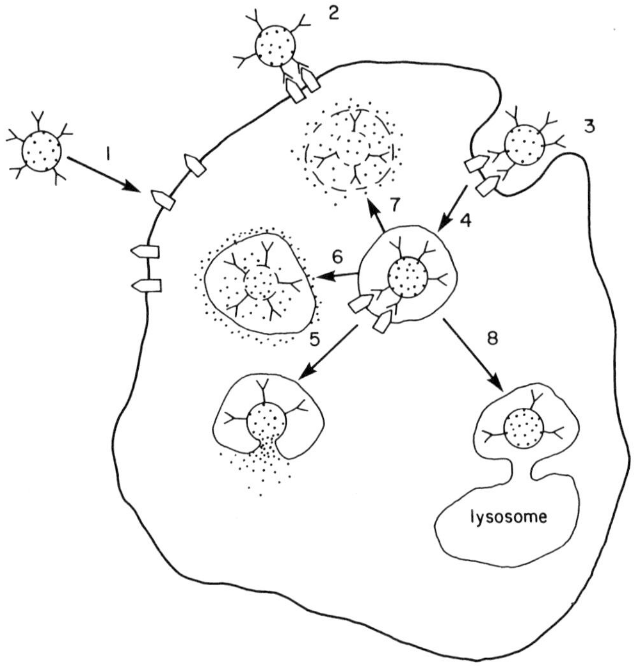

Fig. 4. Probable mechanism of immunoliposome uptake by target cells. 1). Binding of immunoliposome to target, 2&3) movement of bound immunoliposome to coated pits, 4) endocytosis into endosome, 5) fusion with endosome, 6) leakage of contents into endosome, 7) endosome rupture, or 8) delivery of contents to lysosome (Connor and Huang, 1987).

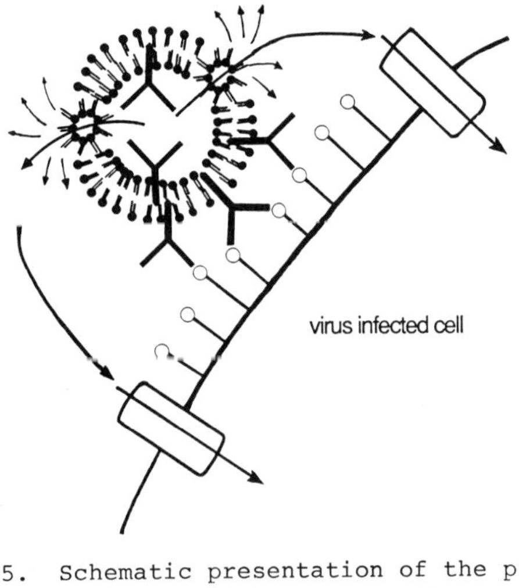

Fig. 5. Schematic presentation of the preferred mechanism of target-sensitive immunoliposome mediated site-specific cytosolic delivery of transportable drugs (Ho et al., 1986b).

(DOPE:cholesterol:oleic acid=4:4:2). Nude mice were inoculated with RDM4 cells by intraperitoneal (i.p.) injection. Seven days after inoculation, liposome-encapsulated DNA or free DNA was injected i.p. into the mice. Twenty-four hours later, 8-Br-cAMP and isobutyl methyl xanthine (MIX) were injected. After five hours, ascites fluid was collected and RDM4 cells separated from peritoneal macrophages by incubation on glass. Mice were sacrificed and the CAT activity in the blood, heart, spleen, lung, stomach, kidney and liver were assayed. We also monitored the liposome distribution by following a ^{3}H-lipid marker.

Mice bearing no tumors but treated with immunoliposomes or tumor bearing mice treated with antibody-free liposomes showed a high degree of liposome accumulation in the spleen. The accumulation of liposomes in the spleen was reduced dramatically in tumor bearing mice treated with immunoliposomes. The accumulation of liposomes in other organs was also affected by the presence of antibody on the liposomes.

Although RDM4 cells were not the only cells which took up significant amounts of pH-sensitive immunoliposomes, these cells were unique in the expression of significant levels of CAT activity. Peritoneal macrophages in the same mice did not express CAT activity despite the fact that they accumulated significant amounts of immunoliposomes. Spleen cells, which also took up a large amount of immunoliposomes, expressed only very low levels of CAT activity. Taken together, these results indicate that CAT expression in our model system is target-specific. The pH-sensitive liposome design was found to be essential for CAT expression in RDM4 cells since pH-insensitive immunoliposomes were unable to yield significant levels of expression.

This recent work has wide ranging implications for gene therapy. pH-sensitive immunoliposomes offer high efficiency of transformation and safety since all components of the system are nontoxic. The system also potentially provides a high degree of specificity by using appropriate antibody for target cells to which immunoliposome can gain access.

F. Mechanism of Delivery

We have shown that pH-sensitive immunoliposomes can efficiently deliver their encapsulated contents to targeted cells. pH-sensitive immunoliposomes are taken up through a receptor-mediated endocytosis mechanism and encounter the acidic pH of the endosome-lysosome system (Fig 4). Cytoplasmic delivery by pH-sensitive immunoliposomes is dependent upon acidification of the endocytic vesicle system since weak bases such as NH_4Cl or chloroquine block the release of liposome contents (Connor and Huang, 1985; Connor and Huang, 1986; Collins and Huang, 1987). The site of release is probably the endosome since the lysosome-sensitive drug araC, DTA and DNA can be efficiently delivered by pH-sensitive immunoliposomes (Connor and Huang, 1986).

Since large macromolecules such as DTA or plasmid DNA cannot cross lipid membrane barriers it is unlikely that leakage of pH-sensitive immunoliposome contents into the endosome can account for the efficient delivery we observed for these macromolecules (Collins and Huang, 1987; Wang and Huang, 1987a; Wang and Huang, 1987b). We have shown that pH-sensitive liposomes become fusion-competent at pH's in the range of those encountered in the endosome (Connor and Huang, 1984). It seems probable that pH-sensitive immunoliposomes translocate their contents into the target cell cytoplasm by fusing with the endosome membrane from within (Fig. 4). Another possibility is that pH-sensitive immunoliposomes induce endosome rupture thereby releasing the encapsulated contents into the cytoplasm. We cannot at present distinguish between liposome-endosome fusion and liposome-induced endosome rupture. Regardless of the actual mechanism, it is clear that pH-sensitive immunoliposomes are able to mediate the targeted cytoplasmic delivery of macromolecules and cytotoxic drugs.

G. In vivo Distribution and Stability (Connor et al., 1986)

The primary goal in the development of immunoliposomes is to establish an effective *in vivo* targeted delivery system. We therefore investigated the biodistribution and *in vivo* stability of pH-sensitive immunoliposomes (DOPE:oleic acid=8:2) to determine the feasibility of this system as an *in vivo* delivery system.

Immunoliposomes and liposomes were prepared to contain ^{125}I-labeled tyraminyl inulin (Sommerman, et al., 1984), a nonmetabolizable aqueous marker, and injected intravenously into either C3H or Balb/c mice. pH-sensitive as well as pH-insensitive immunoliposomes were injected. Lipid accumulation was determined by using hexadecyl [^{3}H] cholestanyl ether. pH-sensitive liposomes accumulated primarily in the lung tissue and was more pronounced in C3H than in Balb/c mice. In addition, antibody-free liposomes showed more lung accumulation than did immunoliposomes. The presence of antibody in the liposome slightly increased the liver uptake of liposomes. pH-insensitive liposomes accumulated primarily in the liver and spleen. Again, the presence of liposomal antibody slightly increased liposome uptake by the liver.

Another phenomenon observed for pH-sensitive liposomes was the separation of the internal aqueous marker from the lipid marker. As mentioned previously, pH-sensitive liposomes accumulated primarily in the lung while the aqueous marker, inulin, accumulated in the kidney and was subsequently excreted. This result suggests that pH-sensitive liposomes are unstable *in vivo*. We confirmed this hypothesis by measuring the plasma induced leakage of calcein from pH-sensitive and pH-insensitive liposomes. While pH-insensitive liposomes were fairly stable in 100% mouse plasma, pH-sensitive liposomes exhibited rapid leakage. The exact mechanism of plasma-induced leakage is probably due to the extraction of oleic acid from liposomes by serum albumin.

From the results obtained above, it is reasonable to conclude that pH-sensitive liposome composed of DOPE and oleic acid is an ineffective tool for drug delivery following intravenous injection. We have shown, however, that the cholesterol containing pH-sensitive immunoliposomes can effectively deliver their encapsulated contents when injected via the i.p. route (Wang and Huang, 1987b). These liposomes are much more stable in serum than the cholesterol-free liposomes (Tsusaki and Huang, unpublished observations). Currently, we are working on immunoliposome systems which are stable in plasma yet retain pH-sensitivity.

III. TARGET-SENSITIVE IMMUNOLIPOSOMES (Ho and Huang, 1985; Ho et al., 1986)

The pH-sensitive immunoliposome system described above exploits the cellular endocytic pathway to achieve targeted drug delivery. While delivery is very efficient for highly endocytic cells, there exist target cells with little or no endocytic activity where pH-sensitive immunoliposomes may not be applicable. To overcome this problem we have developed target-sensitive immunoliposomes which are able to release encapsulated drugs at the surface of the target cell.

Unsaturated PE does not form stable bilayers unless a stabilizing component(s) is added. Among the components used to stabilize PE bilayer are (dinitrophenyl) caproyl-PE (Ho and Huang, 1985), oleic acid (Düzgünes et al., 1985; Huang and Liu, 1985), palmitoyl homocysteine (Connor et al., 1984) and glycophorin A (Ho and Huang, 1985; Taraschi, et al., 1982). For the production of target-sensitive immunoliposomes we stabilized PE bilayers with fatty acid-derivatized antibody prepared as described previously (Huang et al., 1981). The exact stabilization mechanism of the PE bilayer by antibody is unclear. It seems likely that the orientation of the antibody in the bilayer is important for its stabilizing effect. Since derivatization of antibody occurs primarily in the Fc region (Huang et al., 1981) the Fab portion of the antibody molecules are presumably exposed to the aqueous environment. According to hydration theory (Marcevelja and Radiz, 1976), PE does not attract a significant number of water molecules due to its small head group charge density at neutral pH. PE therefore has a dynamic shape of an inverted cone and thus prefers hexagonal phase. PE can be stabilized in a bilayer form by complementary molecules which have a wedge-like dynamic shape. The stabilization of PE bilayer by antibody is probably due to such complementation since the Fab portion of IgG is highly hydrated. Upon binding to the target, the target sensitive immunoliposome is destabilized leading to the release of the liposome contents. The destabilization is probably the result of "contact capping" of the antibody which leaves the bulk of the bilayer destabilized (Fig. 4). Following binding, the liposomes should destablize and release their contents in high concentration at the cell surface. The drugs used must be able to be transported into the cell by an efficient transport system, as is the case for nucleoside analogs.

Our model system explored the ability of target-sensitive immunoliposomes to deliver cytotoxic nucleoside analogs into herpes simplex virus (HSV) infected cells. Target-sensitive immunoliposomes were prepared which were stabilized with fatty acid-derivatized antibody against HSV antigen glycoprotein D. The immunoliposomes contained either araC or acyclovir (ACV). The target cells used in our model system were mouse L929 cells infected with HSV. Target sensitive immunoliposomes were incubated with the HSV-infected cells and cytotoxicity determined by [^{3}H] thymidine incorporation into DNA. Antiviral activity was determined by assaying the virus yield of the culture.

Target sensitive immunoliposomes encapsulating araC which is not viral selective, reduced virus production with a 50% effective dose ED_{50} of ~1.8 ng/ml. At least a 1000 fold higher dose of free araC was required to achieve the same effect. The nonspecific cytoxicity of araC encapsulated in target sensitive immunoliposomes was reduced at least 4 orders of magnitude below that for free drug. Similar but less dramatic results were obtained using ACV instead of araC. From these results it is clear that target-sensitive immunoliposomes enhanced the specific potency of antiviral drugs while decreasing non-specific cytotoxicity to the cell population. We can therefore conclude that our system exhibits target selectivity. The antiviral effect of araC mediated by the target-sensitive immunoliposomes could be blocked by the nucleoside transport inhibitors, p-nitrothiobenzylinosine and dipyridamole. This suggests that target sensitive immunoliposomes release their contents in high concentrations at the cell surface upon binding the target cell. The drugs, in this case nucleoside analogs, are then taken up by the cell through cellular transport system (Fig. 5).

The principle described here should also be applicable to engineer target-sensitive liposomes for a wide variety of target cell population for which specific antigens and antibodies are available. However, delivery is limited to small molecules, since macromolecules are not rapidly transported into cells.

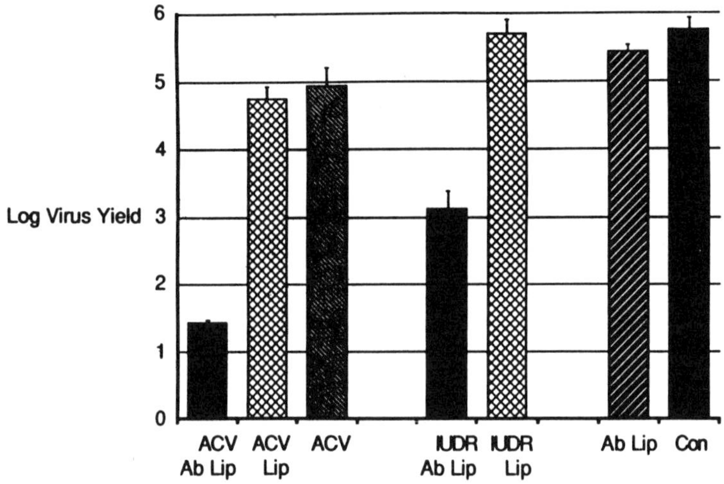

Fig. 6. Yield reduction assay. Cells infected 4 hr previously with HSV-1 were incubated at 4°C for 1 hr with liposomes (Lip) bearing either anti-gD antibody (Ab) or no antibody and containing either IUDR or ACV. After washing, the cells were returned to 37°C until 12 hr Pl when viral yield was measured by titration. Controls included cells exposed to free ACV for 1 hr and cells receiving no treatment (Con.) The results represent the mean of four replicates ± SD (Norley, et al., 1986).

IV. SUSTAINED RELEASE IMMUNOLIPOSOMES (Norley, et al., 1986; Norley et al., 1987).

Despite the obvious potential of immunoliposomes as drug delivery systems, therapeutic applications have been few. The major obstacle is that upon entry into the circulation, liposomes are rapidly taken up by the reticuloendothelial system (RES) and may not reach the target site (Poste, 1983).

We have been recently investigating the use of immunoliposomes in a system where the RES is not a problem, namely the corneal surface. We are especially interested in the development of immunoliposomes for the treatment of ocular herpatic keratitis (OHK) which is a leading cause of blindness in the U.S. (O'Brien, 1984). Currently, OHK is treated by frequent topical application of antiviral drugs to the infected eye (Balfour, 1984). We wanted to design an immunoliposome system which would bind specifically to herpes-infected corneal cells and release the encapsulated antiviral drugs over a long period of time. Here we summarize the results of our *in vitro* model of this system.

Immunoliposomes were prepared with phosphatidylcholine and cholesterol to encapsulate acyclovir (ACV) or iododeoxyuridine (IUDR). The liposomes were targeted to HSV gD by incorporation of fatty acid-derivatized monoclonal anti-HSV-gD into the liposome membrane. IUDR was found to leak out of the liposome much faster (80%/hr at 37°C) than ACV. Tissue culture rabbit corneal cells or isolated mouse cornea were infected with HSV and treated with immunoliposome containing either IUDR or ACV. Virus yield reduction assays were done to evaluate the efficacy of viral inhibition by the drug-loaded immunoliposomes.

With this model system, target antigen (gD) was shown to be expressed by 4 hours post infection. Drugs were shown to be effective up to 5 hours post infection. Using this data, we incubated liposomes containing either IUDR or ACV and targeted with anti-HSV-gD antibody with cells or cornea during this 4-5 hr "window". Incubations were carried out at 4°C to minimize drug release from unbound liposomes. The cells were then washed to remove unbound vesicles and free drugs and then incubated at 37°C for 8 hr before determination of viral yield.

As shown in Fig 6, ACV-containing liposomes were able to dramatically inhibit viral production. Untargeted liposomes or free ACV were ineffective, since they were presumably washed away. IUDR-containing liposomes gave similar though less dramatic results due to fast leakage of the drug from the bound liposomes.

Liposomes loaded with appropriate antiviral drugs and targeted to HSV surface antigens may prove to be an effective treatment for OHK. Current studies are underway to determine the effectiveness of this model system in an *in vivo* rabbit model.

V. CONCLUSIONS AND FUTURE PERSPECTIVES

The model systems described here demonstrate that immunoliposomes have great potential for the targeted cytoplasmic delivery of encapsulated molecules. Several obvious advantages of immunoliposomes can be seen including (1) specific interaction of drugs (2) protection of drugs and macromolecules from degradation (3) low inherent liposomal toxicity

(4) ease of preparation and (5) reduction in nonspecific toxicity of toxic drugs with corresponding increase in target toxicity. The main disadvantages of immunoliposomes are (1) lack of appropriate antigen-antibody pairs for certain potentially desirable targets (2) instability of immunoliposomes *in vivo* and (3) the rapid uptake of immunoliposomes by the RES. We are currently attempting to produce immunoliposome systems which can surmount the problems so that greater drug delivery potential of immunoliposomes can be realized.

REFERENCES

Balfour, H.H., 1984, Acyclovir and other chemotherapy for herpes group viral infections, Ann. Rev. Med., 35:279.

Collins, D. and Huang, L., 1987, Cytotoxicity of DTA fragment to toxin-resistant murine cells delivered by pH-sensitive immunoliposomes, Cancer Res., 47:735.

Connor, J., Yatvin, M., and Huang, L., 1984, pH-Sensitive liposomes: acid induced fusion, Proc. Natl. Acad. Sci. USA, 81:1715.

Connor, J. and Huang, L., 1985, Efficient cytoplasmic delivery of a fluorescent dye by pH-sensitive immunoliposomes, J. Cell Biol. 101:582.

Connor, J. and Huang, L., 1986a, pH-Sensitive immunoliposomes as an efficient and target-specific carrier for antitumor drugs, Cancer Res. 46:3431.

Connor, J., Norley, N. and Huang, L., 1986b, Biodistribution of pH-sensitive liposomes, Biochim. Biophys. Acta., 884:474.

Connor, J. and Huang, L., 1987, Acid-induced fusion of liposomes, in "Cell Fusion," Arthur E. Sowers, ed. Plenum Publishing Corp., New York.

Donovan, J.J., Simon, M.I. and Montal, M., 1985, Requirements for the translocation of diphtheria toxin fragment A across lipid membranes, J. Biol. Chem., 260:8817.

Düzgünes, N., Straubinger, R.M., Baldwin, P.A., Friends, D.S., and Papahadjopoulos, D., 1985, Proton-induced fusion of oleic acid-phosphatidylethanolamine liposomes, Biochemistry, 24:3091.

Ellens, H., Bentz, J., and Szoka, F.C., 1984, pH-Induced destabilization of phosphatidylethanolamine liposomes, Biochemistry, 23:1532.

Gill, D.M. and Pappenheimer, A.M., Jr., 1971, Structure-activity relationships in diphtheria toxin, J. Biol.Chem., 246:1492.

Ho, R.J.Y. and Huang, L., 1985, Interactions of antigen-sensitized liposomes with immobilized antibody: a homogeneous solid-phase immunoliposome assay, J. Immunol., 134:4035.

Ho, R.J.Y., Rouse, B.T., and Huang, L., 1986a, Target sensitive immunoliposomes: preparation and characterization, Biochemistry, 25:5500.

Ho, R.J.Y., Rouse, B.T., and Huang, L., 1986b, Target sensitive immunoliposomes as an efficient drug carrier for antiviral activity, Biochemistry, submitted.

Huang, A., Huang, L., and S.J. Kennel, 1980, Monoclonal antibody covalently coupled with fatty acid, J. Biol. Chem., 255:8015.

Huang, A., Kennel, S.J., and Huang, L., 1981, Immunoliposome labeling: a sensitive and specific method for cell surface labeling, J. Immun. Methods, 46:141.

Huang, A., Kennel, S.J., and Huang, L., 1983, Interactions of immunoliposomes with target cells, J. Biol. Chem., 258:14034.

Huang, L. and Lui, S.S., 1984, Acid induced fusion of liposomes with inner membranes of mitochondria, Biophys. J. 45:72a.

Keen, J.H., Maxfield, F.R., Hardegree, M.C., and Habig, W.H., 1982, Receptor-mediated endocytosis of diphtheria toxin by cells in culture, Proc. Natl. Acad. Sci. USA, 79:2912.

Marcelja, S. and Radiz, M., 1976, Repulsion of interfaces due to boundary water, Chem. Phys. Lett., 42:129.
Norley, S.G., Huang, L., and Rouse, B.T., 1986, Targeting of drug loaded immunoliposomes to herpes simplex virus-infected corneal cells: an effective means of inhibiting virus replication in vitro, J. Immunol., 136:681.
Norley, S.G., Sendele, D., Huang, L., and Rouse, B.T., 1987, Inhibition of herpes simplex virus replication in the mouse cornea by drug containing immunoliposomes, Investigative Ophthalmology and Visual Sciences, in press.
O'Brien, W.J., 1984, Herpetic eye disease in animals as models for therapeutic studies of acute and latent herpes-virus infections, in: "Herpesvirus," Fred Rapp, ed. Alan R. Liss, Inc. New York.
Ohkuma, S. and Poole, B., 1978, Fluorescent probe measurements of the intra lysosomal pH in living cells and the perturbation by various agents, Proc. Natl. Acad. Sci. USA, 80:3334.
Poste, G., 1982, Liposome targeting in vivo: problems and opportunities, Biol. Cell, 47:19.
Rustum, Y.M., Mayhew, E., Szoka, F., and Campbell, J., 1981, Inability of liposome encapsulated 1-β-D-arabinofuranosyl-cytosine nucleotides to overcome drug resistance in L1210 cells, Eur. J. Cancer Clin. Oncol., 17:809.
Sommerman, E.F., Pritchard, P.H., and Cullis, P.R., 1984, ^{125}I labelled inulin: a convenient marker for deposition of liposomal contents in vivo, Biochem. Biophys. Res. Commun., 122:319.
Taraschi, T.F., Van der Steen, A.T.M., DeKruijff, B., Tellier, C., and Verkleij, A.J., 1982, Lectin-receptor interaction in liposomes: evidence that binding of wheat germ agglutinin to glycoprotein phosphatidylethanolamine vesicles induce non-bilayer structure, Biochemistry, 21:5756.
Tycko, B. and Maxfield, F.R., 1982, "Rapid acidification of endocytic vesicles containing 2 macroglobulin, Cell, 28:643.
Uchida, T., Kim, J., Yamaizumi, M., Miyaki, Y., and Okada, Y., 1979, Reconstitution of lipid vesicles associated with HVJ (Sendai virus) spikes, J. Cell Biol., 80;10.
Wang, C.Y. and Huang, L., 1987a, Expression of endogenous gene delivered by pH-sensitive immunoliposomes, submitted.
Wang, C.Y. and Huang, L., 1987b, pH-sensitive immunoliposomes mediate delivery and controlled expression of foreign gene in nude mouse, submitted.
Wynshaw-Boris, A., Lugo, T.G., Short, J.M., Fournier, R.E.K., and Hanson, R.W., 1984, J. Biol. Chem., 259:12161.

OSMOTIC FORCES AND THE FUSION OF BIOMEMBRANES

Jack A. Lucy and Quet F. Ahkong

Department of Biochemistry and Chemistry
Royal Free Hospital School of Medicine (University of London)
Rowland Hill Street, London, NW3 2PF, UK

INTRODUCTION

The possibility that osmotic forces are involved in the molecular mechanisms that govern the fusion of biomembranes has been of wide interest since it was observed that solutions of high osmolarity can inhibit the exocytosis of chromaffin granules in adrenal medullary cells (Hampton and Holz, 1983; Pollard et al., 1984). This work followed that of Finkelstein and his colleagues who had previously demonstrated, in an experimental model for exocytosis, that under appropriate conditions phospholipid vesicles will fuse with a planar phospholipid bilayer under the influence of an osmotic gradient (Cohen et al., 1980). It is, however, less well known that Racker and his co-workers had shown in an earlier interesting paper that unilamellar proteoliposomes containing phosphatidylserine, which ceased to fuse in the presence of calcium ions after the liposomes had grown to diameters of about 100 nm, could be induced to fuse further by an osmotic gradient (internal osmotic pressure higher than external) and form single-walled liposomes with diameters exceeding 1 micron as well as multilayered vesicles (Miller et al., 1976).

In our laboratory we have investigated the artificially-induced fusion of erythrocytes as a model for the fusion-fission reactions of biomembranes over a period of many years and, in early work, we made a number of observations which implicated osmotic phenomena in the membrane fusion process. From our findings we concluded that cell swelling by colloidal osmosis plays an essential role in the fusion of erythrocytes by fusogenic lipids (Ahkong et al., 1973a, 1973b). We then attempted to fuse erythrocytes by causing them to swell in hypotonic media under various experimental conditions (unpublished work with Dr D. Fisher and Dr A.H. Goodall). These experiments were unsuccessful because, at that time, the restraining influence of the protein skeleton of the erythrocyte membrane was not understood. Subsequently, we elucidated the importance of the activation of cellular proteinases and the degradation of the membrane skeleton in the fusion of erythrocytes by several chemicals (Quirk et al., 1978; Ahkong et al., 1980; Lang et al., 1984), and this has enabled us to re-investigate the possible role of osmotic forces in the fusion of erythrocytes. In this paper we discuss recent findings which indicate that osmotic forces are involved in the fusion of erythrocytes by chemical and electrical means. We also discuss a new molecular model that we have proposed for the fusion reactions of biomembranes in which osmotic forces

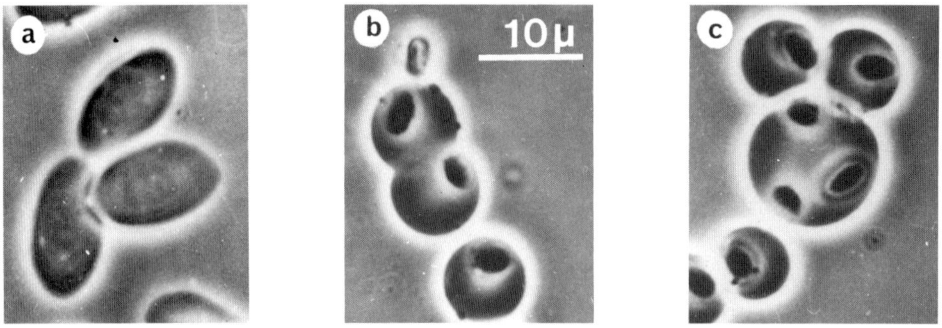

Fig. 1. Light micrographs of hen erythrocytes. (a) The cells are initially oval in appearance. (b) They change shape, becoming swollen and spherical, on treatment with oleoylglycerol. (c) Finally, when calcium ions are also present, multinucleated cells are formed.

are an essential feature; we hope that this model may stimulate work on ion gradients which could be involved in the fusion-fission reactions of biological membranes occurring *in vivo*.

FUSOGENIC LIPIDS AND ERYTHROCYTES

A number of simple lipids are capable of inducing the artificial fusion of hen erythrocytes. The fusogenic lipids include saturated carboxylic acids (containing 10-14 carbon atoms), longer but unsaturated carboxylic acids and their monoesters of glycerol, and retinol. Erythrocytes that are treated with these lipids exhibit colloid osmotic swelling, and they become spherical within a few minutes. When EDTA is present, the swollen cells lyse more readily than they fuse but, when two or more of the spherical cells come together in the presence of calcium ions, they aggregate and fuse before they lyse (Fig. 1) (Ahkong et al., 1973a, Lucy 1973). Fusion between oval, unswollen erythrocytes was never seen in our early work, and every chemical found to induce fusion also caused the cells to become spherical (Ahkong et al., 1973b). Other workers subsequently made similar findings. Thus, Kosower et al., (1977, 1978) reported that, while the fusion of erythrocytes from Leghorn hens by a membrane mobility agent was preceded by the cells becoming round, cells from White Rock hens, which did not fuse under the same conditions, failed to change shape. Maggio et al., (1978) also observed that hen erythrocytes

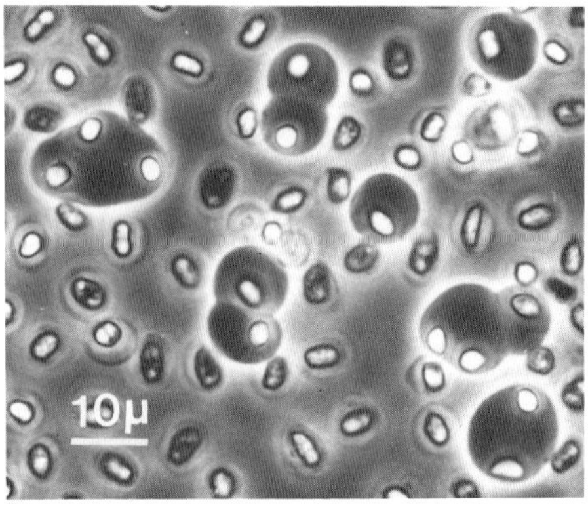

Fig. 2. Swelling, fusion and lysis of hen red blood induced by heating the cells to 50°C.

fused by polysialogangliosides became round after 1.5h., and that bi- and multinucleated cells then appeared after about 3 h. of incubation.

In related work on the effects of heat on hen erythrocytes, we noted that the cells rapidly changed in shape, from oval to spherical, on heating to 48-50°C. Some of the swollen, spherical cells then fused to produce multinucleated cells (Fig. 2) and, as with chemically-induced fusion, lysis of both un-fused and fused cells occurred (Ahkong et al. 1973b). At the time of this work we remarked that a related phenomenon that involves membrane fusion, is the fragmentation of human erythrocytes into small haemoglobin-containing spheres that occurs on heating the cells to 50° C, which was reported earlier by Ponder (1955). Recently it has been demonstrated that, although human erythrocytes swell much less than a perfect osmometer at osmolarities between 200-700 mOsm, their swelling behaviour in this range is approximately ideal after heating to 50° C for 60 min., and it is thought that this is a consequence of the fact that spectrin denatures at this temperature (Heubusch et al., 1985). Looking at our early work in retrospect, it now seems that damage to the membrane skeleton of erythrocytes is probably the feature that is common to heating and exposure to a lipid fusogen, which allows the cells to participate in osmotically-driven membrane fusion reactions.

During an investigation of the fusion of fibroblasts induced by the lipid fusogen, oleylamine, we found that fusion was similarly preceded by rounding of the cells. The fibroblasts then swelled to form large clear spaces immediately beneath the plasma membranes, which were attributed to a rapid influx of water. After 10-15 min. swollen cells that were in contact fused to form polykaryons (Fig. 3). The dividing membranes between the fusing cells "popped" with great speed, apparently having reached a critical point due to mechanical stress or chemical damage (Bruckdorfer et al., 1974).

Although the function of cell swelling in the fusion process was far from clear from these early experiments, it nevertheless seemed that an increased membrane permeability might be associated with changes in membrane structure that are necessary for fusion to occur (Ahkong et al.,

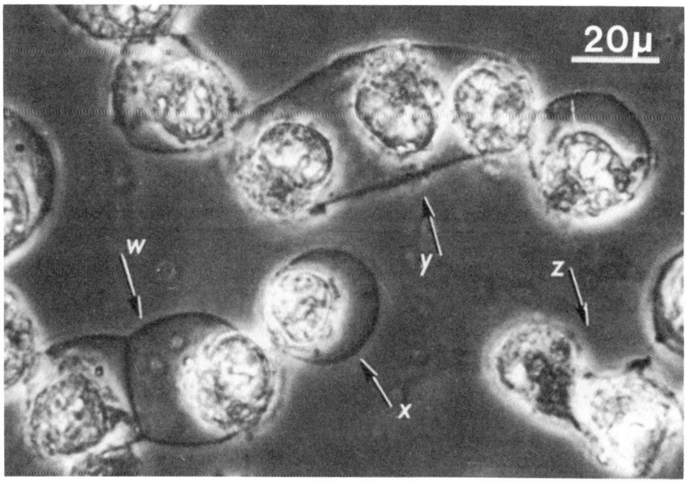

Fig 3. The fusion of mouse A 9 fibroblasts by oleylamine; w and x, rounded, swollen cells in close contact; y, a homokaryon with three nuclei; z, fused cells that are undergoing lysis.

1973a). With hindsight, however, it is apparent that we were unduly influenced in our thinking about the molecular mechanisms involved in membrane fusion by the extent to which cells treated with fusogenic lipids were swollen before they aggregated, and we gave insufficient attention to the possibility that continued swelling after aggregation might be more directly related to the fusion event.

FUSION OF ERYTHROCYTES BY OSMOTIC SHOCK AND IN PROGRESSIVELY DILUTED MEDIA

Since erythrocytes are not induced to fuse by osmotic swelling alone, it was not initially possible to decide whether fusion induced by chemical fusogens is causally related to the osmotic changes seen or whether it is merely accompanied by them. A likely reason for the failure of erythrocytes to fuse when subjected to osmotic swelling alone is that structural changes in the plasma membrane, which are induced by chemical fusogens, do not occur. For example, proteinases are important in chemically-induced cell fusion (Quirk et al., 1978; Thomas et al., 1983; Kosower et al., 1983), and the fusion of erythrocytes by some chemical fusogens involves the proteolysis of ankyrin by a serine protease (Lang et al., 1984) and the proteolysis of band 3 by a calcium-activated proteinase (Lang et al., 1984; Ahkong et al., 1980). Inhibitors of these enzymes inhibit the cell fusion. It also seems that the electrically-induced fusion of erythrocytes in low-ionic strength media involves a perturbation of membrane proteins (Donath and Arndt, 1984).

In view of these various findings, we recently investigated the possibility that osmotic swelling might be capable of driving the fusion of erythrocytes if the restraining membrane skeleton is previously damaged or destroyed by pre-incubating the cells with ionophore A23187 and calcium ions before subjecting them to a hypotonic environment. Incubation of hen erythrocytes with the ionophore and calcium induces a proteolytic breakdown of spectrin and goblin (Thomas et al., 1983), and a release of the intramembranous particles of the plasma membrane from the restraining effects of the membrane skeleton (Vos et al., 1976). Prolonged treatment of the cells in this way or subsequent incubation at 47°C results in cell fusion (Ahkong et al., 1975), which is probably facilitated by a calcium-stimulated phosphodiesterase that yields diacylglycerol (Allan and Michell, 1978) which may be fusogenic (Ahkong et al., 1973a). Under the conditions of our experiments, however, little fusion was apparent after 2 h. at 37°C in the pre-treated cells but, only 5 min. after adding water to dilute the buffered salt solution 2.5-fold, binucleate cells were clearly seen to have been formed by the fusion of previously separate cells (Ahkong and Lucy, 1986). A proportion (10%) of the erythrocytes were labelled with 6-carboxyfluorescein diacetate, which enters cells and is hydrolysed to 6-carboxyfluorescein (Goodall and Johnson, 1982) and, 5 min. after the osmotic shock, fluorescent cells were observed to have fused with non-fluorescent cells to produce weakly fluorescent cells. At the same time, the fluorescent cells fused with one another (Figs. 4a and 4b).

Although human erythrocytes normally show greatly restricted volume changes at osmolarities in the physiological range, they nevertheless swell like a simple osmometer in solutions that are more dilute than 150mM (Heubusch et al., 1985). This provides an alternative way of allowing a change in the osmotic environment to drive the fusion of these cells. To this end, monolayers of human erythrocytes were incubated in an isotonic salt/buffer solution that was progressively diluted, first by 1.5-fold, then by a further 1.7-fold, and finally replaced by water alone. Dextran was present until the final stage to reduce the rate of swelling, and lanthanum ions (0.5mM) were present throughout to inhibit lysis. This procedure yielded very large polycells of irregular shape that retained their haemoglobin, as well as giving rise to lysed single cells and

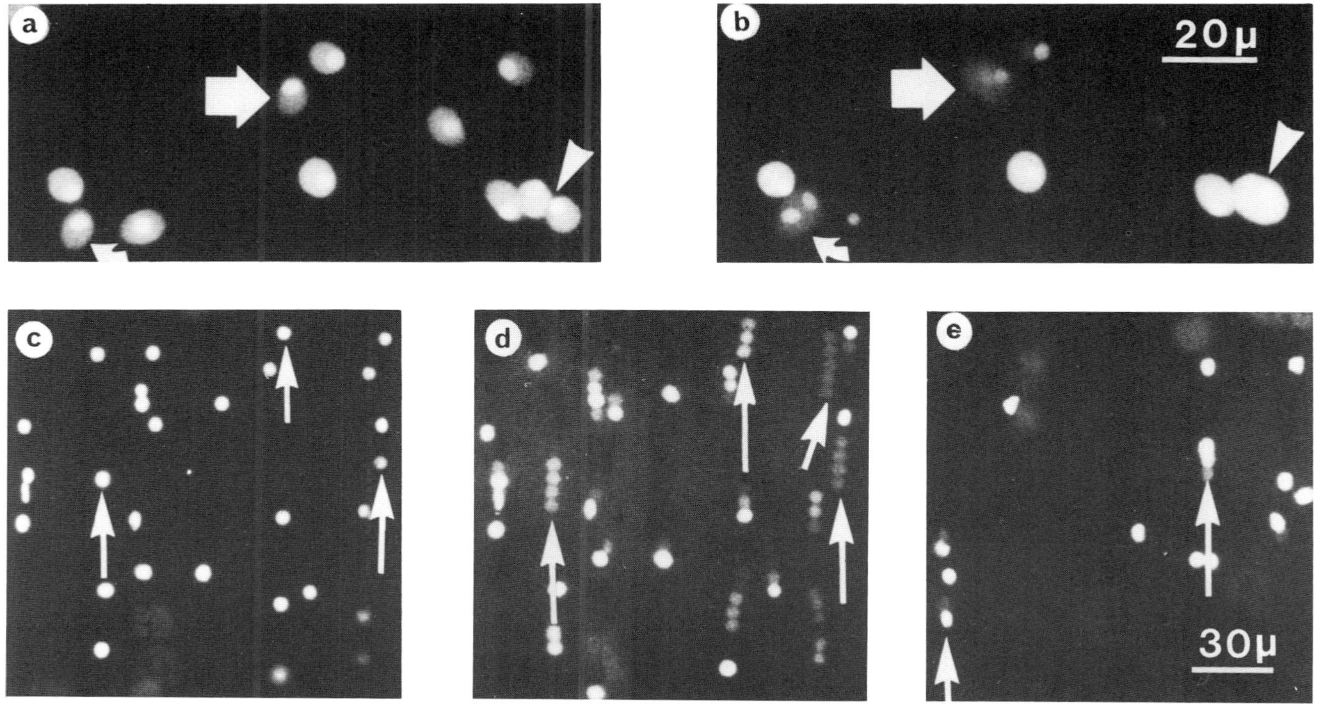

Fig. 4. Erythrocytes visualized by fluorescence microscopy: 0% of the cells were labelled with 6-carboxyfluorescein. (a) Hen erythrocytes with A23187 and Ca^{2+} for 2 h at 37° C. (b) Cells as in (a) but 5 min after diluting the buffer 2.5-fold: the arrowed cells have fused. (c) Human erythrocytes aligned by dielectrophoresis in 150 mM erythritol. (d) Cells as in (c) but 20 s after the breakdown pulse: fluorophore in the arrowed cells has transferred to 3 or 4 unlabelled cells. (e) Cells, in 200 mM erythritol, 15 sec after the pulse; fluorophore has transferred only to cells immediately adjacent to fluorescent (arrowed) cells.

polyghosts (Ahkong and Lucy, 1986). When a proportion of the erythrocytes was labelled with carboxyfluorescein, very large cells of varying fluorescent intensity were formed.

It is therefore apparent that, under appropriate experimental regimes, osmotic swelling is able to induce the fusion of both hen and human erythrocytes.

ELECTRICALLY-INDUCED CELL FUSION OF ERYTHROCYTES AND LYMPHOID CELLS

We have also investigated electrically-induced cell fusion in order to see if there is any evidence for the involvement of osmotic forces in the fusion of cells by this procedure.

Human erythrocytes were labelled with carboxyfluorescein, and a mixture of labelled and unlabelled cells was then suspended in a hypotonic solution of 150 mM erythritol and 10mM histidine which also contained 1 mg/ml of pronase (Zimmermann et al., 1981). After alignment of the cells by dielectrophoresis, a square-wave d.c. pulse of 3.5kV/cm and 20μsec. duration was applied. Unlabelled erythrocytes that were adjacent to fluorescein-labelled cells in the "pearl chains" of aligned cells instantly became fluorescent. Short chains of four or five (and sometimes up to fourteen) fluorescent cells were thus formed (Figs. 4c and 4d), and these chains then gave rise within 3 min. to fused cells that were observable by phase-contrast microscopy. We therefore concluded that the transfer of carboxyfluorescein, which occurs virtually instantaneously with the breakdown pulse (cf., Sowers, 1984), results from the formation by the electrical breakdown pulse of small cytoplasmic bridges between adjacent cells (cf., Zimmermann, 1982).

By contrast to the striking formation of long chains of fluorescent cells with erythrocytes in 150mM erythritol, the fluorophore transferred only to single cells that were immediately adjacent to the fluorescent cells in experiments with less swollen erythrocytes in 200 mM erythritol (Fig. 4e). Chains of fluorescent cells were not formed, and few fused cells were seen by phase-contrast microscopy. In 400 mM erythritol, the instantaneous transfer of fluorescence was neglible, and no fused cells were produced (Ahkong and Lucy, 1986). The degree to which the cells were swollen in a hypotonic medium when the electrical breakdown pulse was applied thus affected the extent to which the cells were fused, and we concluded that electrically-induced cell fusion is governed by a combination of the applied electrical-compressive force and the osmotic force on the cells. It is interesting to note that these observations resemble those made by Zimmerberg et al., (1980) in their studies on osmotically-driven fusion in model membrane systems. They found that phospholipid vesicles that are pre-swollen fuse more readily with a phospholipid bilayer than more flacid vesicles under the same osmotic gradient.

In experiments in our laboratory on the electrofusion of myeloma cells, carboxyfluorescein was transferred from labelled to adjacent unlabelled myeloma cells immediately following exposure to the breakdown pulse and, under appropriate conditions, fluorescent chains of 11-17 cells were formed. Similar observations were made with lymphocytes, and also with mixed populations of myeloma cells and lymphocytes. As with erythrocytes, osmotic effects were found, and transfer of the cytoplasmic label occurred more extensively when myeloma cells and lymphocytes were exposed to breakdown voltages of 1.5- 3.5 kV/cm in 200 mM inositol than when the cells were in 300 mM inositol (Brown et al., 1987). It is also interesting that NS1 myeloma cells in a 280 mM solution of inositol, free from pronase but containing 0.1 mM calcium and 0.5mM magnesium ions

(Vienken and Zimmermann, 1985), fused much less readily than when pronase was present (as monitored by movement of the fluorophore). This pronase-free, inositol medium has been reported to be particularly useful for the preparation of hybridoma cells by electrofusion, and it is therefore conceivable that hybridoma cells are not derived from cells that are fused immediately by the breakdown pulse. Instead, they may result from fusion occurring some time afterwards. The feasibility of this is indicated by the fact that an intercellular transfer of fluorophore has been observed with erythrocytes during the colloid osmotic swelling that occurs several minutes after their permeabilization by a breakdown pulse (Lucy, 1986).

FUSION INDUCED BY POLY(ETHYLENE GLYCOL)

A few years ago it was claimed by Honda et al., (1981) that the fusogenic properties of commercial preparations of PEG are due to impurities and that, although the polymer is able to aggregate human erythrocytes, it has no intrinsic ability to fuse them. Subsequently, in experiments with hen and human erythrocytes, we showed that the fusogenic behaviour of four commercial preparations of PEG were unaffected by purification, and that a preparation like that used by Honda and his colleagues contained contaminating substances which enhanced the fusion of erythrocytes (Smith et al., 1982). Hui et al., (1985) have similarly reported that purified PEG fuses erythrocyte ghosts. Furthermore, not only has recrystallised PEG been found to fuse carrot protoplasts as efficiently as the unpurified polymer, but it has also been reported that a sample of PEG 8000, which was inefficient in fusing erythrocytes with fibroblasts, was fusogenic for carrot protoplasts (Boss, 1983). In addition, Boni et al., (1984) observed that alterations in phospholipid polymorphism are caused by both purified and commercial preparations of PEG, while Parente and Lentz (1986) recently reported that impurities present in commercial PEG had no effect on the rate or extent of the fusion of phospholipid vesicles. Wojcieszyn et al., (1983) nevertheless found that PEG does not fuse erythrocytes with fibroblasts after it has been purified, and these workers therefore attributed the fusion event itself (as distinguished from cell aggregation) to the action of unknown impurities in commercial PEG.

What explanation can be offered for these contradictory reports on the fusogenic behaviour of PEG? We suggest that one possibility is that PEG does not disrupt the membrane skeleton sufficiently in some cases for fusion to occur. It is interesting in this connection that whether or not a lipid fusogen successfully fuses erythrocytes from differing species can depend on whether it effectively activates the endogenous proteinases which degrade the membrane skeleton of the treated cells. Thus proteolysis associated with the fusion of human erythrocytes by chlorpromazine primarily involves activation of an endogenous serine proteinase, while a calpain-like enzyme is involved in the degradation of the membrane skeleton of rat erythrocytes fused by benzyl alcohol; these observations may reflect the relatively high concentration of the endogenous inhibitor of calpain (calpastatin) in human erythrocytes (Lang et al., 1984). Indeed it has recently been shown by Glaser and Kosower (1986) that the membrane mobility agent, A_2C, fuses rat erythrocytes, but not human erythrocytes, because calpain is inhibited by calpastatin in the human cells. PEG, by contrast, seems not to activate endogenous proteinases at all when it fuses human erythrocytes (Schindler et al., 1980; Smith and Palek, 1982). Instead, the polymer may physically disrupt the membrane skeleton by its dehydrating action. It thus seems possible that PEG may not be able to fuse human erythrocytes with fibroblasts (Wojcieszyn et al., 1983) unless a lipid-soluble fusogen is also present, as an impurity, to activate proteinases which degrade the membrane skeletons of one or both cells sufficiently to allow them to fuse. Significantly, when low concentrations of lipid-soluble fusogens (e.g. oleoylglycerol), which activate endogenous

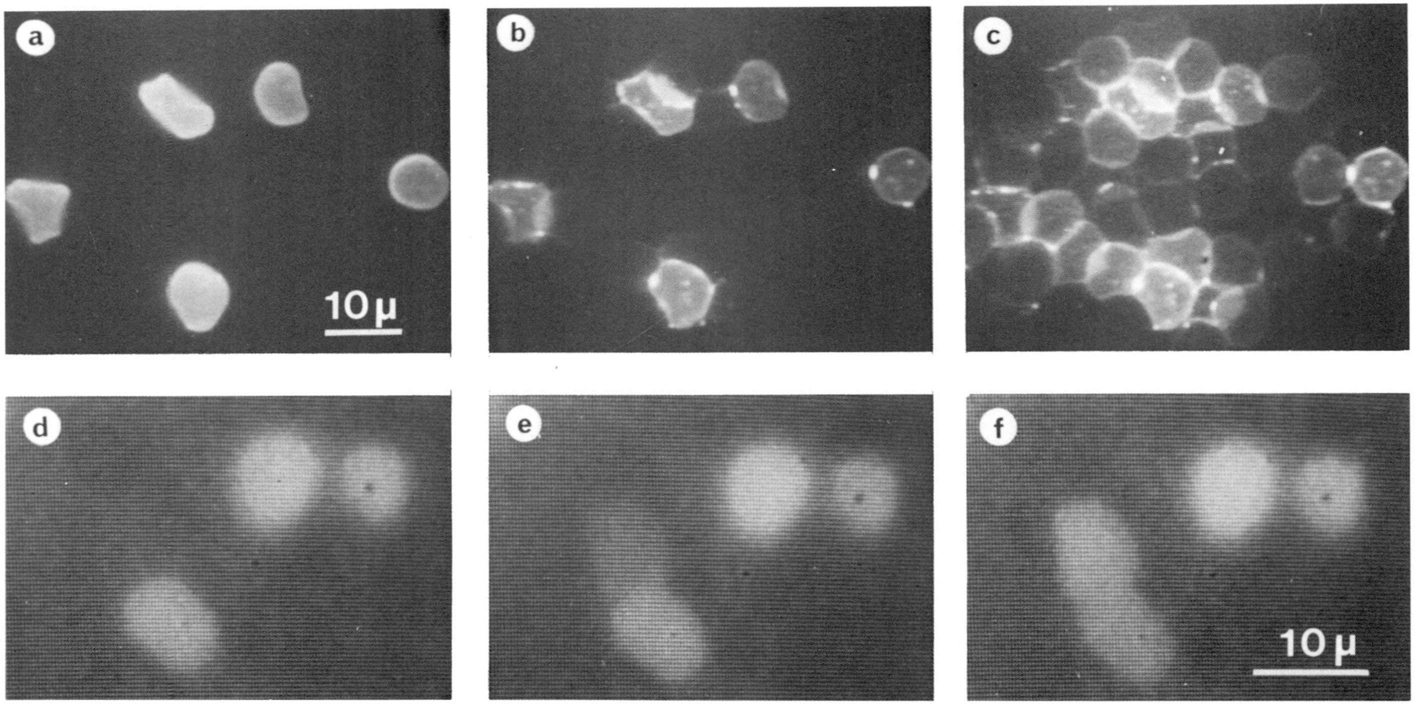

Fig. 5. Human erythrocytes visualized by fluorescence microscopy. (a) 10% of the cells were labelled with the carbocyanine membrane-probe, $diIC_{16}$. (b) Cells as in (a) but 30 sec after replacing buffered saline by 40% PEG 6000: the probe has diffused into adjacent unlabelled cells. (c) Cells as in (b) but 5.5 min later: most cells have become labelled. (d) 10% of the cells were labelled with 6-carboxyfluorescein and are shown after the 40% PEG had been replaced by isotonic buffer. (e and f) Micrographs from a video recording 500 msec and 1 sec after (d) showing a sudden, rapid transfer of cytoplasmic fluorophore to an unlabelled cell.

proteinases in erythrocytes, are mixed with PEG they act synergistically with the polymer to enhance cell fusion (Smith et al., 1982). Huang and Hui (1986) have recently also shown that 0.05% spermidine, 100 mM trichloracetic acid, and 4% ethanol each lower the threshold concentration of PEG (to 25%) required for the fusion of human erythrocytes, apparently by creating zones in the plasma membranes that are free from intramembraneous particles. They have also proposed that the additional PEG (up to 45%) needed for most cell fusion work is likely to be associated with the property of PEG that aggregates and precipitates membrane proteins to expose intramembraneous particle-free areas of membrane.

Is cell fusion induced by PEG driven by osmotic forces? Since close-range, repulsive forces between orientated, surface layers of water prevent phospholipid bilayers from coming into close proximity, withdrawal of water from the polar head groups by PEG enables the bilayers to come close together (Parsegian et al, 1984). Consequently, many studies have been undertaken on the behaviour of phospholipid systems in solutions of PEG, but this work has not shown how its dehydrating action enables the polymer to induce cell fusion rather than simply the very close apposition of the plasma membranes of cells. In experiments with human erythrocytes that were labelled with carboxyfluorescein, we found that fusion of the cells treated with a 40% solution of PEG 6000 begins only when the PEG is subsequently diluted, and that it mostly occurs when the diluted polymer solution is replaced by isotonic buffer. We therefore concluded that, while the dehydration of PEG-treated cells brings them into close contact, it is their rehydration that causes cell fusion (Ahkong and Lucy, 1986).

Wojcieszyn et al., (1983) have observed that a carbocyanine membrane probe spreads from labelled erythrocytes to unlabelled fibroblasts within 1-2 min. in a concentrated solution of PEG. However, because purified PEG did not induce cell fusion in their experiments, they concluded that PEG promotes a unique type of membrane apposition that permits a rapid, intercellular transfer of lipid probes irrespective of its ability to fuse cells. Recently we have investigated the effect of purified PEG on the movement from labelled to unlabelled human erythrocytes of octadecyl rhodamine B chloride, which does not dissociate from membranes either by spontaneous transfer of free monomers through the aqueous phase or by a collision mediated transfer process (Hoekstra et al., 1984). No movement of rhodamine or carbocyanine probes occurred in the absence of PEG but, in the presence of 40% purified PEG, both probes started to diffuse into the membranes of unlabelled cells within 30 sec. (Figs. 5a and 5b). This indicated that continuity between the phospholipid bilayer membranes of adjacent erythrocytes is established almost immediately in concentrated, dehydrating solutions of the polymer, and it was suggested that this involves micro-regions of shared phospholipid bilayer that may be formed from the plasma membranes of cells when they come together as a consequence of the dehydrating action of the polymer. After 6 min., the fluorescent, lipid probes had diffused into the membranes of the majority of the previously unlabelled erythrocytes (Fig. 5c) (Ahkong et al., 1987).

In these experiments, a sensitive video camera enabled us to observe that cell fusion itself, as judged by the movement of cytoplasmic carboxyfluorescein, often occurred suddenly after the PEG solution was replaced by an isotonic buffer. Figs. 5d and 5e show that sufficient carboxyfluorescein was transferred to a recipient cell to render it visible within 500 msec. After a further 500 msec., the donor and recipient cells were approximately equally fluorescent (Fig. 5f). With some erythrocytes, a complete transfer of the fluorophore occurred in 320 msec. Wojcieszyn et al., (1983) have suggested that water soluble proteins do not diffuse from erythrocytes (loaded by hypotonic lysis) into fibroblasts until the PEG is removed because the proteins are immobilized in the presence of the concentrated polymer. It would seem, however that the sudden, rapid transfer of carboxyfluorescein, which we have found when PEG is replaced by an isotonic buffer, is much more likely to result from the rupture of membraneous barriers between the cells than from the return of their cytoplasmic microviscosity to normal. Our finding is also consistent with the fact that cytoplasmic bridges between PEG-treated cells have not been seen until the polymer is diluted and the cells resuspended in an isotonic buffer (Knutton, 1979; Robinson et al., 1979; Krähling, 1981; Ahkong and Lucy, 1986). It has therefore been suggested that a membrane, possibly a shared bilayer, separates the cytoplasms of PEG-treated cells until it is stretched and broken when the replacement of PEG by an isotonic buffer causes water to enter the cells (Ahkong et al., 1987).

The failure of erythrocytes to fuse in the presence of PEG contrasts with the behaviour of phospholipid vesicles, which have been reported to fuse in concentrated solutions of the polymer by several groups of investigators (Boni et al., 1981; Aldwinckle et al., 1982, Parente and Lentz, 1986), and it has been concluded by Parente and Lentz (1986) that dilution and the osmotic swelling that may accompany dilution are not required for the fusion of large phospholipid vesicles by PEG. Consequently, although it is often assumed that the fusion of phospholipid vesicles is a valid model for cell fusion, it is possible that the mechanism by which cells fuse after PEG has been removed differs from that involved when vesicles fuse in its presence. Since fusion of the vesicles is accompanied by the formation of membrane discontinuities that are sufficiently large to allow an extensive release of entrapped, radio-labelled PEG 4000 (used as a marker) to occur, it has been proposed that liposomes fuse at the sites of these dehydration-induced discontinuities in concentrated solutions of PEG (Aldwinckle et al., 1982). Conceivably, stabilizing effects on the phospholipid bilayers of plasma membranes that are exerted by the membrane skeleton, and/or by the gel-like contents of cells (Elgsaeter et al., 1986), may be responsible for preventing this comparatively leaky type of fusion from occurring during the dehydration of PEG-treated erythrocytes.

AN OSMOTIC MOLECULAR MODEL FOR MEMBRANE FUSION

Since our observations indicate that osmotic forces are involved in the fusion of erythrocytes by chemical and electrical means, we have proposed a general molecular model for the fusion of biological membranes that involves osmotic phenomena (Lucy and Ahkong, 1986). The new model contrasts with previous proposals for the molecular mechanisms underlying membrane fusion, which have been based solely on the structures and physical properties of the participating molecules and ions. Another way in which the new model differs from most previous molecular models for fusion is that a phospholipid bilayer, rather than a non-bilayer structure, functions as an intermediate in the fusion process. Thus a central feature of the model is that fusion-fission reactions depend on two perturbed, phospholipid bilayers being merged into a single, shared bilayer, which is subsequently ruptured by osmotic forces.

A first requirement for fusion to occur is that the repulsive forces between two membranes must be overcome. This aspect of the fusion process has been concisely reviewed by Parsegian et al., (1984). For membranes that are 2 to 3 nm apart, the repulsive forces between the bilayers are essentially electrostatic, and they are sensitive to the identity and charge of the lipid polar groups and the ionic composition of the medium. Hydration repulsion dominates the interactions between membranes as they approach more closely together. Both forces may be overcome, for example, by adding calcium ions. Bilayers of phosphatidylserine repel each other by electrostatic and hydration forces but in the presence of calcium ions the bilayers approach so closely that there is virtually no water between them, and it appears that the binding energy of the ions to the polar groups of the phospholipid molecules exceeds the initial energy of hydration of these groups. Similarly PEG, at concentrations of 30-60%, removes some 90% of the water from phospholipid bilayers, enabling them not only to come together but also forcing them into very close proximity.

What happens when two phospholipid bilayers are pushed together with a force that is sufficient to overcome their mutual repulsion? More than ten years ago, Neher (1974) suggested that a new bilayer is formed, and several groups of workers have recently obtained experimental confirmation of this idea. For example, Horn (1984) found that when a very large force was applied to two mica surfaces, each coated with a bilayer of phosphatidylcholine, the mica surfaces pinched-in to a separation that corresponded to a single bilayer. The pinched-in region then rapidly enlarged until the single bilayer occupied an area of 30 microns in diameter. Significantly, the formation of a single bilayer occurred more often with egg phosphatidylcholine than with synthetic dilaurylphosphatidylcholine, and also when a small quantity of hexane was incorporated into the bilayers. Horn concluded that the factors which determine whether fusion will occur are related not so much to the forces between bilayers as to the forces within bilayers that give them their cohesive strength. It is relevant to note that, in the fusion of phospholipid vesicles with a phospholipid bilayer induced by an osmotic gradient, the addition of decane to a "hydrocarbon-free" bilayer increases the rate of fusion (Cohen et al., 1984), and that hydrocarbons can induce erythrocytes to fuse (Ahkong et al., 1974). The formation of a single bilayer has also been observed when bilayers of oleoylglycerol (which contain hexadecane) are pushed together mechanically as water is removed from between them (Fisher and Parker, 1984), and similarly when phospholipid bilayers are pushed together by being inflated (Melikyan et al., 1983).

Relatively little consideration has been given in the past to the possibility that the fusion of biomembranes might involve two phospholipid bilayers merging into a single, shared bilayer because it has been thought that the energy required for the process is much greater than is likely to be available in biological systems. Markin et al., (1984) have estimated that the pressure required for the mutual penetration of two bilayers of neutral phospholipids is approximately 10^9 N/m^2, and they have commented that a 100km column of water would be needed to obtain such a pressure. However, much experimental evidence indicates that fusogenic agents perturb bilayers. For example, lysophosphatidylcholine and oleoylglycerol perturb phospholipid bilayers in model systems and give rise under equilibrium conditions to spherical micelles (Bangham and Horne, 1964) and a hexagonal phase respectively (Howell et al., 1973; Cullis and Hope, 1978). These phase changes have prompted suggestions that spherical micelles (Lucy, 1970), the hexagonal phase (Cullis and Hope, 1978) and inverted micelles (Verkleij et al., 1980) are formed as transient intermediates in the fusion process, but there is no clear evidence that non-bilayer phases are actually involved in the fusion of biological membranes. We therefore suggest that agents which perturb phospholipid bilayers are potentially fusogenic because perturbed bilayers can be merged into a single shared bilayer with a smaller expenditure of energy than unperturbed bilayers. On this basis, whether or not the perturbing agent induces the formation of a non-bilayer phase in model phospholipid systems is irrelevant to membrane fusion.

In our recently proposed model for membrane fusion, we suggested that one way of generating a force that would lead to the formation of a single, shared bilayer is to allow water to enter one or both of the membrane-bound compartments that are concerned in the fusion reaction (Lucy and Ahkong, 1986). In membrane fusion occuring physiologically in biological systems, this movement of water may arise because, for example, osmotically-active particles have accumulated in one or both compartments, or because functionally impermeable membranes, which previously excluded water, have become permeable. Similar movements of water occur in experimental models for membrane fusion, e.g. in the osmotically-driven fusion of phospholipid vesicles with a bilayer, and in the fusion of erythrocytes that are treated with permeabilizing, chemical fusogens. However, in our recent work on the fusion of human erythrocytes by purified PEG that is discussed above, we observed that rhodamine and carbocyanine fluorophores diffused from the membranes of unlabelled to labelled cells within 30 sec. of adding 40% PEG (Ahkong et al., 1987). If this resulted from the formation of micro-regions of shared bilayer between the treated cells it seems that, in this instance, removal of water from between the cells by the dehydrating action of PEG, rather than the entry of water into them, was responsible for the forcible merging of two phospholipid bilayers.

The final event in our proposed model for the fusion-fission reaction of biological membranes is fission of the shared bilayer, which results from stretching of the shared bilayer caused by osmotic swelling. This

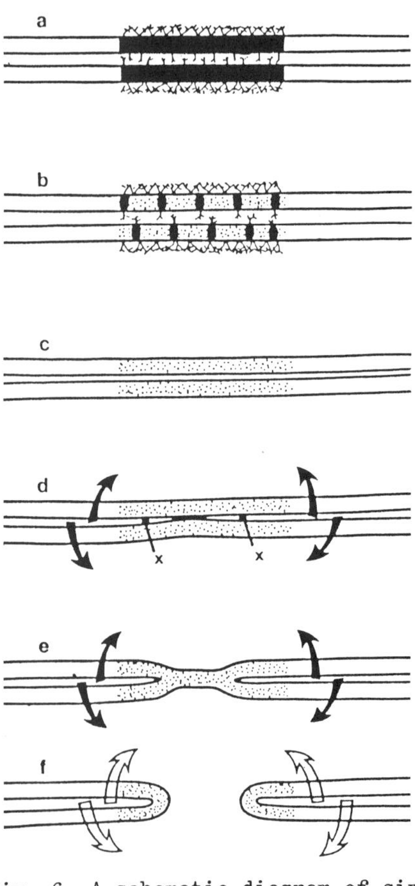

Fig. 6. A schematic diagram of six proposed events for the fusion-fission reactions of biological membranes. (Lucy and Ahkong, 1986).

will give rise to a hole, which is initially of only about 10-15 nm in diameter, but which will widen rapidly as swelling proceeds. The "explosive" movement of carboxyfluorescein from labelled to unlabelled human erythrocytes, which we observed when 40% PEG 6000 was removed and replaced by an isotonic buffer (Figs. 5d-5f), is consistent with an intermediate bilayer being ruptured by membrane stretching that results from the osmotic entry of water. Electron micrographs obtained in this study showed that many cytoplasmic channels linked the fused cells. The smallest channels were about 20 nm in diameter, and they were seen only after the 40% PEG was replaced by the buffer (Ahkong et al., 1987).

Fig. 6 shows, in diagrammatic form, the events that are proposed in our model for the fusion-fission reactions of biological membranes (Lucy and Ahkong, 1986). Thus, Fig. 6a illustrates the initial, close approach of the two membranes, while Fig. 6b is a representation of the perturbation of the phospholipid bilayers at the fusion site that is diagramatically illustrated by a change from solid to hatched shading. In Fig. 6c the membrane proteins are perturbed. The schematic absence of integral membrane proteins, membrane-skeleton proteins and cytoskeletal proteins from the potential fusion site in Fig. 6c is intended to indicate that the lipid bilayers are relatively unrestrained by membrane proteins, and it does not imply that the bilayers are totally free from all polypeptides (eg. fusogenic, viral peptides) at the developing site of fusion. The solid arrows in Fig. 6d represent the entry of water into one, or both (as illustrated) of the two membrane-bound compartments, which causes swelling. The regions marked "x" are indicative of structures or mechanisms that hold the membranes together, adjacent to the fusion site, as swelling proceeds; in cell fusion, adjacent swelling cells that are not held together will lyse separately rather than fuse. In Fig. 6e, the perturbed phospholipid bilayers are forced together at the fusion site, under the influence of the pressure generated by the osmotic gradient, and they form an intermediate, shared bilayer that is common to the two fusing membranes. Finally, in Fig. 6f, the shared bilayer ruptures at one or both of the points where three bilayers meet, and the original bilayers then re-seal. A continued entry of water (open arrows) will cause further swelling and, in the case of cell fusion, will induce the fused cells to lyse (cf. Figs. 2 and 3).

Aspects of this model can be applied to electrically-induced cell fusion. With cells that are aligned by dielectrophoresis, an a.c. field of sufficient strength to flatten adjacent cells against one another has to be applied before exposure to the breakdown pulse (Zimmermann and Vienken, 1982). This presumably brings the apposed plasma membranes close enough together for fusion to occur. There is also some evidence that the breakdown pulse may damage the membrane skeleton (Donath and Arndt, 1984). However, since fusion is virtually instantaneous with the breakdown pulse, the shared bilayer shown in Fig. 6e must be very short-lived. We have therefore suggested that the electrical compressive force, postulated in the electro-mechanical model for membrane breakdown by Zimmermann et al., (1980), firstly causes the phospholipids of the closely appposed bilayers to interdigitate and then, in an effectively continuous process, induces extensive thinning that leads directly to membrane rupture (Lucy and Ahkong, 1986). As discussed above, any applied osmotic or hydrostatic pressure will supplement the electrical compressive force. It is noteworthy that cell fusion generally occurs when the field strength is 1.5-2 times greater than the minimum necessary for membrane breakdown (Zimmermann and Vienken, 1982): this might be a consequence of fusion proceeding via a transient shared bilayer, rather than via the individual breakdown of the membranes of closely apposed cells (Zimmermann, 1982). Interestingly, we have observed that many of the erythrocytes in a given field which remain unfused on exposure to the breakdown pulse are nevertheless permanently attached to their neighbours following exposure to the breakdown pulse.

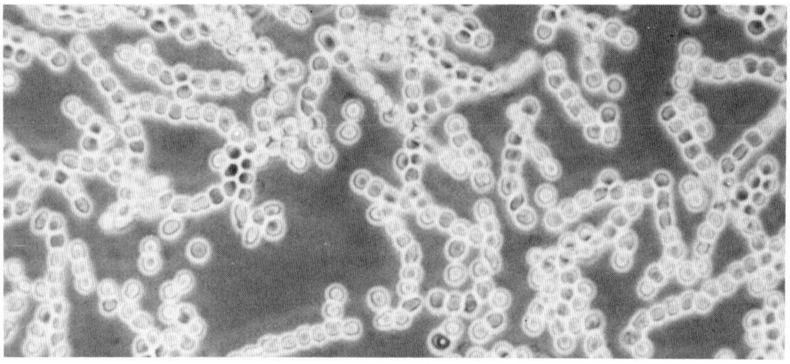

Fig.7. Human erythrocytes, originally aligned by dielectrophoresis, remain attached to one another in the absence of an a.c. field after exposure to a breakdown pulse, and are not dispersed by thermal or mechanical motion.

This is shown by the fact that, when the a.c. field is disconnected, the cells do not separate but remain in chains that are not dispersed by thermal or mechanical motion (Fig. 7). If some erythroctes are pre-labelled with octadecyl rhodamine B chloride, the fluorophore diffuses into the membranes of the unlabelled cells to which they become permanently attached (but not fused) by the breakdown pulse (Ahkong et al, 1987). Hence we suggest that a breakdown pulse, which is just below the threshold that is necessary to induce sufficient membrane thinning for complete fusion, may nevertheless have enough energy to compress the membranes of aligned cells into shared bilayers (Fig. 6e) at the points of very close contact. The linked cells, which will also be permeabilized by the pulse, may then fuse completely if the shared bilayer is subsequently ruptured osmotically (Fig. 6f) but this will not occur until several minutes after breakdown pulse (cf. Lucy, 1986).

CONCLUDING REMARKS

In conclusion, it is pertinent to comment that the availability of an extrinsic source of energy, in the form of an osmotic gradient, offers a possible solution to a fundamental problem that is inherent in all models which postulate that the formation of a non-bilayer organization of phospholipids is responsible for the fusion-fission reactions of biological membranes. Namely, if a non-bilayer intermediate forms spontaneously on treating membranes with a fusogenic agent, what causes the intermediate to rupture in the subsequent fission reaction? In our osmotic model, membrane fusion involves the forced merging of two perturbed phospholipid bilayers into a single, shared bilayer which is then forcibly ruptured, and this riddle therefore does not arise.

The illustrations are reproduced with permission as follows:- Fig.1. from Ahkong et al., (1973a), Fig. 2 from Ahkong et al., (1973b), Fig. 3 from Bruckdorfer al., (1974), Fig. 4 from Ahkong and Lucy (1986), Fig. 5 from Ahkong et al., (1987), and Fig. 6 from Lucy and Ahkong (1986).

REFERENCES

Ahkong, Q. F., Fisher, D., Tampion, W., and Lucy, J. A., 1973a, The fusion of erythrocytes by fatty acids, esters, retinol and α-tocopherol, Biochem. J., 136:147.

Ahkong, Q.F., Cramp, F.C., Fisher, D., Howell, J.I., Tampion, W.,

Verrinder, M., and Lucy, J.A., 1973b, Chemically-induced and thermally-induced cell fusion: lipid-lipid interactions, Nature New Biology, 242:215.

Ahkong, Q.F., Howell, J.I., Tampion, W., and Lucy, J.A., 1974, The fusion of hen erythrocytes by lipid-soluble and surface-active immunological adjuvants, FEBS Lett., 41:206.

Ahkong, Q.F., Tampion, W., and Lucy, J.A., 1975, Promotion of cell fusion by divalent cation ionophores, Nature Lond., 256:208.

Ahkong, Q.F., Botham, G.M., Woodward, A.W., and Lucy, J.A, 1980, Calcium-activated thiol-proteinase activity in the fusion of rat erythrocytes induced by benzyl alcohol, Biochem. J., 192:829.

Ahkong, Q.F., and Lucy, J.A., 1986, Osmotic forces in artificially induced cell fusion, Biochem. Biophys. Acta, 858:206.

Ahkong, Q.F., Desmazes, J-P., Georgescauld, D., and Lucy, J.A., 1987, Unpublished observations.

Aldwinkle, T.J., Ahkong, Q.F., Bangham, A.D., Fisher, D., and Lucy, J.A., 1982, Effects of poly(ethylene glycol) on liposomes and erythrocytes. Permeability changes and membrane fusion, Biochim. Biophys. Acta, 689:548.

Allan, D., and Michell, R.H., 1978, A calcium-activated polyphosphoinositide phosphodiesterase in the plasma membrane of human and rabbit erythrocytes, Biochim. Biophys. Acta, 508:277.

Bangham, A.D., and Horne, R.W., 1964, Negative staining of phospholipids and their structural modification by surface active agents as observed in the electron microscope, J. Mol. Biol., 8:660.

Boni, L.T., Stewart, T.P., Alderfer, J.L., and Hui, S.W., 1981, Lipid-polyethylene glycol interactions: I. Induction of fusion between liposomes, J. Membrane Biol., 62:65.

Boni, L.T., Stewart, T.P., and Hui, S.W., 1984, Alterations in phospholipid polymorphism by polyethylene glycol, J. Membrane Biol., 80:91.

Boss, W.F., 1983, Poly(ethylene glycol)-induced fusion of plant protoplasts, Biochim. Biophys. Acta, 730:111.

Brown, S.M., Ahkong, Q.F., and Lucy, J.A., 1987, Osmotic pressure and the electrofusion of myeloma cells, Biochem. Soc. Trans., 14,1129.

Bruckdorfer, K.R., Cramp, F.C., Goodall, A.H., Verrinder, M., and Lucy, J.A., 1974, Fusion of mouse fibroblasts with oleylamine, J. Cell. Sci., 15:185.

Cohen, F.S., Zimmerberg, J., and Finkelstein, A, 1980, Fusion of phospholipid vesicles with planar phospholipid bilayer membranes. II Incorporation of vesicular membrane marker into the planar membrane, J. Gen. Physiol., 75:251.

Cohen, F.S., Akabas, M.H., Zimmerberg, J., and Finkelstein, A., 1984, Parameters affecting the fusion of unilamellar phospholipid vesicles with planar bilayer membranes, J. Cell. Biol., 98:1054.

Cullis, P.R., and Hope, M.J., 1978, Effects of fusogenic agent on membrane structure of erythrocyte ghosts and the mechanism of membrane fusion, Nature Lond., 271:672.

Donath, E., and Arndt, R., 1984, Electric-field-induced fusion of enzyme treated human red cells: kinetics of intermembrane protein exchange, Gen. Physiol. Biophys., 3:239.

Elgsaeter, A., Stokke, B.T., Mikkelsen, A., and Branton, D., 1986, The molecular basis of erythrocyte shape, Science, 23:1217.

Fisher, L.R., and Parker, N.S., 1984, Osmotic control of bilayer fusion, Biophys. J., 46:253.

Glaser, T., and Kosower, N.S., 1986, Calpain-Calpastatin and fusion. Fusibility of erythrocytes is determined by a protease-protease inhibitor (calpain-calpastatin) balance, FEBS Lett., 206::115.

Goodall,H., and Johnson, M.H., 1982, Use of carboxyfluorescein diacetate to study formation of permeable channels between mouse blastomers, Nature Lond., 295:524.

Hampton, R.Y., and Holz, R.W., 1983, Effects of changes in osmolarity on

the stability and function of cultured chromaffin cells and the possible role of osmotic forces in exocytosis, J. Cell Biol., 96:1082.
Heubusch, P., Jung, C,Y., and Green, F.A., 1985, The osmotic response of human erythrocytes and the membrane cytoskeleton, J. Cell Physiol, 122:266.
Hoekstra, D., de Boer, T., Klappe, K., and Wilschut, J., 1984, Fluorescence method for measuring the kinetics of fusion between biological membranes, Biochemistry, 23:5675.
Honda, K., Maeda, Y., Sasakawa, S., Ohno, H., and Tsuchida, E., 1981, The components contained in polyethylene glycol of commercial grade PEG-6,000) as cell fusogen, Biochem. Biophys. Res. Commun., 101:165.
Horn, R.G., 1984, Direct measurement of the force between two lipid bilayers and observation of their fusion, Biochim. Biophys. Acta, 778::224.
Howell, J.I., Fisher, D., Goodall, A.H., Verrinder, M., and Lucy, J.A., 1973, Interactions of membrane phospholipids with fusogenic lipids, Biochim. Biophys. Acta, 332:1.
Huang, S.K., and Hui, S.W., 1986, Chemical co-treatments and intramembrane particle patching in the poly(ethylene glycol)-induced fusion of turkey and human erythrocytes, Biochim. Biophys. Acta, 860:539.
Hui, S.W., Isac, T., Boni, L.T., and Sen, A., 1985, Action of polyethylene glycol on the fusion of human erythrocyte membranes, J. Membrane Biol., 84:137.
Knutton, S., 1979, Studies of membrane fusion. III. Fusion of erythrocytes with polyethylene glycol, J. Cell. Sci., 36:61.
Kosower, E.M., Kosower, N.S., and Wegman, P., 1977, Membrane mobility agents. IV. The mechanism of particle-cell and cell-cell fusion, Biochim. Biophys. Acta, 471:311.
Kosower, N.S., Wegman, P.O., Neiman, T., and Kosower, E.M., 1978, Membrane mobility agents. V. Genetic variability in the fusibility of hen red cells, Exp. Cell Res., 166:454.
Kosower, N.S., Glaser, T., and Kosower, E.M., 1983, Membrane-mobility agent-promoted fusion of erythrocytes: fusibility is correlated with attack by calcium-activated cytoplasmic proteases on membrane proteins, Proc. Natl. Acad. Sci., USA, 80:7542.
Krähling, H., 1981, Investigations on polyethylene glycol-induced cell fusion - freeze fracture observations Acta histochemica, Suppl., 23:219.
Lang, R.D.A., Wickenden, C., Wynne, J., and Lucy, J.A., 1984, Proteolysis of ankyrin and of band 3 in chemically-induced cell fusion: Ca2+ is not mandatory for fusion, Biochem. J., 218:295.
Lucy, J.A., 1970, The fusion of biological membranes, Nature, 227:814.
Lucy, J.A., 1973, The chemically-induced fusion of cells, in: Membrane-mediated information, Vol. 2, P.W. Kent, ed., Medical and Technical Publishing Co., Lancaster.
Lucy, J.A., 1986, Salient features of artificially induced cell fusion, Biochem. Soc. Trans., 14:241.
Lucy, J.A., and Ahkong, Q.F., 1986, An osmotic model for the fusion of biological membranes, FEBS Lett., 199:1.
Maggio, B., Cumar, F.A., and Caputto, R., 1978, Induction of membrane fusion by polysialogangliosides. FEBS Lett., 90:149.
Markin, V.S., Kozlov, M.M., and Borovjagin, V.L., 1984, On the theory of membrane fusion. The stalk mechanism, Gen. Physiol. Biophys., 5:361.
Melikyan, G.B., Abidor, I.G., Chernomordik, L.V., and Chailakhyan, L.M., 1983, Electrostimulated fusion and fission of bilayer lipid membranes, Biochim. Biophys. Acta, 730:395.
Miller, C., Arvan, P., Telford, J.N., and Racker, E., 1976, Ca++-induced fusion of proteoliposomes: dependence on transmembrane osmotic gradient, J. Membrane Biol., 30:271.
Neher, E., 1974, Asymmetric membranes resulting from the fusion of two black lipid bilayers, Biochim. Biophys. Acta, 373:327.

Parente, R.A., and Lentz, B.R., 1986, Rate and extent of poly(ethylene glycol)-induced large vesicle fusion monitored by bilayer and internal contents mixing, Biochemistry, 25:6678.
Parsegian, V.A., Rand, R.P., and Gingell, D., 1984, Lessons for the study of membrane fusion from membrane interactions in phospholipid systems, in: Cell Fusion, CIBA Symposium No. 103, D. Evered and J. Whelan, eds., Pitman, London.
Pollard, H.B., Pazoles, C.J., Creutz, C.E., Scott, J.H., Zinder, O., and Hotchkiss, A., 1984, An osmotic mechanism for exocytosis from dissociated chromaffin cells, J. Biol. Chem., 259:1114.
Ponder, E. 1955, Red cell structure and its breakdown, Springer-Verlag, Vienna.
Quirk, S.J., Ahkong., Q.F., Botham, G.M., Vos, J., and Lucy, J.A., 1978, Membrane proteins in human erythrocytes during cell fusion induced by oleoylglycerol, Biochem. J. 176:159.
Robinson, J.M., Roos, D.S., Davidson, R.L., and Karnovsky, M.J., 1979, Membrane alterations and other morphological features associated with polyethylene glycol-induced cell fusion, J. Cell Sci., 40:63.
Schindler, M., Koppel, D.E., and Sheetz, M.P., 1980, Modulation of membrane protein lateral mobility by polyphosphates and polyamines, Proc. Natl. Acad. Sci. USA, 77:1457.
Smith, C.L., Ahkong, Q.F., Fisher, D., and Lucy, J.A., 1982, Is purified poly(ethylene glycol) able to induce cell fusion? Biochim. Biophys. Acta, 692:109.
Smith, D.K., and Palek, J., 1982, Modulation of lateral mobility of band 3 in the red cell membrane by oxidative cross-linking of spectrin, Nature Lond., 297:424.
Sowers, A.E., 1984, Characterization of electric field-induced fusion of erythrocyte ghost membranes, J. Cell. Biol., 99:1989.
Thomas, P., Limbrick, A.R., and Allan, D., 1983, Limited breakdown of cytoskeletal proteins by an endogenous protease controls Ca2+-induced membrane fusion events in chicken erythrocytes. Biochim. Biophys. Acta, 730:351.
Vienken, J., and Zimmermann, U., 1985, An improved electrofusion technique for production of mouse hybridoma cells, FEBS Lett., 182:278.
Verkleij, A.J., van Echteld, C.J.A., Gerritsen, W.J., Cullis, P.R., and de Kruijff, B., 1980, The lipidic particle as an intermediate structure in membrane fusion processes and bilayer to hexagonal hII transitions, Biochim. Biophys. Acta, 600:620.
Vos, J., Ahkong, Q.F., Botham, G.M., Quirk, S.J., and Lucy, J.A., 1976, Changes in the distribution of intramembranous particles in hen erythrocytes during cell fusion induced by the bivalent-cation ionophore A23187, Biochem. J., 158:651.
Wojcieszyn, J.W., Schlegel, R.A., Lumley-Sapanski, K., and Jacobson, K.A., 1983, Studies on the mechanism of polyethylene glycol-mediated cell fusion using fluorescent membrane and cytoplasmic probes, J. Cell. Biol., 96:151.
Zimmerberg, J., Cohen, F.S., and Finkelstein, A., 1980, Micromolar Ca2+ stimulates fusion of lipid vesicles with planar bilayers containing a calcium-binding protein, Science, 210:906.
Zimmermann, U., 1982, Electric field-mediated fusion and related electrical phenomena, Biochim. Biophys. Acta, 694:227.
Zimmermann, U., Pilwat, G., Péqueux, A., and Gilles, R., 1980, Electro-mechanical properties of human erythrocyte membranes: the pressure-dependence of potassium permeability, J. Membrane Biol., 54:103.
Zimmermann, U., Pilwat, G., and Richter, H-P., 1981, Electric-field-stimulated fusion: increased field stability of cells induced by pronase, Naturwissenschaften, 68: S 577.
Zimmermann, U., and Vienken, J., Electric field-induced cell-to-cell fusion, J. Membrane Biol., 67:165.

FUSION IN BIOLOGICAL AND MODEL MEMBRANES:

SIMILARITIES AND DIFFERENCES

Joshua Zimmerberg

Physical Sciences Laboratory, Division of Computer Research and Technology, National Institutes of Health
Bethesda, MD 20892

Fusion is an exotic phenomenon, attractive to study, poorly understood (cf. this volume). This chapter will focus on measurements of single fusion events in two systems with similar geometry -- phospholipid vesicles to planar phospholipid membranes and secretory granules to plasma membranes in exocytosis. Large expanses of phospholipid bilayer are joined together in both. Observation of particle-free pre-fusion zones has suggested that the role of protein in cell membrane fusion is to get out of the way, and allow lipid bilayer to contact and fuse (Ornberg & Reese, 1981; Chandler, 1987). This hypothesis suggests that the fundamental molecular event is the same in lipid and biological membrane fusion. However, in examining the details of the fusion reaction on the scale of tens of angstroms, we see major differences between phospholipid bilayer fusion and exocytotic fusion. The stable, hydrated pore which in exocytosis develops between the inside of the vesicle and the extracellular space is not usually detected in bilayer fusion. Osmotic swelling of the vesicle does not precede exocytotic fusion although it appears to drive phospholipid bilayer fusion. Lipid membranes fusion is irreversible; cell membranes seem to occasionally exhibit a reversible step before fusing. These observations suggest that some of the components of cell membranes other than lipids play crucial roles in the control, catalysis and molecular mechanism of cell membrane fusion.

We have come to this conclusion from experiments fusing a sphere to a plane: the geometry found in exocytosis. Our strategy has been to observe exocytosis to test an hypothesis based upon results from model systems. We have demonstrated fusion of phospholipid vesicles to planar phospholipid membranes (Zimmerberg et al, 1980a; Cohen et al, 1980). We have also observed exocytosis, and taken components away from cells until fusion is disrupted (Zimmerberg & Liu, 1987). We hope to add cellular components to lipid bilayers until fusion phenomenology is created. By understanding mechanisms in the defined system and observing the biological system with probes of increased definition, we hope the two approaches will converge.

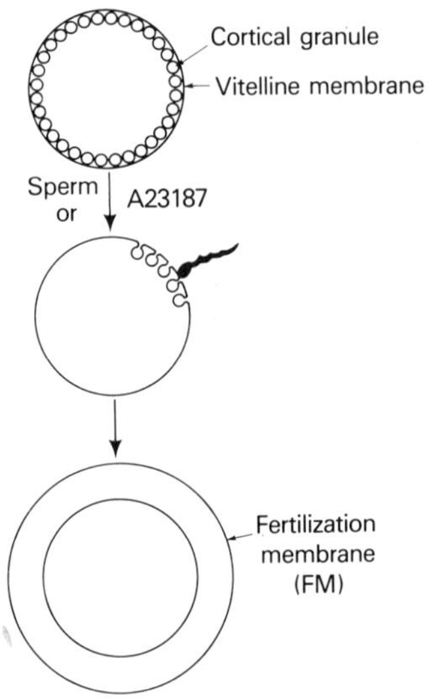

Fig. 1. A cartoon illustrating fertilization of an echinoderm egg. The inner surface of an egg is coated with 510,000 secretory vesicles termed cortical granules. Sperm induces a massive, synchronous exocytosis of these granules. Increasing intracellular calcium by any number of methods, including adding the calcium ionophore A23187 to the bathing sea water causes exocytosis as well. The result of exocytosis is the secretion of a number of hydrolases, peroxidases, glucanase, protease, hyalin and other substances. These in turn modify, elevate, and harden the vitelline envelope to form the fertilization membrane.

SEA URCHIN EGGS

In one system, a series of experiments have demonstrated how little of the cell is needed to see exocytosis. The inner surface of the echinoderm egg is covered with cortical granules. Upon fertilization, intracellular calcium rises and causes granules to fuse to the cell surface (figure 1)(Vacquier, 1981). Physiologically, the exocytotic reaction requires both intact sperm and egg. However, fragments of egg surface -- oolemma with adherent secretory granules -- will react to micromolar calcium in-vitro (Vacquier, 1975; Baker & Whitaker, 1978). This reaction leads to the sine-qua-non of exocytosis: membrane coalescence and vectorial content discharge (figure 2) (Whitaker & Baker, 1983; Zimmerberg et al, 1985). Thus cytoplasm per se is not required for fusion. No specific ionic requirements can be found, i.e. cortical granule exocytosis proceeds in raffinose or sucrose solutions upon the addition of calcium as readily as in solutions having the ionic composition of egg cytoplasm (Liu & Zimmerberg, 1985; Zimmerberg & Liu, 1987). Thus all of the components needed to observe exocytosis triggered by physiologic concentrations of calcium are contained in or adherent to the plasma membrane. Many inhibitors of cytoskeletal function have no effect on exocytosis in this system (Whitaker & Baker, 1983).

Further disruption of the planar cortex, by treating with sulfhydryl reagent, does inhibit exocytosis (Jackson et al, 1985). Washing the membranes with solutions of high ionic strength removes

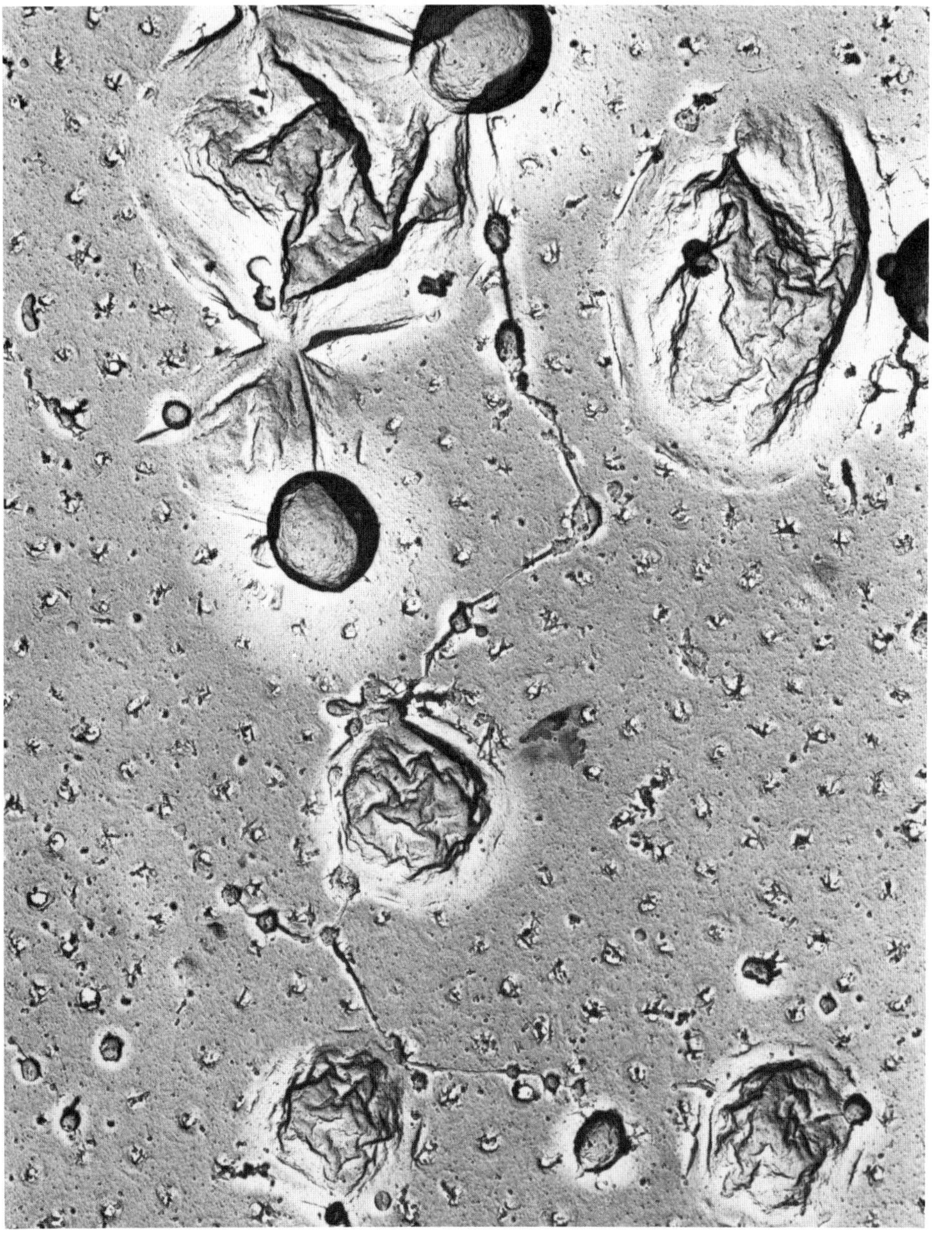

Fig. 2. Single and multiple fusion events. Replica of a fixed, critical point dried, rotary shadowed preparation of isolated planar cortex of S. purpuratus. The cortex was sheared forcibly during preparation to detach granules for clarity. The cortex was then exposed to 17 mM Ca^{++} for 30 seconds. Large domes have formed, continuous with the plasma membrane, as the result of exocytosis and hydration of secretory contents, detaching that region of compound membrane from the vitelline. These domes are visible in high contrast light microscopy. With a full complement of granules, the dome is continuous for each entire cortex and ultimately rips, releasing the contents to the bathing medium.

cortical granules (Kopf et al, 1982). Granules can be added back in low ionic strength solution and caused to fuse with 100 μM calcium, but the physiological response to 1 μM calcium is lost (Crabb & Jackson, 1985). Perhaps one of the proteins conferring calcium affinity is an extrinsic membrane protein, lost at high ionic strength. Such a protein had been observed in some species of sea urchin by Sasaki (Sasaki, 1984). Alternatively, the surface potential close to the calcium binding site may be altered, as with polycations (Crabb & Jackson, 1986; McLaughlin & Whitaker, 1986).

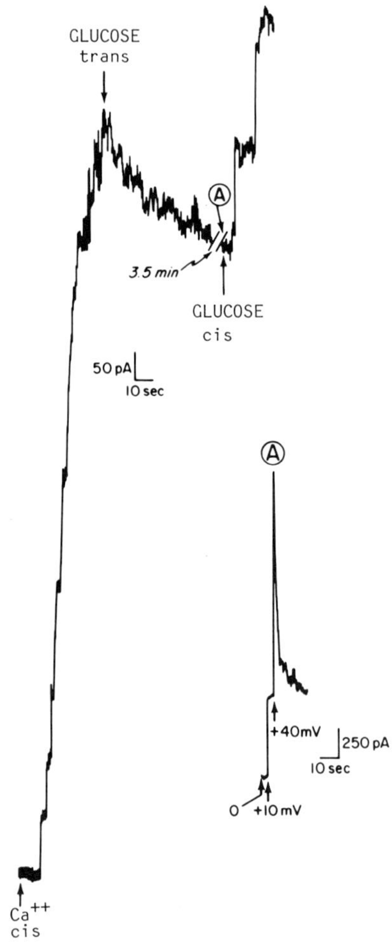

Fig. 3. Demonstration that osmotic gradients affect fusion. A membrane (4.5% asolectin, 5% DPPC) separating symmetrical salt solutions (100 mM NaCl, 10 mM MES, 0.1 mM EDTA, pH 6.0) and clamped at V = 10 mV has VDAC-containing vesicles in the <u>cis</u> compartment at a concentration of ~ 3X $10^8/cm^3$. Upon addition of 40 mM (104 mosM) $CaCl_2$ to the <u>cis</u> compartment, fusion proceeds. Addition of 104 mM glucose to the <u>trans</u> compartment immediately suppresses further fusion. (The small decline in conductance may represent inactivation of channels by decane.) Subsequent addition of 104 mM glucose to the <u>cis</u> compartment causes fusion to resume. The insert at (A) demonstrates that the conductance induced by fusion is characteristic of VDAC; i.e.; it turns off when the voltage is shifted from 10 to 40 mV. (From Cohen et al., 1980)

VESICLE/PLANAR MEMBRANE FUSION

The synthetic approach to studying membrane fusion centers on lipid bilayer membranes. Experimental schemes have been developed to study both the forces between membranes as they are forced together (see Rand and Parsegian, this volume), and to detect fusion of phospholipid membranes. Since this volume is replete with examples of vesicle-vesicle fusion I will discuss only vesicle-planar bilayer fusion, which I have recently reviewed (Zimmerberg, 1987a). Addition of phospholipid vesicles to one side of a planar phospholipid membrane does not lead to spontaneous fusion (Zimmerberg et al, 1980; Cohen et al, 1980; Cohen et al, 1984; Finkelstein et al, 1986). Conditions which lead to intimate adhesion of phospholipid membranes, such as millimolar calcium and negatively charged membranes, lead to a pre-fusion state (Cohen et al, 1984; Akabas et al, 1984). However, neither membrane coalescence nor content transfer is seen. Coalescence and transport of contents <u>are</u> seen when adherent vesicles are osmotically swollen (Cohen et al, 1980; Akabas et al, 1984; Cohen et al, 1982) (figure 3). Thus it is possible to produce membrane fusion with phospholipid bilayers.

OSMOTIC SWELLING IN EXOCYTOSIS

Since addition of calcium to secretory granules adherent to plasma membrane causes fusion, and addition of osmotic swelling to phospholipid vesicles adherent to planar membrane causes fusion, it is possible that calcium is acting to provoke swelling of cortical granules. This swelling can be measured directly in hypertonic milieu which slows the dispersal of the cortical granule core (Zimmerberg et al, 1985; Whitaker & Zimmerberg, 1987; Zimmerberg & Whitaker, 1985). However, in the planar cortex of the sea urchin egg, it is impossible to tell whether the swelling is a direct consequence of calcium action, and leads to fusion, or is a result of fusion (thereby exposing granule contents to sea water) (Green, 1978). This has been the sticking point in a number of systems where swelling is associated with exocytosis and shrinkage inhibits fusion (Finkelstein et al, 1986; Schuel, 1978). In one case membrane fusion was demonstrated without content swelling, but it was not possible to rule out the possibility of a transient swelling (Green, 1982).

We have recently measured secretory granule swelling and membrane fusion simultaneously during exocytosis of mast cells of beige mice (Zimmerberg et al, 1986). In every experiment membrane fusion preceded granule swelling (figures 4,5). Breckenridge and Almers (1987) have independently confirmed these findings. Thus it is impossible that swelling could have caused fusion in that system. Here, then, is a major difference between phospholipid and biological membrane fusion: Whereas we must swell phospholipid vesicles to get fusion, flaccid secretory granules fuse without swelling. Incorporation of channels from native vesicles to planar membranes has become a wide-spread technique for the study of transport, but fusion has not been demonstrated rigorously in those systems (Zimmerberg, 1987a; Cohen, 1986). To date, no protein or lipid combination has been found that, when added to the vesicle/planar membrane system, yields fusion of flaccid vesicles. By comparison, adding calcium-binding protein to planar phospholipid membranes has the effect of reproducing the calcium/magnesium specificity (Zimmerberg et al, 1980b). One then observes fusion at micromolar levels of calcium, but osmotic swelling of the adherent vesicle is still needed (figure 6). Thus other elements are needed for biological fusion, at least in some systems.

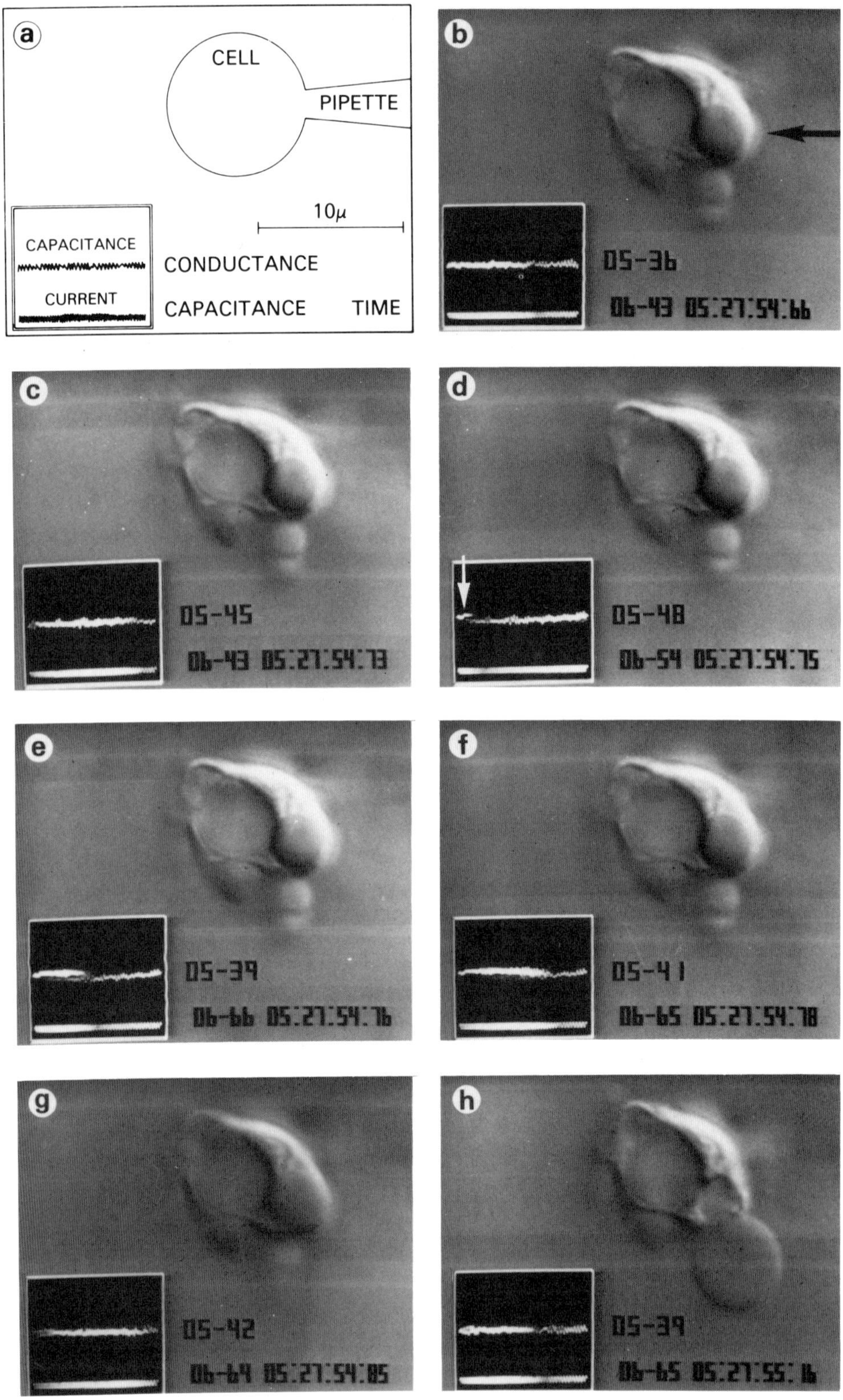
a
CELL
PIPETTE
10μ
CAPACITANCE
CURRENT
CONDUCTANCE
CAPACITANCE
TIME
b
c
d
e
f
g
h

Fig. 4. The increase in capacitance occurs before granule swelling is detected. Fig. 4(a) is a key for the video fields that follow. From left to right, in the lower half of panels (b)-(h), one sees the oscilloscope traces of capacitance, and the eight-digit timer with hr, min, s, and s x 10^{-2} . Above is the cell image with the patch pipette located as in (a); the image was taken with a Dage 65 video camera and stored on tape (Sony VO2850 U-matic video cassette recorder). Sequential frames of a typical experiment are shown. In (b) and (c) no activity occurs, but these fields provide a reference for granule size. The arrow points to the granule. In (d) capacitance changed from 643 to 654, and this jump is seen on the capacitance record of the oscilloscope tracing (arrow). In (e) the capacitance has risen to 666, but no granular swelling can be detected by morphometry. In the succeeding three fields ((f), (g), and (h)) swelling can be detected with no further change in capacitance. Conductance transiently changes during the experiment, but no average change corresponds with the fusion event. Capacitance calibration, 100 = 1 pF. Oscilloscope parameters: 50 ms/trace, persistence of 50 ms. (From Zimmerberg et al., 1987)

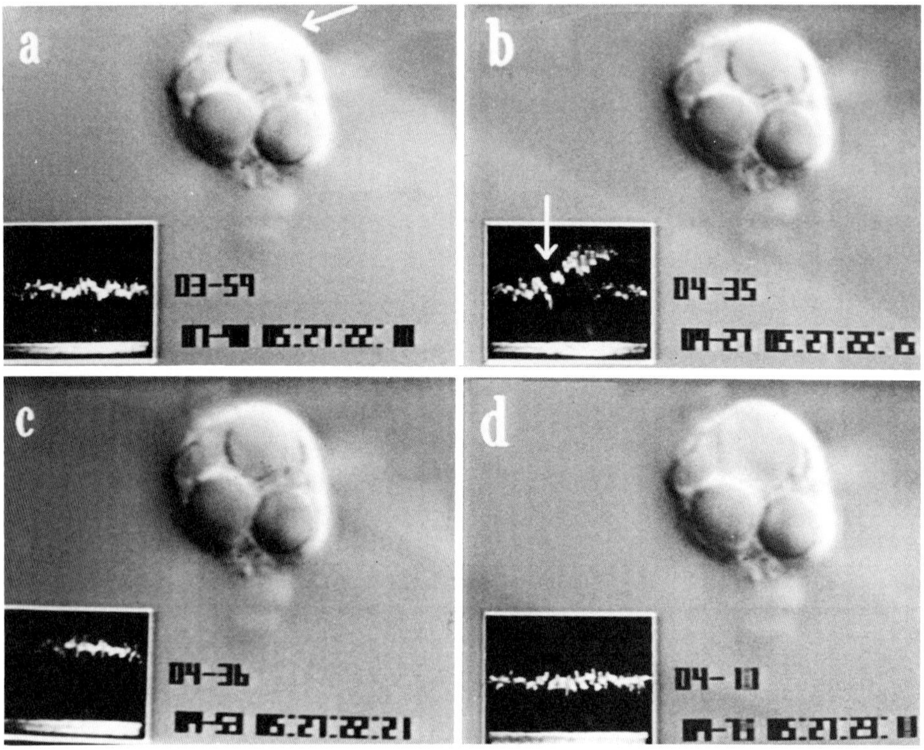

Fig. 5. Membrane fusion precedes granule swelling during exocytosis in beige mouse mast cells. Sequential fields of one experiment are shown. Cells are patch-clamped in the whole-cell configuration and capacitance is measured with a phase-sensitive detector. The cell is simultaneously observed with differential interference contrast and video-enhanced microscopy. The top of the border to the oscilloscope tracing is 10 m; magnification to screen was 13,400. The top oscilloscope tracing (A.C.-coupled, 50 msec/sweep) is capacitance, the bottom is current. The top digits in the middle are conductance and the bottom capacitance, 100 units = 200 femptofarads. Time is at the bottom right; the changing times are seconds and hundredths of seconds. In panel a the resting capacitance of the cell is seen with an arrow at the granule which will soon fuse. In panel b the capacitance change due to fusion of the secretory granule to the plasma membrane is evident (arrow). In panel c the capacitance change is almost finished but there is no vesicular swelling. Swelling is evident at 22:50, and obvious at 23:11, panel d. (From Zimmerberg, 1987a)

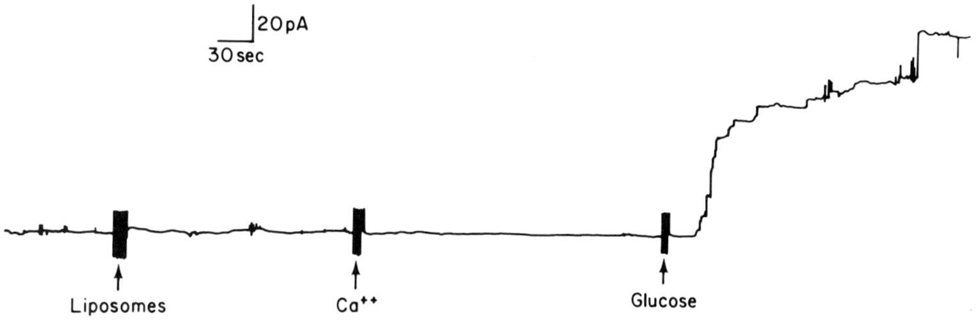

Fig. 6. Demonstration of the requirement of an osmotic gradient for fusion of phospholipid vesicles with a planar bilayer membrane containing calcium-binding protein. The membrane, clamped at V = 10 mV, separates symmetrical solutions (100 mM NaCl, 10 mM MES, 0.1 mM EDTA, pH 6.0). At the first arrow, vesicles (liposomes) are added to the *cis* compartment to a concentration of ~ 3 x 10^8 vesicles per milliliter, and at the second arrow Ca^{2+} is added to the *cis* compartment to a free ion concentration of 10 μM. Note that no fusion events occur in either the 3-minute period prior to Ca^{2+} addition or in the 4-minute period following Ca^{2+} addition. Within 30 seconds after addition of glucose (third arrow) to the *cis* compartment to a concentration of 104 mM, fusion proceeds at a rapid rate. A membrane was formed at room temperature by the brush technique (20) across a 1-mm^2 hole in a Teflon partition; the membrane-forming solution contained 4.5 percent asolectin, 0.5 percent diphytanoyl phosphatidylcholine DPPC), and 0.5 percent calcium-binding protein. (From Zimmerberg et al., 1980b)

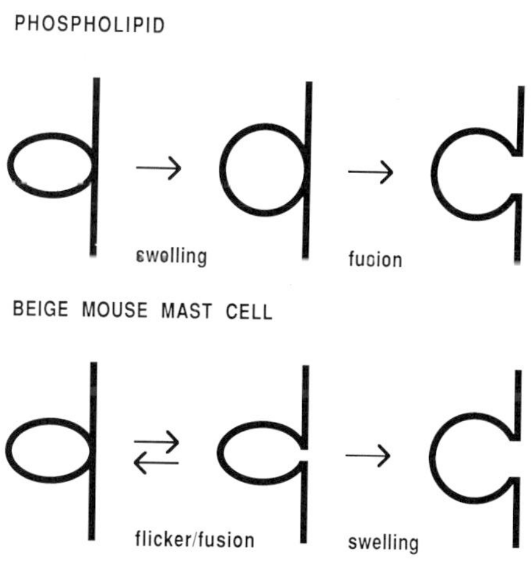

Fig. 7. Differences between artificial and natural membrane fusion. Phospholipid vesicles first adhere to planar membranes, then fuse if swollen to a taut and bursting state. Flaccid histamine-containing secretory granules of the beige mouse mast cell will first fuse to the plasma membrane before swelling is detected, swelling seems to follow exposure of granule contents to the extracellular medium. Flicker fusion is sometimes seen, perhaps a reversible aspect of fusion finalized by swelling.

CONTROL IN BIOLOGICAL FUSION

For what are these components needed? Consider the remaining differences between the fusion phenomena. First, fidelity. While phospholipid fusion to planar bilayers is accompanied by about 50% lysis without fusion (Niles & Cohen, 1987), biological fusion is presumably well controlled to prevent leakage, especially when toxic compounds are safely sequestered within secretory granules. In fact, the word exocytosis stems from the biological advantage of having an intracellular space that is outside of the cell. Many cellular events are under the strict control of feedback systems, often involving phosphorylation of allosteric proteins to change activity.

Second, reversibility. In phospholipid membrane fusion, the fusion event occurs irreversibly in one step. Recent experiments using capacitance to measure the increase in cell surface when a granule adds to that surface through fusion suggest a reversible component to membrane fusion, termed flicker fusion (Fernadez et al, 1984; Zimmerberg et al, 1986).

Third, initial pore formation is instantly wide with phospholipid membranes but is stable at much smaller sizes for measurably longer times in exocytotic fusion (Zimmerberg, 1987b; Zimmerberg et al, 1986).

Some of these differences, role of swelling, fidelity, and reversibility, are summarized in figure 7. The last, pore formation, has only recently been measured (Zimmerberg et al, 1986; Breckenridge and Almers, this volume).

EXOCYTOTIC PORE FORMATION

The basis of the measurement of pore development during exocytosis is seen in figure 8. In order to see the increase in cell capacitance which occurs when a granule fuses to the cell surface, ionic charge must be able to get in and out of the granule from the external medium. The size of the exocytotic pore determines the ease of ionic charge motion. Over a limited range, the amount of the capacitance change detected is a measure of the size of the pore. Since we measure the capacitance as a function of time, we can derive the size of an equivalent pore, assuming bulk conductivity.

We have derived the equivalent circuit for the electrical changes which occur during exocytosis (Zimmerberg et al, 1986). We assume no changes in the series resistance of the pipette and no other changes in cell membrane capacitance and conductance during the course of the fusion event. Since the real and imaginary components are linearly independent, and the only unknown is pore resistance, this gives two sets of data to fit the pore resistance. Thus our measurement of the exocytotic pore is based upon two independent data sets.

We see changes in size equivalent to a pore 100 Å in length varying between 18 and 52 Å in diameter. The timing of pore formation is variable. Some pores open instantly (<3 ms). Others open slowly, over 100 ms. Others open in steps, stable for a while and then open more continuously over many 10's of ms. Equivalent pore formation in phospholipid membrane fusion is much faster, and cannot be detected. We see instantly the full value of channels inserted via fusion of vesicles, not a signal attenuated by a series pore resistance. An equivalent pore would be more than 300 Å in less than 200 ms (Cohen et al, 1980; Zimmerberg, 1987). Thus the detailed motion of membrane components is very different in model and biological membrane fusion.

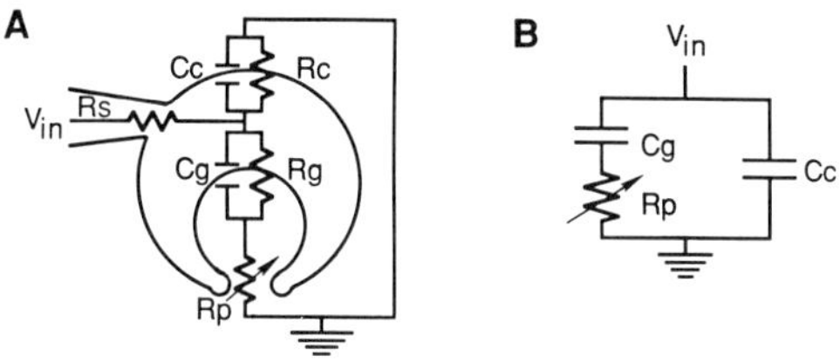

Fig. 8. A. Sketch of the exocytotic pore that has resulted from fusion. Superimposed on the cartoon is the equivalent circuit. R_c and C_c are the resistance and capacitance of the cell membrane, R_g and C_g are the resistance and capacitance of the granule membrane, and R_p is the resistance of the exocytotic pore. The series resistance of the patch electrode is R_s; the command voltage is Vin.

B. The simplified equivalent circuit. Rc and Rc can be neglected for a sine wave of 806 Hz because $R_{c,g} >> 1/2\pi fC_{c,g}$. The change in admittance as a result of fusion is given in complex notation by

$$\Delta Y(\omega) = R_p(\omega C_g)^2/((\omega C_g R_p)^2 + 1) + j(\omega C_g/((\omega C_g R_p)^2 + 1).$$

Note that as a result of fusion, R_p varies from infinity, before fusion, to a small value after fusion. Fusion does not cause any net change in the real part of the admittance. But during the time that the capacitance increases, R_p decreases, and the real part of the admittance increases and then decreases. The maximum possible change (due to varying Rp) in the real component of the admittance equals $\omega C_g/2$, which is one-half the final change of the imaginary component, ωC_g. The experimental conductance transients (real component of the admittance) are in quantitative agreement with this theoretical prediction, lending support to this model. (From Zimmerberg et al., 1987)

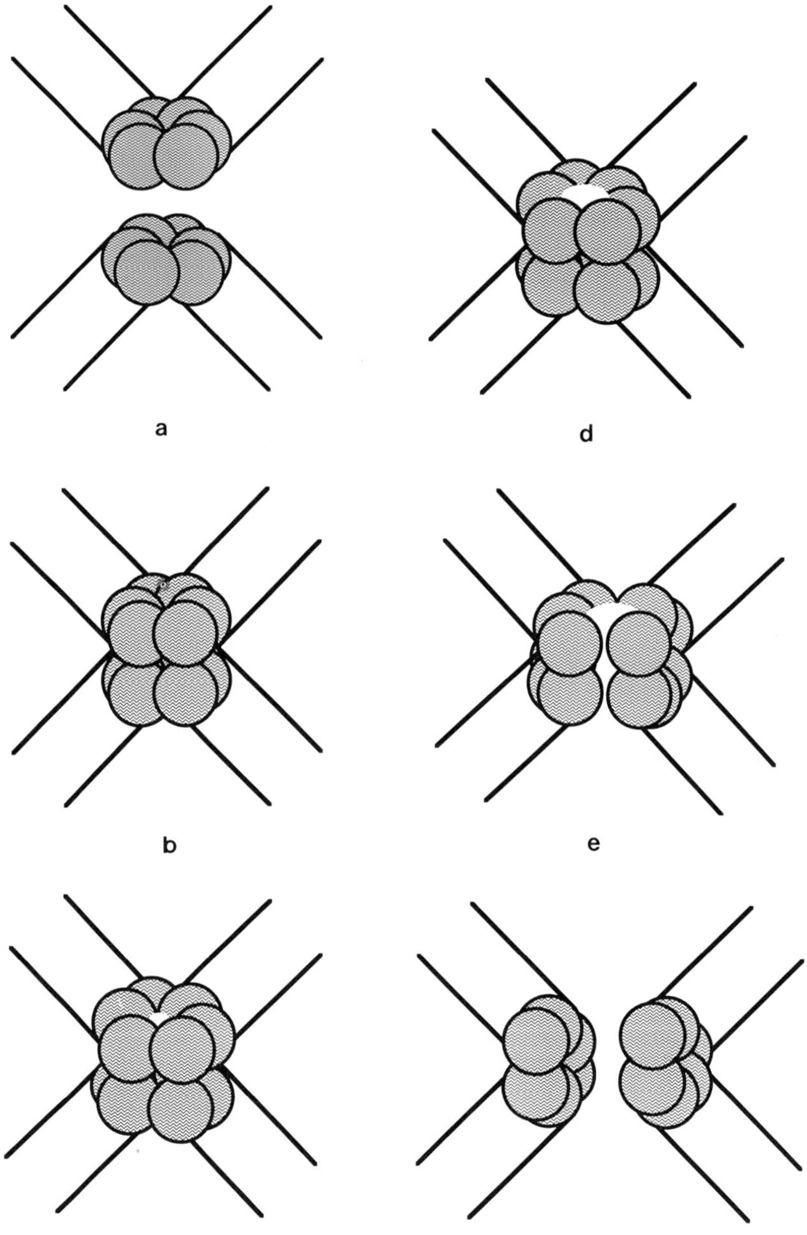

Fig. 9. Protein-pore model of membrane fusion (speculation). Two membranes are initially distinct with happily hydrated surfaces (a). The appropriate biological trigger (e.g. Ca^{++}, GTP-γ-S) causes a conformational change in a putative multi-subunit fusion protein causing approach and adhesion because of attractive forces between complementary surfaces (b). Other portions of the protein hydrate in response to the biological trigger, forming a hydrated pore between the top and bottom surfaces (c). This pore widens (d) and can exist stably as it widens (e), or shut again (flicker) (c). The subunits ultimately rearrange and the new hydrated surfaces separate (f), completing fusion.

CONVERGENCE?

Having described these differences between fusion of phospholipid vesicles to planar membranes and fusion of secretory granules to plasma membrane, we return to the question of the role of the other membrane components. We have no data on the role of gangliosides and sterols, except that their incorporation into planar bilayers does not affect the basic response. Adhesion, dehydration, and osmotic swelling of the vesicle is needed for its fusion to a planar mosaic membrane. Protein may be thought of as playing a crucial role in exocytotic fusion. Perhaps protein can act as a wedge, providing the separation of phospholipids effected in model membranes with osmotic swelling. Perhaps it acts as a phospholipase, creating lipids which need to separate from the lamellar phase (see D Siegel, this volume). Perhaps it is the protein itself which first fuses and separates, only involving lipid at a much later stage (figure 9). By reconstituting proteins from biological membranes into planar and vesicular phospholipid membranes, we hope to have a set of defined membranes which fuse with the kinetics seen physiologically and which will not need osmotic swelling for fusion.

REFERENCES

Akabas, M.H., Cohen, F.S., and Finkelstein, A., 1984, Separation of the osmotically driven fusion event from vesicle-planar membrane attachment in a model system for exocytosis, J. Cell Biol., 98:1063.

Baker, P.F., and Whitaker, M.J., 1978, Influence of ATP and calcium on the cortical reaction in sea urchin eggs. a, Nature (Lond.), 276:513.

Breckenridge, L.J., and Almers, W., 1987, Final steps in Exocytosis observed in a cell with giant secretory granules, Proceedings of the National Academy of Sciences, 84:1945.

Chandler, D.E., 1987, Exocytosis and endocytosis: membrane fusion events captured in rapidly frozen cells, Curr. Top. Membr. Trans., in press.

Cohen, F.S., 1986, Fusion of liposomes to planar bilayers, in: " Ion Channel Reconstitution" Miller, C. ed., Plenum Press, New York.

Cohen, F.S., Zimmerberg, J., and Finkelstein, A., 1980, Fusion of phospholipid vesicles with planar phospholipid bilayer membranes. II. Incorporation of a vesicular membrane marker into the planar membrane, Journal of General Physiology, 75:251.

Cohen, F.S., Akabas, M.H., and Finkelstein, A., 1982, Osmotic swelling of phospholipid vesicles causes them to fuse with a planar phospholipid bilayer membranes, Science, 217:458.

Cohen, F.S., Akabas, M.H., Zimmerberg, J., and Finkelstein, A., 1984, Parameters affecting the fusion of unilamellar phospholipid vesicles with planar bilayer membranes, J. Cell Biol., 98:1054.

Crabb, J.H., and Jackson, R.C., 1985, In-vitro reconstitution of exocytosis from plasma membrane and isolated secretory vesicles, J. Cell Biol., 101:2263.

Crabb, J.H., and Jackson, R.C., 1986, Polycation inhibition of exocytosis from sea urchin egg cortex, J. Membrane Biol., 91:85.

Fernadez, M., Neher, E., and Gomperts, B.D., 1984, Capacitance measurements reveal stepwise fusion events in degranulating mast cells, Nature (London), 312:453.

Finkelstein, A., Zimmerberg, J., and Cohen, F.S., 1986, Osmotic swelling of vesicles: its role in the fusion of vesicles with planar phospholipid bilayer membranes and it possible role in exocytosis, Annual Review of Physiology, 48:163.

Green, D.P.L., 1978, The induction of the acrosome reaction in guinea-pig sperm by the divalent metal caption ionophore A23187, J. Cell Sci., 32:137.

Green, D.P.L., 1982, The course of the acrosome reaction in guinea-pig sperm, J. Cell Sci., 54:161.

Jackson, R.C., Ward, K.K., and Haggerty, J.G., 1985, Mild proteolytic digestion restores exocytotic activity to N-ethymaleimide-inactivated cell surface complex from sea urchin eggs, J. Cell. Biol., 101:6.

Kopf, G.S., Moy, G.W., and Vacquier, V.D., 1982, Isolation and characterization of sea urchin egg cortical granules, J. Cell Biol, 95:924.

Liu, J., and Zimmerberg, J., 1985, Calcium-stimulated secretion in sea urchin eggs: exocytosis proceeds in vitro in the absence of other ions, Biological Bulletin, 169:562.

McLaughlin, S., and Whitaker, M., 1986, A Clue to the location of the Ca-binding site(s) responsible for exocytosis and phosphotidylinositol 4,5-bisphosphate (PIP2) Hydrolysis in an isolated plasma membrane preparation from sea urchin eggs, Biophysical Journal, 49:133a.

Niles, W.D., and Cohen, F.S., 1987, Video fluorescence microscopy studies of phospholipid vesicle fusion with a planar phospholipid membrane. Nature of membrane - membrane interactions and detection of contents release, J. Gen. Phys., in press.

Ornberg, R.L., and Reese, T.S., 1981, Beginning of exocytosis captured by rapid-freezing of Limulus amoebocytes. J. Cell Biol., 90:40.

Sasaki, H., 1984, Modulation of calcium sensitivity by a specific cortical protein during sea urchin egg cortical vesicle exocytosis, Dev. Biol., 101:125.

Schuel, H., 1978, Secretory functions of egg cortical granules in fertilization and development: a critical review, Gamete Res., 1:299.

Vacquier, V.D., 1975, The isolation of intact cortical granules from sea urchin eggs: calcium ions trigger granule discharge, Dev. Biol., 43:62.

Vacquier, V.D., 1981, Dynamic changes of the egg cortex, Dev. Biol., 84:1.

Whitaker, M.J., and Baker, P.F., 1983, Calcium-dependent exocytosis in an in vitro secretory granule plasma membrane preparation from sea urchin eggs and the effects of some inhibitors of cytoskeletal function, Proc. R. Soc. London B, 218:397.

Whitaker, M., and Zimmerberg, J., 1987, Inhibition of exocytosis in sea urchin eggs by polymer solutions, Journal of Physiology,:in press.

Zimmerberg, J., 1987a, Fusion of phospholipid vesicles to planar phospholipid membranes, in: "Cellular Membrane Fusion" Wilschut. J., and Hoekstra, D. ed., Marcel Dekker, New York

Zimmerberg, J., 1987b, Simultaneous electrical and optical measurements of membrane fusion during exocytosis of beige mouse mast cells, Bioscience Reports, in press.

Zimmerberg, J., and Liu, J., 1987, Ionic and permeability requirements for exocytosis in vitro in sea urchin eggs, J. Membrane Biology, :in press.

Zimmerberg, J., and Whitaker, M., 1985, Irreversible swelling of secretory granules during exocytosis caused by calcium, Nature, 315:581.

Zimmerberg, J., Cohen, F.S., and Finkelstein, A., 1980a, Fusion of phospholipid vesicles with planar phospholipid bilayer membranes. I. Discharge of vesicular contents across the planar membrane, Journal of General Physiology, 75:251.

Zimmerberg, J., Cohen, F.S., and Finkelstein, A., 1980b, Micromolar concentrations of calcium stimulate fusion of phospholipid vesicles with planar phospholipid bilayer membranes when a calcium binding protein is present in the planar membrane, Science, 210:906.

Zimmerberg, J., Sardet, C., and Epel, D., 1985, Exocytosis of sea urchin egg cortical vesicles in vitro is retarded by hyperosmotic sucrose: kinetics of fusion monitored by quantitative light-scattering microscopy, Journal of Cell Biology, 101:2398.

Zimmerberg, J., Curran, M., Cohen, F.S., and Brodwick, M., 1986, Simultaneous electrical and optical measurements show that membrane fusion precedes secretory granule swelling during exocytosis of beige mouse mast cells, PNAS, 84:1585.

EARLY STEPS IN THE EXOCYTOSIS OF SECRETORY VESICLES IN MAST CELLS

W. Almers and L.J. Breckenridge

Department of Physiology and Biophysics
University of Washington Medical School
Seattle, Washington

REGULATION OF SECRETION IN MAST CELLS

Under physiologic conditions, mast cells are stimulated to secrete ("degranulate") when a multivalent antigen simultaneously binds two or more IgE antibodies that, in turn, are bound to IgE receptors in the mast cell membrane. The antigen-antibody binding apparently brings IgE-receptors into close proximity so that they form dimers or oligomers. Since the IgE receptor is a membrane-spanning protein, this oligomerization communicates itself to the cytoplasmic face of the membrane, and thereby initiates a chain of events that ultimately leads to degranulation (Ishizaka & Ishizaka, 1984). Any maneuver leading to oligomerization or clustering of IgE receptors induces degranulation. Probably the first effect of IgE-receptor oligomerization is to activate a phospholipase-C, probably via GTP binding to a G-protein (Gomperts, 1986).

An increase in cytosolic [Ca^{2+}] (White et al., 1984) has long been thought to be a crucial event in stimulus-secretion coupling (e.g. Forman et al., 1973). In single mast cells loaded with the Ca-indicator Fura-2, stimulation with antigen is seen to cause an explosive increase in cytosolic [Ca^{2+}], from a resting level of about 150 nM to several micromolar (Almers & Neher, 1985). [Ca^{2+}] then declines again, in some cells completely and in others to a level of about 1 μM that is often maintained for minutes (Neher & Almers, 1986). Hence in some cells, the increase in cytosolic [Ca^{2+}] has a transient and a maintained component. Earlier work (e.g. Mazurek et al., 1984) has suggested that the increase in cytosolic [Ca^{2+}] is produced by Ca-influx through a Ca channel activated by the antigen-antibody-IgE-receptor complex , but this view could not be confirmed in measurements on single cells. Patch-clamp studies have failed to reveal an antigen-induced Ca current (Lindau & Fernan dez, 1986a,b), and full-size transient increases in cytosolic [Ca^{2+}] as well as degranulation are produced by antigen even in the absence of external Ca (Neher & Almers, 1986). Evidently, the transient component is due to release from internal stores, probably as a result of inositoltrisphosphate liberation. Nonetheless, Ca-influx through a transmembrane channel may well contribute to the frequently observed maintained component of [Ca^{2+}] increase. The cell membranes of both neutrophils (von Tscharner et al., 1986) and mast cells (Lindau & Fernandez, 1986a) carry nonselective cation channels that are activated by cytosolic [Ca^{2+}] and, at least in neutrophils, are permeable to Ca. In mast cells, these channels carry inward current when they are activated by cytosolic [Ca^{2+}] (Almers & Neher, 1987) and contribute to the increase in membrane conductance that often accompanies degranulation (Lindau & Fernandez, 1986). It is possible that similar channels mediate the depolarization-inhibited Ca-influx in rat basophilic leukemia cells (Mohr & Fewtrell, 1987). Ca influx through such channels may well be important for degranulation by maintaining a high cytosolic [Ca^{2+}] after the transient release from internal stores has run its course. However, there is no evidence that these channels are coupled directly to the antigen-IgE-receptor complex; their activation by cytosolic Ca^{2+} would be a fairly remote consequence of antigen binding.

Cytosolic [Ca^{2+}] is not the only variable regulating secretion in mast cells or other granulocytes. In fact, large [Ca^{2+}] increases may be observed without degranulation, and external application of the (presumably membrane-impermeant) secretagogue 48/80 may, under some conditions, induce degranulation without an intervening increase in cytosolic [Ca^{2+}] (Neher & Almers, 1986). Probably, degranulation requires, in addition, the phosphorylation of an unknown substrate by protein kinase-C, a membrane-associated enzyme activated by diacylglycerol, the other product of phosphoinositide breakdown (Nishizuka et al., 1984). Furthermore, recent studies on single cells have provided clear evidence that the secretion-controlling machinery must first be sensitized before it can respond to cytosolic [Ca^{2+}], and that a G-protein is required in this sensitization (Neher, 1987). Probably, both phosphorylation and Ca binding are early steps in exocytosis (see also Gomperts, 1986).

MONITORING EXOCYTOSIS OF SINGLE VESICLES BY CAPACITANCE MEASUREMENTS

Whenever a vesicle fuses with the cell membrane, the cell membrane area must increase. Since all biological membranes have a capacitance of about 1 $\mu F/cm^2$, the area of the membrane surrounding a single cell may be assayed by measuring the cell membrane capacitance, C_m. A large advance was made when Neher & Marty (1982) combined gigaseal recording with the use of a lock-in amplifier to measure C_m changes at femtofarad resolution. With this method, they observed C_m changes representing the fusion of single secretory vesicles with the cell membrane. Fig. 1 shows a recording of C_m versus time, made with a lock-in amplifier on a rat peritoneal mast cell. The trace in Fig. 1 starts just after a firepolished glass micropipette, filled with a solution designed to mimick the cytoplasm, had been sealed onto the mast cell. The membrane patch beneath the pipette orifice was then ruptured by a pulse of suction (first arrow) so that the pipette interior became continuous with the cytosol ("whole cell" configuration, Hamill et al., 1982). Now the cell could be voltage clamped. At the second arrow, a sine wave (800 Hz) was supplied both to the command input of the voltage clamp and to the lock-in amplifier; from this time onwards the output of the lock-in amplifier tracks C_m.

Besides ATP, NaCl, traces of EGTA and buffer, the pipette solution contained 20 μM GTPgammaS, a non-hydrolyzable analog of GTP. GTPgammaS irreversibly activates G-proteins and stimulates secretion in mast cells (Fernandez et al., 1984) even in the absence of antigen-IgE-receptor complex formation. Within 1 min after patch rupture, small stepwise increments in capacitance appeared, became increasingly frequent and eventually led to an explosive, 2-4 fold increase in C_m (off scale in Fig. 1). Each stepwise increment results from the fusion of a single secretory vesicle with the cell membrane (Fernandez et al., 1984). Superimposed on the steps, there was also a slower decline that showed no indication of being stepwise at the resolution of this method. This decline was induced by the increase in cytosolic [Ca^{++}] (Almers & Neher, 1987) that accompanies mast cell activation and may represent a burst of Ca-activated pinocytosis.

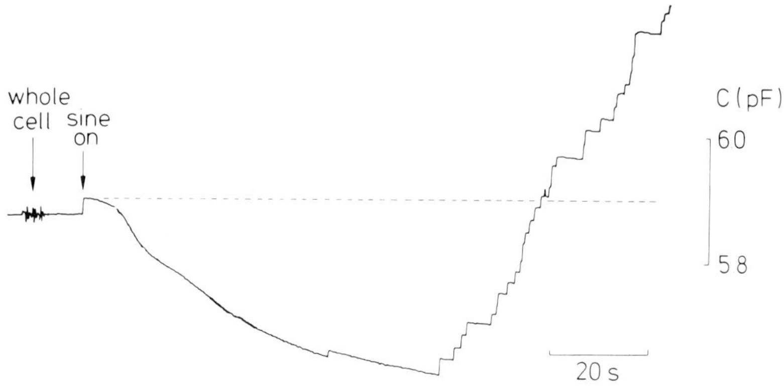

Fig. 1. C_m during the early phase of degranulation in a rat peritoneal mast cell. C_m was measured as the component of the sinusoidal current whose phase leads the voltage sinusoid by 90° (Neher & Marty, 1982). From Almers & Neher, 1987.

DOES MECHANICAL STRESS INITIATE MEMBRANE FUSION IN EXOCYTOSIS?

Exocytosis involves membrane fusion, an event that takes place in nearly every cell. Experiments as in Fig. 1 tell us when exocytosis of single vesicles occurs, but not how.

Actually, biological membranes are notoriously disinclined to fuse. For example, many cells are connected by gap junctions and hence their membranes approach each other within molecular dimensions. Yet they never fuse. A resting mast cell is so tightly packed full of secretory granules that many granules touch each other or the cell membrane, yet fusion and exocytosis happen only when the cell is stimulated. The myelin sheath of myelinated nerve is another example with its indefinitely stable, extensive and extremely close apposition of adjacent Schwann cell membranes. The innate resistance of biological membranes to fusion is a direct consequence of the architecture of the lipid bilayer (see, e.g., Rand, 1981). While bilayers (and biological membranes) are encouraged by van der Waals attractive forces to stick together, a small separation between them is always maintained. The reasons are (1) electric repulsion provided by the small amount of charged lipids that is usually present in biological membranes, and, more importantly, (2) the "hydration force" that tends to retain a 2-3 nm thick layer of water molecules covering the highly hydrophilic lipid head groups (see Rand & Parsegian, 1986; Ohki et al., this volume).

Nonetheless, lipid vesicles can be made to fuse, both with each other (e.g. Bearer et al., 1982) and with planar lipid bilayers (e.g. Cohen et al., 1984; Akabas et al., 1984). It is thought that during bilayer fusion, lipids transiently assume non-bilayer conformations, such as "hexagonal phase" or "inverted micellar intermediates". Such structures may have been observed in fusing lipid vesicles (e.g. Hui et al., 1981). The force that creates such nonbilayer conformations is thought to be the lateral stress on the bilayer that results either from osmotic pressure gradients (e.g. Akabas et al., 1984) or when originally spherical vesicles are forced to increase their surface-to-volume ratio because their membranes adhere to that of other vesicles and try to form large planar areas of close contact (see, e.g., Rand

& Parsegian, 1986). A possibly close relationship between the fusion of pure lipid bilayers and that of biological membranes was suggested by the observation that in many systems, exocytosis is associated with granule swelling. However, the need for swelling has not been clearly established (Finkelstein et al., 1986), and may not exist in chromaffin cells (Holz, 1986).

The point may be directly tested in beige mice, a murine mutant where mast cells have secretory granules so abnormally large that, under a light microscope, they may be seen individually to swell during exocytosis (Curran et al., 1984). Zimmerberg et al. (1987) and ourselves (Breckenridge & Almers, 1987) have made simultaneous video recordings and C_m measurements on degranulating mast cells of beige mice. Stepwise increases in C_m were observed to accompany the swelling of individual granules. Fig. 2 shows a brief episode of degranulation in a single cell, during which one secretory granule underwent exocytosis. It plots C_m against time (continuous line) as well as the diameter of one of the granules (dots), measured on polaroid photographs taken of successive video frames. In this particular experiment the pipette solution had twice the tonicity of the external fluid and of the cytosol. This caused the granule to shrink during the first 30 s of whole-cell contact. Later the granule swelled again, but the swelling was preceded by the C_m step. Clearly, C_m measurements report fusion long before the first signs of vesicle swelling.

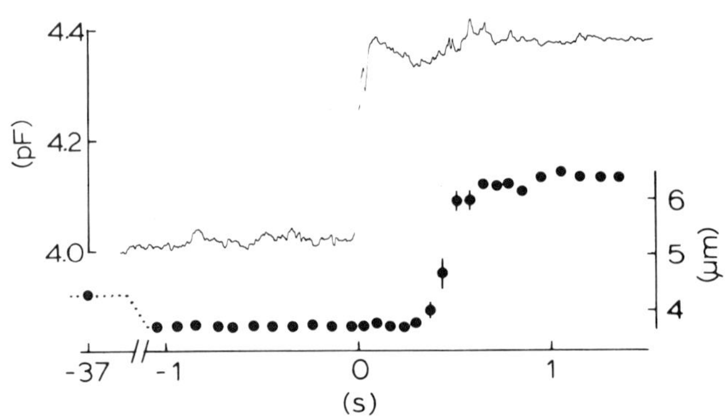

Fig. 2. Beige mouse mast cell. C_m (upper trace) during exocytosis of the vesicle whose diameter is indicated by the dots (lower trace). Degranulation triggered by 20 μM GTPgammaS in the pipette. Experimental procedures and analysis as in Figs. 4, 5 of Breckenridge & Almers (1987a), except that 300 mM sucrose was added to the solution in the pipette to increase the tonicity of the pipette solution to about twice normal.

It could be argued that a very slight swelling may have occurred before the C_m step but was missed because of the limited optical resolution. If so, however, the swelling was so small that it did not even take up the slack introduced in the vesicle membrane by previous osmotic shrinkage. Hence membrane fusion during exocytosis in mast cells is not driven by stress on the vesicle membrane.

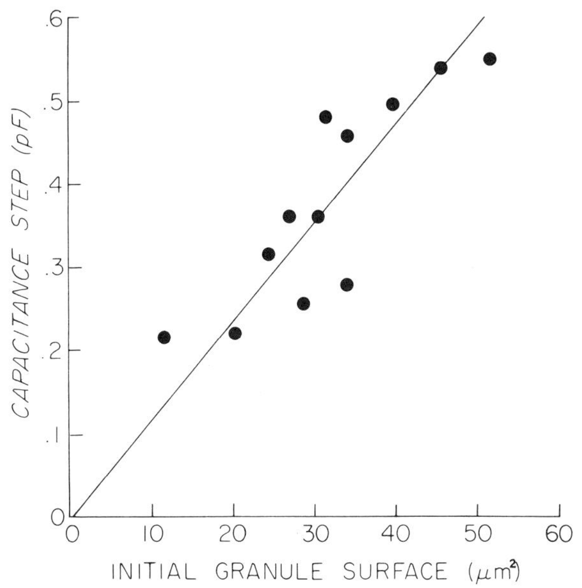

Fig. 3. Final amplitude of C_m steps, as in Fig. 2, plotted against the surface area of the vesicle that was observed to swell immediately afterwards. The surface area was calculated from the diameter of the vesicle measured before swelling started. Experimental procedures and analysis as in Fig. 2 except that the pipette solution was isotonic to avoid vesicle shrinkage.

Fig. 3 provides additional evidence that the C_m steps really represent exocytosis of the vesicles that were subsequently observed to swell. These experiments were done with an isotonic solution in the pipette, so that granule shrinkage was avoided. Fig. 3 shows that the C_m step at the moment of exocytosis is strongly correlated with the vesicle membrane surface, calculated from the initial diameter of the exocytosing mast cell granule. The correlation coefficient was 0.87, and the chance of observing this correlation by accident is less than 0.001. The regression line through the origin had a slope of 1.17 $\mu F/cm^2$, in agreement with the capacitance universally observed in biological membranes.

That the C_m step precedes vesicle swelling and can occur even in flaccid vesicles indicates that exocytosis in mast cells is initiated by mechanisms very different from those that initiate the fusion of pure lipid vesicles with planar lipid bilayers. Forces other than osmotic gradients or mechanical stress must drive the early steps in membrane fusion.

CAPACITANCE FLICKER

Possible clues about an early step came from the curious behavior of some degranulating mast cells. While results as in Fig. 1 are observed about 80% of the time, Fig. 4 represents, on a faster time scale, a less common response to identical experimental conditions. The first C_m step in this cell (arrow) was not maintained; instead, C_m returns to the original level, remains there for 15 s and then increases again, this time permanently (see also Fernandez et al., 1984). Also, the first step shows a "ragged" appear-

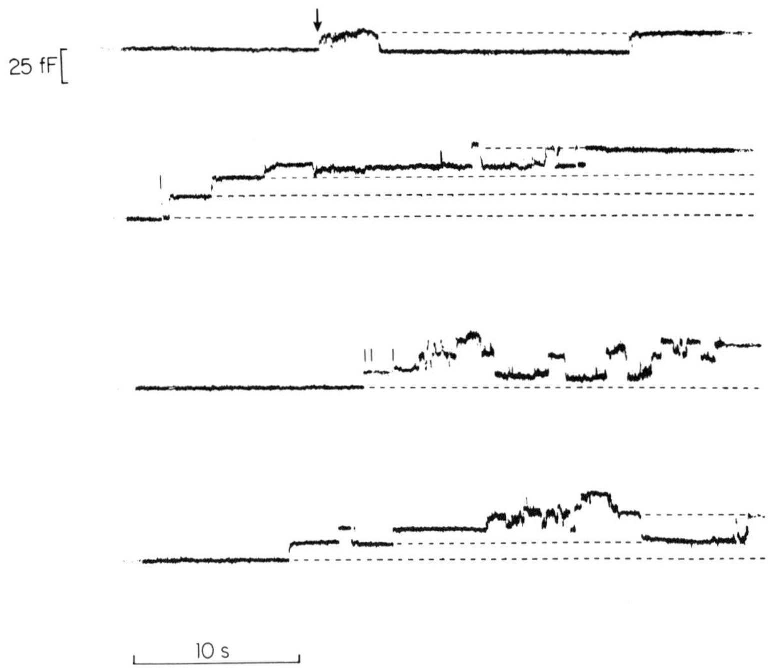

Fig. 4. Rat peritoneal mast cell. Conditions as in Fig. 1, except that the pipette contained 150 mM K-glutamate in place of NaCl, as well as 1 mM ATP and 0.5 mM $MgCl_2$. The four traces shown form a continuous record. Dashed lines indicate capacitance levels; the dashed line in the first trace represents the same capacitance as the lowermost dashed line in the second trace; the uppermost dashed line in the second trace represents the same capacitance as the lowermost dashed line in the third trace. Small vertical lines indicate transitions that were observed on the oscilloscope, but were too brief to reproduce on this chart recording. The downward slope of the traces is due to a Ca-dependent, gradual C_m decline that may represent pinocytosis. From Almers & Neher, unpublished.

ance, familiar in single-channel recordings where it signifies a flickering between two levels that is too fast to be temporally resolved. Rapid and repeated increases and decreases in C_m were observed frequently also in later sections of this experiment. When mast cells of beige mice degranulate, there is in each cell nearly always at least one vesicle that shows a similar behavior, and contributes an increment in capacitance that increases and decreases repeatedly, sometimes disappearing entirely before reappearing and becoming stably established. Recalling that C_m steps report only the formation of an electrical connection between the vesicle lumen and the cell exterior, one concludes that the electric connection, and hence early stages of membrane fusion, are reversible and need not represent the (presumably irreversible) coalescence of vesicle with cell membrane.

Capacitance flicker is strongly reminiscent of the current flicker observed in single ion channels. Just as the rapid appearance and disappearance of current through a single channel reports the rapid opening and closing of that channel, so does the rapid appearance and disappearance of vesicle capacitance represent fluctuations, or even the rapid making and breaking, of an electric connection between the inside of the vesicle and the cell exterior. "C_m flicker" strongly suggests that the first connection is a narrow aqueous pore that, in some vesicles, may open and close rapidly and repeatedly before giving way to a complete and presumably irreversible fusion of the two adjacent membranes. We have also monitored the loss of a fluorescent dye from single vesicles induced to fuse with the cell membrane. While C_m steps showing no "flicker" lead immediately to release of the dye, such release does not occur during "flicker" (Breckenridge & Almers, 1987). Evidently no significant secretion takes place during flicker. The finding confirms that during flicker the electric connection between vesicle lumen and cell interior is a narrow channel.

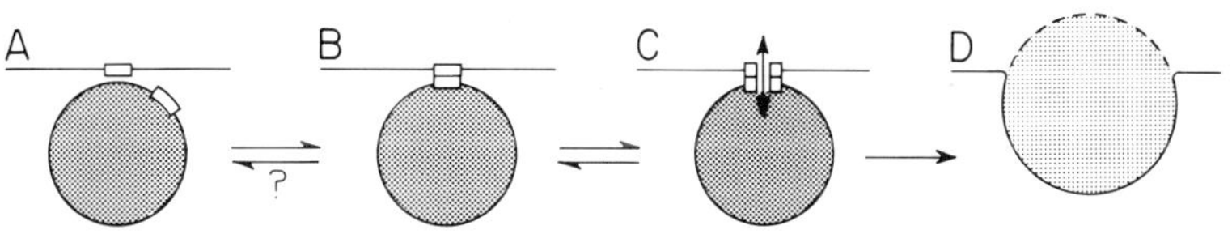

Fig. 5. Hyopthetical steps in membrane fusion during exocytosis. From Breckenridge & Almers, 1987a.

Fig. 5 illustrates a hypothesis that would be consistent with our observations. The vesicle (A) first binds to the cell membrane, possibly via specific sites (B). As in virus infection, this could be accomplished with a single membrane protein located in the membrane of either the vesicle or cell, or both. The binding protein(s) may well contain regulatory site(s) that control exocytosis, either by binding Ca^{++} or by being the substrate of a kinase or phosphatase. Upon binding, cell and vesicle membranes come in close apposition. Next comes a crucial step in exocytosis, which may also be the one that is regulated by Ca binding or (de)phosphorylation: the formation of an aqueous channel which we term the fusion pore (C). The channel will pass electric current and small solutes, but is evidently narrow enough to retard the escape of quinacrine from the vesicle (Breckenridge & Almers, 1987a). A step increase in C_m is expected to accompany the opening, and a step decrease the closing of a fusion pore. Hence fusion pores may cause C_m "flicker" by opening and closing repeatedly before giving way to a wider and more permanent opening (D). It is not known what forces drive the transition from (C) to (D). Whether the fusion pore is proteinaceous (as implied in Fig. 4) or lipidic is unknown.

THE CONDUCTANCE OF THE FUSION PORE

It might seem that the most straightforward way to measure the conductance of the fusion pore would be to patch-clamp individual vesicles and characterize the ion channels in the vesicle membrane. The vesicles of beige mouse mast cells are large enough to be patch-clamped, and preliminary experiments have shown this to be worthwhile as well as possible. For detecting fusion pores, however, this approach has the problem that it is hard to tell whether an ion channel thus observed are involved in forming the fusion pore. Indeed, none of the ion channels in the vesicle membrane may serve this purpose, because they span only the vesicle membrane while the fusion pore must span two membranes (the membranes of both vesicle and cell). Hence we took a different approach (Breckenridge & Almers, 1987b).

Suppose the potentials of cell and vesicle membrane are different. This is easily achieved, since the cell membrane potential may be set to any desired value by voltage clamping. The moment the fusion pore opens and the first electrical connection between cell exterior and vesicle interior is formed, a capacitative current transient must flow through the fusion pore to equalize the membrane potentials of vesicle and cell membranes. In mast cells of beige mice, the step increases in C_m are indeed preceded by current transients with a mean reversal potential (E_r) of -76 mV. Fig. 6 shows an example of a current transient recorded at 30 mV, along with its time integral. The transients have these properties: (a) They decline monotonically and (2) carry a charge Q that is linearly related to the final amplitude (ΔC) of the C_m step. (3) The ratio $Q/\Delta C$ depends linearly on holding potential with a slope of one, and intercepts the abscissa at E_r = -76 mV. (4) FCCP, a H^+-ionophore, shifts E_r to 0 mV. (1)-(4) are strong evidence that the current transients serve to charge the vesicle membrane through the fusion pore. E_r gives the membrane potential of the vesicle, 76 mV lumen positive to cytosol. The initial conductance of the fusion pore, g, is obtained either from the amplitude (I_o) of the transient,

$$g = I_o * \Delta C/Q$$

or from the time constant of decline (tau):

$$g = tau/\Delta C$$

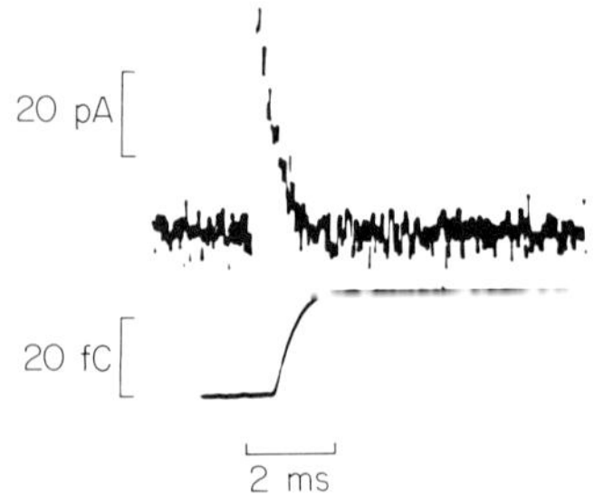

Fig. 6. Current transient through the fusion pore of an exocytosing vesicle (upper trace) and its time integral (lower trace). Cell membrane potential held at 30 mV, experimental conditions as in Fig. 3. Beige mouse mast cell. After Breckenridge & Almers, 1987b.

Estimates by the two methods usually are in good agreement. The mean value for g is 230 pS, independent of the cell membrane potential. This conductance corresponds to a probable pore diameter of < 2 nm (see, e.g., Hille, 1984).

The further time course of g was reconstructed from the electrical admittance (800 Hz voltage sinusoid) contributed by the fusing vesicle. Real and imaginary components of the admittance were sampled at 800 Hz with a new, synchronous-integration lock-in amplifier. We find that g grows to about 1 nS within 2-3 ms, as if the fusion pore rapidly dilated. The dilated fusion pore may persist for hundreds of milliseconds and can repeatedly close and open again, or expand further to complete membrane fusion. Evidently the fusion pore that constricts or dilates (or opens or shuts) during flicker has a much higher conductance than the first fusion contact.

How rapidly could a small molecule dissolved in a vesicle be released through a flickering pore? The rate constant, a, for the loss of a substance with diffusion coefficient D through a pore of length l and radius r may be calculated by (Hille, 1984):

$$a = \pi * r^2 * D / (1 + \pi * r/2)$$

If concentration gradients within the vesicle may be neglected, the substance would be lost exponentially with time constant given by the ratio V/a where V is the volume of the vesicle. For $D = 4 * 10^{-6} cm^2/s$ (midway between values for sucrose and ATP) and $l = 15$ nm and $r = 2$ nm, pore dimensions that would give a pore conductance of about 0.7 nS (eqn. 8-1 of Hille, 1984), the time constant becomes 120 s for a typical, large 4 μm diameter vesicle. This is imperceptibly slow on the time scale at which we measured loss of dye from exocytosing vesicles (Breckenridge & Almers, 1987a). However, the typical 0.8 μm diameter vesicle found in peritoneal mast cells from rats and normal mice could discharge the substance with a time constant of only about 1 s.

CONCLUSIONS

We have investigated the exocytosis of single secretory vesicles in peritoneal mast cells of rats and beige mice. Our data, as well as those of Zimmerberg et al. (1987) demonstrate that membrane fusion, or at least the early events in this process, do not depend on applying mechanical stress to the vesicle membrane. Although the granule matrix filling mast cell vesicles clearly swells when exposed to extracellular solution, such swelling is a consequence rather than the cause of exocytosis. Possibly, the only physiological role of granule swelling is to extrude the granule from the vesicle cavity once membrane fusion has occurred. Osmotically shrunken vesicles are capable of exocytosis also in chromaffin cells (Holz & Senter, 1986). There is not a single instance where a requirement of membrane stretch in exocytosis has been clearly established. Hence we are inclined to abandon theories that attribute a central role in the initiation of exocytosis to osmotic gradients.

Instead, the electrical measurements reviewed here have shown that one of the earliest steps (and perhaps the first step) in membrane fusion is the sudden formation of a 230 pS ion conductance pathway that connects the lumen of the vesicle with the cell exterior. Such a large conductance can not be explained by any structure that requires an ion to cross a region of low dielectric constant, such as the hydrocarbon core of a lipid bilayer. In particular, it is hard to see how structures such as "inverted micellar intermediates" or "hexagonal phases" (see, e.g., Rand & Parsegian, 1986) could provide such an efficient pathway for ion flux. The 230 pS conductance must be due to the formation of an aqueous pore. Such a structure is likely to share many properties with ion channels.

Interestingly, the initial conductance of the fusion pore (230 pS) is of the same order of magnitude as the conductance of a single gap junction channel (160 pS; Veenstra & de Haan, 1986), the only other aqueous pore known to span two membranes. However, there are important differences between fusion pores and gap junction channels. Whereas a gap junction channel slowly forms as a stable structure that may survive for the life of the cell, the fusion pore must form within seconds (or even milliseconds in neurons) and within milliseconds give way to a wider structure. Most importantly, gap junctions do not lead to the fusion of adjacent membranes.

While the biochemical nature of the fusion pore is unknown, it is tempting to suggest that it is formed from transmembrane proteins. One might imagine that it results from an aggregation of membrane protein subunits that form an annulus with a channel in the middle. The aggregation must be loose, and allow the rapid incorporation of lipids or of a variable number of additional subunits so that the pore may dilate rapidly. The fusion pore may in this regard be analogous to the channel formed by the antibiotic alamethicin, which is similarly thought to form channels of fluctuating diameter by aggregation of a variable number of subunits (e.g. Boheim, 1974). But how does the fusion pore disassemble and leave behind a seamless tube of lipid bilayer connecting the vesicle lumen with the extracellular space? Understanding this process will likely require entirely new concepts that have no parallel in our present understanding of ordinary ion channels.

ACKNOWLEDGEMENT

This work was supported by NIH grant AR 17803.

REFERENCES

Akabas, M. H., Cohen, F. S., and Finkelstein, A., 1984, Separation of the osmotically driven fusion event from vesicle-planar membrane attachment in a model system for exocytosis, J. Cell Biol., 98:1063.

Almers, W., and Neher, E., 1985, The Ca-signal from fura-2 loaded mast cells depends strongly on the method of dye loading, FEBS Letters, 192:13.

Almers, W., and Neher, E., 1987, Gradual and stepwise changes in the membrane capacitance of rat peritoneal mast cells, J. Physiol. (Lond.), 386:205.

Bearer, E. L., Duzgunes, N. Friend, D. S., Papahadjopoulos, D., 1982, Fusion of phospholipid vesicles arrested by quick freezing. The question of lipidic particles as intermediates in membrane fusion, Biochim. Biophys. Acta, 693:93.

Boheim, G., 1974, Statistical analysis of alamethicin channels in black lipid membranes, J. Membrane Biol., 19:277.

Breckenridge, L. J., and Almers, W., 1987a, Final steps in exocytosis observed in a cell with giant secretory granules, Proc. Natl. Acad. Sci U.S.A., 84:1945.

Breckenridge, L. J., and Almers, W., 1987b, Currents through the fusion pore that forms during exocytosis of a secretory vesicle, Soc. Neurosci. Abst., Vol. 13.

Cohen, F. S., Akabas, M. H., Zimmerberg, J., and Finkelstein, A., 1984, Parameters affecting the fusion of unilamellar phospholipid vesicles with planar bilayer membranes, J. Cell Biol., 98:1054.

Curran, M. J., Brodwick, M.S., and Edwards, C., 1984, Direct visualization of exocytosis in mast cells, Biophys. J., 45:170a.

Fernandez, J. M., Neher, E., and Gomperts, B. D., 1984, Capacitance measurements reveal stepwise fusion events in degranulating mast cells, Nature 312:453.

Finkelstein, A., Zimmerberg, J., Cohen, F. S., 1986, Osmotic swelling of vesicles: Its role in the fusion of vesicles with planar phospholipid bilayer membranes and its possible role in exocytosis, Ann. Rev. Physiol., 48:163.
Forman, J. C., Mongar, J. L., and Gomperts, B. D., 1973, Calcium ionophores and movement of calcium ions following the physiological stimulus to a secretory process, Nature, 245:249.
Gomperts, B. D., 1986, Calcium shares the limelight in stimulus-secretion coupling, Trends Biochem. Sci., 11:290.
Hamill, O. P., Marty, A., Neher, E., Sakmann, B., and Sigworth, F. J., 1981, Improved patch-clamp techniques for high resolution current recording from cells and cell-free membrane patches, Pflügers Arch., 391:85.
Hille, B., 1984, "Ionic Channels in Excitable Membranes," Sinauer Associates, Sunderland, Mass.
Holz, R. W., 1986, The role of osmotic forces in exocytosis from adrenal chromaffin cells, Ann. Rev. Physiol., 48:175.
Holz, R. W., and Senter, R. A., 1986, The effects of osmolarity and ionic strength on secretion from adrenal chromaffin cells permeabilized with digitonin, J. Neurochem., 46:1835.
Hui, S. W., Stewart, T. P., and Boni, L. T., 1981, Membrane fusion through point defects in bilayers, Science, 212:921.
Ishizaka, T., and Ishizaka, K., 1984, Activation of mast cells for mediator release through IgE receptors, Prog. Allergy, 34:188.
Lindau, M., and Fernandez, J. M., 1986a, IgE-mediated degranulation of mast cells does not require the opening of ion channels, Nature, 319:150.
Lindau, M., and Fernandez, J. M., 1986b, A patch-clamp study of histamine-secreting cells, J. Gen. Physiol., 88:349.
Mazurek, N., Schindler, H., Schuerholz, T. H., and Pecht, I., 1984, The cromolyn binding protein constitutes the Ca channel of basophils opening upon immunological stimulus, Proc. Natl. Acad. Sci. U.S.A., 81:6841.
Mohr, F. C., and Fewtrell, C., 1987, Depolarization of rat basophilic leukemia cells inhibits calcium uptake and exocytosis, J. Cell Biol., 104:783.
Neher, E., 1987, Stimulation by GTP-gamma-S renders secretion in mast cells sensitive towards free intracellular calcium, Biophys. J., 51:433a.
Neher, E., and Almers, W., 1986, Fast calcium transients in rat peritoneal mast cells are not sufficient to trigger exocytosis, Europ. Molec. Biol. Organiz. J., 5:51.
Neher, E., and Marty, A., 1982, Discrete changes of cell membrane capacitance observed under conditions of enhanced secretion in bovine adrenal chromaffin cells, Proc. Natl. Acad. Sci. U.S.A., 79:6712.
Nishizuka, Y., Takai, Y., Kishimoto, A., Kikkawa, U., and Kaibuchi, K., 1984, Phospholipid turnover in hormone action, Rec. Prog. Hormone Res., 40:301.
Ohki, S., 1988, Surface tension, hydration energy and membrane fusion, in: "Molecular Mechanisms of Membrane Fusion," Plenum Press, New York.
Rand, R. P., 1981, Interacting phospholipid bilayers: Measured forces and induced structural changes, Ann. Rev. Biophys. Bioeng., 10:277.
Rand, R. P., and Parsegian, V. A., 1986, Mimicry and mechanism in phospholipid models of membrane fusion, Ann. Rev. Physiol., 48:201.
Veenstra, R. D., and DeHaan, R. L., 1986, Measurement of single channel currents from cardiac gap junctions, Science, 233:972.
von Tscharner, V., Prod'hom, B., Baggiolini, M., and Reuter, H., 1986, Ion channels in human neutrophils activated by a rise in free cytosolic calcium concentration, Nature, 324:369.
White, J.R., Ishizaka, T., Ishizaka, K., and Sha'afi, R., 1984, Direct demonstration of increased intracellular concentration of free calcium as measured by quin-2 in stimulated rat peritoneal mast cell, Proc. Natl. Acad. Sci. U.S.A., 81:3978.

Zimmerberg, J., Curran, M., Cohen, F. S., and Brodwick, M., 1987, Simultaneous electrical and optical measurements show that membrane fusion precedes secretory granule swelling during exocytosis of beige mouse mast cells, Proc. Natl. Acad. Sci. U.S.A., 84:1585.

ELECTROFUSION AND ELECTROTRANSFECTION OF CELLS

Ulrich Zimmermann

Lehrstuhl für Biotechnologie
Universität Würzburg
FRG

INTRODUCTION

Biotechnology is one of the new key technologies of the present day. Modern biotechnology is expected to solve the central problems of the human race such as sufficient supplies of energy, raw materials, food and efficient treatment of diseases as well as the degradation of oil contaminations and the removal of pollution from drinking water.

The technical possibilty of changing the properties of cells at the genetic level (genetic engineering) has led (or can lead) to the development of microbial strains with increased, changed or novel metabolic activities, to the production of microbial or animal strains for the manufacture of pharmaceuticals (e.g. monoclonal antibodies) and to the breeding of new plants which give higher yields or are resistant to diseases or brackish water. In principle, the genetic information of an organism can be changed in vitro by two complementary techniques: somatic hybridization with the aid of cell fusion and direct gene (plasmid) transfer. The conventional fusion and gene transfer techniques which mainly use chemicals or inactivated virus are not always very efficient and are partly founded on an empirical basis. The development of electrical methods for in vitro cell fusion and for DNA (and protein) transfer across biological membranes in my laboratory in the last decade (1-7) may well represent an interesting alternative to the convention methods since they allow the process of fusion and gene transfer to be monitored by optical means and to be predicted by physical laws. The large scale application of these methods has thus moved into the realms of possibility (1-7).

Electrofusion of cells and electrically induced incorporation of low molecular weight substances and macromolecules into cells (so-called electropermeabilisation) is based on the reversible electrical breakdown of the cell membrane which leads to a high state of permeability of the cell membrane of controlable duration.

Reversible electrical breakdown of cell membranes was discovered in 1973 (8,9) and contradicted the then current school of thought that high electric fields irreversibly destroyed cells (10). A number of scientists (11,12) initially disputed the existence of such electric field efffects in biological membranes and only later confirmed (13,14) that it is possible to make the cell membrane permeable for prolonged periods of time under reversible conditions, in order to manipulate the intracellular pool of cells by exchange with macromolecules.

A historical overview of the development of these field methods has been reviewed (2). A number of further review articles have been published on this subject (3-7), so the author of this article will concentrate most atttention to recent developments in the fundamentals of the methods.

FUNDAMENTAL ASPECTS OF ELECTROFUSION AND ELECTROPERMEABILIZATION

Both electrofusion and electrotransfection (or more generally electropermeabilization if entrapment of proteins, drugs etc. are included) are based on the reversible increase in membrane permeability in response to a local electrical breakdown of the membrane in the presence of an external field (1-9). Therefore, the same equipment can be used for both applications, i.e. fusion and gene (or drug or protein) transfer through biological membranes followed by intracellular entrapment (7). The difference between the two applications is only that electrofusion requires membrane contact between at least two cells, whereas electropermeabilization is performed on freely suspended cells at relative low suspension densities. The establishment of membrane contact very often needs further equipment and different media composition than electropermeabilization (6,7). One exception is seen in cell suspensions of very high density. When the average diameter between the cells is small application of the field pulse leading to membrane breakdown is sufficient to give fusion (5). The very first electrofusion of cells utilised this approach (15). Membrane contact between cells at lower suspension densities can be achieved in a controlled manner by physical forces (4-6), such as by application of inhomogeneous alternating electric fields (leading to dielectrophoresis and subsequently cell chain formation), by ultrasound, by mechanical forces or by inhomogenous magnetic fields (provided that the cellular membrane was labelled with small magnetic particles).

Electrical breakdown occurs in freely suspended or attached cells under two conditions: 1) exposure to a field pulse of high intensity (kV/cm range), depending on cell radius, temperature and pulse length and 2) exposure to very short duration (nano- to microsecond range), depending on cell radius, the electrical properties of the biological membrane and on the resistivities of the intracellular and extracellular environment. In response to the external field pulse a potential difference is built up across the membrane capacitor due to charge separation. If the breakdown voltage of the membrane is exceeded at a certain electric field strength, the membrane will locally break through leading to areas of high permeability. The life span of this stage of high permeability depends on the temperature; at 4°C it can last up to 15 to 30 min (depending on the applied field strength and on the species used), whereas at 20°C and, in particular, at 37°C the permeabilized areas reseal very quickly under restoration of the original membrane resistance and impermeability. The term ´reversible electrical breakdown´ refers to the primary event and reveals nothing about the molecular mechanism leading to breakdown which has still to be elucidated. A number of models have been proposed and there is one among them which postulates that electrical breakdown leads to local instabilities in the membrane by way of electromechanical compression and electric-field induced tension, so that ´pores´ are able to form (6). Although the pore model is certainly a very lucid one, we have to point out that the molecular interpretation of electrical breakdown in terms of pores is not proven. On the contrary, evidence provided by Zimmermann and coworkers (4-6) that organelles and whole cells can be sequestered by electrical means across the membrane of the host cell under appropriate field conditions, seems to rule out the formation at least of very large and stable pores. Therefore, the term ´electroporation´, introduced by Neumann et al. (16) to the term 'reversible electrical breakdown', is very appealing, but may not be totally accurate under the circumstances. In fact, it may well be misleading since other permeability changes in the membrane structure generated by the breakdown pulse are not taken into account (e.g. flip-flop of phosoplid molecules, structural defects etc., see 6). Therefore, the author of this article would regard the use of the term electroporation as inappropriate - at least for the time being. The term electropermeabilization or electroinjection, on the other hand, are quite acceptable if one places more emphasis on the consequence of reversible electrical breakdown. The increased permeability of the membrane is, after all, the significant event as far as biological applications are concerned.

ELECTROTRANSFECTION

The electropermeabilization technique is becoming increasingly popular for gene transfer in a variety of eukaryotic cell types (2,16-21). The appeal of the technique lies in its relative simplicity compared to conventional techniques, its greater degree of control and reproducibility, and the avoidance of chemical reagents which are often lethal to the cells. Moreover, transfection frequency is considerably higher compared with conventional techniques provided that the field conditions are carefully selected in the light of the theory (1-6) underlying the mechanism of electrical breakdown. However, this latter aspect has been overlooked by several authors with the consequence that empirical protocols for electrotransfection have been published which resemble more an art than real science and are partly contradictionary to the boundary conditions of reversible electrical breakdown (1-7).

Stopper and Zimmermann (see e.g.21,22) have re-investigated in the last two years the field and cell parameters which control the yield of transfections and reported (at least for mammalian cells) transfection frequencies of the order of 0.1 to 3% in the presence of very low concentrations of DNA in the external medium. The results of these authors can be summarized as follows:

1. Field pulse application has to be performed at 4°C confirming previous results of my laboratory (1-7) that the incorporation of foreign molecules and particles at low temperature is preferable to that at higher temperatures. Transfection yield and viability of L-cells and myeloma cells were much better at 4°C as compared to 23°C or 37°C. Field pulse application below 4°C (20) has to be avoided because of irreversible deterioration of the membrane at this temperature under electric field conditions (1).

2. The temperature dependence of the breakdown voltage has to be taken into account if the pulses are injected into a cooled suspension. Work on lipid bilayer membranes and on the giant algal cells of <u>Valonia utricularis</u> have shown that the breakdown voltage of a unit membrane at 37°C is about 0.5 V, at 20°C is about 1 V and at 4°C is about 2 V (if appropriate pulse lengths are used, see e.g. (2,6)). From this it can be easily calculated that the field strength of the breakdown pulse must be at least 10 kV/cm in order to achieve reversible electrical breakdown of the cell membranes at 4°C. In electrofusion experiments, which are normally - but not always- performed at room temperature the field strength has to be correspondingly smaller (roughly by a factor of 2, Ref. 6,7).

3. Injection of 5 to 10 consecutive field pulses into the cell suspension leads to much higher transfection frequencies than application of only one or two pulses (Fig. 1). Most importantly, the time interval between the pulses has to be in the order of 1 min. In contrast, shorter time intervals, such as

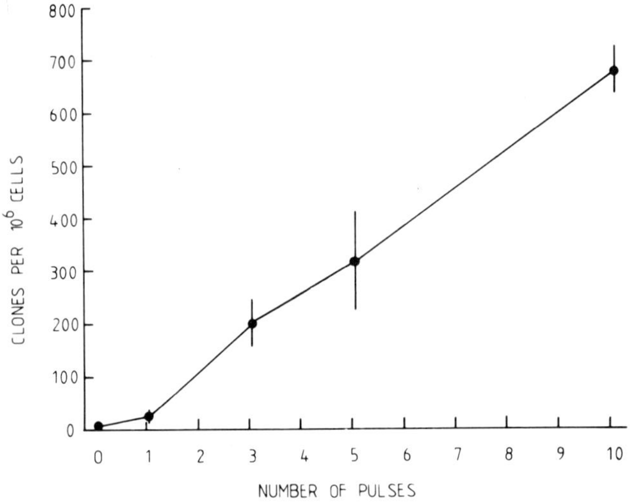

Fig. 1. The relationship between transfection yield and the number of applied pulses at 4°C. The pulse duration was 5 μs and the time interval between the pulses was 1 min. Field application at 4°C was achieved using the capacitor discharge method (6). The field strength was 10 kV/cm. The mouse L-cells were pretreated with dispase and suspended in an isoosmostic solution containing 30 mM KCl. The DNA per cell ratio was 3.3 pg/cell. The results are presented as means ± SD. Data taken from (22).

a couple of seconds (16,17,20), resulted in a decrease of transfection yield if pulse trains are applied.

The higher yield in response to a pulse train (if the appropriate time interval is adjusted) can be explained theoretically by the membrane potential distribution of spherical cells in electric fields (1-7). In response to a single pulse the potential built up across the untreated membrane is greatest at membrane sites oriented in the field direction and progressivly decreases towards zero at sites located perpendicular to the field vector. These locations perpendicular to the field direction can never be electropermealized. The situation is further complicated by studies which show that asymmetric permeabilization of the two hemispheres of the cell can occur due to the intrinsic membrane potential (6). Thermal-convectional, diffusional, and rotational events between the pulses with a 1 min interval should therefore be expected to expose these relatively unpermeabilized areas of the membrane to the electric field in response to successive pulsing. Yield enhancement in this way therefore probably reflects a greater degree of membrane permeabilization which thus improves the chance of DNA uptake occurring. The requirement of a time interval of 1 min between the pulses is further essential because this time period seems to be sufficient at 4°C for membrane resealing to occur to an extent that a breakdown potential can be built up across the membrane after application of the first breakdown pulse. When the time interval between the pulses is of the order of seconds the resealing process at 4°C is incomplete (1,2,6) with the ´pores´ remaining open thus leading to a short circuit through the cell interior instead of building up a breakdown potential across the membrane. Under these conditions the current flow through the cell interior will lead to secondary, adverse side effects (such as local heating, microelectrophoresis, osmotic processes and interactions of the field with cellular components) which, as shown previously (1-6), lead to cell death. Therefore a train of 5 to 10 pulses only yield a high transfection frequency if the appropriate time interval is chosen (22).

4. The duration of a single pulse of the pulse train should be not longer than 5 to 10 μs (for eggs, which have a much larger radius than normal cells, the pulse duration has to be extended correspondingly because of the longer relaxation time of the membrane charging process). This statement only holds for conductive solutions in which electric field-mediated plasmid transfer is performed. For electrofusion experiments which are normally carried out in weakly conductive solutions (1-7), the pulse duration has to be extended to about 20 to 40 μs because of the linear relationship between the relaxation time and the external resistivity.

5. The composition of the medium, in which the cells are suspended during field application is of great importance and decides about the success of transfection.The medium should contain about 30 mM KCl (21,22) because of the following reasons. The presence of KCl minimizes the release of intracellular potassium ions into the external medium during the state of high permeability. Considerable loss of intracellular potassium ions affects cell viability because the activity of many enzymes is stimulated by the presence of potassium ions. Furthermore, the presence of KCl in the external medium is important because of the asymmetry of breakdown (for details, see 6). Isosmolarity of the medium has to be adjusted by the addition of appropriate amounts of low-molecular weight sugars such as inositol or sorbitol. Addition of high molecular weight compounds have to be avoided. For example, in the presence of about 5 mM inulin the transfection rate is considerably decreased. A possible explanation for failure of transfection in the presence of inulin and other compounds of comparable molecular weight arises from the observation that under normal conditions the diameter of the cells almost doubled after resealing in response to the application of 5 to 10 field pulses. The flow of water into the permeabilized cell may be coupled with a DNA flow that caused the yield increase. This increase in volume can be explained by the colloid osmotic pressure theory (2).

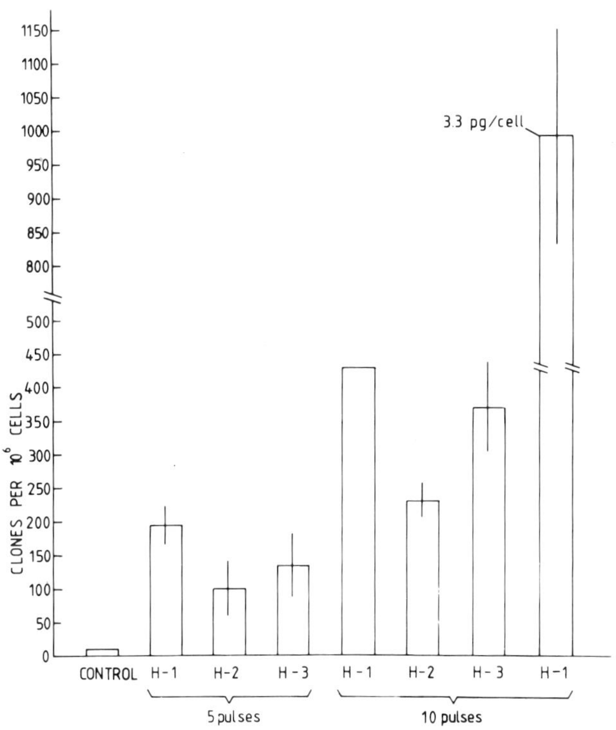

Fig. 2. The effect on yield of transfecting different subpopulations (H-1, H-2 and H-3) of mouse L-cells with a DNA per cell ratio of 1.0 pg/cell and in one case (where indicated) 3.3 pg/cell; otherwise the field conditions were identical to those described in Fig. 1. The control cells were not pulsed and the DNA concentration was the same as that used for treated cells. The results are presentedd as means ± SD. Data taken from (22).

6. Not only the biophysical parameters have to be taken into account but also the biological parameters have to be considered. Electrical pulsing of early exponentially cells resulted in more than a doubling of transfection yield. This was found to correlate with the size distribution of the different cell populations and probably with the state of growth cycle. Of particular importance is the subpopulation problem (and the related changes of the biological parameters of the cells when cultivated over long periods). The yield of stably transfected L-cells was dependent on the subpopulation selected from the bulk culture and used in the experiments although the growth and biophysical conditions were held constant (see Fig.2). The variations between the subpopulations may be due to either different membrane permeability and/or DNA uptake, integration, and expression. Certainly, the characterization of subpopulations (for example by means of size distribution measurements, see section below) is important in the future for establishing high yielding cell lines.

These considerations may help to transfer the protocols developed by Stopper and Zimmermann (21,22) for attached and suspended mammalian cell lines to other cell systems.

ELECTROFUSION

The many reports published in the last years on electrofusion demonstrate that this technique can be applied to all living cells, eggs, liposomes and planar lipid bilayer membranes and combinations of them (1-7). The universality of the electrofusion technique results from the fact that both electrical breakdown and the physical procedures leading to the required membrane contact reflect characteristic features of lipid-protein and lipid bilayer membranes. There is also now considerable evidence from many laboratories that electrofusion leads to high yields of viable hybrids.

A body of literature is available (6,7), showing electric-field-mediated hybridization of bacteria cells, yeast cells, mammalian cells and plant protoplasts. Most importantly, the production of hybridoma cells could be enhanced by electrofusion of murine myeloma cells with lymphocytes (6,7). Results obtained up to 1986 were reviewed several times by the author under different aspects and the interested reader is referred to this literature (1-7). In particular, these reviews summarize the equipment and media constraints which have to be taken into account for successful electrofusion. In this section the author will, therefore, restrict himself to a discussion of more recent developements in his laboratory on this subject matter.

Even though electrofusion is controlled by vectorial and controlable forces it is not possible to develop a generalized protocol for a specific cell system due to the inherent variations of cell cultures and - in the case of yeast and plant cells - of protoplast preparation, as already mentioned above for electrotransfection. Experiments performed with cells of two yeast strains carrying different genetic markers and plasmids (23) have shown that the yield of hybrids can considerably vary despite of constant and optimal field conditions. Current work were, therefore, focused on the developments of methods for predicting pulse application outcomes as a function of cellular phyisology and morphology. Progress was achieved by Broda et al. (23) in the prediction of hybrid yield prior to electrofusion by measuring and characterizing the size distributions of the two fusion partners. It was assumed that this parameter would play an important role in hybrid yield because the size influences the dielectrophoretic positioning of the cells as well as the electric breakdown of the membrane. Theory (2,6) shows that optimal fusion would be obtained if all cells in a suspension subjected to electrofusion had the same volume. However, size measurements in a hydrodynamically focussing particle analyzer indicated that there were distributions which, depending on the strain and the culture, preparation and pre-incubation conditions, may be positively or negatively skewed and/or bimodal.

Figs. 3 and 4 each show histograms of volume distributions for yeast protoplasts of the Saccharomyces cerevisiae strain AH22 (pADH 040-2) and AH215 which were selected by Broda et al.(23) as model systems. When protoplasts of AH22 and AH215 exhibiting volume distributions as shown in Fig. 3 were fused very high yields of hybrids were obtained. Much lower yields were seen when the preparations exhibited the volume distributions shown in Fig.4. Comparision of both figures leads to the definition of the parameters characterizing the distribution (Fig.5) and, in turn, to the definition of a protoplast index´ which correlates well with the expected hybrid yield.

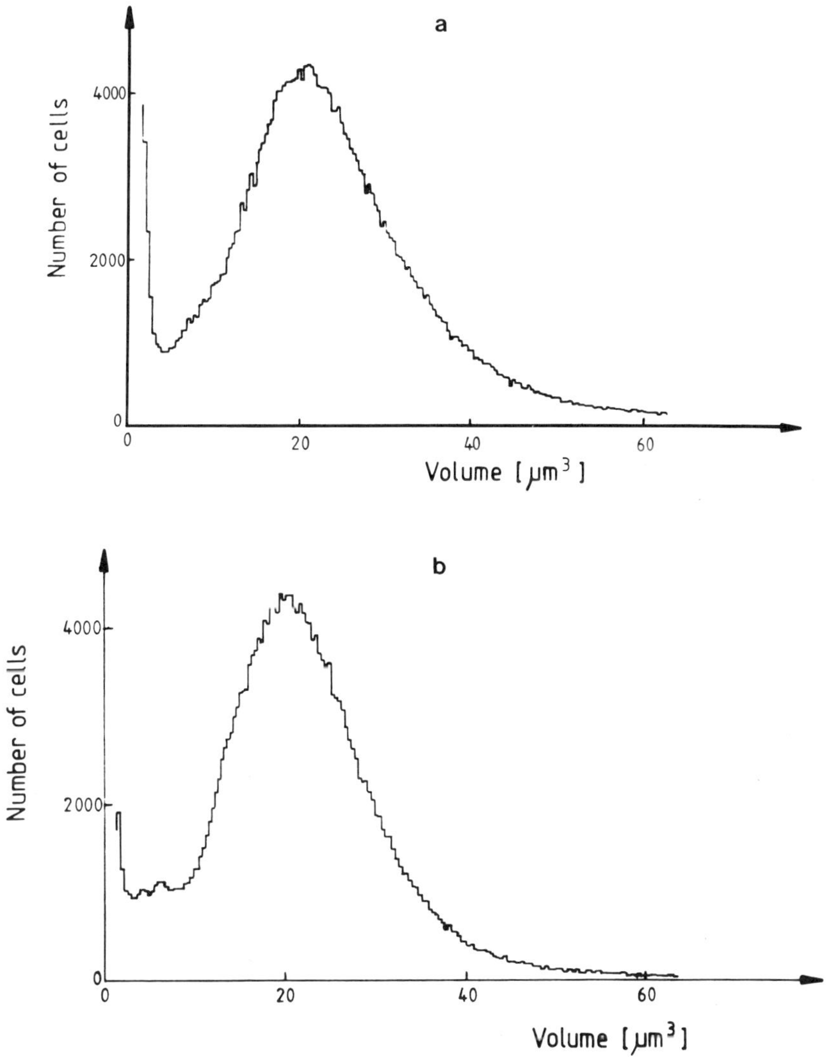

Fig 3. Volume distributions of protoplasts of the two yeast strains AH22 (pADH040-2) (a) and AH215 (b). Both volume distributions show a small amount of debris and undersized protoplasts, a small half-width and the same modal volume. The fusion of protoplasts from these preparations led to a very high yield of hybrids (protoplast index 0). Data taken from (23).

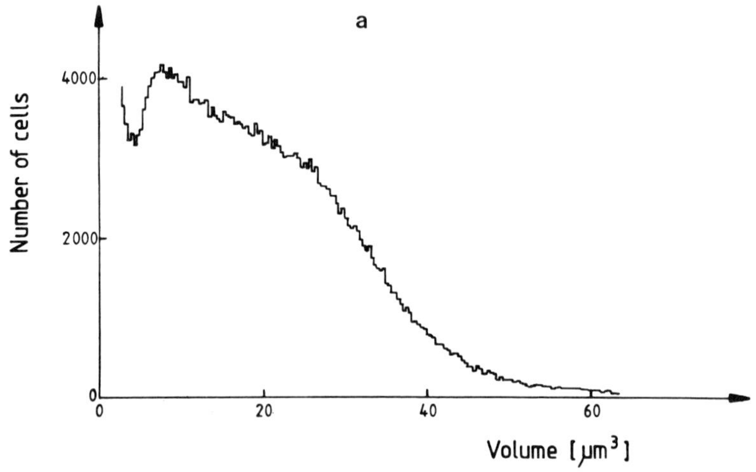

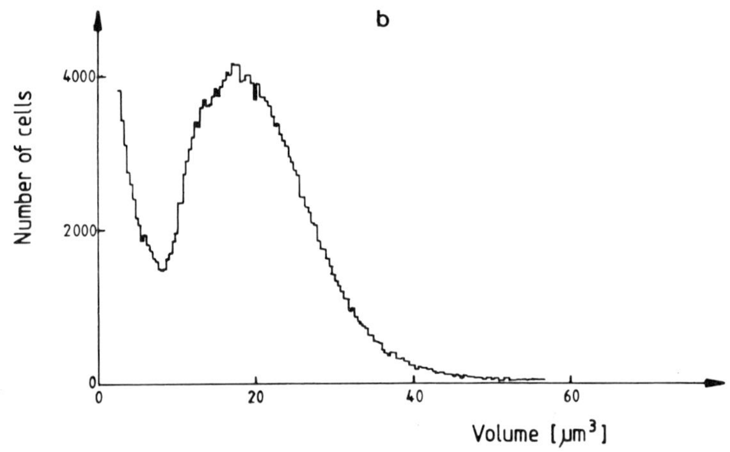

Fig. 4. In contrast to Fig.3 these volume distributions show a large amount of small protoplasts and a marked skewness to the right, especially for the strain AH22 (a). The modal volume (8 μm^3) of the strain AH22 is different from the optimal volume of about 20 μm^3. The protoplast index of these two fusion partners is 6. Data taken from (23).

All the distributions exhibited a ´bimodality´ which is particularly prominent for strain AH22 in Fig. 4. The first peak is the result of varying amounts of debris and tiny cells present in the protoplast population of each strain and of varying amount of superimposed electronic noise. The second peak is attributed to the intact protoplasts of average size. The right-hand skewness of the size distribution is due to ´abnormal´ large cells. Debris and tiny cells obviously disturb electrofusion. The presence or absence of these debris and cells can be described by three factors; these are the modal volume, the ´debris-to-cell ratio´ and the ´modal volume + half-width´ (Fig.5). The ´debris-to-cell ratio´ and the ´modal volume + half width´ take into account the relative proportions of undersized protoplasts (and of debris) and of very large cells in the populations, respectively. The ´protoplast index´ is the sum of these factors determined for the size distribution of each fusion partner and assumes values between 0 (Fig. 3) and 8 (Fig. 4). The index is calculated according to standard methods in statistics. Fig. 6 shows that there is a very good correlation between hybrid yield and the protoplast index. In particular, it could be shown that the protoplast index of fresh protoplast preparations with a very high index (leading to a poor hybrid yield) could be improved by incubation of protoplasts in a special, growth-limiting medium at 29°C.

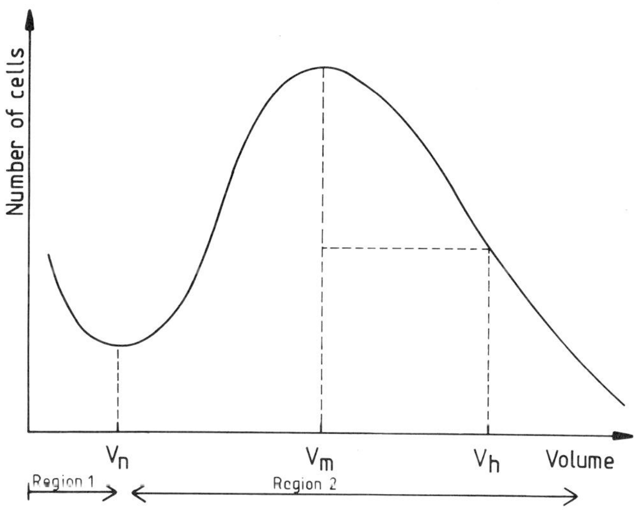

Fig. 5. Scheme of the volume distribution to explain the parameters which are necessary to calculate the protoplast index. According to Figs. 3 and 4, the volume distribution is composed of the volume distribution of the protoplasts (region 2) and the distribution which is due to debris, tiny protoplasts and electronic noise (region 1). The amplitude of the volume, V_n, at the local minimum between regions 1 and 2 is a measure of the number of very small particles. V_m is the modal volume and V_h is the ´modal volume + half-width´, which is a measure of the skewness of the curve indicating the proportion of larger protoplasts. Data taken from (23).

Pre-incubation, however, had no effect on yield - or even decreased yield - if the protoplast index of the fusion partners was optimal just after preparation of the protoplasts.

The introduction of the ´protoplast index´ is certainly the first quantitative approach to overcome the problems of biological variability by means of the measurements of characteristic biophysical parameters of the cell. The implementation of this procedure was also recently demonstrated for the optimization of the yield of hybridoma cells by electrofusion. These investigations and a more extensive study of the processes leading to the variation in shape of the size distributions of the two yeast strains carrying both either the plasmids or no plasmids revealed another, very interesting aspect of this method (24). The pronounced right-hand skewness of the size distributions (related to the 'unusual' large cells) could be correlated with an increase in plasmid copy number because the plasmid codes for a protein (β-lactamase) which is accumulated in the cell. Due to the colloid-osmotic pressure exerted by these proteins water is taken up resulting in a measurable increase in size. In agreement with this interpretation Broda et al. (24) found that the volume distributions of plasmid free and plasmid carrying strains show similar shapes if the cells were cultured in a complex nutrition medium. However, in minimal media with

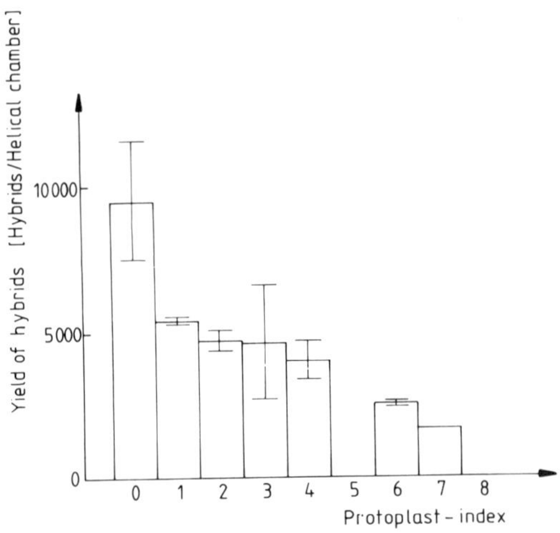

Fig. 6. The strong correlation between hybrid yield and protoplast index calculated from the volume distributions of both fusion partners indicates that the shape and the modal volume of the volume distribution of each fusion partner strongly affects the hybrid yield. Most protoplast preparations yield a protoplast index number in the range of 1 - 4; index numbers 5 and 8 have not been observed.

increasing selection pressure for maintenance of the plasmid the differences between the shapes of the volume distributions of plasmid carrying and plasmid free cells increased. This electronic, on-line measurements of the size distributions of plasmid containing cells may be, therefore, a useful method in the choice of appropriate nutrition and selection media. Furthermore, this method can in principle be applied to the determination of the ratio of plasmid free and plasmid containing cells in a growing population provided that marked differences between the volume distributions of these two types of cells exist. Such measurements are of fundamental interest in studies of plasmid stability and segregation properties - in particular in combination with electrotransfection of cells.

NUCLEAR MEMBRANE FUSION IN ELECTROFUSED CELLS

A further requirement for enhancement of hybrid yield in electrofusion is the better understanding and improvement of the fusion of nuclei which is certainly beside membrane resealing the most important step in the establishment of a hybrid once the plasma membranes of the fusion partners have been intermingled. This aspect of electrofusion has not received much attention in the past although Zimmermann et al. (3) reported evidence that fusion of nuclear membrane occurred some time after electrofusion of the plasmalemma of dielectrophoretically aligned, unsynchronized erythroleukaemic cells. In contrast, in chemically or virally induced fusion, nuclear membrane fusion has been reported only seldom and - if at all- only after synchronisation by colcemide treatment or by centrifugation or by GTP addition. Bertsche et al. (25) studied the fusion of nuclei in electrofused cells using staining procedures and DNA flow cytometry. Homogeneous and heterogeneous electrofusion of Ehrlich ascites tumor cells, Muntjac cells and V 79-S181 cells were performed in salt-balanced solutions at low temperature. The experiments showed that incubation of the cells exposed to the field pulse in fusion media for about 2 h was required to complete cell and, in particular, nuclear membrane fusion. Under optimum electrofusion conditions it was found that half of the fused cells (about 50 to 70%) underwent nuclear membrane fusion. The process of nuclear membrane fusion after application of the breakdown pulse could be divided into four phases: 1) cell contact associated with flattening of the adjacent plasmalemma regions and simultaneous repositioning of the nuclei into the polar (i.e. membrane contact) regions; 2) attachment of the nuclei; 3) enlargement of the contact region between the attached nuclei and 4) ´engulfment´ of the smaller nucleus by the larger one. Evidence was presented by these authors that the repositioning of the nuclei into the contact zone between attached, unsynchronized cells resulted from intracellular dielectrophoresis occuring during cell alignment.

Experimentally it was found, that the frequency of the alignment field had an influence on the cell and nuclear membrane fusion. Fig. 7 shows homogeneous fusion of Ehrlich ascites tumor and Muntjac cells as a function of the alignment frequency. It is evident that both plasma membrane (cell) and nuclear membrane fusion of Ehrlich ascites tumor cells had a maximum value at 1 to 2 MHz and decreased towards higher frequencies in the same manner. A similar behaviour was observed for fusion of Muntjac cells with the exception that the maximum of the two fusion processes occurred between 3 and 4 MHz. Analysis of the stained cells under the microscope indicated that at the optimum frequencies accumulation and attachment of the nuclei of adjacent cells in a chain assumed the maximum value. Taking into account the relaxation time of the charging process of the plasma membrane theory (6) predicts that above about 1 MHz (depending on cell species and the electrical properties of the membrane and the media) the plasma membrane capacity no longer shielded the cell interior from participation in the conduction process. Above this cut-off frequency a potential difference can be built up across the nuclear membrane leading to repositioning of the nuclei into the contact zone of the plasmalemma of two attached cells. At higher frequencies the cut-off frequency of the nuclear membrane is reached with the consequence that the nuclear membrane also becomes more conductive. Under these conditions the lowered potential difference across the nuclear membrane is no longer more sufficient to achieve intracellular dielectrophoresis. This intracellular dielectrophoresis apparently facilitates fusion of nuclei once intermingling of the plasma membranes has occurred. Increase in nuclear membrane fusion should increase hybrid yield. Indeed, as shown in preliminary experiments (Neil and Zimmermann, unpublished data) the yield of hybridoma by electrofusion could be considerably enhanced when instead of 1 MHz (6) a frequency of the alignment field of 4 MHz was applied for 30 s before pulse application.

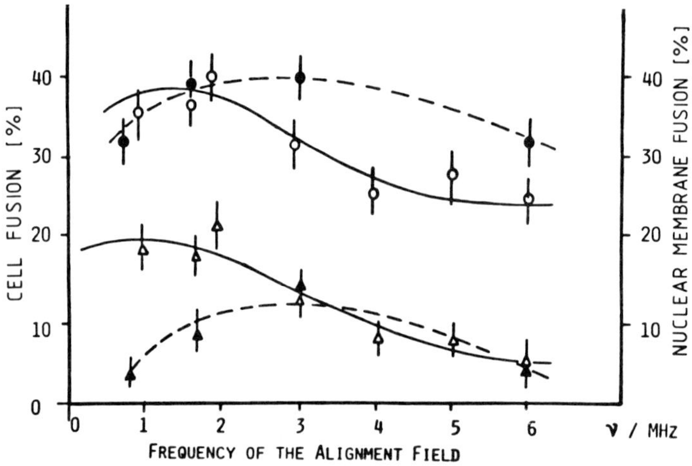

Fig. 7. Cell fusion and nuclear membrane fusion as a function of the frequency of the alignment field (field conditions: strength of the alternating field, 220 V/cm, duration time 30 s, followed by 2 x 3 field pulses of 5 kV/cm strength and 15 µs duration). Cells were suspended in salt-balanced solutions during field application and kept for 2 h in these solutions after application of the fusion pulses. Nuclear membrane fusion was measured by Giemsa staining.
The circles refer to cell fusion and the triangles to nuclear membrane fusion. The open symbols represent fusion between Ehrlich ascites tumor cells and the closed symbols between Muntjac cells. Data taken from (25).

This inherent feature of electrofusion is obviously one of the main reasons why this physical technique is demonstrably superior to the conventional techniques (which are based on random fusion) and also paves the way for the development of new efficient protocols for fusion of other cell types.

CONCLUSIONS

The purpose of this article was to identify some major contributing factors in electropermeabilization and electrofusion. The foregoing considerations should assist in establishing the optimum conditions as they apply to each targeted cell type. The advent of new instruments for determining of more biophysical parameters of cells and for detecting cellular and subcellular events during the entire field procedures will doubtless contribute valuable insight into system control. It is hoped that optimization of these procedures will also allow to select stable electrofusion- and electrotransfection-adapted subclones or subpopulations and to design automated large scale production of transfected cells and new hybrid cell lines.

ACKNOWLEDGEMENTS

This work was supported by grants of the DFVLR (No. 01QV3541) and of the SFB 176 (Deutsche Forschungsgemeinschaft).

REFERENCES

1. Zimmermann, U., Scheurich, P., Pilwat, G. and Benz, R., 1981, Cells with manipulated functions: New perspectives for cell biology, medicine and technology, Angewandte Chemie (Int. Ed. Engl.) 20:325
2. Zimmermann, U., 1982, Electric field-mediated fusion and related electrical phenomena, Biochim. Biophys. Acta 694:227
3. Zimmermann, U., Vienken, J., Pilwat, G. and Arnold, W.M., 1984, Electrofusion of cells, in: Cell fusion, D.Everid and J.Whelan, eds., Pitman, London.
4. Zimmermann, U., Vienken, J. and Pilwat, G., 1984, Electrofusion of cells, in: Investigative microtechniques in medicine and biology, Vol. 1, J. Chajen and L. Bitensky, eds., Marcel Dekker, New York.
5. Zimmermann, U.,Vienken, J., Halfmann, J. and Emeis, C.C.,1985, Electrofusion: a novel hybridization technique, in: Advances in biotechnological processes, Vol. 4, A. Mizrahi and A.L. van Wezel, eds., Alan R. Liss Inc., New York.
6. Zimmermann, U., 1986, Electrical breakdown, electropermeabilization and electrofusion, Rev. Physiol. Biochem. Pharmacol. 105:175.
7. Zimmermann, U. and Urnovitz, H., 1987, Principles of electrofusion and electropermeabilization, Meth. Enzymol.,in press.
8. Zimmermann, U., Schulz, J. and Pilwat, G., 1973, Transcellular ion flow in E.coli B and electrical sizing of bacteria, Biophys. J. 13:1005.
9. Zimmermann, U., Pilwat, G. and Riemann, F., 1974, Reversible electrical breakdown of cell membranes, Z. Naturforsch. 29c:304.
10. Schwan, H.P.,1985, Historical Review, state of the art, open problems, in: Interactions between electromagnetic fields and cells, A. Chiabrera, C. Nicolini and H.P. Schwan, eds., Plenum Publishing Corp.
11. Neumann, E. and Rosenheck, K., 1972, Permeability changes induced by electric impulses in vesicular membranes, J. Membrane Biol. 10:279.
12. Tsong, T.Y. and Kingsley, E., 1975, Hemolysis of human erythrocytes induced by a rapid temperature jump, J. Biol. Chem. 250:786.
13. Lindner, P., Neumann, E. and Rosenheck, K., 1977, Kinetics of permeability changes induced by electric impulses in chromaffin granules, J. Membrane Biol. 32:231.
14. Kinosita, K. and Tsong, T.Y., 1978, Survival of sucrose-loaded erythrocytes in the circulation, Nature 272:258.
15. Zimmermann, U. and Pilwat, G., 1978, Electric field-induced effects in cell membranes, Abstracts of the Sixth International Biophysics Congress, Kyoto, Japan.
16. Neumann, E., Schaeffer-Ridder, M., Wang, Y. and Hofschneider, P., 1982, Gene transfer into mouse lyoma cells by electroporation in high electric fields, EMBO J. 1:841.
17. Karube, I., Tamiya, E. and Matsouka, H., 1985, Transformation of Saccharomyces cerevisiae spheroplasts by high electric pulse, FEBS Lett. 182:90.
18. Fromm, M.E., Taylor, L.P. and Walbot, V.,1986, Stable transformation of maize after gene transfer by electroporation, Nature 319:791.
19. Auer, D., Brandner, G. and Bodemer, W. 1976, Dielectric breakdown of the red blood cell membrane and uptake of SV40 DNA and mammalian RNA, Naturwissenschaften 63:391.
20. Potter, H., Weir, L. and Leder, P., 1984, Enhancer-dependent expression of human k immunoglobulin genes introduced into mouse pre-B lymphocytes by electroporation, Proc. Natl. Acad. Sci. USA, 81:7161.
21. Stopper, H., Zimmermann, U. and Wecker, E., 1985, High yields of DNA transfer into mouse L-cells by electropermeabilization Z. Naturforsch. 402c:133.
22. Stopper, H., Jones, H. and Zimmermann, U., 1986, Large scale transfection of mouse L-cells by electropermeabilization, Biochim. Biophys. Acta, in press.

23. Broda, H.G., Schnettler, R. and Zimmermann, U., 1987, Parameters controlling yeast hybrid yield in electrofusion: the relevance of pre-incubation and the skewness of the size distributions of both fusion partners, Biochim. Biophys. Acta, in press.
24. Broda, H.G., Schnettler, R. and Zimmermann, U., 1987, Monitoring of plasmid-gene expression in yeast cell populations by size distribution measurements, FEBS Lett. in press.
25. Bertsche, U., Mader, A. and Zimmermann, U., 1987, Nuclear membrane fusion in electrofused mammalian cells, Biochim. Biophys. Acta, in press.

PRE-SELECTION OF B-LYMPHOCYTES BY ANTIGEN FOR FUSION TO MYELOMA CELLS BY PULSED ELECTRIC FIELD (PEF) METHOD

Tian Yow Tsong[1], Masahiro Tomita[1], and Mathew M. S. Lo[2]

[1]Department of Biological Chemistry
The Johns Hopkins University School of Medicine
Baltimore, Maryland 21205, U.S.A.

[2]National Institute on Drug Abuse
Addiction Research Center
4940 Eastern Avenue
Baltimore, Maryland 21224, U.S.A.

I. Introduction

Cells treated with pulsed electric field (PEF) tend to fuse together when brought into close contact by various means. The original discovery of this phenomenon is hard to trace because of the lack of documentation. Early findings of cells fused into large aggregates or mass by PEF were not fully appreciated and, therefore, few serious efforts were made to investigate the underlying biophysical mechanisms causing cell membrane fusion. It is quite possible that the value of electrofusion in biology and medicine were not fully realized by pioneering physicists. Evidently, several investigators and their coworkers continued with this research and are responsible for the recent resurrection of pulsed electric field method to induce cell fusion (Zimmermann et al., 1985; Neumann et al., 1980;Tsong, 1983;Teissie et al., 1982;Lo et al., 1984). Their championing of the potential applications of this technology in genetics and agriculture, and their experiments have contributed to recent interest paid by molecular and cell biologists alike to the PEF method (see e.g. chapters by Zimmermann, Sowers, Hui and Lucy). Another related area of current excitement, namely, the electric perforation of cell membranes for loading gene material, DNA, into cells has also helped to gain the popularity of the PEF method (Neumann et al., 1982; Potter et al., 1984; Tsong, 1983). Cell fusion using PEF is relatively simple. However, like other methods of cell fusion, fused cells are not always viable or able to proliferate unless certain precautions are taken to prevent the colloidal swelling and the subsequent lysis of cells after the PEF treatment. The advantage offered by PEF cell fusion over the use of chemical reagents, such as polyethylene glycol (PEG), is that the fusion efficiency is higher, and the toxicity of chemicals can be avoided. Additionally, certain cell types (e.g. plant cells and human cells) are usually not easily fused using PEG method. Another distinctive advantage of PEF method, to which this chapter is devoted, is that "molecule specific selection", or "biospecific selection", of cells can be effectively employed in conjunction with fusion. In contrast, biospecific selection is unlikely to succeed when used in conjunction with chemical fusogens because these chemicals would interfere with molecular recognition process of cell membrane receptors.

II. Electroporation And Electrofusion

General Considerations: The two phenomena are closely related, and in fact, "electric-perforation", or "electropermealization", or "electroporation" of cell membranes is perhaps responsible for the fusion of cells. Thus, certain knowledge on electroporation phenomenon should help one to understand mechanisms of electrofusion and to design experiments using PEF method. Cells suspended in isotonic medium when exposed to a field strength exceeding the critical breakdown potential of membranes have their permeability modified. Cell membranes facing towards the two electrodes becomes permeable to ions and molecules of various sizes which are otherwise impermeable in the normal state. The phenomenon was initially described as "dielectric breakdown" or "electric breakdown" of cell membranes, a terminology which implies a global breakdown in the membrane. To my knowledge, the first systematic study of this phenomenon was by Sale and Hamilton (1967, 1968). Later, this phenomenon was studied by several laboratories, notably, that of Neumann's, Zimmermann's, and our own. However, these studies can not be construed as a direct extension of Sale and Hamilton's work, because of various seemingly unrelated experimental circumstances leading to their findings. It is now generally agreed that electroporation of cell membranes is a local phenomenon, except in extreme conditions such as those used by Charlmers et al. (1985). In an attempt to load clinically useful enzymes into red cell ghosts by PEF method, Charlmers and coworkers have used PEF of 20 kV/cm or higher of millisecond duration. The result was a total rupture and fragmentation of red cells and ghosts, as any experienced investigator would have expected. Since electrofusion of cells occur close to the onset of cell killing, caution must be taken to avoid using extreme conditions (Teissie et al., 1982; Lo et al., 1984). PEF DNA transfection occurs at a slightly higher electric field intensity, namely around the mid-point of the cell killing. Another difference of the two methods is that PEF cell fusion requires, that cells to be fused, are in tight contact either at the time of, or shortly after (Sowers, 1986) the PEF treatment. This cell adhesion, of course, should be avoided in the PEF DNA transfection experiment, in order to prevent undesired fusion of cells.

Colloidal Lysis of Cells: Not every type of cell can survive electroporation of membranes. The most common finding is that after the electroporation, cells begin to swell and their shape becomes rounded (Kinosita & Tsong, 1977a,b, 1978). Since electroporation of membranes either is the cause or is an associated event of PEF cell fusion (Teissie et al., 1982; Sowers, 1986), it is not surprising to find that fused heterokaryon or hybridoma cells appear very often badly swollen. Unless these cells, whether single or heterokaryon, can repair their damaged membranes, they usually lyse after a period of incubation. The swelling phenomenon, caused by the colloidal osmotic pressure of cytoplasmic macromolecules, can be prevented by adding 25-35 mM of solutes which are impermeable to the perforated membranes of the PEF treated cells (Kinosita & Tsong, 1977a,b,c, 1978; Serpersu et al., 1985). The procedure is outlined in Fig. 1, which illustrates how an electroperforated cell membrane can reseal and survive under such conditions. These observations suggest that cells after electrofusion are preferably suspended in a "resealing medium" and kept at a temperature near the growth temperature so that the PEF treated cells and fused heterokaryons can heal their damaged membranes (Kinosita & Tsong, 1977b, 1978; Teissie et al., 1982; Tsong, 1983). We have used "pulsing medium" which consists of 5 mM phosphate buffer and isotonic sucrose, and "healing medium" which consists of 2 mM adenine, 10 mM glucose, 1 mg/ml BSA, 94 mM Na^+, 5 mM K^+, 20-35 mM stachyose (or an oligosaccharide), 1 mM Ca^{++}, 1 mM Mg^{++}, 20 mM HEPES, at pH 7.2 (Kinosita & Tsong, 1977b, 1978; Serpersu et al., 1985; Tsong, 1983).

Non-selective Cell Adhesion for Electrofusion: There are basically two types of procedures which can bring cells to adhere to one another for electrofusion, namely the mechanical and the chemical methods. Mechanical methods include:

1) Pelleting of cells by sedimentation followed by incomplete

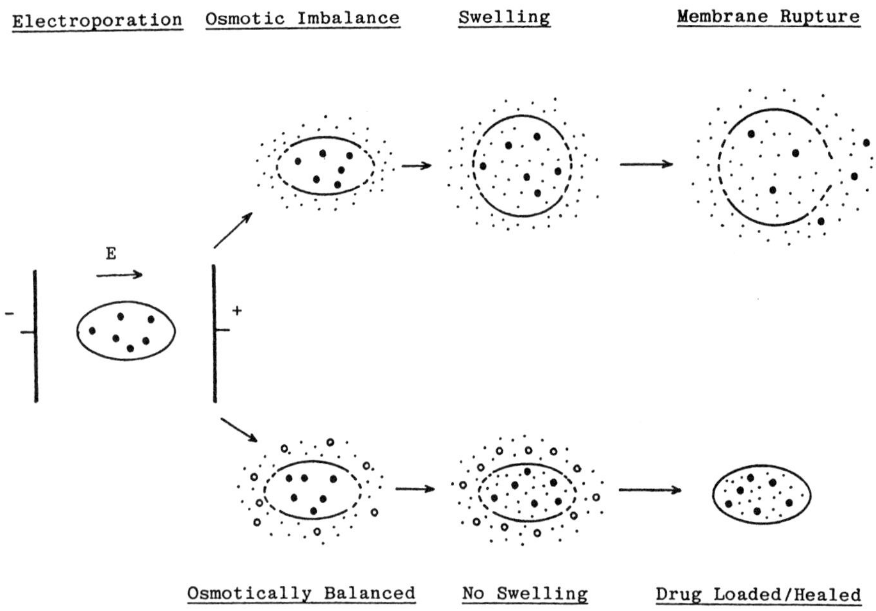

Fig. 1 Colloidal osmotic lysis and resealing of an electrically perforated cell. Upper path: When the cell membrane is electrically implanted with micropores, the membrane becomes permeable to ions and small probe molecules (in dots). These pores are small enough to block passage of cytoplasmic macromolecules (filled circles). As the result, the colloidal osmotic pressure of the cytoplasmic macromolecular content leads to influx of water. The cell swells, and eventually the membrane ruptures because of over swelling. Lower path: If on the other hand, molecules larger than the size of electric pores (open cicles) are added to external medium to counter balance the colloidal osmotic pressure of the cytoplasmic macromolecules, cell swelling can be prevented. Under such conditions, electrically perforated membrane will heal, at 37°C. Before complete resealing, exogenous molecules are loaded. Membrane-resealed cell now carries exogenous molecules, e.g. DNA in its cytoplasm. This cell behaves like a normal cell, except for the difference the effect of the loaded exogenous molecules.

dispersion into a suspension for electrofusion. This method is simple to perform in any biological laboratory, and the fusion efficiency is as high as any mechanical method in practice. The disadvantage is that there is little control over the size of cell aggregates to fuse and what kind of cells to fuse.

2) Micromanipulation- if only a few cells need to be processed, and if cells are relatively large, they can be micromanipulated into contact for electrofusion. In this case, cell selection by morphology may be partially achieved.

3) Dielectrophoresis- cells exposed to a megahertz a.c. field of 10-50 V/cm tend to aggregate to form "pearl chain" or "grape bundle" (Pohl, 1978; Zimmermann et al., 1985). When kilovolt PEF is then applied to these cell aggregates they fuse. By a careful adjustment of the a.c. field, it is possible to control the size of aggregates. The method perhaps has more

visual appeal, and also has some theoretical interest, however, in terms of practicality in electrofusion it is no more efficient or selective than any other mechanical or chemical method.

Non-selective chemical methods include use of agglutinins and lectins to induce cell aggregations for electrofusion. We have tested concanavalin A and other agglutinins for PEF fusion experiments (unpublished). The optimal concentration of these reagents to induce small size cell aggregations is sometimes difficult to determine, but once it is done, the method is clean and efficient. Different cell types also respond to different agglutinins, and some knowledge on these specific reagents is required. There are other factors which investigators have found useful, e.g. pre-treatment of cell surface by proteolytic enzyme such as pronase or by removing to certain extent the carbohydrate coverage of the cell surface. None of these observations is systematically explored, however. Their effect is also unknown.

III. Pre-Selection of B-Lymphocytes for Electrofusion

Monoclonal Antibody Technology: The most significant difference between the polyclonal and monoclonal antibody technologies is the screening of a single hybridoma in the former for propagation (Kohler & Milstein, 1975; Galfre & Milstein, 1981). Thus, antibody prepared is derived from one cell, or monoclonal. B-Lymphocytes harvested from spleen are polyclonal, i.e. they are a mixed population of cells which express, on their membrane surface, receptors that recognize a large variety of antigens. Even for the subpopulation which would produce antibody against a particular antigen, each hybridoma produces antibody against a specific part, or epitope, of the antigen molecule. Antibody prepared from the monoclonal technology, in principle, recognizes only one single epitope of the antigen. The importance of the monoclonal antibody technique in medical sciences is without question, though it is not the subject of this chapter. What one should recognize, however, is that until now no cell fusion methods discussed above (random fusion) are capable of producing monoclonal cell lines without going through a tedious screening of the hybridoma cells produced in these experiments. Obviously, to produce monoclonal cell lines, one must preselect B-lymphocytes that are destined to produce antibody against an antigen of choice for fusion to myeloma cells and disregard all other lymphocytes which are irrelevant for this purpose.

Design Using Biotin-Avidin Bridge: Our initial experiments using non-selective chemical reagents to elicit cell adhesion for PEF fusion resulted in vastly improved yield of hybridoma cells. For example, by using 1 μg/ml pronase, 20 μg/ml concanavalin A, or both combined, the improvement in frequency of fused hybridoma was, respectively, 1000, 50000 and 100000 folds over PEF treatment without adding these reagents (Lo et al., 1984; Conrad et al., 1987). This frequency was, respectively, 100, 5000, and 10000 fold higher compared to our control experiment using PEG method, under similar conditions, but without the help of pronase or concanavalin A. The improved frequency of cell fusion using PEF would permit hybridoma to rare antigen more likely to be found. However, to screen the vast number of hybridoma cells produced by PEF could pose an insurmountable problem for most biological laboratories. To circumvent the problem, we designed a means to pre-select B-lymphocytes for adhesion to myeloma cells before PEF fusion.

The design (Lo et al., 1984) is shown in Fig. 2. In this scheme, antigen for which monoclonal antibody was to be raised was used to select those lymphocytes which would be destined to produce the desired antibody. The strong affinity of avidin for biotin was used to bring cell adhesion. The basic protocol is given as the following.

1) Antigen was first chemically cross-linked to avidin. Avidin has four binding sites for biotin, of which at least one must be preserved during the cross linking process.

2) To achieve this goal, avidin was first bound to iminobiotin sepharose which was pre-treated with a cross linking reagent 1,5-difluoro-2,4-dinitrobenzene (DFDNB). After the crosslinking of antigen to avidin, the antigen-avidin complex was then eluted out from the iminobiotin column by a pH 3.8 buffer.

3) Myeloma cell surface was coated with N-hydroxysuccinimide (NHS) biotin.

4) The antigen-avidin complex was added to spleen cells and after a short incubation, the suspension was then mixed with a suspension of NHS-biotin coated myeloma.

5) The total concentration of cells was kept at about 1 x 10^6/ml, i.e. one tenth of that normally used, to avoid non-specific cell adhesion. Under these conditions, those lymphocytes with bound antigen-avidin complex on membranes should bind to myeloma, and all other lymphocytes should not. PEF of suitable combinations (e.g. 4 pulses at 2 - 4 kV/cm, 5 us) was then applied to the suspension. After the PEF treatment, cells were incubated in a healing medium discussed above at 37°C for about 30 minutes.

6) All other procedures followed are standard procedures for screening of hybridoma and cloning of monoclonal cells.

Ideally done, these procedures should produce only hybridomas which would make antibody against the antigen used in the selection of lymphocytes. Thus, the additional screening step for hybridoma, as is normally done using PEG method may be omitted. The range of affinity of antibody to the antigen may also be specified by adjusting the concentration

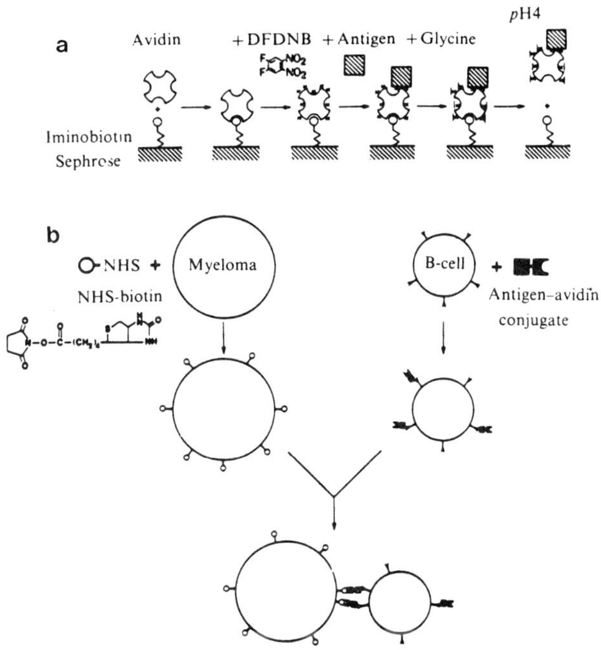

Fig. 2 Pre-selection of B lymphocytes for electrofusion to myelomas using avidin-biotin bridge. a) An iminobiotin sepharose column is used to crosslink an antigen to avidin. b) Myeloma cell surface is coated with biotin, using NHS-biotin, and the surface of relevant B cells is decorated with avidin through binding of the antigen-avidin conjugate to the surface receptor. When these two populations of cells are mixed, only relevant B cells can adhere to myeloma to form cell doublets. Subsequent electrofusion of the mixture will produce only hybridomas that are destined to secrete antibodies against the antigen used in the selection of B cells.

of antigen used in the lymphocyte selection process, i.e. steps (4) and (5). Although we have many experiments in which these expectations were met, in most other experiments hybridoma prepared did not always produce desired antibodies and normal selection procedure for hybridoma should not be completely eliminated. The less than totally clean selection of B-lymphocytes by the antigen-avidin complex, perhaps, was due to non-specific interaction of either avidin or the antigen with the lymphocytes membrane. This being the case, the method outlined here has definite advantage over the dielectrophoresis-PEF fusion and the conventional PEG fusion.

Another benefit of the pre-selection of lymphocytes for PEF

fusion is that the time required for immunization is much shortened. Typically, rabbit or mouse after immunization with an antigen will go through several stages of immune response. The primary response occurring one week after antigen administration produces predominantly IgM type that combine weakly with the antigen. After the primary response a recessed period is observed, which is followed by a secondary response 4-8 weeks later. Antibodies produced during the secondary response period are mainly IgG type which has higher affinity for the antigen. This division of response time is arbitrary, and even in an early stage of immune response there are lymphocytes expressing on their membrane surfaces antigen receptor of IgG type, although these cells are relatively rare. With specific recognition procedure it would be possible to pick up even these rare B-cells for PEF fusion to myeloma cells. Such, indeed, was the case, and we have succeeded in obtaining hybridoma cells at an early state of immune response which produced monoclonal antibodies of IgG type against angiotensin converting enzyme, enkephalin convertase and other proteins. The affinity of these monoclonal antibodies was typically in the range of 10^8 to 10^9. Saving of time and rare antigen was significant.

IV. Further Development of the Molecular Selection Procedure

The significance of the biotin-avidin bridging procedure outlined above goes beyond the methodology itself. The concept of molecular selection of cells for PEF fusion or for other kinds of biological research should broaden the possibility and the perspective of other areas of research where such selection procedures are beneficial, for example, targeting of cells for drug delivery or for radioactive irradiation. There are limitless variations with which such molecule and biospecific recognition can be accomplished. This does not mean that there are no useful rules for the design of new procedures. Let us consider some practical aspects in hybridoma technology. It is essential that any method designed should be applicable to a wide variety of antigens, with least modification required for each antigen. The toxicity of chemicals should be small. After treatment, e.g. coating of myeloma cell surface with biotin or antigen, these cells should be able to survive and to function normally. An ideal method is such that any investigator of moderate laboratory experience should be able to use it, and with reasonable expectation of success, i.e. of getting desired hybridoma cells. And the essence of bioselection should not get lost in the course of an experiment.

Obviously, the biotin-avidin bridging technique discussed above is far from meeting these ideal situations. We have been testing other combinations. Some are listed below.

1) Substitute streptavidin for avidin in the above design. Streptavidin possesses the same high affinity binding as avidin to biotin but it lacks the high background in avidin system. Avidin is highly positively charged at a neutral pH and tends to binds to DNA and negatively charged lipids. Streptavidin has an isoelectric point near neutrality and does not bind non-specifically.

2) Most difficulties we experienced came from the use of iminobiotin Sepharose column. Here the recovery of avidin-antigen complex ranged from a few % to less than 0.1% as checked by using ^{125}I-labeled antigen. The reliability was unassured. If one can eliminate the column step and still achieve molecule specific selection, it would be a substantial improvement over the original design. Let us consider three possibilities.

The first one is to coat myeloma, directly, with an antigen. The antigen coated myeloma will select B cells for adhesion. PEF fusion can then be applied. Coating may be achieved by nonspecific binding of antigen to myeloma cell surface, or by chemical cross linking to be discussed below. This design has some problem. For each antigen, one would need to find the right conditions for coating it to myeloma cells and still preserve the viability of the modified myeloma cells. The problem may not be severe though, and only experiments can tell.

Second, one can prepare F_{ab} of anti-myeloma antibodies. These F_{ab} fragments should be quite stable for storage. An antigen can be easily chemical cross linked to the F_{ab}. After mixing myeloma with the F_{ab}-antigen complex, the myeloma cells will be now coated with F_{ab}-antigen complex. These cells when mixed with a lymphocyte preparation, will select correct B-cells to adhere to for electrofusion. Once optimal conditions for cross linking an antigen to anti myeloma F_{ab} is worked out, the method would have general applicability.

The third method is potentially the most convenient. An antigen is chemically cross linked to a biotin instead of an avidin as in previous design. Myeloma cells are also coated with biotin. First, antigen-biotin complex is mixed with B-cells. Antigen will bind to correct B-cells. At this step, concentration of antigen-biotin complex and B cells can be adjusted to select monoclonal antibody of desired affinity. The B cells-antigen mixture is then mixed with excess amount of myeloma. In the step followed, a suitable amount of avidin is added to the myeloma/B cell mixture. Potentially, three types of cell doublets should form beside unpaired B and myeloma cells still in the suspension. These cell doublets are B/B, B/myeloma, and myeloma/myeloma doublets. Since excess myeloma cells are present, initially, only B/myeloma and myeloma/myeloma will predominate. After PEF fusion and HAT election, only myeloma/B cell hybridoma will survive. In principle any of these survived cells should secrete monoclonal antibody against the antigen.

For chemical cross linking, several reagents have been tested. They are summarized below:

1) Attachment of the antigen to imino Sephadex-immobilized avidin with homobifunctional cross-linkers, e.g. 1,5-difluoro-2,4-dinitrobenzene (DFDNB).

2) Attachment of the antigen to avidin with heterobifunctional cross linker, such as, maleimidobenzoyl-N-hydroxysuccinimide ester (MBS) and N-succinimidyl 3-(2-pyridyldithio)propionate (SPDP), to avidin in solution or on solid phase.

3) Chemical activation of the antigen to N-hydroxysuccinimide derivative and subsequent reaction with avidin.

4) Biotinylation of the antigen with NHS-biotin, and subsequent binding to avidin.

Peptides and proteins of greater than 100 amino acid residues are often derivatized by using methods 1 and 2, and that of smaller sizes are derivatized by methods 2 and 4. Haptens such as carbohydrates and steroids have been successfully derivatized with method 3. Membrane associated antigens may be biotinylated directly in situ and then solubilized with

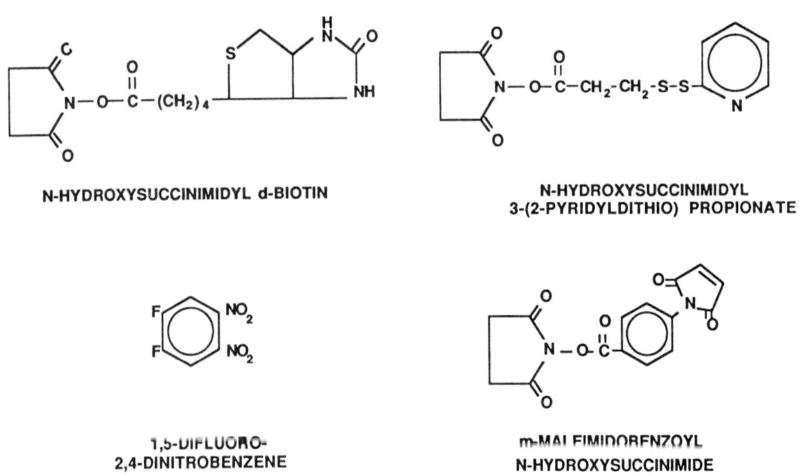

Fig. 3 Chemical structures of bifunctional crosslinkers used in our experiments.

detergents. In principle, immunization process does not require purified antigens, although in the lymphocyte selection process the use of purified antigen will alleviate, to a great extent, the tedious screening usually required later. The chemical structure of these cross linkers are shown in Fig. 3.

One important consideration in the selection of a chemical crosslinker is whether the crosslinker would alter the antigenic property of the antigen (Lo & Tsong, 1987). Modification of basic residues such as lysin may adversely affect the solubility of the antigen. Charged residues are also commonly found in antigenic epitopes (Geyson et al., 1987). Such being the case, the method (2) which uses MBS for crosslinking (Lui et al., 1979; Youle & Neville, 1980) have definite advantage. MBS is first conjugated to avidin through lysin residues. This provides a substantial reduction in the isoelectric point of avidin from pI of 10 to a more neutral value, thus, reducing the nonspecific binding usually experienced with unmodified avidin. The avidin-MBS conjugate prepared in this manner reacts with an antigen, which has been pre-treated with 10 mM dithiothreitol, through free -SH groups. After 20 hours, cysteine is added to block any remaining unreacted maleimide groups. The conjugate can be used without further purification (Tomita, Lo and Tsong, in preparation). When purification is required, it can be done by using ion-exchange chromatography, or by gel filtration, depending on the nature of the antigen used.

V. Case Studies

Several antigens have been tested using different biospecific chemical crosslinkings for PEF fusion. For rat lung angiotensin converting enzyme (ACE), enkephalin convertase and a nine-amino acid peptide, bradykinin, conjugated to human albumin the first method, i.e. the method using iminobiotin/sepharose immobilized avidin, was used. In the case of ACE, it was shown that the immunization process could be greatly shortened. For example, in one experiment, a C57BL/6 mouse received two intraperitoneal injections of pure ACE (2 µg) five days apart. Three days later, the spleen cells were harvested for electrofusion to biotinylated myeloma cells. Eleven wells, against 144 plated wells, showed cell growth and all these wells produced antibodies to ACE. Typically, we use two weekly intraperitoneal injections of pure or crude antigens before harvesting the spleen cells. Careful examination of wells containing growing hybridoma cells in several experiments showed only a single growing colony in each well. Similar results were obtained with enkepharin convertase and bradykinin/albumin conjugate.

The fourth method, i.e. the use of NHS-biotin to conjugate with antigens, was assessed using peptide antigens, human parathyroid hormone (PTH) and insulin. For PTH, 53 hybrid colonies were obtained. Only 5 showed strong immunoreactivity and 4 showed weak reactivity. In contrast, anti-insulin antibodies were detected in 4 out of 12 growing hybrid colonies. Three showed weak immunoreactivity (10^{-7} to 10^{-8} M) and one showed high binding affinity (10^{-9} M). The elevated non-specific hybridoma formation compared to other examples just mentioned may be due to the weak immune response these peptide hormones elicit. Other cross linking methods have also been applied to several antigens. A systematic study will be done to assess the efficiency of each method and to compare the strength and weakness of different methods.

VI. Perspective

Electroporation and electrofusion of cell membranes are two related phenomena generally regarded artificial in making and are not part of the natural process of cells. The importance of studying these phenomena is because of the popular applications of PEF method in DNA transfection and cell fusion. However, the interaction of a cell membrane with electric fields may have physiological relevance. The interaction is Coulombic. The field strengths used in the electroporation and the electrofusion may be greater or close to the range of the normal transmembrane electric field (Tsong, 1983). For example, a transmembrane potential of 10 - 250 mV, normally recorded for cells is equivalent to a transmembrane electric field of 20 - 500 kV/cm if the thickness of the lipid bilayer is taken to be 5 nm. This is an enormous electric field, and it is sustained with constant energy input from the cell, through the electron transport chain, the photo-oxidative reaction, or the hydrolysis of ATP. Thus, transmembrane electric fields are maintained to serve many essential functions of cells (Tsong & Astumian, 1986, 1987, 1988). In many cases, the opening/closing of channel proteins have been shown to depend on the transmembrane electric potential. Nerve impulses and conductions are familiar examples. Study of PEF induced cellular phenomena should be able to contribute to the understanding of the effect of *in vivo* electric fields on cell functions (Tsong & Astumian, 1986, 1987, 1988).

Acknowledgement This work was supported in part by NIH grant GM28795 to T.Y.T. We thank our past and present colleagues for their contributions.

References

Charlmers, R.A. 1985. Comparison and potential of hypo-osmotic and iso-osmotic erythrocyte ghosts and carrier erythrocytes as drug and enzyme carriers. Biblthca haemat., 51: 15-24.

Conrad., M.K., Lo., M.M.S., Tsong, T.Y., and Snyder, S.H. 1987. In: "Cell Fusion", Sowers, A., Ed., Plenum Publishing Corp., New York.

Galfre, G., and Milstein, C. 1981. Production of monoclonal antibodies: Strategies and procedures. Methods in Enzymol., 73: 3-46.

Geysen, H.M., Taimer, J.A., Kodda, S.J., Mason, T.J., Alexander, H., Getzoff, E.D., and Lerner, R.A. 1987. Chemistry of antibody binding to a protein. Science, 235: 1184-1190.

Kinosita, K., Jr., and Tsong, T.Y. 1977a. Hemolysis of human erythrocytes by transient electric field. Proc. Natl. Acad. Sci. USA, 74: 1923-1927.

Kinosita, K., Jr., and Tsong, T.Y. 1977b. Formation and resealing of pores of controlled sizes in human erythrocyte membrane. Nature (London), 268: 438-441.

Kinosita, K., Jr., and Tsong, T.Y. 1977c. Voltage induced pore formation and hemolysis of human erythrocytes. Biochim. Biophys. Acta, 471: 227-242.

Kinosita, K., Jr., and Tsong, T.Y. 1978. Survival of sucrose-loaded erythrocytes in circulation. Nature (London), 272: 258-260.

Kohler, G., and Milstein, C. 1975. Continuous cultures of fused cells secreting antibody of predefined specificity. Nature (London), 256: 495-497.

Lui, F.T., Finnecker, M., Hamaoka, T., and Katz, D. 1979. New procedures for preparation and isolation of conjugates of proteins and a synthetic copolymer of D-amino acids and immunochemical characterization of such polymer. Biochemistry, 18: 690-697.

Lo, M.M.S., and Tsong, T.Y. 1987. Producing monoclonal antibodies by electrofusion. In press in "Electroporation And Electrofusion in Cell Biology", Eds. Neumann, E., Sowers, A., and Jordan, C., Plenum Publishing Corp., New York.

Lo, M.M.S., Tsong, T.Y., Conrad, M.K., Strittmatter, S.M., Hester, L.D., and Snyder, S.H. 1984. Monoclonal antibody production by receptor-mediated electrically induced cell fusion. Nature (London), 310: 792-794.

Neumann, E., Schaefer-Ridder, M., Wang, Y., and Hofschneider, P.H. 1982. Gene transfer into mouse lyoma cells by electroporation in high electric fields. EMBO J., 1: 841-845.

Pohl, H.A. 1978. "Dielectrophoresis", Cambridge University Press, Cambridge.

Potter, H., Weir, L., and Leder, P. (1984). Enhancer-dependent expression of human kappa immunoglobulin genes introduced into mouse pre-B lymphocytes by electroporation. Proc. Natl. Acad. Sci. USA, 81: 7161-7165.

Sale, A.J.H., and Hamilton, W.A. 1968. Effects of high electric fields on microorganisms. I. Killing of bacteria and yeasts. Biochim. Biophys. Acta, 118: 781-788.

Sale, A.J.H., and Hamilton, W.A. 1968. Effects of high electric fields on microorganisms. III. Lysis of erythrocytes and protoplasts. Biochim. Biophys. Acta, 163: 37-43.

Serpersu, E.H., Kinosita, K., Jr., and Tsong, T.Y. 1985. Reversible and irreversible modification of erythrocyte membrane permeability by electric field. Biochim. Biophys. Acta, 812: 779-785.

Sowers, A. 1986. A long-lived fusogenic state is induced in erythrocyte ghosts by electric pulses. J. Cell Biol., 103: 1358-1362.

Teissie, J., Knutson, V.P., Tsong, T.Y. and Lane, M.D. 1982. Electric pulse-induced fusion of 3T3 cells in monolayer culture. Science, 216: 537-538.

Tsong, T. Y. 1983. Voltage modulation of membrane permeability and energy utilization in cells. Biosci. Reports, 3: 487-505.

Tsong, T.Y., and Astumian, R.D. 1986. Absorption and conversion of electric field energy by membrane bound ATPases. Bioelectrochem. Bioenerg. 15: 457-476.

Tsong, T.Y., and Astumian, R.D. 1987. Electroconformational coupling and membrane protein function. Prog. Biophys. Molec. Biol., In press.

Tsong, T.Y., and Astumian, R.D. 1988. Electroconformational coupling: How membrane-bound ATPase transduces energy from dynamic electric fields. Ann. Rev. Physiol., In press.

Youle, R.J., and Neville, D.M., Jr. 1980. Anti-Thy 1.2 monoclonal antibody linked to ricin is a potent cell-type-specific toxin. Proc. Natl. Acad. Sci. USA, 77: 5483-5486.

Zimmermann, U., Vienken, J., Halfmann, J., and Emeis, C.C. 1985. Electrofusion: A novel hybridization technique. Adv. Biotech. Proc., 4: 79-150.

THE MECHANISM OF ERYTHROCYTE GHOST FUSION BY ELECTRIC FIELD PULSES

Arthur E. Sowers and Veena Kapoor

Biomedical Research and Development
American Red Cross
Rockville, Maryland 20855

SIGNIFICANCE OF ELECTROFUSION

The electric field method of fusing membranes is significant for the general study of membrane fusion mechanisms because of the following unique possibilities:

i) the condition which triggers fusion can be presented and withdrawn within a millisecond or less,
ii) the effects of membrane-membrane contact can be separated from fusion-inducing stimulus,
iii) the effects of chemical conditions can be separated from the fusion-inducing stimulus,
iv) fusion can be induced simultaneously in many membranes,
v) very high fusion yields are possible,
vi) there is preliminary evidence that at least two mechanisms may be involved.

These special characteristics permit the design of unique experiments to, for example, coordinate the induction of fusion and quick freeze samples for analysis by freeze-fracture electron microscopy. The high fusion yields permit efficient data collection in real time data from single cells.

Our present view is that one of the two mechanisms stated in vi) above may have a relatively trivial foundation, but the other mechanism has interesting characteristics which may be significant for the general study of fusion. During fusion membrane components rearrangements may simply be due to self-organizing forces which come into play after a grossly disorganizing disturbance. Such disturbances can originate from laser pulses (Schierenberg, 1987) or microsurgical movements (Diacumakos, 1973) and thus have a trivial foundation. Since rather specific experimental and pulse conditions are needed to achieve fusion of membranes by an electric field pulse, the movements of bilayer molecules during the fusion event may occur in ways which are possibly relevant to other fusion mechanisms

A better understanding of electrically-induced membrane fusion (electrofusion) will require answers to two questions: First, to what extent is the pulse-induced permeabilization (=leakage) involved in the fusion mechanism? Second, if it is not involved, then what membrane condition, property, or structural feature predisposes membranes to fuse? Answers to these questions may not only resolve the question of the electrofusion mechanism but also may permit a new way to study membrane stability and membrane fusion.

KNOWN FACTORS IN METHODOLOGY

Most of what has been written about electrofusion has involved the applications and phenomenology of electrofusion (for previous reviews see Zimmermann, 1982; Berg, 1982; Berg et al., 1983; Pohl et al., 1984; Zimmermann et al., 1985; Zimmermann, 1986; Bates et al., 1987). The state of our knowledge of the electrofusion mechanism in erythrocyte membranes has been reviewed in Sowers and Kapoor (1987a; 1987b; 1988) and Stenger and Hui (1988). Some aspects of the mechanism are also covered in Dimitrov and Jain (1984). The subject of electrofusion is also covered in reviews by Hoffmann and Evans (1986) and Hoffmann (1988) although in both cases the instrumentation aspects are emphasized. All of the reviews contain extensive bibliographies of the literature. Other aspects of electrofusion will be covered in a future monograph devoted to electroporation and electrofusion (Neumann et al., 1988). Although the subject of this chapter is the erythrocyte membrane, reference to other membrane systems will be made as they are relevant to the discussion.

The general requirements for electrofusion are: i) a means for bringing membranes into close contact, ii) a means for creating a strong electric field pulse in the medium, and iii) the requirement that the electric field direction at the point where the membranes are in close contact is perpendicular to the plane of the membranes. These requirements call for a special chamber to contain the membranes and have electrodes which permit an electrical circuit between the suspension and an electronic pulse generator.

While the condition of close membrane-membrane contact can be met by aggregation-inducing chemical conditions (eg. Conrad et al., 1987), the use of dielectrophoresis can induce the membranes to form pearl chains and therefore result in contact between virtually every membrane in a suspension within a few seconds. This formation can be held indefinitely and is reversible. Dielectrophoresis is induced by passing a low strength alternating current through the medium containing the membranes. The random positions of the suspended membranes changes to a pattern of pearl chains shortly after the current is turned on. The membranes return to random positions by Brownian motion after the current is turned off. It is a fortunate quirk of nature that the direction of the pearl-chain alignment is such that the same electrodes in simple chamber designs can be used to carry both the alternating current and the fusion-inducing pulse.

Reports indicate that the specifications of the electric field pulse can be critical. Depending on conditions, a single pulse can result in a significant fusion yield. However, in both cases, additional pulses can increase the fusion yield until the fusion yield is saturated. In some laboratories fusion is induced with a pulse having a square waveform. It has been reported that fusion is induced with similar strengths but that the pulse width (=length) must be in the 10-100 usec range.

Our studies use erythrocyte plasma membranes and a pulse with an exponentially-decaying waveform. These pulses have peak amplitudes that generate a field strength of 500 - 700 V/mm in the medium. The pulse decay half-time must be about 0.6-1.2 msec for maximum fusion yield. As a general rule there is a tradeoff between pulse field strength and decay half-time for a given fusion yield (Fig. 1). This is consistent with other reports. We have also found that fusion yield reproducibility can be highly dependent on pulse characteristics. In other words, a 1-2% change in fusion yield can be detected for a 1-2% change in pulse voltage or decay half-time within the ranges for these two characteristics. Preliminary data (Sowers and Kapoor, 1988) indicate that square waveform pulses of about the same amplitude, but in the range of 20-200 usec in width, also induce fusion yields comparable to exponentially-decaying pulses, but less lumens are formed. Definitive experiments on the qualitative and quantitative effects of the two waveforms, however, have not yet been done. A number of other possibly relevant membrane structures are also induced by pulses and are also discussed elsewhere (Sowers and Kapoor, 1988). Lastly, the fidelity of the waveform of at least the exponentially-decaying pulse appears to be critical since minor changes monitored on a storage screen oscilloscope which were caused by electrical malfunctions were always accompanied by a loss in reproducibility or fusion yield.

In the normal electrofusion protocol the electric pulse is applied while the membranes are already in close contact (Zimmermann, 1982; Bates et al., 1987). In other words this contact is brought about before the pulse treatment is applied. However we discovered that fusion of human erythrocyte ghosts was also possible when pulses were applied to membranes before they came into contact (Sowers, 1983b; Sowers, 1984; Sowers, 1985b; Sowers, 1986a, Sowers and Kapoor, 1987a). This observation led us to recognize that a long-lived fusogenic state was induced in the membrane. This observation has been subsequently confirmed by Bates (Bates et al., 1987). A similar observation appears to have also been reported but not recognized as being due to a possible new fusion mechanism (Zimmermann et al., 1984). Long-lived structural changes have been previously reported in membranes by NMR (Stulen, 1981) and permeability (Chernomordik, 1985) but not related to a biologically relevant membrane function such as

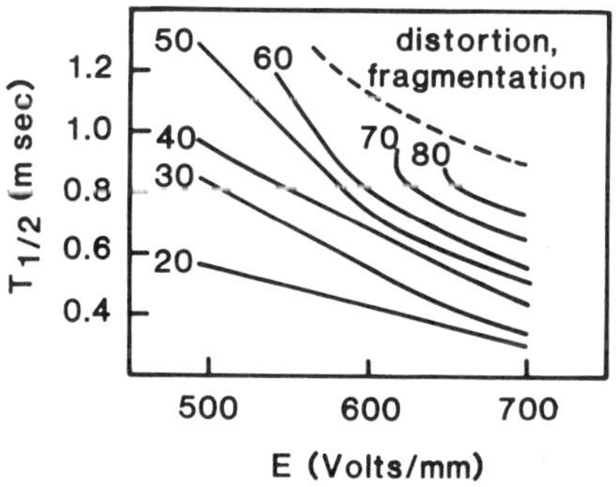

Fig. 1. Dependence of fusion yield in human erythrocyte ghosts on pulse field strength, E, and pulse decay half-time, $T_{1/2}$. One pulse is applied to membranes while they are held in pearl chains by dielectrophoresis. Isofusion contours are shown with fusion yields in percent (range 20-80%). Buffer is sodium phosphate (60 mM, pH = 8.5). Assay is conducted at 22-24°C. Assays use DiI (see text) as fluorescent label. Fusion does not occur for high voltage long pulses (upper right corner) but is instead replaced by a variety of pulse-induced morphological changes (see Sowers and Kapoor, 1988, for further details).

fusion. Permeability induced by an electric pulse also represents a membrane structural alteration and can also be long-lived under certain conditions (Serpersu et al., 1985; Schwister and Deuticke, 1985). Electric field pulses can also induce a lateral structural reorganization in membranes (Sowers and Hackenbrock, 1981). It should be pointed out that the above discussion includes phospholipid-containing artificial membranes in vesicular form (Buschle et al., 1982). The presence of proteins, a structural asymmetry, and extramembrane elements (cytoskeleton and glycocalyx) could fundamentally alter the phenomena we wish to study. A recent freeze-fracture electron microscopy study of electrically-induced fusion revealed at least two and possibly three pulse-induced alterations at the ultrastructural level which may also be relevant to the fusion mechanism (Stenger and Hui, 1986). An additional significant aspect of this study is that the existence of these structures was tracked in time until they decayed out of existence 10-100 sec after the pulse.

PROTOCOL MISCONCEPTIONS

Our experience with erythrocyte ghosts and mitochondrial inner membranes (Sowers, 1983a) have led us to question a number of previously stated requirements (see reviews) for the induction of fusion by electric field pulses. Our contrary findings are as follows.

Phase light optics is not sufficient to study fusion (Sowers, 1984). We have found that fusion products may or may not develop lumens (Fig. 2) thus showing that use of a light microscope equipped with only phase optics capability is not sufficient to study fusion events. Indeed the ratio of lumen producing to non-lumen producing fusion products can be altered with, for example, glycerol. Fusion however can be unmistakably scored if fluorescent labels are used. A fluorescent lipid-soluble label added to the membrane or a fluorescent water-soluble label resealed into the cytoplasmic compartment will move to an adjacent unlabeled membrane thus demonstrating a connection between either the unfused membranes or communication between the unfused cytoplasmic compartments. Our studies routinely use a fluorescent lipid analog, 1,1',-dihexadecyl-3,3,3",3'-tetra-methylindo carbocyanine perchlorate (DiI), which is commonly used in photobleaching studies. It has a high quantum efficiency, is resistant to bleaching, is not fusogenic, and can be used with standard FITC filter sets. We have found that measurements of lateral diffusion coefficients for DiI using pulse-induced fusion are comparable to measurements using photobleaching methods (Sowers, 1985a). We use a ratio of about 14 unlabeled membranes to one labeled membrane. This permits a practical tradeoff between efficient data collection and undermeasurement of fusion because fusion between two close-spaced membranes which are labeled will not be counted. We calculate fusion yield as the ratio: number of events in which the fluorescence moves from a labeled membrane to at least one unlabeled membrane divided by the sum of those events plus the number of labeled membranes which show no movement of fluorescent label. The resulting fraction, which ranges from 0.0 to 1.0, is multiplied by 100 to give percent.

High frequency alternating current is not absolutely necessary (Sowers, 1983; 1984). Alternating current in the hundred KHz or the MHz range has been previously emphasized as necessary for dielectrophoresis. Our experiments utilize alternating current stepped down by a transformer from the utility power lines. Use of high frequency alternating current does appear to minimize electrochemical processes at the electrodes and is useful for chamber designs in which membranes are close to or come into contact with electrodes (Rosenberg et al., 1965).

Chamber geometries need not produce a bulk non-homogeneous electric field either for dielectrophoresis or fusion (Sowers, 1984). Our most often used chamber has rectangular symmetry and places the electrodes far enough away that electrochemical activity (Rosenberg et al., 1965) in the vicinity of the electrodes is highly unlikely to influence processes in the area where membranes are observed (Sowers, 1983a; 1984). Other workers have also observed that homogeneous electric fields will induce dielectrophoresis (Bates et al., 1987; Miles and Hochmuth, 1987; Zachrisson and Bornman, 1984) and a theoretical foundation has been published (Sauer, 1983). The chamber is disposable and replacement of the electrodes which eventually become corroded is inexpensive and requires little effort.

Membrane suspension media need not have low buffer strengths for dielectrophoresis (Sowers, 1984; 1986a). We have universally used 60 Hz alternating current to induce dielectrophoresis in human erythrocyte ghosts, osmotically-swollen mitochondrial inner membranes in phosphate buffers up to 60 mM, in chicken erythrocyte ghosts in ammonium acetate buffers up to 120 mM in strength, and in human white blood cells in the RMPI 1640 cell culture medium. We have used not only our uniform field chamber (Sowers, 1984) but also our non-uniform field chamber (Sowers, 1986b) for these experiments.

Complex and expensive instrumentation is not needed (Sowers, 1984; Sowers and Kapoor, 1988). In the early history of many research areas a large fraction of physical and biophysical experiments involves equipment which is custom built in the laboratory. Reports describing in considerable detail instrumentation assembled by students for electrofusion studies may also be found (Mischke et al., 1986; Miles and Hochmuth, 1987). Literature is available to assist in both the mechanical (Moore et al., 1983) as well as classical (Malmstadt et al., 1962) and modern (Malmstadt et al., 1981) approaches to electronic design. Both fundamental and practical details are well covered in these references. It should be kept in mind, however, that care must be used since lethal high voltages may be present.

STRENGTHS AND WEAKNESSES OF PRESENT HYPOTHETICAL MECHANISM

The current hypothetical mechanism is derived from what has been learned in the 1970s about electric pulse-induced membrane permeabilization although no strong evidence actually supporting a relationship between this permeabilization and fusion has ever been presented. However, the hypothesis has features which are easy to visualize and test in experiments (see below).

Pilwat et al. (1981) proposed a three stage mechanism in which the electric field pulse induces a pair of equal diameter electropores in close spaced membranes (Fig. 2). As a metastable intermediate structure the pores and dispersed lipid molecules recondense into a fused membrane end product rather than back to the two unfused membranes. A major shortcoming of this hypothesis is that the gap between the two membranes is large compared to size of pores and must be spanned with a lipid-containing bridge through an unspecified process. To get around this shortcoming, Dimitrov and Jain (1984) proposed a four stage mechanism in which the membrane immediately around each electropore develops a flare or bend which thus permits the free edge of a pore on one membrane to touch at least a part of a free edge on a second membrane to make a fused membrane. Both the Pilwat et al. (1981) and the Dimitrov and Jain (1984) versions of this hypothesis deal only with the protocol in which membranes are already in close contact at the time the pulse is applied. Neither hypothesis predicts that it should be possible to fuse membranes if the pulses are applied before, instead of during, membrane contact. This is because the

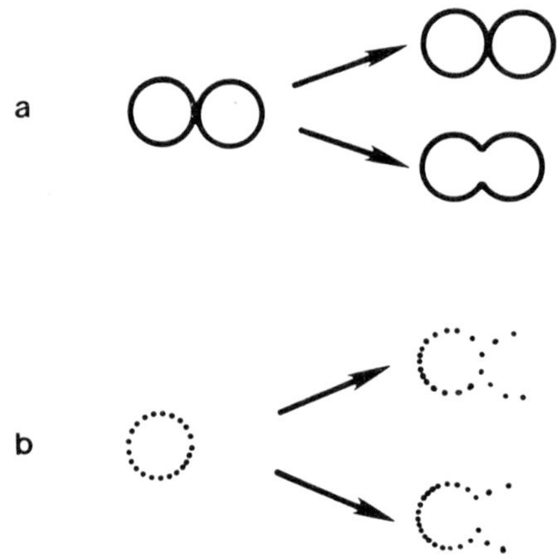

Fig. 2. Fusion as shown by phase and fluorescence light microscopy. a) Phase: fusion products show both nonlumen-producing (top) and lumen-producing (bottom) morphologies; b) fluorescence: same membranes as in a) show fluorescence moving to unlabeled membrane by lateral diffusion if left member is labeled with DiI (see text and Sowers, 1984; 1985).

concentric pore pair is a critical element which is induced as a single entity. Additional discussion may be found elsewhere (Sowers and Kapoor, 1987a; 1987b; 1988). In comparison to the four step Dimitrov and Jain (1984) hypothesis, Rand and Parsegian (1986) have pointed out that the molecular mechanism of phospholipid membrane fusion may involve up to six stages. These are: i) stable membrane apposition, ii) triggering, iii) contact, iv) focused destabilization, v) membrane coalescence, and vi) restabilization.

The above hypotheses deserve some comment. Although fusion and pore formation (permeabilization) are observed to occur above similar if not identical thresholds in pulse field strength, it does not necessarily follow that a causal relationship must exist between pore induction and fusion. The relationship between leakage and fusion in other membrane fusion studies is contentious (see, for example, Siegel, 1987; Szoka, 1987; and Düzgüneş et al., 1987, for reviews). Indeed, electric field pulses may induce many unrelated properties, structures, or other changes (see below) in the membrane and it is conceivable that one class of these entities may not interact with or be related to members of another class (see Sowers and Kapoor, 1988). Our thinking has centered on the operational idea that the electric field pulse induces a change in the membrane (the fusogenic state) which predisposes it to fuse. It may be due to the presence of multiple discrete entities (such as electropores) or the presence of an alteration which is continuously and homogeneously distributed within, or near the bilayer, and exists regardless of volume element size in the membrane. We have sometimes referred to the fusogenic state in terms of fusogenic sites (Sowers, 1986a). Both terms are constructs useful in the design of experiments and for operational discussion of results. Indeed, even when the CF protocol (see below) is used, the term <u>fusogenic state</u> may be reasonable in describing the pulse-induced change in the membrane even though the change is not long-lived.

The holes in the membrane which are induced by an electric field pulse can be probed as molecules similar to or smaller in effective diameter pass through (for reviews see: Tsong, 1983; Neumann, 1984; Knight and Scrutton, 1986). It is possible to model this transport in terms of a diffusion process (Sowers and Lieber, 1986). The generation of these pores has been often compared to dielectric breakdown but could also result from the generation of internal stresses through a piezeoelectric or electrostriction effect (Von Hippel, 1958), transmission of forces due to conformational changes (Porschke, 1985), or initiation of highly complex sequences such as the gel-collapse phenomenon (Tanaka, 1982) or ferroelectric effects (Lovinger, 1983; Goodby, 1986).

Experimental papers have reported large and sometimes highly reversible permeabilization increases in response to membrane hyper-polarization above fairly distinct transmembrane voltage thresholds. The phenomenon which leads to the permeability increase has been referred to as dielectric breakdown, a term borrowed from physics. These thresholds vary from one membrane to another but are generally two-three or more fold higher than metabolic membrane potentials. The reversibility of the permeability may be complete or incomplete and can be rapid or slow (see reviews and Sowers and Lieber, 1986). Electropores which reseal at slow rates are thus an additional pulse-induced structural entity which is long-lived. The membrane potential induced in a spherical-shaped membrane (Fig. 3) by an external electric field can be calculated as follows from an equation derived on rigorous physical principles (Fricke, 1953).

$$V = 1.5\ E\ r\ (\cos\theta) \qquad \text{eqn (1)}$$

where V = induced transmembrane voltage (in volts)
E = external electric field strength (in volts/mm)
r = radius of spherical-shaped membrane (in mm)
θ = angle between radius vector to a point on the sphere and the direction of the electric field (Fig. 3)

The validity of this equation has recently been confirmed for induced transmembrane voltages at least up to those generated by metabolic processes (Gross et al., 1986). This is important because it suggests that if external fields exceed the threshold for breakdown at the poles, then breakdown will occur at locations more distant from the poles (Fig. 4). It also suggests that the induced membrane potential may be different in the vicinity of the electropores and at locations between the electropores. These potentials could differentially affect membrane properties. This equation will also be utilized in a discussion below in connection with a question relating to characteristics of the fusion mechanism. Detailed information about the relationship between diameters, numbers, lifetimes, and locations of induced pores and the characteristics of the electric pulse which induces the pores are not, however, well understood. It has also been reported that low voltage long duration pulses will also lead to permeabilization (Sukharev et al., 1985).

RESULTS OF EXPERIMENTAL STUDIES ON ERYTHROCYTE GHOSTS

Our studies have attempted to both characterize the electrofusion phenomenon and test certain hypotheses concerning certain simple predictions. Although it appears to be far too early to elucidate the finer details of the electrofusion mechanism, certain observations nevertheless suggest that significant new knowledge may be learned. The facts and their implications are as follows:

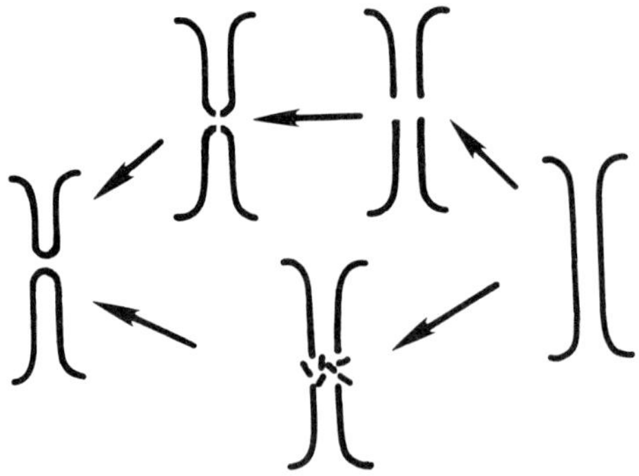

Fig. 3. The electropore hypothesis for the electrofusion mechanism. Left: close-spaced but unfused membranes. Right: fused membrane with a small lumen. Middle: the upper path (adapted from Pilwat et al., 1981) shows that a pulse induces a pair of equal-sized and concentric electropores as a single intact entity (a cloud of lipid molecules is located on axis passing through both electropores); the lower path (according to Dimitrov and Jain, 1984) shows a concentric pair of pores, as in upper path, but the membrane bends or flares off in the immediate vicinity of pores so that the pore edges can bridge the distance between the separate membranes. The field direction is left-right and in plane of the page.

Two protocols will lead to fusion (Sowers, 1983b; 1984; 1985b, 1986a, 1986b, 1987). We have found that pulses will fuse human erythrocyte ghosts regardless of whether the pulses are applied before or after the close membrane-membrane contact is induced. To distinguish between these two protocols we refer to whether pulses are applied to the membranes before or

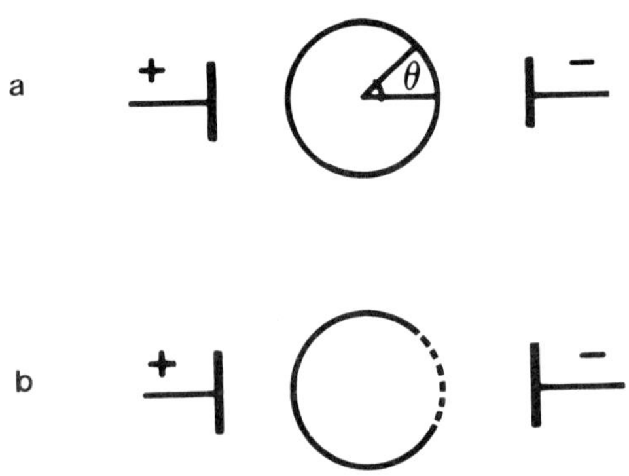

Fig. 4. Spherical-shaped membrane in an electric field.
a) Transmembrane voltage, V, at a given point on the sphere will be related to the bulk field strength, E, the membrane radius, r, and the cosine of the angle, θ, between the field direction and the line from the center of the sphere to the given point according to $V = 1.5Er(\cos\theta)$; b) as observed by low light level video microscopy a single pulse (700 V/mm, 1.2 msec) induces a peak permeability during the following 200 msec interval primarily on the cathod-facing hemisphere of human erythrocyte ghosts in 20 mM sodium phosphate buffer, pH 8.5 (see Sowers and Lieber, 1986).

after they are brought into and held in contact as the Pulse First (PF) or the Contact First (CF) protocol, respectively. The CF protocol is the conventional protocol and is universally discussed in other reviews.

The fact that fusion results from use of the PF protocol shows that a long-lived fusogenic change is induced in the membrane (Preliminary studies using the same media as for human erythrocyte ghosts have not revealed such a long-lived, pulse-inducible state in either chicken embryo erythrocyte ghosts, adult chicken erythrocyte ghosts, or osmotically-swollen mitochondrial inner membranes). The two protocols have qualitatively different pulse treatment requirements. High fusion yields can occur from just one pulse if the CF protocol is used (Fig. 1). When the PF protocol is used, then high fusion yields always require many pulses (Sowers, 1986a; Sowers and Kapoor, 1987a; 1987b; 1988).

Fusion yields are relatively pH-independent but strongly proportional to ionic strength. This is true for human erythrocyte ghosts over the pH range 6.0- 9.0 using the CF protocol and sodium phosphate buffer over the range 20-60 mM, and for chicken erythrocyte ghosts at high ionic strength (Fig. 5). This is consistent with a previous report of a relative pH independence of fusion yield in CHO cells (Blangero and Teissie, 1985). However, the same authors observed a decrease in fusion yield with increasing ionic strength.

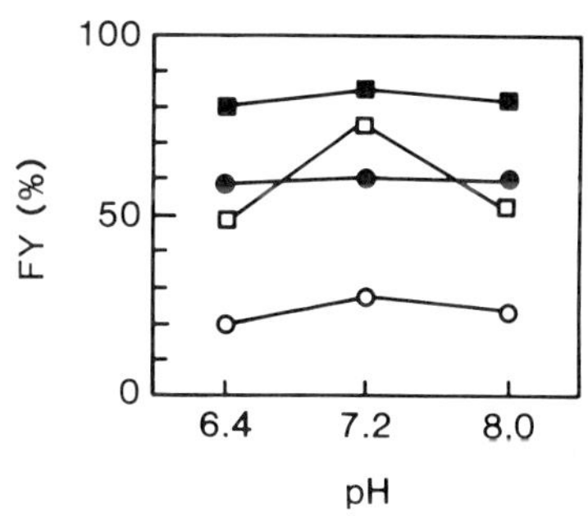

Fig. 5. Ionic strength and pH dependence of fusion yield (FY) in chicken embryo erythrocyte ghosts in response to single pulse ($T_{1/2}$ = 1.2 msec) applied to membranes held in pearl chains at the time of the pulse. Label is DiI. Buffer is ammonium acetate (including 1 mM $MgCl_2$). Buffer strength is 20 mM (open symbols) or 120 mM (solid symbols). Pulse voltage is 800 V/mm (squares) and 400 V/mm (circles).

Electropore numbers, diameters, and lifetimes do not correlate with fusion yields when ionic strength is changed (Sowers, 1986a; 1986b; Sowers and Lieber, 1986; Sowers and Kapoor, 1987). It is conceivable from the hypotheses presented above that conditions favoring increased numbers, diameters, or lifetimes of electropores would lead to increased fusion yields. Using suspensions of erythrocyte ghost membranes into which fluorescent soluble molecules were resealed, we found that an increase in ionic strength led to a decrease in pulse-induced loss of the label. This is in general agreement with other published papers on the effect of buffer strength on electropore induction (see reviews). However, when the same membranes were aligned into pearl chains to induce fusion by the PF protocol, then measured fusion yields went down. This is opposite to what would be expected if electropore number, diameter, or lifetime had a simple relationship with the fusion mechanism. In fact, when the PF protocol is used the ionic strength dependent fusion yields do not correlate with residual pore diameter, which does not appear to change with ionic strength. It is not known what mechanical property of the membrane controls peak and residual electropore diameter, but hemolytic pores show a diameter inversely proportional to ionic strength (Lieber and Steck, 1982a; 1982b). If electropore characteristics are the same regardless of whether membranes are in close contact at the time of the pulse (the CF protocol), then fusion yields still change in a direction opposite to what is expected when ionic strength is changed and argues against a simple relationship between pores and the fusion mechanism.

Residual pores are not likely to be involved in the fusion mechanism when the PF protocol is used (Sowers and Kapoor, 1987b). It is possible that residual electropores can be discrete fusogenic sites if it is hypothesized that a residual pore on one membrane covers a residual pore on a second membrane when the two membranes come into contact. This possibility would thus assemble two electropores into the concentric pore-pair intermediate structure and become yet a new modification of the original Pilwat et al. (1981) hypothesis. Preliminary calculations indicate that probability of assembly of concentric pore pairs when two membranes come into contact may be too low to account for observed fusion yields. The assumptions made in these calculations are being revised, however, and future refinements may lead to a different conclusion.

A shape change nearly always takes place before fusion occurs (Sowers, 1983b; 1984; Sowers and Kapoor, 1988). Human erythrocyte ghosts in >20 mM NaPi buffer at pH 8.5 have an imperfect spherical shape. Upon pulse treatment, they progressively become spherical with greater numbers of pulses. When < 20 mM, the ghost membranes are already spherical and do not change shape upon pulse treatment unless the pulse voltage or decay half-time is too high or too long, respectively. Membranes in contact at the time of pulse application sometimes form lumens. Intact erythrocytes (discocytes or echinocytes) always first become echinocytes and then spherocytes before fusion. Glycerol inhibits lumen formation but not shape change. Sucrose inhibits shape change but not fusion. Shape change in ghosts always involves stomatocyte forms with concave side facing the negative electrode at an intermediate point before the final spherocyte shape is achieved. The relevance of this observation to the fusion mechanism is unknown.

The long-lived fusogenic alteration has very low or zero lateral mobility (Sowers, 1986b). It thus may be in some way linked to the cytoskeleton or to otherwise immobilized integral proteins. This also indicates that if electropores are fusogenic sites, then they can not be entirely made of lipids and be totally surrounded by lipids.

A strong pulse induces an unequal amount of permeabilization in the two hemispheres facing the electrodes (Sowers and Lieber,1986). If one 700 V/mm, 1.2 msec decay half-time pulse applied to a 20 mM sodium phosphate buffer (pH = 8.5) containing suspended human erythrocyte ghosts in which are water soluble fluorescent markers then the induced pores will develop peak diameters of at least 17 nm within 100 msec or less and then reseal within about 200 msec to a residual diameter of 1 nm and then stay at that diameter for at least 20-30 seconds and possibly many minutes. The large short-lived pores have either larger sizes, numbers, or lifetimes on the hemisphere facing the negative pole (Fig. 4) than on the other hemisphere. Whether this involves one, two, or all three pore characteristics (diameter, lifetime, number) can not be answered at this time. It is also possible that two classes of electropores are induced: i) large completely resealing, and ii) small, incompletely resealing. However, this can not be determined from the present data.

The electropores evidently do not coalesce into fewer but larger pores because the loss of soluble labels from the labeled cytoplasmic compartment was never observed to accelerate at any time beyond the 200 msec interval after the pulse. Because all pulse-induced efflux of soluble labels stops after about 200 msec, it is unlikely that the short-lived large pores fragment into long-lived but smaller pores unless they are smaller than about 1 nm in diameter.

If the above observed permeability difference on the two hemispheres also occurs when the membranes are aligned by dielectrophoresis into close contact, then the electropore hypothesis may have to account for pore pairs which are unequal in diameter. It also raises the possibility that in either protocol, the anode-facing and cathode-facing hemispheres participate differently in the fusion mechanism.

Electric field pulses of a lower field strength but longer duration have been observed by freeze-fracture electron microscopy to induce tether formation in mitochondrial inner membranes (Sowers and Kapoor, 1988) and further supports the notion that the two hemispheres experience different forces during the pulse treatment.

Does fusogenic alteration have an all-or-none or a continuously variable character? Since membrane fusion events are a probablistic all-or-nothing process at the level of individual membrane fusion partners fusion must be measured in a relatively large population of membranes. Brownian motion, thermal motion, membrane-membrane proximity, and other fluctuations are probably involved in poorly understood and poorly measureable ways. Since fusion yields increase as the strength of an electric field pulse is increased, two questions arise: i) Does the fusogenic membrane state have a variable character? ii) Is it related to how far the electric field is above the threshold for inducing fusion? A third related question is also suggested: Can one membrane be more fusogenic than the partner it fuses with? The approach to these three questions involves the induce membrane potential as predicted by eqn (1) from the external electric field. Whether eqn (1) accurately predicts the induced membrane potential above the threshold for inducing permeabilization is not yet known. However, it is possible that membrane potential in areas of the membrane which are between the electropores continues to increase. If these areas contribute to the fusogenic membrane alteration, then fusion yield measurements between two dissimilar radius membranes should give two kinds of results (Fig. 6). Three combinations of fusion partners with different diameteres are needed: Large + Large, small + large, and small + small. From eqn (1), fusion yield should be greater for a given pulse strength with large + large fusion partners than with small + small fusion partners because the threshold membrane potential for fusion will be reached with a lower external field strength. The question is what will happen with large + small fusion partners? It is possible that the fusion mechanism will require above threshold induced membrane potentials on both membranes (Fig. 6a) or just one membrane (Fig. 6b). Our attempt to obtain information to answer this question utilized the CF protocol with 5-day chicken embryo erythrocyte ghosts, CEEG, (4.9 um ave. radius) and adult chicken erythrocyte ghosts, ACEG, (3.7 um ave. radius) as fusion partners (Chan, 1977). The results are shown for fusion yields

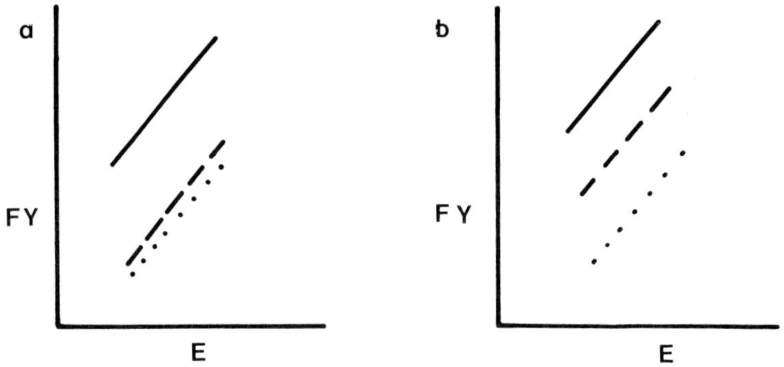

Fig. 6. Possible outcomes of field strength(E)-dependent fusion yield (FY) measurements between similar diameter and dissimilar diameter fusion partners: Large + large (solid), large + small (dashed), and small + small (dotted). a) If fusion mechanism requires both membranes to have electropores, or other fusogenic alteration which is induced above a threshold in E, then the fusion yield for large + small partners would be the same as for small + small; b) if fusion mechanism involves variable or different contributions from partner membranes then fusion yields from mixed diameter partners will be intermediate between large + large and small + small. Assumption is that fusogenic alteration is related to V (see text).

obtained at buffer strengths of 20 mM (Fig. 7a) and 120 mM (Fig. 7b). At low ionic strength it would appear that both membranes participate together in the fusion process, but at 120 mM only the fusion threshold for induction of the fusogenic state on the large membrane appears to be necessary for fusion to occur. It is also possible that biological factors may play a significant role since chicken embryo erythrocyte plasma membranes undergo a change in at least membrane protein composition before becoming adult chicken ertthrocytes (Chan, 1977). It is also clear that ionic strength plays a major role in controling not only fusion yield, but possibly also the fusion mechanism.

A second way to examine the question of whether the two fusion partner membranes participate equally or unequally is to follow the direction in which the fluorescent label laterally diffuses following the application of a pulse (Fig. 8). The observation of a differential pulse-induced permeabilization on the two hemispheres (Sowers and Lieber, 1986) leads to the question of whether differential induction of the fusogenic state also occurs. This possibility was examined using the same combination of large and small diameter membranes except that the scoring of the fusion events was placed into three categories. The fusion yields were measured for the two buffer strengths and shown in Fig. 9. The data show that at low ionic strength and low pulse field strength ($\leq$500 V/mm), lateral diffusion moves preferentially towards the negative electrode. This indeed suggests that more fusogenic sites or a higher fusogenicity is generated in one hemisphere. However, at higher pulse field strengths the movement is in both directions, suggesting that a second fusion mechanism may be superposed upon the first mechanism. At high ionic strength, bidirectional lateral diffusion <u>always</u> dominates. This suggests that at low ionic strength and low pulse field strength, fusogenicity is generated preferentially in only one hemisphere of one of the fusion partner membranes. At higher ionic strength, both membranes participate equally.

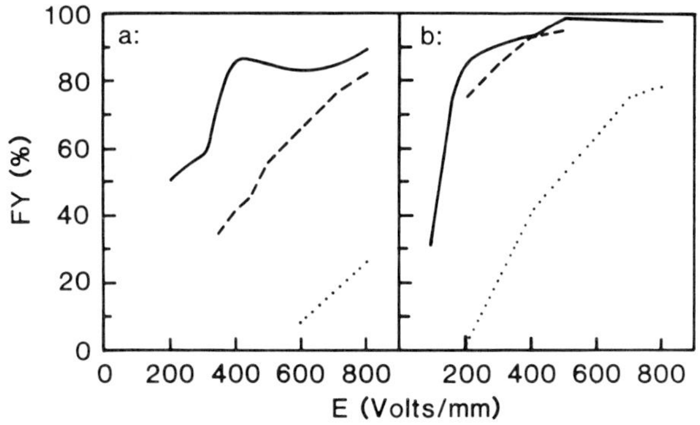

Fig. 7. Measured fusion yields (FY) as a function of pulse voltage when fusion partners are adult chicken erythrocyte ghosts (ACEG) and DiI-labeled chicken embryo erythrocyte ghosts (CEEG) in 20 mM (a) and 120 mM (b) ammonium acetate buffer (containing 1 mM $MgCl_2$) at pH 7.4 (compare with Fig. 6 and see text). Partners are CEEG +CEEG (solid), ACEG + CEEG (dashed), and ACEG + ACEG (dotted). Membranes are held in contact by dielectrophoresis and fusion assayed after a single pulse ($T_{1/2}$ = 1.2 msec).

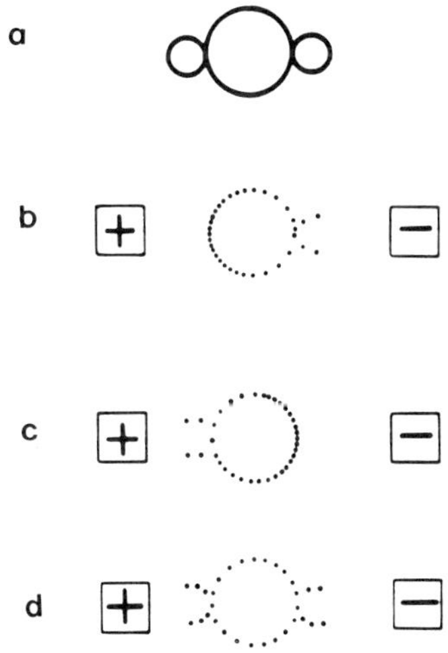

Fig. 8. Expected fusion outcomes from application of single pulse to pearl chain of membranes which include a DiI-labeled chicken embryo erythrocyte ghost flanked by two unlabeled adult chicken erythrocyte ghosts. a) view in phase optics, b-d) view in fluorescence optics: b) lateral diffusion of fluorescent label towards negative electrode, c) lateral diffusion of fluorescent label towards positive electrode, and d) lateral diffusion of fluorescent label towards both electrodes.

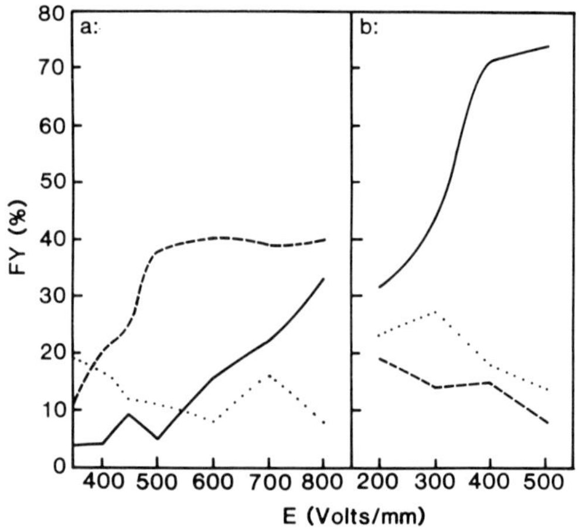

Fig. 9. Direction of DiI migration following application of a single pulse ($T_{1/2}$ = 1.2 msec) to induce fusion of DiI-labeled chicken embryo erythrocyte ghost with one or both flanking unlabeled adult chicken erythrocyte ghosts in 20 mM (a) or 120 mM (b) $MgCl_2$-containing ammonium acetate buffer (compare with Fig. 8) after application of a single pulse. Migration direction: towards negative electrode only (dashes); towards positive electrode only (dots); both directions (solid). See Fig. 8. Fusion yield assay counts three direction possibilities only from alignments containing a large DiI-labeled membrane flanked by two small unlabeled membranes.

SUMMARY

The data suggest that fusion induced in human erythrocyte ghosts may proceed by two mechanisms (Sowers and Kapoor, 1987). The first mechanism, a consequence of the use of the CF protocol, may involve a trivial scenario in which a gross disturbance of the membrane creates holes large enough to be a substantial fraction of a membrane thickness and that self-organizing processes re-stabilize the disturbed membrane into a fused membrane. The second mechanism is clearly revealed if the PF protocol is used and appears to come about as a result of a more subtle pulse-induced change in the membrane. If electropores are involved, then they are very small (residual) and could be on only one of the two hemispheres on each membrane if they are formed from incompletely resealed peak pores. Residual pores are such a small fraction of the membrane thickness as to require a substantial geometrical reconstruction to be involved in the fusion mechanism. It is possible that electropores are not involved in the fusion mechanism if the PF protocol is involved. If this proves true, then a hitherto unrecognized membrane property or condition may be responsible and its elucidation may significantly benefit our understanding of membrane fusion in general. The data also suggest that the fusion process may involve aspects which are not equally shared between the two membranes.

Acknowledgement — This is contribution No. 747 from the American Red Cross laboratories. We thank S. Rossi for typing.

REFERENCES

Bates, G., Saunders, J., and Sowers, A.E., 1987, Electrofusion: principles and applications, in: "Cell Fusion," A.E. Sowers, ed., Plenum Press, New York.

Berg, H., Bauer, E., Berg, D., Forster, W., Hamann, M., Jacob, H.-E., Kurischko, A., Muhlig, P., and Weber, H., 1983, Cell fusion by electric fields, Studia Biophys., 94:93.

Berg, H., 1982, Molecular biological implications of electric-field effects, Studia Biophys., 90:169.

Blangero, C., and Teissie, J., 1985, Ionic modulation of electrically induced fusion of mammalian cells, J. Memb. Biol., 86:247.

Buschle, R., Ringsdorf, H., and Zimmermann, U., 1982, Electric field-induced fusion of large liposomes from natural and polymerizable lipids, FEBS Let., 150:38.

Chan, L.-N., 1977, Changes in the composition of plasma membrane proteins during differentiation of embryonic chick erythroid cell, Proc. Nat'l. Acad. Sci. USA, 74:1062.

Chernomordik, L.V., Sukharev, I.G., Abidor, I.G., 1985, Long-living defects in BLM after reversible electrical breakdown, Biologicheskie Membrany, 2:87.

Conrad, M.K., Lo, M.M.S., Tsong, T.Y., and Snyder, S.H., 1987, Bioselective cell-cell fusion for antibody production, in: "Cell Fusion," A.E. Sowers, ed., Plenum Press, New York.

Diacumakos, E.G., 1973, Microsurgically fused human somatic cell hybrids: analysis and cloning, Proc. Nat'l. Acad. Sci., USA, 69:2959.

Dimitrov, D.S., and Jain, R.K., 1984, Membrane stability, Biochim. Biophys. Acta, 779:437.

Düzgüneş, N., Hong, K., Baldwin, P.A., Bentz, J., Nir, Shlomo, and Papahadjopoulos, D., 1987, Fusion of phospholipid vesicles induced by divalent cations and protons: modulation by phase transitions, free fatty acids, monovalent cations, and polyamines, in: "Cell Fusion," A.E. Sowers, ed., Plenum Press, New York.

Fricke, H., 1953, The electric permittivity of a dilute suspension of membrane covered ellipsoids, J. Appl. Phys., 24:644.

Goodby, J.W., 1986, Optical activity and ferroelectricity in liquid crystals, Science, 231:350.

Gross, D., Loew, L.M., and Webb, W., 1986, Optical imaging of cell membrane potential changes by applied electric fields, Biophys. J., 50:339.

Hofmann, G.A., and Evans, G.A., 1986, Electronic genetic-physical and biological aspects of cellular electromanipulation, IEEE Eng. Med. Biol. Mag., 5:6.

Hofmann, G.A., 1988, Physical and Electronic Aspects of Electro Cell Manipulation, in: "Electroporation and Electrofusion in Cell Biology," E. Neumann, A.E. Sowers, and C.A. Jordan, eds., Plenum Press, New York.

Knight, D.E., and Scrutton, M.C., 1986, Gaining access to the cytosol: The technique and some applications of electropermeabiliziation, Biochem. J., 234:497.

Lieber, M.R., and Steck, T.L., 1982a, A description of the holes in human erythroctye membrane ghosts, J. Biol. Chem., 257:11651.

Lieber, M.R., and Steck, T.L., 1982b, Dynamics of the holes in human erythroctye membrane ghosts, J. Biol. Chem., 257:11660.

Lo, M.M.S., Tsong, T.Y., Conrad, M.K., Strittmatter, S.M., Hester, L.D., and Snyder, S.H., 1984, Monoclonal antibody production by receptor-mediated electrically-induced cell fusion, Nature (London), 310:792.

Lovinger, A.J., 1983, Ferroelectric polymers, Science, 220:1115.

Malmstadt, H.V., Enke, C.G., and Toren, E.C., 1963, "Electronics for scientists," W. A. Benjamin, New York.

Malmstadt, H.V., Enke, C.G., and Crouch, S.R., 1981, "Electronics and Instrumentation for Scientists," Benjamin/Cummings, Menlo Park.

Melikyan, G.B., Abidor, I.G., Chernomordik, L.V., and Chailakhyan, L.M., 1983, Electrostimulated fusion and fission of bilayer lipid membranes, Biochim. Biophys. Acta, 730:395.

Miles, D.M., and Hochmuth, R.M., 1987, Micromanipulation and elastic response of electrically fused red cells, in: "Cell Fusion," A.E. Sowers, ed., Plenum Press, New York.

Mischke, S., Saunders, J.A., and Owens, L., 1986, A versatile low-cost apparatus for cell electrofusion and other physiological treatments, J. Biochem. Biophys. Meth., 13:65.

Moore, J.H., Davis, C.C., and Coplan, M.A., 1983, "Building Scientific Apparatus," Addison-Wesley, London.

Neumann, E., 1984, Electric gene transfer into culture cells, Bioelectrochem. Bioenerget., 13:219.

Neumann, E., Sowers, A.E., and Jordan, C.A., 1988, "Electroporation and Electrofusion in Cell Biology," Plenum Press, New York.

Neumann, E., Schaefer-Ridder, M., Wang, Y., and Hofschneider, P.H., 1982, Gene transfer into mouse lyoma cells by electroporation in high electric fields, EMBO J., 1:841.

Pilwat, G., Richter, H.-P., and Zimmermann, U., 1981, Giant culture cells by electric field-induced fusion, FEBS Let, 133:169.

Pohl, H.A., 1978, "Dielectrophoresis," Cambridge Univ. Press, London.

Pohl, H.A., Pollock, K., and Rivera, H., 1984, The electrofusion of cells, Int. J. Quant. Chem: Quant. Biol. Symp., 11:327.

Porschke, D., 1985, Effects of electric fields on biopolymers, Ann. Rev. Phys. Chem., 36:159.

Rand, R.P., and Parsegian, V.A., 1986, Mimicry and mechanism in phospholipid models of membrane fusion, Ann. Rev. Physiol., 48:201.

Rosenberg, B., Van Camp, L., and Krigas, T., 1965, Inhibition of cell division in Escherichia coli by electrolysis products from a platinum electrode, Nature (London), 205:698.

Sauer, F.A., 1983, Forces on suspended particles in the electromagnetic field, in: "Coherent Excitations in Biological Systems," H. Frohlich, and F. Kremer, eds., Springer Verlag, New York.

Schierenberg, E., 1987, Laser-induced cell fusion, in: "Cell Fusion," A.E. Sowers, Ed., Plenum Press, New York.

Schwister, K., and Deuticke, B., 1985, Formation and properties of aqueous leaks induced in human erythrocytes by electrical breakdown, Biochim. Biophys. Acta, 816:332.

Serpersu, E.H., Kinosita Jr., K., and Tsong, T.Y., 1985, Reversible and irreversible modification of erythrocyte membrane permeability by electric field, Biochim. Biophys. Acta, 812:779.

Siegel, D.P., 1987, Membrane-membrane interactions via intermediates in lamellar-to-inverted hexagonal phase transitions, in: "Cell Fusion," A.E. Sowers, ed., Plenum Press, New York.

Sowers, A.E., 1986a, A long-lived fusogenic state is induced in erythrocyte ghosts by electric pulses, J. Cell Biol., 102:1358.

Sowers, A.E., 1986b, Long-lived fusogenic membrane sites induced by electric field pulses are not free to diffuse laterally in the plane of the membrane, Biophys. J., 49:132a.

Sowers, A.E., 1985a, Movement of a fluorescent lipid label from a labeled erythrocyte membrane to an unlabeled erythrocyte membrane following electric field-induced fusion, Biophys. J., 47:519.

Sowers, A.E., 1985b, Electric field-induced membrane fusion in erythrocyte ghosts: evidence that pulses induce a long-lived fusogenic state and that fusion may not involve pore formation. Biophys. J., 47:171a.

Sowers, A.E., 1984, Characterization of electric field-induced fusion in erythrocyte ghost membranes, J. Cell Biol., 99:1989.

Sowers, A.E., 1983a, Fusion of mitochondrial inner membranes by electric fields produces inside out vesicles: visualization by freeze-fracture electron microscopy, Biochim. Biophys. Acta, 735:426.

Sowers, A.E., 1983b, Red cell and red cell ghost membrane shape changes accompanying the application of electric fields for inducing fusion, J. Cell Biol., 97:179a.

Sowers, A.E., and Hackenbrock, C.R., 1981, Rates of lateral diffusion of intramembrane particles: Measurements by electrophoretic displacement and rerandomization, Proc. Natl. Acad. Sci., USA 78:6246.

Sowers, A.E., and Kapoor, V., 1987a, The electrofusion mechanism in erythrocyte ghosts, in: "Cell Fusion," A.E. Sowers, ed., Plenum Press, New York..

Sowers, A.E., and Kapoor, V., 1987b, Fusogenic membrane alterations induced by electric field pulses, in: "Mechanistic Approaches to Interactions of Electromagnetc Fields with Living Systems," M. Blank and E. Findl, eds., Plenum Press, New York.

Sowers, A.E., and Kapoor, V., 1988, The Mechanism of electrically-induced fusion in erythrocyte membranes, in: "Electroporation and Electrofusion in Cell Biology," E. Neumann, A.E. Sowers, and C.A. Jordan, eds., Plenum Press, New York.

Sowers, A.E., and Lieber, M.L., 1986, Electropores in individual erythrocyte ghosts: diameters, lifetimes, numbers, and locations, FEBS Let., 205:179.

Stenger, D.A., and Hui, S.W., 1986, Kinetics of ultrastructural changes during electrically induced fusion of human erythrocytes, J. Memb. Biol., 93:43.

Stenger, D.A., and Hui, S.W., 1988, Electron microscopy of electrofusion, in: "Electroporation and Electrofusion in Cell Biology," E. Neumann, A.E. Sowers, and C.A. Jordan, eds., Plenum Press, New York.

Stulen, G., 1981, Electric field effects on lipid membrane structure, Biochim. Biophys. Acta, 640:621.

Sukharev, S.I., Popov, S.V., Chernomordik, L.V., and Abidor, I.G., 1985, A patch-clamp study of electrical breakdown of cell membranes, Biologicheskie Membrany, 2:77.

Szoka, F.C., 1987, Lipid vesicles: model systems to study membrane-membrane destabilization and fusion, in: "Cell Fusion," A.E. Sowers, ed., Plenum Press, New York.

Tanaka, T., Nishio, I., Sun, S.-T., and Ueno-Nishio, S., 1982, Collapse of gels in an electric field, Science, 218:467.

Tsong, T.Y., 1983, Voltage modulation of membrane permeability and energy utilization in cells, Bioscience Reports, 3:487.

Von Hippel, A., 1958, Dielectrics, in: "Handbook of Physics," E.U. Condon and H. Odishaw, eds., McGraw-Hill, New York.

Zachrisson, A., and Bornman, C.H., 1984, Application of electric field fusion in plant tissue cutlure, Physiol. Plant., 61:314.

Zimmermann, U., 1982, Electric field-mediated fusion and related electrical phenomena, Biochim. Biophys. Acta, 694:227.

Zimmermann, U., 1986, Electrical breakdown, electropermeabilization and electro fusion, Rev. Physiol. Biochem. Pharmacol., 105:176.

Zimmermann, U., Buchner, K.-H., and Arnold, W.M., 1984, Electrofusion of cells: recent developments and relevance for evolution. in: "Charge and Field Effects in Biosystems," M.J. Allen and P.N.R. Usherwood, eds., Abacus Press, McComb.

Zimmermann, U., Vienken, J., Halfmann, J., and Emeis, C.C., 1985, Electrofusion: A novel hybridization technique, in: "Advances in Biotechnological Processes," vol. 4, A. Mizrahi and A.L. van Wezel, eds., Alan R. Liss, New York.

WATER-MEDIATED EFFECTS OF PEG ON MEMBRANE PROPERTIES AND FUSION

Klaus Arnold, Andreas Herrmann[2], Klaus Gawrisch[1] and Lothar Pratsch[2]

Institute of Biophysics and [1]Department of Physics of the Karl Marx University, Leipzig, G.D.R
[2]Department of Biology of the Humboldt University Berlin, G.D.R.

INTRODUCTION

Poly(ethylene glycol) (PEG) has met wide applications in biology and medicine. PEG is used as a fractional precipitating agent for protein purification and isolation, DNA and protein crystallization and for the cryopreservation of biological material. PEG of a molecular weight of about 6000 is commonly used to induce fusion of liposomes as well as of a wide variety of plant protoplasts and animal cells.

PEG induced cell-cell fusion is now a standard technique in biological, medical and biotechnological practice. However, the interaction of PEG with biological membranes and the mechanisms of membrane and cell fusion are not well understood. The fusogenic activity of PEG has been related to the ability of PEG to alter properties of the lipid phase of membranes, i.e. to lower the surface potential (Maggio et al. 1976), to induce the formation of nonlamellar structures within the membrane (Boni et al. 1981b), to raise the transition temperature of lipids (Tilcock and Fisher 1979), to enhance membrane permeability (Ahkong et al. 1975, Maggio et al. 1976, Aldwinckle et al. 1982), to induce the aggregation of intramembraneous particles (IMP) in cell membranes (Knutton 1979, Krähling et al. 1978, Hui et al. 1985) and to be bound to membranes (Boni et al. 1984a).

In this paper, new evidence is given that PEG is excluded from the membrane surface. This result supports the hypothesis that PEG exerts its influence on membrane systems in an indirect manner due to a change of the physicochemical properties of the aqueous phase (Arnold et al. 1983, Herrmann et al. 1983a, Arnold et al. 1986a). The influence of the changed physicochemical properties of the aqueous phase on the bilayer approach in the lamellar phase, on phase transitions and phase separations, partition coefficients, ion interaction and the organization of membrane proteins is investigated. The relevance of these results for the aggregation and fusion of membranes is discussed.

INFLUENCE OF PEG ON THE PHYSICOCHEMICAL PROPERTIES OF AQUEOUS SOLUTIONS

Hydration of PEG and Volume Exclusion

PEG is one of the simplest synthetic polymers having a high solubility in water. The regular alteration of ether oxygens and ethylene groups of the PEG molecule results in a significant perturbation of the structure of water. The new structure is a complex network consisting of both water and polymer molecules as structural elements (Mc Pherson 1985). The effects of PEG on the structure of water correspond to the effects of molecules which belong to the structure makers (Das and Singhal, 1981). From the chemical point of view the ethylene oxide group can interact with two water molecules on the basis of hydrogen bonding between the water molecules and the ether oxygens (Molynex 1975). But hydration is not a precisely defined quantity and therefore different hydration numbers were determined using different methods (Foster et al. 1984).

The quantity of unbound water in solutions of PEG was determined by differential scanning calorimetry (Tilcock and Fisher 1982). The lowest concentration of polymer at which all the water was bound was dependent on the molecular weight of PEG and ranged from 59.5 wt% for PEG_{200} to 48.1 wt% for PEG_{6000}. These data correspond to a water binding ability of each monomer unit from 1.8 molecules for PEG_{200} to 2.7 molecules for PEG_{6000}. Other authors found that no free water exists in an aqueous solution of more than 38 wt% PEG (Baran 1972, Blow et al. 1978). It has also been demonstrated that PEG may exercise a structuring effect on several layers of water around the polymer. This can involve up to 16 water molecules per monomer unit. This means that all the water would be structured in solutions of more than 13 wt% PEG (Baran et al. 1972).

The restructuring of the water causes an exclusion as well as a promotion of phase separation of macromolecules, particularly proteins. This property is called volume exclusion principle. The proteins are sterically excluded from regions of the water occupied by the inert PEG and they are thus concentrated until their solubility is exceeded and precipitation occurs (Atha and Ingham 1981). The volume exclusion principle was employed to explain the aggregation of sonicated vesicles (Tilcock and Fisher 1982).

Dielectric Properties

Measurements of the static dielectric constants of PEG in the liquid state give values in the order of 10 to 15 with the lower value for those with the higher molecular weight. A drastic decrease of the dielectric constant from $\epsilon \simeq 80$ for water was confirmed on addition of PEG (Arnold et al. 1985a, Foster et al. 1984, Masszi et al. 1986). In Fig. 1 the dielectric constant ϵ measured at a frequency of 7 MHz is given as a function of the PEG concentration and the molecular weight of the polymer. Linear extrapolation of the concentration dependence of the dielectric constant to 100 wt% gives the values of pure PEG. A linear dependence of ϵ on the volume fraction of PEG in a 0.01 M $MnSO_4$ solution was observed by Masszi et al. (1986). From their measurements a stronger decrease of ϵ in the range of low concentrations can be inferred as compared to the data given in Fig. 1. The concentration dependence of the dielectric constant on PEG's of different molecular weights are very similar in the range of higher molecular weights (>1000) indicating that the effect of the polymer depends mainly on the monomer concentration.

In order to understand the effect of a solvent in a solution the polarity is a more useful parameter than the dielectric constant because the dielectric constant is a macroscopic quantity and polarity describes the internal polar properties. Changes in the solvent polarity detected by shifts of the maximum of fluorescence and changes of the quantum yield of ANS after addition of PEG are explained by a reduced rotational mobility of water molecules due to binding to PEG and by a reduced polarity of the surrounding medium of the fluorescence molecule. In Fig. 1 the isotropic hyperfine splitting of the small paramagnetic molecule TEMPO as a function of the PEG concentration is given (Herrmann et al. 1983a). A linear relationship between A_{iso} and the PEG concentration was found. According to the studies of Briere et al. (1965) these results suggest that the polarity of the aqueous solution is reduced upon increasing the PEG concentration.

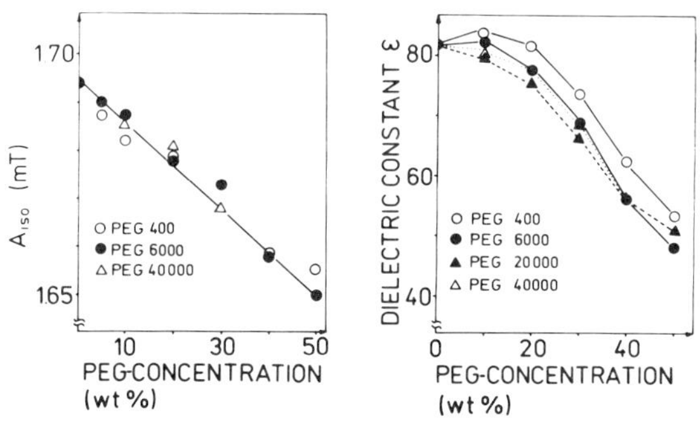

Fig. 1. The isotropic hyperfine splitting A_{iso} of TEMPO and the dielectric constant ϵ of aqueous PEG solutions as a function of PEG concentration (data were partly taken from Herrmann et al. (1983a) and Arnold et al. (1985a)).

The influence of the changed polarity on the partition of molecules between the membrane and the external phase will be discussed in one of the following chapters. Besides this property the dielectric constant of the medium is known to influence the stability of colloids because it changes the electrostatic repulsion. The ion-induced coagulation of particles exhibits an ϵ^3 dependence of the critical coagulation concentration (c.c.c.) on the electrolyte concentration (Overbeek 1982). This behaviour could be of interest for the ion-induced aggregation and fusion of liposomes and cells in the presence of PEG (Hoekstra 1982). At higher concentrations of PEG the dielectric constant of the solvent could be sufficiently low to favour ion-pair formation and the neutralization of ionized groups. More experimental data are necessary to estimate the importance of these factors for the fusion process.

Osmotic Pressure of Aqueous PEG Solutions

All published values of osmotic pressures of aqueous PEG solutions are much higher than predicted by ideal solution theory. Moreover, there are significant differences between osmotic pressures of comparable aqueous PEG solutions reported in the literature. For example, the values reported for a solution of 30 wt% PEG_{6000} in water at 298 K differ from each other by up to a factor of two (Chirife and Fontan 1980, Hui et al. 1985, Herrmann et al. 1985). The results may be influenced by the distribution of molecular weights of PEG depending on the source.

We determined the osmotic properties of aqueous PEG solutions by water vapour pressure measurements relative to the water vapour pressure over pure water at the same temperature. In accordance to Chirife and Fontan (1980) the water activities could be well described by the following formula

$$a_w = X_1 * \exp(-K * X_2^2) \qquad (1)$$

X_1 and X_2 are the mole fractions of water and PEG in the solution. K is a constant which depends on the molecular weight of PEG and the temperature. The experimentally determined parameters K are given in Table 1. The osmotic pressure π is connected with the activity of water molecules by the following formula

$$\pi = - R_oT/V_w * \ln a_w \qquad (2)$$

R_o - absolute gas constant; T - temperature in (K); V_w - molar volume of water in solution.

The osmotic pressures of PEG's of different molecular weight are given in Fig. 2. At a concentration of 50 wt% PEG osmotic pressures were observed ranging between 120 atm (PEG_{6000}) and 200 atm (PEG_{150}). It should be mentioned that the osmotic pressures of ^2H_2O solutions containing the same mole fraction of water coincide with osmotic pressures of H_2O solutions.

Table 1. Experimentally determined K-values for the different PEG's used at 298 K.

Molecular weight of PEG	Assumed average molecular weight	K
20,000 - 17,000 (Serva)	18,500	72,500
7,500 - 6,000 (Serva)	6,750	10,800
420 - 380 (Serva)	400	28.5
150 (Merck)	150	2.8

STRUCTURAL CHANGES IN PHOSPHOLIPID/WATER DISPERSIONS

Dehydration Effects of PEG on Lipid Membranes

^{2}H-NMR of water at membrane surfaces Hydration properties of multilamellar lipid water dispersions can be investigated by ^{2}H-NMR measurements if the lipids are hydrated with ^2H_2O. Water molecules near the membrane surface show a quadrupole splitting. Because of fast exchange between the water in layers far from the membrane surface with the water at the membrane surface the quadrupole splitting decreases with increasing water concentration in the sample. Quantitative studies of lecithin bilayer hydration by ^{2}H-NMR were reported by Finer and Darke (1974). In this paper a model for the quantitative description of lipid membrane hydration was given. It is assumed that at the membrane surface different hydration layers characterized by different effective quadrupole splittings of water exist. All water molecules in one layer exhibit equal splitting . The next layer is only occupied if the layers before are completely filled with a definite number of molecules. The rate of fast exchange of water molecules between all filled layers is in the order of 10^{-11} s. The quadrupole splitting can be quantitatively described by the number of water molecules of each hydration shell and their specific quadrupole splitting. The dependence of the quadrupole splitting on the reciprocal ratio of water molecules per lipid molecule is characterized by a series of straight lines with different slopes. From the curves the number of water molecules per shell and the specific quadrupole splitting of every hydration shell can be deduced (Gawrisch et al. 1985).

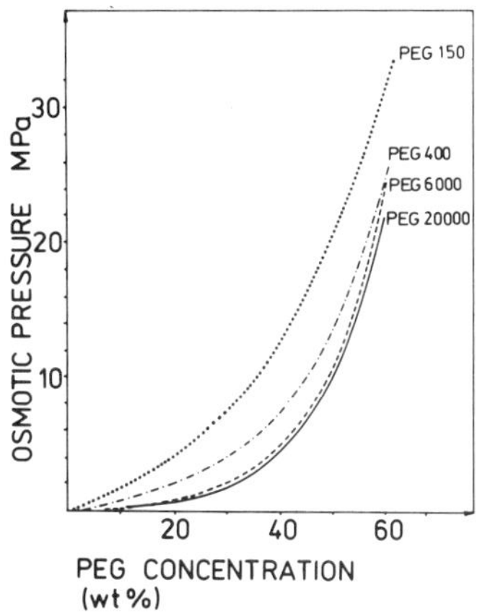

Fig. 2. Osmotic pressures of aqueous PEG solutions as a function of PEG concentration. (For clarity experimental points are not given)

Characteristic ^{2}H-NMR spectra of phospholipid water dispersions are given in Fig. 3a for low water content and in Fig. 3c for high water content (excess of water). On the addition of low amounts of water up to about 23 water molecules per phospholipid all water is arranged between the lamellae and a quadrupole split signal results as shown in Fig. 3a. The spectrum for high water content consists of two components. The narrow intense signal at the centre is assigned to the free water. The broad second component originates from the quadrupole-split signal of the interlamellar water. The spectrum given in Fig. 3b belongs to a phospholipid dispersion with an excess of water after addition of PEG. The quadrupole splitting of the second component is increased compared to the spectrum at the same water content in the absence of the polymer. This increase of the quadrupole splitting indicates a decrease of the amount of water incorporated between the lamellae. The values measured are independent of the experimental conditions whether there is direct contact between PEG and the lecithin or contact only through a dialysis membrane nonpermeable to PEG molecules. From this result we may conclude that the PEG molecules of higher molecular weight will not enter the water layer between lipid bilayers.

The real number of water molecules per lipid incorporated between the bilayers is given in Fig. 4. These values were calculated via the measured quadrupole coupling constants as discussed above. At a fusogenic concentration of 50 wt% PEG approximately 8 water molecules per phospholipid molecule are included between the bilayers. If the area per lipid molecule and the density of water between lipid bilayers are known the thickness of the water layer between lipid bilayers can be calculated.

Membrane dehydration is obviously driven by the high osmotic pressure of the PEG water solution. The distribution of water molecules between the lipid and the PEG phase can be explained by the concept of chemical potential differences of water between different phases. Near a hydrophilic membrane surface the chemical potential of water is decreased in comparison to free water (Rand, 1981). In the PEG water solution the chemical potential of water is decreased due to the interaction with PEG. Differences in the chemical potentials of water molecules between different phases lead to a redistribution of the water until the potentials in both phases are equal. The higher the concentration of PEG in solution the thinner the water layer between lipid bilayers.

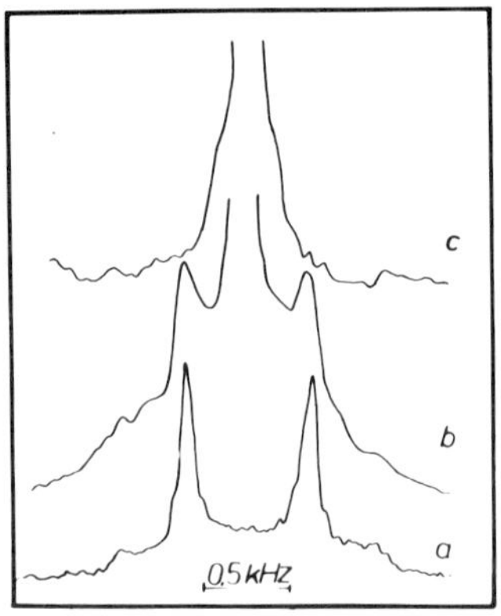

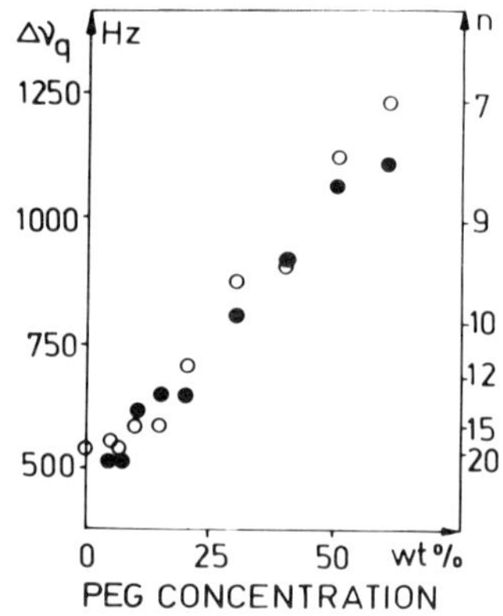

Fig. 3. ^{2}H-NMR spectra of DPPC/^2H_2O dispersions at T = 334 K for samples with (a) 14.3 and (c) 46.5 water molecules per phospholipid molecule. Spectrum (b) was obtained on the addition of 47 wt.% PEG_{400} (Data taken from Arnold et al. (1983)).

Fig. 4. ^{2}H-NMR quadrupole splitting $\Delta\nu_q$ of deuterated water as a function of the concentration of PEG_{20000} for an EYL/^2H_2O dispersion in excess of water. The number of water molecules **n** incorporated between the bilayers is given on the right ordinate. PEG was directly added (o) or separated from the dispersion by a dialysis membrane (●).

Forces between lipid bilayers. With the help of hydrophilic polymers which do not penetrate into the water layer between lipid bilayers repulsive forces between the bilayers can be measured (Rand 1981, LeNeveu et al. 1976). If the attractive forces between bilayers can be neglected in comparison to the repulsive forces the osmotic pressure of the surrounding PEG water solution equals the repulsive pressure between lipid bilayers. Using the ^{2}H-NMR method and the osmotic pressures of PEG the repulsive force between bilayer membranes was determined. An exponential decay of the repulsive force was found as observed before by means of small angle X-ray scattering.

The quantitative agreement of the forces determined by use of PEG and ^{2}H NMR with those obtained by use of dextran and scattering methods (Rand, 1981) is satisfactory. Therefore, from an analysis of experimental results in terms of forces between bilayers we have additional evidence that PEG molecules of higher molecular weight will not enter the water layer between the lipid bilayers. Significant deviations from force parameters between lecithin bilayers reported in the literature were only obtained for PEG of molecular weights lower than 400. At these molecular weights the PEG molecules become able to penetrate into the water layer between lecithin bilayers. But nevertheless, the concentration of PEG in that layer is still lower than the concentration in the surrounding solution, which also results in an osmotic compression of the bilayer system. The low concentration of these PEG molecules in the water layer between the lipid bilayers is therefore an indication for the absence of a specific interaction between lipid and PEG molecules.

The results given above strongly indicate that the PEG-induced aggregation of lipid membranes is caused by the high osmotic pressure of PEG water solutions together with a reduced solubility of PEG molecules in the water layer near to the membrane surface. From measurements with lipid model membranes we have no evidence for a direct interaction between lipid and PEG molecules often discussed in literature.

Effect of Membrane-bound Poly(ethylene oxide)chains on Bilayer-bilayer Interaction

Large molecules adsorbed on or bound to a particle surface can lead to a repulsive force between the particles (for a review see Overbeek (1982)). The following experiments will demonstrate that poly(ethylene oxide) chains anchored in phospholipid bilayers also increase the bilayer-bilayer repulsion (Arnold et al. 1986b).

Poly(ethylene oxide) surfactants with different numbers n of ethylene oxide groups and with a nonyl-phenyl alkyl chain were used (Fig. 5). As shown by NMR experiments these surfactants are incorporated into the bilayer system with the ethylene oxide part extending into the water layer and the alkylphenyl group imbedded in the phospholipid bilayer (Klose et al. 1986). In Fig. 5 the repeat spacing d measured for different surfactants is represented as a function of the number of the ethylene oxide groups at a temperature below and above the phase transition temperature T_u. The surfactants and DPPC molecules were used in a relatively low molar ratio of 1:30. An increase of the concentration is limited because of the ability of the system to form nonbilayer arrangements, i.e. micelles at higher surfactant concentrations.

It can be concluded that the bilayer separation is increased on addition of these surfactants. An additional repulsive force occurs which depends on polymer concentration and length. Below the phase transition temperature the repeat spacing has a smaller dependence on the chain length of the polymer part due to a decreased solubility of the surfactant in the crystalline DPPC bilayer. The interaction of such polymer-coated surfaces was described by a theory given by Bell et al. (1984). A sufficiently good agreement of the experimental data with this theory was found (Arnold et al. 1986b).

The important conclusion from these experiments is that the strong aggregation of liposomes and cells observed after addition of PEG cannot be explained by a bridging mechanism. The existence of the ethylene oxide chains in the phospholipid surface would lead to an increase of the repulsive force. For inducing aggregation the PEG molecules must be separated from the contact region of the bilayers. The influence of such surfactant molecules on hemolysis and fusion in the presence of high concentrations of PEG was examined in regard to their potential as promoters of cell fusion (Honda et al. 1981a). The fusion activity is reduced for surfactants with longer poly(ethylene oxide) chains. This could be caused by the inhibiting effect of the surfactants on the aggregation process.

Phase Transitions

An increase of the temperature of the gel-liquid crystalline transition of pure phospholipid dispersions on addition of PEG was observed by use of DSC, NMR, ESR, freeze-fracture electron microscopy and X-ray diffraction (Tilcock and Fisher 1979, Arnold et al. 1983, Herrmann et al. 1983a). The main transition is broadened, indicating that the cooperativity of the phase transition is decreased. This behaviour is a general property of all structure makers (Das and Singhal 1981). According to the ^{2}H-NMR measurements given above the amount of water bound to the phospholipid headgroups is reduced on addition of PEG. It was shown that the shift of the phase transition temperature correlates very well with the amount of water bound to the headgroups. These results are consistent with those for less hydrated bilayers (Arnold et al. 1983). Freeze-fracture electron microscopy revealed numerous morphological features, some of which are known to occur in lipids at low hydration (Boni et al. 1984b). Therefore it can be concluded that the increase of the phase transition temperature is a direct result of PEG-induced dehydration. It is not necessary to postulate a specific interaction of PEG with the phospholipid molecules.

A dehydration of the phospholipid dispersion results in a decrease of the mobility of the phospholipid molecule for the fatty acid region as well as for the headgroups (Volke et al. 1982). The molecular area of the molecules decreases and the lateral pressure increases. The dehydrating action of PEG on phospholipid dispersions should consequently lead to similar changes of the membrane fluidity as observed for membranes at low water content. The order parameter of model membranes (Surewicz 1983, Herrmann et al. 1983a, Arnold et al. 1986a) and of biological membranes (Surewicz 1983, Arnold et al. 1986a, Gawrisch 1986, Boss 1983) is increased on addition of PEG.

The increase of the order parameter of spin probes in biological membranes is partly caused by an influence of PEG on the organization of proteins (Gawrisch 1986). The mobility of the phosphate group is not influenced by PEG (Arnold et al. 1983). This can be explained by the low sensitivity of the mobility of the phosphate group to the decrease of the water content of phospholipid dispersions. The ^{31}P-NMR spectrum is only influenced by water concentrations of lower than 5 water molecules per headgroup. Such a dehydration is observed for PEG concentrations larger than 50 wt% as demonstrated above. ^{31}P-NMR was also used to investigate the influence of PEG on the lateral mobility of phospholipids in membranes (Gawrisch et al. 1986). A significant change of the lateral diffusion rate of phospholipids in erythrocyte vesicle membranes was not observed.

In many bilayer systems composed of a mixture of lipids a segregation of lipid species into separate coexisting lamellar phases was observed on the mutual approach of membranes. It was shown that sphingomyelin and lecithin mixtures separate into different lamellar phases if the water is decreased (cited in Rand 1981). The interpretation of these data is difficult because it cannot be established that the lateral segregation occurs within the bilayers. The action of PEG on mixtures in multibilayer systems should be comparable to the effects of the lowering of the water content. Mixtures of dipalmitoyl phosphatidylcholine (DPPC) and dipalmitoyl phosphatidylethanolamine (DPPE) were used to study the influence of PEG on the lateral distribution of the components. In Fig. 6 phase diagrams are given for this mixture before and after the addition of PEG_{400} (Gawrisch 1986).

The phase diagram of the sample without PEG exhibits a phase separation of the components in the gel state as was described before (Arnold et al. 1981, Petrov et al. 1982). Addition of 60 wt% PEG increases the phase transition temperatures of the pure components and therefore the whole phase diagram is shifted to higher temperatures. However, the behaviour of the liquidus curve is also changed indicating a phase separation in the liquid crystalline state as well. The different hydrations of the headgroups of DPPC and DPPE could be the reason for the phase separation under the influence of the dehydrating action of PEG. This result is of great importance for the aggregation and fusion process of membranes because it demonstrates that the lateral distribution of phospholipids can be changed under the influence of PEG, also in the liquid crystalline state. This could result in a maximization of the interactions of bilayers as postulated by Rand (1981).

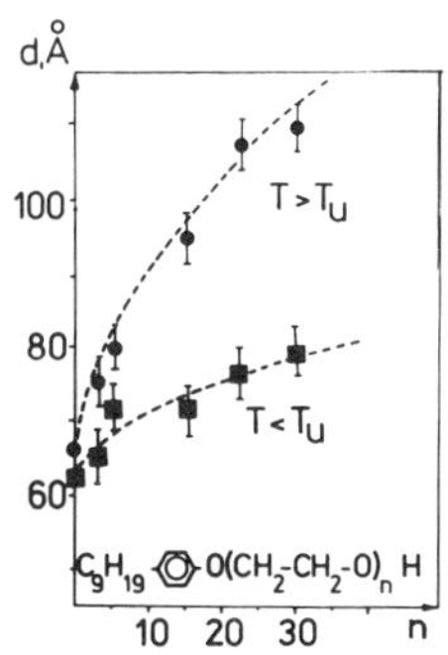

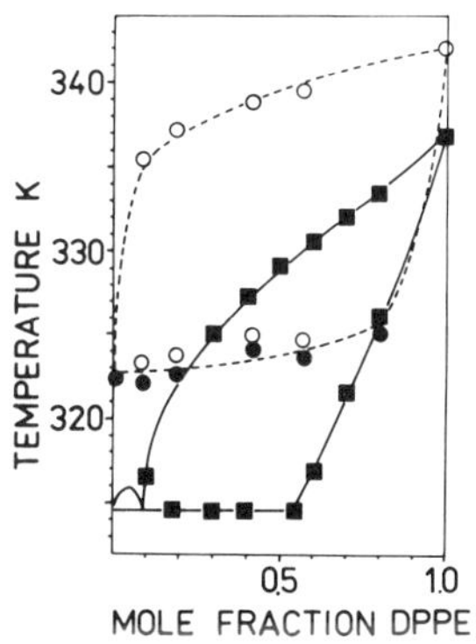

Fig. 5. Changes of the repeat spacing **d** determined from X-ray experiments of a lamellar DPPC/ water dispersion as a function of the chain length of the poly(ethylene oxide) surfactant added in a molar ratio of 1:30. The structure of the molecule is given in the figure. (Data taken from Arnold et al. (1986b)).

Fig. 6. Phase diagrams of DPPC/DPPE mixtures with an excess of water before (■) and after (○) the addition of 60 wt% PEG_{400}.

The formation of nonlamellar phases as a mechanism of membrane fusion is still under discussion (Cullis and Hope 1978, Verkleij et al. 1979). The nonlamellar hexagonal II phase formed by phospholipids can be understood on the basis of the shape and interaction of phospholipids. When the hydrophilic volume of the molecule decreases relative to the hydrophobic volume a hexagonal phase can be formed because the molecules cannot pack on a plane, and a curved surface is preferred. Dispersions from unsaturated phosphatidylethanolamines are examples of systems which have transitions from the bilayer to the hexagonal phase. Boni et al. (1984b) found an increase of the transition temperature from the bilayer to the hexagonal phase and an influence on the formation of lipidic particles on addition of PEG was observed for soybean PE. The repeat spacings of the lamellar and the hexagonal phase are reduced. Different defects in the bilayer occur which could be of interest for the fusogenic activity of PEG. Most of the effects may be explained by the dehydrating action of PEG which reduces the hydrophilic volume and increases the phospholipid packing. However, PEG should have some unique properties since other nonfusogenic agents with comparable dehydrating action are not able to produce similar results (Boni et al. 1984b). This unique property of PEG may result from the influence of PEG on the polarity of the aqueous phase (Arnold et al. 1983) as discussed above.

Partition Coefficients and Membrane Stability

The stability of the membrane is mainly caused by the "hydrophobic effect" which leads to a squeezing of the hydrocarbon chains of amphiphilic molecules out of the water phase. The decreased polarity of the external phase in the presence of PEG should have a drastic influence on the hydrophilic-hydrophobic balance at membrane surfaces. These alterations were considered to induce a change in the partition of molecules between the membrane and the external phase and the conformation of membrane molecules could be altered due to an influence on the arrangement of polar and unpolar groups in the surface. It would be expected that a decrease in the strength of the hydrophobic interaction and a destabilization of the membrane occurs (Arnold et al. 1985a,b).

The influence of PEG on the partition coefficient of a charged spin probe with a long hydrophobic chain was quantitatively studied in different membrane systems (Arnold et al. 1986a). As the polymer concentration increases the partition coefficient decreases exponentially. The chemical potential difference of this spin probe derived from the partition coefficient decreases linearly by about 10 kJ mol^{-1} for PEG concentrations between 0 and 50 wt.%. These measurements reflect a lower thermodynamic stability of the membrane. This is in accordance with other observations showing that spin labelled fatty acids (Herrmann et al. 1983a, Boss 1983) and the fluorescence probes ANS and pyrene (Arnold et al. 1983) leave the membrane on addition of PEG.

Such destabilizations of membranes were found for liposomes in the presence of high PEG concentrations and they were discussed as detergent-like properties of PEG (Saez et al. 1982). The promotion of lipid exchange between multilamellar vesicles at higher concentrations of PEG described by Tilcock and Fisher (1982) could also be caused by this property because the concentration of molecules in the aqueous phase is increased. It is assumed that both the lower polarity of the external phase and the dehydration of the membrane contribute to the decreased difference of the chemical potential. The self-assembling of phospholipids is influenced in such a manner that a minimum aggregate size occurs at about ϵ=33. From other studies it was concluded that the lowest value for the formation of a bilayer structure from phospholipids in polar solvents is about ϵ=35 (Tonkong et al. 1982). McGrath (1981) discussed the influence of membrane dehydration on the chemical potential of the membrane components. It was found that the chemical potential increases and therefore "labile" membrane components can leave the membrane.

Furthermore, the changed chemical potential of one membrane component can lead to a change of the chemical potential of any other membrane component and this is connected with alterations of mechanical and thermodynamical properties of the membrane. Besides desorption of membrane molecules a modified chemical potential of membrane components can therefore drive other processes as phase separation of phospholipids, changes of lateral organization of proteins, formation of defects and microvesicles (McGrath 1981). Such processes were observed at high concentrations of PEG and are assumed to play a crucial role in fusion. Besides a strong aggregation of the membranes a certain degree of bilayer destabilization should be also necessary to preform the membrane for the process of fusion. The experiments demonstrate that PEG makes the membrane surface less polar or more hydrophobic. Ohki (1982, 1985) has recently published a theory for membrane fusion where the increased hydrophobicity of the membrane surface is a main driving force of membrane fusion. It is important to state that these effects can occur without any direct interaction between PEG and the membrane.

Exclusion of PEG from Membrane Surfaces

It has recently been established that PEG is excluded from the surface of water-soluble proteins. It appears that there is a water layer around the proteins which is impermeable to PEG. Its solvent structure is perturbed as compared to the bulk. The protein surface becomes preferentially hydrated (Lee and Lee 1981). Such a preferential hydration is an expression of the difference between the interaction of water and PEG with the protein. Addition of PEG raises the chemical potential of the protein driving the phase separation of polymer and protein. This idea will be valid too when PEG is not completely excluded from the surface and the concentration of the polymer near the surface is only reduced relative to the bulk concentration.

This concept of PEG exclusion from the protein surface should also be applicable to membrane surfaces. A partial or complete exclusion of PEG depending on the molecular weight from the water layer of multilamellar systems was described above. However, experiments performed to characterize the interaction of PEG with the membrane have given contrary results. An interaction with the liposome surface was found by use of tritiated PEG (Boni et al. 1984a). Using spin-labelled PEG Boss (1983) failed to establish a direct interaction between PEG and membrane lipids from plant protoplasts. Many other papers were published concluding from changes of the dynamical and structural properties of membrane components in the presence of PEG that PEG molecules interact directly with the membrane. As we have demonstrated above such influences of PEG on the membrane can also result from an indirect interaction of PEG with the membrane due to an alteration of the properties of the water phase.

Measurements of the electrophoretic mobility of charged liposomes and erythrocytes in the presence of PEG support the idea that PEG has a lower concentration near the membrane surface than in the bulk phase (Arnold et al. 1987). EPR measurements of the surface potential of these systems in the presence of PEG using a positively charged spin probe did not give an increase of the surface potential in the presence of PEG. However, the electrophoretic mobility of erythrocytes and liposomes decreases much less than expected from the change of the viscosity of the suspending medium suggesting an increase of zeta-potential on addition of PEG (Data are given in Fig. 7.). According to Bäumler and Donath (1987) this contradiction can be explained by introducing a distance-dependent viscosity. It is assumed that the concentration of the polymer near the membrane surface is much lower than in the bulk phase. This distance-dependent concentration is accompanied by a viscosity profile with a lower viscosity near the membrane surface relative to the viscosity in the bulk phase. The electrophoretic data are very well described by an exponential profile of the viscosity (Bäumler and Donath, 1987).

The electrophoretic measurements support the conclusion that PEG is excluded from the membrane surface. To summarize, many of the above conclusions about the PEG interaction with proteins can also be applied to membranes. This result is of basic importance for the interpretation of the action of PEG on membranes and the molecular mechanisms of the fusogenic activity of PEG. As we have discussed above many of the effects of PEG on membranes have to be explained by an indirect action mediated by the changed properties of the water phase.

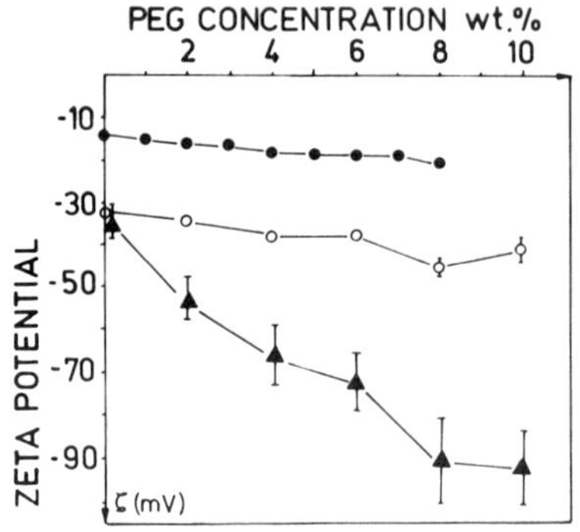

Fig. 7. The electrophoretic zeta-potential of erythrocytes at an ionic strength of 0.01 M NaCl (o) and 0.14 M NaCl (●) and charged liposomes (▲) prepared from a mixture of DPPC and DPPA as a function of the concentration of PEG.

The comparable influence of PEG and dextran on the electrophoretic mobility does not mean that the molecular causes for this property must be equal. Differences of the interaction of these polymers with vesicles were found in a two-phase system of dextran and PEG (Eriksson and Albertsson 1978). The exclusion of dextran from the erythrocyte surface could result from steric contributions because polymers of very high molecular weight are used. The diameter of such polymers is larger than the thickness of the glycocalyx. For the relatively small PEG molecules such an effect does not play a role and the thermodynamically unfavoured interaction of PEG with a polar surface should be important as found for proteins. This difference could possibly explain the different actions of PEG and dextran on membrane systems. On addition of PEG a decrease of the surface potential of lipid monolayers by several hundred millivolts was described by Maggio et al. (1976). Since this drop of the surface potential was not found for liposomes and erythrocytes, we have to assume that the change of the surface potential observed for monolayers results mainly from changes in the orientations of molecular dipoles in the surface.

Ion Binding to Membrane Surfaces

Many studies have indicated that Ca^{2+} is almost universally involved in cell fusion. Synergistic effects of PEG and Ca^{2+} in the fusion of PS-containing vesicles were described (Hoekstra 1982). The content of Ca^{2+} in the fusion medium affects the efficiency of the fusion procedure. The influence of ions on membrane systems is very complex. Ion binding, ionic screening, structural changes, the formation of nonbilayer structures and an influence of bilayer interaction on ion binding were observed. How these processes are influenced by PEG cannot be answered at the present stage of research.

The binding of paramagnetic Pr^{3+} and Mn^{2+} ions to lecithin liposomes was studied by use of NMR and ESR (Gawrisch 1986). From the shift of the NMR signals it was found that addition of 30 wt% PEG increases the surface concentration of ions about threefold. Increased ion binding seems to be a general feature of the action of PEG on neutral phospholipid bilayers. A reason for the increased binding could be the lower solubility of ions in the external phase due to their decreased polarity.

The findings may give an explanation for the enhanced Ca^{2+} permeability of liposomes and erythrocytes in the presence of the polymer (Blow et al. 1979, Aldwinckle et al. 1982). The raised concentration of the ions at the membrane surface taken together with a membrane destabilization should result in an enhanced intracellular ion concentration. Certainly, more experimental investigations are necessary to explain the influence of PEG on the properties of ions at membrane surfaces.

INFLUENCE OF PEG ON THE STRUCTURE AND ORGANIZATION OF MEMBRANE PROTEINS

Generally, the role of membrane proteins in membrane fusion processes is not well understood, and is discussed mainly in the light of the following points: (a) the role of the lateral organization of integral membrane proteins, in particular of intramembraneous particles (IMP) during and after fusion; (b) alterations of the physical state of the cytoskeleton which could be important for the lateral organization of integral membrane proteins; (c) proteolysis enhancing the lateral mobility of proteins and/or producing short polypeptide chains which can act as membrane perturbers (Lucy 1984).

The lateral organization of IMP in PEG induced fusion was investigated on different biological membranes. In partiuclar, the following questions were studied:
(a) Is IMP clustering induced directly by PEG or by the aggregation of membranes mediated by PEG? This is consistent with the question whether the IMP aggregation is observed only in contact regions or over the whole cell membrane. (b) Is the IMP aggregation reversible? (c) Is a disturbance of the cytoskeleton necessary for the reorganization of IMP?

The erythrocyte membrane has been largely used as a model system for investigating the effect of PEG on the organization of membrane proteins. We studied structural (lateral reorganization) and dynamical alterations within the glycocalyx in the presence and after removal of PEG using freeze-fracture electron microscopy and the ESR spin labelling technique (Pratsch et al.,1987). The molecular motion in the membrane surface was studied by (a) covalently labelling sialic acid residues of glycoproteins and glycolipids and by (b) incorporating a charged spin probe into the erythrocyte membrane which is sensitive to structural changes directly on the outer membrane surface (Pratsch et al. 1987, Herrmann et al. 1986). Although, cell aggregation was observed at a low concentration of PEG_{6000} at 7 wt% PEG, no evidence for a direct interaction between PEG and membrane surface was found since the lateral organization of the IMP's and glycocalyx structure was not affected.

In the presence of 45 wt% PEG an intense clustering of IMP's and the formation of particle-free areas were established only at sites of cell-cell contact in agreement with the results of Knutton (1979). After removal of PEG a random particle distribution was reestablished (see also Hui et al. 1985, Krähling 1980, Knutton 1979), but the particle consisted of small patches of aggregated proteins (Pratsch et al. 1987). This is confirmed by the irreversible changes in the spectra of the glycocalyx sensitive spin probes. Spin labelled sialic acid residues exhibit a more restricted motion after removal of PEG. Also the spectra of spin labelled SH-groups of erythrocyte membrane proteins - mainly of the cytoskeleton protein spectrin - were irreversibly altered by PEG (Herrmann et al. 1983b, Surewicz 1983).

Knutton (1979) assumed that the mechanism of IMP aggregation is an electrostatic displacement since it was observed only in contact regions. In agreement with Wojcieszyn et al. (1983) we suggest that the IMP aggregation is caused either by dehydrating lipids and thus affecting fluidity or by directly promoting protein aggregation through dehydration. Due to the lateral reorganization of the membrane unstable phase boundaries may be created which nucleate the fusion process.

Investigating resealed ghosts of human erythrocytes Hui et al. (1985) revealed that IMP-patching preceded cell-cell contact and occurred without direct PEG interaction. This was underlined by the observation that the aggregation of glycophorin reconstituted multilamellar vesicles occurs also in those bilayers to which PEG had no access. As possible reasons they discussed the increase of local electrolyte concentration, the excluded volume effect as well as the altered lipid structure. An influence of the cytoskeleton on the IMP organization was not found. IMP clustering after addition of PEG was also established for other mammalian cell membranes (Krähling 1980, Robinson et al. 1979).

In summary, an IMP aggregation was observed in different cell membranes after addition of PEG whereas the extent of aggregation seemed to be dependent on the cells and their specific membranes as well as on experimental conditions. The presence of PEG alone seems to be sufficient for IMP aggregation whereas a direct interaction between PEG and membrane proteins has not to be assumed. On intact cell membranes the aggregation of IMP is reversible to a certain extent. The cytoskeleton is obviously not involved in IMP aggregation. Taking into account that no evidence for a direct interaction between PEG and membrane proteins was found the following reasons have to be considered as reponsible for IMP aggregation:

(1) Electrostatic repulsion as well as steric interaction caused by the increased segment density due to penetration of the glycocalyx of adjacent cells can lead to IMP aggregation.

(2) Dehydration of lipids as well as of proteins can cause phase separations due to changes in the chemical potential of these components, for instance, a mismatch of the corresponding hydrophobic parts of lipids and proteins can result in protein aggregation (Mouritsen and Bloom 1984). The hydrophobic thickness of the lipid bilayer can be increased by an enhanced packing density.

(3) The physical state of membrane lipids with respect to the fluid transition temperature is known to affect clustering and dispersion of integral membrane proteins and IMP (Devaux and Seigneuret 1985). As pointed out the physical state of membrane lipids depends also on the degree of hydration.

(4) Due to the lowered dielectric constant a clustering of the hydrophilic parts of integral membrane proteins can be favoured.

(5) An alternative explanation can be the excluded volume mechanism which seems to be valid for the precipitation of proteins in aqueous solutions of PEG. In those systems it was found that PEG is excluded from the protein domain ((Atha and Ingham 1981). This results in a phase separation (Lee and Lee 1981).

EFFECT OF PEG ON INTRACELLULAR MICROVISCOSITY

We measured the intracelluar microviscosity of red blood cells in the presence of PEG by ESR measuring the rotational correlation time of a permeable small spin label (TEMPONE) (Herrmann et al. 1985, Herrmann and Müller 1985). These rotational correlation times as a measure of the intracellular microviscosity were compared with those measured in red blood cells suspended in NaCl or sucrose containing media of known osmotic pressure. In this way it was possible to estimate the effective osmotic pressure of PEG solutions stressing the cells. Since different data have been reported on the osmotic pressure π of aqueous PEG solutions (see above) probably due to difficulties arising in measuring π, this method yields additional, independent data of π.

The internal microviscosity of red blood cells is dramatically enhanced by adding PEG to the suspension medium. The rotational correlation time of TEMPONE located in the cytoplasm increases threefold by raising the concentration of PEG_{6000} from 0 wt% to 30 wt%. This increase is caused by a considerable shrinkage of the cells. We estimated a relative cell volume of 0.5 at 30 wt% PEG_{6000} which is approximately the limit of cell shrinkage. The effective osmotic pressure of PEG solutions estimated is in good agreement with that obtained by water vapour measurements.

For cells remaining intact after washing out PEG, the intracellular microviscosity resembles that of the control cells.

SUMMARY

Studies of the electrophoretic mobility of liposomes and erythrocytes demonstrate that the concentration of PEG near the membrane surface is reduced compared to the bulk phase. Therefore an indirect action of PEG on membrane systems due to a change of the physicochemical properties of the aqueous phase is postulated. The influence of PEG's of different molecular weight on the dielectric constant, the polarity and the osmotic pressure is studied. The strong aggregation of liposomes and cells in the presence of PEG is interpreted as a mutual approach of membranes due to an osmotic compression because PEG is excluded from the contact region of the membranes. PEG chains fixed to the membrane surface have the opposite effect and increase the bilayer-bilayer separation.

The partition coefficients of amphiphilic molecules are decreased in the presence of PEG which indicates a lowering of the thermodynamic stability of the membrane as a result of the changed physicochemical properties of the water phase. Such bilayer destabilization could play a crucial role in the fusion process. Most of the actions of PEG on the phase behaviour of phospholipid systems are comparable to the effects of the decrease of the water content. Shifts of the phase transition temperature, increase of the order parameter and lateral phase separations are described. The aggregation of intramembraneous particles observed on addition of PEG could also be connected with the high ability of PEG to dehydrate membranes.

Since there is no evidence for a direct interaction between PEG and membranes we conclude that the fusogenic activity of PEG arises from the altered physicochemical properties of water which lead to a strong aggregation of membranes, a destabilization of the membrane, changes of the membrane structure and osmotic stress on cells. It is important that these effects can occur without any direct interaction between PEG and the membrane.

ACKNOWLEDGMENT

We thank Dr. E. Donath for helpful discussions.

REFERENCES

Ahkong, Q. F., Fisher, D., Tampion, W., and Lucy, J. A., 1975, Mechanisms of cell fusion , Nature (London), 253: 194-195.

Ahkong, Q. F., and Lucy, J. A., 1986, Osmotic forces in artifically induced cell fusion, Biochim. Biophys. Acta, 858: 206-216.

Atha, D. H., and Ingham, K. C., 1981, Mechanism of precipitation of proteins by PEG, J. Biol. Chem., 256: 12108-12117.

Aldwinckle, T. J., Ahkong, Q. F., Bangham, A. D., Fisher, D., and Lucy, J. A., 1982, Effects of polyethylene glycol on liposomes and erythrocyte permeability changes and membrane fusion, Biochim. Biophys. Acta, 689: 548-560.

Arnold, K., Lösche, A., and Gawrisch, K., 1981, ^{31}P-NMR investigations of phase separation in phosphatidylcholine/phosphatidylethanolamine mixtures, Biochim. Biophys. Acta, 645:143-148.

Arnold, K., Pratsch, L., and Gawrisch, K., 1983, Effect of poly(ethylene glycol) on phospholipid hydration and polarity of the external phase, Biochim. Biophys. Acta, 728: 121-128.

Arnold, K., Herrmann, A., Pratsch, L., and Gawrisch, K., 1985a, The dielectric properties of aqueous solutions of PEG and their influence on membrane structure, Biochim. Biophys. Acta, 815: 515-518.

Arnold, K., Herrmann, A., Gawrisch, K., and Pratsch, L., 1985 b, Mechanisms of PEG-induced fusion, studia biophysica, 110: 135-141.

Arnold, K., Borin, M., and Azizova, O., 1986 a, Influence of PEG on the partition of a charged spin probe, Coll. Polymer. Sci., 264: 248-253.

Arnold, K., Lvov, Y. M., Szögyi, M., and Györgyi, S., 1986 b, Effect of poly(ethylene oxide)-containing surfactants on membrane-membrane interaction, studia biophysica, 113: 7-14.

Arnold, K., Zschörnig, O., Herold, W. and Barthel, D., 1987, Exclusion of PEG from membrane surfaces, Mol. Cryst. Liqu. Cryst., submitted.

Bäumler, H. and Donath, E., 1987, Does dextran indeed significantly increase the surface potential of human red blood cells, studia biophysica, in press.

Baran, A. A., Solomentseva, I. M., Mank, V. V., and Kurilenko, O. D., 1972, Role of the solution factor in stabilizing disperse systems containing water soluble polymers, Dokl. Akad. Nauk. USSR, 207: 363-366.

Bell, G. I., Dembo, M. and Bongrand, P., 1984, Cell adhesion, competition between nonspecific repulsion on specific binding, Biophys. J., 45: 1051-1064.

Blow, A. M. J., Botham, G. M., Fisher, D., Goodall, A. H., Tilcock, C. P., and Lucy, J. A., 1978, Water and calcium ions in cell fusion induced by PEG, FEBS lett., 94: 305-310.

Blow, A. M. J., Botham, G. M., and Lucy, J. A., 1979, Calcium ions and cell fusion, Biochem. J., 182: 555-563.

Boni, L. T., Stewart, T. P., Alderfer, J. L., and Hui, S. W., 1981 a, Lipid-PEG interactions: (I) Induction of fusion between liposomes, J. Membrane Biol., 62: 65-70.

Boni, L. T., Stewart, T. P., Alderfer, J. L., and Hui, S. W., 1981 b, Lipid-PEG interactions: (II) Formation of defects in bilayers, J. Membrane Biol., 62: 71-77.

Boni, L. T., Hah, J. S., Hui, S. W., Mukherjee, P., Ho, J. T., and Jung, C. Y., 1984 a, Aggregation and fusion of unilamellar vesicles by poly(ethylene glycol), Biochim. Biophys. Acta, 775: 409-418.

Boni, L. T., Stewart, T.P., and Hui, S. W., 1984 b, Alterations in phospholipid polymorphism by PEG, J. Membrane Biol., 80: 91-104.

Boss, W. F., 1983, Poly(ethylene glycol)-induced fusion of plant protoplasts. A spin-label study, Biochim. Biophys. Acta, 730: 111-118.

Briere, R., Lemaire, H., and Rassat, A., 1965, Nitroxdes XV: Synthèse et etude de radicaux libres stables piperidiniques et pyrrolidinique Bull. Soc. Chim. Fr., 11: 3273-3283.

Chirife, J., and Fontan, C. F., 1980, A study of the water activity lowering behaviour of polyethylene glycols in the intermediate moisture range, J. Food Sci., 45: 1717-1719.
Cullis, P. R., and Hope, M. J., 1978, Effects of fusogenic agent on membrane structure of erythrocyte ghosts and the mechanism of membrane fusion, Nature (London), 271: 672-674.
Das, S. and Singhal, G., 1981, Effects of change of water structure on the phase transition of liposomes of DPL, Int. J. Quantum Chem., 20: 495.
Devaux, P., and Seigneuret, M., 1985, Specifity of lipid-protein interactions as determined by spectroscopic techniques, Biochim. Biophys. Acta, 822: 63-125.
Eriksson, E and Albertsson, P.-A., 1978, The effect of lipid composition on the partition of liposomes in aqueous two-phase systems, Biochim. Biophys. Acta, 507: 425-432
Finer, E. G., and Darke, A., 1974, Phospholipid hydration studied by deuteron magnetic resonance spectroscopy, Chem. Phys. Lipids, 12: 1-16.
Foster, K. R., Cheever, E., Leonhard, J. B., and Blum, F. D., 1984, Transport properties of polymer solutions, Biophys. J., 45: 975-984.
Gawrisch, K., Richter, W., Möps, A., Balgavy, P., Arnold, K. and Klose, G. 1985, The influence of water concentration on the structure of egg yolk phospholipid/water dispersions, studia biophysica, 108: 5-16.
Gawrisch, K., Stibenz, D., Möps, A., Arnold, K., Linss, W. and Halbhuber, K.-J., 1986, The rate of lateral diffusion of phospholipids in erythrocyte microvesicles, Biochi. Biophys. Acta, 856: 443-447.
Gawrisch, K., 1986, Molekulare Mechanismen und Membranveränderungen bei der durch Polyethylenglykol induzierten Zellfusion, Thesis B, Leipzig.
Herrmann, A., Pratsch, L., Arnold, K., and Laßmann, G., 1983a, Effect of PEG on the polarity of aqueous solutions and on the structure of vesicle membranes, Biochim. Biophys. Acta, 733: 87-94.
Herrmann, A., Arnold, K., Pratsch, L., and Laßmann, G., 1983b, Influence of polyethylene glycol on the structure of the erythrocyte membrane: An ESR study, Biomed. Biochim. Acta, 9:1151-1155.
Herrmann, A., Arnold, K., and Pratsch, L., 1985, The effect of osmotic pressure of aqueous PEG solutions on red blood cells, Bioscience Rep., 5:689-696.
Herrmann, A., and Müller, P., 1985, Correlation of the internal microviscosity of human erythrocytes to the cell volume and the viscosity of hemoglobin solutions, Biochim. Biophys. Acta,885: 80-87.
Herrmann, A., Laßmann, G., Groth, T., Donath, E. and Hillebrecht, B., 1986 Conformational alterations within the glycocalyx of erythrocyte membranes studied by spin labeling, Biochim. Biophys. Acta, 861,111-121.
Hoekstra, D., 1982, Role of lipid phase separation and membrane hydration in phospholipid vesicle fusion, Biochemistry, 21: 2833-2840.
Honda, K., Maeda, Y., Sasakawa, S., Ohno, H., and Tsuchida, E., 1981 a, Activities of cell fusion and lysis of the hybrid type of chemical fusogens (I) structure and function of the promotor of cell fusion, Biochem. Biophys. Res. Commun., 100: 442-447.
Hui, S. W., Isac, T., Boni, L. T., and Sen, A., 1985, Action of polyethylene glycol on the fusion of human erythrocyte membranes. J. Membrane Biol., 84: 137-146.
Klose, G., Brückner, S., Bezzabotnov, V. Yu., Borbely, S. and Ostanevich, Yu. M., 1986, Hydration and swelling of the total lipid fraction of egg yolk lecithin and the influence of some additives studied by small angle neutron scattering, Chem. Phys. Lipids, 41, 293-307.
Knutton, S., 1979, Studies of membrane fusion: III. Fusion of erythrocytes with PEG, J. Cell Sci., 36: 61-72.
Krähling, H., 1980, Untersuchungen zum Mechanismus der PEG-induzierten Zellfusion, Thesis, Stuttgart.
Krähling, H., Schinkewitz, U., Barker,, A., and Hülser, D. F., 1978, Electronmicroscopical and electrophysiological investigations on PEG-induced cell fusion, Cytobiol., 17: 51-61.

Lee, J. C., and Lee, L. L. Y., 1981, Preferential solvent interactions between proteins and PEG, J. Biol. Chem., 256: 625-631.

Le Neveu, D. M., Rand, R. P., and Parsegian, V. A., 1976, Measurement of forces between lecithin bilayers, Nature (London), 259, 601-603.

Lucy, J. A., 1984, Fusogenic mechanisms, in: "Cell fusion," Ciba Foundation symposium 103, Pitmans Books, London.

Maggio, B., Ahkong, Q. F., and Lucy, J. A., 1976, PEG, surface potential and cell fusion, Biochem. J., 158: 647-650.

Masszi, G., Koszorous, L., and Lakatos, T., 1986, Investigation of hydration of macromolecules. III. Study of PEG homologues by microwave measurements, Arch. Biochem. Biophys. Hung., 21: 263-281.

McGrath, J. J., 1981, Thermodynamic modelling of membrane damage, in: "Effect of Low Temperatures on Biological Membranes", G. J. Morris and A., Clarke, eds., Academic Press, London.

McPherson, A., 1985, Use of PEG in the crstallization of macromolecules, in: "Methods of Enzymology", Vol. 114, Academic Press, New York.

Molynex, P., 1975, Synthetic polymers, in: "Water, a comprehensive treatise"; F. Franks, ed., Vol. 4, Plenum Publishing Corp., New York, pp. 569-757.

Mouritsen, O. G., and Bloom, M., 1984, Matress model of lipid-protein interactions in membranes, Biophys. J., 46: 142-153.

Ohki, S., 1982, A mechanism of divalent ion-induced PS membrane fusion, Biochim. Biophys. Acta, 689: 1-11.

Ohki, S., 1985, Membrane fusion: Theory and Experiments, studia biophysica, 110: 95-104.

Overbeek, J. Th. G., 1982, Strong and weak points in the interpretation of colloid stability, Adv. Coll. Interf. Sci., 16: 17-30.

Petrov, A. G., Gawrisch, K., Brecesinski, G., Klose, G., and Möps, A., 1982, Optical detection of phase transitions in simple and mixed lipid-water dispersions, Biochim. Biophys. Acta, 690: 1-7.

Pratsch, L., Schwede, I., Meyer, H.W., Herrmann, A. and Laßmann, G., 1987 PEG-mediated rearrangements in erythrocyte membranes, Biochim. Biophys. Acta, submitted.

Rand, R. P., 1981, Interacting phospholipid bilayers: Measured forces and induced structural changes, Ann. Rev. Biophys. Bioeng., 10: 277-314.

Robinson, J. M., Ross, D. S., Davidson, R. L., and Karnovsky, M. J., 1979, Membrane alterations and other morphological features associated with PEG-induced cell fusion, J. Cell Sci., 40: 63-75.

Saèz, R., Alonso, A., Villena, A., and Goni, F. M. 1982, Detergent like properties of PEG's in relation to model membranes, FEBS-Lett., 137: 323-327.

Surewicz, W. K., 1983, ESR study on the mechanism of PEG-membrane interaction, FEBS lett., 151: 228-231.

Tilcock, C. P. S., and Fisher, D., 1979, Interaction of phospholipid membranes with PEG, Biochim. Biophys. Acta, 577: 53-61.

Tilcock, C. P. S., and Fisher, D., 1982, The interaction of phospholipid membranes with poly(ethylene glycol). Vesicle aggregation and lipid exchange, Biochim. Biophys. Acta, 688: 645-652.

Tonkonog, L. A., Vasilenko, I. A. and Serebrennikova, G. A., 1982, Formation of aggregates of phospholipids in nonaqueous polar solutions, Dokl. Akad. Nauk USSR, 267: 402-405.

Verkleij, A. J., Mombers, C., Gerritsen, W. J., Leunissen-Bijvelt, L., and Cullis, P. R., 1979, Fusion of phospholipid vesicles in association with the appearence of lipidic particles as visualized by freeze-fracturing, Biochim. Biophys. Acta, 555: 358-361.

Volke, F., Arnold, K., and Gawrisch, K., 1982, The effect of hydration on the mobility of phospholipids in the gel state, Chem. Phys. Lipids, 31: 179-189.

Wojcieszyn, J. W., Schlegel, R. A., Lumley-Sapanski, K., and Jacobson, K. A., 1983, Studies on the mechanism of PEG-mediated cell fusion using fluorescent membrane and cytoplasmic probes, J. Cell Biol., 96: 151-159.

CONTROL OF CELL MEMBRANE FUSION BY LIPID COMPOSITION

David S. Roos

Department of Biological Sciences
Stanford University
Stanford, CA 94305

INTRODUCTION AND OVERVIEW

The fusion of biological membranes is an integral component of many physiological processes, occurring within and between cells in numerous situations of interest to such diverse fields as cell biology, developmental physiology, and medical pathology (reviewed in Poste and Nicholson, 1978; Lucy, 1982; Sowers, 1987). Observations of multinucleated cells in vertebrate tissues were first made by medical pathologists as early as the mid-19th century (reviewed by Langerhans, 1868), and cell fusion has continued to intrigue morphologists ever since, yet the precise molecular mechanisms governing the fusion of biological membranes remain unclear. In recent years, cell and membrane fusion techniques have been used extensively for the advancement of modern cell genetics (Davidson, 1977). Further development and refinement of applications for the biotechnology industry has lent increasing impetus to investigations into the biophysical and biochemical bases of cell fusion and other membrane interactions (de St. Groth and Scheidegger, 1980; Furusawa, 1980; Zimmermann, 1982; Gregoriades, 1984; Schlegel and Lieber, 1987).

The experiments reviewed below describe a genetic approach to the elucidation of membrane fusion mechanisms, based on examining the susceptibility or resistance of mutant cells to the polyene chemical fusogen polyethylene glycol (PEG*). Biochemical and physiological investigation of a series of PEG fusion-resistant cell lines (Roos and Davidson, 1980) has demonstrated a correlation between cell membrane lipid composition (especially acyl-chain saturation) and the fusion response of a given cell culture (Roos and Choppin, 1985a). In further studies, careful manipulation of the lipid composition of these cells has allowed nearly complete control over the fusion phenotype (Roos and Choppin, 1985b), directly demonstrating the importance

---

**Abbreviations used in this report:* Clone 1D, a 5-bromodeoxyuridine resistant mouse fibroblast cell line LMTK$^-$ Clone 1D (Kit *et al.*, 1963); F_8, F_{16}, F_{40}, fusion-resistant cell lines isolated from Clone 1D by 8, 16, or 40 cycles of polyethylene glycol-induced fusion (Roos and Davidson, 1980); MHV, murine hepatitis virus, strain A59; PEG, polyethylene glycol; S/P ratio, ratio of saturated to polyunsaturated fatty acyl chains; VSV, vesicular stomatitis virus, Indiana serotype.

of membrane lipids in regulating cell fusion, and providing a useful approach to the control of cell fusion in various technological applications (Roos *et al.*, 1987).

Initially, this project focused on cell fusion induced by polyethylene glycol, because of its convenience of use, reproducibility, absence of extraneous biological material, and importance to the development of fusion technology. Nevertheless, the results obtained with PEG apply equally well to other forms of membrane fusion, including virus-induced fusion and spontaneously occurring fusion. These findings suggest similarities in certain structural aspects, if not the precise mechanisms, of a variety of forms of membrane fusion. While the original studies were carried out in a particular system of cultured mouse fibroblasts, our experimental observations have now been found to apply to other cell systems as well (Roos *et al.*, 1987). Further studies, briefly outlined but beyond the scope of this report, have taken advantage of the availability of the cell lines described here to demonstrate the importance of lipid composition to such diverse membrane-associated phenomena as immune recognition of tumor cells, metastatic spread of neoplastic disease, and the determination of virus host range.

ISOLATION AND CHARACTERIZATION OF FUSION-ALTERED CELL LINES

PEG fusion-resistant mutants

Individual cultured cell lines generally exhibit a characteristic response to a given fusogen such as PEG, but comparisons between cell lines have revealed considerable differences in the potential range of fusion response (Davidson and Gerald, 1976; Blow *et al.*, 1978; Wang *et al.*, 1982). Consideration of these observations suggested both the importance of working with a single, well-defined, clonal cell line, and the feasibility of isolating mutants which differ in their response to fusogenic treatment. The murine fibroblast cell line designated Clone 1D (Kit *et al.*, 1963) has been extensively characterized with respect to its PEG fusion response (Davidson *et al.*, 1976; Robinson *et al.*, 1979) and provided a suitable parental strain for the studies described below. Although several different techniques for isolating PEG fusion-resistant mutants were attempted, the most successful (if laborious) approach involved repeated cycles of treatment with fusogenic concentrations of PEG (50% by weight) followed by outgrowth of the remaining unfused cells (Roos and Davidson, 1980). This repetitive protocol produced a series of cell lines designated F_1, F_2, F_3, etc., where the subscripted number indicates the number of cycles of selection required to produce each cell line.

The susceptibility of these cell lines to PEG-induced fusion is illustrated in Figure 1. Typically, >90% of the cells in PEG-treated Clone 1D cultures fuse into large, multinucleated syncytia. However, increasingly selected cell lines are progressively less susceptible to fusion: only ~70% of the nuclei in F_8 cultures are ultimately found in fused cells, F_{16} cells exhibit ~50% fusion, and the F_{40} mutants are characterized by <20% fusion. Thus, we have succeeded in isolating highly PEG fusion-resistant mutants from the original, fusible Clone 1D cell line. It has not, however, proved possible to generate mutants which are wholly resistant to the fusion-inducing effects of PEG (despite considerable effort in this direction), a finding which was suggested to result from insurmountable constraints governing viable cell architecture (Roos *et al.*, 1983). The nature of these constraints can now be understood at the biochemical level, as will be discussed below.

Each of the PEG fusion-resistant cell lines is stable, maintaining its characteristic fusion response even when carried in culture for several months without PEG treatment or other selection (Roos *et al.*, 1983). Each

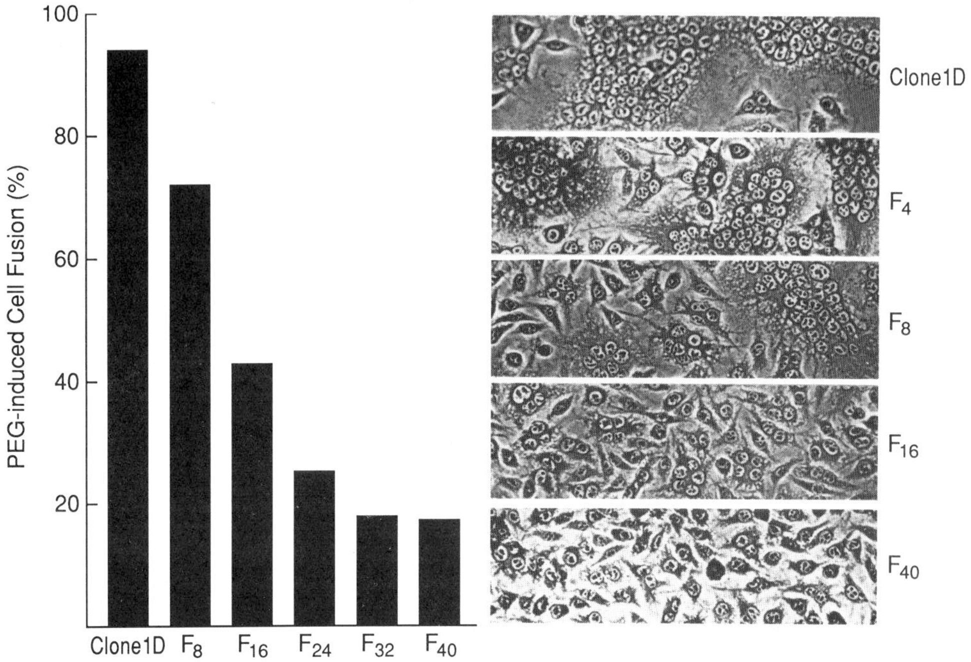

Figure 1. Response of Clone 1D and fusion-altered cell lines to PEG treatment. Confluent cultures of the parental cell line Clone 1D and PEG-resistant mutants were treated for 1 min with 50% PEG-1000 as previously described (Roos and Davidson, 1980) and fixed and stained for microscopy. Fusion response was scored as the percentage of nuclei found in fused cells of any size. PEG fusion-resistant cell lines F_4, F_8, ..., F_{40} were selected by repeated cycles of PEG treatment followed by outgrowth of the remaining unfused cells; subscripts indicate the number of cycles of selections required to produce each cell line. The selection protocol employed produced a series of stable, homogeneous cell lines each of which is slightly more resistant to PEG-induced fusion than its predecessor, as discussed in the text. Reproduced, with permission, from Roos *et al.* (1983) and Roos and Choppin (1985a).

cell line also appears to be homogeneous in its fusion response; examination of clonal cultures isolated from several different cell lines of intermediate susceptibility to PEG demonstrates that each of these lines consists of a population of cells which are uniformly reduced in fusibility (Roos and Davidson, 1980). The F_8 cell line, for example, is composed of individual cells each of which is approximately one-half as likely to fuse as are the parental Clone 1D cells*, a finding which contrasts sharply with expectations for an intermediate stage in the selection of classical genetic mutants, where we might predict a mixed population containing fusible parental cells enriched with unfusible mutants. This observation has interesting implications for the genetic basis of fusion resistance.

---

*Combining the overall fusion response for a given cell culture (e.g. data from Figure 1) with observations of cell density, the degree of cell spreading, etc., it is possible to model the fusion response mathematically (Röhme and Thorburn, 1981; D.S. Roos and B. Singer, unpublished analyses). Applying such models yields an underlying probability of fusion for a single pair of adjacent Clone 1D cells ≈ 45%, following a standard PEG treatment protocol. (The overall fusion response for Clone 1D is much higher [cf. Figure 1], because each cell contacts several adjacent cells, fusion with any one of which is sufficient to be scored as a fused cell.) The calculated probability that a closely apposed pair of F_8 cells will fuse is 27%, and the underlying fusion probabilities for F_{16} and F_{40} cells are 12% and 4%, respectively.

PEG-fusion mutants are unusually susceptible to virus-induced fusion

Having isolated the PEG fusion-resistant cell lines described above, it was of interest to characterize their response to other fusogenic treatments as well. Most enveloped viruses are capable of causing the fusion of biological membranes, as membrane fusion is a necessary step in their infectious pathway (Choppin and Scheid, 1980; White *et al.,* 1983; Spear, 1987). Under appropriate conditions, many of these viruses are also capable of inducing fusion between cells. Several of the PEG-resistant cells lines have been examined for their susceptibility to fusion following treatment with various enveloped viruses, as shown in Table I.

Somewhat surprisingly, increasingly PEG-resistant cells are increasingly *susceptible* to fusion induced by a variety of enveloped viruses which cause fusion under different conditions (D.S. Roos, C. Stevenson, K.V. Holmes, and P.W. Choppin, manuscript submitted for publication). Vesicular stomatitis virus (VSV) readily infects these cells and causes fusion both by bridging adjacent cells with virus particles which subsequently fuse with each cell, and through the action of viral fusion proteins inserted in the surface membranes of infected cells. Although VSV-infected Clone 1D and F_{40} cells produce similar titers of virus (not shown), Clone 1D cells are markedly less susceptible to VSV-induced fusion. Murine hepatitis virus (MHV) infects both Clone 1D and F_{40} cells and is capable of fusing both cell types, but the overall extent of fusion is much higher in F_{40} cultures, at least in part because Clone 1D cells are unable to produce infectious virions. Sendai virus is unable to infect either cell line, but can cause fusion upon the addition of high titers of inactivated virus (sometimes called fusion from without; Bratt and Gallaher, 1969). Under these conditions as well, more extensive fusion was seen in F_{40} cells than in the Clone 1D cell line. Similar observations of increased susceptibility to virus-induced fusion in PEG-resistant cell lines have been made for many other viruses, including influenza, SV5, and visna virus (an ovine retrovirus).

Although the inverse correlation between chemically-induced and virus-induced fusion was unexpected, it is notable that the PEG-resistant cell lines, selected without any intentional reference to other charteristics, are nevertheless altered with respect to another membrane-associated phenomenon, namely their susceptibility to virus-induced fusion. Across the graded series of increasingly PEG fusion-resistant mutants, the progressive nature of fusion alteration holds up for all viruses tested (cf. F_{16} data in Table I). The

Table I. Fusion Response of PEG-Resistant Cell Lines

	Percent fusion following indicated treatment				
Cell Line	PEG	VSV*	Sendai†	MHV‡	none (spontaneous fusion)
Clone 1D	96	8 (+)§	6 (–)	22 (–)	2
F_{16}	58	ND	17 (–)	45 (+)	4
F_{40}	24	20 (+)	20 (–)	64 (+)	5

* Indiana serotype; cell fusion following brief exposure to acid pH (White *et al.*, 1981).
† Z-strain; fused 'from without' using inactivated virus (Holmes and Choppin, 1966).
‡ Strain A59; fusion carried out according to Holmes *et al.* (1981).
§ +/– indicates the ability/inability to produce infectious virions.

inverse relationship between PEG and virus response may explain the observation that BHK-21 cells, which have frequently been used for studies on virus-induced fusion because of their extreme sensitivity to many fusogenic viruses (Holmes and Choppin, 1966; Compans *et al.*, 1966, Helenius *et al.*, 1980; White *et al.*, 1981), are among the most highly PEG fusion-resistant cell lines we have encountered (Wang *et al.*, 1982).

Spontaneous fusion has also been examined in these cell lines (Roos and Choppin, 1985b), and also changes progressively across the series of increasingly fusion-altered mutants, as shown in the rightmost column of Table I. As found for virus-induced fusion, spontaneous fusion is inversely correlated with PEG fusion response in these cell lines. Although the overall level of spontaneous fusion is low in both Clone 1D and F_{40} cultures, differences between these cell lines have been reproducibly observed over several years of study and are statistically highly significant ($P < 0.001$).

BIOCHEMICAL ANALYSIS OF FUSION-ALTERED CELL LINES

The availability of closely related cell lines which differ so dramatically in their response to diverse fusogenic treatments provided an opportunity to explore the biochemical basis of the fusion alterations observed in these cells. As an added advantage, the existence of a graded series of stable mutants intermediate between Clone 1D and F_{40} has permitted the search for biochemical alterations to be narrowed, focusing on differences which are progressive across the entire series of cell lines isolated.

Protein analysis of fusion-altered cells

Initial efforts towards elucidating biochemical differences associated with the altered fusion response in these cells concentrated on the analysis of their protein composition (Roos, 1985). A small number of differences have been detected in two-dimensional electrophoretic separations of proteins isolated from Clone and F_{40} cells, including a series of high molecular weight membrane glycoproteins which are increasingly abundant in increasingly fusion-altered cells. Differences have also been seen in protein glycosylation and growth inhibition by certain lectins. Overall, however, protein differences between Clone 1D and F_{40} are slight, and no functional properties are known for those differences which have been found. While antibodies can be prepared with slight specificity for either Clone 1D or F_{40} cells, these antibodies have no clear effect on the fusion response. Any direct association of membrane proteins with fusion is further called into question by experiments on the fusion of protease-treated cells: Clone 1D cells remained more highly susceptible to PEG-induced fusion than the F_{40} mutants even after exposure to a crude mixture of pancreatic proteases, a treatment which degraded the high molecular weight proteins described above, eliminated cell-specific antibody staining, and greatly diminished lectin binding*.

While specific proteins must surely be involved at some level in generating the contrasting fusion responses of Clone 1D and F_{40} cells, differences in the lipid composition of these cell lines appear to be centrally involved in mediating the fusion alterations described above. An extensive analysis of the lipid composition of Clone 1D and F_{40} cells has previously been published (Roos and Choppin, 1985a), and will be summarized here only briefly.

---

*Fusion of protease-treated cells was performed in suspension or according to the 'pancake fusion' technique (O'Malley and Davidson, 1977).

Lipid composition is directly correlated with the fusion-altered phenotype

Several aspects of the lipid composition of these cell lines fulfill the requirement of progressive change across the series of increasingly PEG fusion-resistant mutants, as summarized in Table II. Although these data are derived from biochemical analyses of whole cells, qualitatively similar results are obtained from membrane preparations (Roos, 1985).

The parental cell line, Clone 1D, which is highly susceptible to PEG-induced fusion, contains approximately equal amounts of saturated and polyunsaturated fatty acyl chains (S/P ratio ≈ 1.0), but the extent of fatty acid saturation increases progressively in increasingly PEG-resistant cells (S/P ratio ≈ 4.0 in F_{40} cells). This change in acyl chain composition results from a general increase in the prevalence of all saturated species (palmitate, stearate, etc.) and a decline in the prevalence of most polyunsaturates (linoleic acid, arachidonate, etc.) (Roos and Choppin, 1985a). Fusion-altered cells also contain high levels of an unusual ether-linked neutral lipid (1-alkyl, 2,3-diacyl glycerol; Snyder, 1972; Mangold and Paltauf, 1983). Neutral ether lipids (excluding plasmalogens) are virtually absent from Clone 1D cells, but increase in prominence to rank second only to cholesterol as the major neutral lipid in F_{40} cells. The decreased ganglioside content in fusion-altered cells is intriguing, particularly because of its presence in membranes, but variations in ganglioside composition dependent on cell density and culture conditions overwhelm the differences between cell lines (Roos *et al.*, 1987).

Acyl chain saturation and ether lipid content are both closely correlated with susceptibility to fusion across the series of mutant cell lines under study. In fact, these phenomena may be metabolically linked (Roos and Choppin, 1985a): ether-lipids are unusual among neutral lipids in being preferentially localized in cell membranes (Pfleger *et al.*, 1968; Wood *et al.*, 1970; Schroeder and Vagelos, 1976; Friedberg and Halpert, 1978), carrying with them the highly saturated acyl chains characteristic of neutral lipids (Spector *et al.*, 1981).

FUNCTIONAL CONTROL OF CELL FUSION BY LIPID COMPOSITION

Although the biochemical analyses of these cells reveal strong correlations between lipid composition and fusion response, such correlations are

Table II. Lipid Composition of Fusion-altered Cell Lines*

	Cell Line			
	Clone 1D	F_8	F_{16}	F_{40}
PEG-induced fusion	95	68	46	19
Acyl chain saturation (S/P ratio)	1.0	2.8	3.3	4.0
Neutral ether-linked lipids (relative content)	1	12	23	36
Triglycerides (relative content)	1	4	6	8
Total gangliosides (relative content)	4	3	3	1

* For details of lipid analysis see Roos and Choppin (1985a).

fundamentally circumstantial. To evaluate the functional relevance of these correlations, the lipid content of Clone 1D cells and its fusion-altered derivatives has been modified in several ways. Through the use of such 'lipid-alteration' experiments it has been possible to directly demonstrate the importance of lipid composition in controlling cell membrane fusion.

Alteration of acyl-chain saturation completely controls PEG-induced fusion

The degree of lipid saturation in all of these cell lines is remarkably constant under normal culture conditions, but saturation can be experimentally controlled by growth in medium containing high concentrations of various fatty acids (presented as sodium salts, conjugated to serum albumin, or by more elaborate procedures), as shown in the upper panel of Table III. When Clone 1D cells were grown for 18 hours in the presence of 10^{-4} molar saturated fatty acid, the S/P ratio increased to >3:1, similar to the highly saturated levels found in untreated F_{40} cultures. Growth in polyunsaturated fatty acids suppressed the acyl chain saturation of F_{40} cells to levels to levels at or below those normally found in the parental cell line Clone 1D.

The effects of these changes on the fusion response of Clone 1D and F_{40} cells is illustrated in Figure 2. As already seen, control cultures of Clone 1D cells are highly susceptible to PEG-induced fusion (panel 2*A*), while little fusion occurs in similarly treated F_{40} cells (panel 2*B*). However, when the fatty acid saturation of Clone 1D cultures was altered to resemble F_{40} by growth in high concentrations of saturated fatty acids (in this case nonadecanoic acid), the fusion response was abolished (panel 2*C*). Conversely, susceptibility to PEG-induced fusion was restored in F_{40} cells grown in polyunsaturated fatty acid (such as linolenic acid; panel 2*D*). Quantitatively, these treatments were capable of completely controlling PEG fusion, as shown in the second panel of Table III. A time course and dose response curves for fatty acid supplementation and PEG-induced fusion have previously been published (Roos and Choppin, 1985b).

Additional experiments have probed the biological limits of the PEG fusion response in these cells. Untreated Clone 1D cells are relatively high in polyunsaturated acyl chain content. However, when grown in the presence of linolenic acid (or other polyunsaturates), the S/P ratio of these cells was suppressed even below control levels (Table III, top panel). Accordingly, the cells in these cultures became even more highly sensitive to PEG-induced fusion than usual. As untreated Clone 1D cells normally fuse virtually to completion following a standard treatment with 50% PEG (cf. Figure 1), the enhanced fusion of these cells was detected as a decline in the concentration of PEG required to stimulate fusion (Roos and Choppin, 1985b). By contrast, F_{40} cells could not be rendered more highly saturated than their usual state. Growth in high concentrations of saturated fatty acid was highly toxic to the F_{40} mutants, suggesting that these cells have reached the physiological limit of PEG fusion resistance: higher acyl chain saturation, which might be expected to produce a completely PEG-resistant line, simply kills the cells.

Control of virus-induced fusion and spontaneous fusion by fatty acid supplements

Virus induced fusion of Clone 1D cells and its PEG-resistant derivatives is inversely correlated with chemically mediated fusion, as already seen (Table I). Nevertheless, the same lipid supplement experiments which allowed control over the PEG fusion response also altered the susceptibility of these cells to virus-induced fusion (lower half of Figure 2; third panel of Table III). Murine hepatitis virus, the most fusogenic virus yet tested in this system, fuses F_{40} cells extensively while producing relatively little fusion of Clone 1D cells (panels 2*F* and 2*E*, respectively). Appropriate lipid supplements, however, were able to render Clone 1D cells susceptible to MHV fusion and F_{40} cells MHV-resistant (panels 2*G* and 2*H*). Thus modification of cellular lipid chain saturation by growth in fatty acid supplemented medium is able to control the response of these cell lines to both chemical and viral fusogens in parallel, despite the fact that these two classes of fusogens exert opposite effects on the cells. Spontaneously occurring fusion, another phenomenon inversely related to the PEG fusion response, can also be controlled by fatty acid supplements, as seen in the bottom panel of Table III.

Table III. Control of Cell Fusion by Modification of Fatty Acid Composition

Acyl chain saturation (S/P ratio)

Cell Line	Lipid Supplement*: polyunsaturated FA	control	saturated FA
Clone 1D	0.31	0.93 →	3.02
F_{40}	0.54 ←	2.93	2.80

PEG-induced fusion (%)†

Cell Line	Lipid Supplement*: polyunsaturated FA	control	saturated FA
Clone 1D	91	93 →	17
F_{40}	89 ←	20	[13]‡

MHV-induced fusion (%)†

Cell Line	Lipid Supplement*: polyunsaturated FA	control	saturated FA
Clone 1D	15	24 →	61
F_{40}	9 ←	56	[70]‡

Spontaneous fusion (%)†

Cell Line	Lipid Supplement*: polyunsaturated FA	control	saturated FA
Clone 1D	<2	2 →	15
F_{40}	<2 ←	5	[20]‡

* Fatty acid supplements were 10^{-4} molar linolenic acid (18:3) or nonadecanoic acid (19:0), supplied as conjugates to delipidated serum albumin (~1:1 ratio of FA to BSA).
† Spontaneous fusion subtracted to yield PEG-specific or virus-specific fusion.
‡ Quantitation unreliable due to high toxicity in F_{40} cells treated with saturated fatty acid.

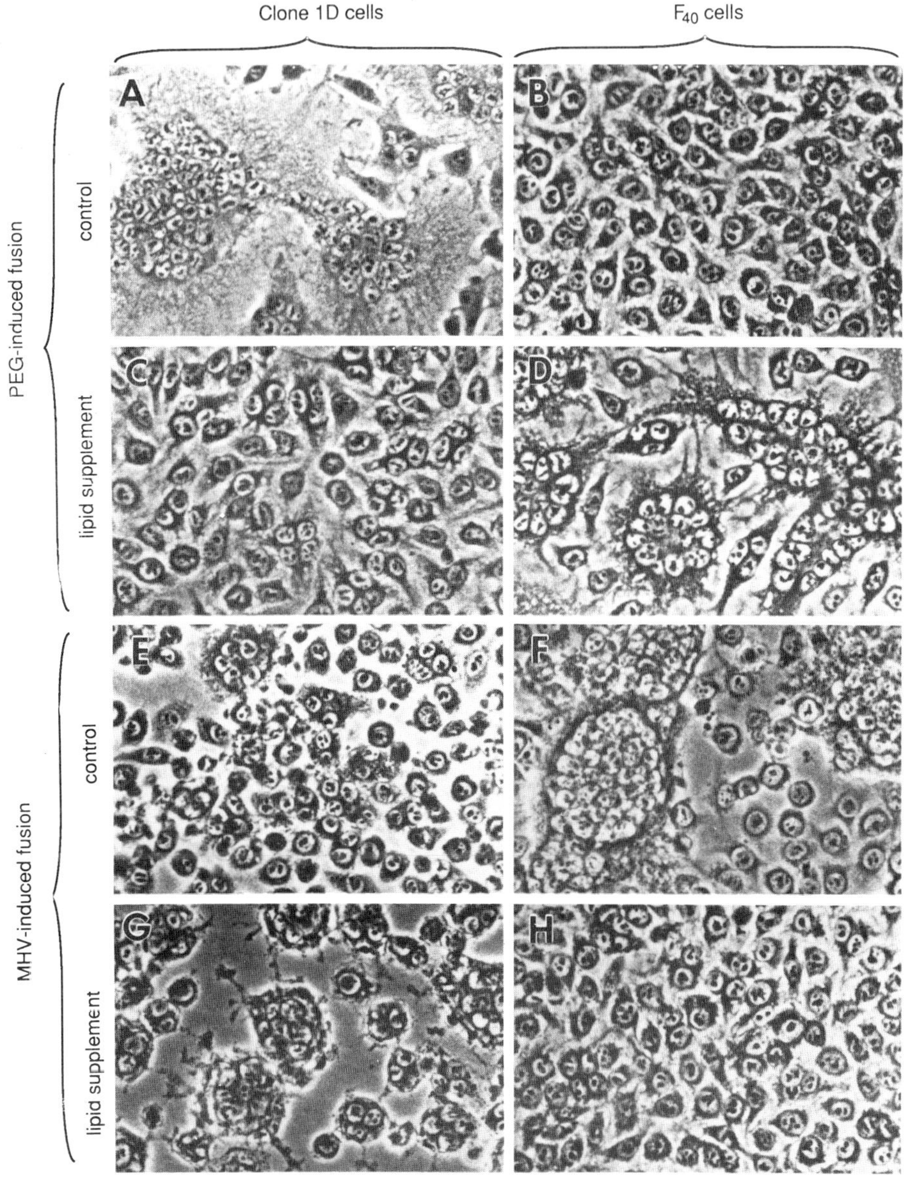

Figure 2. Control of cell fusion by lipid alteration. Confluent cultures of Clone 1D cells (panels *A, C, E,* and *G*) or F_{40} cells (panels *B, D, F,* and *H*) were grown for 18 hr in either control medium (panels *A, B, E,* and *F*) or medium supplemented with exogenous fatty acids conjugated to serum albumin (panels *C, D, G,* and *H*). Supplements were 10^{-4} molar nonadecanoic acid (19:0) for Clone 1D, and 10^{-4} molar linolenic acid (18:3) for F_{40}. Cultures were then treated with polyethylene glycol (PEG; panels *A-D*) or infected with murine hepatitis virus (MHV; panels *E-H*). See Roos and Choppin (1985b) for further experimental details.

Control cultures of Clone 1D are highly sensitive to PEG-induced fusion (panel *A*) but resistant to virus-induced fusion (panel *E*). F_{40} cells exhibit the reverse phenotype: they are PEG-resistant but sensitive to virus-induced fusion (panels *B* and *F*, respectively). Growth of F_{40} cells in high concentrations of polyunsaturated fatty acid, however, corrects the altered fusion response in these mutants, rendering them sensitive to PEG and virus-resistant (panels *D* and *H*). Conversely, increasing the acyl chain saturation of Clone 1D cells mimics the PEG-resistant/virus-fusible phenotype characteristic of F_{40} (panels *C* and *G*). In sum, both PEG-induced and virus-induced fusion can be completely controlled in this system, through the appropriate addition of lipid supplements to the growth medium.

Lipid composition and fusion response of tumor explants

In addition to the fatty acid alterations found in F_{40} cells, neutral ether-lipid content is elevated in these mutants, as noted above. While supplementing the growth medium with various fatty acids provides an effective method for controlling both acyl chain saturation and cell fusion, these experiments also result in parallel changes in ether-lipid concentration (Roos, 1985). Careful fluctuation analysis of a variety of lipid supplement experiments points to chain saturation, rather than ether-lipid content, as the principle factor in regulating fusion response, but more direct evidence for this conclusion comes from an unexpected source.

Following the observation of others that ether-linked lipids are frequently associated with a wide variety of animal tumors (Bollinger, 1967; Wood and Snyder, 1967; Snyder, 1972; Lin *et al.,* 1978), we have used the series of cell lines described here to study the effects of lipid composition on tumor formation in mice (Roos and Choppin, 1984). For reasons known, perhaps, only to Mnemusyne*, growth of these cell lines as tumors produces tissue containing high levels of ether-lipid (similar to F_{40} cells grown in culture), while retaining a low ratio of saturated to polyunsaturated fatty acyl chains (similar to Clone 1D). Thus the prevailing conditions of *in vivo* growth in mice have fortuitously separated the phenomena of fatty acid saturation and ether-lipid content more effectively than yet achieved by any experimental program *in vitro.* Table IV presents the results of biochemical analyses and tests for fusion response of cultured explants derived from these tumors.

Tumors arising from either Clone 1D or F_{40} cells exhibit the low S/P ratio characteristic of Clone 1D cells, coupled with the high ether-lipid content characteristic of F_{40}. Although it is not possible to directly assess susceptibility to PEG-mediated fusion in tumor tissue, numerous tumors derived from Clone 1D and F_{40} cell lines have been removed from their hosts, disrupted and filtered to produce single cell suspensions, and returned to culture for further analysis (Roos, 1985). In culture, these cells initially retained their high ether-lipid and low acyl-chain saturation, and all tumor explants (regardless of the cell line originally injected to generate the tumor) fused readily following PEG treatment -- as expected from the Clone 1D-like fatty acid saturation, but contrary to the F_{40}-like ether lipid content. As the tumor explants were maintained in culture, each individual culture reverted to the initial lipid composition and fusion phenotype of the cell line originally injected: Clone 1D-derived tumor explants lost their ether-lipid with little effect on fusion response, while F_{40}-derived tumor cells returned to their characteristic high degree of acyl chain saturation and lost their sensitivity to PEG. The combination of these results clearly links fusion response with acyl chain saturation rather than ether-lipid content.

IMPLICATIONS OF THESE STUDIES

Applications of cell and membrane fusion

The experiments described above demonstrate a functional association between lipid composition and cell membrane fusion. Through the use of defined media supplements it has been possible to completely control the response of these cells to a variety of fusogenic treatments. Although for any given cell line the extent of fusion induced by chemical agents (such as PEG)

*The muse of mice, a little-known illegitimate daughter of Mnemosyne (goddess of memory and mother of the Muses by Zeus).

and enveloped viruses (such as MHV) are inversely correlated, appropriate lipid supplement experiments were able to control both phenomena at once. Thus, for example, parallel cultures of an intermediate fusion-altered cell line such as F_{16} can be made to mimic the PEG-sensitive/virus-resistant phenotype of Clone 1D by growth in high concentrations of polyunsaturated fatty acid, or to resemble the virus-sensitive/PEG-resistant character of F_{40} cells, solely by growth in saturated fatty acid.

The ability to control the extent of membrane fusion virtually at will has numerous potential applications in the design of experimental protocols involving cell hybridization or other forms of membrane fusion (Roos *et al.*, 1987). Even where the direct manipulation of fusion through lipid alteration may be undesirable or infeasible, the relationships between different types of fusion which have been elucidated above may provide useful clues to the selection of an appropriate fusogen. Careful studies on the extent of fusion between several of these cell lines which differ in their inherent responsiveness to PEG supplement existing models of cell fusion (Röhme and Thorburn, 1981), and provide a rational basis for maximizing desirable interactions in cell hybridization procedures (D.S. Roos, R.L. Davidson, and P.W. Choppin, manuscript in preparation).

Membrane fusion mechanisms

Several aspects of cellular lipid composition are associated with these fusion-altered cell lines, but the degree of acyl chain saturation and ether-lipid content are most closely correlated with the observed differences in fusion phenotype. As previously noted (Roos and Choppin, 1985a), these two phenomena may be metabolically linked, and it is therefore not surprising

Table IV. Lipid Composition and Fusion Response of Tumor Explants

Cell/Tissue sample*		FA Saturation (S/P ratio)	Ether Lipid	PEG Fusion Response (%)
Clone 1D	Control cells (prior to injection)	1.0 ± 0.1†	–	94 ± 5†
	Tumor tissue	1.1	+++	ND
	Tumor explants cultured <2 days	ND	++	74
	Explants after 5 days in culture	1.1	–	81
	Explants after 14 days in culture	1.2	–	91
F_{40}	Control cells (prior to injection)	4.0 ± 1.2†	++	21 ± 8†
	Tumor tissue	1.1	+++	ND
	Tumor explants cultured <2 days	ND	++	70
	Explants after 5 days in culture	1.9	++	23
	Explants after 14 days in culture	3.6	++	19

* 10^6 Clone 1D or F_{40} cells were injected intramuscularly into 6-week-old irradiated C3HeB/FeJ mice, and tumors were removed for culture six weeks post-injection. Each data point is the average of results from at least three independent tumors. ND, not done.

† ± standard deviation.

that they have proven to be quite refractory to independent manipulation. Table V presents a qualitative summary of the relevant results from a variety of experiments which have produced alterations in fatty acyl chain saturation, ether-lipid content, or both. Supplementation of the culture medium with fatty acids varying in their degree of saturation and conformational stereochemistry has allowed substantial control over acyl chain composition with lesser, but significant, variation in ether-lipid content. In these experiments, PEG fusion response can be dramatically altered (cf. Table III and Figure 2), although it is somewhat surprising that both *cis* and *trans* poly-

Table V. Cell Fusion and Lipid Composition: Summary of Lipid Alteration Experiments*

	Treatment	Ether Lipid	FA saturation	PEG Fusion
Control				
Clone 1D	none	–	–	+
F_{40}	"	+	+	–
Fatty Acid Supplements				
Clone 1D	+19:0	+/–	+	–
F_{40}	+18:3	–/+	–	+
Trans Fatty Acids				
Clone 1D	+18:2*trans*	–	– (but S+t/P ↑)	+
F_{40}	" "	–/+	– (but fluidity low)	+
Fatty Alcohol Supplements				
Clone 1D	+FAOH (18:2)	+	–/+	+/–
F_{40}	" "	+	+	–
Temperature Shift				
Clone 1D	40°C	–	+/–	+
F_{40}	29°C	–/+	–	–/+
Tumorigenesis				
Clone 1D	tumor explants	+ → –	–	+
F_{40}	" "	+	– → +	+ → –

* See text for descriptions of experimental protocols.

unsaturates serve to stimulate fusion (Roos and Choppin, 1985b). Fatty alcohols, important precursors in ether-lipid biosynthesis (Hajra, 1983; Brown and Snyder, 1983; Friedberg *et al.*, 1983), have also been used as a media supplement to increase ether-lipid levels (according to the procedures of Cabot and Snyder, 1980). Although ether-lipid production could be substantially stimulated in Clone 1D cells, no effect was observed on PEG-induced fusion beyond the level expected from accompanying changes in fatty acid saturation. Another experimental approach to the alteration of acyl chain saturation involves growing cells at different temperatures (Ferguson *et al.*, 1975). The most useful experiments, however, emerged from the analysis of lipid composition and fusion response in cells derived from mouse tumors, as discussed above and in Table IV. Overall, PEG fusion response is most directly correlated with fatty acid saturation. Virus-induced fusion has only been examined in a limited number of these lipid alteration experiments, but wherever tested, virus fusion has been found to be inversely correlated with susceptibility to the fusogenic effects of PEG.

It is tempting, based on theoretical considerations and work on artificial systems, to base models for cell fusion on altered membrane fluidity (Ahkong *et al.*, 1973; Kosower *et al.*, 1975; Papahadjopoulos *et al.*, 1979; Chaudhury and Ohki, 1981; Cullis *et al.*, 1983). The central role of acyl-chain saturation in controlling fusion response superficially suggests that increased membrane fluidity increases susceptibility to PEG-induced fusion, while stiffer membranes predispose cells to virus-induced fusion. This interpretation is dfficult to reconcile, however, with the observation that both *cis* and *trans* polyunsaturates serve to promote PEG-induced fusion (Roos and Choppin, 1985b), despite the fact that *trans* polyunsaturated acyl chains stiffen the surface membranes of these cells as effectively as do saturated lipid supplements (unpublished data). It seems more likely that the role played by fatty acyl chains in controlling fusion is an indirect one, perhaps exerting an effect through interactions with membrane proteins. Such crucial controlling roles for lipids in regulating protein function have recently come to light in a number of systems (Sandermann, 1978; Cockroft and Gomperts, 1979; Benveniste and Vargaftig, 1983; Nishizuka, 1984; Low *et al.*, 1986), and may also apply to cellular and viral proteins which induce membrane fusion.

Regardless of the precise biochemical basis (or bases) of cell fusion, the system described above provides a useful approach to the elucidation of whatever factors may be involved. It may be possible to use specific lipid species, known from these studies to relate to the control of cell fusion, as affinity reagents to isolate further fusion factors. Alternatively, taking advantage of the genetic basis of fusion alterations in these somatic cell mutants, it may be possible to directly search for the DNA sequences responsible for the fusion-altered phenotype of F_{40} cells.

Use of membrane fusion mutants to study other membrane-associated phenomena

The cell lines described in this report have proved to be useful in a number of studies unrelated to cell fusion (unpublished observations, and Roos and Choppin, 1984). Although the initial intent of these experiments was to select PEG fusion-resistant mutants, the result was to isolate cell lines which are progressively altered in several aspects of their membrane biochemistry. This panel of closely related cell lines can now be used to study other membrane-associated phenomena in addition to PEG-induced fusion. A partial listing of membrane interactions which preliminary studies have shown to vary across the series of increasingly PEG-resistant cell lines includes: fusion induced by various enveloped viruses, chemicals other than PEG, and spontaneously occurring fusion; pinocytosis and membrane recycling; viral infectivity, maturation, and transformation; bacterial and mycoplasmal invasion and pathogenesis; tumorigenicity, metastatic spread *in vivo* and in model systems; and immune recognition both *in vivo* and *in vitro*. As the cell lines described above are stable, homogeneous, and now partially characterized genetically and biochemically, it may be possible to extend the analysis of these phenomena to link particular aspects of their structure with these various functions, similar to the lipid-alteration studies which have so clearly linked acyl-chain saturation with fusion response.

ACKNOWLEDGEMENTS. I would like to express my thanks to Dr.s Purnell W. Choppin (The Rockefeller University, New York City, New York, and Howard Hughes Medical Institute, Bethesda, Maryland), Richard L. Davidson (Center for Genetics, University of Illinois Medical Center, Chicago, Illinois), and Kathryn V. Holmes (Department of Pathology, Uniformed Services University of the Health Sciences, Bethesda, Maryland) for their collaboration, advice, and support during the course of the studies described, and to Karen Winkel and Linda Wilkinson for able and enthusiastic technical assistance.

REFERENCES

Ahkong, Q.F., D. Fisher, W. Tampion, and J.A. Lucy, 1973, The fusion of erythrocytes by fatty acids, esters, retinol, and α-tocopherol, *Biochem. J.* **136:**147-155.

Benveniste, J., and B.B. Vargaftig, 1983, Platelet-activating factor: an ether lipid with biological activity, *In:* "Ether Lipids: Biochemical and Biomedical Aspects," H. Mangold and F. Paltauf, editors, pp. 355-376, Academic Press, New York.

Blow, A.M.J., A.J. Botham, D. Fisher, A.H. Goodall, C.P.S. Tilcock, and J.A. Lucy, 1978, Water and calcium ions in cell fusion induced by poly(ethylene glycol), *FEBS Lett.* **94:**305-310.

Bollinger, J.N., 1967, The isolation and tentative identification of diacylglyceryl ethers from the Walker 256 carcinoma of the rat and a human lymphosarcoma, *Lipids* **2:**143-148.

Bratt, M.A., and W.R. Gallaher, 1969, Preliminary analysis of the requirements for fusion from within and fusion from without by Newcastle disease virus, *Proc. Nat'l Acad. Sci. USA* **64:**536-543.

Brown, A.J., and F. Snyder, 1983, The method of alkyldihydroxyacetone-P synthase, *J. Biol. Chem.* **258:**4184-4189.

Cabot, M.C., and F. Snyder, 1980, Manipulation of alkyklglycerolipid levels in cultured cells: fatty alcohol versus alkylglycerol supplements, *Biochim. Biophys. Acta* **617:**410-418.

Chaudhury, M.K., and S. Ohki, 1981, Correlation between membrane expansion and temperature-induced membrane fusion, *Biochim. Biophys. Acta* **642:**365-374.

Choppin, P.W., and A. Scheid, 1980, The role of viral glycoproteins in adsorption, penetration, and pathogenicity of viruses, *Rev. Infect. Dis.* **2:**40-61.

Cockroft, S., and B.D. Gomperts, 1979, Evidence for a role of phosphatidylinositol turnover in stimulus-secretion coupling, *Biochem. J.* **178:**681-687.

Compans, R.W., K.V. Holmes, S. Dales, and P.W. Choppin, 1966, An electron microscopic study of moderate and virulent virus-cell interactions of the parainfluenza virus SV5, *Virology* **30:**411-426.

Cullis, P.R., B. de Kruijff, M.J. Hope, A.J. Verkleij, R. Nayar, S.B. Farren, C. Tilcock, T.D. Madden, and M.B. Bally, 1983, Structural properties of lipids and their functional roles in biological membranes, *In:* "Membrane Fluidity in Biology," R.C. Aloia, editor, pp. 40-83, Academic Press, New York.

Davidson, R.L., 1977, Genetics of cultured mammalian cells, as studied by somatic cell hybridization, *Nat'l Cancer Inst. Monograph* **48:**21-30.

Davidson, R.L., and P.S. Gerald, 1976, Improved techniques for the induction of mammalian somatic cell hybridization by polyethylene glycol, *Somatic Cell Genet.* **2:**165-176.

Davidson, R.L., K.A. O'Malley, and T.B. Wheeler, 1976, Polyethylene glycol-induced mammalian cell hybridization: effect of polyethylene glycol molecular weight and concentration, *Somatic Cell Genet.* **2:**271-280.

de St. Groth, S.F., and D. Scheidegger, 1980, Production of monoclonal antibodies: strategy and tactics, *J. Immunol. Methods* **35:**1-21.

Ferguson, K.A., M. Glaser, W.H. Bayer, and P.R. Vagelos, 1975, Alteration of fatty acid composition of LM cells by lipid supplementation and temperature, *Biochem.* **14:**146-151.

Friedberg, S.J., and M. Halpert, 1978, Ehrlich ascites tumor cell membranes: an abnormality in ether lipid content, *J. Lipid Res.* **19:**57-64

Friedberg, S.J., S.T. Weintraub, M.R. Singer, and R.C. Greene, 1983, The mechanism of ether bond formation in O-alkyl lipid synthesis in Ehrlich ascites tumor, *J. Biol. Chem.* **258:**136-142.

Furusawa, M., 1980, Cellular microinjection by cell fusion: technique and applications in biology and medicine, *Int. Rev. Cytol.* **62:**29-67.

Gregoriades, G., editor, 1984, "Liposome Technology," CRC Press, Boca Raton, Florida.

Hajra, A.K., 1983, Biosynthesis of O-alkylglycerol ether lipids, *In:* "Ether Lipids: Biochemical and Biomedical Aspects," H. Mangold and F. Paltauf, editors, pp. 85-106, Academic Press, New York.

Helenius, A., J. Kartenbeck, K. Simons, and E. Fries, 1980, On the entry of Semliki Forest virus into BHK-21 cells, *J. Cell Biol.* **84:**404-420.

Holmes, K.V., and P.W. Choppin, 1966, On the role of the response of the cell membrane in determining virus virulence. Contrasting effects of the parainfluenza virus SV5 in two cell types, *J. Exp. Med.* **124**:501-520.

Holmes, K.V., E.W. Doller, and L.A. Sturman, 1981, Tunicamycin resistant glycolsylation of a coronavirus glycoprotein: demonstration of a novel type of viral glycoprotein, *Virology* **115:**334-344.

Kit, S., D.R. Dubbs, L.J. Piekarski, and T.C. Hsu, 1963, Deletion of thymidine kinase activity from L cells resistant to bromodeoxyuridine, *Exp. Cell Res.* **31:**297-312.

Kosower, N.S., E.M. Kosower, and P. Wegman, 1975, Membrane mobility agents II. Active promoters of cell fusion, *Biochim. Biophys. Acta* **401:**530-534.

Langerhans, T., 1868, Über Riesenzellen mit wandständigen Kernen in Tuberkeln und die fibröse Form des Tuberkels, *Virchow's Arch. Path. Anat. Physiol.* **42**:382-404.

Lin, H.J., F.C.S. Ho, and C.L.H. Lee, 1978, Abnormal distribution of O-alkyl groups in the neutral glycerolipids from human hepatocellular carcinomas, *Cancer Res.* **38:**946-949.

Low, M.G., M.A.J. Ferguson, A.H. Futerman, and I. Silman, 1986, Covalently attached phophatidylinositol as a hydrophobic anchor for membrane proteins, *Trends Biochem. Sci.* **11:**212-215.

Lucy, J.A., 1982, Biomembrane fusion, *In:* "Biological Membranes," Vol. 4 (D. Chapman, editor, pp. 367-415, Academic Press, London.

Mangold, H, and F. Paltauf, editors, 1983, "Ether Lipids: Biochemical and Biomedical Aspects," Academic Press, New York.

Nishizuka, Y., 1984, The role of protein kinase C in cell surface signal transduction and tumor promotion, *Nature (Lond.)* **308:**693-698.

O'Malley, K.A., and R.L. Davidson, 1977, A new dimension in suspension fusion techniques with polyethylene glycol, *Somatic Cell Genet.* **3:**441-448.

Papahadjopoulos, D., G. Poste, and W.J. Vail, 1979, Studies on membrane fusion with natural and model membranes, *Methods Membr. Biol.* **10:**1-121.

Pfleger, R.C., N.G. Anderson, and F. Snyder, 1968, Lipid class composition and fatty acid composition of rat liver plasma membranes isolated by zonal centrifugation, *Biochemistry* **7:**2826-2833.

Poste, G., and G.L. Nicholson, editors, 1978, "Membrane Fusion," Cell Surface Reviews, Vol. 5, Elsevier/North Holland, Amsterdam.

Robinson, J.M., D.S. Roos, R.L. Davidson, and M.L. Karnovsky, 1979, Membrane alterations and other morphological features associated with polyethylene glycol-induced cell fusion, *J. Cell Sci.* **40:**63-75.

Röhme, D., and D. Thorburn, 1981, Quantitative cell fusion: derivation and application of theoretical models, *Somatic Cell Genet.* **7:**43-57.

Roos, D.S., 1985, "Membrane Fusion, Lipid Composition, and Tumorigenicity of Cultured Cells," Ph.D. thesis, pp. 201+xviii, The Rockefeller University, New York.

Roos, D.S., and P.W. Choppin, 1984, Tumorigenicity of cell lines with altered lipid composition, *Proc. Nat'l Acad. Sci. USA* **81:**7622-7626.

Roos, D.S., and P.W. Choppin, 1985a, Biochemical studies on cell fusion I. Lipid composition of fusion-resistant cells, *J. Cell Biol.* **101:**1578-1590.

Roos, D.S., and P.W. Choppin, 1985b, Biochemical studies on cell fusion II. Control of fusion response by lipid alteration, *J. Cell Biol.* **101:**1591-1598.

Roos, D.S., and R.L. Davidson, 1980, Isolation of mouse cell lines resistant to the fusion-inducing effect of polyethylene glycol, *Somatic Cell Genet.* **6:**381-390.

Roos, D.S., J.M. Robinson, and R.L. Davidson, 1983, Cell fusion and intramembrane particle distribution in polyethylene glycol-resistant cells, *J. Cell Biol.* **97:**909-917.

Roos, D.S., R.L. Davidson, and P.W. Choppin, 1987, Control of cell fusion in polyethylene glycol-resistant cell mutants: applications to fusion technology, *In:* "Cell Fusion," A.E. Sowers, editor, pp. 123-144, Plenum Press, New York.

Sandermann, H., 1978, Regulation of membrane enzymes by lipids, *Biochim. Biophys. Acta* **515:**209-237.

Schroeder, F., and P.R. Vagelos, 1976, Effects of phospholipid base analogs on the subcellular membrane ether composition of suspension cultured LM cells, *Biochim. Biophys. Acta* **441:**239-254.

Schlegel, R.A., and M.R. Lieber, 1987, Microinjection of culture cells via fusion with loaded erythrocytes, *In:* "Cell Fusion," A.E. Sowers, editor, pp. 457-478, Plenum Press, New York.

Snyder, F., editor, 1972, "Ether Lipids: Chemistry and Biology," Academic Press, New York.

Sowers, A.E., editor, 1987, Cell Fusion, Plenum Press, New York.

Spear, P.G., 1987, Virus-induced cell fusion, *In:* "Cell Fusion," A.E. Sowers, editor, pp. 3-32, Plenum Press, New York.

Spector, A.A., S.N. Mathur, T.L. Kaduce, and B.T. Hyman, 1981, Lipid nutrition and metabolism of cultured mammalian cells, *Prog. Lipid Res.* **19:**155-186.

Wang, E.W., D.S. Roos, M.H. Heggeness, and P.W. Choppin, 1982, Function of cytoplasmic fibers in syncytia, *Cold Spring Harbor Symp. Quant. Biol.* **46:**997-1012.

White, J., M. Matlin, and A. Helenius, 1981, Cell fusion by Semliki Forest, influenza, and vesicular stomatitis virus, *J. Cell Biol.* **89:**674-679.

White, J., M. Kielian, and A. Helenius, 1983, Membrane fusion proteins of enveloped animal viruses, *Quart. Rev. Biophys.* **16:**151-195.

Wood, R., and F. Snyder, 1967, Characterization and identification of glyceryl ether diesters present in tumor cells, *J. Lipid Res.* **8:**494-500.

Wood, R., N.G. Anderson, and D.C. Swartzendruber, 1970, Characterization of lipids isolated from membranous material, *Arch. Biochem. Biophys.* **14:**190-197.

Zimmermann, U., 1982, Electric field-mediated fusion and related electrical phenomena, *Biochim. Biophys. Acta* **694:**227-277.

ROLE OF PHOSPHOLIPID ASYMMETRY IN CELLULAR MEMBRANE FUSION

Robert A. Schlegel and Patrick Williamson

Department of Molecular and Cell Biology
The Pennsylvania State University
University Park, PA 16802

Department of Biology
Amherst College
Amherst, MA 01002

INTRODUCTION

Because of its relative simplicity and the wealth of information available on its membrane, the erythrocyte has served as a useful model for studies of the mechanisms of cellular and membrane fusion. Insight has been sought most often from scrutinizing intercellular fusion systems, while relatively little attention has been given to intracellular fusion and to what might be learned from evaluating conditions conducive to this type of fusion. Similarly, alterations in the protein portion of the membrane which might play a role in fusion have been given far more attention than the role which alterations in phospholipids, particularly their arrangement, might play in erythrocyte fusion. In this paper, we explore phospholipid spacing as a potential determinant of intracellular erythrocyte membrane fusion. In particular, we postulate that erythrocyte membranes with tightly-packed phospholipids are fusion-incompetent, whereas those with loosely-packed phospholipids are fusion-competent; and that packing is controlled by the maintenance or loss of the normal asymmetric distribution of phospholipids in the erythrocyte membrane.

PHOSPHOLIPID ASYMMETRY AND ITS MAINTENANCE

One of the well-documented properties of the erythrocyte membrane is the asymmetric distribution of phospholipids between the inner and outer leaflets of the bilayer: phosphatidyl choline (PC) and sphingomyelin are found primarily in the outer leaflet, whereas phosphatidyl ethanolamine (PE) and phosphatidyl serine (PS) are, respectively, largely or exclusively restricted to the inner leaflet (Op den Kamp, 1979). Because PE and PS contain a disproportionate share of the unsaturated sidechains present in the membrane (Williams et al., 1966), this asymmetry in headgroup location implies an asymmetry of fatty acyl sidechain distribution which imparts distinct characteristics to the two leaflets: the inner leaflet is less tightly packed and more fluid than the outer (Tanaka and Ohnishi, 1976; Williamson et al., 1982; Seigneuret et al., 1984; Morrot et al., 1986).

The fluorescent lipophilic probe merocyanine 540 (MC540) is relatively excluded from tightly-packed hemileaflets in comparison to loosely-packed ones (Williamson et al., 1983). Because the dye molecule contains a permanent, localized charge, it cannot penetrate beyond the exterior leaflet when externally applied (Schlegel et al., 1980; Williamson et al., 1982). These properties give it the power to discern the difference in packing between the two leaflets of the erythrocyte. Only small amounts of MC540 bind to normal erythrocytes because of the tightly-packed character of their exterior leaflet. If erythrocytes lose their phospholipid asymmetry, however, previously interior phospholipids appear on the surface, as lipids randomize between the two leaflets, resulting in a loosely-packed exterior leaflet which binds much larger amounts of dye (Williamson et al., 1982, 1985). Increased dye binding can be assessed either qualitatively by fluorescence microscopy (Schlegel et al., 1987b), quantitatively by determining actual amounts of dye bound by a population of cells (McEvoy et al., 1986; Schlegel et al., 1987b) or semi-quantitatively at the single cell level by flow cytometry (Schlegel et al., 1987a).

Both MC540 studies and classical biochemical procedures have contributed to a body of evidence implicating the cytoskeletal protein spectrin in the maintenance of erythrocyte phospholipid asymmetry. When the network of spectrin which underlies the erythrocyte membrane is missing or perturbed, whether for genetic reasons or as a result of experimental manipulation, phospholipid asymmetry is lost (Haest and Deuticke, 1976; Haest et al., 1978; Williamson et al., 1982; Kumar and Gupta, 1983; Reed et al., 1985; Choe et al., 1985, 1986). These data have suggested an hypothesis that PE and PS are selectively retained in the inner leaflet through electrostatic interactions with the spectrin/cytoskeletal network, specific for those lipids whose headgroup carries at least a partial net negative charge. A competing hypothesis is that a headgroup-specific, ATP-dependent phospholipid pump is responsible for both establishing and maintaining phospholipid asymmetry (Seigneuret and Devaux, 1984). However, the kinetics of this activity suggest that the pump is actually a bidirectional catalyst of lipid flip/flop which itself could not maintain lipid asymmetry, but which could facilitate expression of shifts in phospholipid equilibrium distributions (Williamson et al., 1987).

VESICULATION

A naturally-occurring biological experiment provides a first clue as to the potential role of spectrin and maintenance of phospholipid asymmetry in the control of membrane fusion. Mice with the sph/sph genotype suffer from severe anemia. The spherocytic erythrocytes found in their circulation contain essentially no spectrin (Greenquist et al., 1978), and in contrast to erythrocytes from normal littermates, they have lost the normal asymmetry in phospholipid distribution; they also stain with MC540 (Williamson et al., 1982). Interestingly, when cultures of these spherocytes are viewed microscopically, they display frequent blebs and attached vesicles; and the isolated membranes show a strong tendency toward fragmentation and vesiculation, a characteristic they share with normal erythrocyte membranes after spectrin has been extracted with very low ionic strength buffers. Intramembrane fusion of internal surfaces can account for all of these observations, suggesting that loss of spectrin and asymmetry are conducive to fusion. If loss of lipid asymmetry does indeed produce the capacity for fusion, the external leaflet of these cells might be expected to be as fusable as the internal leaflet. Although no direct evidence of this capacity is available, it is provocative that in spite of the tendency of these cells to fragment and vesiculate, the average cell volume of the spherocytes is larger than that of normal control erythrocytes, with a very broad distribution of sizes, consistent with spontaneous fusion of spherocytes with themselves and with their vesiculated fragments.

Another example of a genetic disorder in which loss of asymmetry is accompanied by vesiculation is sickle cell disease. Deoxygenation/ reoxygenation of sickle erythrocytes in vitro results in the release of spicules which pinch off to form small vesicles and chains of vesicles (Allan et al., 1982). Similar spicules and vesicles are present in the circulation of sickle cell patients (Allan et al., 1982; Chilcote et al., 1982; Dumaswala et al., 1982). These elements are nearly devoid of spectrin (Allan et al., 1982) and have lost their phospholipid asymmetry, assessed either biochemically or by staining with MC540 (Choe et al., 1986). The ability to visualize spicules while they are still attached to the cell has allowed an examination of the disposition of the spicule membrane before detachment from the cell. When MC540 is used to assess the disposition of lipids, the attached spicules are found to bind no more dye than the remainder of the membrane (Choe et al., 1986). This result would suggest that the spicules contain spectrin. In fact, immunofluorescence analysis of deoxygenated cells using anti-spectrin antibody revealed the presence of the cytoskeletal in attached spicules, extending even to the tip (Williamson, Costigan, and Schlegel, unpublished). This result suggests that mere mechanical extrusion of the membrane will not produce spectrin-free/symmetric vesicles. As implied by the characteristics of the released vesicles, detachment of the spectrin network from the membrane, with a presumptive loss of lipid asymmetry, is also required before fusion and vesiculation can take place.

In addition to these genetic disorders which predispose erythrocyte membranes to vesiculation, normal erythrocytes can be manipulated in a number of ways to induce vesiculation experimentally, thus providing systems where further analyses of the relationship between membrane structure and fusion capacity may be accessible to study. When erythrocytes are incubated for several hours in the absence of an energy source, they shed small vesicles (Lutz et al., 1977). Exfoliation of similar vesicles results when Ca^{2+} levels are elevated in erythrocytes using a Ca^{2+} ionophore (Allan and Thomas, 1981). In both of these cases, the vesicles are spectrin-free, and in both cases the vesicles have lost their asymmetry as assessed either biochemically or with MC540 (Williamson et al., 1982; Choe et al., 1985).

Collectively, these results suggest that separation of the membrane bilayer from the spectrin cytomembrane network, with an attendant loss of lipid asymmetry, is a common antecedent to the intracellular fusion events which result in vesiculation. However, since spectrin/membrane separation and loss of lipid asymmetry are always associated in these vesiculation systems, the independent requirement of each can not be ascertained. In other words, it may be that dissociation of spectrin is both necessary and sufficient, irrespective of loss of asymmetry, for fusion to occur--that dissociation of spectrin removes a physical barrier which prevents already fusion-competent leaflets from contacting one another.

ENDOCYTOSIS

When vesiculation is internal rather than external, the process is termed endocytosis. When this occurs, the surface which fuses is the external leaflet of the bilayer and therefore is free of spectrin. Studies of endocytosis thus offer the possibility of addressing the importance of asymmetry in controlling fusion in the absence of an impediment to contact of fusing partners posed by spectrin. Irrespective of whether internal leaflets are constitutively fusion-competent or become so upon loss of asymmetry, external leaflets might be expected to become fusion-competent upon loss of asymmetry through introduction of interior phospholipids in the first case, or through acquiring the same composition as the fusion-competent inner leaflet in the latter case.

Endocytosis can be induced in neonatal human reticulocytes and erythrocytes by concanavalin A (Scheckman and Singer, 1976). Indirect immunoferritin labeling of thin sections reveals that invaginations and endocytic vesicles do not stain for spectrin, whereas the adjacent membrane stains heavily (Tokuyasu et al., 1979). These results suggest that limited portions of the membrane may lose their asymmetry converting that portion of the external leaflet to a fusion-competent condition. If so, in a membrane in which both kinds of lipid distribution are present, areas of endocytosis should be marked by a looser lipid packing and corresponding increase in binding of MC540. Because reticulocytes and erythrocytes from adults do not possess this endocytic capacity, it being removed during enucleation (see below), the immediate precursors of reticulocytes, enucleating erythroblasts, were studied.

When enucleating erythroblasts are examined by immunofluorescence for the presence of spectrin, the portion of the cell which contains the nucleus is not stained, whereas the incipient reticulocyte portion is (Geiduschek and Singer, 1979). In contrast, staining with MC540 showed the reverse pattern with the incipient reticulocyte not staining and the nucleus-containing portion of the bilobulate cell staining (Schlegel et al., 1982). These results suggest that the enucleating erythroblast is dichotomous with respect to transbilayer distribution of phospholipids with a lipid-symmetric and a lipid-asymmetric portion coexisting within one cell. Remarkably, this segregation is apparently accompanied by a segregation of those regions of the membrane which are specialized for endocytosis, since the bulk of the clathrin-coated regions of the membrane appear to segregate in the portion of the membrane which stains with MC540 (Marshall et al., 1984), providing circumstantial evidence for a link between the absence of membrane lipid asymmetry and active membrane fusion of the extracellular leaflet of the bilayer.

To examine this relationship more closely, mouse erythroleukemia cells (MEL) were used because of their ready availability and ease of manipulation. When MEL were stained with MC540, they displayed a patchy distribution of fluorescence superimposed upon the ring fluorescence characteristic of general membrane staining. When they were induced to differentiate, the stainable patches were collected into one region of the membrane, simulating the events which precede enucleation in mouse erythroblasts (Schlegel et al., 1980, 1982). If uninduced cells were incubated at 37°C following staining, some of the dye appeared to be internalized as judged by a change in the focal plane in which it could be found. To confirm this observation, a water-soluble, impermeant quencher of MC540 (Williamson et al., 1982, 1985) was used to extinguish external fluorescence. As can be seen in Figure 1, addition of quencher removed the patchy ring staining observed on unquenched cells (arrow, panel b), but left undimmed concentrations of dye distributed irregularly throughout the cell (panel d). This pattern of staining was completely distinct from that produced by dye permeating the plasma membrane, observed either in dead cells or cells permeabilized with detergent, where staining was very much brighter, and included intense fluorescence from the nuclear membrane. One such cell can be seen in the upper left of panel d, although the details of its staining pattern are obscured by the overexposure required to reveal the pattern observed in the remaining, living cells.

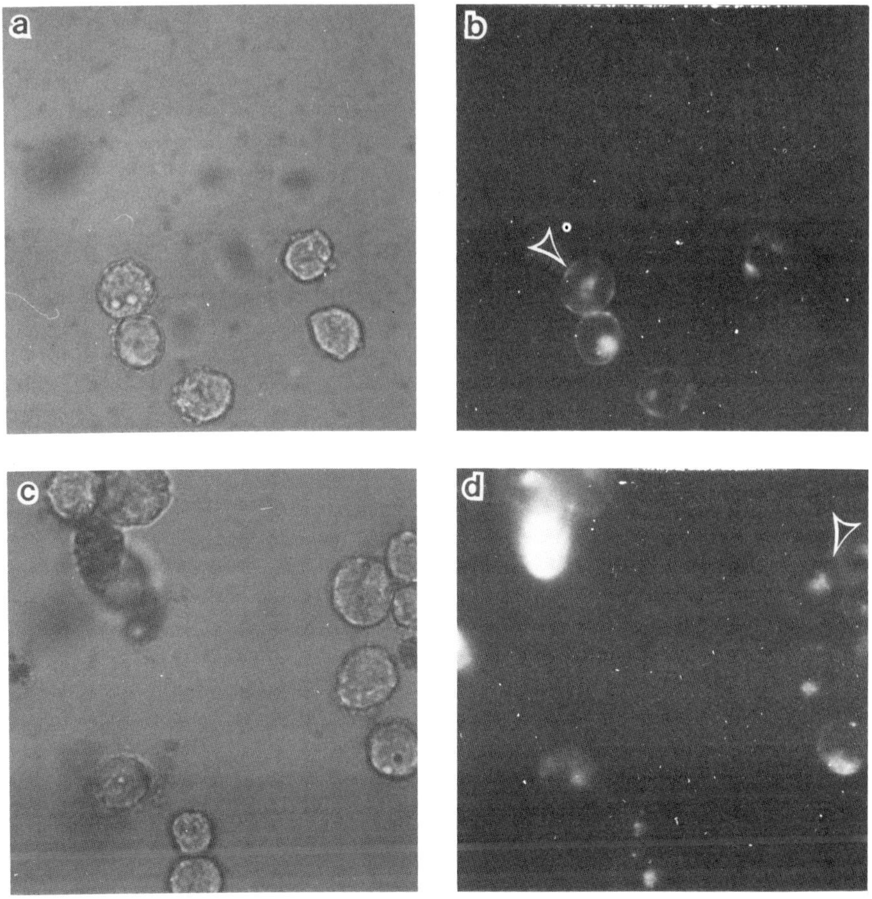

Figure 1. MC540 staining of MEL cells. A superinducible subclone of the Friend murine erythroleukemia cell line 745-PC-4-7 (Volloch and Housman, 1982; a kind gift of D. Housman) was itself subcloned and used within 3 days of isolation. Cells were stained with 4 ug/ml of MC540 in phosphate-buffered saline (PBS) containing 1% fetal bovine serum (FBS), incubated for 5 minutes at 37°C, and washed twice in PBS + 1% FBS. Following an additional 5-minute incubation at 37°C, quencher (diS-$C_3N^+(CH_3)_3$-5) was added to the cells shown in panels c and d at a final concentration of 50 ug/ml. Cells were photographed using Kodak recording film as described elsewhere (Williamson et al., 1982; Schlegel et al., 1987b). The arrow in panel b notes the ordinary ring staining observed whenever these cells are stained in the absence of quencher. This distinct ring staining is absent from the cells in panel d. The arrow in panel d notes a cell whose external fluorescence has been quenched, containing a single region into which dye has begun to accumulate. The survival of this region in the presence of quencher indicates its internal location. With prolonged incubation, this region becomes more prominent and distinct, and can be seen in a progressively larger fraction of the stained cell population.

When cells were incubated for longer times following staining, dye became concentrated in a single, fairly large structure; a typical example of this staining which persists following application of quencher is indicated by the arrow in panel d. Double staining with MC540 and the nuclear stain Hoechst 33258 indicated that this structure was not within the nucleus, but rather was located in the cytoplasm within a cleft or indentation in the nucleus, considered to be the Golgi region of the cell (Novikoff and Goldfischer, 1961).

These results suggest that regions of the MEL plasma membrane marked by MC540 are internalized by endocytosis and traffic to the Golgi complex. The MEL system has the disadvantage, however, that the spherical shape of the cell, combined with its small amount of cytoplasm, make visualization of the endocytic process difficult. To provide better viewing conditions and at the same time test the generality of these observations in other non-erythroid cells, attached, spread, and actively endocytosing macrophages were stained with MC540. As seen in Figure 2, these cells fluoresced with a punctate pattern of staining in which the nucleus was clearly discernible by its near absence of fluorescence compared with the cytoplasm. As was the case with MEL cells, when the impermeability of the plasma membrane was compromised, as indicated by loss of the ability to exclude trypan blue, a totally different pattern of staining was seen. The nuclear membrane stained brightly indicating that the punctate pattern was not a consequence of permeation of dye through the plasma membrane.

Internalization of stained membrane might be simply a consequence of the fact that MC540 binds to the membrane bilayer, and this bilayer by definition cycles through endocytic vesicles. Were this the case, any non-

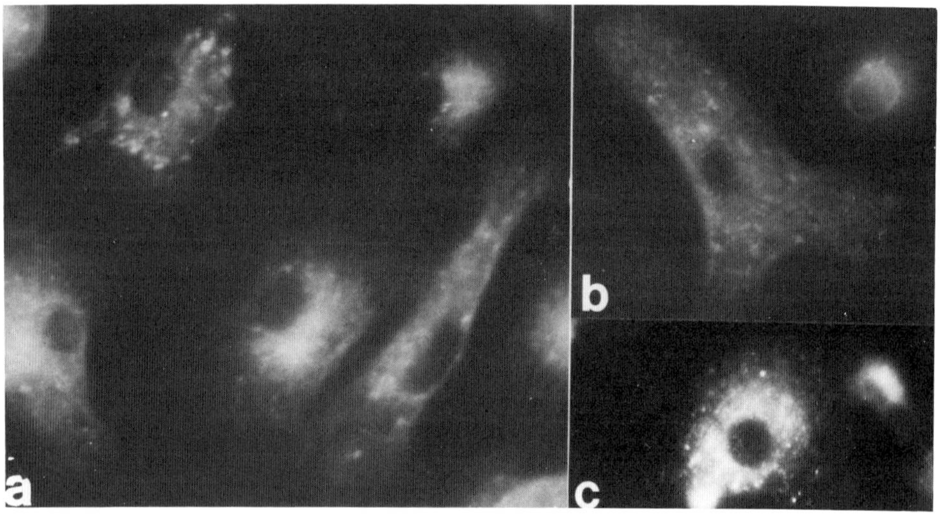

Figure 2. MC540 staining of mouse macrophages. Primary splenic mouse macrophages spread on glass coverslips were incubated at 37°C for 15 minutes in buffer containing 10 ug/ml of MC540 and 2% fetal bovine serum, and washed twice. Coverslips were wet-mounted and photographed as described elsewhere (Schlegel et al., 1987b). In the upper right-hand corner of panel b, a cell can be seen which was unable to exclude trypan blue indicating loss of membrane integrity. The nuclear membrane is stained, the periphery of the cell is faintly visible as a halo of fluorescence and the cytoplasm is devoid of punctate staining.

specific membrane stain should produce a pattern similar to that observed with MC540. To test this possibility, macrophages were stained with the membrane probe diI-C_{18} (3). Under conditions where MC540 produced a punctate pattern of staining, this probe exhibited weak ring fluorescence typical of general surface-membrane staining (data not shown). This observation implies that binding to the membrane bilayer is insufficient to result in dye internalization into vesicles as observed with MC540, and therefore suggests that MC540 stains only selected regions of the plasma membrane preferentially involved in the extracellular leaflet fusion event that gives rise to endocytic vesicles.

Such behavior is restricted neither to erythrocytes nor to other blood cells since mouse fibroblasts and Chinese hamster ovary cells exhibit similar punctate patterns when stained with MC540 (unpublished). Perhaps one function of the clathrin network apposed to the endocytosing membrane is the exclusion of spectrin or spectrin-like molecules to allow a fusion-competent external membrane leaflet to develope. If a transbilayer redistribution of phospholipids is involved in the acquisition of fusion-competence in coated pits in these various types of cells, MC540 may be acting as a marker of symmetric membrane in a surrounding of asymmetric membrane. Regardless, MC540 is unique in allowing visualization of the constitutive endocytosis universal to nearly all cells.

INTERCELLULAR FUSION

If localized loss of lipid asymmetry produces foci of fusion-competence in the external leaflet, then a generalized loss of asymmetry over the entire cell should convert the entire cell surface from fusion-incompetent to fusion-competent, thus enhancing intercellular fusion. It must be emphasized that fusion-competence does not mean that such membranes spontaneously fuse. It is generally agreed that membrane fusion is a multistep process, and that initial events, such as very close apposition of bilayers, must precede actual bilayer disruption and reformation. Thus, a fusogenic stimulus may be required to initiate or induce fusion, but a competent membrane leaflet is necessary for the stimulus to be effective. Therefore, erythrocytes with symmetric membranes cannot be expected to spontaneously fuse with one another, but inducing them to fuse should be facilitated. Recently, methods for producing lipid-symmetric erythrocyte ghosts under controlled conditions (Williamson et al., 1985) offer the prospect of examining the role of transbilayer lipid distribution in intercellular fusion, by testing whether symmetric ghosts are more susceptible to fusion than asymmetric ones.

As of yet, we know of no direct test of this prediction, but there is some information in the literature which is pertinent. Erythrocytes can be fused using electric field pulses (Sowers and Kapoor, 1987; Sowers, this volume). Although dielectrophoreis induced by alternating current followed by a direct current pulse has most often been used to induce electrofusion, erythrocyte fusion can also be induced by reversing the order of these two treatments, even when significant time intervals separate the pulse from the subsequent dielectrophoresis (Sowers, 1986). This evidence for a long-lived fusogenic state of erythrocytes is of particular interest when viewed in the context of a separate report which indicates that following dielectric breakdown of erythrocytes, more PE becomes accessible on the external leaflet (Dressler et al., 1983). Are the fusogenic sites induced by direct current pulses actually sites in the membrane where phospholipid asymmetry has been lost? Such an hypothesis could be tested in several ways. Pulsed erythrocytes could be probed with MC540 for localized regions of lipid-symmetric membrane detected by enhanced staining. Another route would be to determine whether symmetric erythrocyte ghosts (Williamson et al, 1985) are fusible upon dielectrophoresis.

When polyethylene glycol (PEG) is used to induce fusion of erythrocytes, treatments with agents such as spermidine and trichoroacetic acid have been shown to enhance fusion (Huang and Hui, 1986). Such treatments also induce IMP clustering in the presence of PEG, whereas other treatments effect neither clustering nor fusion. To test whether active agents were inducing a loss of asymmetry, and that the consequent fluidization of the outer leaflet was allowing clustering of IMP particles creating IMP-free regions requisite for fusion, erythrocytes treated as described were stained with MC540. No increase in binding could be detected by flow cytometry (Tullius, Williamson and Schlegel, unpublished). It would still be of interest, however, to determine whether symmetric ghosts are more easily fused than their asymmetric counterparts, i.e., whether fluidization of the outer leaflet upon loss of asymmetry facilitates IMP clustering conducive to fusion.

Lucy and Ahkong (1986) have postulated that a critical step in Sendai-virus induced intercellular fusion of erythrocytes is a permeabilization of the cells which results in colloid-osmotic swelling (and under some conditions, hemolysis). Such a permeabilization might have additional consequences as well: during hemolysis and resealing, erythrocytes do not maintain phospholipid asymmetry unless certain precautions are taken (Williamson et al., 1985). Permeabilization might therefore serve the dual role of driving fusion osmotically and at the same time generating a lipid-symmetric membrane with a fusion-competent exterior leaflet. Indeed, when erythrocytes are fused by lipid-soluble fusogens, colloid-osmotic lysis dominates fusion in the absence of Ca^{2+}, the presence of which is conducive to loss of asymmetry (Williamson et al., 1985).

MECHANISMS

How does loss of asymmetry convert a fusion-incompetent to a fusion-competent membrane? More specifically, what critical change is effected by introducing inner leaflet phospholipids into the outer leaflet? There are several possibilities, namely: 1) the presence of anionic phospholipids, particularly PS and/or PE, 2) an increase in fluidity/spacing of lipid molecules, or 3) an increase in the hydrophobicity of the hemileaflet.

Because PS is a relatively abundant molecule which is normally restricted exclusively to the inner leaflet, this single species of phospholipid could be responsible for determining whether a leaflet is fusion-competent or not. For example, Ca^{2+} induces a lateral phase separation of PS (Papahadjopoulos et al., 1977), and might thereby induce fusion in PS-containing membranes, as occurs during Ca^{2+}-induced vesiculation. Ca^{2+} might also induce fusion by a different mechanism. Inner monolayer lipids of erythrocytes adopt the hexagonal II phase in the presence of Ca^{2+}, whereas the outer monolayer lipids are inert in this regard (Hope and Cullis, 1979). Should the hexagonal II phase, or the closely-related inverted micelle structure, be an essential intermediate in the fusion process, as has been suggested (Siegel, 1986), the appearance of susceptible lipids in the outer leaflet which occurs as asymmetry is lost might be responsible for its conversion to a fusion-competent state.

As indicated above, if lateral mobility of integral membrane proteins were restricted by the close packing of the outer leaflet, relief of this constraint upon loss of asymmetry and fluidization/loosening of lipid packing might facilitate IMP clustering. As pointed out, there is a strong correlation between IMP clustering and the ability to fuse erythrocytes with PEG (Boni and Hui, 1987).

Finally, might this alteration in packing be more directly responsible for conferring fusion-competence? It is generally agreed that an essential element in the fusion process is the very close approach of apposed bilayers (Lucy and Ahkong, 1986). A wider spacing of lipid headgroups might result in an increase in the exposure of the hydrophobic core of the membrane to solvent. In fact, when the partitioning of lipid-symmetric and lipid-asymmetric erythrocytes in an aqueous two polymer system was compared, symmetric cells were found to be more hydrophobic than asymmetric ones (McEvoy et al., 1986), as would be expected if the hydrophobic core of the bilayer were more accessible to solvent. This increase in hydrophobicity may well be responsible for the enhanced interactions of symmetric erythrocytes with other cells (Schlegel et al., 1985b; McEvoy et al., 1986). It may also facilitate a closer approach of bilayer fusion partners to each other or reduce the energy barrier to rearrangement of the two bilayers at the point of membrane fusion itself. Indeed, lipid-soluble fusogens such as unsaturated fatty acids (Ahkong et al., 1973) and simple hydrocarbons (Ahkong et al., 1974) may serve the dual role of increasing lipid spacing upon intercalation into the membrane (the latter by inserting into the hydrophobic membrane core, increasing the distance between phospholipid headgroups) as well as perturbing the membrane sufficiently to induce fusion.

SUMMARY AND IMPLICATIONS

Explanations for the phenomenon of phospholipid asymmetry have tended toward the teleological. However, evidence is now accumulating which indicates that inner leaflet phospholipids are confined to the interior so that the cell is able to present a non-interactive surface to reticuloendothelial cells (Schlegel and Williamson, 1987) and to blood components (Zwaal et al., 1977). We suggest that fusion-incompetent be added to the lexicon of non-interactive terminology. To reiterate, membrane leaflets with tightly-packed phospholipids, such as the outer leaflet of erythrocytes, are postulated to be fusion-incompetent; those with loosely-packed lipids, such as the inner leaflet of erythrocytes, are fusion-competent. Fusion between the fusion-competent cytoplasmic surfaces of erythrocytes is prevented by the cytomembrane spectrin network, which is also involved in maintaining asymmetry. When this barrier is removed, the cytoplasmic surfaces of the erythrocyte can make contact and fuse, resulting in vesiculation. If membrane phospholipid asymmetry is lost, the external leaflet becomes disordered and fusion-competent. If the loss is localized, apposition and fusion of external surfaces results in endocytosis. A more generalized loss of asymmetry abolishes the non-interactive nature of the cell surface, producing fusion-competence, and lending the cell to the ultimate interaction, intercellular fusion.

Evidence for an asymmetric distribution of phospholipids in the plasma membranes of cells other than erythrocytes appears sporadically in the literature but is not yet nearly as conclusive as it is for erythrocytes. Nonetheless, the evidence presented here on endocytosis in a variety cells lends weight to arguments that distinct lipid domains can exist within the plasma membrane of cells other than erythrocytes (Schlegel et al., 1985a). It is possible that these domains contain the same lipid composition as the remainder of the membrane but are distinct by virtue of a different transbilayer distribution of phospholipids. Therefore, studies of lipid asymmetry and its maintenance in erythrocytes may have much broader implications in cellular physiology than previously imagined.

ACKNOWLEDGEMENTS

We thank Leslie McEvoy for her help in preparing the manuscript. R.A.S. is an Established Investigator of the American Heart Association.

REFERENCES

Ahkong, Q.F., Fisher, D., Tampion, W., and Lucy, J.A., 1973, The fusion of erythrocytes by fatty acids, esters, retinol and α-tocophorol, Biochem. J., 136:147.

Ahkong, Q.F., Howell, J.T., Tampion, W., and Lucy, J.A., 1974, The fusion of hen erythrocytes by lipid-soluble and surface-active immunological adjuvants, FEBS Lett., 41:206.

Allan, D., Limbrick, A.R., Thomas, P., and Westerman, M.P., 1982, Release of spectrin-free spicules on reoxygenation of sickled erythrocytes, Nature, 295:612.

Allan, D., and Thomas, P., 1981, Ca^{+2}-induced biochemical changes in human erythrocytes and their relation to microvesiculation, Biochem. J., 198:433.

Boni, L.T., and Hui, S.W., 1987, The mechanism of polyethylene glycol-induced fusion in model membranes, in: "Cell Fusion," A.E. Sowers, ed., Plenum, New York.

Chilcote, R.R., Jones, B., and Damper, C., 1982, Membrane vesiculation in HbSS sickle cell anemia, Blood, 60:44a.

Choe, H.-R., Willaimson, P., and Schlegel, R.A., 1985, Disruption of phospholipid asymmetry in erythrocyte vesicles deficient in spectrin, Cell Biol. Intnl. Rep., 9:597.

Choe, H.-R., Schlegel, R.A., Rubin, E., Williamson, P., and Westerman, P., 1986, Alteration of red cell membrane organization in sickle cell anaemia, Brit. J. Haematol., 63:761.

Dressler, V., Schwister, K., Haest, C.W.M., and Deuticke, B., 1983, Dielectric breakdown of the erythrocyte membrane enhances transbilayer mobility of phospholipids, Biochim. Biophys. Acta, 732:304.

Dumaswala, U., Bryan, D.J., and Greenwalt, T.J., 1982, Studies of the mechanism of membrane changes in sickle cell anemia, Blood, 60:45a.

Geiduschek, J.B., and Singer, S.J., 1979, Molecular changes in the membranes of mouse erythroid cells accompanying differentiation, Cell, 16:149.

Greenquist, A.C., Shohet, S.B., and Bernstein, S.E., 1978, Marked reduction of spectrin in hereditary spherocytosis in the common house mouse, *Blood*, 51:1149.

Haest, C.W.M., and Deuticke, B., 1976, Possible relationships between membrane proteins and phopholipid asymmetry in the human erythrocyte membrane, *Biochim. Biophys. Acta*, 436:353.

Haest, C.W.M., Plasa, G., Kamp, D., and Deuticke, B., 1978, Spectrin as a stabilizer of the phospholipid asymmetry in the human erythrocyte membrane, *Biochim. Biophys. Acta*, 509:21.

Hope, M.J., and Cullis, P.R., 1979, The bilayer stability of inner monolayer lipids from the human erythrocyte, *FEBS Lett.*, 107:323.

Huang, S.K., and Hui, S.W., 1986, Chemical co-treatments and intraparticle patching in the poly(ethylene glycol)-induced fusion of turkey and human erythrocytes, *Biochim. Biochys. Acta*, 860:539.

Kumar,A., and Gupta, C.M., 1983, Red cell membrane abnormalities in chronic myelogenous leukemia, *Nature*, 303:632.

Lucy, J.A., and Ahkong, Q.F., 1986, An osmotic model for the fusion of biological membranes, *FEBS Lett.*, 199:1.

Lutz, H.U., Lui, S.-C., and Palek, J., 1977, Release of spectrin-free vesicles from human erythrocytes during ATP depletion, *J. Cell Biol.*, 73:548.

Novikoff, A.B., and Goldfischer, S., 1961, Nucleosidediphosphatase activity in the Golgi apparatus and its usefulness for cytological studies, *Proc. Natl. Acad. Sci. USA*, 47:802.

Marshall, L.M., Thureson-Klein, A., and Hunt, R.C., 1984, Exclusion of erythrocyte-specific membrane proteins from clathrin-coated pits during differentiation of human erythroleukemia cells, *J. Cell Biol.*, 98:2055.

McEvoy, L., Williamson, P., and Schlegel, R.A., 1986, Membrane phopholipid asymmetry as a determinant of erythrocyte recognition by macrophages, *Proc. Natl. Acad. Sci. USA*, 83:3311.

Morrot, G., Cribier, S., Devaux, P.F., Gelderwerth, D., Davoust, J., Bureau, J.F., Fellmann, P., Herve, P., and Frilley, B., 1986, Asymmetric lateral mobility of phospholipids in the human erythrocyte membrane, *Proc. Natl. Acad. Sci. USA*, 83:6853.

Op den Kamp, J.A.F., 1979, Lipid asymmetry in membranes, *Ann. Rev. Biochem.*, 48:47.

Papahadjopoulos, D.W., Vail, J., Newton, C., Nir, S., Jacobson, K., Poste, G., and Lazo, R., 1977, Studies of membrane fusion. III. The role of calcium-induced phase changes, *Biochim. Biophys. Acta*, 465:579.

Reed, J.A., Kough, R.H., Williamson, P.L., and Schlegel, R.A., 1985, Merocyanine 540 recognizes membrane abnormalities of erythrocytes in chronic myelogenous leukemia, *Cell Biol. Intnl. Rep.*, 9:43.

Schekman, R., and Singer, S.J., 1976, Clustering and endocytosis of membrane receptors can be induced in mature erythrocytes of neonatal but not adult humans, *Proc. Natl. Acad. Sci. USA*, 73:4075.

Schlegel, R.A., Phelps, B.M., Waggoner, A., Terada, L., and Williamson, P., 1980, Binding of merocyanine 540 to normal and leukemic erythroid cells, Cell, 20:321.

Schlegel, R.A., Phelps, B.M., Cofer, G.P., and Williamson, P., 1982, Enucleation eliminates a differentiation-specific marker from normal and leukemic murine erythroid cells, Exptl. Cell Res., 139:321.

Schlegel, R.A., Lumley-Sapanski, K., and Williamson, P., 1985a, Insertion of lipid domains into plasma membranes by fusion with erythrocytes, Biochim. Biophys. Acta, 846:234.

Schlegel, R.A., Prendergast, T.W., and Williamson, P., 1985b, Membrane phospholipid asymmetry as a factor in erythrocyte-endothelial cell interactions, J. Cell Physiol., 123:215.

Schlegel, R.A., and Williamson, P., 1987, Membrane phospholipid organization as a determinant of blood cell-reticuloendothelial cell interactions, J. Cell Physiol., in press.

Schlegel, R.A., McEvoy, L., Weiser, M., and Williamson, P., 1987a, Phospholipid organization as a determinant of red cell recognition by the reticuloendothelial system, Advances in the Biosciences, in press.

Schlegel, R.A., Reed, J.A., McEvoy, L., Algarin, L., and Williamson, P., 1987b, Maintenance and loss of membrane phospholipid asymmetry by loaded red cells and its effects on cellular interactions, Meths. Enzymol. 149, in press.

Seigneuret, M., and Devaux, P.F., 1984, ATP-dependent asymmetric distribution of spin-labled phospholipids in the eryhtrocyte membrane: relation to shape changes. Proc. Natl. Acad. Sci. USA, 81:3751.

Seigneuret, M., Zachowski, A., Hermann, A., and Devaux, P.F., 1984, Asymmetric lipid fluidity in human erythrocyte membrane: new spin label evidence, Biochemistry, 23:4271.

Siegel, D.P., 1986, Inverted micellar intermediates and the transition between lamellar, cubic, and inverted hexagonal lipid phases. II Implications for membrane interactions and membrane fusion, Biochem. J., 49:1171.

Sowers, A.E., 1986, A long-lived fusogenic state induced in erythrocyte ghosts by electric pulses, J. Cell Biol., 102:1358.

Sowers, A.E., and Kapoor, V., 1987, The electrofusion mechanism in erythrocyte ghosts, in: "Cell Fusion," A.E. Sowers, ed., Plenum, New York.

Tanaka, K.-I., and Ohnishi, S.I., 1976, Heterogeneity in the fluidity of intact erythrocyte membranes and its homogenization upon hemolysis, Biochim. Biophys. Acta, 426:218.

Tokuyasu, K.T., Schekman, R., and Singer, S.J., 1979, Domains of receptor mobility and endocytosis in the membranes of neonatal human erythrocytes and reticulocytes are deficient in spectrin. J. Cell Biol., 80:481.

Volloch, V., and Housman, D., 1982, Terminal differentiation of murine erythroleukemia cells: physical stabilization of end-stage cells, J. Cell Biol., 93:390.

Williams, J.H., Kuchmak, M., and Witter, R., 1966, Fatty acids in phospholipids isolated from human red cells, Lipids, 1:391.

Williamson, P., Bateman, J., Kozarsky, K., Mattocks, K., Hermanowicz, N., Choe, H.-R., and Schlegel, R.A., 1982, Involvement of spectrin in the maintenance of membrane phase state asymmetry, Cell, 30:725.

Williamson, P., Mattocks, K., and Schlegel, R.A., 1983, Merocyanine 540: a fluorescent probe sensitive to lipid packing, Biochim. Biophys. Acta, 732:387.

Williamson, P., Algarin, L., Bateman, J., Choe, H.-R., and Schlegel, R.A., 1985, Phospholipid asymmetry in human erythrocyte ghosts, J. Cell. Physiol., 123:209.

Williamson, P., Antia, R., and Schlegel, R.A., 1987, Maintenance of membrane phospholipid asymmetry: lipid-cytoskeletal interactions or lipid pump?, submitted.

Zwaal, R.F.A., Comfurius, P., and van Deenan, L.L.M., 1977, Membrane asymmetry and blood coagulation, Nature, 268:358.

ULTRASTRUCTURAL STUDIES OF THE KINETICS OF FUSION

S. W. Hui, D. A. Stenger and S. K. Huang

Biophysics Department
Roswell Park Memorial Institute
Buffalo, NY 14263

INTRODUCTION

There have been many studies of the mechanisms of membrane fusion, using different physical and biochemical techniques. A large number of these works propose models depicting the structural changes during the fusion event. Most of these models are based on indirect evidence from physical measurements, and are not directly related to structural parameters. While models are very helpful in describing molecular events, the actual spatial arrangements of structural features during fusion events must be derived from visual data. Therefore, imaging studies at high spatial and temporal resolution are indispensable in the study of the mechanisms of membrane fusion.

Imaging at high spatial and temporal resolution simultaneously is not a trivial problem. Video recording of light microscope images at the TV rate (1/30 sec) is a good start. It has a distinct advantage of allowing continuous, single sample recording for a long period, thereby following the sequence of fusion events unambiguously. Much valuable information has been obtained by this method (Rand et al., 1985). However, the light microscope is limited in spatial resolution, and the temporal resolution is limited by the recording rate, which is in turn limited by the signal to noise ratio at a given light level. Electron microscopy increases the spatial resolution by 3 orders of magnitude, but traditionally has required long specimen fixation times. Only with the advance of rapid freezing techniques can events at given time points be preserved. The state-of-art rapid freezing technique enables the quenching of the specimen in less than a ms, and is accurate to ms in timing the initiation of freezing (Heuser and Reese, 1981). In combination of the high resolution of the electron microscope, the rapid freezing method offers a means to achieve both high temporal and spatial resolution in imaging. The disadvantage is that, like most electron microscopic studies, one cannot follow the same specimen through a sequence of events. The temporal order of structural events must then be reconstructed, using statistical approaches.

Arresting an intermediate stage of membrane fusion poses an additional problem. Most observed fusion phenomena occur over a period of time after the application of the fusion stimulus. A single fusion event may be initiated at any time point during this period, and last for only a short interval. The cumulative rate of individual events is then expressed as the "kinetics" of fusion. Let the number of individual fusion events initiated at time t be x(t). This function represents the fusion rate of most bulk fusion assays. Assume that each individual event lasts for a period of T, within which an intermediate structure is detectable, and that the sample is frozen "instantaneously" within a time interval f. The probability P(t) of arresting a fusion intermediate during a fusion event initiated at time t=a may be estimated as:

$$P(a) = \int_{a-T}^{a+f} x(t)\,dt \,/\, \int_0^{\infty} x(t)\,dt. \qquad (1)$$

The numerator represents the number of on-going events which either initiate or last until the freeze quenching time f. Here we simply assume that x(t) is independent of temperature, an unjustfied claim for the sake of simplicity. The freezing time f is limited by the time resolution requirement, and is further limited by the need to vitrify ice for morphological preservation. In reality, the time f can be regarded as the time taken to reduce the temperature enough to stop any fusion events. In a typical experiment, f is in the order of 1 ms. It now becomes obvious that the best chance to arrest an intermediate stage of fusion is to freeze-quench the sample at a time when x(t) is at or near a maximum. Therefore, a kinetic fusion assay of the bulk sample is a valuable guide for ultrastructural studies.

For freeze fracture studies, the chance that a fracture plane cutting through a fusing membrane constitutes an additional probability which must be multiplied to P(t).

We have applied fluorescence assays for contents mixing or lipid mixing to determine the kinetics of fusion in a number of membrane systems. From the resulting kinetic curves, we selected the optimal time points to freeze quench samples for ultrastrucural study by freeze fracture electron microscopy. The results are discussed in the following sections.

CALCIUM-INDUCED FUSION OF PHOSPHOLIPID VESICLES

The fusion of phospholipid vesicles containing anionic phospholipids may be triggered by calcium. This phenomenon is well studied, and the kinetics have been analyzed using the mass action theory (Nir et al., 1983). We have chosen a system which may be induced to fuse after the addition of 4 mM calcium. The large unilamellar vesicles (LUV), are prepared by the reverse evaporation method (Szoka et al. 1982). The vesicles contain mixtures of bovine brain phosphatidylserine (PS) and transacylated phosphatidylethanolamine (PE) (from egg phosphatidylcholine) at a 1:3 ratio. The fusion kinetics of this system have been measured experimentally (Duzgunes et al., 1981). However, a higher sample concentration than that used for their experiment is needed for freeze quenching experiments to increase the likelihood of observing vesicles in the freeze fracture plane. We performed the fluorescence assay for the slightly higher lipid concentration of 0.315 mM, to be comparable to the electron microscopy experiment. The standard Tb/DPA fusion assay (Duzgunes et al., 1981) was used. Two ratios of Tb or DPA containing vesicles were used to estimate the aggregation and fusion rate coefficients as explained by Nir (1983). The fluorescence and leakage curves in terms of the percentage of maximum fluorescence, are given in Fig. 1. The fluorescence curve, after correction for leakage, represents the cumulative fusion events. The rate of fusion x(t) is the time derivative of this curve. For 0.315 mM of lipid, the maximal rate occurred within the first 60 s following the addition of calcium.

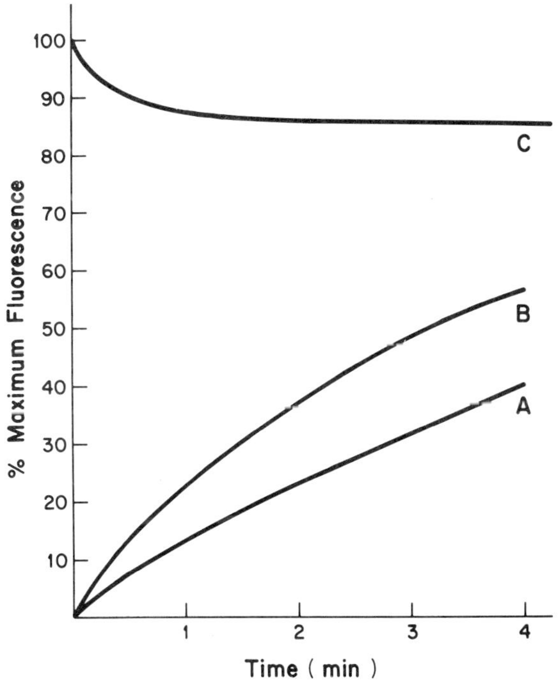

Fig. 1. Fusion of PS/PE vesicles as indicated by the fluorescence of Tb and dipicolinic acid (DPA) at ratios 1:1 (A), or 1:9 (B) after the addition of 4 mM of $CaCl_2$. Leakage of Tb/DPA mixture is indicated by (C).

For electron microscopic study, the lipid suspension was mixed with calcium on the specimen holder and freeze-quenched in liquid propane after a given time. For longer time points, calcium was injected and rapidly mixed with the lipid vesicles for a chosen period. The sample was then placed in a copper sandwich for freeze-quenching. Freeze fracture electron micrographs in Fig. 2 show a variety of morphologies. Since the sample morphology is not homogeneous, one can only derive the sequence of events indirectly by the frequency of occurrence of various structural features. One obvious fact is that the vesicles aggregated immediately after the addition of calcium (Fig.

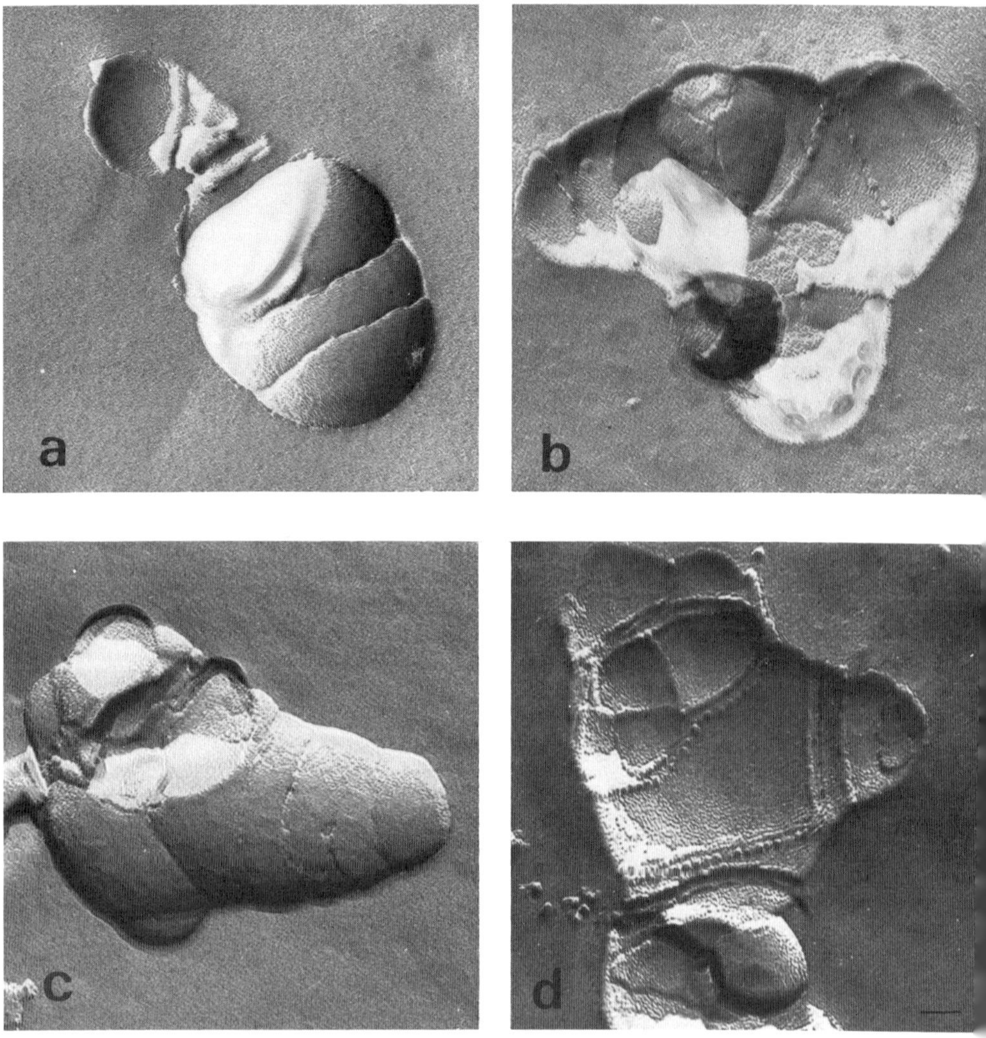

Fig. 2. Freeze fracture electron micrographs of PS/PE vesicles frozen-quenched at (a) 2 s, (b) 37 s, (c) 64 s, and (d) 154 s after the addition of 4 mM of $CaCl_2$. B=100nm.

2a). The barrier between aggregated vesicles can be seen breaking into disjoined, bead-like structures in 37 s (Fig. 2b), and continue to disintegrate at 1 min. after the addition of calcium (Fig. 2c). The remaining partition materials between fusing vesicles seem to transformed into strings of beady or tubular remnants in 2.5 minutes (Fig. 2d). These features are not frequently observed between the 0 and 37 s time points when the fusion rate is the highest, thus they may be insignificant by-products rather than features representing a necessary intermediate stage of membrane fusion. We did not detect an unusual increase of any types of structural features at or near the connecting areas between aggregated vesicles during the peak fusion time. The occasional observation of lipidic particle-like features may not be representative of fusion intermediates, in agreement with the observation by Rand et al., (1985).

Using equation 1 as a guide, we may estimate the probability of observing short lived (T<f) intermediates in this system. Taking the freezing time f to be 1 ms, the percent fusion rate (%/s taken as the derivative of the fluorescence curve) at 0.5 minute to be 0.003/sec, and the probability of fracturing through a contact area between two vesicles to be 0.1, the total P(t) is about $3x10^{-7}$. If each vesicle undergoes one round of fusion only, then the chance of observing intermediate states of fusion while two vesicles are fusing is 1 in 3 million. This is based on the estimate that T= or < f. If T > f, then the chance of seeing an intermediate is better. Therefore the absence of features pertaining to fusion intermediates in freeze fracture electron micrographs does not necessary lead to the conclusion that fusion intermediates do not exist in the system. At such a low probability level, a search for ultrastructural evidence for fusion intermediates is subject to much artifact.

Although the rapid freezing technique does not reveal any structural intermediate pertaining to the fusion mechanism in this system, it does provide information on the aggregation and fusion in the microscopic sense. The mass action model, which has successfully described the fusion reactions of many systems (Nir et al., 1985), stipulates the formation of dimers, trimers .....oligomers as the result of aggregation. These predictions are difficult to test by bulk measurements. Freeze fracture electron microscopy provides an ideal way to test this theory. At each given instant of freeze-quenching, the percentage of vesicles in the dimer, trimer .....oligomer forms can be measured. The degree of fusion can be estimated from the increase of the mean vesicle size. This is done statistically by measuring 4000 electron microscopic images of vesicles. The dominance of oligimers over dimers at the expense of monomers is as predicted (fig. 3), and the size increase also agrees with that predicted by mass action modelling (Hui et al., to be published), using the aggregation and fusion rate constants derived independently from the fluorescence measurement (fig. 1). This is the first verification of the mass action model of fusion at the microscopic scale.

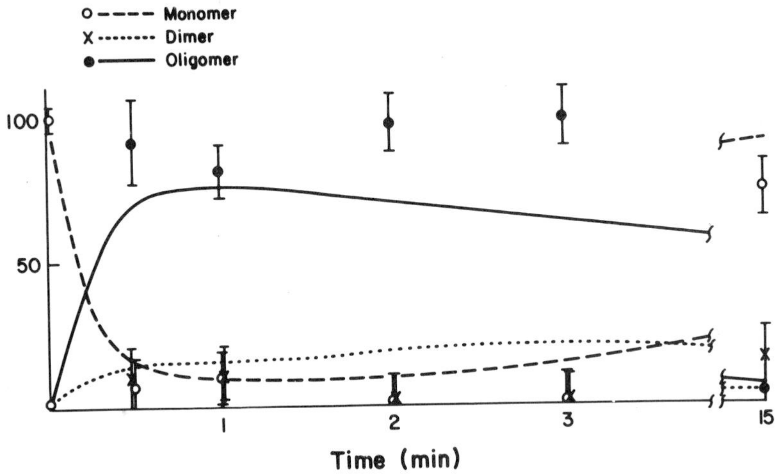

Fig. 3. Percentage of vesicle aggregation forms after calcium addition. Curves represent values predicted by the mass action model.

ELECTROFUSION OF ERYTHROCYTES

Electrofusion provides an ideal condition for the detection of fusion intermediates because the fusion events are more synchronized. As a consequence of synchronization, the entire process is shortened, and the probability of arresting them can be very high if the timing is chosen correctly.

To determine the kinetics of fusion by fluorescence contents mixing assays, we have constructed a special front face fluorometer curvette with built-in electrodes. Samples are sealed within the gap between the pair of stainless steel electrodes and placed in the light path of a fluorometer. Human erythrocyte ghost membranes were loaded with Tb or DPA as described by Hoekstra et al. (1985) to be used for quantitating contents mixing as a measure of fusion. The fusion protocol is similar to that described by Zimmermann (1982). Briefly, the loaded ghosts, in 1:1 mixture, were prealigned by dielectrophoresis in a 1 MHz electric field of the strength 500 V/cm. A short (15 - 50 μs) square pulse at 2.5-5.0 kV/cm was applied to induce fusion between aligned cell ghosts. Fusion was measured as the increase of fluorescence upon the mixing of Tb and DPA, whereas leakage was measured as the decrease of fluorescence from vesicles loaded with mixed Tb+DPA and suspended in a EDTA-containing medium.

Fluorescence increase was detected immediately after the pulse application. Within 2 seconds the fluorescence reached its peak value, then declined slowly over a period of tens of seconds (Fig. 4). An analysis of the fluorescence data is given elsewhere (Stenger and Hui, to be published).

The maximum rate of fusion occurs within the first second, and reaches a percentage rate of 0.5/s. Taking again the freezing time f to be 1 millisecond, and the fracturing probability of 0.1, we arrive at a $P(t)=5x10^{-5}$. The probability is still too low to allow detection by freeze-quenching unless T>>f. If there are 10 features per fusing cell and the T for these features is in the order of 0.1 second, then P(t) becomes 0.05 or 5% which should allow experimental observation.

For freeze quenching the samples, a rapid freezing copper sandwich device was modified to double as electrodes. The two opposite copper plates were spaced apart by thin strips of insulating material. Pronase-treated human erythrocytes suspended in isotonic sucrose solution were aligned by dielectrophoresis as described, and were induced to fuse by a single 15 µs pulse of 5kV/cm field strength. The timing between the pulse and the freezing was controlled to within 0.1 second. At least 40 micrographs were taken at each time point and the features on them were analysed.

At the 0.1s time point, the intermembrane aqueous spacing is reduced from 15 nm during dielectrophoresis to less then 2 nm at some points. There are many patches on the cell surface which appear torn from the surrounding membrane (Fig. 5a). These discontinuous areas (DA) are seen on about 20 % of the fractured cells. Intramembraneous particles (IMP) are less dense, and often absent from many DA. At 2 seconds after the pulse, which corresponds to the maximum of fluorescence, few DA are observed, and are replaced by lipidic-particle-like point defects (PD) as the predominent membrane abnormality. These PD seem to be confined to the flattened areas of the membrane which are

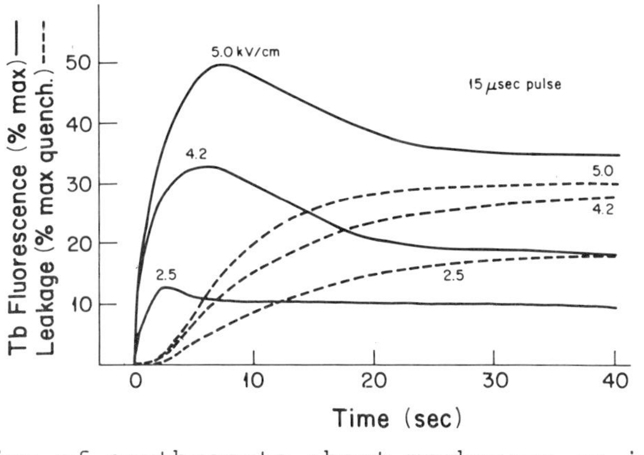

Fig. 4. Fusion of erythrocyte ghost membranes as indicated by the fluorescence of mixed contents of Tb- or DPA-containing ghosts after the application of an electric pulse.

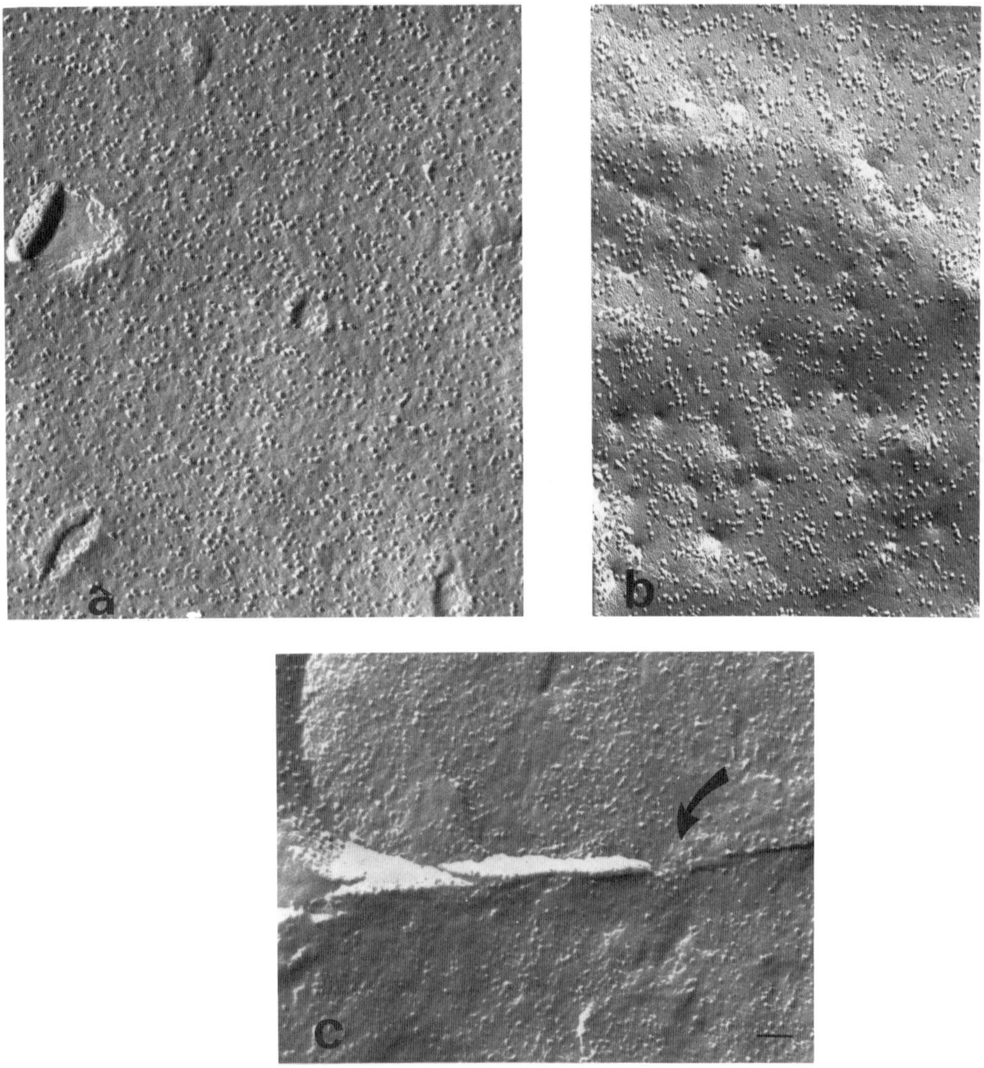

Fig. 5. Freeze fracture electron micrographs of erythrocytes frozen-quenched at (a) 0.1 s, (b) 2 s, and (c) 10 s after the electric pulse. Arrow indicates a cytoplasmic bridge in a cross-fractured plane. B=100 nm.

likely to be the contact zones between two adjacent cells (Fig. 5b). The intercellular distance at this time point returns to the prefusion value of 15 nm, except at the attachment sites which adjoin two fused cells. At 10 seconds after the pulse, most of the PD disappear, and cytoplasmic connections between a few fused cells become apparent. These cytoplasmic connections or lumens become well established between pairs of fusion partners in 60 seconds (Fig. 5c). Apart from these connections, the rest of the cells are indistinguishable from the control population.

The dominant feature which appear at the maximum fusion time points are DA. A statistical presentation of the appearence of these features at given time points are presented (Stenger and Hui, 1986). The frequency of appearence in 20% of cells indicates that their life time, T, is much longer than f. However, they are unlikely to represent permeable sites to the Tb/DPA complex since leakage is rather minor at this early time point (Fig. 4). Most of these DA disappear at a latter time. A cartoon model has been proposed (Stenger and Hui, 1986) which depicts some of these DA are precursors for fusion, leading to a subsequent cytoplasmic bridging. However, not all DA are expected to have this fate, in fact most of them just disappear without leading to any fusion.

The study of electrofusion events give us a handle on capturing fusion events. What we see are longer lived features associated with fusion, but not necessarily as mandatory intermediate features. The short living ones, if any, may still elude detection by freeze quenching due to low probability.

FUSION INDUCED BY POLYETHYLENE GLYCOL (PEG)

Among many membrane fusion systems, the PEG-induced fusion enjoys the most attention with regard to ultrastructural study (see review by Hui and Boni, 1987). It is known that PEG causes the aggregation of IMP, and the degree of the induced IMP aggregation is directly related to the susceptibility of the membrane to PEG-induced fusion (Roos and Choppin, 1986). Dr. Roos has described this in an earlier chapter. We shall focus on the capturing of the fusion intermediates of this system by rapid freezing. We will also discuss the mechanism of IMP aggregation in relation to the fusion efficiency.

Lipid vesicles fuse as soon as a fusogenic concentration of PEG is added to the suspension. This can be monitored by light scattering as well as by NMR (Boni et al., 1984a), and more recently by fluorescence life time measurement (Parente and Lentz, 1986). According to the latter paper, the highest fusion rate occurs at about 1 to 2 minutes of PEG treatment depending on the concentration of PEG. The percentage rate is approximately 0.005/sec at low PEG concentrations (see chapter by Lentz in this book), and may increase by an order of magnitude at high concentrations. The arresting probability is therefore comparable to that of calcium induced fusion. At the earliest time point of 1 second after the mixing of 50% PEG with egg PC SUV, fusion is seen to take place, although no unusual structural features attributable to fusion intermediate is observed (Boni et al., 1981). The increasingly larger vesicles soon become multilamellar (fig. 6a) in 5 minutes, most likely through the mechanism of defect formation (fig. 6b). These defects are common and are well characterized in dehydrated lipids (Boni et al., 1984b). Figure 6b shows confocal domains that connect the inner layers to the outer ones, thus relax the constrain of lamination.

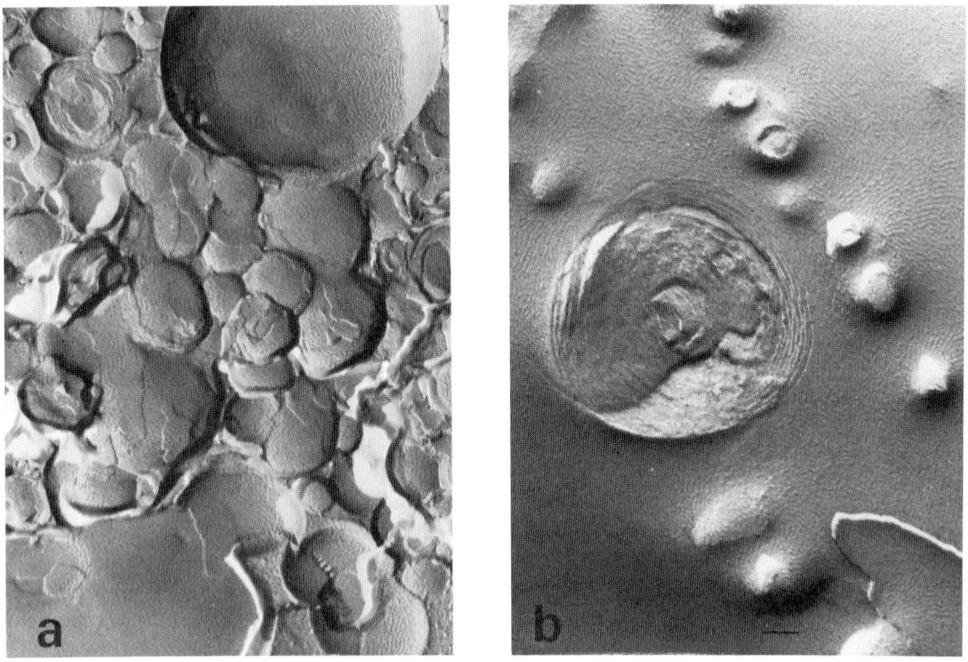

Fig. 6. Freeze fracture electron micrographs of (a)"unilamellar" vesicles and (b) multilamellar vesicles of egg PC after the addition of PEG 8000 to 50% (w/v). Bar=100 nm.

The formation of multilamellar products is apparently restricted to lipid vesicles, and never occurs in the fusion of biomembranes. Furthermore, cell fusion does not occur during the PEG treatment, although lipid exchange may take place between fusion partners (Wojceszyn et al., 1983). The opening of lumens and the free exchange of cytoplasm between fusion partners are initiated only during the dilution step after the PEG treatment. We therefore decided to investigate the structural changes at the dilution step.

We first examined a simple system, the cytoskeleton-free vesicles budded from erythrocytes by repeated calcium/EDTA treatment. This way the cytoskeletal effect is excluded. The vesicles show a random distribution of IMP on their fracture faces before the treatment of PEG. At 6% or more PEG, the IMP become aggregated regardless of contact between vesicles. If contacts occur, the contact areas are always free of IMP (Fig. 7a). Intact erythrocytes behave similarly except that the threshold concentration of PEG is higher for similar effects to be observable (Hui et al., 1985). The effects are reversible up to 12% of PEG.

The capability of a fusogen to induce IMP aggregation was found to parallel their fusion efficiency (Hui el al., 1985). Based on the evidence from this and other laboratories with regard to the intimate relation between IMP aggregation and membrane fusion, we performed an experiment to induce IMP aggregation by other means than

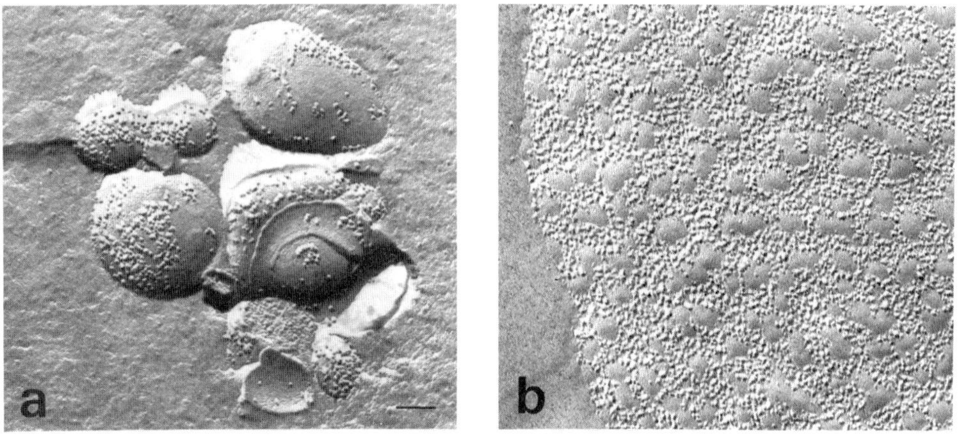

Fig. 7. Freeze fracture electron micrographs of (a) cytoskeleton free erythrocyte vesicles in 6% PEG, and (b) turkey erythrocyte after the treatment of 10 μg/ml of concanavalin A. Bar=100 nm.

PEG, in order to test if such a process would facilitate PEG-induced fusion. When a number of membrane aggregating or perturbing agents were tested individually, those agents causing IMP aggregation and exposing large areas of IMP-free areas in the membranes were found to be effective fusogens (Huang and Hui, 1986). Figure 7b shows a turkey erythrocyte treated with concanavalin A. Circular areas free of IMP are seen over the entire cell. The appearance of these IMP-free areas reduces the threshold of PEG needed for fusion, to as low as 5%. Thus the creation of IMP free patches is the single, most important factor in PEG-induced fusion, which is believed to be centered on the lipid bilayer. This is in contrast to protein-induced fusions where a single or a specific group of protein is often observed at the center of the fusion site.

DISCUSSION AND CONCLUSION

The recording of structural changes during a fusion event by electron microscopy is not a simple task. Although the advance of the rapid freezing technique enables us to arrest instantaneouly any features in the kinetic process, the probability of freezing and revealing such features depends on a number of other factors. The most important factor is the rate of fusion which determines the frequency of individual fusion event occurring per unit time. Since the freezing time is relatively short, the fraction of fusion events initiated during that time is small, unless the remains of previously initiated, long-lived fusion events are added to the probability. Thus the second most important factor is the life time of fusion intermediates. The longer the life time, the easier it is to visualize the structure. In some cases, the life time is infinitely long (the intermediate is stable), then the detection becomes a certainty. Paradoxically, such structures can not be defined as fusion intermediates any more. There are various reports ascerting that those stable structures are equivalent to fusion intermediates, but the association is at best circumstantial. The true intermediate structure must vanish after the fusion is completed, and an incomplete fusion is not strictly a fusion.

Even for the most synchronized fusion system like the electrofusion, the highest percentage fusion rate is only 0.5/s as measured by fluorescence methods. A freezing rate of 1 millisecond reduces the probability to 0.0005/s unless the life time of fusion is much longer than a ms. The DA feature that we observed is such a long lived intermediate or by-product. For slower fusion rates such as in the calcium or PEG-induced fusions, the probability is even lower for detecting such long lived features. Thus the rarity of observed features does not imply that fusion intermediates do not exist, nor that these features are indistinguishable from the normal membrane structure.

We can not conclude whether there is a common fusion mechanism for different types of membrane fusion based on ultrastructure studies alone. At this stage it is mainly due to the lack of information. We know that there is at least an alternative mechanism involved in protein-induced fusion, i.e. one that centers on a visible protein at the center of the fusion zone. I have deliberately skiped over the area of protein and virus induced fusion. This topic has been a subject of much discussion (Pinto da Silva, et al., 1980; Bachi et al., 1973), and will be dealt with in detail in other talks of this symposium.

Besides revealing fusion intermediates, the rapid freezing technique also contributes to clarifying a number of pre-requisites or post mortum factors associated with fusion. The importance of prior vesicle aggregation and protein clearance from potential fusion sites are good examples. In addition, we have supplied microscopic evidence for the mass action model of fusion. We have provided a structural rationale for the potencies of various fusogens. And we have developed a model for electric fusion entirely based on microscopic evidence.

Electron microscopy will continue to contribute to the elucidation of the molecular mechnism of membrane fusion. Without such support from structural study, much room will be left for speculation.

ACKNOWLEDGEMENT

This work is supported by grants GM28120 and GM30969 from the NIH. Parts of the results presented here are from collaborations with Drs. Nir, Boni, Ohki and Yeagle. Discussions with Drs. Bentz and Zimmermann are appreciated.

REFERENCE

Bachi, T., Aguet, M. and Howe, G., 1973. Fusion of erythrocytes by Sendai virus studied by immuno-freeze-etching, J. Virol., 11:1004.

Boni, L.T., Stewart, T.P., Alderfer, J.L. and Hui, S.W., 1981. Lipid polyethylene glycol interaction: I: Induction of fusion between liposomes. J. Memb. Biol., 62:65.

Boni, L.T., Hah, J.S., Hui, S.W., Mukherjee, P., Ho, J.T. and Jung, C.Y., 1984a. Aggregation and fusion of unilamellar vesicles by poly(ethylene glycol). Biochim. Biophys. Acta, 775:409.

Boni, L.T., Stewart, T.P. and Hui, S.W., 1984b. Alterations in phospholipid polymorphism by polyethylene glycol. J. Membrane Biol., 80:91

Duzgunes, N., Wilschut, J., Fraley, R. and Papahadjopoulos, D., 1981. Studies on the mechanism of membrane fusion: role of head-group composition on calcium- and magnesium-induced fusion of mixed phospholipid vesicles. Biochim. Biophys. Acta, 642:182.

Heuser, J.E. and Reese, T.S., 1981. Structural changes after transmitter release at the frog neuromuscular junction. J. Cell Biol. 88:564.

Hoekstra,D., Wilshut, J. and Schorphof, G., 1985. Fusion of erythrocyte ghosts induced by calcium phosphate. Eur. J. Biochem. 146:131.

Huang, S.K. and Hui, S.W., 1986. Chemical co-treatments and intramembrane particle patching in the poly(ethylene glycol)-induced fusion of turkey and human erythrocytes. Biochim. Biophys. Acta, 860:539.

Hui, S.W. and Boni, L.T., 1987. Membrane fusion induced by polyethylene glycol: dehydration and structural defects. In: "Cellular Membrane Fusion" (Wilshut, J. and Hoekstra. D., eds., Marcel Dekker, New York), in press.

Hui, S.W., Isac, T., Boni, L.T. and Sen, A., 1985. Action of polyethylene glycol on the fusion of human erythrocyte membranes. J. Membrane Biol. 84:137.

Nir, S., Bentz, J., Wilschut, and Duzgunes N., 1983. Aggregation of phospholipid vesicles. Progress in Surface Sciece, 13:1.

Parente, R.A. and Lentz, B.R., 1986. Rate and extent of poly(ethylene glycol)-induced large vesicle fusion monitored by bilayer and internal contents mixing. Biochem. 25:6678.

Pinto Da Silva, P.P., Shimizu, K. and Parkison, C., 1980. Fusion of human erythyrocytes induced by Sendai virus: freeze-fracture aspects. J. Cell Sci., 43:419.

Rand, R.P., Kachar, B. and Reese, T.S.,1985. Dynamic morphology of interacting phosphatidylserine vesicles. Biophys. J. 47:483.

Roos, D.S. and Choppin, P.W., 1985. Biochemical studies on cell fusion. J. Cell Biol. 101:1578.

Stenger, D.A. and Hui, S.W., 1986. Kinetics of ultrastructural changes during electrically induced fusion of human erythrocytes. J. Membrane Biol. 93:43.

Szoka, F., Olson, F., Heath, T., Vail, W., Mayhew, E. and Paphadjopoulos, D., 1980. Preparation of unilamellar liposomes of intermediate size (0.1-0.2 μm) by a combination of reverse phase evaporation and extrusion through polycarbonate membranes. Biochim. Biophys. Acta, 601:559.

Wojcieszyn, J.W., Schlegel, R.A., Lumley-Sapanski, K. and Jacobson, K.A., 1983. Studies on the mechanism of polyethylene glycol-mediated cell fusion using fluorescent membrane and cytoplasmic probes. J. Cell Biol. 96:151.

Zimmermann, U., 1982. Electric field-mediated fusion and related electrical phenomena. Biochim. Biophys. Acta, 694:227.

pH TRIGGERED SYNTHETIC PEPTIDES: MODELS FOR VIRAL FUSION SEQUENCES

Francis C. Szoka, Jr., Nanda K. Subbarao and Roberta A. Parente

Department of Pharmacy and Pharmaceutical Chemistry
School of Pharmacy, University of California
San Francisco, CA 94143

INTRODUCTION

Membrane fusion is an essential element of enveloped viral infectivity. Efforts to understand the molecular details of this process have focused on specific lipid interactions and more recently on protein-lipid interactions[1]. Enveloped viruses like influenza, are internalized into endosomes where the low pH of this compartment catalyzes a conformational change in the hemagglutinin protein rendering it fusion competent[2]. Although the role of specific amino acid sequences at the amino terminal of the hemagglutinin have been implicated in this process[3], the molecular details remain obscure.

Our approach to studying this problem was to design a peptide *de novo*, in which specific alterations in the amino acid sequence could be made to investigate the role of secondary structure and hydrophobicity in the membrane destabilizing and fusing properties of peptides[4]. Previous studies have demonstrated that many membrane interactive proteins or peptides are organized in an amphipathic helix[5]. Consequently, a sequence was selected that would be water soluble and aperiodic at pH 7.4 but would undergo a transition to an amphipathic α helix when the pH was lowered(Figure 1). Constituent amino acids were chosen based on their preference to adopt helical conformations[6] and the hydrophobicity of the side chains[7]. Glutamic acid with a pK_a value of about 5, was selected to provide pH sensitivity since it has been previously shown that polyglutamic acid undergoes a conformational change to an α helix upon lowering the pH to 5.0[8]. The 30 amino acid peptide that emerged from this process was designated GALA after the major repeat unit, Glu-Ala-Leu-Ala, contained in the sequence (WEAALAEALAEALAEHLAEALAEALEALAA)[4] (figure 2).

GALA undergoes a pH dependent aperiodic to α helix transition with an apparent pK_a of 5.7. At low pH it associates with lipid bilayers and induces release of contents from large unilamellar vesicles and fusion of small unilamellar vesicles.

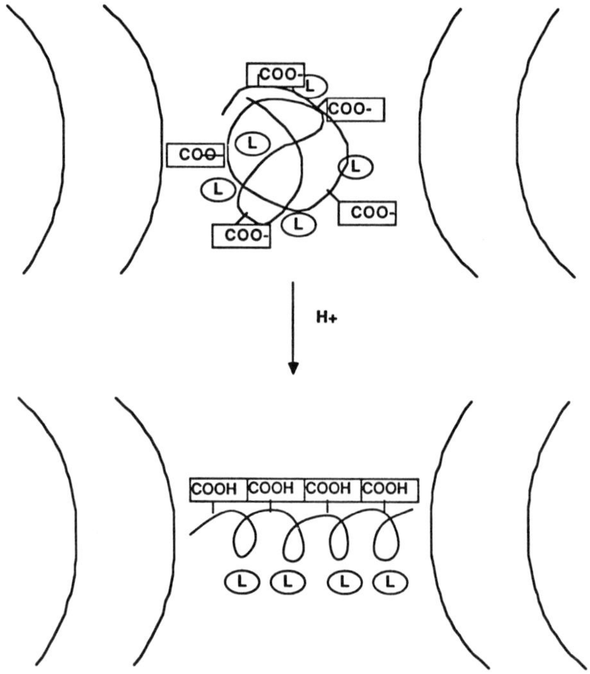

Figure 1. A water Soluble Peptide Fusogen that Undergoes a pH Triggered Conformational Change Exposing a Membrane Perturbing Amphipathic Face.

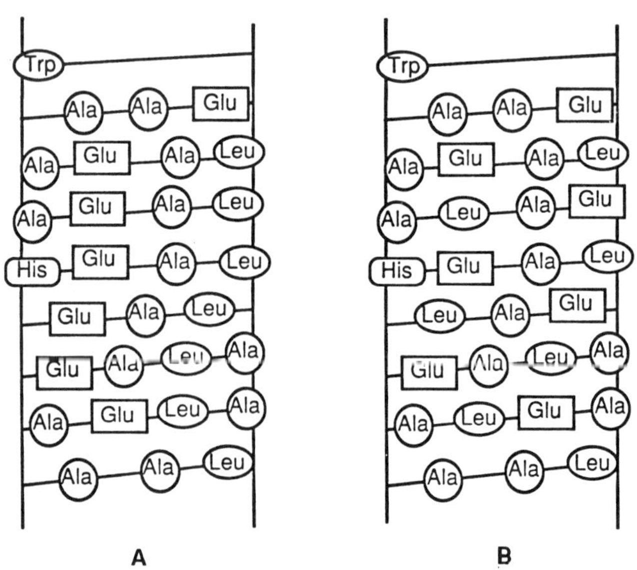

Figure 2. Helical Grid Representation of GALA (A) and LAGA (B).

MATERIALS AND METHODS

Phospholipids were obtained from Avanti Polar Lipids (Birmingham, AL). Fluorescent probes were purchased from Molecular Probes (Eugene, OR).

Synthesis of Peptides

Peptides were synthesized using solid phase methods on a Merrifield resin and purified by reverse phase HPLC (Dynamax, C_{18} column) using a linear gradient of 0.1% trifluoroacetic acid in acetonitrile and water. Peptides were analyzed on a Vydac C_{18} reverse phase column. Purity was greater than 99% when monitored at 220 nm. Details of the synthesis, purification and characterization have been published[4].

Circular Dichroism (CD)Measurements

The CD spectra of the peptides were scanned using a Jasco 500 A spectropolarimeter attached to an IBM PC. All spectra were obtained in a sample chamber flushed with nitrogen at 23° C. The spectra were recorded in 1 mm path length quartz optical cells at 0.05 mg GALA per mL. Ellipicity values are expressed as degrees centimeter^2 per decimole.

Preparation of Lipid Vesicles : Characterization of Leakage of Contents and Fusion of Bilayers

Large unilamellar vesicles (LUV) and small unilamellar vesicles (SUV) were prepared and characterized as previously described[4,9]. Fluorescence assays of lipid mixing[10] and contents leakage[11] were performed on a SPEX fluorolog photon counting instrument (Edison, NJ).

RESULTS AND DISCUSSION

The sequence of GALA is represented on a helical grid diagram in figure 2A. The acidic glutamate residues are aligned on one face of the helix while the hydrophobic leucine residues are aligned on the opposite face. Tryptophan and histidine residues are included in the sequence as fluorescence and NMR probes. In addition, the histidine residue provides a convenient moiety for radiolabelling the peptide with ^{125}I. An analog of GALA designated LAGA has the same amino acid composition but with a placement of the residues which disrupts the amphipathic face(figure 2B).

The hydrophobicity profilc of a seven residue window of GALA at pH 7.5 is compared to the hydrophobicity profile of the consensus sequence of the amino terminus obtained from various influenza virus hemagglutinins[12] in figure 3A. Values for the residue hydrophobicities were taken from Fauchere and Pliska[7]. Employing this hydrophobicity scale permits the calculation of the effect of lowering pH on the hydrophobicity of GALA (figure 3B). As the pH is reduced resulting in the protonation of the Glu residues, the hydrophobicity of GALA becomes similar to that of the consensus sequence. At both pH values the hydrophobicity profiles of LAGA resemble that of GALA.

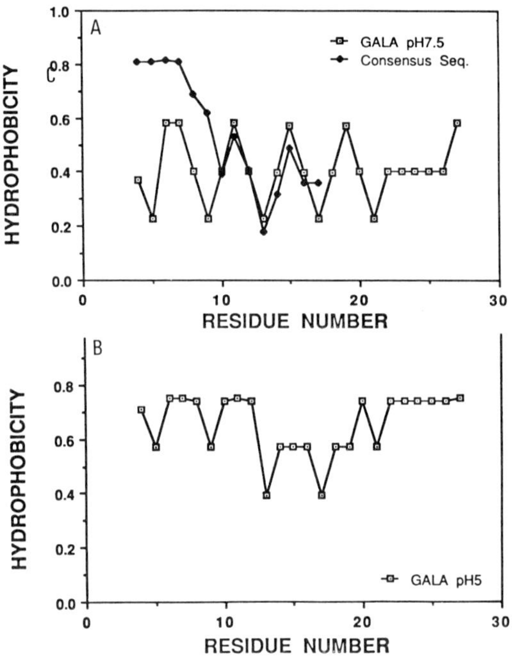

Figure 3. Seven Residue Hydrophobicity Plots of GALA and the Consensus Sequence of the Amino terminus of Influenza Virus at pH 7.5 (A) and GALA at pH 5.0 (B).

As indicated by CD the conformation of GALA changed from one of a predominant aperiodic structure at neutral pH to one with an appreciable quantity of α helix at pH 5.0 (figure 4).

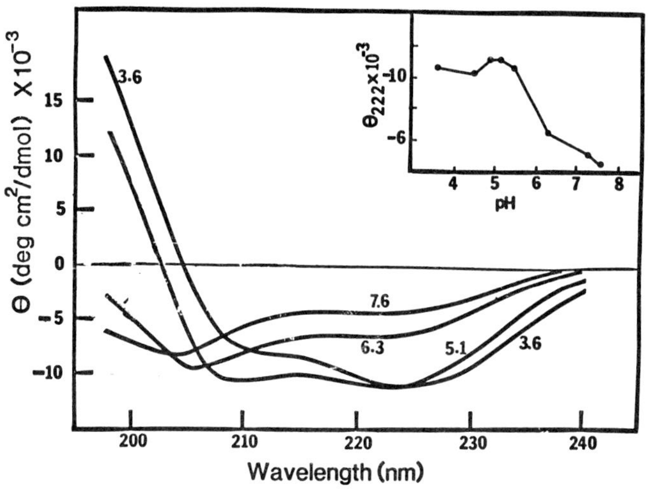

Figure 4. Circular Dichroism spectra of GALA at Varying pHs (from reference 4, with permission).

GALA also acquired an appreciable amount of α helical content in the presence of high concentrations of monovalent cations and at 5-20 mM concentration range of divalent cations.

The sensitivity of tryptophan to the polarity of its environment was used to demonstrate an interaction between GALA and egg phosphatidylcholine lipid vesicles (EPC LUV) at low pH. The fluorescence emission spectrum of GALA at pH 7.5 had a maximum at 360 nm. At pH 5.0 this maximum shifted to 350 nm in the absence of lipid vesicles which could be due to an altered conformation of the peptide or due to a different aggregation state of the peptide. At pH 7.5, in the presence of EPC LUV the fluorescence intensity and position of the maximum were unaffected. However at pH 5.0, in the presence of EPC LUV, the fluorescence maximum showed a blue shift to 335 nm. The fluorescence changes observed with GALA upon lowering the pH were completely reversible when the pH was increased to 7.5. Peptide induced leakage of vesicle contents was monitored by dequenching of the fluorophore aminonapthalene-1,2,3-trisulfonic acid[11] (ANTS). This probe has previously been shown to remain encapsulated in vesicles at low pH. Typical leakage curves are presented in Figure 5 contrasting the results for a lipid/peptide ratio of 500/1 at pH 5.0 and 7.5. Leakage was a function of the pH and lipid to peptide ratio. Complete leakage of vesicle contents in 10 min was observed down to a lipid/peptide ratio of 2500/1. In general, leakage was significant at lipid/peptide ratios greater than 500/1 only below pH 5.7 (Figure 6). An important observation was that LAGA or truncated versions of GALA containing the repeat unit $(EALA)_n$, n=1-4 did not induce any significant leakage from EPC LUV at pH 5.0.

Since results from CD indicated that divalent cations can promote α helical content in GALA at pH 7.5, the ability of GALA to promote leakage under these conditions was also investigated. Divalent cations induced leakage with the following potency: $Ca^{2+}>Zn^{2+}>Mg^{2+}$. However the most potent divalent cation induced significantly less leakage than protons at pH 5.0.

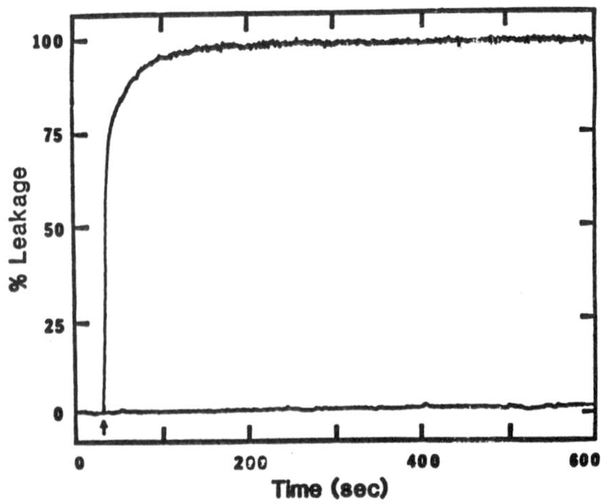

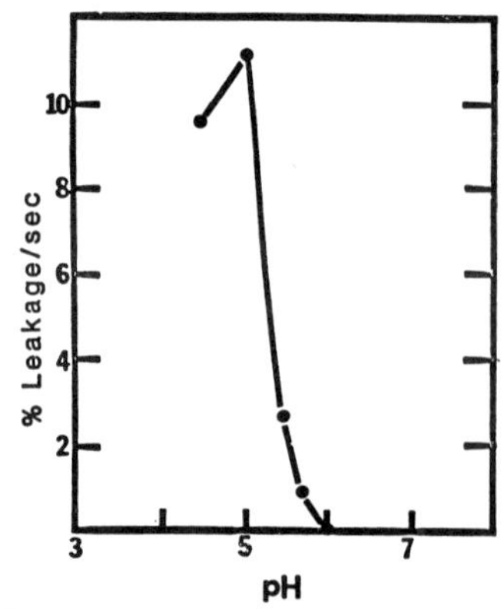

Figure 5 (left)Percent Leakage of Contents from EPC LUV at a Lipid/Peptide Ratio of 500/1. Upper Trace pH 5.0 buffer, lower trace pH 7.5 buffer. (from reference 4, with permission.)

Figure 6 (right)Initial Rate of Leakage as a Function of pH at a lipid/peptide Ratio of 500/1. (from reference 4, with permission.)

Additional studies demonstrated that leakage from the liposomes at low pH was an "all or none " event and that water soluble molecules with a molecular weight greater than about 700 were not released from the vesicles by the action of GALA at low pH.

When EPC LUV with a diameter greater than 100 nm were the target membrane, GALA induced leakage of vesicle contents but not fusion of the bilayer. These latter experiments employed lipid mixing experiments and changes in vesicles diameter as measured by laser light scattering and were performed at lipid/peptide ratios as low as 50/1.

The results were considerably different when EPC SUV prepared by sonication techniques with a diameter of 50 nm were used. In this case GALA, at a lipid/peptide ratio of 100/1 induced a low pH dependent lipid mixing. This lipid mixing was correlated with an increase in vesicle size to about 75 nm. The change in size was not reversed by increasing the pH to 7.5. Lipid mixing was not observed at pH 7.5 in this system. Neither LAGA nor the truncated versions of GALA were able to induce lipid mixing or promote an increase in vesicle diameter under comparable lipid/peptide molar ratios at low pH in either LUV or SUV.

The original goal was to design a water soluble pH triggered peptide fusogen as a model for viral fusogenic sequences. The concept is illustrated in Figure 1. Clearly we have achieved some of the original objectives. In GALA we have been able to balance the hydrophobicity of the peptide so that at pH 7.5, it is water soluble and does not interact with phosphatidylcholine bilayers. However when the pH is reduced GALA undergoes a random coil to α helix transition that exposes a hydrophobic face. In this conformation GALA is able to partition into the bilayer, induce release of encapsulated low molecular weight compounds and to catalyze the fusion of SUV but not LUV.

The absence of any similar effects to GALA using the truncated peptides, which do not have any appreciable secondary structure but have similar hydrophobicity profiles to GALA, indicate that a minimal length is required for the GALA induced membrane destabilization. Results with LAGA, a peptide with the same amino acid composition, length, hydrophobicity profile and a similar helical content at pH 5.0 as GALA indicate that the topology of the residues is important for the membrane destabilization; as measured by leakage and fusion.

These results encourage us to believe that the basic GALA sequence can be modified to obtain more quantitative correlates between peptide sequence and fusogenic ability.

The absence of fusogenic activity of GALA against larger diameter vesicles is similar to results from many other laboratories where fusion was reported only with SUV[13]. Small unilamellar vesicles have high radii of curvature and are poised to fuse, and in fact do so under appropriate conditions even in the absence of peptides[14]. To catalyze fusion of LUV, it appears that close apposition and bilayer destabilization are more difficult to achieve[15]. In the case of Influenza virus, the fusion proteins must be anchored to the bilayer surface in order to catalyze fusion[16]. Whether this is due to the state of aggregation of the fusion protein or the ability of the membrane anchored protein to simultaneously destabilize both bilayers is not known. The availability of synthetic peptides such as GALA should permit this question to be addressed in a rigorous fashion.

ACKNOWLEDGEMENTS

We thank Profs. D. Papahadjopoulos and F. Yang for the generous use of instruments and Drs. L. Nadasdi and K. Pongracz for the synthesis of LAGA and the truncated versions of GALA. This work was supported by NIH GM29514 (FCS) and a Damon Runyon-Walter Winchell Fellowship DRG-907 (RAP).

REFERENCES

1. Szoka, F.C., Lipid Vesicles: Model Systems to Study Membrane-Membrane Destabilization and Fusion. <u>in</u>:Cell Fusion, A.E. Sowers, ed., Plenum Press, New York, 1987.
2. White, J., Kielian, M. & Helenius, A., Membrane Fusion Proteins of Enveloped Animal Viruses, Q. Rev. Biophys. 16:151 (1983).
3. Gething, M.J., Doms, R.W., York, D. & White, J.,Studies on the Mechanism of Membrane Fusion: Site-specific Mutagenesis of the Hemagglutinin of Influenza Virus, J. Cell Biol. 120:11(1986).
4. Subbarao, N. K., Parente, R.A., Szoka, jr., F.C., Nadasdi, L. and Pongracz, K., pH-Dependent Bilayer Destabilization by an Amphipathic Peptide, Biochemistry 26:2964(1987).
5. Kaiser, E.T. and Kezdy, F.J., Amphiphilic Secondary Structure: Design of Peptide Hormones, Science 223: 249 (1984).
6. Chou, P.Y. and Fasman, G.D., Conformational Parameters for Amino Acids in Helical, β Sheet and Random Coil Regions Calculated from Proteins, Biochemistry 13: 211 (1974).
7. Fauchere, J. and Pliska, V., Hydrophobic Parameters of Amino-acid Side Chains From the Partitioning of N-acetyl-amino amides, Eur. J. Med. Chem., 18:369(1983).
8. Tiffany, M.L. and Krimm, S., Circular Dichroism of the "Random" Polypeptide Chain, Biopolymers 8:347(1969).
9. Szoka, F.C. and Papahadjopoulos, D., Comparative Properties and Methods of Preparation of Lipid Vesicles (Liposomes). Ann. Rev. Biophys. Bioeng. 9:467(1980).
10. Struck, D.K., Hoekstra, D., & Pagano, R.E.,Use of Resonance Energy Transfer to Monitor Membrane Fusion, Biochemistry 20:4093(1981).
11. Ellens, H., Bentz, J. & Szoka, F.C.,pH Induced Destabilization of Phosphatidylethanolamine Containing Liposomes: Role of Bilayer Contact, Biochemistry 23:1532(1984).
12. Lamb, R.A., The Influenza Virus RNA Segments and Their Encoded Proteins, <u>in</u> :Genetics of Influenza Viruses, P. Palese and D.W. Kingsbury, eds., Sprinder-Verlag, Berlin (1983).
13. Hong, K., Duzgunes, N., Meers, P.R. & Papahadjopoulos, D., Protein Modulation of Liposome Fusion, <u>in</u> :Cell Fusion, A.E. Sowers, ed.,Plenum Press, NY(1987).
14. Wong, M. and Thompson, T.E., Aggregation of Dipalmitoylphosphatidylcholine Vesicles, Biochemistry 21:4133(1982).
15. Ohki, S., Physicochemical Factors Underlying Lipid Membrane Fusion, <u>in</u> :Cell Fusion, A.E. Sowers, ed., Plenum Press,NY(1987).
16. White, J., Helenius, A. and Gething, M.-J., Haemagglutinin of Influenza Virus Expressed from a Cloned Gene Promotes Membrane Fusion, Nature 300:658(1982).

MEMBRANE FUSION IN MODEL SYSTEMS FOR EXOCYTOSIS: CHARACTERIZATION OF CHROMAFFIN GRANULE FUSION MEDIATED BY SYNEXIN AND CALELECTRIN

William J. Zaks and Carl E. Creutz

Department of Pharmacology and the Programs in Biophysics
Neuroscience and Cell and Molecular Biology
University of Virginia, Charlottesville

SUMMARY

Two sensitive fluorescence lipid-mixing assays were used to study the fusion of isolated chromaffin granules. Intermixing of membrane lipids was found to be the rate-limiting step in the fusion process. The threshold Ca^{2+} concentration for fusion was 1 mM in the absence of mediator protein, but 10 μM in the presence of synexin or 32K calelectrin. Cis-unsaturated free fatty acids increased both the magnitude of the fusion response and its sensitivity to Ca^{2+} to within the micromolar range. Fusion required an intact, native chromaffin granule and was inhibited by hypertonic medium. High concentrations of synexin or calelectrin also inhibited granule fusion, but not granule aggregation. The properties of these proteins suggest they may play a specific role in promoting membrane fusion in exocytosis.

INTRODUCTION

The final event in the exocytotic process may be considered to be the fusion of the secretory granule with the plasma membrane, or, during compound exocytosis, with the membrane of another secretory granule. The fusion of synexin-aggregated chromaffin granules by arachidonic acid to form large vesicles has been proposed as a model for the granule fusion events occurring during compound exocytosis in stimulated chromaffin cells (1). However, a detailed analysis of fusion in this system is lacking. For example, it is unclear whether synexin has a specific role in the fusion process, per se, or functions solely by promoting chromaffin granule aggregation. Also, a function for calelectrin, a 32KDa protein with chromaffin granule aggregating activity, in this system has yet to be established. Furthermore, the influence of the granule core contents on the fusion process remains poorly defined. In this report an attempt has been made to address these and other issues using improved assays for measuring chromaffin granule fusion.

Evaluation of the application of fluorescence - based fusion assays to chromaffin granule fusion

Multiple granule fusion events can be visualized in the phase microscope as the formation of large vesicular structures after addition of *cis*-unsaturated fatty acids to synexin-aggregated chromaffin granules (1), however, only a semiquantitative measure of fusion may be obtained from an analysis of the size or frequency of these structures. The decrease in turbidity of a suspension of aggregated granules upon addition of fatty acid is consistent with formation of these large poorly refractive structures, but also occurs during granule lysis and might not detect fusion events unassociated with such post-fusion changes in granule structure. In order to circumvent the limitations of these approaches to study fusion in this system, the possibility of applying a quantitative method for continuously measuring fusion was investigated. Initial studies focused on the assay described by Hoekstra et. al. (2) which is based on the relief of fluorescence self-quenching of octadecylrhodamine B (R18) following fusion of nonlabelled membranes with membranes labelled with a self-quenching concentration of R18. Conditions are maintained so that there is a proportionality between the efficiency of self-quenching and the probe surface density, ensuring that the fluorescence increase is proportional to the extent of probe dilution due to fusion. Figure 1 shows the time course of fluorescence changes during a representative chromaffin granule fusion experiment using the R18 assay. (A detailed description of experimental procedures is included in Appendix I). Part A of the curve is the fluorescence signal obtained from a dilute suspension of chromaffin granules (5-8 μg granule protein/ml in a ratio of 1 labelled:4 nonlabelled) and synexin in 0.1 mM EGTA. This fluorescence signal remains constant for the duration of the experiment (1200 sec) in the absence of further additions, indicating the lack of spontaneous transfer of the probe between membranes. Part B shows the effect of addition of Ca^{2+} buffer to bring the free Ca^{2+} concentration to 1 mM. The magnitude of the rapid decrease in fluorescence is both Ca^{2+} and synexin (or 32K calelectrin) concentration dependent. A similar, although smaller rapid decrease in fluoresence is observed upon addition of 1 mM Ca^{2+} in the absence of granule aggregating protein. A similar decrease in fluorescence was reported when Ca^{2+} was added to chromaffin granules labelled with rhodamine-PE (3). This phenomenon may be due to a direct interaction of Ca^{2+} with the rhodamine molecules in the plane of the membrane. In the presence of 32K calelectrin (or synexin), the decrease in fluorescence varies as a function of protein concentration. The protein concentration dependence of this phenomenon agrees well with the

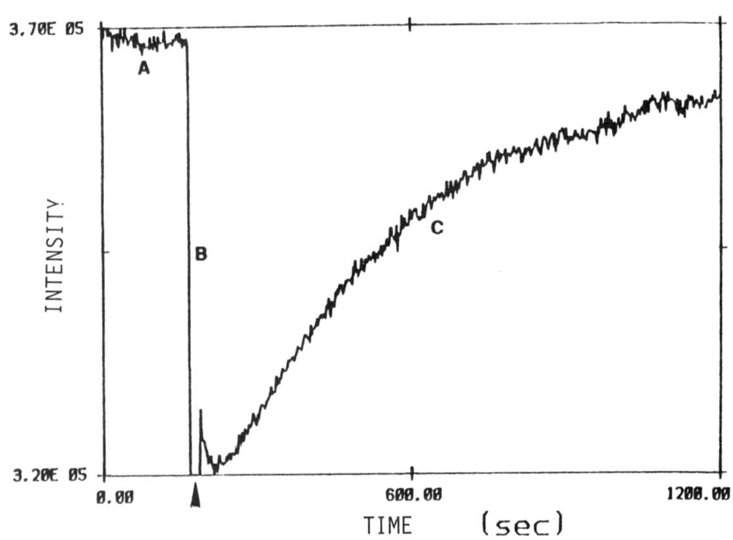

Figure 1. Time course of octadecylrhodamine (R18) fluorescence during a representative chromaffin granule fusion experiment. R18 labelled granules were mixed with a 4-fold excess of unlabelled granules (final A540 = 0.03) in the presence of 0.5 μg/ml oleic acid and 10 μg/ml synexin, as described in appendix I. The fluorescence emission intensity at 590 nm (excitation at 560 nm) was monitored for 3 min and then a pCa = 3.0 calcium buffer was added at the time indicated by the arrow.

protein concentration dependence of binding ^{125}I-32K calelectrin to the membrane (4). In addition, the Ca^{2+} concentrations required for the fluorescence changes in the presence of 32K calelectrin also agree with those necessary for 32K calelectrin binding to membranes (4). These data suggest that protein binding to the membrane surface is responsible for the rapid fluorescence quenching.

The slow increase in fluorescence observed in part C of Fig. 1 is believed to represent fusion. The quantitative measure of the process used in this report is the difference in fluorescence between the time of minimum fluorescence and some later time, usually 700 sec. However, a detailed characterization of the R18 assay in the chromaffin granule system revealed behavior different from that predicted on the basis of the theory applied to liposome fusion. For example, the relative fluorescence change associated with fusion was observed to vary in a linear manner with increasing fraction of labelled granules (Fig. 2), and not in the manner described by the expression, $F = x(1-x)$, where x is the fraction of labelled granules, as predicted by theory (see Appendix II for derivation). In addition, the fluorescence change associated with fusion varied linearly with increasing concentration of incorporated R18 (Fig. 3), and not as the square of the concentration as predicted (see Appendix II for derivatin). The experimentally observed behavior might be due to a nonhomogeneous distribution of R18 in the chromaffin granule membrane such that the majority resides on the outer leaflet of the bilayer. During fusion a fraction of the R18 might cross to the inner leaflet of the membrane where it might experience a marked increase in quantum yield due to interaction with granule

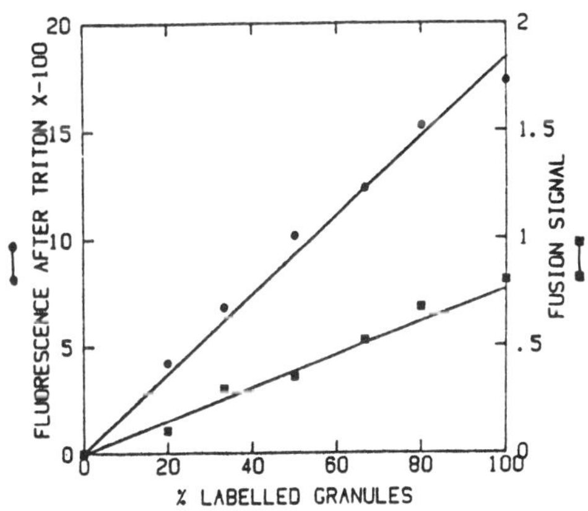

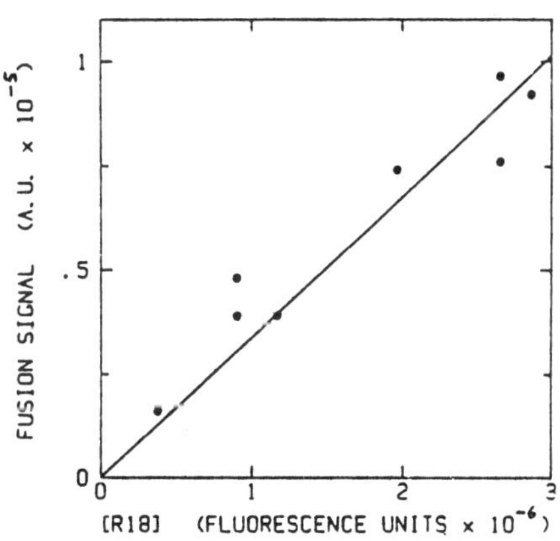

Figure 2. (LEFT) Dependence of the fusion signal on the fraction of labelled chromaffin granules. R18 labelled granules were mixed with nonlabelled granules in varying proportions, keeping the total granule concentration constant. Each reaction mixture contained 1.0 μg/ml 32K calelectrin. Fusion was initiated by the addition of pCa = 3.0 buffer and was recorded as the increase in fluorescence after 700 seconds. After each fusion experiment was completed 1% triton X-100 was added and the fluorescence at 590 nm recorded.

Figure 3. (RIGHT) Dependence of the fusion signal on the concentration of R18 in labelled granules. Granules labelled with a 4-fold excess of nonlabelled granules in the presence of 1 μg/ml 32K calelectrin. Fusion was initiated by addition of pCa = 3.0 buffer. The concentration of R18 in the labelled vesicles was determined fluorometrically at the end of fusion assay after dissolution in 1% triton.

core contents. However, at present we have no evidence for such a mechanism.

In view of the nonclassical behavior of the R18 fluorescence assay in the chromaffin granule system, it is difficult to unambiguously assign the observed fluorescence changes to actual fusion events. Because of this difficulty we felt it necessary to corroborate the data obtained with the R18 assay with that obtained by a second assay, the NBD/Rho resonance energy transfer assay described by Struck et. al. (5) and recently applied to the chromaffin granule system by Morris and Bradley (3). This assay is based on the relief of resonance energy transfer between the two non-exchangeable fluorescent phosphatidylethanolamine derivatives, N-(7-nitro-2,1,3-benzoxadiazol-4-yl) phosphatidylethanolamine and N-(lissamine Rhodamine B sulfonyl) phosphatidylethanolamine as they dilute from a labelled membrane bilayer into a nonlabelled one during membrane fusion. In this case the efficiency of resonance energy transfer is directly proportional to the acceptor (N-Rho-PE) surface density and independent of donor concentration. As shown in Fig. 4 the addition of Ca^{2+} to a suspension of chromaffin granules in a ratio of 1 dual-labelled:4 nonlabelled, resulted in a decrease in resonance energy transfer characterized by a decrease in acceptor fluorescence. The time course of the decrease in acceptor fluorescence agrees well with that of the increase in R18 fluorescence shown in Fig. 1. Although the effects of varying the probe surface densities as well as the ratio of labelled to nonlabelled granules on the fluorescence signal associated with fusion were not examined in detail for the NBD/Rho assay, it was observed that fluorescence signal was detected when labelled granules alone were used, a result consistent with theory. Equipped with the R18 and NBD/Rho fusion assays, a detailed study of chromaffin granule fusion was undertaken.

Granule Concentration Dependence of Fusion

The rate of fusion of initially separate granules is dependent on granule granule interaction and is therefore expected to be second order with respect to granule concentration. A lower thanb second order rate (if observed) would imply that the frequency of granule-granule collisions is not the limiting factor for the rate of intermixing of membrane lipids. In Fig. 5

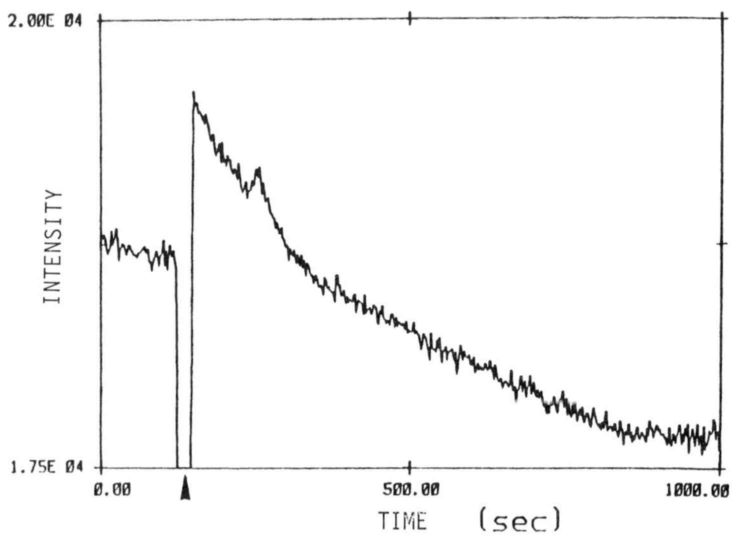

Figure 4. Time course of fluorescence changes accompanying fusion of NBD/Rhodamine labelled chromaffin granules. Dual-labelled granules were mixed with a 4-fold excess of unlabelled granules (final A540 = 0.3) in the presence of 10 μg/ml 32K calelectrin as described in appendix I. pCa = 3.0 calcium buffer was added at the time indicated by the arrow.

the rate of chromaffin granule fusion, at 1 mM free Ca^{2+} and a fixed ratio of 32K calelectrin to chromaffin granules, is plotted against the concentration of chromaffin granules (R18 labelled:nonlabelled = 1:4). Over the range of granule concentrations used, the rate of fusion is directly proportional to granule concentration and hence first order with respect to granules. Therefore, the fusion process itself, and not aggregation, is rate limiting under these conditions. A similar first order process was observed for granule fusion promoted by high concentrations of Ca^{2+} in the absence of 32K calelectrin (not shown). The latter result is in agreement with that of Morris and Bradley, who, using the Rho/NBD RET assay to study Ca^{2+}-dependent fusion of chromaffin granules in the absence of a mediator protein, found that fusion was several fold slower than the aggregation induced by Ca^{2+} (3).

Structural Requirements for Fusion

The addition of nonlabelled lysed chromaffin granules to a suspension of intact R18 labelled granules leads to a decrease in the fluorescence change associated with granule fusion (Fig. 6). The dependence of the inhibition on the concentration of lysed granules is similar to that calculated on the basis of complete lack of fusion between intact and lysed granules. If labelled granules only fused amongst themselves in aggregates consisting of both labelled-intact and nonlabelled lysed granules, one can calculate from the probability of granule interaction that the observed fusion signal should should decrease as the inverse function of x+1 where x is the nonlabelled granule concentration expressed as a multiple of the concentration of labelled granules (see Appendix III for derivation). This relationship, plotted as closed triangles in Fig. 6 agrees well with the experimental data plotted as closed circles. Unfortunately, since this experiment was not repeated with the NBD/Rho assay, it is unclear whether the lack of fusion signal is due to failure of the granules to fuse or failure of the R18 assay to report fusion. The latter possibility is raised due to the possible mechanism suggested earlier by which the R18 assay detects fusion, i.e. transbilayer movement of R18 to a fluorescence enhancing environment. This environment

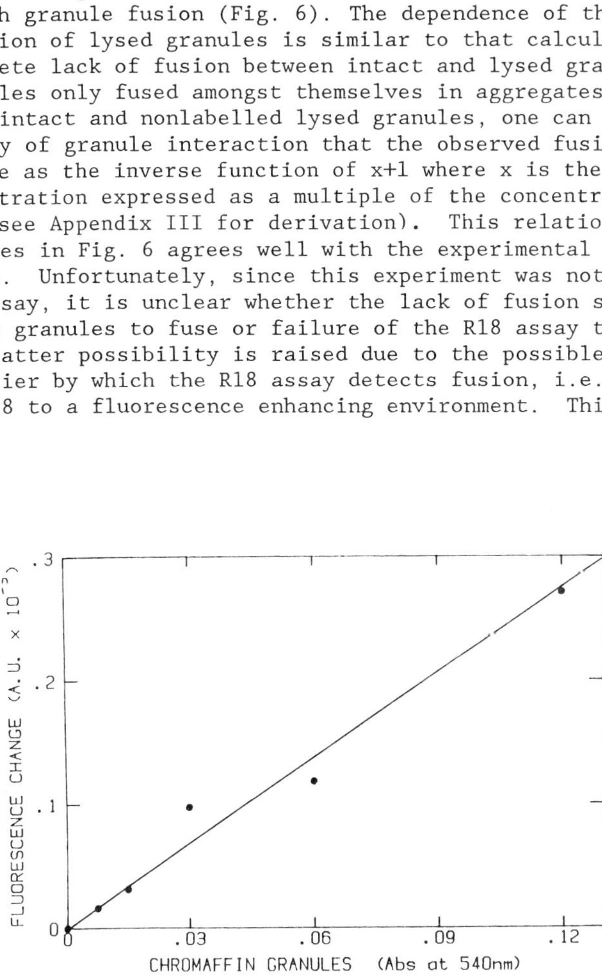

Figure 5. The rate of fusion is a linear function of chromaffin granule concentration. Increasing amounts of R18 labelled granules were mixed with a 4-fold excess of unlabelled granules to produce a series of samples with total granule concentration varying from 0 to 0.12 absorbance units at 540 nm. These samples were incubated with a constant proportion of 32K calelectrin (33 μg/ml per absorbance unit at 540 nm). Fusion was initiated by addition of pCa = 3.0 buffer.

would be significantly altered if one or both fusing granules were lysed. The R18 assay also fails to detect fusion between lysed labelled and lysed nonlabelled granules. This result is corroborated by the study of Morris and Bradley, in which, using the NBD/Rho assay, fusion was not detected between lysed chromaffin granules (3), suggesting that the apparent lack of fusion may not be an artifact of the R18 assay. Why do intact granules fuse while lysed granules do not? To answer this question it is appropriate to examine the differences between these two states of the organelle. Perhaps the most obvious difference is the lack of core contents of the lysed granules. The nature of these contents have been studied in great detail and their composition is well defined, even though the nature of the structure formed by the contents is still poorly understood (6). The total concentration of core constituents is approximately 0.9M yet chromaffin granules are stable in isotonic solution, and have only slightly hypertonic internal osmotic pressure (7). However, conditions have been described where these core contents were able to manifest their latent osmotic strength. Since it has been speculated that osmotic swelling may play an important role in fusion reactions of biological membranes (11), the role of osmotic forces in chromaffin granule fusion was investigated.

Osmotic Effects on the Fusion Process

Addition of cis-unsaturated free fatty acids such as arachidonic acid to granules aggregated by synexin causes the formation of large vesicles as the result of fusion of hundreds of individual granules (1). The observation that formation of these vesicles was inhibited if the osmolarity of the medium bathing the aggregates was increased (1), was consistent with other observations showing inhibition of membrane fusion reactions (9) and exocytosis in several cell types (10) by hypertonic solutions. Since lysed chromaffin granules would presumably lack a mechanism for producing an osmotic stress on their membranes, lack of fusion in this system would be consistent with the osmotic theory of fusion. To test this theory further

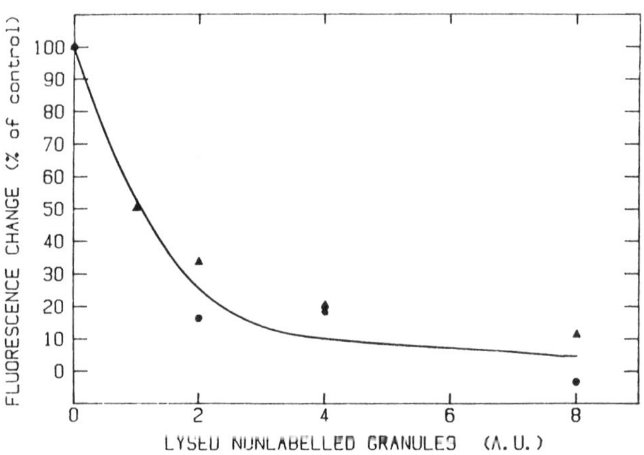

Figure 6. Inhibition of chromaffin granule fusion by lysed granule membranes. R18 labelled, intact granules (A540 = 0.006) were incubated with increasing concentrations of hypotonically lysed, unlabelled granule membranes. Fusion was initiated by addition of $CaCl_2$ to 10 mM final concentration, in the absence of aggregating protein or free fatty acid. The concentration of unlabelled membranes is expressed as a multiple of the labelled granule concentration. Experimental data is plotted as closed circles. The closed triangles represents the fusion response predicted on the basis of complete lack of fusion between intact and lysed granules.

the effect of osmotic changes on the fusion of intact chromaffin granules was investigated. When intact granules were placed in a medium made hypertonic by sucrose or stachyose, a slow (2-5 minutes), approximately log-linear decrease in R18 fluorescence with time was observed prior to addition of Ca^{2+}. The total degree of quenching was dependent on the concentration of the osmoticant, but even at the highest solute concentrations was of limited magnitude; when R18 was initially 40% quenched, 540 mM solute increased quenching by an additional 7.5% compared with 13% as a result of 32K calelectrin binding. Interestingly, the magnitude of the protein-induced quenching decreased as the magnitude of the osmoticant-induced quenching increased, although the two effects were not additive. Intact chromaffin granules behave as ideal osmometers and shrink when placed in hypertonic solutions (7). Changes in membrane structure occurring as a result of granule shrinkage may increase the effective R18 surface density so as to produce these fluorescence changes. Similar slow osmoticant-induced fluorescence changes were observed for NBD/Rho dual-labelled chromaffin granules. In this case the acceptor fluorescence increased, a result consistent with a decreased effective surface area. Such a change in probe surface density would be expected to increase the fluorescence change associated with fusion. However, in view of the anomalous behavior of the R18 assay in this system, no correction was made for this effect when calculating the fusion response. In view of the small magnitude of these effects it is unlikely they appreciably alter the fluorescence changes associated with fusion. The fluorescence change associated with fusion of intact chromaffin granules initiated by Ca^{2+} addition (after the osmotically induced fluorescence change had ceased) was inhibited by increasing hypertonicity, as shown in Fig. 7. The degree of inhibition was variable, however, (eg. compare fig. 7 and fig.8) and it also appeared that higher osmolarities were required to inhibit fusion of "fresh" chromaffin granules than those which had been permitted to "age" for several hours in 0.3 M sucrose at 4°C. Furthermore, the fusion response observed under approximately isotonic conditions progressively decreased in magnitude as the

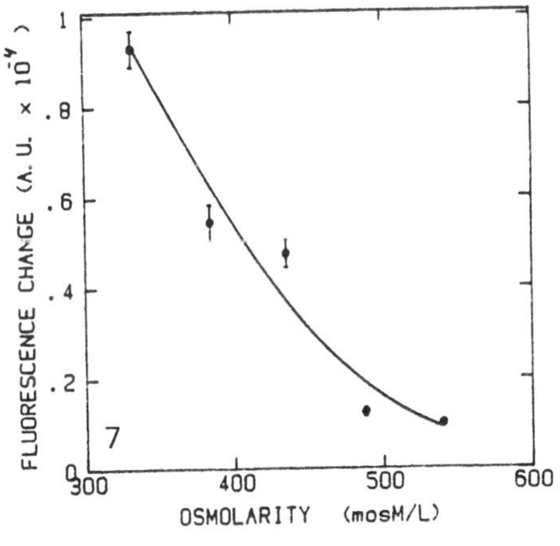

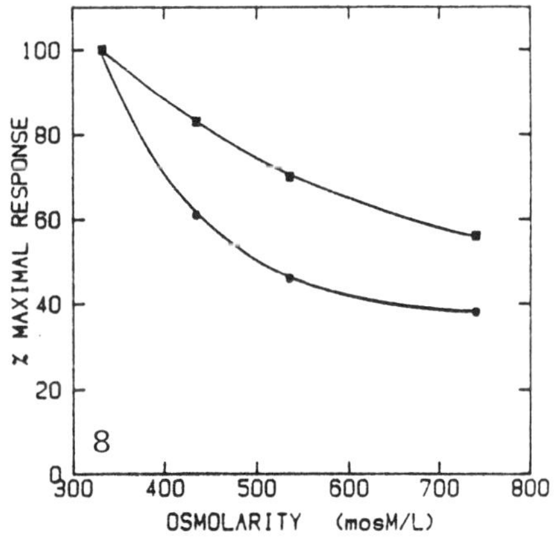

Figure 7. (LEFT) Osmotic supression of chromaffin granule fusion, as monitored using the R18 fluorescence assay. Osmolarity was increased by the addition of stachyose. Other conditions = pCa 3.0; 0.5 μg/ml oleic acid; and 1.5 μg/ml 32K calelectrin. Total granule concentration = 0.03 A540 units.

Figure 8. (RIGHT) Osmotic suppression of chromaffin granule fusion, monitored using the R18 assay (closed circles) or the NBD/Rhodamine energy transfer assay (closed squares). Osmolarity was increased by the addition of stachyose. Other conditions: pCa 3.0; and 13 μg/ml 32K calelectrin. Total granule concentration = 0.3 A540 units.

granules aged. Morris and Bradley also observed that the fusion reaction between chromaffin granules decreased as the granules aged, and also that granules dehydrated during preparation on sucrose gradients did not fuse as well as those prepared in isotonic media (3). These changes are consistent with the loss of osmotically active core contents from the granules under these conditions. To insure that the apparent osmotic suppression of fusion observed with the R18 assay was not an artifact, the experiment was repeated using the NBD/Rho resonance energy transfer assay. As shown in Fig. 8, this assay gives results qualitatively similar to those obtained with the R18 assay. These decreased fusion responses did not appear to be due to a decrease in chromaffin granule aggregation as a result of the increased viscosity of the sucrose or stachyose containing solutions because the increase in acceptor fluorescence following contact of chromaffin granules singly labelled with NBD-PE and Rho-PE was unaffected by the osmoticants (not shown).

Effect of Free Fatty Acids on the Fusion Process

When chromaffin granules are aggregated by synexin, granule fusion appears to be a rare event as detected by either EM or phase microscopy (1). Fusion in this system appears to require the addition of a *cis*-unsaturated free fatty acid (1). We therefore examined the effects of free fatty acids on the fusion process as revealed by the fluorescence fusion assays. As shown in Fig. 9 the R18 assay reports that high concentrations of oleic acid markedly increased the rate and extent of granule fusion promoted by either synexin or 32K calelectrin. Fatty acids also promoted fusion at 1-20 mM Ca^{2+} alone, suggesting that their action is not specific for the protein-mediated process. Effective concentrations of oleic acid, expressed relative to total chromaffin granule lipid, were similar to those required to promote vesicle formation in synexin-induced aggregates of chromaffin granules (1). These concentrations of oleic acid were also similar to those required to lower the Ca^{2+} concentrations necessary for 32K calelectrin binding to membranes and 32K calelectrin-induced granule aggregation (4). The relative specificity of different fatty acids was as follows: *cis*-unsaturated fatty acids such as

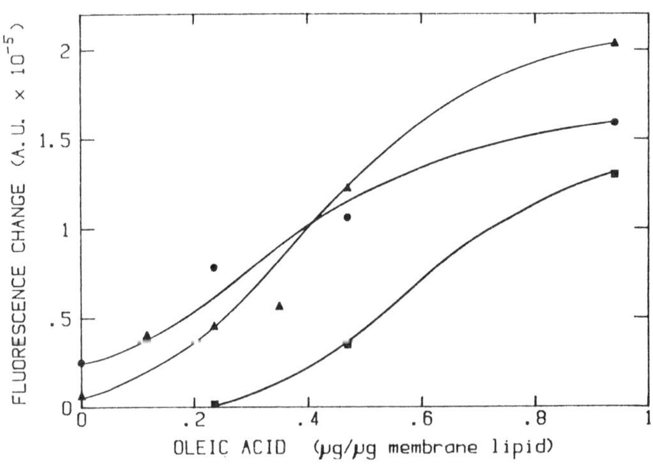

Figure 9. Dependence of chromaffin granule fusion on oleic acid concentration. R18 labelled granules were mixed with a 4-fold excess of unlabelled granules (final A540 = 0.03) in the presence of varying concentrations of oleic acid and either no aggregating protein (squared), 0.84 μg/ml 32K calelectrin (circles) or 2.8 μg/ml synexin (triangles). Fusion was initiated by the addition of pCa 3.0 buffer.

oleic acid, or arachidonic acid were most effective, while long chain saturated fatty acids such as palmitic acid and the methyl ester of oleic acid were ineffective. Of intermediate effectiveness were the <u>trans</u>-unsaturated fatty acid, elaidic acid, and the medium chain saturated fatty acid, myristic acid. The specificity of this process appeared similar for fusion promoted either by synexin or 32K calelectrin.

Effects of Ca^{2+}-binding Proteins on Granule Fusion

Ideally, an analysis of the effect of Ca^{2+} binding proteins on chromaffin granule fusion must deal with the effect of Ca^{2+} and protein binding on the quenching of the fluorescent membrane probes and the influence this will have on the fluorescence change associated with fusion. Protein binding to R18 labelled chromaffin granules increases rhodamine quenching (from 14% to 37% in the limit of infinite probe dilution) and decreases the slope of the fluorescence quenching curve (data not shown). However, since it is unknown how these changes affect the fluorescence change associated with fusion, no correction for these effects has been made in the analysis of the following data. In light of these uncertainities, it must be emphasized that these fusion assays provide only a qualitative picture of the effect of Ca^{2+} and protein on the fusion response in this system.

Chromaffin granule fusion as measured by the R18 assay occurs at high Ca^{2+} concentrations in the absence of granule aggregating protein as shown in Fig. 10. Similar Ca^{2+} requirements for granule fusion have been observed using the NBD/Rho resonance energy transfer assay (3) or electron microscopy (12). The Ca^{2+} requirements for granule fusion measured by both the R18 and NBD/Rho assays are much lower in the presence of 32K calelectrin or synexin as shown in Fig 11. These Ca^{2+} requirements are similar to those for granule aggregation by these proteins (4). Since <u>cis</u>-unsaturated free fatty acids increase the apparent Ca^{2+} affinity for activation of granule aggregation, we also examined the effect of oleic acid on the fusion of chromaffin granules induced by 32K calelectrin. As shown in Fig. 12 oleic acid (0.05 mg/mg granule lipid) appeared to lower the Ca^{2+} requirements for granule fusion in addition to increasing the maximum level of fusion observed.

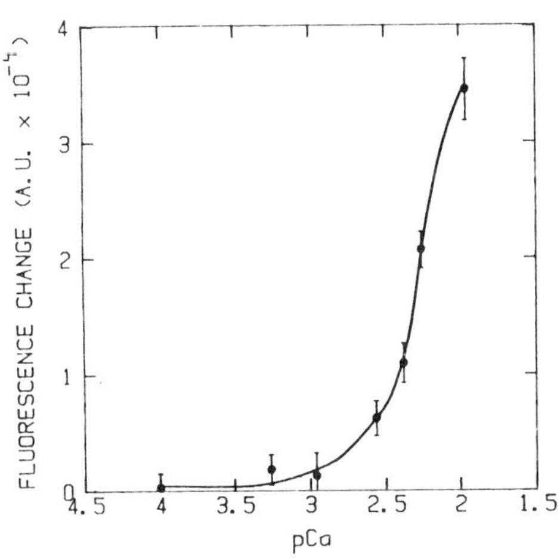

Figure 10. Calcium dependence of chromaffin granule fusion in the absence of an aggregating protein. R18 fluorescence assay was used in the presence of 0.5 μg/ml oleic acid. Total granule concentration = 0.03 A540 units.

When the protein concentration dependence for chromaffin granule fusion was examined, we observed two interesting features (Fig. 13). First, granule fusion required similar, if not identical, amounts of protein as protein binding and granule aggregation. The relationship between binding, as detected by R18 quenching, and fusion is shown for synexin (fig. 13). Second, fusion was inhibited at high protein concentrations. This inhibitory effect was also seen using the NBD/Rho assay, as shown for 32K calelectrin. The most obvious mechanism to explain this inhibitory effect is physical separation of the bilayers due to the presence of protein in the

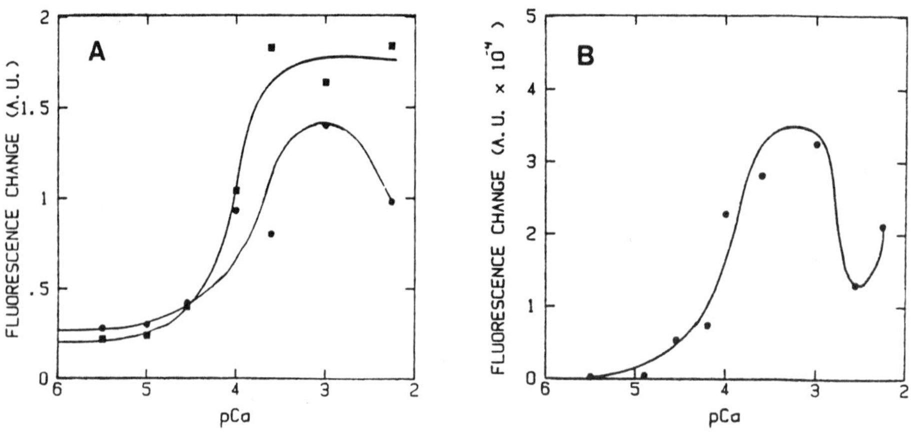

Figure 11. Calcium dependence of chromaffin granule fusion in the presence of 32K calelectrin or synexin. The R18 assay (circles, granule A540 = 0.03) or the NBD/rhodamine assay (squares, granule A540 = 0.3) was used with 1 μg/ml 32K calelectrin (circles, part A), 10 μg/ml 32K calelectrin (squares), or 6.5 μg/ml synexin (part B).

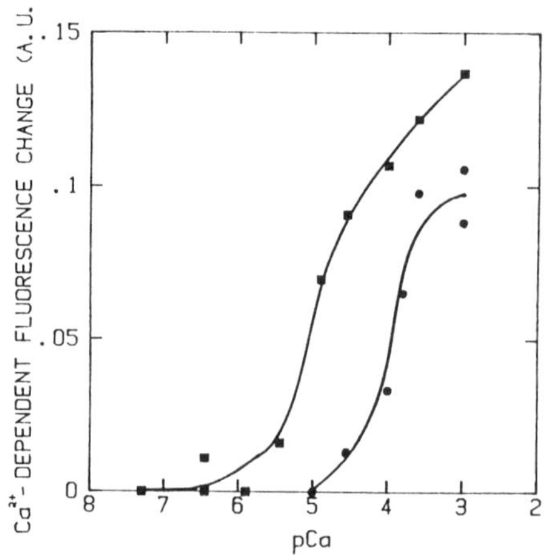

Figure 12. Effect of oleic acid on the calcium dependence of chromaffin granule fusion induced by 32K calelectrin. R18 assay was used in the presence of 1 μg/ml calelectrin and the absence (circles) or presence (squares) of 1 μg/ml oleic acid. Total granule concentration = 0.03 A540 units.

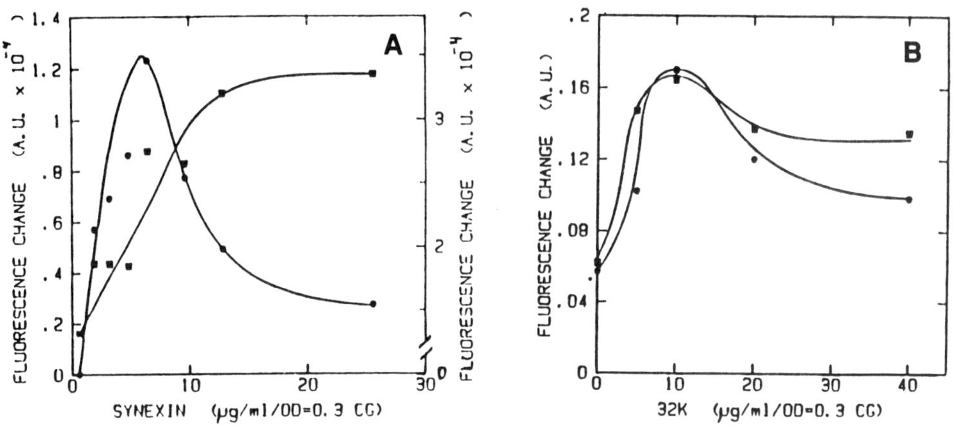

Figure 13. Dependence of chromaffin granule fusion on the concentration of synexin (A) or 32K calelectrin (B). R18 assay (circles, part A and part B) or NBD/rhodamine assay (squares, part B) was applied in the presence of pCa 3.0 buffer and the absence of exogenous fatty acids. In part A (squares) the magnitude of the initial rapid decrease in fluorescence associated with protein binding to the membrane is also shown.

intermembrane space. This phenomenon occurs when the maximal binding capacity of the membrane is approached and as aggregation reaches its zenith. Since less than one-tenth of the surface area of the granule is covered by 32K calelectrin or synexin under these conditions, it is difficult to envisage how bilayer interaction would be prevented unless the protein self-aggregated to form a "patch" of molecules situated at the zone of intermembrane contact.

DISCUSSION

We have observed that chromaffin granule fusion requires an intact native chromaffin granule. Granules lysed by hypoosmotic shock failed to fuse. Although the mechanism for this selectivity is unclear, we also observed that fusion was inhibited by increased osmolarity. It is believed that the mechanism for osmotic sensitivity of membrane fusion in some systems is inhibition of osmotically induced vesicle swelling (11). This swelling event has been shown to facilitate fusion of vesicles to planar phospholipid bilayers (9). Since chromaffin granules have potentially hyperosmotic core contents, these organelles may be capable of applying an osmotic force to their membranes. If the above theory of osmotic pressure sensitivity applies to this system, it would suggest that an osmotic force on the granule membrane due to the granule core contents plays a role in chromaffin granule fusion. We may speculate that membrane fusion has two requirements. First the membranes must be brought into close apposition by the adhesive forces of the granule aggregating proteins. Second, the apposed membranes must be physically forced together with sufficient energy to bring about membrane reorganization. For the latter process to occur, both membranes must be resistent to movement away from the opposing membrane. The internal osmotic pressure could provide this resistive force. If this idea is correct, it would suggest that exocytosis would be localized to sites of plasma membrane stress, and would fail to occur at sites of easily deformable membrane. Changes in membrane tension could potentially be locally regulated in the cell by gel-sol transformations as well as by interactions with cytoskeletal elements.

Most models for membrane fusion require some perturbatio of the membrane phospholipids for fusion to occur (11). This perturbation may take the form of an increase in tension due to pressure imbalance as postulated above, or it may result from physiochemical interactions. The observed ability of *cis*-unsaturated free fatty acids to promote fusion may be the result of the second mechanism. The fatty acids decrease the polarization of diphenylhexatriene in biological membranes consistent with a "disordering" of the bilayer (13). This increase in fluidity would presumably increase the probability of lipid interactions between apposed membranes.

The chromaffin granule aggregating proteins were observed to promote chromaffin granule fusion at > 10 μM Ca^{2+} in the absence of exogenous free fatty acids. In the presence of high concentrations of fatty acids, significant fusion was observed between 1-10 μM Ca^{2+}, well within the physiological range of intracellular Ca^{2+} concentration in stimulated cells. The Ca^{2+} requirements for promoting chromaffin granule aggregation and fusion were similar, both appearing biphasic over a range of Ca^{2+} concentration from pCa 5 to pCa 2. One phase of activity from pCa 5 to pCa 3.5 was stimulating for both granule aggregation and fusion. However, while the second phase of activity from pCa 3.5 to pCa 2 promoted additional granule aggregation, it appeared to inhibit fusion as shown in Fig. 11. These data suggest that the proteins modulate fusion by mechanisms other than by simple granule aggregation. Interestingly, fusion was observed to be inhibited at high protein concentration, although aggregation was not (4). Furthermore, the amount of protein required for inhibition of fusion varied as a function of Ca^{2+}, and this does not appear to be explained by different amounts of protein bound to the chromaffin granule membrane. This suggests that some functional property of the protein, other than the one regulating binding, is responsible for inhibition of fusion. One possibility is Ca^{2+} dependent self-association, such that protein polymers form at the sites of membrane contact and physically prevent the close membrane contact necessary for fusion.

In summary, the fusion of chromaffin granules mediated by Ca^{2+}, protein, and *cis*-unsaturated free fatty acid appears to be an excellent model for Ca^{2+}-dependent exocytosis in general. Figure 14 shows in cartoon form the relationships of these components in chromaffin granule fusion. The critical component in this system is the granule aggregating protein. It is reasonable to assume this model system can be applied to study secretion from non-chromaffin cells simply by substituting other secretory granules for the chromaffin granules, since the aggregating proteins have the ability to recognize a wide variety of biological membranes. Because of this property they are ideal candidates for a role in generalized secretory processes. Furthermore, fusion in this system shares two important features with exocytosis: 1) the effective Ca^{2+} concentration is in the micromolar range in the presence of free fatty acids. Interactions with regions of specific lipid composition may conceivably decrease this Ca^{2+} requirement further. 2) fusion is inhibited by hypertonicity. This model also provides a possible molecular basis for other secretory phenomena: 1) differential secretion from one or more distinct populations of granules may be a consequence of nonhomogenous distributions of different aggregating protein, or differences in the lipid composition of the different granule types. 2) the limited extent of exocytosis observed in secreting cells, even in the presence of elevated Ca^{2+} levels, may be a consequence of the "autoinhibition" due to the recruitment and binding of large amounts of membrane-binding protein over a prolonged period of stimulation. The model also predicts that fusion may be regulated by local changes in state of tension of the plasma membrane, perhaps regulated by membrane-cytoskeleton interactions or cortical sol-gel transformations.

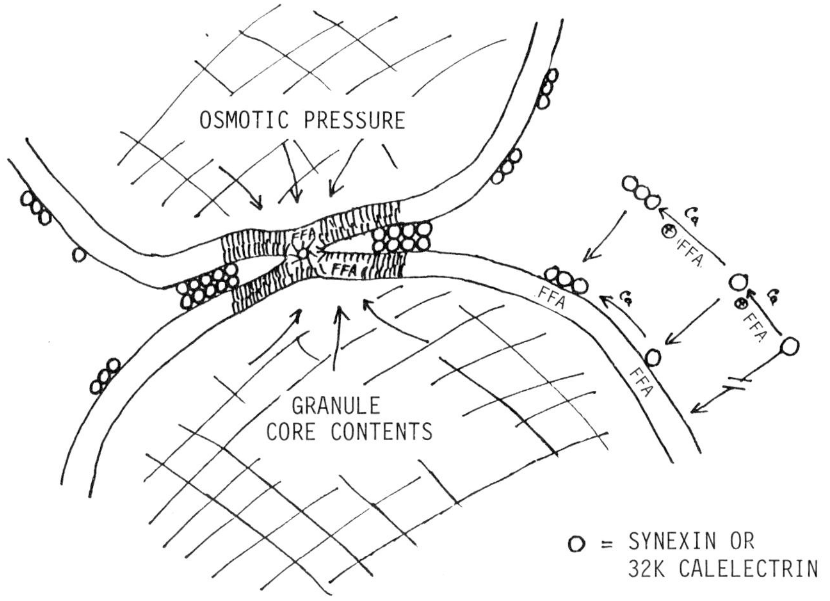

Figure 14. Model for chromaffin granule aggregation and fusion. Granule aggregating protein (synexin or 32K calelectrin), in the presence of elevated intracellular Ca^{2+} and free fatty acids (FFA), binds to chromaffin granule membranes and polymerizes, causing granule aggregation. The granule core contents provide an internal osmotic pressure which forces the opposing membranes together into a nonbilayer structure. Free fatty acids may further destabilize the membranes at the point of interaction leading to fusion.

APPENDIX I: Methods

NBD/Rho Resonance Energy Transfer Lipid Mixing Assay

Chromaffin granules were labelled with fluorescent lipids as described (3). Small unilamellar vesicles (NBD-PE/Rho-PE 1:1) prepared by sonication were incubated with freshly isolated chromaffin granules (10 μg probe/mg granule protein) in 0.3 M sucrose for approximately 12 hrs at 4°C. The suspensions were centrifuged at 12000 xg for 10 min at 4°C and the pellets were washed twice by resuspension in 2 ml of ice-cold 0.3 M sucrose to remove unincorporated probe. Chromaffin granule fusion assays were performed on nonstirred samples at 22°C in 0.5 cm cuvettes containing 0.5 ml of chromaffin granules (1 labelled:4 nonlabelled; OD 540 = 0.3), 0.24 M sucrose, 30 mM KCl 0.1 mM EGTA, 2.5 mM chelator (HEDTA, ADA), 40 mM Hepes-NaOH, pH 7.0, 0-15 mM $CaCl_2$, and varying concentrations of free fatty acids, stachyose, added sucrose, and 32K calelectrin or synexin. The samples were excited at 460 nm and fluorescence was recorded at 590 nm using a Schott 515 cut-off filter to reduce scattering artifacts. Baseline fluorescence in the presence of 0.1 mM EGTA was allowed to stabilize prior to starting the reaction by addition of Ca^{2+} buffer.

R18 Fluorescence Dequenching Lipid Mixing Assay

Freshly isolated chromaffin granules were labelled with octadecylrhodamine B (R18) as described (2). A concentrated ethanolic solution of R18 (1 mg/ml) was added dropwise to a vortexed suspension of chromaffin granules in 0.3 M sucrose. The amount of added R18 was calculated to be 3 to 5 mol% of the total granule membrane phospholipid (450 nmol/mg protein (6)), and care was taken to keep the total ethanol concentration below 1%. The mixture was incubated in the dark for 1 hr at 22°C and nonincorporated R18 was removed by 2 cycles of centrifugation (12000 xg for 10 min at 4°C) and resuspension in 2.0 ml of ice-cold 0.3 M sucrose. Chromaffin granule fusion was measured on a nonstirred sample at 22°C in a 0.5 cm cuvette containing 0.24M sucrose, 30 mM KCl, 0.1 mM EGTA, 2.5 mM chelator, 40 mM Hepes-NaOH, pH 7.0, and varying concentrations of chromaffin granules, $CaCl_2$, free fatty acids, stachyose, added sucrose, 32K calelectrin or synexin. The samples were excited at 560 nm and fluorescence was recorded at 590 nm. Most experiments were done at a low optical density (0.03) to reduce light scattering artifacts. Baseline fluorescence in the presence of 0.1 mM EGTA was allowed to stabilize prior to initiating fusion by addition of Ca^{2+}-buffer.

Proteins

Synexin (4) and 32K Calelectrin (14) were isolated from bovine liver as previously described.

Fluorescence measurements

Uncorrected fluorescence spectra and time-dependent measurements were made on a Spex Fluorolog 2 Model 111C spectrofluorometer using the ratio mode to reduce signal variability due to lamp instability.

Calcium Measurements

Ca^{2+} buffers were employed to maintain medium Ca^{2+} concentrations at precisely defined levels. A series of chelators (EDTA, HEDTA and ADA) were employed as described by Durham (15) and the pCa was checked with a Ca^{2+} selective electrode (Radiometer) standardized with solutions of known Ca^{2+} activity. The values of pCa reported here therefore reflect Ca^{2+} activities, rather than Ca^{2+} concentrations.

APPENDIX II: Anticipated behavior of the R18 dequenching fluorescence assay.

Dependence of Fusion Signal on Labelled: Nonlabelled Granule Ratio

Consider a mixed population of labelled and nonlabelled vesicles where the surface density, D, of the probe is constant, but the fraction of labelled vesicles, x, is variable. Multiple fusion events in this system will ultimately result (theoretically) in the formation of one large hybrid vesicle. The degree to which the probe surface density will change during this process is a function of the fraction of labelled vesicles. The surface density after fusion, D_f, is given by the expression, $D_f = xD_i$, where D_i is the initial surface density of the labelled vesicles. Since the efficiency of quenching, $Q = (F_T - F)/F_T$, is adjusted to be directly proportional to D, we may write

$$x = Q_f/Q_i$$

where Q_i and Q_f are the pre- and post-fusion levels of probe self-quenching, and F_T is the fluorescence at infinite probe dilution, i.e. in the presence of triton X-100. Hence,

$$\Delta Q = Q_f - Q_i = Q_i\ (x-1)$$

Since the fluorescence change associated with fusion, $\Delta F = -\Delta Q\ F_T$, we have,

$$\Delta F = F_T Q_i\ (1-x)$$

Since $F_T = kx$, where k is a constant of proportionality, and Q_i is independent of x, we obtain,

$$\Delta F = k'x(1-x)$$

where $k' = Q_i k$. Q.E.D.

<u>Dependence of Fusion Signal on R18 Concentration</u>

Consider a mixed population of labelled and nonlabelled vesicles where the fraction of labelled vesicles is constant, but the probe surface density is variable. Proceeding as above, we may write

$$\Delta F = F_T\ Q_i\ (1-x)$$

In this system, however, x is constant, and both F_T and Q_i are directly proportional to D_i, which in turn is directly proportional to c, the concentration of membrane associated probe. Hence we may write

$$\Delta F = kc^2$$

where k is a constant of proportionality. Q.E.D.

APPENDIX III: Predicted Influence of Lysed Membranes on Chromaffin Granule Fusion

The process of chromaffin granule fusion is modelled as the sum of granule dimerization reactions. Since the extent of fusion observed experimentally is limited, higher order reactions are neglected. In a uniform population of labelled granules, G, a fluorescence increase is associated with the fusion of granule dimers, GG. If lysed membranes, M, are then added to the system and prevent fusion, then no fluorescence increase is associated with the formation of granule-membrane dimers, GM, or membrane-membrane dimers, MM. In effect, the formation of the granule-membrane dimers, GM, reduces the number of granule dimers, GG, that can form, thus reducing total fluorescence. Using the binomial theorem we can calculate the distribution of dimer types that form as the granules and membranes interact:

Granule-granule dimers GG	Granule-membrane dimers GM	Membrane-membrane dimers MM
p^2	$2pq$	q^2

where p = fraction of granules in the mixed population, and q = fraction of lysed membranes in the mixed population. If the granule concentration is taken as 1, and the added membrane concentration as x, then $p = \frac{1}{1+x}$ and $q = \frac{x}{1+x}$. Since the fluorescence increase seen will be proportional to the number of granule-granule dimers that form, we may write ΔF = (a constant) x (fraction of total granules or membranes present as granule-granule dimers) x (total concentration of granules plus membranes) $= k \times p^2 \times (1+x) = \frac{k}{(1+x)} = \frac{100}{(1+x)}$, after normalization of $\Delta F = 100$ in the absence of membranes $(x = o)$.

This expression was used to calculate the theoretical points shown in figure 6.

REFERENCES

1. Creutz, C.E. J. Cell Biol. 91:247-256 (1981).
2. Hoekstra, D., de Boer, T., Klappe, K. and Wilschut, J. Biochemistry 23:5675-5681 (1984).
3. Morris, S.J. and Bradley, D. Biochemistry 23:4642-4650 (1984).
4. Zaks, W.J. and Creutz, C.E. J. Biol. Chem., submitted (1987).
5. Struck, D.K., Hoekstra, D. and Pagano, R.E. Biochemistry 20:4093-4099 (1981).
6. Winkler, H. and Westhead, E. Neuroscience 5:1803-1823 (1980).
7. Sudhof, T.C. Biochem. Biophys. Acta 684:27-39 (1982).
8. Reference deleted in proof.
9. Finkelstein, A., Zimmerberg, J. and Cohen, F.S. Ann. Rev. Physiol. 48:163-174 (1986).
10. Pollard, H.B., Pazoles, C.J. and Creutz, C.E. Recent Progress in Hormone Research 37:299-332 (1981).
11. Lucy, J.A. and Ahkong, Q.F. FEBS Lett. 199:1-11 (1986).
12. Morris, S.J., Costello, M.J., Robertson, J.D., Sudhof, T.C., Odenwald, W.F. and Haynes, D.H., J. Auton. Nerv. Syst. 7:19-33 (1983).
13. Karnovsky, M.J., Kleinfeld, A.M., Hoover, R.L. and Klausner, R.D. J. Cell Biol. 94:1-6 (1982).
14. Creutz, C.E., Zaks, W.J., Hamman, H.C., Crane, S., Martin, W.H,., Gould, K.L., Oddie, K.M. and Parsons, S.J. J. Biol. Chem. 262: 1860-1868 (1987).
15. Durham, A.C.H. Cell Calcium 4:33-46 (1983).

Acknowledgements

This study was supported by research grants from the N.I.H. (AM/DK 33151) and the N.S.F. (PCM8206453) and was done during the tenure of an Established Investigatorship (to C.E.C.) from the American Heart Association with funds contributed in part by the Virginia affiliate. W.J.Z. was supported by N.I.H. Medical Scientist Training Grant GM07055.

SYNEXIN, CALCIUM AND THE HYDROPHOBIC BRIDGE HYPOTHESIS FOR MEMBRANE FUSION

Harvey B. Pollard, Eduardo Rojas, A. Lee Burns and Claudio Parra*

Laboratory of Cell Biology and Genetics, National Institutes of Diabetes, and Digestive Diseases National Institutes of Health, Bethesda, MD (USA)
*Department of Physiology and Biophysics, Faculty of Medicine, University of Chile, Santiago, Chile

INTRODUCTION

Synexin is a calcium binding protein of 47,000da which causes chromaffin granules (1-5) and chromaffin granule ghosts (6), as well as acidic liposomes (7,8) to aggregate and fuse. In addition to binding to granule membranes and liposomes, synexin also binds selectively to the inner aspect of chromaffin cell plasma membranes (9). Furthermore, synexin dependent fusion structures of chromaffin granules appear quite similar to the large vacuolar, compound exocytotic structures seen in secreting chromaffin cells (3-5). These facts have lead us to anticipate that synexin might be involved in fusion processes occurring during exocytosis (5). We have therefore devoted substantial efforts to trying to understand how synexin indeed fuses these biological and artificial membranes.

We have now begun to learn some details of the fusion process by comparing synexin dependent fusion of intact chromaffin granules with the same process using chromaffin granule ghosts (6). The two processes are not exactly identical, possibly because the physical and chemical states of the membranes are quite different. However, the ghosts give us the advantage of being able to follow kinetically both membrane mixing (using octadecyl rhodamine) and volume mixing (using FITC-Dextran loaded ghosts) simultaneously. From studies on synexin driven fusion of ghost membranes we have learned that membrane mixing is 2-5 times faster than volume mixing during the fusion process. Therefore, the membrane mixing step may precede the volume mixing step during synexin-driven fusion.

CALCIUM-ACTIVATED SYNEXIN BINDS TO IMMOBILIZED PHOSPHOLIPID MONOLAYERS

The mechanism by which synexin may cause membranes to fuse may involve specific interaction of synexin with acidic phospholipids (8). For example, calcium-activated synexin can cause some types of acidic phospholipid liposomes to fuse. However, liposomes formed from phosphatidylinositol only aggregate. Therefore, using the liposome technology it was difficult to distinguish differential lipid affinity from the subsequent fusion process. To help resolve this problem we recently developed a solid phase technique for examining the interaction of synexin with different phospholipids (10).

This technique involved placing different phospholipids onto 96-well microtitration plates, constructed from a quite hydrophobic plastic. After drying away the methanol solvent, we added buffer, presumably allowing the phospholipids to orient on the plate with hydrophobic tails towards the plastic surface. We found that 125-[I]-labelled synexin bound to acidic phospholipids such as phosphatidylserine in a calcium dependent manner. However, labelled synexin did not bind to phosphatidylcholine or to neutral lipids such as triglyceride and cholesterol. Furthermore, synexin binds to different acidic or neutral lipids in a specific order, as summarized in Table I.

Table I. Calcium-Dependent Binding of 125-[I]-Synexin to Phospholipids

PHOSPHOLIPID SPECIES	PHOSPHOLIPID FOR 1/2 MAXIMUM BINDING, PMOL/WELL
Phosphatidylethanolomine (PE)	29
Phosphatidylserine (PS)	36
Phosphatidic acid (PA)	90
Phosphatidylinositol (PI)	400
Phosphatidylcholine (PC)	>>2000

These data were abstracted from reference 10

A [SYNEXIN-CALCIUM] COMPLEX BINDS TO THE PHOSPHOLIPID MONOLAYER

These data thus indicated that synexin could bind to specific acidic phospholipids in a calcium dependent manner, and that the mechanism of fusion induced by synexin could involve such an interaction. However, the details of this binding process remained obscure. For example, did synexin bind to calcium, and thus become competent to interact with specific phospholipids? Alternatively, calcium itself could modify the membrane and make molecules on it competent to bind synexin. As a final alternative, both processes could occur simultaneously. To resolve this issue we turned to a dynamic, radiometric method for measuring the interaction of synexin with phosphatidylserine monolayers at the air-water interface (11).

As shown in Figure 1, a flow radiation detector was placed above a dish containing sucrose-histidine buffer, pH 6.0 (1), CaCl (200 uM), 45-[Ca], and synexin where appropriate. An abrupt increase in counts at the surface was initiated by addition of a phosphatidylserine monolayer, placed on the buffer surface by addition of small amounts of known masses of phospholipids dissolved in chloroform. This is illustrated in the left and right drawings in the following graph. For the 45 - [Ca] case, the detector measures the radioactivity in a volume including the top 660 microns. From the data in Figure 1 it is apparent that 45-[Ca] activity is acquired by the surface layer immediately and does not change over the 90 minute course of the experiment.

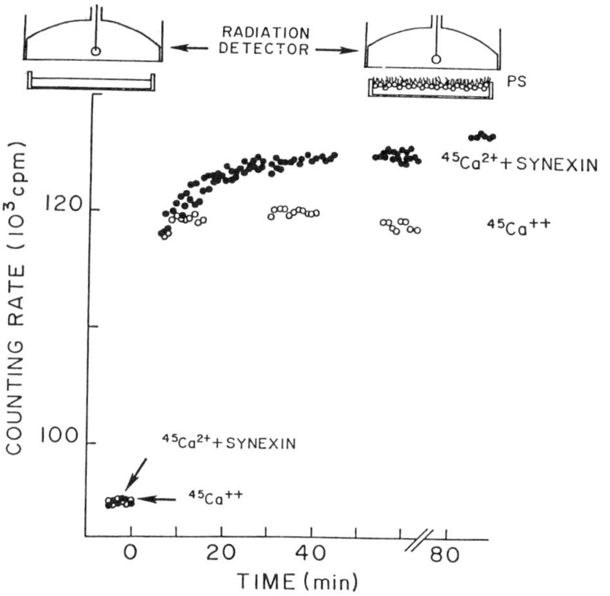

Figure 1. Influence of synexin (ca. 18ug/5ml) on recruitment of 45-[Ca] (200 uM) to phosphatidylserine (PS) monolayer. Drawings above the graph represent geometry of the detector. Open circles (O) represent calcium alone. Closed circles (●) represent synexin + calcium. Counts include all isotopes from the interface to 660 microns below the interface. All surfaces were coated with bovine serum albumin (1 mg/ml) dissolved in buffer prior to the experiment, and we ascertained that 14-[C]-BSA was inert to all experimental manipulations described here.

By contrast, if synexin is mixed with the 45-[Ca], more activity is acquired at the surface, and it takes more than one hour to approach equilibrium. Taken at face value these data would tend to support the concept that synexin binds the 45-[Ca] and brings it to the phospholipid monolayer at the air-water interface. In support of this concept is the fact that in the absence of phospholipid synexin did not cause any increase in radioactivity at the surface. More direct evidence comes from parallel studies on the surface tension of monolayers of phosphatidyl serine, to which synexin and calcium were added. Indeed, as shown in Table II, synexin (18ug/5ml buffer) in the presence of 200 uM Ca^{++} raised the surface tension of the monolayer from 32.1 dyne/cm to 36.4

dyne/cm. Neither synexin alone nor calcium alone was able to effect this change. One reasonable, qualitative interpretation of these data is that calcium caused synexin to leave the subphase and intercalate into the monolayer. Consistent with this interpretation is the fact that under the same conditions synexin activity could no longer be detected in the subphase after calcium was added (data not shown). We will return to an analysis of other surface pressure data in Table II after verifying the dependability and meaning of the increase in the calcium signal when synexin is added.

Since we know that synexin binds two molecules of calcium/mole of synexin monomer (2), it is possible to measure the number of calcium molecules brought to the surface by synexin, and to derive from that the minimum number of synexin molecules in the monolayer. To do this we need only to know the specific activity of the 45-[Ca], and to verify that the absorption of calcium to the surface follows some sort of predictable isotherm. We can not use the Langmuire isotherm, since it is valid only for uncharged species. Taking account of the effects of charged substances on the surface free energy is, in fact, quite complicated.

However, there is an empirical equation, called the Temkin isotherm (11), which states that for charged substances the number of absorbed particles/unit area is proportional to the log of the concentration of the charged species (in this case, calcium) in the subphase. To verify that our experimental data could be represented by this equation we varied the concentration of calcium in the subphase, using two different phospholipids, phosphatidylserine and phosphatidylinositol, and using two different surface pressures, corresponding to 40 and 60 $Å^2$/phospholipid.

As shown in Figure 2, the concentration in the subphase was varied between 10 uM (pCa 5) and 1000 uM (pCa3), and the number of calcium ions absorbed to the monolayer estimated from the specific activity. From our knowledge of the exact number of phospholipid molecules added to the surface, we were also able to calculate the ratio of calcium to phospholipid for each condition, as illustrated on the right vertical axis. Over the entire range of calcium concentrations, the expected linear relationship was observed. Furthermore, as also expected, raising the surface concentration of phospholipids from 60 $Å^2$/phospholipid to 40 $Å^2$/phospholipid raised the slope of the adsorption isotherm. With more phospholipids on the surface more calcium was bound.

We can now, with some assurance, compute the number of calcium molecules recruited to the surface upon the addition of calcium, as depicted in Figure 1. In the absence of synexin, 2.67×10^{15} molecules of calcium become recruited to the monolayer. Upon adding 18ug synexin, an additional 0.67×10^{15} calcium molecules are recruited to the monolayer. If each synexin molecule binds 2 molecules of calcium, then the calculated amount of synexin absorbed to the surface would be 0.33×10^{15} or 26ug. The difference between 18 and 26ug may be based on the possibility that less than 2 calcium ions were bound/synexin monomer, or that the difference is simply experimental error.

Further, though still indirect, evidence that the 45-[Ca] activity reaching the interface is borne by synexin can be obtained by measuring the apparent diffusion constant of the labelled species. We do not know the diffusion constant of calcium-activated synexin as yet. However, if we make the assumption that it is linearly proportional to the diffusion constant of the somewhat larger bovine serum albumin molecule (MW=65,000; D= 5.94 x 10^{-7}cm/sec), we could compute the time synexin would take to come to equilibrium with the surface. The appropriate equation is $\tau = \Delta X^2/D$ where X= 0.065cm and D=c.8.21 x 10^{-7}cm/sec. The time to equilibrium would then be 1.4 hours, a value approximately correct simply from inspection of the time course in Figure 1. We can

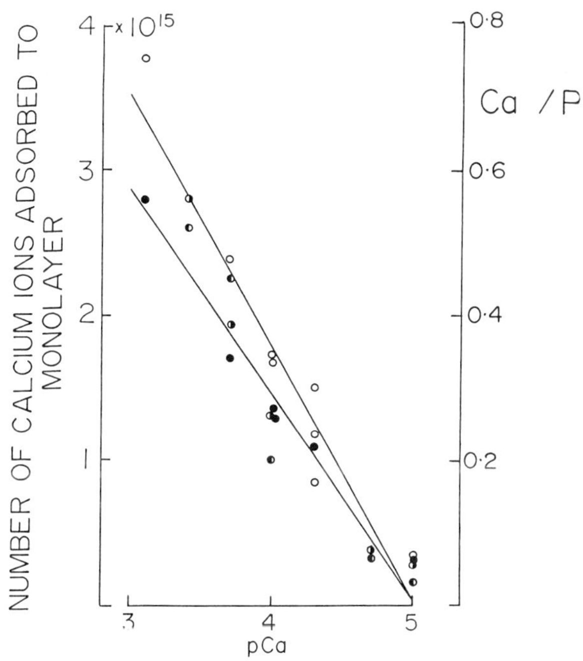

Figure 2. Temkin isotherms of 45-[Ca] binding to the phosphatidylserine and phosphatidylinositol monolayers as a function of calcium concentration in the subphase. The higher slope is 40 Å^2/PL, while the lower slope is 60 Å^2/PL. The symbols, (0) and (●) represent PS and PI, respectively at the higher surface concentration, while (◐) and (◑) represent PS and PI at the lower surface concentration.

therefore conclude from all these data that most of the additional calcium bound to the PS monolayer was probably brought there by synexin. However, we can not exclude the possibility that the ambient calcium modified the surface of the monolayer in such way as to make it more hospitable to the synexin-calcium complexes.

We can now return to further analysis of the data in Table II. The top panel in the Table contains data for a phosphatidylserine (PS) bilayer, and shows that synexin, in the presence of calcium, can elevate the surface tension of the bilayer by ca. 4 dynes/cm. Synexin may have therefore been able to insert into the monolayer, a conclusion consistent with data in Figure 1. However, when the monolayer is made of phosphatidylinositol (PI), synexin actually causes, if anything, a reduction in the surface tension. One possible explanation may lie in the fact that while calcium-activated synexin can cause PS liposomes to fuse, it only causes PI liposomes to aggregate (8). This fact was mentioned in the first part of this paper without explanation. But, with the monolayer data in hand it may be that synexin can not penetrate past the charged headgroup of PI once it binds. Therefore one could anticipate that synexin might only crosslink (or aggregate) such membranes but not fuse them. We will return later to the question of whether penetration of synexin into the bilayer is somehow involved in the mechanism of synexin-driven membrane fusion.

SYNEXIN PRIMARY SEQUENCE INDICATES AN "INTRINSIC MEMBRANE PROTEIN" TYPE STRUCTURE

When one considers how synexin might cause two membranes to fuse, two different mechanisms come to mind. In one case synexin might just bind to and crosslink the fusing membranes, allowing yet a second process to complete the fusion process. Alternatively, synexin might mediate fusion by an entirely synexin-specific process, perhaps involving intimate penetration of the synexin into both membranes. A hint that the latter mechanisms might be a real possibility has come from our initial studies on the sequences of parts of the synexin molecule (10).

For these experiments we prepared cDNA libraries from synexin-rich tissues in a lambda-gt11 expression vector system. We then searched these libraries for clones making proteins reactive with one or more of a variety of anti-synexin monoclonal antibodies and one polyclonal antibody. From studies on these partial clones we determined that the structure of synexin appeared to consist of lengthy segments of hydrophobic amino acids, punctuated by shorter segments of charged and neutral amino acids. Such a structure is typical of many "intrinsic membrane proteins," such as the acetycholine receptor (see ref 10 for detailed discussion), and the opposite of most "water soluble proteins."

However, synexin, in the absence of calcium, is quite water-soluble. This must mean that under low calcium conditions the protein must fold in such a way as to hide its hydrophobic sequences. But in the presence of calcium synexin becomes quite hydrophobic and even polymerizes (2). This must mean that under these conditions synexin folds in such manner as to hide its hydrophilic sequences. Preliminary spectroscopic data is also consistent with substantial conformational changes occurring upon addition of calcium to synexin. For this reason we have adopted the very general two-dimensional model for synexin conformation depicted in Figure 3, and the model of polymerization driven by hydrophobic forces depicted in Figure 4.

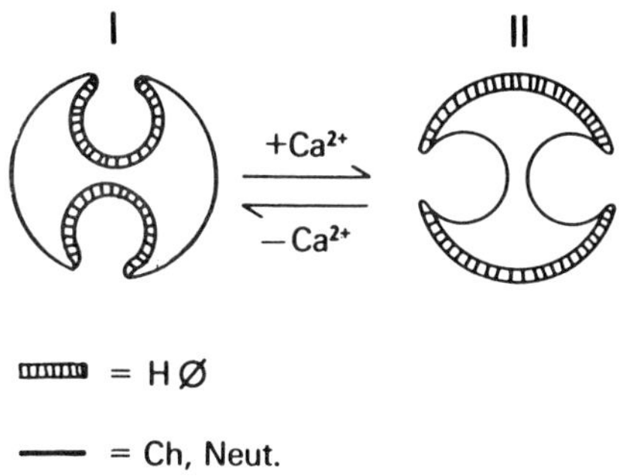

Figure 3. Two-dimensional model for calcium-dependent conformational transitions in synexin. This figure is reproduced from that in ref. 10.

Table II. Effect of Calcium and Synexin on the Surface Tension of PS and PI Monolayers

EXPERIMENT	SURFACE TENSION DYNES/CM
PHOSPHATIDYLSERINE	
BUFFER ALONE	63.0
PS MONOLAYER (60 Å^2/PL)	31.5
+200uM Ca^{++}	32.1
+200uM Ca^{++}, 18 ug SYNEXIN	36.4
PHOSPHATIDYLINOSITOL	
BUFFER ALONE	63.0
PI MONOLAYER (60 Å^2/PL)	33.0
+200uM Ca^{++}	ND
+200uM Ca^{++}, 18 ug SYNEXIN	27.9

These data were collected from the samples analyzed as in Figure 1, after close approach to equilibrium (1.5 hours). ND means not determined; PL means phospholipid molecule. Surface tension was measured by the Du Noüy tensiometer.

A.

B.

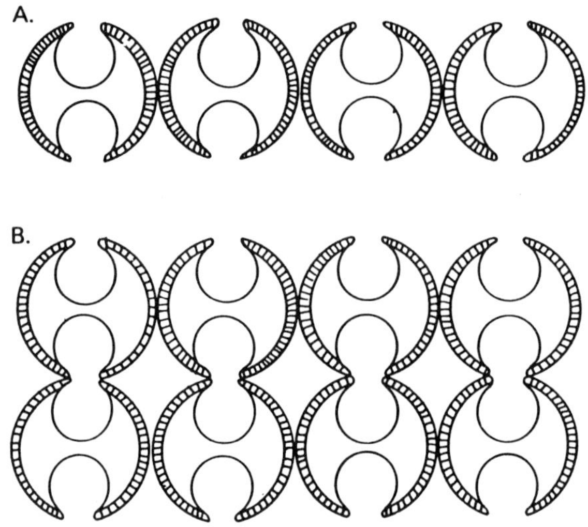

Figure 4. In the presence of calcium synexin polymerizes into rods, 100Å x 50Å x 50Å (A). These rods themselves polymerize side by side to form more extended structures (B). The drawing is based on that in ref. 10.

MEMBRANE CAPACITANCE STUDIES INDICATE THAT SYNEXIN ENTERS THE BILAYER

The surface tension studies summarized in Table II indicate that synexin could penetrate into a phospholipid monolayer. Furthermore, just like a true intrinsic membrane protein, the sequencing studies indicate that synexin could perhaps adopt a conformation that would permit it to enter into the interior of a bilayer. However, neither of these studies demonstrate that this actually occurs in bilayer membranes. In fact, there are only a few experimental approaches that could clearly distinguish between synexin binding to a membrane and synexin inserting into the membrane. One such approach is to study the effect of calcium-activated synexin on membrane capacitance, a method which we have recently exploited with some success (12).

For capacitance measurements, a phosphatidylserine bilayer can be easily prepared at the tip of a patch pipet, using the now classical double dip methods. As illustrated in Figure 5 the bilayer in such a pipet (drawing on the left) can be represented by an equivalent circuit in which the membrane is a capacitor. Upon application of a voltage pulse current flows in and places a charge, Q, on the capacitor. The capacitance of the membrane is just the ratio, Q/V, which can be easily measured from the current which flows into the capacitor over the time course of the voltage pulse. In the case of a simple membrane that contains no mobile charges and does not conduct charge the slope of the curve relating Q and V (lower drawing, Figure 5) is linear.

However, if the membrane contains a dipole, such as synexin or other charged species, within the bilayer, the capacitance is substantially changed. The reason is as follows, and is illustrated in Figure 6. The capacitance of a capacitor, or membrane, is given by the equation in the lower part of the figure. In the case of a bilayer at the tip of a patch pipet, the area (A) and the thickness

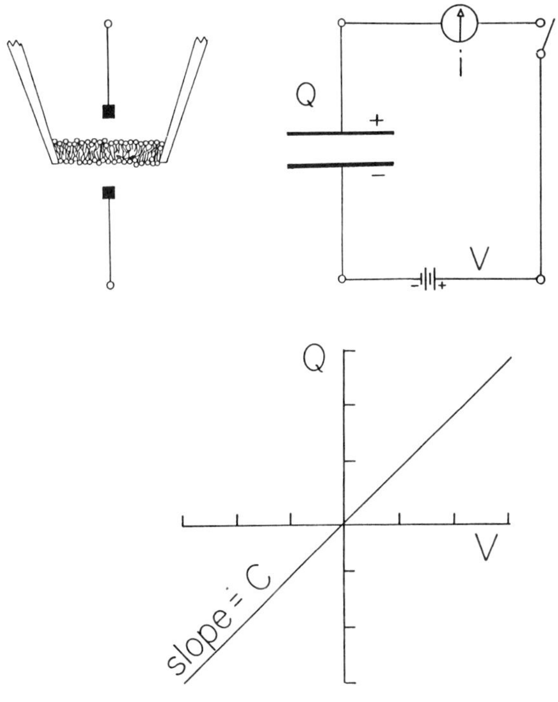

Figure 5. Techniques for measuring capacitance of a bilayer at the tip of a patch pipet. Drawing on the left represents a phosphatidylserine bilayer at the tip of the pipet, while the objects on either side are the electrodes. The equivalent circuit on the right shows how current (I), flowing into the capacitor as a consequence of voltage (V), results in accumulation of charge (Q) on the membrane. The graph below shows the linear relationship between Q and V, for a simple, membrane, the slope being the capacitance (C).

(d) are relatively fixed. However, if a dipole such as synexin were to be placed in the otherwise low dielectric interior of the membrane, the apparent dielectric constant would rise. This then is the key to using capacitance changes to distinguish protein binding from protein insertion to a membrane. Binding would change neither A nor d nor $\varepsilon/\varepsilon_0$. By contrast inserting a protein would raise the apparent dielectric constant.

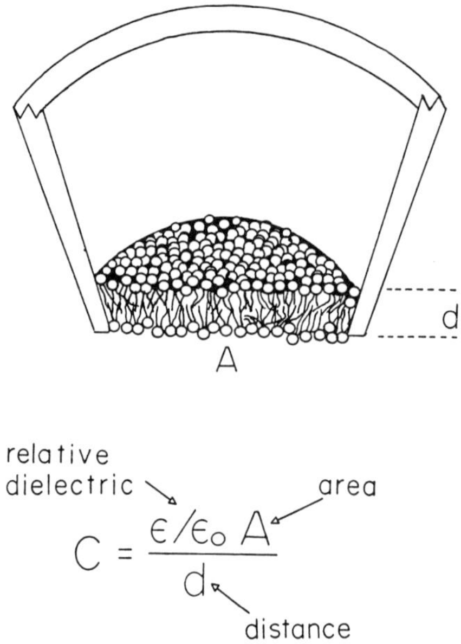

Figure 6. Geometry of the bilayer formed at the tip of the patch pipet, and the equation for capacitance (C) of the membrane. ε_0 is the permittivity of free space while $\varepsilon/\varepsilon_0$ is the relative dielectric constant of the bilayer. A is the area of the bilayer (about 1 square micron), and d is the thickness of the bilayer (about 40Å).

Some representative data from a capacity measurement experiment are shown in Figure 7. Upon application of a pulse of 50 mV to a simple membrane (Figure 7 A), current flows into the bilayer and then ceases to flow as the membrane becomes charged (Figure 7 B, "Mb"). However, if the synexin has been added to the membrane more current flows, and it flows for a longer time (Figure 7 B, "Mb + syn"). Since charge accumulated is the time integral of the current, increased area under the trace means greater charge. The difference between the current flowing into the membrane alone and the membrane supplemented with synexin is the synexin-specific current, and is shown in Figure 7C. From this one can compute the synexin-specific charge, and therefore the increase in capacitance due to synexin. Upon addition of synexin the membrane capacitance rises from ca. 20 fFarad to ca. 200 fFarad. The data also reveal that these changes in membrane capacitance occur only if the medium also contains calcium, and that no changes occur if the medium instead contains up to 50mM magnesium or barium. We have concluded from these and other data that calcium-activated synexin is able to penetrate into the bilayer of PS membranes, and that once inside it synexin is able to move within the confines of the membrane in response to voltage gradients. The latter conclusion is based on voltage dependence of the steady-state distribution of synexin-specific mobile charge, where the apparent effective valence of synexin is -1.46 electronic charges, and where a displacement of 16.7 mV in membrane potential induces a e-fold change in the charge transferred at the midpoint potential (21 mV).

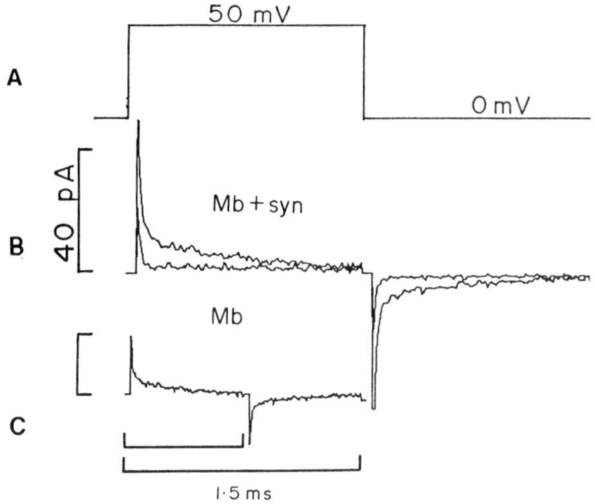

Figure 7. Protocol to measure the synexin-specific capacitance of a phosphatidylserine bilayer formed at the tip of a patch pipet. A. A +50 mV pulse was placed across the bilayer. B. Currents flowing into a plain membrane (labelled "Mb") and a synexin-impregnated membrane (labelled "Mb + syn") were measured. C. The difference between the two traces in B were calculated and taken to represent synexin-specific current transients. (Data are taken from Ref. 12)

We consider these capacitance data to be convincing as to the fact that calcium-activated synexin molecules do penetrate into the low dielectric substance of the bilayer. However, our search for additional compelling evidence recently lead us to observe that in the phosphatidylserine bilayer system synexin also exhibits calcium channel activity. This means that synexin not only can enter the membrane but also can penetrate across the membrane. This observation has had important consequences for our thoughts about how synexin might mediate membrane fusion, as will become apparent below.

To illustrate the evidence for synexin channel activity we show in Figure 8 the experimental arrangement and an example of data showing unitary conductance changes. The experimental setup involves two pipets, one containing a bilayer (on the right) and the other containing synexin. To initiate the experiment we applied synexin to the bilayer by pressure injection. In order to detect synexin channels, some modest changes in the ionic gradients in the bath and pipet had to be made, compared to conditions needed to detect capacitance changes. However, as in the case of synexin-induced capacitance changes, synexin channel activity was absolutely dependent upon calcium in the medium.

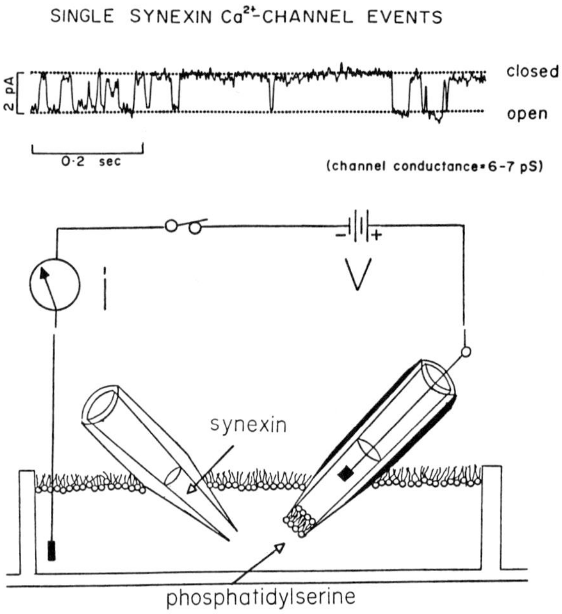

Figure 8. Detection of synexin calcium channels in phosphatidylserine bilayer membranes. A bilayer was prepared by the double dip method, as for capacitance measurements, in the pipet on the right. Purified synexin dissolved in bath buffer was then applied to bilayer using microscopic control and pressure injection. With the appropriate electrochemical gradients, unitary conductance events could be readily observed. In this case these events had a magnitude of 6-7 pS, at V (holding) = 0 mV.

"HYDROPHOBIC BRIDGE" HYPOTHESIS FOR SYNEXIN-DRIVEN MEMBRANE FUSION

In most forms of membrane fusion the fusing membranes join by eventually having their initially juxtaposed (cis) leaflets become co-linear, and then by having their contra-lateral (trans) leaflets also become co-linear. The process is a stereochemical mystery, but our recent studies on synexin have lead us to make a proposal for the mechanism of synexin-driven fusion which we have termed the "hydrophobic bridge" hypothesis. The basis aspects of this model are summarized in Figure 9, which we have slightly modified from a previous rendition (10). The basic idea, which we shall develop in brief, is that hydrophobic synexin polymers enter both fusing membranes and provide a "bridge" over which lipids on the adjacent membrane surfaces may cross and mix. The advantage of this hypothesis is that it is primarily based on the known properties of synexin and on the biological and natural membranes with which synexin actually promotes fusion.

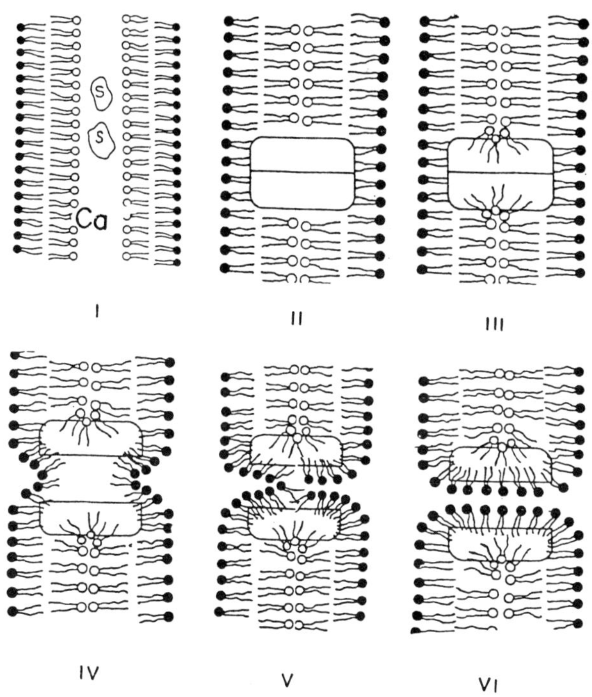

Figure 9. Hydrophobic bridge hypothesis for synexin-driven membrane fusion. (I): Two membranes are poised to fuse. Open circles represent acidic phospholipids on the cytoplasmic (cis) leaflet of a membrane; closed circles represent phospholipids on the opposing (trans) leaflet. Calcium causes synexin monomers (labelled "S") to polymerize. (II): Calcium-activated synexin polymer binds to the acidic phospholipids on both cis surfaces, penetrates into both bilayers and crosslinks the membranes. (III): Phospholipids cross the "hydrophobic bridge" supplied by synexin, thereby allowing the cis leaflets of the fusing membrane to mix. (IV): Synexin polymer dissociates in the low dielectric environment, providing a hydrophobic pathway for the trans leaflets of the fusing membrane to mix. (IV): Synexin polymer dissociates in the low dielectric environment, providing a hydrophobic pathway for the trans leaflets to approach. (V): Trans leaflets complete their reorientation, leaving (VI): synexin in the bilayer of the newly fused membranes. Synexin may then leave the membrane when the calcium level drops, but we have little information as yet on that process. (This sequence is a modification of the model described in reference 10).

As iterated above, each step in this model is based on known properties of synexin; and so far as we know, under physiologic conditions synexin exhibits little affinity for membranes without first being activated by calcium. For that reason, in resting cells synexin is a soluble, cytosolic protein and we depict it in this way in step I (Figure 9). We also know that calcium activated synexin has affinity for acidic phospholipids, that it can penetrate into phosphatidylserine monolayers, and that it can even span bilayers. For this reason in step II (Figure 9) we depict synexin penetrating both cis leaflets and crosslinking them. In studies on fusing chromaffin granule membrane ghosts we learned (6) that membrane mixing (step III, Figure 9) to occur at an early time. This membrane mixing occurs as the phospholipids on the cis leaflets cross the hydrophobic bridge provided by synexin, and is the key event in our model. It may also be apparent that in step III synexin becomes the armature of a new bilayer, a reasonable structure considering the fact that synexin does exactly this in our model systems.

The next steps, involving contact and fusion of the trans bilayers, are the most speculative. However, they are still based on known properties of synexin. For example, we know that synexin polymers tend to become destabilized in low dielectric media. In addition, the number of synexin dipoles in a phosphatidylserine bilayer might be in excess of what could be expected should polymeric synexin remain associated. Furthermore, the voltage sensitivity of the capacitative current indicates that synexin dipoles can move within the confines of the bilayer. Therefore, dissociation of the synexin polymer within the bilayer, or perhaps some energetically equivalent step, might occur. This would provide additional hydrophobic surfaces for the trans bilayer to enter upon (Step IV, Figure 9), and upon which to reorient into new, complete and asymmetric bilayers (steps V and VI, Figure 9). However, whether by this mechanism or by some yet to be determined analogous process, the movement and mixing of the trans monolayers probably occurs after the mixing of the cis monolayers. This is implicit in the experiments showing that membrane mixing occurs faster than volume mixing when synexin causes fusion of the chromaffin granule ghosts.

Following fusion, synexin may depart from the membrane. We can assume this based on the fact that in tissues and dissociated cells most synexin is "soluble." Furthermore, calcium-dependent binding of synexin to intact chromaffin granules, phospholipid monolayers, and inside-out chromaffin cell plasma membranes is substantially reversible. Nonetheless some synexin remains behind under these circumstances. Therefore, whether the residual synexin molecules are those which took part in fusion events remains to be determined.

Finally, we might add parenthetically that the capacitance changes and channel activity we have observed with pure phosphatidylserine bilayers can also be observed when synexin is applied to inside-out patches from chromaffin cell plasma membranes. This means that the various properties ascribed to synexin in model membranes can, in practice, also obtain *in vivo*.

CONCLUSIONS

Synexin is quite an unusual protein being water soluble in the absence of calcium, and membrane-soluble upon exposure to calcium. Once in the membrane it becomes mobile within the bilayer, and can even span the bilayer, connecting both sides. Molecular biology studies provide at least one explanation, since the partial sequences now at hand indicate that synexin is structured like many intrinsic membrane proteins. The game that synexin plays is to hide this structure when in its water soluble condition, presumably by means of a conformational strategy. While such a strategy is also used by a variety of bacterial toxins (eg, diphtheria toxin, colicin I), synexin is the only protein we know which does this for positive reasons within normally functioning cells. Perhaps others will soon be found.

Synexin is also unusual in its fidelity to calcium as the mediator of its' conformational state. The fact that synexin is also a calcium channel, and displays the same fidelity to calcium in terms of ion selectivity, may indicate that control of conformation and of conductance may be regulated in the same sites. This may mean that conformational changes, membrane insertion, channel function and membrane fusion may be inexorably entwined in this molecule.

While synexin may prove to be a prototype for fusion mechanisms in other biological systems, many other proteins bearing some apparent operational resemblances to synexin have been recently reported (13). We anticipate that some of them may prove to have fusion properties, and that the hydrophobic bridge mechanism may be a common denominator for their action.

REFERENCES

1. Creutz, C.E., Pazoles, C.J. and Pollard, H.B. (1978) J. Biol. Chem., 253:2858-2866.
2. Creutz, C.E., Pazoles, C.J. and Pollard, H.B. (1979) J. Biol. Chem., 254:553-558.
3. Creutz, C.E. and Pollard, H.B. (1981) Biophys. J., 37:119-120.
4. Creutz, C.E., J. Cell Biol., 91:247-256.
5. Pollard, H.B., Creutz, C.E., Fowler, V.M., Scott, J.H. and Pazoles, C.J., (1982) Cold Spring Harbor Symposium on Quant. Biol., 46:819-833.
6. Stutzin, A. (1986) F.E.B.S. Lett., 197:274-280.
7. Hong, K., Duzgunes, N. and Papahadjopoulos, D. (1981) J. Biol. Chem.,256:3641-3644.
8. Hong, K., Duzgunes, N. and Papahadjopoulos, D. (1982) Biophys. J., 37:297-305.
9. Scott, J.H., Creutz, C.E. and Pollard, H.B. (1985) F.E.B.S. Lett., 180:17-21.
10 Pollard, H.B., Rojas, E. and Burns, A.L. (1987) Ann. N.Y. Acad. Sci.. 493:524-551.
11. Davies, J.T. and Rideal, E.K., Interfacial Phenomena Academic Press (NY & Longon) 1961.
12. Rojas, E., and Pollard, H.B. (1987 F.E.B.S. Lett. in press.
13. Pollard, H.B. and others (1985) Vitamins and Hormones 42:109-195.

ACKNOWLEDGEMENTS

The authors wish to thank Ms. Diane Johnson-Seaton for her help in preparation and testing of synexin, and Mrs. Delma Tyler for preparing the manuscript for publication. The authors also gratefully acknowledge the NIDDK-Cystic Fibrosis Foundation Joint Fellowship Program for support of part of this work.

MOLECULAR MECHANISM OF PROTEIN-MEDIATED LOW PH-INDUCED MEMBRANE FUSIONS

Shun-ichi Ohnishi and Masayuki Murata

Department of Biophysics, Faculty of Science
Kyoto University
Sakyo-ku, Kyoto 606, Japan

INTRODUCTION

Membrane fusion and splitting play an essential role in endocytosis, transport of proteins and other materials to cellular membrane structures, and exocytosis. They also constitute an essential initial and final steps in the infection of enveloped viruses. Fusion of virus envelope with cell membranes has been most extensively studied and a wealth of informations has been accumulated (for example, White et al.,1983; Ohnishi,1987). HVJ (hemagglutinating virus of Japan) and other paramyxoviruses can fuse with cell membranes at neutral as well as acidic pH's. Other viruses, however, can fuse only at acidic pH. The different pH characteristics differentiates the cell entry mechanism for these viruses. While paramyxoviruses can transfer their genome into cytosol by fusion with cell surface membrane, the other viruses will do this by fusion in endosomes after endocytosis when the intraendosomal pH becomes lower than 6.

Envelope fusions are mediated by glycoproteins situated in the virus envelope. These fusion proteins contain a hydrophobic segment at the N-terminus or internally (see Table 1). The segment consists of 20 or more hydrophobic amino acids. That of virus proteins which have fusion activity only at acidic pH contains 2 or three acidic resiues in most cases. These hydrophobic segments are assumed to play a direct role in the fusion reaction because of the hydrophobicity or the sequence homology (Gething et al., 1978).

Low pH-induced fusion activity of influenza virus has been extensively studied since its discovery (Maeda & Ohnishi,1980; Huang et al.,1981; Lennard & Miller,1981). The three dimensional structure of the virus fusion protein HA (hemagglutinin) has been determined by X-ray analysis (Wilson et al.,1981) and the conformational change induced at acidic pH has been studied by various techniques (Skehel et al.,1982; Sato et al.,1983; Ruigrok et al.,1986). The amino acid sequence of the fusogenic hydrophobic segment at the N-terminus of HA2 subunit has been determined for many strains of A and B type viruses as well as influenza C virus. Variants and mutants with altered pH dependence of fusion have been isolated and studied. Low pH is required for the conformational change of HA by which the fusogenic segment may approach and interact with the target membranes. We have proposed that protonation of the acidic residues in the fusogenic segment may also be required for the interaction leading to fusion (Maeda & Ohnishi,1980; Ohnishi,1985; 1987).

In order to investigate the role of the putative fusogenic segment and the proposed requirement of protonation of the acidic residues directly, and also to obtain molecular insight into the protein-mediated membrane fusion, we have studied fusion in a simplified system, fusion of neutral phospholipid (egg yolk PC (phosphatidylcholine)) vesicles induced by hydrophobic peptides. Fusion was assayed using spin-labeled PC (Maeda et al.,1975; Kuroda et al.,1985). This method is based on the increase in the ESR peak height on fusion of membranes with those containing spin-labeled PC owing to dilution of spin labels. Other assays based on the internal content mixing (Wilschut et al.,1980) and dequenching of octadecyl Rhodamine fluorescence (Hoekstra et al.,1984), and also direct morphological observation were made.

Table I. Some examples of hydrophobic segment in virus fusion proteins.

Hemagglutinating virus of Japan (HVJ) or Sendai virus	
F1	FFGAVIGTIALGVATSAQITAGIALAEAR-
Influenza virus HA2	
A/PR/8/34	GLFGAIAGFIEGGWTGMIDGWYGYH-
B/LEE/40	GFFGAIAGFLEGGWEGMIAGWHGYT-
C/CAL/18	IFGIDDLIIGLLFVAIVETGIGGYLLGSR-
Semliki Forest virus	
E1	-KVYTGVYPFMWGGAYCFCDSENTQLSEAYVDR-
Mouse mammary tumor virus	
gp 36	FVAAIILGISALIAIITSFAVATTALVK-

Two different kinds of peptides were studied for the fusion activity. One is a 20 amino acid peptide with the same sequence as that of the HA2 N-terminal segment of influenza virus (HA peptide) (see Fig.1A). This peptide showed a low pH-induced fusion activity similar to that of the parent virus (Murata et al., 1987a). Another one is bee venon melittin, a 26 amino acid peptide with a N-terminal hydrophobic segment (19 hydrophobic and 1 positively charged residues) and a C-terminal positively charged segment (Fig.1Ba). With this peptide, we attempted to modify its fusion activity to be acid sensitive by converting the amino groups into negatively charged carboxyl groups. For this purpose, we succinylated all the four lysyl amino groups in the peptide by the reaction with succinic anhydride (Fig.1Bb). Also we prepared trypsinized succinylated melittin which lacks the C-terminal 4 residues. The succinylated melittin showed a low pH-induced fusion activity in contrast to intact melittin which was active in a wide range of pH (Murata et al.,1987). We have also studied protonation process of the carboxyl groups in succinylated melittin by ^{13}C NMR spectroscopy and showed directly that the protonation parallels the fusion activation. The NMR data also suggest dynamic nature of the interaction of the peptide with lipid bilayer (Murata et al.,1987).

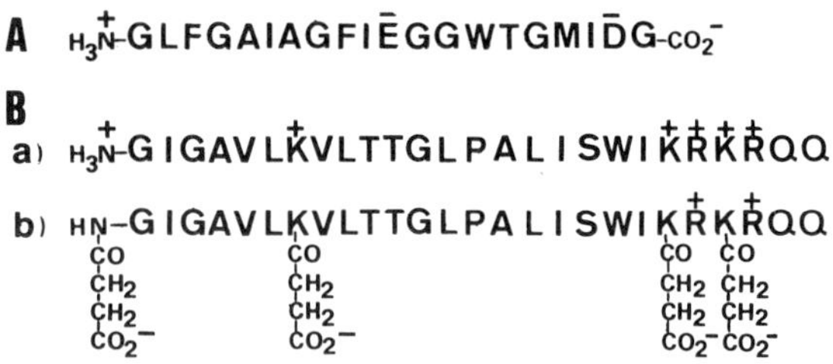

Fig.1. Peptides used for the present fusion study. A) A 20 amino acid peptide with same sequence as that of HA2 N-terminal segment of influenza virus A/PR/8/34 hemagglutinin (HA peptide). B) Bee venom melittin (a) and succinylated melittin (b).

FUSION ACTIVITY OF SUCCINYLATED MELITTIN IN COMPARISON WITH THAT OF INTACT MELITTIN

pH Dependence of Fusion

Both intact and succinylated melittins caused a rapid (within 1 min) and efficient fusion (Fig.2) but showed a great contrast in the pH characteristics (Fig.3). While intact melittin caused fusion at neutral as well as acidic pH's, succinylated melittin had fusion activity only at acidic pH. Intact melittin has a broad pH dependence, with a tendency of decreasing fusion efficiency at lower pH, succinylated melittin has a sharp rise in the activity at pH 5.2 (Fig.3). Light scattering of the vesicle suspension in the presence of succinylated melittin did not increase at pH values higher than 5.2 but increased at the lower pH values. This indicates that the peptide does not cause aggregation of vesicles in the pH range where it does not cause fusion.

The fusion efficiency of intact and succinylated melittins was dependent on the amount of peptide relative to phospholipid. It increased with the increase in the relative amount of peptide (see Fig.2 for example).

Trypsinized succinylated melittin did not cause fusion at acidic as well as neutral pH's. The removal of the four C-terminal residues from succinylated melittin therefore abolished the fusion activity. Light scattering measurements indicated no aggregation of vesicles at pH values higher than 5.2 but aggregation at the lower pH's, though to weaker extent than succinylated melittin.

Rapid Switching On and Off of Fusion by Shifting pH

Fusion initiated by succinylated melittin at acidic pH can be stopped immediately by shifting the pH to neutral and restarted again by restoring the pH. The switching on and off can be done repeatedly. The final level of fusion was the same as that without changing the pH to neutral.

Protonation of Carboxyl Groups in Succinylated Melittin

In order to follow the protonation process, succinylated melittin with ^{13}C-labeled carboxyl and carbonyl groups was synthesized by reaction with 1,4-^{13}C-succinic anhydride (see chemical formula in Fig.4). The ^{13}C

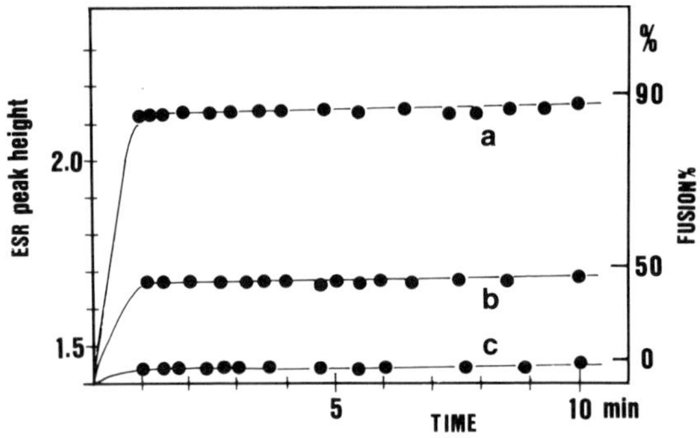

Fig.2. Fusion of PC vesicles induced by succinylated melittin at acidic pH. Succinylated melittin at 52 uM (a), 7.9 μM (b), or none (c) was added to a mixture of egg PC sonicated vesicles (3.6 mM) and those containing 5% spin-labeled PC (1.2 mM) at pH 5.0 at room temperature (23°C). The normalized ESR peak height was plotted against time after the addition. The degree of fusion was estimated from the normalized peak height with reference to a control curve (see Kuroda et al.,1985). The normalized peak height for egg PC vesicles containing 1.25 % spin-labeled PC was taken as 100% fusion; 100% fusion corresponds to the complete mixing of vesicle lipids in the mixture. (Taken from Murata et al.,1987).

NMR spectrum showed 6 peaks both in the presence and absence of vesicles. These peaks moved upfield with the decrease in pH (Fig.4). The titration curve can be fitted satisfactorily by the equation with a single pK_a value

$$pH = pK_a + \log[(\delta - \delta_{HA})/(\delta_A - \delta)]$$

The full line curve for the peaks 5 and 6 for the carboxyl carbons in Fig.4 was drawn by the equation using the best fit parameter value for pK_a. The pK_a value as well as the peak positions for the acid and neutral forms are given in Table II. Several important conclusions can be drawn from these data.

1) The pK_a value in the presence of vesicles (5.2) agrees with the half maximal pH of fusion (5.15). The parallelism between fusion and protonation strongly suggests induction of the fusion activity by protonation of the acidic residues in the hydrophobic peptide.

2) The pK_a value in the presence of vesicles (5.2) was higher than that in the absence of vesicles (4.8). The higher shift indicates stabilization of the protonated form in the presence of vesicles. A probable model for the stabilization, though not proved, is entrance of the neutralized hydrophobic segment into the hydrocarbon layer of the lipid bilayer (see Fig.5Bc).

3) The NMR peak for each carboxyl and carbonyl groups always consists of a single line in the pH titration. It did not split into two lines corresponding to the acid and neutral forms with change in the relative intensity on changing pH. This indicates a rapid exchange between the protonated and deprotonated forms, faster than the NMR time

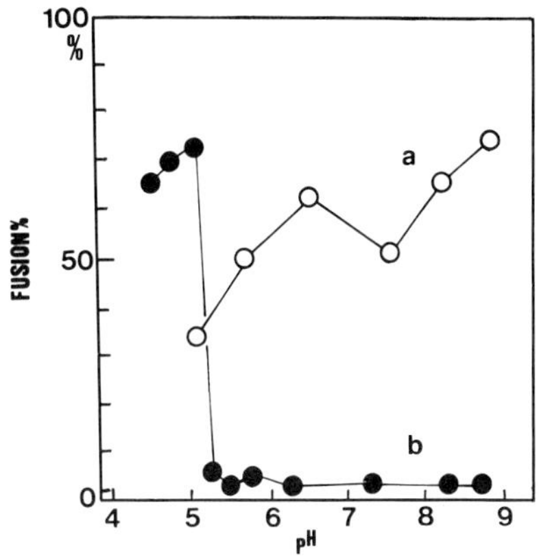

Fig.3. pH dependence of fusion of PC vesicles induced by intact (a) and succinylated (b) melittin. Fusion was assayed in a system of egg PC vesicles (3.6 μM), those containing 5% spin-labeled PC (1.2 mM), and melittin or succinylated melittin (31 μM). The degree of fusion after 10 min was estimated from the normalized ESR peak height. (Taken in part from Murata et al.,1987).

Table II. The ^{13}C NMR peak position and the pK_a value for the carboxyl group in ^{13}C-labeled succinylated melittin in the presence and absence of egg PC vesicles. The chemical shift value of the peak 6 with reference to 2,2,4,4-tetradeutero-3-(trimethylsilyl) propanesulfonic acid and the difference between the deprotonated δ_A and the protonated δ_{HA} forms $\Delta\nu$ are given.

Vesicles	Chemical shift, ppm		$\Delta\nu$, Hz	pK_a
	δ_A	δ_{HA}		
+	183.75	180.75	301.2	5.20
-	184.00	181.00	301.2	4.83

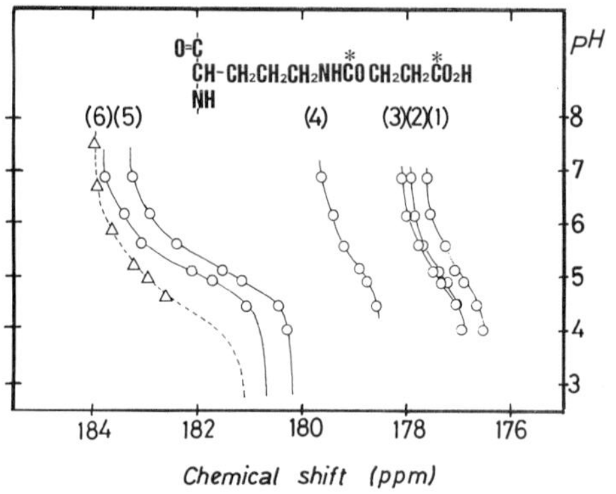

Fig.4. Change with pH in the ^{13}C NMR peak position of ^{13}C-labeled succinylated melittin in the presence of egg yolk PC vesicles (○). Data points for the peak 6 in the absence of vesicles (△) are also shown for comparison. The ^{13}C labels are on the carboxyl and carbonyl carbons indicated by asterisk in the formula. (Taken in part from Murata et al.,1987).

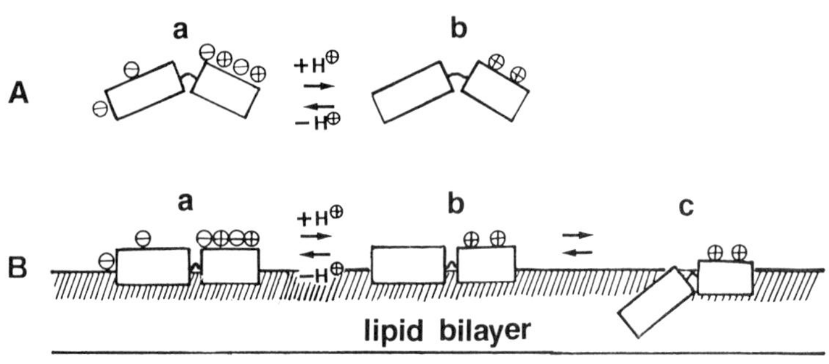

Fig.5. A model for rapid protonation and deprotonation of the carboxyl groups in succinylated melittin in the absence (A) and presence (B) of lipid bilayer membrane. In (B), the protonated hydrophobic segment is assumed to be stabilized inside the bilayer (c). The ^{13}C NMR data suggest fast exchange (faster than 3 ms) between (a) and (b) in (A), and between (a), (b), and (c) in (B). (Taken from Murata et al.,1987).

scale $(\Delta\nu)^{-1} \sim 3$ ms (see Table II). In the absence of vesicles, this simply means a rapid protonation-deprotonation reaction (Fig.5A). In the presence of vesicles, the rapid exchange should include the stabilized position for the protonated form. If the latter were inside the lipid bilayer as discussed above, the exchange of the hydrophobic segment on the surface of bilayer and the protonated hydrophobic segment inside bilayer should also be fast (Fig.5B). Interaction of the protonated hydrophobic segment with bilayer membranes may thus be dynamic. Such dynamic nature can very well explain the observed rapid switching on and off of fusion reaction on shifting the pH to acid and neutral, respectively.

FUSION ACTIVITY OF A 20 AMINO ACID PEPTIDE WITH THE SAME SEQUENCE AS THAT OF HA2 N-TERMINAL SEGMENT

This peptide (HA peptide) was synthesized by the Merrifield solid phase method and studied for the fusion activity. The peptide caused a rapid and efficient fusion at acidic pH but not at neutral pH. The fusion reaction reached a saturation level within 5 min. The fusion efficiency depended on the amount of the peptide relative to phospholipid, increasing with the increase in the relative amount. The pH dependence of fusion is shown in Fig.6. The fusion started to occur at pH 6.0 and reached a maximum at pH 5.0. This pH dependence is similar to that of the parent virus (Murata et al.,1987a).

The fusion initiated at acidic pH can be stopped immediately on shifting the pH to neutral and can be restarted again on restoring the pH, as in the case of succinylated melittin (Murata et al.,submitted).

The fusion reaction is probably induced by protonation of the acidic residues in the peptide, though not yet directly measured. In this simplified system, there are no other titratable groups both in the phospholipid and the fusogen in that pH range. Interaction of the protonated peptide with lipid bilayer can also be dynamic as in the case of succinylated melittin.

These results indicate that the putative fusogenic segment in HA has fusion activity at acidic pH induced probably by protonation of the acidic residues.

DISCUSSION

Fucion Mechanism

Our results show that hydrophobic peptides of 20 or more amino acids can cause fusion of neutral phospholipid vesicles. The presence of a few acidic residues in the peptide modifies its fusion activity to be active at acidic pH. These acidic residues are negatively charged at neutral pH but neutralized by spontaneous protonation at acidic pH because of the pK_a value of ca 5 in the presence of lipid membranes. The neutralization will facilitate entrance of the hydrophobic peptide into the hydrocarbon core of the bilayer. The charges in the peptide should be neutralized for its entrance because of a very large work required to bring a charge into hydrocaron layer. This neutralization will occur spontaneously at acidic pH but require work at neutral pH. We can estimate the work from the pK_a value, amounting ~ 16 kJ per mole residue. Therefore, the free energy for transfer of hydrophobic peptide with two such acidic residues into hydrocarbon layer would be ca 32 kJ/mole lower at acidic pH than at neutral pH (Ohnishi,1985;1987).

We do not yet know how the peptide-lipid interaction leads to fusion. The hydrophobic peptide entered into lipid bilayer would disturb the surrounding lipid molecules. A possibility is to create micelle-like structure which can mediate fusion (Yamada & Ohnishi,1986; Ohnishi,1987). Non-bilayer structures such as cylinder and inverted-micelle have been observed and suggested to be an intermediate in fusion (Cullis & Hope,1978). If the interaction were dynamic, the disturbance should also be dynamic.

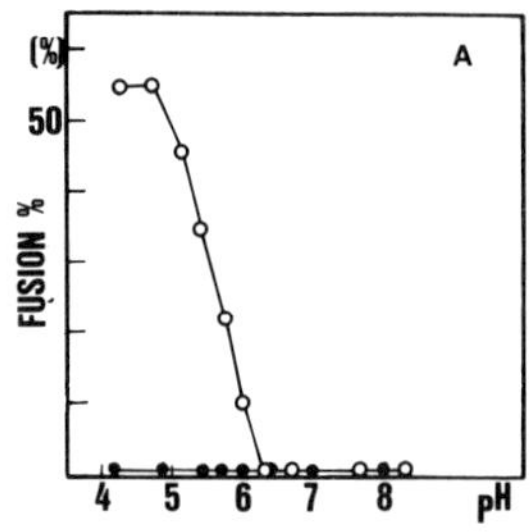

Fig.6. pH dependence of fusion of PC vesicles induced by HA peptide. Peptide (68 μM) was added to a mixture of sonicated egg PC vesicles (3 mM) and those containing 10% spin-labeled PC (1 mM) at various pH values and the ESR spectrum was measured at room temperature (23° C) (○). The fusion % was estimated from the normalized ESR peak height. (●) represents data in the absence of peptide.

Implications to virus membrane fusion mechanism

When influenza virus binds to the cell surface receptor (sialoglycoproteins or sialoglycolipids), the virus envelope and the cell membrane are still far (ca 10 nm) apart. On lowering pH, some acidic residues, His, Asp, Glu, of HA will be protonated to cause conformational change. The conformational change would make the two membranes come close and the HA2 N-terminal hydrophobic segment approach and interact with the target cell membrane lipid bilayer. Protonation of the acidic residues in the hydrophobic segment may also be required for the interaction as discussed above. Lipid bilayer domain as the target has been suggested (Bächi et al.,1973; Yoshimura et al.,1985; Tsuji & Ohnishi,1986).

Low pH for the virus fusion activation may therefore be required for the conformational change and protonation of the hydrophobic segment. Both processes occur either at the same or different pH ranges. If the pH ranges are different, then the process occurring at lower pH would determine the overall pH profile of fusion. As an example, low pH fusion activity of mouse mammary tumor virus can be explained as follows. This virus has a low pH-induced fusion activity but its fusion protein gp 36 does not contain acidic residues in the N-terminal hydrophobic segment (Table I). In this case, the conformational change of gp 36 would occur at acidic pH. Therefore, the hydrophobic segment can be exposed to approach target membranes only at acidic pH although it can interact with membranes both at neutral and acidic pH.

Recently, influenza virus HA mutants and variants have been isolated and studied for modification in the fusion activity (For example, Daniels et al,1985; Gething et al.,1986). It should be noted, however, in these mutation studies that the modified pH characteristics of fusion can be caused by change in the pH dependence of the conformational change and/or change in the pH for neutralization of the hydrophobic segment. Even mutations in the HA2 N-terminal segment can affect pH for the conformational change as well as pH for its protonation.

ACKNOWLEDGEMENTS

We are indebted to Kuniaki Nagayama, Sho Takahashi, and Yuki Sugahara for their cooperations and discussions.

REFERENCES

Bachi,T.,Aguet,M.,& Howe,C.(1973). Fusion of erythrocytes by Sendai virus studied by immuno-freeze-etching. J.Virol. 11, 1004-1012.

Cullis,P.R.,& Hope,M.J.(1978). Effects of fusogenic agent on membrane structure of erythrocyte ghosts and the mechanism of membrane fusion. Nature 271, 672-674.

Daniels,R.S.,Downie,J.C.,Hay,A.J.,Knossow,M.,Skehel,J.J.,Wang, M.L.,& Wiley,D.C.(1985). Fusion mutants of the influenza virus hemagglutinin glycoprotein. Cell 40, 431-439.

Gething,M.-J.,White,J.M.,& Waterfield,M.D.(1978). Purification of the fusion protein of Sendai virus: Analysis of NH_2-terminal sequence generated during precursor activation. Proc.Natl.Acad.Sci.US 75, 2737-2740.

Gething,M.-J.,Doms,R.W.,York,D.,& White,J.(1986). Studies on the mechanism of membrane fusion: Site-specific mutagenesis of the heamgglutinin of influenza virus. J.Cell Biol. 102, 11-23.

Hoekstra,D.,de Boer,T.,Klappe,K.,& Wilschut,J.(1984). Fluorescence method for measuring the kinetics of fusion between biological membranes. Biochemistry 23,5675-5681.

Huang,R.T.C.,Rott,R.,& Klenk,H.-D.(1981). Infleunza viruses cause hemolysis and fusion of cells. Virology 110, 243-247.

Kuroda,K.,Kawasaki,K.,& Ohnishi,S.(1985). Kinetic analysis of fusion of hemagglutinating virus of Japan with erythrocyte membrane using spin-labeled phosphatidylcholine. Biochemistry 24, 4624-4629.

Lenard,J.,& Miller,D.K.(1981). pH-dependent hemolysis by influenza, Semliki Forest virus, and Sendai virus. Virology 110, 479-482.

Maeda,T.,Asano,A.,Ohki,K.,Okada,Y.,& Ohnishi,S.(1975). A spin label study on fusion of red blood cells induced by hemagglutinating virus of Japan. Biochemistry 14,3736-3741.

Maeda,T.,& Ohnishi,S.(1980). Activation of influenza virus by acidic media causes hemolysis and fusion of erythrocytes FEBS Lett. 122, 283-287.

Murata,M., Nagayama,K., & Ohnishi,S.(1987). Membrane Fusion Activity of Succinylated Melittin Is Triggered by Protonation of Its Carboxyl Groups. Biochemistry 26, 4056-4062.

Murata,M., Sugahara,Y.,Takahasgi,S., & Ohnishi,S. pH-Dependent Membrane Fusion Activity of a Twenty Amino Acid Peptide of Influenza Virus Hemagglutinin. J.Biochem. 102, No.4 (1987a).

Ohnishi,S.(1985). Membrane fusion: Mechanism and role in cellular response to enveloped virus. in Biomolecules Electronic Aspects (Nagata,C.,Hatano,M.,Tanaka,J.,& Suzuki,H.Eds), pp.227-252, Japan Scientific Societies Press, Tokyo, & Elsevier, Amsterdam.

Ohnishi,S.(1987). Membrane Fusion Activity and Mechanism of Enveloped Viruses. in Membrane Fusion in Transport, Development, and Viral Infection (Duzgunes,N., & Bronner,F. Eds), in press, Academic Press, New York.

Ruigrok,R.W.H.,Wrigley,N.G.,Calder,L.J.,Cusack,S.,Wharton,S.A Brown,E.B.,& Skehel,J.J.(1986). Electron microscopy of low pH structure of influenza virus haemagglutinin. EMBO J. 5, 41-49.

Sato,S.B.,Kawasaki,K.,& Ohnishi,S.(1983). Hemolytic activity of influenza virus hemagglutinin glycoproteins activated in mildly acidic environments. Proc.Natl.Acad.Sci.US. 80, 3153-3157.

Skehel,J.J.,Bayley,P.M.,Brown,E.B.,Martin,S.R.,Waterfield,M.D White,J.M.,Wilson,I.A.,& Wiley,D.C.(1982). Changes in the conformation of infleunza virus haemagglutinin at the pH optimum of virus-mediated membrane fusion. Proc. Natl.Acad.Sci.US. 79, 968-972.

Tsuji,A.,& Ohnishi,S.(1986). Restriction of the lateral motion of Band 3 in the erythrocyte membrane by the cytoskeletal network: Dependence on spectrin association state. Biochemistry 25, 6133-6139.

White,J.,Kielian,M.,& Helenius,A.(1983). Membrane fusion proteins of enveloped animal viruses. Quart.Rev.Biophys. 16, 151-195.

Wilschut,J.,Duzgunes,N.,Fraley,R., & Papahadjopoulos,D.(1980). Studies on the mechanism of membrane fusion: Kinetics of calcium ion induced fusion of phosphatidylserine vesicles followed by a new assay for mixing of aqueous vesicle contents. Biochemistry 19, 6011-6021.

Wilson,I.A.,Skehel,J.J.,& Wiley,D.C.(1981). Structure of the haemagglutinin membrane glycoprotein of influenza virus at 3 A resolution. Nature 289, 366-373.

Yamada,S.,& Ohnishi,S.(1986). Vesicular stomatitis virus binds and fuses with phospholipid domain in target cell membranes. Biochemistry 25, 3703-3708.

Yoshimura,A.,Yamashina,S.,& Ohnishi,S.(1985). Mobilization and aggregation of integral membrane proteins in erythro cytes induced by interaction with influenza virus at acidic pH. Exp.Cell Res. 160, 126-137.

pH-DEPENDENT FUSION OF VESICULAR STOMATITIS VIRUS WITH CELLS: STUDIES OF MECHANISM BASED ON AN ALLOSTERIC MODEL

Robert Blumenthal, Anu Puri, Anne Walter and Ofer Eidelman

Section on Membrane Structure and Function, LMMB
National Cancer Institute, NIH
Bethesda, Md.

INTRODUCTION

Most enveloped animal viruses enter cells by receptor-mediated endocytosis through coated pits (White et al, 1983). In the endocytic vesicle rapid acidification occurs, a process which triggers fusion of the viral membrane with that of the endosome, and the nucleocapsid is ejected into the cytosol. Some viruses, such as Sendai, fuse directly with the plasma membrane at neutral pH. The viruses which enter via the endocytic pathway can be made to fuse directly with the plasma membrane by lowering the pH of the medium. In fact, a common assay for viral fusion is counting cell-cell fusion events after inducing viral-cell fusion by exposing the culture to low pH for a short time. In order to study mechanisms of viral membrane fusion it is necessary to measure the fusion of intact virions with target membranes directly. In our laboratory we are focussing on the study of mechanisms of entry of Vesicular Stomatitis Virus (VSV). The envelope of VSV consists of a bilayer membrane with a single type of spike glycoprotein, the G protein, which contains all the functions needed to induce membrane fusion (Florkiewicz and Rose, 1984). However, the structural domains responsible for those functions have not been characterized.

We have studied fusion of VSV with Vero cells (Blumenthal et al, 1987) using the fusion assay which is based on self-quenching of octadecyl rhodamine (R18) when it is inserted into the intact virus (Hoekstra et al, 1984). Relief of R18 self-quenching occurs as result of fusion of virus with target membrane (Fig.1). We examined the kinetics, extent, temperature dependence, effect of osmotic strength of the medium, effect of neutralizing antibodies and the reversibility of the fusion reaction. The quantitative real-time spectroscopic assay for viral-cell fusion enables further detailed examination of hypotheses regarding possible mechanisms of viral spike glycoprotein-mediated membrane fusion.

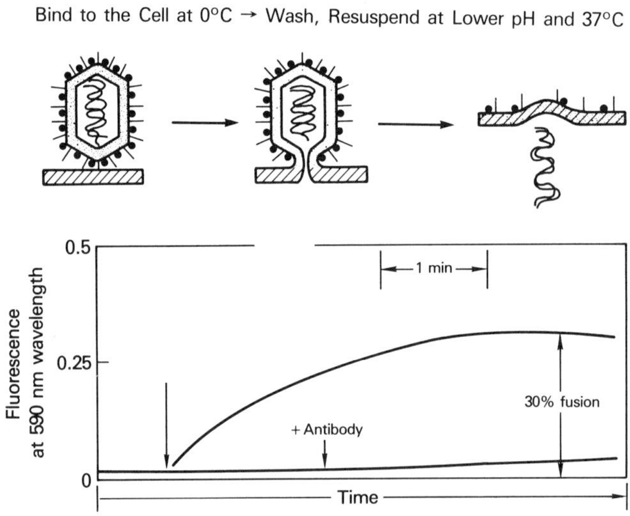

Fig. 1. Virus-cell fusion detected by fluorescence dequenching. The top cartoon is drawn to indicate the spreading of the lipid probe over a larger surface area after fusion, which is measured by fluorescence dequenching. The lipid mixing is presumably concomitant with the injection of the nucleocapsid into the cells' cytosol. The bottom cartoon is drawn to indicate the kinetic pattern. The R18VSV-Vero complex is added to the cuvette at pH 7.4 and 37°C at 0 time, and fusion is triggered after 1 min (arrow) by adding a small amount of low pH buffer to the cuvette. The figure also shows a control experiment with neutralizing antibody.

Recently, such a hypothesis was formulated (Blumenthal, 1987b) based on presumed conformational transitions and cooperative properties of viral spike glycoproteins. In this paper we will describe the model in more detail and discuss how it fits some of the experimental data we have gathered on fusion of VSV with Vero cells.

ALLOSTERIC MODEL FOR VIRAL PROTEIN-MEDIATED FUSION

Fig. 2 shows two possible models to describe how viral proteins might interact with the target membrane to induce the lipid rearrangements required for fusion. The models drawn are fanciful representations to facilitate discussion, rather than specific proposals for the mechanism of fusion. On the basis of the high protein density found in viral envelopes, it is reasonable to assume that the virus binds to the target membrane at many points. The nature of the binding might be electrostatic, specific site-site recognition, or penetration into the acyl chain region of the target membrane. As a result of the multiple binding and presumably following a protein conformational change, the two membranes could be physically deformed and thus brought into close contact. This latter event might precipitate fusion between the two membranes either by overcoming the repulsion forces (steric, electrostatic, hydration), or by inducing an excessive radius of curvature (Blumenthal, 1987a). A number of contributions to this symposium deal extensively with those forces between lipid bilayers (Evans, this volume; Rand, this volume), and with possible lipid structures as intermediates in the catastrophic rearrangements that occur when two separate bilayers are melded into a single bilayer (Cullis, this volume; Siegel, this volume; Lucy, this volume).

In the proposed allosteric model (Blumenthal, 1987b), the viral spike glycoproteins are considered as regulatory proteins involved in catalysis of fusion. Initially, the model does not deal with the specific mechanism of catalysis of fusion, but rather with consequences of ligand-protein interactions on the regulation of fusion. An analogy may be drawn to allosteric enzymes discussed by Monod, Wyman and Changeux (1965), where not the mechanism of catalysis, but the effects of regulatory ligands, conformational changes and/or cooperativity, on the functioning of the protein was considered. The allosteric model is not dependent on mechanism, but analysis of experimental observations according to the model will provide constraints on any proposed mechanism of fusion mediated by regulatory proteins.

Conformational changes and subunit interactions have been studied in viral spike glycoproteins. The best-characterized fusion protein is the hemagglutinin HA glycoprotein of influenza virus. Based on the 3-dimensional structure of the ectodomain of influenza virus and on physico-chemical studies of pH-dependent conformational changes of that

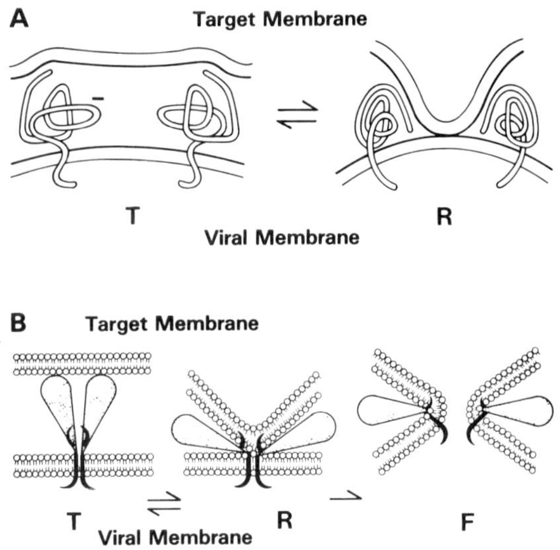

Fig. 2. Hypothetical mechanisms of membrane fusion mediated by viral fusion proteins (VSV G protein (A) and the influenza virus hemagglutinin (B)). The viral spike glycoprotein oligomers are proposed to bind with the target membrane. Ligand (H^+) binding to the oligomers shifts their conformation from the T state to the R state in such a way that the viral and target membrane are brought in close proximity and destabilized in a manner that leads to fusion (the F state). Only two of the n possible subunits of the oligomer are shown; the subunits are assumed to be linked by quarternary interactions (not shown). See text for further explanation.

domain, the following picture has emerged (Doms and Helenius, this volume): The protein is a trimeric rod-shaped molecule 135 Å in length, consisting of a stem and three globular, highly folded domains (stalks) at the top. The globular domains are composed of the HA1 chain, and since they contain the binding sites for sialic acid, they provide the crosslinking feature. The stem contains the hydrophobic HA2 N-terminal sequences implicated in fusion (fusion peptide). At low pH, the stalks open up and expose the fusion peptide, which then interacts with the target membrane, presumably destabilizing it, and thereby inducing fusion.

On the basis of what is known about the influenza virus HA protein, it seems reasonable to think of the induction of fusion in terms of an allosteric model with the following assumptions. The viral spike glycoproteins are assumed to be arranged as oligomers, consisting of a number (n) of subunits linked through quarternary interactions. In Fig. 2 only two of the n subunits are drawn as separate entities. This oligomer is capable of undergoing a "concerted" conformational change from what is conventionally known as the "tense" or T state (inactive) to the "relaxed" or R state which is active (Monod, Wyman and Changeux, 1965). Each subunit in the oligomer contains a regulatory site for a ligand (H^+) with a dissociation constant K_d. Once the protein has undergone a conformational transition from the T state to the R state, the fusion process is catalysed leading to the melding of the two linked membranes into the F state. Although all of the spike proteins of a particular virion can undergo the conformational change, only those involved in the virus-target interaction are relevant. Indeed, according to the model shown in Fig. 2, the transition of a single oligomer is considered sufficient to precipitate the fusion event.

In the experiments with the R18-labeled virus, a single fusion event presumably causes the movement of about 1500 R18 molecules from the viral membrane to the target membrane where they diffuse over a larger surface area. A single fusion event is induced by the transition of an oligomer in the virus-target complex from T to R to F (Fig. 2). The R18 probe monitors fusion in a way that each single event results in a quantal increase in fluorescence. The total fluorescence increase is equal to the fluorescence increase per fusion event times the number of virions fused. This situation is similar to ion channel opening which involves the conformational transition of channel protein oligomers (Changeux et al, 1984) resulting in movement of about 1000 ions/msec through the channel which is measured as a single channel conductance (Ehrenstein et al, 1974).

DERIVATION

We are considering an "exclusive" model, i.e. the T state does not induce fusion or bind the ligand (H^+), although it does bind to the target membrane. The equilibrium constant for the conformational change between T and R states of the oligomer is L. In the calculation L=100 (see below), i.e. in the absence of ligand the oligomer is predominantly in the T state and inactive. Binding the ligand shifts the conformational equilibrium towards the active form thereby enabling the fusion process to occur. The process is cooperative in that binding to one subunit stabilizes the active conformation of n subunits. The transitions of the model are characterized by the scheme shown in Fig. 3. The forward and backward rate constants for the conformational change are k_1 and k_{-1} respectively, the rate constant for fusion induced by the R_i state is k_{fi}, and the forward and backward rate constants for ligand (H^+) binding are k_a and k_d respectively.

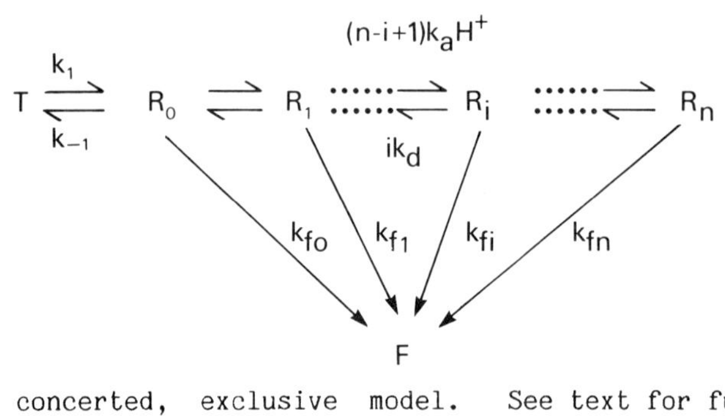

Fig. 3. A concerted, exclusive model. See text for further explanation of the symbols.

We assume that the protonation reaction is instantaneous, so that

$$(n-i+1)\ k_a H^+ R_{i-1} = i\ k_d R_i \quad (1a)$$

$$\text{or}\quad R_i = ((n-i+1)/i) R_{i-1}\ \alpha \quad (1b)$$

where $\alpha = H^+/K_d$ and $K_d = k_d/k_a$ is the dissociation constant for the protonation site.

Iteration of eqn (1b) yields the relationship:

$$R_i = \binom{n}{i} \alpha^i R_o \qquad (2)$$

The sum of all R states (S) is then given by:

$$S = \sum_{i=0}^{n} R_i = \sum_{i=0}^{n} \binom{n}{i} \alpha^i R_o = (1+\alpha)^n R_o \qquad (3)$$

By substituting for R_i according to eqns (2) and (3) we can simplify the kinetic scheme shown in Fig. 3 to:

$$T \underset{k_{-1}/(1+\alpha)^n}{\overset{k_1}{\rightleftarrows}} S \xrightarrow[\beta k_f]{} F \qquad (4)$$

If all the R states are fusion-competent ($k_{fo}=k_{f1}=\ldots k_{fi}=\ldots k_{fn}=k_f$), as the original model (Blumenthal, 1987b) then $\beta=1$. Assuming that k_f is the rate limiting step, a simple solution for rate constant for fusion was derived (Blumenthal, 1987b).

$$\text{rel. fusion rate const.} = \frac{(1+\alpha)^n}{L+(1+\alpha)^n} \qquad (5)$$

where $L=k_{-1}/k_1$. This expression is very similar to that derived by Monod, Wyman and Changeux, 1965) for a concerted model for allosteric enzymes. However, recent data on pH activation of VSV indicate that only the protonated R states are fusion-competent and that the conformational change is rate limiting (see below). As a first approximation, we assume that only R_n is fusion competent. If we set $k_{fi}=0$ for all R states except R_n ($k_{fn}=k_f$), then it can be shown that:

$$\beta = \frac{\alpha^n}{(1+\alpha)} n \qquad (6)$$

A set of 3 differential equations can be written expressing the changes in the fraction of bound oligmer in the T, S and F (fusion) states as a function of time according to the scheme shown in eqn (5). Solution of the 3 differential equations yields the following expression for % fusion as a function of time for a given α (Blumenthal, 1987b):

$$\% \text{ fusion} = 100 \left(1 + \frac{\lambda_3 e^{-\lambda_2 \tau} - \lambda_2 e^{-\lambda_3 \tau}}{\lambda_2 - \lambda_3}\right) \qquad (7)$$

where $\tau = k_1 t$ (dimensionless time). The rate constants are given by:

$$\lambda_{2,3} = (1+\kappa_1+\kappa_2)(1 \pm [1 - 4\kappa_2/(1+\kappa_1+\kappa_2)^2]^{1/2})/2 \qquad (8)$$

where

$\kappa_1 = k_{-1}/(k_1(1+\alpha)^n) = L/(1+\alpha)^n$ and

$\kappa_2 = \beta k_f/k_1$.

The % fusion can be calculated as a function of time according to eqn (7) for a given n, L, k_f/k_1 and α. To calculate the relative fusion rate as a function of ligand concentration, we set $\tau=1$ or $t=1/k_1$ in eqn (7). It can be shown that at saturation ($\alpha \to \infty$, i.e. at very low pH) 63% fusion has taken place at $\tau=1$. The relative fusion rate is then obtained by setting $\tau=1$ in eqn (8) and dividing by 63 (yielding a factor of 1.58):

$$\text{rel fusion rate const.} = 1.58 \left(1 + \frac{\lambda_3 e^{-\lambda_2} - \lambda_2 e^{-\lambda_3}}{\lambda_2 - \lambda_3}\right) \qquad (9)$$

EXPERIMENTS WITH VSV AND VERO CELLS

VSV was labeled with octadecyl rhodamine (R18) as follows: A solution of R18 in ethanol (1 mg/ml) was injected into a suspension of VSV while vortex mixing, and the mixture was held at room temperature for 15 min. Excess R18 was removed by passing the incubation mixture through a Sephadex G-75 mini column. Virus recovery was 60 -75 %. R18VSV was preincubated with Vero cells at 4°C and unbound virus was removed. Small aliquots of R18VSV-cell complexes were added to 2 ml of buffer prewarmed at 37°C. Subsequently the pH was lowered by adding 20-50 µl of a low pH buffer (see

Fig. 1). Percent fusion was calculated according to eq. (10):

$$\% \text{ FUSION} = \frac{F - F_0}{F_t - F_0} \; 100 \qquad (10)$$

F_0 and F are the fluorescence units at time 0 and at a given time point, and F_t is the fluorescence after disruption of virus and cells by detergent. F_t is divided by 1.56, a correction factor to account for the difference in quantum efficiency of R18 between plasma membrane and triton micelle (Blumenthal et al, 1987).

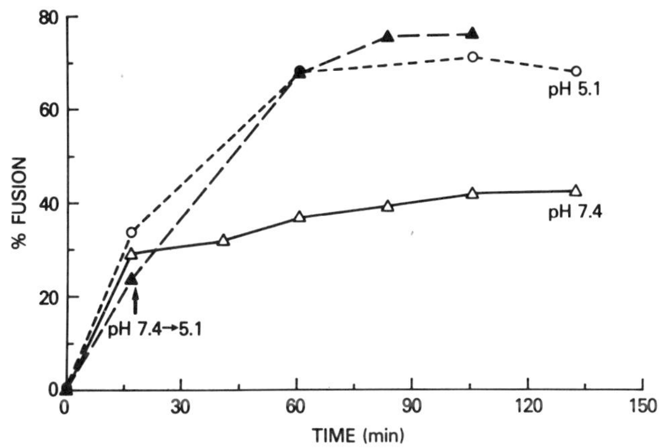

Fig. 4: Kinetics of fusion of R18-labeled VSV with Vero cells. R18VSV was preincubated with Vero cells at 4^oC, pH 7.4, for 60 min. The cells were washed by centrifugation. Fifty microliters of the R18VSV-Vero complex were injected into a prewarmed cuvette at 37^oC, pH 7.4. One hundred seconds after injection of R18 VSV-Vero, the pH in the medium was changed by adding 20-100 µl 1 M MES. Percent fusion was calculated from fluorescence dequenching values according to eqn (10). (○---○), fusion at pH 5.1 (with the plasma membrane); (△—△), fusion at pH 7.4 (with the endosome membrane after endocytosis); (▲---▲), fusion at pH 7.4 was switched to pH 5.1 after 20 min (arrow).

The time course of fluorescence change over 2 hrs is shown in Fig. 4. At pH 5.1 the fluorescence signal increased with an approximately exponential time course reaching, after 15 min, a value which corresponds to 25% fusion.

The fusion continued to increase gradually after those 15 min and reached about 75% after 2 hrs. Fig. 4 also shows a similar experiment at pH 7.4. The degree of fusion after 15 min was about the same as at pH 5.1, but the rate attenuated considerably after about 20 min. Based on the initial kinetics of R18 dequenching, effects of inhibitors, temperature dependence, and dependence on osmotic pressure, we have identified the dequenching observed at pH 7.4 as fusion of R18VSV with the endosome membrane after entry of the virus via the endocytic pathway (Blumenthal et al, 1987). The observation in Fig. 4 that fusion via the endocytic pathway (at pH 7.4) diminished after about 20 min, might be due to the fact that the cells are depleting their metabolic energy resources or even dying. However, the remaining virions which had not entered that pathway and were still on the plasma membrane could be made to fuse with the plasma membrane by lowering the pH to 5.1 (Fig. 4).

In addition to revealing the two pathways of viral entry, the R18 dequenching enabled us to make the following observations about fusion of VSV with the plasma membrane (Blumenthal et al, 1987). i) the pH profile of Vero-VSV fusion at the plasma membrane as measured by the dequenching method, paralleled that observed for VSV-induced cell-cell fusion, ii) fusion was blocked by adding neutralizing antibody to the Vero-VSV complexes, iii) pH-dependent fusion occured with fragments and with intact cells at similar rates indicating that pH-, voltage-, or osmotic gradients are not required for viral fusion, and iv) activation of the fusion process by lowering the pH is reversible, in that fusion is arrested by raising the pH back to 7.4.

These experiments enabled us to measure rates of fusion of R18VSV with Vero cells as a function of pH. We measured the fluorescence increase in the first 15 min for fusion reactions which were completed in an hour (see Fig. 4). In Fig. 5, the "rate constants" (i.e % fusion in 15 min) are plotted as a function of pH. They are normalized to the fastest "rate constant" at pH 5.0.

COMPARISON OF THE MODEL WITH EXPERIMENTAL DATA

Equation (9) contains too many adjustable parameters to fit to the few data points shown in Figure 5. However, based on our activation experiments (see below) we estimate L and k_f/k_1 to be 100 and 150 respectively. To examine the effect of cooperativity we set a value for n and adjust the K_d to get the best fit.

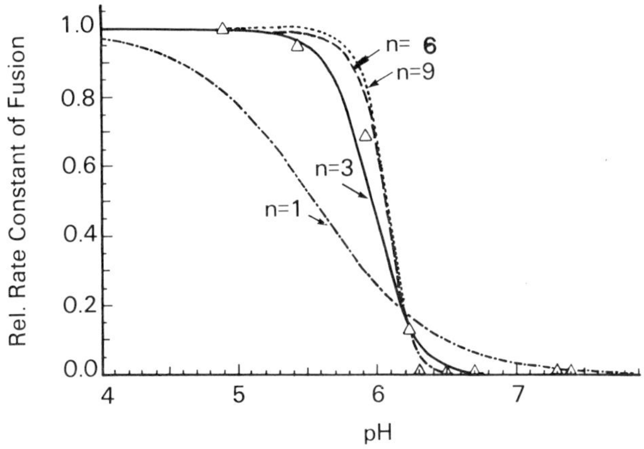

Fig 5. pH-dependent fusion of Vesicular Stomatitis virus with Vero cells. Triangles are fusion rate constants (Blumenthal et al, 1987), normalized to the fastest rate constant at pH 5. The smooth curves are generated from eqn (9) using the SAAM computer program (Berman et al, 1983) for different values of n, setting L=100 and k_f/k_1=150. To fit the curves the pK values had to be adjusted to 5.17, 5.85, 6.05, and 6.21 for n values of 1, 3, 6 and 9 respectively.

Fig. 5 shows the curves, calculated according to eqn (9), for different values of n, leaving L and k_f/k_1 as above. A non-cooperative behavior (n=1) clearly does not fit the data, while the slope at n=3 also appears to be too shallow. On the other hand, it is not possible to determine the difference between n=6 and n=9 from those data. Fitting the data to eqn (9) involved adjusting the pK of the active group. In case of n=9, the best-fitting pK value was 6.22, and for n=6 a pK of 6.05 gave the best fit. Therefore, if independent evidence were available for the apparent pK of the group involved in the conformational change, then a determination of the cooperativity of the oligomers involved in the conformational change would be feasible.

The parameter often considered in studies of viral fusion is the threshold pH for the induction of fusion. Influenza virus mutants exhibit different pH thresholds (Doms and Helenius, this volume). With VSV the threshold was different when cell fusion was observed in cells infected by VSV compared to cells transfected with the plasmid containing the cloned cDNA expressing the G protein (Florkiewicz and Rose, 1984). In the allosteric model, the threshold depends not only on the pK of the groups involved in the conformational change, but also on other factors, e.g. the initial R/T ratio or L. This is illustrated in Fig. 6.

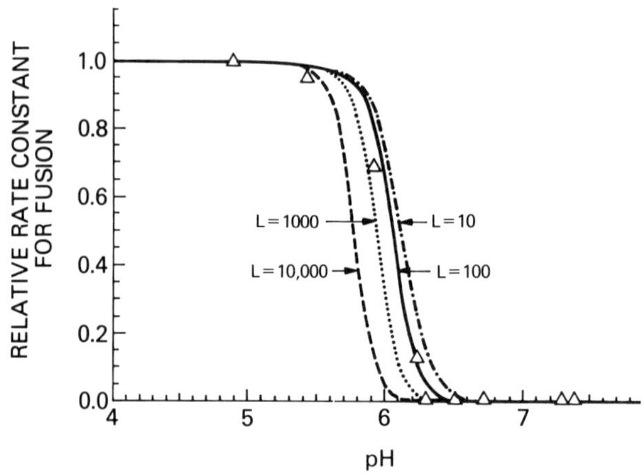

Fig. 6. pH-dependent fusion of Vesicular Stomatitis virus with cells. See legend to Fig. 5 for explanation of data points and generation of the curves. Values of k_f/k_1, n and pK were set to 150, 6 and 6.05 respectively, and curves were generated for different L's.

We fix the parameters $k_f/k_1=150$, n=6 and pK=6.05, and examine how the apparent threshold varies with L. As shown in Fig. 6, the apparent threshold moves to a lower pH with increasing L. The difference in threshold between VSV-infected and transfected cells could reflect an interaction between the G protein oligomers and another VSV protein, for example the matrix (M) protein. Perhaps the M protein affects L by interacting with allosteric sites on the G protein. Moreover, presence of the M protein could increase n by clustering the G protein.

The paucity of data points in Figs. 5 and 6 does not allow a precise determination of the parameters of the allosteric model. However, the model provides a framework for further studies on the changes in kinetics of fusion brought about by manipulating the viral spike glycoproteins into the different conformations in the absence of targets. The effect of manipulating the VSV G protein by pretreatment at low pH is presented in the next section. The results provide further constraints on the parameters, which determine the model.

ACTIVATION OF VSV G BY PRETREATMENT AT LOW pH

During a study aimed at developing methods to distinguish biologically relevant fusion mediated by VSV G protein from non-biological fusion, and to study mechanisms of viral fusion by manipulation of the viral spike glycoproteins, we preincubated virus alone at different pH values and temperatures, or pretreated virus alone with different chemical agents and enzymes (Puri et al, 1987). We then examined the effect of those preincubation conditions on rates and extents of fusion via the two pathways mentioned above. In the case of influenza virus, pretreatment at low pH causes inactivation of fusion (see Nir et al, this volume). To our surprise, we obtained completely opposite results with VSV in that fusion was enhanced after pretreatment at low pH. This observation fits very well into the framework of the allosteric model. By pretreatment in the absence of target membranes we obtained further insights into the T → R transition, and were able to estimate the relationship between the rate constants for fusion k_f and for the conformational change k_1.

Fig. 7 outlines the the protocol used to modify the virus. R18VSV was exposed to pH 5.5 at 0.2-0.6 mg/ml viral protein at a given temperature for 0.25-2.0 hr. We found that viral aggregates were formed by this treatment; they were removed by low speed centrifugation. The R18VSV in the supernatant was then incubated for 30 min with Vero cells at pH 7.4 and 4^oC. After removal of unbound virus, Vero-VSV complexes were suspended in buffer at pH 7.4. After 1 min, the pH was lowered by addition of acid and fusion was followed by fluorescence dequenching.

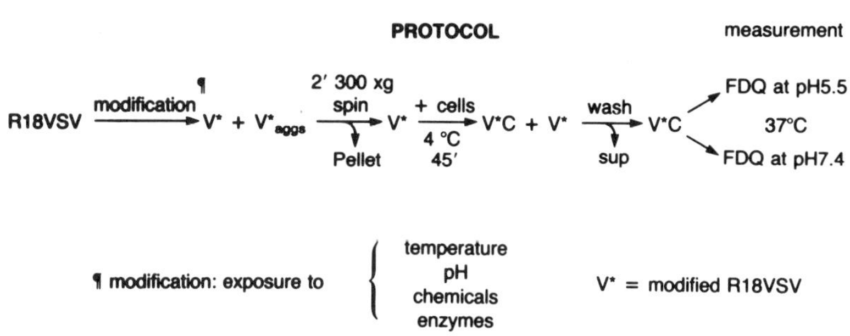

Fig. 7. Protocol for modification of VSV. See text for further explanation.

Fig. 8 shows that fusion of pretreated virus initially proceeds at a much faster rate than that of the untreated virus. After about 400 sec the fast process levels off to about 40 % fusion. Subsequently, fusion proceeds at a much slower rate to about 100 % (data not shown).

The pattern of fusion of pH-activated virus can be interpreted in the framework of the allosteric model: Preincubation at low pH in the absence of target causes the transition to the protonated R state. When the pH is brought back to 7.4 at $4^{o}C$, a portion of the viral spike protein oligomers will be in the T state and the remainder will be in the R_o (unprotonated) state. Binding to target membranes at pH 7.4 and $4^{o}C$ will "freeze" the oligomers in the T $\leftrightarrow$ R equilibrium states, which were initially produced by pretreatment at low pH in the absence of targets. Consistent with this hypothesis is the observation that preincubation of VSV -Vero complexes at pH 5.5 and $4^{o}C$ does not result in subsequent enhancement of the fusion rate at $37^{o}C$ (data not shown). This indicates that the conformational transitions of the oligomer in the bound state are very slow at $4^{o}C$. To examine whether the oligomers can be brought back to the T state in the absence of target membranes, we incubated activated virus at pH 7.4 and $4^{o}C$ without target membranes and monitored the subsequent fusion rate after incubation with cells. Those incubation conditions resulted in the same fusion rate as that of untreated virus (data not shown). This indicates that the activation process can be reversed, i.e. the oligomers can be brought back to the T state in the absence of target membranes.

Since its oligomers are predominantly in the T state, untreated virus does not fuse with the plasma membrane at pH 7.4 and $37^{o}C$. However, neither did the low pH-treated virus fuse with the plasma membrane at $37^{o}C$ and pH 7.4. Since a large portion of oligomers of the low pH-treated virus are in the R_o state, this observation indicates that the R_o form is inactive (k_{fo}=0). The pretreated virus only fused by lowering the pH of the VSV-Vero complex (Fig. 8), indicating that only the protonated R state is fusion competent. Since a portion of the oligomers had already been brought to the R state by the pretreatment, the only kinetic barrier to fusion is the R $\rightarrow$ F transition, which has a rate constant k_f. The slower rate seen after the R state has been depleted, is determined by the conformational transition rate (k_1). The ratio between k_f and k_1, estimated from the slopes of the fast and slow rates, is about 150. We use that number in our simulations of the model (see preceeding section). The high k_f/k_1 ratio indicates that the conformational transition is rate limiting for the oligomer in the absence of pre-treatment.

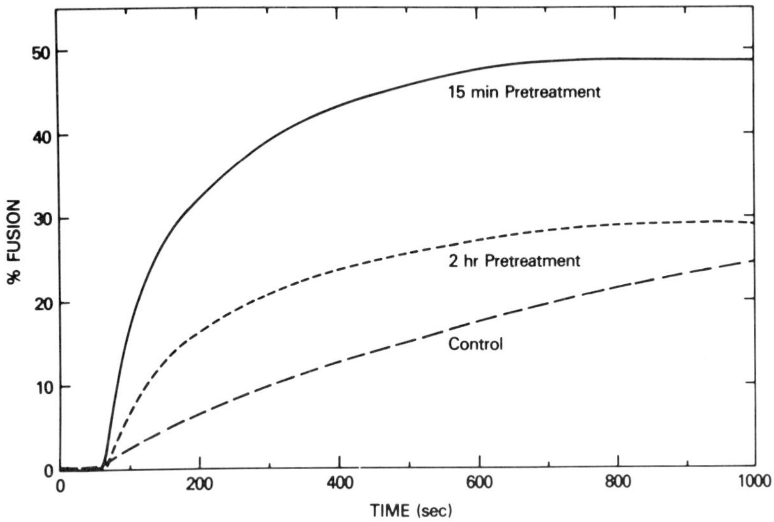

Fig. 8: Fusion of VSV with Vero cells after pretreatment at low pH. R18VSV was exposed to pH 5.5 for 0.25-2.0 hr and larger aggregates were removed by low speed centrifugation. The supernatant was bound to the cells and kinetics of fusion was followed. Fusion with Vero cells was measured at pH 5.2 after pretreatment of R18VSV at pH 5.5 for 15 min (——), or 120 min (·----·). The control R18VSV was pretreated at pH 7.4 for 15 min (– – –).

According to the model, the preincubation only changed the initial conditions, not the intrinsic rate constants. The ratio between the amplitudes of the fast and slow components are a measure of the initial $T \leftrightarrow R$ equilibrium. Since the amplitude of the fast component in the untreated virus is less than 1 % of the amplitude of the slow component, we estimate the equilibrium constant L to be 100 in our calculations (see preceding section).

We had previously shown that the process of triggering fusion is reversible in that fusion activated at pH 5.5 can be arrested by reversing the pH to 7.4. Since the conformational change is rate-limiting, pH neutralization arrests fusion by increasing the rate of the back reaction of the conformational change. Moreover, those oligomers which had been committed to the R state will not induce fusion at neutral pH, since, according to the model, fusion requires both the conformational change <u>and</u> the low pH (i.e. protonation of the R state).

Our observation that VSV fusion is activated by low pH treatment is different from that seen with influenza virus, which is inactivated by low pH treatment (Nir et al, this volume). Assuming that the inactivation of influenza virus is not due to aggregation of virus, we are led to propose another, desensitized, state in the allosteric model in the case of the HA protein. Such a desensitized state has also been proposed in an allosteric model for the acetylcholine receptor (Changeux et al, 1984). According to Nir et al (this volume) the desensitized state of HA involves self-aggregation of HA proteins, which have exposed their hydrophobic portion. VSV does not have any obvious long hydrophobic stretches in its extracytoplasmic non-membrane spanning sequence, and therefore might not be denesitized that easily. However, the observation in Fig. 8, that the amplitude of the activated state is lower after pretreatment for 2 hrs than for 15 min might indicate that the VSV G protein oligomers undergo a transition to a desensitized state after very long times of incubation at the low pH.

CONCLUSIONS

With the accurate means of monitoring fusion between intact virions and cells we are able to test hypotheses regarding the role of viral glycoproteins in mediating membrane fusion at the molecular level. We have proposed the allosteric model as a working hypothesis to analyse data on fusion mediated by viral spike glycoproteins. Our data on activation of VSV spike glycoproteins by pretreatment at low pH reveals two aspects of the model which were not apparent in the original formulation (Blumenthal, 1987b): 1) only the protonated R state is fusion competent, and 2) the conformational transition is the rate-limiting step in the fusion process.

Although fusion proteins involved in exocytosis have not been characterized, we propose that they are allosteric proteins similar to viral spike glycoproteins, with regulatory sites for Ca^{2+} rather than for H^{+}. However, other regulatory sites appear to be sensitive to GTP analogues (Barrowman et al, 1986). It will be very interesting to analyse data on Ca^{2+} and GTP-dependence of exocytosis in terms of this model.

REFERENCES

Barrowman, M.M., Cockcroft, S., and Gomperts, B.D., 1986 Nature 319: 504

Berman, M., Beltz, W.F., Greif, P.C., Chabay, R. and Boston, R.C, 1983, "Consam User's Guide", National Institutes of Health, Bethesda, Md.

Blumenthal, R. (1987a) Current Topics in Membranes and Transport 29: 203-254.

Blumenthal, R., 1987b, Cell Biophys. in press

Blumenthal, R., Bali-Puri, A., Walter, A., Covell, D. and Eidelman O., 1987, J. Biol. Chem. in press

Changeux, J-P., Devillers-Thiery, A. and Chemoulli, P , 1984, Science 225: 1335

Ehrenstein, G., Blumenthal, R., Latorre, R., and Lecar, H., 1974, J. Gen Physiol 63: 707.

Florkiewicz, R.Z. and Rose, J.K., 1984, Science 225: 721.

Hoekstra, D., de Boer, T., Klappe, K. and Wilschut, J., 1984, Biochemistry 23: 5675

Monod, J., Wyman, J., and Changeux, J-P. 1965, J. Mol. Biol. 12: 88

Puri, A., Eidelman, O., Walter, A. and Blumenthal, R., 1987, in preparation.

White, J., Kielian, M. and Helenius, A. 1983, Q. Rev. Biophys. 16: 151.

PROPERTIES OF A VIRAL FUSION PROTEIN

Robert W. Doms and Ari Helenius

Department of Cell Biology
Yale University School of Medicine
333 Cedar Street, New Haven
CT 06510

Abstract The mechanisms by which proteins mediate biological membrane fusion are not clearly understood. We have analyzed acid-induced changes in influenza hemagglutinin (HA), a viral membrane fusion factor, and correlated them with its acid-triggered fusion activity. Our results, some of which are reported here for the first time, indicate that the extramembraneous domain of HA acquires amphiphilic properties when converted to the fusion active form by mild low pH treatment. We show that each HA subunit acquires the ability to bind approximately 10 additional molecules of Triton X-100. A fusion inactive precursor of HA, termed HAO, does not acquire this property. The trimeric ectodomain fragment of HA tends, moreover, to dissociate after acid treatment into dimers and trimers. A model for the acid-induced changes in HA and the mechanism of fusion are discussed.

INTRODUCTION

Viral membrane fusion proteins mediate the fusion of the viral membrane with a host cell target membrane during virus entry (see White et al. 1983). The means by which they accomplish this is of considerable interest for understanding viral pathology and protein-mediated membrane fusion in general. The fusion factors are, as a rule, glycoproteins associated with the viral membrane via hydrophobic transmembraneous anchor peptides. A large, polar, glycosylated domain (the ectodomain) protrudes from the viral envelope forming a spike projection. A smaller polar domain (the endodomain, usually C-terminal) is located inside the internal virus compartment. All fusion proteins analyzed in detail so far have proven to be oligomeric, often consisting of homotrimers (Wiley et al. 1977, Fuller 1987, Vogel et al. 1986, Doms et al. 1987).

Of the viral fusion proteins studied, the hemagglutinin (HA) of influenza virus is the best characterized. HA is a homotrimer with a total molecular weight of 224kD (Wilson et al. 1981). Each subunit consists of two disulfide-linked glycopolypeptides, HA1 and HA2, of which the latter constitutes the transmembraneous subunit. HA1, which is entirely outside the membrane, is responsible for binding the virus to sialic acid residues on the surface of the host cell. The ectodomain of HA can be released from the viral membrane in its entirety by bromelain digestion. The resulting trimeric, water-soluble fragment, termed BHA, has been crystallized and its structure determined to high resolution by X-ray studies (Brand and Skehel 1972; Wilson et al. 1981). The structure shows that the HA1 glycopolypeptides form three independent globular domains at the distal end of the molecule; the tip of the spike. Each HA1 contains the four major antigenic epitopes to which virus neutralizing antibodies are directed as well as the sialic acid binding site which mediates attachment of influenza to the plasma membrane of the host cell (Wiley et al. 1981). HA2, located in the stem of the spike, has two noteworthy structural features; a 76Å long alpha helix which forms a coiled-coil with the helices of the two adjoining subunits (many of the non-covalent interactions which stabilize the trimer are localized on or close to this domain), and a hydrophobic amino terminus implicated in viral infectivity and in membrane fusion.

After attachment to the plasma membrane via the sialic acid binding sites in HA1 the virus is internalized by receptor-mediated endocytosis (Matlin et al. 1981) and delivered to endosomes. In these mildly acidic prelysosomal organelles the protein undergoes an acid induced, irreversible conformational change. The low pH form of HA catalyzes fusion between the viral and endosomal membranes (for reviews see White et al. 1983, Yoshimura and Ohnishi, 1984; Doms et al. 1987). The viral nucleocapsid thereby gains entry to the host cell cytoplasm and eventually the nucleus where replication occurs (Herz et al. 1981).

The significance of the HA2 amino terminus was first noted in 1976 when Klenk and co-workers observed that HA is initially synthesized as a single polypeptide precursor, HAO, which is proteolytically processed to HA at a point late in biosynthesis. They noted that virions containing the precursor became infectious only after the activating cleavage (Klenk et al. 1975). The amino terminus of HA2 created by the cleavage is hydrophobic and highly conserved. In addition, it bears homology with the amino terminus of the Sendai F protein (Gething et al. 1978). These observations, plus the fact that site directed mutagenesis of the amino terminus has deleterious effects on fusion activity (Gething et al. 1986), strongly indicate that this region plays a critical role in membrane fusion. We have suggested that it interacts hydrophobically with the target membrane and helps to bring the two membranes into close proximity (Doms et al. 1985, Doms et al. 1987).

In this paper we describe some previously unpublished experiments on the hydrophobic moiety in the ectodomain of HA and HA0, and the dissociation of HA's ectodomain in acid. In the discussion, these data and other information obtained on the HA, are used to update and review our model for the acid induced conformational change and fusion activity.

MATERIAL AND METHODS

Preparation of virus, BHA, and anchor minus HA and HAO The X:31 strain of influenza virus was propagated in embryonated eggs and isolated as described (Doms et al. 1985). Preparation of the water soluble ectodomain of HA, termed BHA, has also been described (Doms et al. 1985). Briefly, virus was digested with bromelain overnight in a reducing environment. Viral cores were pelleted by centrifugation and the BHA containing supernatant concentrated by Amicon filtration. BHA was then iodinated by the method of Fracker and Speck (1978) and further purified on a continuous sucrose gradient as described below.

Anchor minus HAO was collected from the media of CV-1 cells transfected with an SV40 late replacement vector containing a truncated HA gene lacking the sequences encoding the transmembrane and cytoplasmic domains (Gething and Sambrook, 1982). Anchor minus HA0 (A^-HAO) labeled with ^{35}S-methionine was isolated by ricin affinity chromatography and sucrose density gradient centrifugation basically as described by Boulay et al. (1987). Anchor minus HA (A^-HA) was obtained from A^-HAO by trypsin digestion also as described (Boulay et al. 1987).

Liposome binding Liposomes were prepared from phosphatidylcholine, phosphatidylethanolamine, sphingomyelin, cholesterol, and phosphatidic acid in molar ratios of 1:1:1:1.5:0.3 after the method of White and Helenius (1980). BHA, A^-HA, and A^-HAO were incubated with liposomes for 30 minutes, 37°C at the indicated pH, reneutralized, and the liposomes separated from unbound protein by bouyant density centrifugation as described by Doms et al. (1985).

Velocity gradient sedimentation BHA in 300ul of 100mM NaCl, 20mM MES, 30mM Tris (MNT buffer) containing 0.1% TX100 was placed atop a continuous 5%-25%

sucrose gradient (w/v) also in MNT/0.1% TX100 buffer. The samples were centrifuged in a SW40 rotor for 16 hours at 40,000 RPM, 20°C. The gradients were fractionated from the bottom, and the amount of ^{125}I or ^{35}S activity in each fraction determined. Sucrose concentrations were determined by using a refractometer. Proteins with known $s_{20,w}$ values were centrifuged in parallel gradients, and their positions determined by protein assay. The $s_{20,w}$ values of BHA and HA were determined by the methods of both McEwen (1967) and Martin and Ames (1961).

Sedimentation equilibrium centrifugation. Sedimentation equilibrium centrifugation was carried out in a Beckman Airfuge according to a modification of the procedure first described by Bothwell et al. (1978). ^{125}I-BHA in a volume of 120ul was centrifuged in 220ul cellulose propionate tubes of the A-100/30 rotor at 4°C for 22 hours. Speed was monitored at least 6 times during the course of each run with a General Radio Co. Type 1513-A Stroboscope. The solution contained 0.1% TX-100, 100mM NaCl, 5mM $NaPO_4$ and the pH was 7.4. BSA (5mg/ml) was included to reduce convective mixing upon deceleration. Solution densities were varied by including varying proportions of $D_2{}^{18}O$, D_2O and H_2O. By employing a variety of densities it is possible to determine the molecular weight of an unknown particle with increased accuracy, both because large particles may be, in effect, "floated" into the range where the method is most accurate, and because the partial specific volume may be determined directly. Further, the contributions of bound detergent to particle mass may be blanked by estimating the molecular weight at a solution density which matches the bouyant density of the detergent being used (Reynolds and Tanford, 1976).

Following centrifugation, the fractional concentration of ^{125}I-BHA from the top 40ul of the tube was determined. The log of fractional concentration varies as a linear function of a protein's reduced molecular weight:

$$\frac{Mr(1-\bar{v}D)}{2RT} \times w^2$$

where Mr is the true molecular weight, w^2 is the rotational velocity, $\bar{v}$ is the partial specific volume, D is the solution density and R and T have their usual meanings. The reduced molecular weight of BHA was computed by comparison with an empirically derived standard curve prepared for the A-100/30 rotor with well characterized reference proteins. The partial specific volume of TX-100 was taken to be 0.99 (Meunier et al. 1972).

Determination of TX100 binding. The amount of TX100 bound to BHA following acid treatment was determined by a modification of the technique of Simons et al. (1973). BHA (300ug protien) was incubated at pH 4.8 for 30 minutes, 37°C in MNT buffer with 0.1% ^{3}H-TX100. The solution was neutralized, then passed over a ricin affinity column. The column was washed with MNT buffer containing ^{3}H-TX100. The protein was eluted with 0.2M galactose in the same buffer. Fractions were monitored for both ^{3}H-TX100 and for protein by the method of Bradford (1972). An aliquot of the peak fraction was hydrolyzed and subjected to amino acid analysis to obtain an accurate protein concentration. The amount of detergent bound was determined by subtracting the base line ^{3}H value from the peak protein fractions.

RESULTS

The influenza HA undergoes an irreversible conformational change under the mildly acidic conditions which are required for influenza-membrane fusion. Although incapable of catalyzing fusion itself, the isolated ectodomain of HA (BHA) has been quite useful as a tool in the analysis of the conformational change. In previous studies Skehel et al. (1982) and Doms et al. (1985) have demonstrated that BHA, which in its neutral pH form is water-soluble, becomes amphiphilic when exposed to pH < 5.3. It binds to liposomes and to nonionic detergents, and in the absence of other amphiphiles it aggregates to form "protein micelles" (see Fig.1). Thus, its solubility properties resemble those of integral membrane proteins after acid treatment.

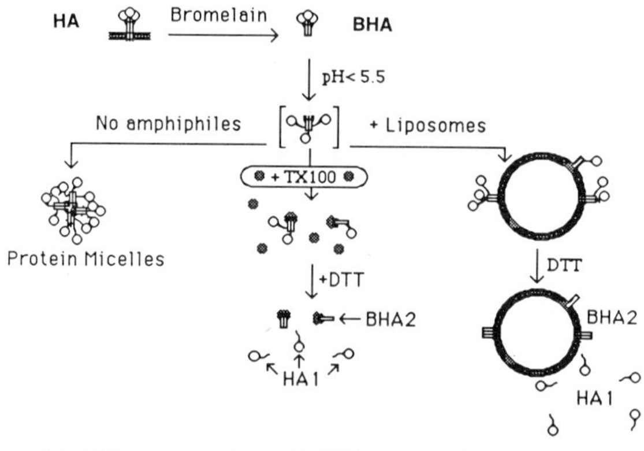

Figure 1 Amphiphilic properties of BHA. BHA is generated from HA by bromelain digestion. Acid treatment induces a conformational change in BHA which exposes a previously hidden hydrophobic moiety. In the absence of amphiphiles such as lipids or detergents, the acid BHA rapidly aggregates to protein micelles (Skehel et al. 1982). In TX100 solution it associates with detergent, and if liposomes are present it associates with them (Skehel et al. 1982, Doms et al. 1985). Partial dissociation of the trimers may take place (Nestorowicz et a. 1985, Doms and Helenius 1986). Reduction of the acid form of BHA with dithiothreitol (DTT) leads to the dissociation of HA1 and BHA2 (Graves et al. 1983). After reduction the HA2 chains remain associated with the detergent and liposomes (Doms et al. 1985).

In this study we have extended our analysis to include two other soluble, anchor-free forms of the hemagglutinin to further analyze the responses of the isolated ectodomain to acidity and to define differences between HA and HA0 which could explain why one is fusion active and the other not . The additional anchor-free forms were: **1)** *Anchor minus HA0* (A^-HA0), which was obtained by expressing a cDNA clone of HA which lacked the sequences encoding the transmembrane and endodomains (Gething and Sambrook, 1982). A late replacement SV40 vector was used. **2)** *Anchor minus HA* (A^-HA) was derived from A^-HA0 by trypsin activation.

Like HA0, the A^- HA0 is synthesized in the endoplasmic reticulum. However, since it lacks a membrane anchor, it is subsequently secreted into the medium as a soluble protein (Gething and Sambrook 1982). Sucrose velocity gradient centrifugation showed that when A^-HA0 was produced from a vector containing the HA gene of the Japan strain of influenza, 70% of the secreted A^-HA0 was monomeric ($4S_{20,W}$) and 30% trimeric ($9S_{20,W}$). In the X-31 strain we have previously observed a lower fraction of trimers (Copeland et al. 1986). The trimeric Japan A^-HA0 recovered from the sucrose gradients could, like authentic HA0, be activated to A^-HA by trypsin digestion after which it was functionally indistinguishable from BHA.

Liposome binding We have previously shown that BHA binds to liposomes of different composition in a pH dependent manner (Doms et al. 1985). To determine whether ^{35}S-methionine labeled A^-HA or A^-HAO behave similarly, we incubated them with liposomes containing phosphatidylcholine, sphingomyelin, phosphatidylethanolamine, cholesterol and phosphatidic acid (1:1:1:1.5:0.3) for 30 minutes in different pH buffers. The samples were reneutralized by the addition of base, made 40% w/w with sucrose and subjected to flotation in bouyant sucrose density gradients (Doms et al. 1985). As shown in Figure 2, it was found that A^-HA bound to liposomes with the same pH dependence as BHA prepared from isolated virus. Binding could not be reversed by the addition of high salt or chaotropic agents, which suggested that attachment occurred by hydrophobic interactions. Thus, A^-HA proved to accurately reflect the properties of BHA.

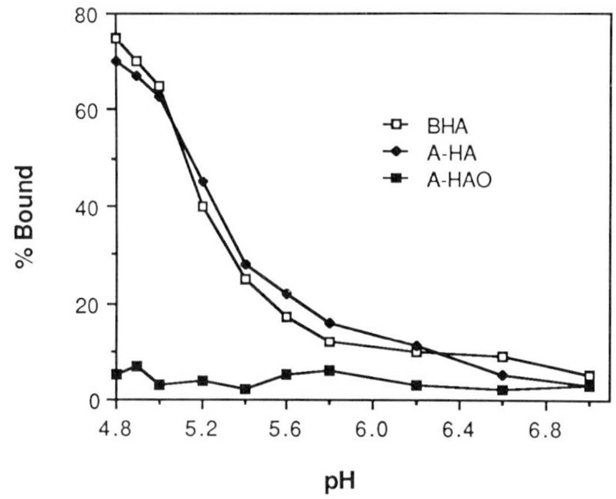

Figure 2 Binding of BHA, A$^-$HA and A$^-$HAO to liposomes. Protein was incubated with liposomes for 30 minutes, 37°C at the indicated pH. The amount of liposome associated protein was determined by a simple flotation assay as described in the Methods section.

A$^-$HAO, on the other hand, failed to bind liposomes under any of the conditions tested (Fig. 2). The reason for this was not that the protein failed to undergo an acid dependent change (it became for instance totally digestible with trypsin after acid treatment) but rather that it did not expose a hydrophobic moiety that could mediate the attachment (see also Boulay et al. 1987).

Detergent partitioning. Integral membrane proteins form complexes with nonionic detergent (Helenius and Simons, 1975). As a consequence they tend to partition into the detergent phase during Triton X-114 phase separation (Bordier 1981). We incubated BHA, A$^-$HA and A$^-$HAO at pH 7 or pH 5.0 prior to TX114 phase separation. In agreement with their water solubility, the neutral pH forms of BHA, A$^-$HA and A$^-$HAO were recovered in the aqueous phase (Figure 3). After acid treatment, however, A$^-$HAO remained in the aqeuous phase while A$^-$HA was recovered in the detergent phase. This result confirmed that A$^-$HA becomes amphiphilic when acidified and A$^-$HAO does not. The difference between A$^-$HA and A$^-$ HA0 confirmed that the HA0 does not expose a hydrophobic moiety when exposed to acid pH. This probably explains why it is not fusion competent. We have shown elsewhere that the main difference between HA and HA0 is not in pH sensitivity as such; they both undergo irreversible changes at acid pH (Boulay et al. 1987).

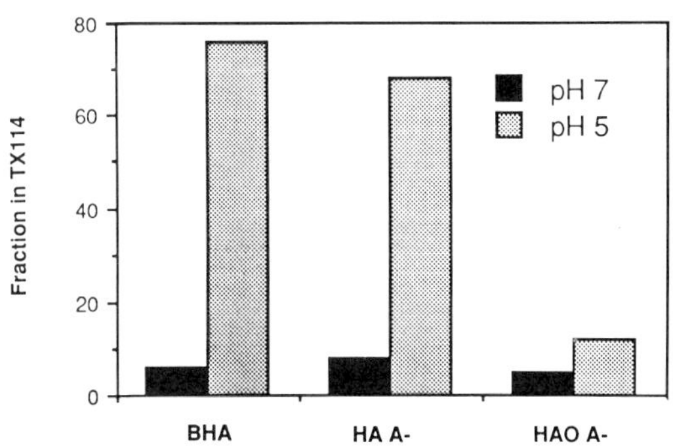

Figure 3 Triton X114 partitioning. BHA, A$^-$HA and A$^-$HA0 were incubated at pH 7 or pH 5 prior to phase separation in TX114 (Bordier 1981). The fraction of the protein recovered in the detergent phase is indicated.

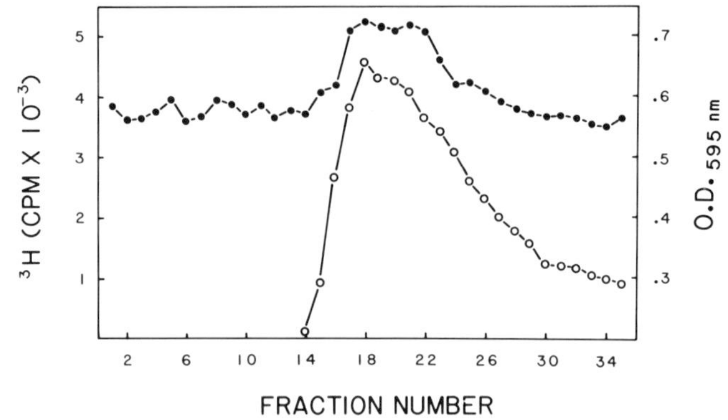

Figure 4 Binding of Triton X-100 to acid BHA. Acid treated BHA was bound to a ricin affinity column which was equilibrated with ^{3}H-TX100. The BHA was eluted with 0.2M galactose/^{3}H-TX100 and aliquots of each fraction taken for protein assay. The amount of detergent bound was determined by subtracting the base line ^{3}H value from the peak protein fractions.

Detergent binding to acid treated BHA While previous studies and the results above indicated that acidified A⁻HA and BHA bound nonionic detergent (Skehel et al. 1982, Doms et al. 1985, Boulay et al. 1987), the actual amount of detergent bound was not determined. Integral membrane proteins usually bind about 70-120 molecules of TX100, i.e. the equivalent of about one micelle (Helenius and Simons 1975).

To quantitate TX100 binding, 0.5 mg of BHA was mixed with 0.1% ^{3}H-TX100 for 30 minutes at pH 4.8. The solution was neutralized by the addition of base and passed over a ricin-Sepharose column. The column was equilibrated with ^{3}H-TX100. The BHA, bound to the lectin by its terminal galactose residues, was eluted in 0.2M galactose with 0.1% ^{3}H-TX100. As shown in Figure 4, a peak of ^{3}H-TX100 eluted with the protein. The exact amount of protein in the peak fractions was determined by amino acid analysis and the amount of ^{3}H-TX100 estimated from the radioactivity above the 0.1% background. We found that 0.09mg of TX100 was bound per mg of acidified BHA which is equivalent to 30 detergent molecules per BHA trimer. The partial specific volume of the BHA-TX100 complexes calculated from these values was 0.75. The binding values were somewhat variable in the three determinations preformed, but, in every instance, there was less bound TX100 than usually observed for integral membrane proteins.

The relatively small amount of TX100 bound suggested that the binding site(s) for detergent were either small or that they were relatively inaccessible. Given that the complexes between detergents and amphiphilic proteins usually have the character of mixed micelles, and that the size of the intramembraneous portion therefore does not directly correlate with the amount of detergent binding, we favor the second option. It is likely that the presence of obstructing protein elements close to the hydrophobic moiety could sterically prevent the formation of a larger micellar detergent accumulation with a bulky surface layer of hydrated polyoxyethylene.

Sedimentation analysis of acid BHA We found, as have others (Skehel et al. 1982, Nesterowicz et al. 1985), that the neutral form of BHA sediments as a homogenous particle with $s_{20,w}$ = 8.9 (Fig. 5a), and that it dissociates after acid treatment (Nestrowicz et al. 1985). The acidified BHA was recovered in either two (7.7S and 4.3S) or three peaks (11S, 7.7S and 4.3S) depending on the concentration of sample and the pH used for acidification (Fig. 5b,c). All three were observed when relatively large amounts of BHA (20 ug/ml) were acidifed to pH 4.8 and applied to the gradients (Figure 5b and Doms and Helenius 1986). The three peaks contained intact HA1 and BHA2 polypeptide chains as judged by SDS-PAGE. When trace amounts of ^{125}I-BHA were used, only the 7.7S and 4.3S particles were present (Fig. 5c). The relative distribution of radioactivity in the three peaks as a

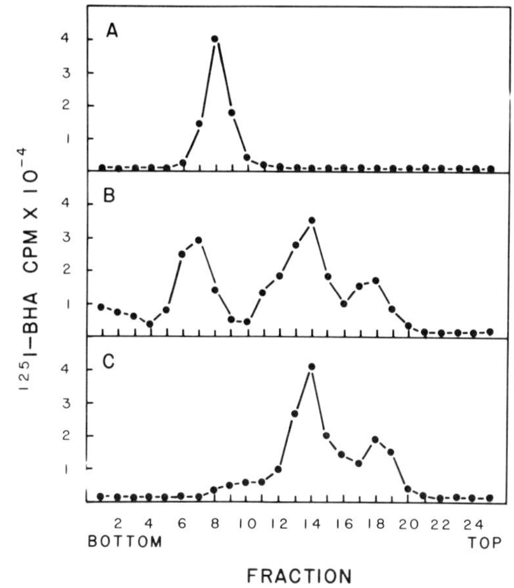

Figure 5 Sedimentation of BHA at pH 7.0 and pH 4.8. BHA was centrifuged at pH 7.0 (A) or after acid treatment for 15 minutes (B). The concentration of BHA was 20ug/ml protein. The protein was neutralized and centrifuged on a continuous sucrose gradient containing TX100.

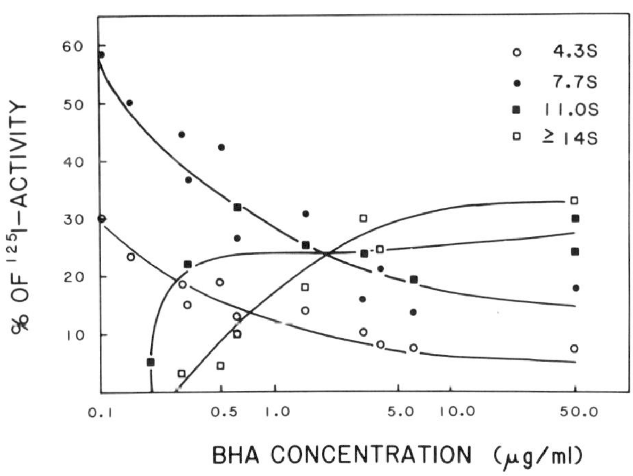

Figure 6 Effects of protein concentration on BHA dissociation. BHA at the indicated concentration was incubated at pH 4.8 for 15 minutes at 37°C, neutralized, and centrifuged as above. Following fractionation, the 11S, 7.7S, and 4.3S peaks were identified, and the fraction of BHA in each determined. The proportion of BHA sedimenting to the bottom of the tube was also determined.

function of BHA concentration is shown in Figure 6. It is seen that the degree of dissociation is highly concentration dependent. The sedimentation of BHA was the same in several types of non-ionic detergents. It was not affected by the inclusion of 0.5M salt in the gradients or by reneutralization after the acid treatment.

The pattern of dissociation did not correspond to that observed by Nestorowicz et al. (1985) who only observed one dissociation product. In our hands the BHA gave two dissociated forms. The ratio between the forms was highly sensitive to the conditions used and the protein concentration. We were able to reproduce a result similar to that of Nestorowicz et al. (1985) when the pH of acidification was 4.8 instead of 5.2.

Sedimentation equilibrium analysis on BHA Given that all three peaks contained intact, nondissociated HA1 and BHA2 and that the 11S material corresponded to the sedimentation of trimeric BHA, we suspected that the 7.7 S peak represented dimers and the 4.3S peak monomers. To obtain more definite molecular weight information, we performed sedimentation equilibrium analysis on isolated 7.7S particles in a Beckman airfuge (Agnew, W.S. in preparation). Aliquots of the 7.7S BHA were mixed with various amounts of H_2O, D_2O, and $D_2{}^{18}O$ and centrifuged in the airfuge until equilibrium was reached. By measuring the fractional depletion of the protein from the upper third of the gradient, the molecular weight for the complete complex was computed to be 115kD ($\pm$25kD). We assumed that aproximately 13kD was due to bound detergent, giving a value for the protein moiety of 102kD. This value falls between those expected for monomeric (69kD) and dimeric (138kD) BHA. It was, however, likely that the 7.7S complexes had partially dissociated during the long centrifugation. When aliquots of the protein recovered after centrifugation were re-centrifuged on a sucrose density gradient, approximately half of the BHA sedimented in the 7.7S peak and half in the 4.3S peak (Figure 7). In light of this result, we interpreted the102kD molecular weight estimate as approximating an average molecular weight of BHA in the mixture. This indicated that the 4.3S particles were, indeed, BHA monomers and the 7.7S particles dimers.

The main conclusion from these data, and the results of Nestorowicz et al.(1985), is that BHA tends to dissociate at acid pH to monomers and dimers. We have previously shown that trimers of intact HA and HA0 are, in contrast, stably associated under the same conditions (Doms and Helenius 1986). The hydrophobic C-terminal membrane anchors, which are lacking in BHA, thus appear to confer stability to HA. They apparently help keep the trimer together even under conditions that open up the ectodomain. The difference in stability of HA and BHA trimers can also be demonstrated using sodium dodecyl sulfate; HA is only partially dissociated by this denaturing detergent (Doms and Helenius 1986) whereas BHA is fully dissociated (not shown).

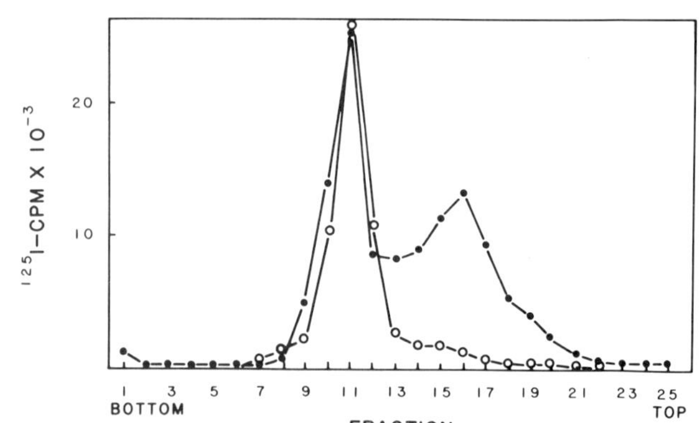

Figure 7 Dissociation of 7.7S BHA following sedimentation equilibrium centrifugation. Pooled fractions of 7.7S BHA from the gradient shown in Figure 5B were dialyzed against MNT buffer and recentrifuged on a sucrose density gradient (-○-). The remainder of the 7.7S BHA was subjected to sedimentation equilibrium sedimentation after which aliquots were recentrifuged on sucrose density gradients to see if dissociation had occurred (-●-).

DISCUSSION

The molecular mechanism by which influenza HA - or any other protein - catalyzes membrane fusion is not known. There is, however, sufficient information about the HA to allow consideration of alternative models that can be subjected to experimental verification. The model that we favor has been described in some detail in previous publications (White et al. 1983, Doms et al. 1985, 1987). We will review it here in the light of the most recent experimental data including the results described above.

Dissociation of the ectodomain. The HA exists as an extended, trimeric structure in the viral membrane. The trimer is stabilized by non-covalent interactions (hydrogen bonds, electrostatic and hydrophobic interactions) primarily between HA2 polypeptides in the stem region (Daniels et al. 1985). The data suggest that these, as well as the few interactions observed between the HA1 domains at the top of the molecule, are extensively disrupted by acid treatment. The evidence for this is as follows:

1. *The association of the HA1 subunits to the rest of the trimer is severed* Following acid treatment and reduction of the interchain disulfide bond, the HA1 subunits are readily dissociated from the complex in monomeric form (Graves et al. 1983, Doms et al. 1985). The contacts between HA1 and HA2 subunits are thus greatly weakened by acid treatment, and it seems possible that the HA1 chains remain attached to the rest of the spike primarily by the interchain disulfide as indicated in Figure 8. Consistent with this model are the observations from many laboratories that new antigenic epitopes in the HA1 and HA2 subunits are exposed after acid treatment (Boulay et al. 1987, White and Wilson, submitted) and that other epitopes at the molecule's tip, which are located close to the intersubunit contacts (epitopes B and D), are modified or lost (Webster et al. 1983, Yewdell et al. 1983, Daniels et al. 1983, Jackson and Nestorowicz, 1985, Copeland et al. 1986). Negative staining electron microscopy has also suggested that the ectodomain of HA dissociates after acid treatment (Ruigrok et al. 1986, Doms and Helenius 1986).

2. *Interactions between the ectodomains of HA2 are affected* The evidence for this is indirect. *First,* BHA - the isolated ectodomain fragment - dissociates to form dimers and monomers when exposed to pH values that trigger fusion in intact HA (Nesterowicz et al. 1985, Doms and Helenius 1986, Fig. 5). *Second,* fusion variants which lack one or more of the residues involved in interchain electrostatic bonds between HA2 subunits undergo acid conformational changes at a higher pH than the wt HA (Daniels et al. 1985, Rott et al. 1985, Doms et al. 1986). This suggests that not only do these salt bridges provide stability to the trimer, but they dissociate during the acid induced conformational change. One of the salt bridges in question is located at the very base of the ectodomain close to the interchain disulfide, the others are in the mid portion of the stem (Wilson et al. 1981, Daniels et al. 1985, Doms et al. 1986).

3. *A major change takes place at the base of the trimer* That the changes are not restricted to the top domains is also shown by the exposure of new antigenic binding sites relatively close to the molecule's base. The C-terminus of HA1 and the N-terminus of HA2 are both exposed to their corresponding antipeptide antibodies (White and Wilson, submitted).

4. *HA remains trimeric* Hydrodynamic data and electron microscopy (Ruigrok et al. 1985, Doms and Helenius, 1986) have shown that in spite of extensive opening up of the ectodomain, intact HA molecules remain trimeric after acid treatment (Doms and Helenius, 1986). The contrast with BHA, which tends to dissociate after acid treatment, suggests that the transmembrane domain of HA plays an important role in stabilizing the acid form of the HA trimer.

Exposure of a hydrophobic domain Another well documented change which correlates with fusion activity is the exposure of a hydrophobic moiety in the ectodomain. Acidified BHA aggregates in aqueous solution, binds to liposomes, and associates with nondenaturing detergent (Skehel et al. 1982, Doms et al. 1985). Intact HA, which binds detergent at neutral pH to its transmembrane anchor binds additional detergent following acid treatment judging by its partitioning in Triton X114 (Boulay et al. 1987). HA0 and A^{-}HA0 - both fusion incompetent - undergo acid conformational changes but they do not expose a hydrophobic moiety (Boulay et al. 1987 and Figure 3).

The hydrophobic moiety exposed in acid HA and BHA resides in the HA2 subunit (Skehel et al. 1982, Doms et al. 1985). Recent labeling studies with photoactivatable phospholipids have shown that the BHA2 chain of acid treated BHA interacts with the liposome membrane (Harter et al., personal communication) but it is not yet clear which peptide moieties are inserted. It is, however, almost certain that the amino terminus of HA2 is involved. It clearly becomes exposed in the acid conformation of the activated HA, and it is the most notably hydrophobic sequence in the entire ectodomain. Further, mutations in this peptide have striking effects on HA's fusion activity (Gething et al. 1986a,b). While the N-terminus is clearly exposed, as indicated by the black domains in Figure 8, we cannot rule out the possibility that other regions of HA2 (i.e. the central alpha helix of HA2) contribute to the hydrophobic moiety. Spectroscopic data from Skehel et al. (1982) have shown that there is no major change in overall alpha helix content of HA after acid treatment. This means that the long central alpha helices in HA2 probably remain intact, but it does not mean that they necessarily remain in unaltered configurations with respect to each other. It will be very interesting to know whether these helices somehow interact with the target membrane as well.

Our present view of the acid induced conformational change in HA is shown schematically in Figure 8. In this working model we have tried to integrate the various experimental observations: the dissociation of the ectodomain, exposure of the N-terminae of the three HA2 subunits, and sterical constraints around the new hydrophobic moieties that limit detergent binding.

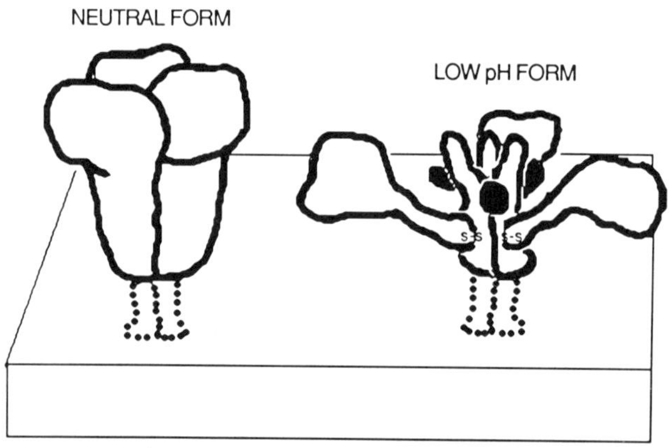

Figure 8 Model of the neutral and acid conformations of HA. At acid pH, the trimeric structure is maintained by the transmembrane anchors even though subunit interactions in the ectodomain are largely disrupted. HA1 is associated with HA2 primarily by a disulfide bond at the base of the molecule. The hydrophobic amino terminae of HA2 are colored black.

How does acid HA mediate membrane fusion? Even in this relatively well-characterized system the mechanism of protein-mediated membrane fusion remains elusive. There are, however, four general points worth making which may throw some light on the problem:

First, for fusion to be possible the acidic form of HA must in some way bring the two membranes together and overcome the hydration force at a focal site of interaction (for discussion of the hydration force see Rand, 1981). The hydration force constitutes the main obstacle against any membrane fusion reaction.

Second, by eliminating the water between the two membranes the HA may indirectly induce a change in lipid configuration and/or composition at the site of contact (for discussion see Rand, 1981), and this could lead to fusion. HA could, in addition, directly perturb the lipid bilayer in a way that would favor the joining of the two bilayers.

Third, as far as we know, the fusion reaction is very tight, it does not involve transient membrane lysis or rupture (White et al. 1982, Young et al. 1983). We would therefore suggest that fusion of the bilayers may occur in two steps with the external leaflets fusing first and the internal leaflets later. In this way the focal site of fusion remains tight; at every moment two intact leaflets will be present to provide a sealed membrane barrier between the internal volumes of the fusing membrane compartments and the bulk solution.

Fourth, it is possible that the fusion activity requires more than just a single HA trimer. Indirect evidence from our laboratory suggests that fusion may depend on cooperative interactions between several HA spikes (Doms et al. 1985, Doxsey et al. 1985). On the other hand, recent target inactivation analyses have indicated that the target size for inactivation of hemolysis activity corresponds to a single HA monomer (Gibson et al. 1986).

According to our present thinking, fusion is initiated by the ectodomains interacting hydrophobically with the target membrane. Since BHA bound to liposomes can be removed by base treatment (Doms et al. 1985) and since the N-terminal hydrophobic sequence is relatively short (the first charged residues occur at positions 11 and 14), we think that the interaction may only involve the external bilayer leaflet of the target membrane. A relatively superficial attachment with the target membrane is also suggested by our observation that BHA binds considerably less detergent than do most integral membrane proteins. It is possible that the interaction is mediated by helical peptide segments interacting in the plane of the target membrane. Insertion of polypeptide moieties into the external bilayer leaflet may disrupt lipid packing and/or the hydration shell to such an extent that coalescence of the viral and target membranes ensues.

Can HA-induced fusion occur at neutral pH? Influenza viruses fuse with detectable efficiency with highly negatively charged liposomes in buffers of neutral pH (Stegmann et al. 1986, Haywood and Boyer, 1985). This has been shown with vesicles of cardiolipin, phosphatidylserine and vesicles containing gangliosides. On the basis of these data it has been suggested that the acid-induced conformational change in HA may not always be required, and that acidity may not play any role in influenza virus entry (Haywood and Boyer, 1985). The fact that fusion with biological membranes always requires acidification and that the cell biological results strongly favour a role for acidic endosomes strongly argue that the fusion observed in neutral pH media is an artefact due to the highly charged target membranes used (Stegmann et al. 1986).

A likely explanation, not yet properly considered by investigators in the field, is that the pH close to the surface of highly negatively charged membranes can be considerably lower than in the bulk solution and thus induce fusion by the normal acid-induced pathway at an apparently neutral pH. The Guoy-Chapman electrical double layer theory predicts the pH close to a highly negatively charged membrane surface will be considerably more acidic than the bulk solution. The higher the charge of the membrane and the lower the concentration of Na^+ in the medium, the lower the surface pH (Schnitzer, in press). It can easily be a pH unit lower than bulk pH. Whether any conditions exist where such effects play a major role in biologically relevant fusion events remains to be seen. It is clear, however, that even in the case of erythrocyte membranes, which are very negatively charged, low pH in the bulk medium is required to induce fusion activity (Stegmann et al. 1986).

It is of interest, in this context, that many soluble proteins and membrane proteins with no known biological fusion function become fusogenic when denatured at acidic pH (Doms et al. 1987). Again fusion is most effective with negatively charged lipids. This phenomenon suggests a role for surface denaturation as a mediator of fusion reactions and it emphasizes the need for caution in interpreting fusion data in non-physiological conditions.

Protein catalyzed membrane fusion clearly poses a complicated, though not intractable problem. A multidiciplinary approach will be required to fully understand the process. In particular, biophysical techniques will be needed to complement the extensive biochemical and genetic approaches taken thus far. Hopefully, by examining well characterized biologically significant fusion factors such as the influenza HA, general themes applicable to more general cellular fusion reactions will become evident.

ACKNOWLEDGEMENTS

We wish to thank Dr. Bill Agnew of the Department of Physiologiy, Yale School of Medicine, for advice on the sedimentation equilibrium experiments. This work was supported by a grant from NIH (18599) and the Medical Scientist Training Program.

REFERENCES

Bordier, C. (1981) Phase separation of integral membrane protiens in Triton X-114 solution. J. Biol. Chem. 256:1604-1607.

Bothwell, M.A., G.J. Howlett and H.K. Schachman (1978) A sedimentation equilibrium method for determining molecular weights of proteins with a tabletop high speed air turbine centrifuge. J. Biol. Chem. 253:2073-2077.

Boulay, F., R.W. Doms, I. Wilson and A. Helenius (in press) The influenza hemagglutinin precursor as an acid sensitve probe of the biosynthetic pathway.

Bradford, M.M. (1972) A rapid and sensitive method for the quantification of microgram quantities of protein utilizing the principle of protein dye binding. Anal. Biochem. 72:248-254.

Brand, C.M. and J.J. Skehel (1972) Crystalline antigen from the influenza virus envelope. Nat. (London) New Biol., 238:145-147.

Copeland, C.S., R.W. Doms, E.M. Bolzau, R.G. Webster and A. Helenius (1986) Assembly of influenza hemagglutinin trimers and its role in intracellular transport. J. Cell Biol. 103:1179-1191.

Daniels, R.S., J.C. Downie, A.J. Hay, M. Knossow, J.J. Skehel, M.L. Wang, and D.C. Wiley (1985) Fusion mutants of the influenza virus hemagglutinin glycoprotein. Cell, 40:431-439.

Daniels, R.S., A.R. Douglas, J.J. Skehel and D.C. Wiley (1983) Analyses of the antigenicity of influenza hemagglutinin at the pH optimum for virus-mediated membrane fusion. J. Gen. Virol. 64:1657-1662.

Doms, R.W., D.S. Keller, A. Helenius and W.E. Balch (1987) Assembly and transport of VSV G protein. submitted.

Doms, R.W., J. White, F. Boulay and A. Helenius (in press) Influenza virus hemagglutinin and membrane fusion. in Cellular Membrane Fusion, eds. J. Wilschut and D. Hoekstra. Marcell Dekker, New York.

Doms, R.W., M.-J. Gething, J. Henneberry, J. White, and A.H. Helenius (1986) A variant influenza virus hemagglutinin that induces fusion at elevated pH. J. Virol., 57:603-613.

Doms, R.W. and A. Helenius (1986) Quaternary structure of influenza virus hemagglutnin after acid treatment. J. Virol. 60:833-839.

Doms, R.W., A.H. Helenius, and J.M. White (1985) Membrane fusion activity of the influenza virus hemagglutinin: the low pH-induced conformational change. J. Biol. Chem., 260:2973-2981.

Doxsey, S., J. Sambrook, A. Helenius and J. White (1985) An efficient method for introducing macromolecules into living cells. J. Cell Biol. 101:19-27.

Fraker, P.J. and J.C. Speck, Jr. (1978) Protein and cell membrane iodinations with a sparingly soluble chloroamide, 1,3,4,6-tetrachloro-3a,6a- diphenyl-glycoluracil. Biochem. Biophys. Res. Commun., 80:849-857.

Fuller, S.D. (1987) The T=4 envelope of Sinbis virus is organized by interactions with a complementary T=3 capsid. Cell 48:923-934.

Gething, M.-J., R.W. Doms, D. York, and J. White (1986a) Studies on the mechanism of membrane fusion: Site-specific mutagenesis of the hemagglutinin of influenza virus. J. Cell Biol., 102:11-23.

Gething, M.-J., R.W. Doms, J. White and A. Helenius (1986b) Studies on the mechanism of membrane fusion. in Protein Engineering: Applications in Science, Medicine and Industry. Ed. M. Inouye and R. Sarma. pp. 193-215. Academic Press, New York.

Gething, M.-J. and J. Sambrook (1982) Construction of influenza hemagglutinin genes that code for intracellular and secreted forms of the protein. Nature 300:598-603.

Gething, M.-J., J.M. White and M.D. Waterfield (1978) Purification of the fusion protein of Sendai virus: analysis of the NH_2-terminal sequence generated during precursor activation. Proc. Natl. Acad. Sci. USA 75:2737-2740.

Gibson, S., C.Y. Jung, M. Takahashi and J. Lenard (1986) Radiation inactivation analysis of influenza virus reveals different target sizes for fusion, leakage, and neuraminidase activities. Biochem. 25:6264-6268.
Graves, P.N., J.L. Schulman, J.F. Young and P. Palese (1983) Preparation of influenza virus subviral particles lacking the HA1 subunit of HA: unmasking of cross-reactive HA2 determinants. Virology 126:106-116.
Haywood, A.M. and B.P. Boyer (1985) Fusion of influenza virus membranes with liposomes at pH 7.5. Proc. Natl. Acad. Sci. 82:4611-4615.
Helenius, A. and K. Simons (1975) Solubilization of membranes by detergents. Biochim. Biophys. ACTA 415:29-79.
Herz, C., E. Stavnezer, R.M. Krug and T. Gurney (1981) Influenza virus, an RNA virus, synthesizes its messenger RNA in the nucleus of infected cells. Cell 26:391-400.
Jackson, D.C. and A. Nestorowicz (1985) Antigenic determinants of influenza virus hemagglutinin. XI. Conformational changes detected by monoclonal antibodies. Virol. 145:72-83.
Martin, R.G. and B.N. Ames (1961) A method for determining the sedimentation behavior of enzymes: application to protein mixtures. J. Biol. Chem. 236:1372-1379.
Matlin, K.S., H. Reggio, A. Helenius and K. Simons (1981) Infectious entry pathway of influenza virus in a canine kidney cell line. J. Cell Biol. 91:601-613.
McEwen, C.R. (1967) Tables for estimating sedimentation through linear concentration gradients of sucrose solution. Anal. Biochem. 20:114-149.
Meunier, J.C., R.W. Olsen, and J.P. Changeux (1972) Studies on the cholinergic receptor from *Electrophorus Electricus*. Effect of detergents on some hydrodynamic properties of the receptor protein in solution. FEBS Lett. 24:63-68.
Nestorowicz, A., G. Laver and D.C. Jackson (1985) Antigenic determinants of influenza virus hemagglutinin. X. A comparison of the physical and antigenic properties of monomeric and trimeric forms. J. Gen. Virol. 66:1687- 1695.
Rand, R. (1981) Interacting phospholipid bilayers: measured forces and induced structural changes. Ann. Rev. Biophys. Bioeng. 10:277-314.
Reynolds, J.A. and C. Tanford (1976) Determination of molecular weight of the protein moiety in protein-detergent complexes without direct knowledge of detergent binding. Proc. Natl. Acad. Sci. 73:4467-4470.
Rott, R., M. Orlich, H.-D. Klenk, M.L. Wang, J.J. Skehel and D.C. Wiley (1984) Studies on the adaptation of influenza viruses to MDCK cells. EMBO J. 3:3329-3332.
Ruigrok, R.W.H., N.G. Wrigley, L.J. Calder, S. Cusack, S.A. Wharton, E.B. Brown and J.J. Skehel (1986) Electron microscopy of the low pH structure of influenza virus hemagglutinin. EMBO J. 5:41-49.
Simons, K., A. Helenius and H. Garoff (1973) Solubilization of the membrane proteins from Semliki Forest virus with Triton X-100. J. Mol. Biol. 80:119-133.
Skehel, J.J., P.M. Bayley, E.B. Brown, S.R. Martin, M.D. Waterfield, J.M. White, I.A. Wilson and D.C. Wiley (1982) Changes in the conformation of influenza hemagglutinin at the pH optimum of virus mediated membrane fusion. Proc. Natl. Acad. Sci., USA 79:968-972.
Stegmann, T., D. Hoekstra and J. Wilschut (1986) Fusion activity of influenza virus. A comparison between artificial and biological target membrane vesicles. J. Biol. Chem. 261:10966-10969.
Vogel, R.H., S.W. Provencher, C.-M. Bonsdorff, M. Adrian and J. Dubochet (1986) Envelope structure of Semliki Forest virus reconstructed from cryo-electron micrographs. Nature 320:533-535.
Webster, R.G., L.E. Brown and D.C. Jackson (1983) Changes in the antigenicity of the hemagglutinin molecule of H3 influenza virus at acidic pH. Virol. 126:587-599.
White, J., M. Kielian and A. Helenius (1983) Membrane fusion proteins of enveloped animal viruses. Q. Rev. Biophys. 16:151-195.
White, J., J. Kartenbeck and A. Helenius (1982) Membrane fusion activity of influenza virus. EMBO J. 1:217-222.
White, J. and A. Helenius (1980) pH-dependent fusion between the Semliki Forest virus membrane and liposomes. Proc. Natl. Acad. Sci. 77:3273-3277.
Wiley, D.C., I.A. Wilson and J.J. Skehel (1981) Structural identification of the antibody binding sites of Hong Kong influenza haemagglutinin and their involvement in antigenic variation. Nature 289:373-378.

Wiley, D.C., J.J. Skehel and M. Waterfield (1977) Evidence from studies with a cross-linking reagent that the hemagglutinin of influenza virus is a trimer. Virol. 79:446-448.
Wilson, I.A., J.J. Skehel and D.C. Wiley (1981) Structure of the hemagglutinin membrane glycoprotein of influenza virus at 3A resolution. Nature (London) 289:366-373.
Yewdell, J.W., W. Gerhard and T. Bachi (1983) Monoclonal anti-hemagglutinin antibodies detect irreversible antigenic alterations that coincide with the acid activation of influenza virus A/PR/8/34-mediated hemolysis. J. Virol. 48:239-248.
Yoshimura, A. and S.-I. Ohnishi (1984) Uncoating of influenza virus in endosomes. J. Virol. 51:497-504.
Young, J. D.-E., G.P.H. Young, Z.A. Cohn and J. Lenard (1983) Interaction of enveloped animal viruses with planar bilayer membranes: Observations on Sendai, Influenza, Vesicular Stomatitis, and Semliki Forest viruses. Virology 128:186-194.

PARAMETERS AFFECTING THE FUSION OF VIRUSES WITH ARTIFICIAL AND BIOLOGICAL MEMBRANES

Dick Hoekstra, Karin Klappe, Toon Stegmann and Shlomo Nir*

Laboratory of Physiological Chemistry, University of Groningen
Bloemsingel 10, 9712 KZ Groningen, The Netherlands
*The Hebrew University of Jerusalem, Rehovot, Israel

INTRODUCTION

Membrane fusion constitutes an essential element in the overall mechanism by which enveloped viruses infect cells. During fusion, the viral nucleocapsid is introduced into the cytoplasm, representing a prerequisite for viral replication. The cellular site of entry depends on the family to which a virus belongs (Fig. 1). Paramyxoviruses, such as Sendai virus or Newcastle disease virus, but also measles viridae, fuse directly with the plasma membrane at neutral pH. Most other viruses, such as toga viruses (e.g. Semliki Forest Virus) and orthomyxoviruses (e.g. influenza viruses) are fusing with the endosomal membrane 'from within', after being internalized by receptor-mediated endocytosis. In these cases, fusion is triggered by the mild acidic pH, generated in the endosomal compartment. It has been well established that fusion between a viral membrane and a cellular target membrane occurs (1,2) and that it provides a convenient mechanism that allows the nucleocapsid to circumvent two potential membrane barriers. It is much less clear, however, how the molecular factors that are involved in fusion, are bringing about the actual membrane merging. By characterizing the parameters that affect the fusion activity of viruses, attempts are made to understand the mechanism of this event. Viral membrane glycoproteins are intimately involved in the overall process that leads to the intracellular delivery of the viral genomes. For Sendai virus, the fusion process is a two-step reaction, consisting of a binding step and the actual fusion event. Binding is mediated by the binding protein HN, which contains hemagglutination and neuraminidase activity. The fusion process itself is accomplished by the F protein, which consists of two disulfide-bonded polypeptides, F_1 and F_2. The F protein is derived from an inactive precursor, F_0. The latter is activated upon its proteolytic cleavage - without a loss of polypeptide fragments - by a host cell enzyme. For influenza virus, both the binding and fusion activity are confined to one viral spike glycoprotein, the hemagglutinin, HA. In its active form it also consists of two disulfide-linked subunits, HA_1 and HA_2. The binding activity is located on HA_1, while the fusion activity resides on the HA_2 polypeptide. As for Sendai virus, influenza also has two types of spike proteins, the other being NA. This protein contains neuraminidase activity but does not seem to be involved in the entry mechanism.

As membrane fusion plays a ubiquitous role in the biological life of a cell, a detailed knowledge of the mechanism of fusion of viruses might well

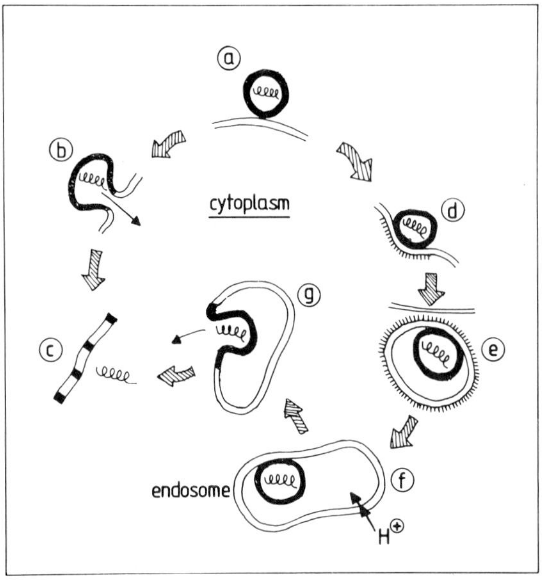

Fig. 1. Mechanisms of entry of enveloped viruses into cells. Some families of viruses release their nucleocapsid into the host cell following the fusion of their envelope with the plasma membrane at neutral pH (a-c). Many others (d-g) are first taken up into the cells by means of receptor-mediated endocytosis (d, e). After reaching the endosomal compartment (f), their envelopes subsequently fuse with the endosomal membrane "from within", triggered by the low pH that is generated in this organelle. As a result, the nucleocapsid gains access to the cytoplasm (g). After fusion, the components of the viral and target membrane (i.e. plasma membrane or endosomal membrane, c) randomize by lateral diffusion.

provide the framework for clarifying fusion mechanisms that mediate numerous intracellular processes. Thus, apart from the significance of revealing the fusogenic properties of these particles per se, viruses and their membranes, being relatively simple in composition, also serve, in a broader context, as a valuable 'model system'.

Much of our current knowledge regarding membrane fusion mechanisms has been derived from studies on the fusion of artificial membranes, such as liposomes (3,4). Key elements, facilitating the recognition of molecular factors involved in the fusion of such systems were the ability (i) to monitor continuously the kinetics of fusion, (ii) to quantitate fusion and (iii) to model fusion by kinetic analysis that allowed to discriminate between the aggregation step and the fusion reaction per se (5). It would appear highly advantageous, therefore, to adapt similar experimental and theoretical approaches to systems involving the fusion of biological membranes. Furthermore, although fusion between liposomes does not reflect the fusion of biological membranes, the validity of carrying out such work relies particularly on the ability to recognize certain trends, that might be essential to fusion at biological conditions. This will be illustrated by briefly summarizing some of our work on the interaction of viruses with biological and liposomal membranes.

PROCEDURES TO MEASURE VIRUS FUSION ACTIVITY

At appropriate conditions, viruses can induce the fusion between cells. Hence, cell-cell fusion seems to be a convenient measure for viral fusion activity. Yet, it should be recognized that viral envelope-cell fusion and cell-cell fusion are separate events while the first can occur without proceeding to the latter. For example, it has been shown that reconstituted Sendai virus envelopes (RSVE) can fuse with human erythrocytes and various murine cell lines as efficiently as native Sendai viruses, i.e. similar amounts of viral membrane glycoproteins are inserted into the target cell membranes. Yet, the extent of cell-cell fusion, as induced by RSVE is much lower than that observed for the intact virus. (6, van Bleek and Hoekstra, unpublished observation). Moreover, cell-cell fusion consists of two, kinetically different consecutive steps, i.e. local membrane fusion of apposed membranes followed by osmotically driven cell swelling. Only the latter step, which is secondary to the actual fusion reaction, can be discerned while furthermore, quantitative information concerning the number of virus particles that fuse with the cells, is not obtained. Neither is this the case, when determining the hemolytic properties of viruses, which, like cell-cell fusion, is as such a convenient measure for the biological activity of viruses. Hemolysis is also secondary to fusion and is thought to result from the insertion of partially "damaged" viral envelopes into the erythrocyte membrane (7). A substantial number of viruses fuse, prior to the occurrence of hemolysis or of alterations in the permeability of cultured cells, thus reflecting a gradual acquisition of a threshold level of membrane damage (8). This also explains the inability to monitor virus-induced fusion of erythrocyte ghosts by means of an aqueous contents mixing assay, as described previously (9). It appears that most of the probes, entrapped within the aqueous volume of the ghosts to register fusion, is released prior to the occurrence of significant ghost-ghost fusion (Hoekstra, unpublished observation).

The desirability to obtain quantitative and kinetic data on membrane fusion, including that of viruses, in order to dissect the molecular mechanisms underlying this event, has culminated into the development of a variety of assays. They included among others, assays based on the use of radioactive (10) or spin-labeled markers (11,12). More recently a number of fluorescent assays haven been introduced (8,13-15) and they offer the great advantage to monitor continuously the fusion process, to detect fusion in a very sensitive manner and to reveal the fate of molecules during or after fusion. Above all, these techniques are relatively easy to apply. The principles of the assays are based on either resonance energy transfer (RET) or fluorescence quenching (Fig. 2). The RET assay (16) involves the incorporation of two fluorescent phospholipid derivatives in the target membrane, usually a liposome. Alternatively, it is also possible to insert the probes into reconstituted viral envelopes, by addition of the fluorophores to the detergent extract, containing the viral membrane components, prior to starting the dialysis procedure (17). Upon exciting the donor, the energy from a photon absorbed by the donor is transferred to the acceptor, the excitation spectrum of which overlaps with the emission spectrum of the donor. As a result, the donor fluorescence, monitored at its emission wavelength, will be quenched. Upon fusion of the labeled vesicles with a virus, lipid mixing and hence dilution of the probes occurs. The energy transfer efficiency decreases and therefore, the donor fluorescence increases. The latter can be conveniently monitored as a measure of fusion, provided that lipid exchange can be excluded. When incorporated into membranes at a relatively high concentration, certain fluorescent lipid(-like) derivatives may display a concentration-dependent fluorescence selfquenching (8,13,15). Upon fusion with non-labeled membranes, mixing of lipids will result in probe dilution and a concomitant increase of fluorescence is seen reflecting the kinetics of fusion, i.e. when appropiate control experiments have excluded the possibility of lipid transfer. In

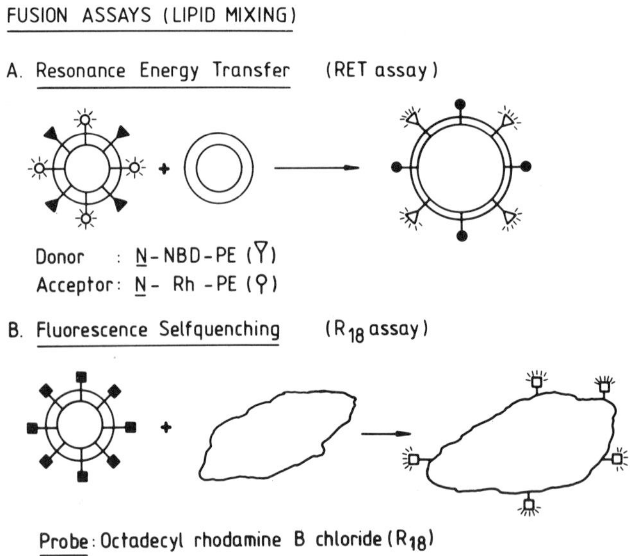

Fig. 2. Schematic representation of the principles of fluorescence assays used to monitor the fusion of viruses with artificial and biological membranes. A. RET assay. When exciting the donor, the acceptor will fluoresce. Upon fusion, energy transfer is abolished when the probes dilute. In this case, donor fluorescence is detectable at its emission wavelength. (for details see text and ref. 16) B. R_{18} assay. Membranes are labeled with a selfquenching concentration of R_{18}. Upon dilution, the selfquenching is relieved and the increase in fluorescence reflects a measure of fusion (for details see ref.13).

contrast to the probes commonly used in the RET assay (N-(7-nitro-2,1,3-benzoxadiazol-4-yl)phosphatidylethanolamine, N-NBD-PE, and N-(lissamine Rhodamine B sulfonyl)phosphatidylethanolamine, N-Rh-PE), others such as octadecylrhodamine B chloride (R_{18}; 13) and pyrene sulfonyl phosphatidylethanolamine (PSPE; 8) can be readily incorporated (by exogenous addition) into native biological membranes at concentrations where selfquenching occurs. Unfortunately, both probes are not homogeneously distributed within the lateral plane of the viral bilayer (8; Hoekstra et al., unpublished observation). Whereas in liposomes the selfquenching of R_{18} increases proportionally with the probe concentration, labeling of the virus with relatively low concentrations already displays an unexpectedly high degree of selfquenching. This would suggest a preferential association of lipid probes, such as R_{18} or PSPE with distinct domains in the viral membrane. When fluorescently-tagged virions, labelled with a low (ca. 1 mol%) concentration of R_{18}, are treated with a protein structure-modifying reagent such as dithiotreitol (DTT) or trypsin, a (partial) redistribution of the probe occurs as reflected by an increase in fluorescence. Such an effect is no longer apparent at higher concentrations ($\gg$ 3 mol%). Whether viral proteins directly affect the lateral distribution of these probes remains to be established but it is interesting to note that a viral protein from vesicular stomatitus virus has been reported to affect similarly the lateral organization of lipid probes in the bilayer (18). Yet, in spite of this drawback, the versatile use of the R_{18} assay has become apparent in numerous studies. By comparison with other, independent techniques (see below), it can be inferred that this approach can properly reflect the fusion of viruses with both biological and artificial membranes, thus constituting an important tool in clarifying the mechanism of fusion of viruses.

FUSION OF VIRUSES WITH BIOLOGICAL MEMBRANES

A key element in any fusion assay that relies on the mixing of lipids is the necessity to exclude that mixing occurs via a monomeric exchange process. The results that were obtained in this respect for the R_{18}-assay indicate that after its insertion the probe does not significantly dissociate from viral, liposomal or erythrocyte membranes (13,19,20). Furthermore, the presence of the probe (3-4 mol %, with respect to total viral phospholipids) within the viral membranes does not interfere with the hemolytic properties of viruses (19). Also, the kinetics of fusion between viruses and liposomes are indistinguishable when monitored with either the R_{18} or RET assay (13,17). Note in this regard, that the concentration of the RET probes , located in the liposmal bilayer, is much lower than that of R_{18}, which is inserted into the viral membrane.

To characterize the parameters that affect binding and fusion of Sendai virus, the interaction of R_{18}-labelled virions with both erythrocytes (19,21,22) and cultured cells (van Bleek and Hoekstra, submitted) has been studied in considerable detail. Binding of Sendai virus to biological target membranes is a temperature-independent phenomenon. The stability of viral attachment is sensitive, however, to a shift in temperature: a fraction of the virions (ca. 30%), attached at 4°C, rapidly (half time = 2.5 min) elutes from the cell surface at 37°C. The elution can be attributed to a spontaneous temperature-induced release, rather than to viral neuraminidase activity (21). Still 25% of the cell-associated virus fraction dissociated from the cell surface in the presence of the neuraminidase inhibitor 2-deoxy-2,3-dehydro-N-acetylneuraminic acid. Although the viral neuraminidase activity, contained in the HN protein, appears to be capable of cleaving sialic acid from its receptor side, it remains to be determined, therefore, to what extent desialylation is actually related to viral elution. Obviously, the fraction of the total viral dose that binds to the cells is dependent on the target membrane concentration. On the other hand, the number of virus particles that can maximally bind per cell is limited. Erythrocyte ghosts yield an average of ca. 1200 Sendai virions (21,23) while, for example, as many as 3000 virions can bind per mouse spleen cell (van Bleek and Hoekstra, submitted).

As treatment of the virus with the reducing agent dithiothreitol diminishes the extent of viral binding (19,20), it would appear that for proper (electrostatic) interaction of the HN binding protein with target membrane receptors, certain structural requirements for functional expression of the protein have to be met as well. Interestingly, binding of the virus is not affected by alterations in the pH, i.e., within the range of 5.0-9.0.

Relative to the fusion reaction of viruses that fuse at mild acidic pH, the fusion of Sendai virus is a slow and inefficient process. Figure 3 illustrates the initial fusion event between the virus and erythrocyte ghosts as target membranes, monitored with the R_{18} assay. The fusion reaction may proceed for several hours and the final extent of fusion depends on the number of virus particles added per ghost (see below). In contrast to the binding, the fusion of Sendai virus appears to be a much more strictly controlled process. The fusion activity, mediated by a separate membrane glycoprotein, F, displays a strict temperature dependence, i.e. it only becomes apparent at a temperature of ca. 23°C or above. Alterations in the physical state of both the viral or target membrane, as determined by fluorescence polarization measurements, do not take place around this temperature. Neither can changes in the kinetics and extent of viral binding or temperature-induced conformational changes in the F protein

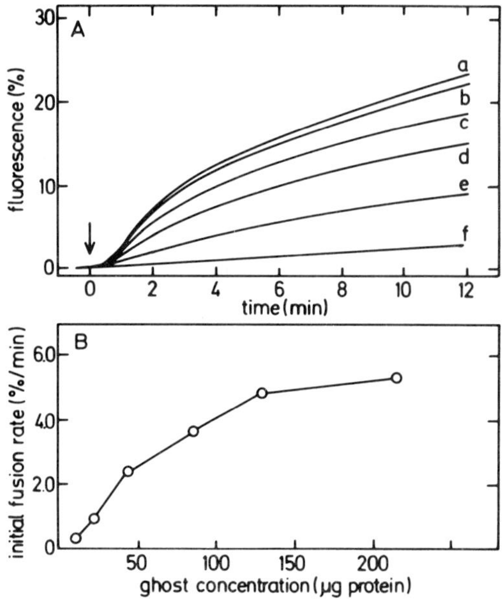

Fig. 3. Fusion of Sendai virus with erythrocyte ghosts.
A. Various concentrations of ghosts (216(a), 130(b), 86(c), 43(d), 22(c) and 11(f) μg of protein in a final volume of 2 ml incubation medium) were added to R_{18}-labelled Sendai virus (arrow) and the relief of fluorescence selfquenching was monitored continuously in a fluorometer. Note the presence of the lag phase prior to the onset of the fluorescence increase.
B. From the curves in figure A, the initial fusion rates were calculated from the slope of the steepest part of the fluorescence tracings. The rates were plotted as a function of the ghost concentration.

explain the existence of a fusion threshold temperature (19). Rather, the rapid increase in the kinetics of fusion, occurring between 23°C and 37°C, overlaps remarkably well with the drastic increase in the rotational mobility of the viral proteins, that takes place within the same temperature range (24). As binding is not affected by temperature, immobile complexes of HN molecules, being located at the focal sites of fusion, may restrain the F protein molecules from coming into direct contact, which is thought to be necessary for triggering fusion (1), with the target membrane at temperatures below 23°C. In this way, the HN protein could thus play a regulatory role in the overall process that leads to fusion. With influenza virus, a similar phenomenon has been noted (25). Also in this case, a threshold temperature for fusion of ca. 20°C is seen. A considerable increase in the mobility of the hemagglutinin, HA (which contains both the binding and fusion activity), occurs above this temperature. It has been proposed that this increase is necessary for allowing the molecular rearrangements which accompany the fusion event to take place (25). These rearrangements involve a low-pH induced conformational change of the HA molecule, which causes the exposure of the previously buried, hydrophobic N-terminal amino acid segment of the HA_2 subunit. A similar hydrophobic sequence of amino acids is already exposed (at neutral pH) at the N-terminal region of the F_1 subunit which together with a disulfide-linked F_2 subunit constitutes the fusion protein of Sendai virus. It is thought that these hydrophobic segments are directly involved in triggering the actual fusion reaction by penetrating into the target membrane, thus causing local perturbations which lead to merging of the apposed membranes (cf. Fig. 8). Whereas low pH triggers the fusion of influenza virus, it impedes the fusion of Sendai virus. With both erythrocyte ghosts and various murine cell lines, the latter virus displays a pH optimum for fusion around pH 7.5 (19). Mild acidic or alkaline conditions irreversibly inactivate the viral fusion protein, whereas binding is not impaired. This inactivation is due to a pH-induced conformational change in the protein (26), implying that a proper conformation of the F protein is, as for influenza, essential to its functional activity. This is further emphasized by the observation that a disintegration of the disulfide-bond between the F_1 and F_2 polypeptide, by treatment of the virus with reducing agents, abolishes the fusion activity. It turns out (21) that the F protein (i.e. fusion) is much more sensitive to such treatments than the HN protein (binding).

An interesting difference between Sendai virus and influenza virus is, that the binding and fusion function of Sendai are located on different proteins, HN and F, respectively, while for influenza both functions are contained in one protein, HA. This difference may well be reflected by the differences in the initial kinetics of fusion with erythrocyte membranes (cf. Fig. 3). For Sendai virus, a lag phase (20-30 sec) in fusion is consistently apparent, irrespective of whether viruses are allowed to attach at 4°C (19). For influenza virus a short lag phase (ca. 6 sec) is only seen when the virus is added to ghosts at 37°C (22). Furthermore, it is also intriguing that an incubation of Sendai virus *per se* for 10 min at pH 7.4 and 37°C (i.e. the optimal conditions for fusion) does not alter the fusion capacity of the virus, whereas a similar incubation of influenza at its fusion optimum (37°C, at pH 5.2) inactivates the fusion activity as revealed by subsequent addition to ghosts. The latter has been attributed to an irreversible aggregation of HA in the viral membrane (22,25). This instantaneous inactivation can be partially prevented when the incubation of the virus at this pH is carried out in the presence of a target membrane (22,25) at a temperature below 20°C (where binding takes place but no fusion). Apparently, at those conditions the HA molecules, due to their interaction with target membrane receptors, become 'fixed', such that they are becoming protected against selfaggregation and hence, irreversible inactivation. The latter still occurs, however, when incubating the virus per se at these conditions. These considerations lead us to suggest that in case of Sendai virus, the role of HN may well extend beyond that of simply conveying binding activity to the virus. Rather, the regulatory role of HN as suggested above may also apply to regulating, possibly indirectly, the lateral distribution of F in the plane of the bilayer. These aspects, the validity of which clearly requires future work, may thus be reflected by the persistent presence of a lag phase prior to fusion and the protection against inactivation as a result of selfaggregation at 37°C and pH 7.4.

The kinetics of fusion of Sendai virus with erythrocyte membranes is affected by the density of attached viral particles (21). At relatively low densities the relationship between the initial fusion rate and particle density is linear. However, it deviates form linearity at higher densities ($>$ 170 viruses bound per ghost), suggesting that adjacent particles affect one another with respect to their ability to fuse. Furthermore, in light of the maximal binding capacity per ghost (ca. 1200, see above), these observations also indicate that not all binding sites convey equal fusion susceptibility to the viruses. In fact, our results indicate that, as for binding, fusion is limited, i.e. only some 200 virus particles maximally can fuse per ghost, representing 15-20% of the attached virions at saturable binding conditions. Higher percentages (50-60%) were found for various murine cell-lines, implying that the exact number may be cell type-dependent. On the other hand, at non-saturable binding conditions the number of attached virions that eventually fuse with ghosts is dependent on particle density. At a relatively low input multiplicity, i.e. less than 100 virus particles per ghost, essentially all virions fuse, implying that the percentage of viral dose that fuses can be as high as 90% (23).

The potential presence of specific sites for the fusion of viruses at cell surfaces, as inferred from the distinct difference of binding versus fusion capacity, has recently also been suggested by an analysis of the interaction of reconstituted Sendai virus envelopes with cultured cells with FPR techniques (27). It was demonstrated that the fraction of cell-associated RSVE which fuses with the cell varies, depending on the location on the cell surface. In the context of this study (27), it is finally relevant to note that a comparison of the various fluorescence dequenching methods, including the R_{18}-assay, with the FPR technique on RSVE-erythrocyte fusion revealed that under identical experimental conditions, essentially identical results were obtained by all methods.

FUSION WITH LIPOSOMES: A COMPARISON WITH BIOLOGICAL MEMBRANES

Because of their simplicity and the ease of manipulating their composition, liposomes or lipid vesicles may serve as an attractive model system to reveal fundamental aspects of virus fusion activity. In addition, because of the availibility of the appropiate techniques, the results of such studies can be directly correlated to those obtained with biological target membranes. Various viruses, e.g. Sendai virus, influenza virus and vesicular stomatitus virus, avidly fuse with pure negatively charged vesicles (cf. Figs. 6 and 7) consisting of CL or PS (14,28-30). This would suggest, at least in the case of Sendai and influenza virus, that a specific sialic acid-containing receptor molecule is not required for initial virus-target membrane interaction. However, the characteristics of fusion of influenza virus with ghosts deviate from those of fusion with CL liposomes (22). This is reflected, for example, by the observation that fusion with CL liposomes does not exhibit a strict dependence on the low pH-induced conformation change in the HA molecule (31). Furthermore, virus-CL vesicle fusion is triggered immediately, i.e. without a lag phase in fusion as commonly seen when ghosts are mixed with viruses at 37°C. The artificial nature of fusion between viruses and pure acidic phospholipid vesicles is even more pronounced in the case of Sendai virus. As for influenza virus, fusion of Sendai virus is triggered immediately when mixed with CL (cf. Fig. 7) or PS vesicles at neutral pH (28). However, both the rate and extent of fusion dramatically increase with decreasing pH and optimal activity is seen around pH 4.0. Analysis of the virus-liposome mixtures, incubated at pH 5.0 and 7.4 on sucrose density gradients (Fig. 4) demonstrated that fusion had actually occurred at both conditions and that lipid dilution does not result from a monomeric transfer of the fluorescent probe, reporting the fusion event. The initial kinetics of the low pH-induced fusion reaction are ca. 3-fold faster than those observed at neutral pH. However, in contrast to fusion at neutral pH, fusion at acidic pH is essentially complete and involves all viral and liposomal particles present in the incubation mixture. At both neutral and acidic pH (Fig. 5) virus-liposome fusion products display a capacity to fuse subsequently only with liposomes but not with virions. In passing, these experiments, in conjunction with the density gradient analysis, also indicate that in spite of a non-homogeneous distribution of the fluorophore in the viral membrane, a significant interference with the kinetics of fusion does not become apparent and complete randomization occurs upon merging of viral and target membrane. This was also reflected by a comparison between RET and R_{18} assay (13,17). Obviously, the occurrence of fusion between Sendai virus and liposomes at acidic pH is in sharp contrast to the assumed biological fusion activity which, with ghosts or cultured cells as targets, is optimal around pH 7.5 while it becomes irreversibly inactivated at acidic pH. At acidic pH and with CL vesicles as target membranes, parameters such as trypsinization and DTT treatment only partially affect the ability of Sendai virus to fuse with the vesicles, thus further emphasizing the non-specific fusogenic behavior with such liposomes.

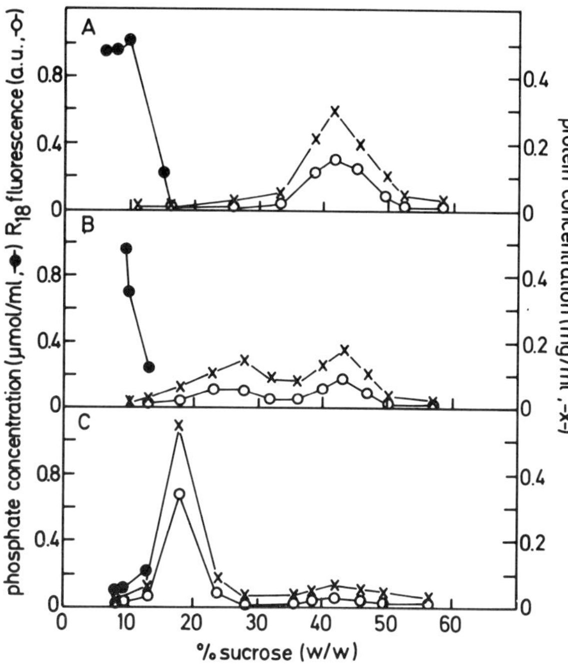

Fig. 4. Density gradient analysis of virus-liposome fusion products. R_{18}-labeled Sendai viruses were fused with CL vesicles at pH 7.4 (panel B) or pH 5.0 (panel C) for 15 min at 37°C, The fusion mixtures were analysed on continuous sucrose gradients (10-60%). The combined results of liposomes and viruses, run on separate gradients, are shown in panel A. Phosphate (●) was determined to trace the position of the liposomes on the gradient, and protein (x) to mark the virus.

Evidence has been obtained which demonstrates that at low pH both F and HN, when separately reconstituted into lipid vesicles, can trigger membrane fusion (32, Hoekstra and Klappe, unpublished observation). On the other hand, zwitterionic liposomes containing the ganglioside $GD1_a$ display a fusion behavior with both Sendai and influenza virus which, in many respects, is similar to that seen for erythrocyte ghosts (22,28). It is concluded that depending on the pH and the liposomal lipid composition, different mechanisms are operational in the fusion process between viruses and liposomes. Hence, fusion with liposomes can, at least partly, be accomplished in a manner which may not bear physiological significance.

Interestingly, with both CL liposomes and ghosts, the fusion reaction of influenza virus can in a reversible manner be switched on and off by pH shifts from mild acidic to neutral and vice versa. These observations indicate that the low pH-induced irreversible conformational change in the HA molecule does not in itself suffice to activate the fusion activity. Rather, protonation of the acidic amino acids contained within the hydrophobic segment of the HA protein appears to be equally important.

In studies concerning the fusion of liposomes per se, it has been well established that the ultimate barrier for membrane fusion is constituted by strong repulsive hydration forces due to the water molecules associated with the lipid headgroups (33,34). Elimination of this barrier is pertinent to the occurrence of fusion in such systems. Studies carried out with lipid vesicles of various composition suggested that the interaction of virus with target membranes may also be governed by hydrophobic interactions and impeded by repulsive hydration forces. For example, the inclusion of PC in

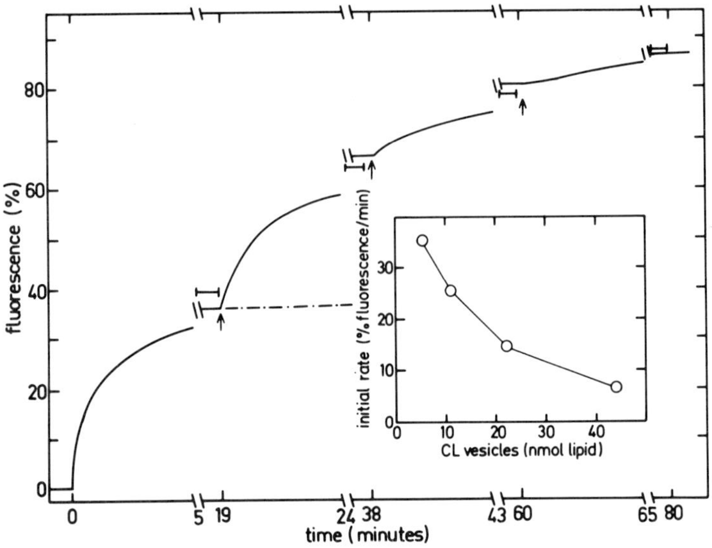

Fig. 5. Low pH-induced fusion between Sendai virus and CL LUV; the occurrence of multiple fusion events.
At time zero, equal amounts of viral lipids and CL liposomes were mixed, after a preincubation of the virus at pH 5.0 at 37°C. Fusion was monitored and after its completion, an additional amount of CL vesicles, equal to the total amount of lipid of the first addition, was added (arrow). This was repeated twice. The bars represent the calculated level of fluorescence, that would be reached upon complete randomization of the lipids. Note that after each addition, fusion involves virtually all virus and liposomal particles present in the incubation mixture. Apparently, one virus particle can fuse with several liposomes and fusion products, at least those obtained after one round of fusion, do not fuse with native viruses (—·—·—) The insert shows the decrease in the initial fusion rate after the sequential additions of liposomes to the virus-liposome fusion products. For experimental details, see reference 28.

acidic phospholipid vesicles strongly inhibits virus-liposome fusion (14, 28; Fig. 7). Replacement of PC by PE, however, sustains the fusion reaction, an effect which is likely to be attributed to the distinct state of hydration of both lipids. Moreover, the presence of small amounts of the dehydrating agent polethylene glycol enhances the rate and extent of fusion of Sendai virus with liposomes (35). Interestingly, also with erythrocyte target membranes, the fusion of Sendai virus is strongly accellarated in the presence of low (non-fusogenic) amounts of PEG (Hoekstra and Klappe, unpublished observation). Besides the state of membrane hydration, variations in lipid packing density may also affect the fusion susceptibility of viruses with liposomes, similarly as reported for protein induced-fusion of liposomes (36). Thus, the observation that the fusion susceptibility of Sendai virus with CL vesicles is higher than that with PS and zwitterionic GD_{1a}-containing vesicles (in that order) may reflect increasing constraints in the ability of the viral proteins to penetrate into the lipid matrix as the packing density of the bilayer increases. This notion is supported by experiments on protein-induced fusion of lipid

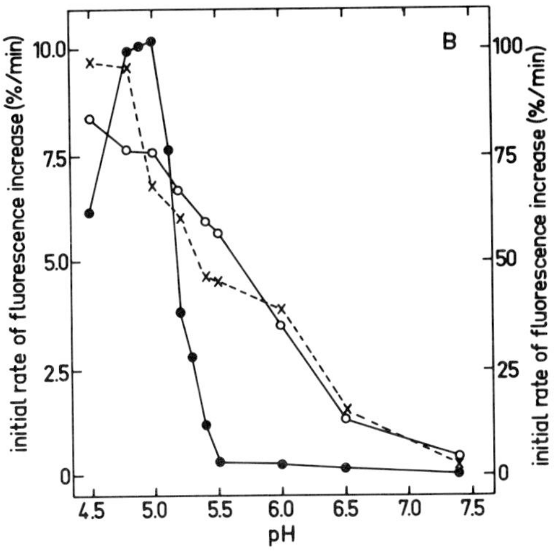

Fig. 6. Fusion of influenza virus (X-47 strain) with liposomes of various compositions as a function of pH. R_{18}-labeled viruses were injected into a cuvette containing the liposomes, suspended in a medium of the desired pH. Fusion was monitored continuously and the initial fusion rates were calculated and plotted as a function of the pH. Fusion was carried out with liposomes (LUV) consisting of CL (o—o, R_{18} assay and x-x, RET assay; right-hand scale) or PC/PE/ Gangliosides (•—•, left-hand scale). Note the considerable differences in the initial rates, depending on the lipid composition. Except for quantitative differences the pH-dependent fusion pattern for influenza virus-erythrocyte ghosts fusion very closely resembles that seen for the ganglioside-containing liposomes.

vesicles of the same composition in which fusion was correlated to changes of the intrinsic fluorescence of the protein, that was taken as an indirect measure for protein penetration. Thus, it was observed that the initiation of both lysozyme- and α-lactalbumin-induced fusion commensurates with an abrupt enhancement of the intrinsic fluorescence (Hoekstra et al., unpublished observation). However, assuming that a hydrophobically mediated perturbation of the bilayer triggers the fusion of viruses and taking into account that at least part of the fusion reaction with CL- and PS-vesicles is due to a non-specific fusion process, it would appear that this type of interaction is less subtle in liposomal than in biological systems. Nevertheless, the results obtained on virus-liposome fusion would be consistent with the hypothesis (Fig. 8) that penetration of hydrophobic segments of the viral fusion proteins represents a key element in triggering fusion activity.

Finally, the assays, that are used to monitor fusion in a continuous manner, follow essentially the overall fusion reaction, which involves an initial binding step of adjacent membranes and the actual fusion reaction per se. However, in conjuntion with a theoretical model, which is based on a

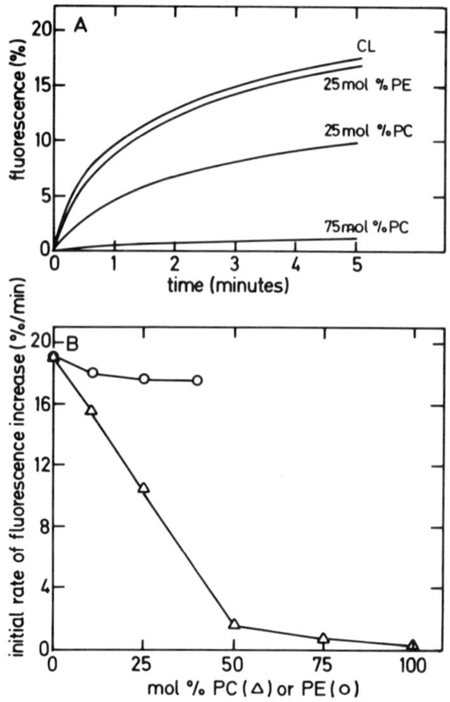

Fig. 7. Effect of the incorporation of PC and PE into CL liposomes on Sendai virus fusion activity.
A. The fusion of Sendai virus with liposomes, (LUV), consisting of CL and various amounts of PC or PE was monitored continuously. The temperature was 37°C and the medium pH 7.4. Note the absence of a lag phase in fusion upon addition of the liposomes to the virus-containing cuvette (cf. fig. 3). The initial fusion rates were calculated as described in fig. 3 and plotted (B) as a function of the mol% of PC or PE.

simple mass action kinetic model (5), the overall fusion reaction can be analyzed in terms of a second-order process of membrane adhesion followed by the first order fusion reaction itself. By determining the aggregation rate constant, a reflection is obtained of the balance of forces that mediates close apposition while the fusion rate constant provides a direct measure of the significance of parameters affecting the molecular events involved in the fusogenic destabilization of apposed bilayers. Such an analysis reveals, for example, that both PEG treatment and pH affect the rate constant of fusion rather that of aggregation (35). Application of these analyses, as described elsewhere in this volume (see Nir et al.), in conjunction with the experimental approaches as described here, will be important tools to further clarify the structural and functional features of the viral proteins that allow enveloped virions to exploit membrane fusion as the initial step towards viral replication.

ACKNOWLEDGMENTS

Mrs Lineke Klap is gratefully acknowledged for expert help in preparing the manuscript. Part of this work was carried out under the auspices of the Netherlands Foundation for Chemical Research (SON) with financial support from the Netherlands Organization for the Advancement of Pure Research (ZWO).

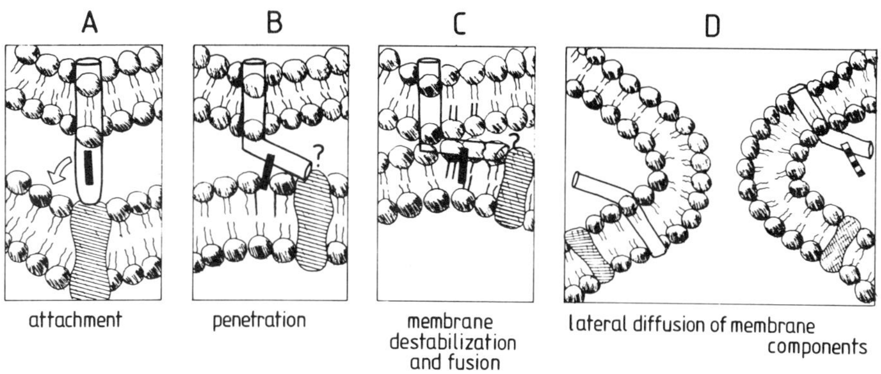

Fig. 8. Hypothetical mechanism of the fusion of enveloped viruses with target membranes. The various, sequential stages that can be discerned in the overall process that leads to fusion between a viral membrane (top) and a target membrane (bottom) are: (i) attachment of the virus, to either a sialic acid-containing glycolipid or glycoprotein receptor (A); (ii) penetration of the hydrophobic amino-acid domain, located in the fusion protein (bar), into the target membrane (B); (iii) membrane destabilization and fusion (C) and (iv) randomization of viral and target membrane components (D). For influenza virus, a low pH- induced conformational change in the HA structure is necessary in order to expose the hydrophobic domain of the HA_2 subunit (A,B). In case of Sendai virus, the hydrophobic segment in the F_1 subunit of the F protein is already exposed at neutral pH and penetrates into the target membrane at that pH without the need for a conformational change (B). Note that binding of Sendai virus is accomplished by a separate binding protein (not shown) which subsequently enables the F protein to interact with the target membrane. Penetration of the hydrophobic segments (C) presumably causes a hydrophobic dehydration which originates in the tight interactions between the hydrophobic segments of the viral fusion proteins and the hydrophobic interior of the bilayer. As a result of this interaction, the apposed membranes will be "pulled" together and interfacial water will be expelled causing a local dehydration at the interbilayer contact sites which, in turn, will give rise to lipid packing defects. The latter act as the focal sites for fusion.

REFERENCES

1. White, J., Kilian, M. and Helenius, A. (1983) Q. Rev. Biophys. 16, 151-195.
2. Lenard, J. and Millar, D.K. (1983) in Receptor-Mediated Endocytosis and Processing (Cuatrecasas, P. and Roth, T. Eds), pp. 121-138, Chapman and Hall, London.
3. Düzgünes, N. (1985) Subcell. Biochem. 11, 195-286.
4. Wilschut, J. and Hoekstra, D. (1986) Chem. Phys. Lipids 40, 145-166.
5. Nir, S. Bentz, J. and Portis, A. (1980) Adv. Chem. Ser. No. 188, 75-106.
6. Henis, Y.I. and Gutman, O. (1987) Biochemistry 26, 812-819.
7. Hosaka, Y. and Shimizo, K. (1977) Cell Surf. Rev. 2, 129-155.
8. Micklem, K.J., Nyaruwe, A. and Pasternak, C.A. (1985) Molec. Cell, Biochem. 66, 163-173.
9. Hoekstra, D., Wilschut, J. and Scherphof, G. (1985) Eur. J. Biochem. 146, 131-140.
10. Haywood, A.M. and Boyer, B.P. (1982) Biochemistry 21, 6041-6046.
11. Maeda, T., Kawasaki, K. and Ohnishi, S. (1981) Proc. Natl. Acad. Sci. U.S.A. 78, 4133-4137.
12. Lyles, D.S. and Landsberger, F.R. (1979) Biochemistry 18, 5088-5095.
13. Hoekstra, D., De Boer, T. Klappe, K. and Wilschut, J. (1984) Biochemistry 23, 5675-5681.
14. Stegmann, T. Hoekstra, D. Scherphof, G. and Wilschut, J. (1985) Biochemistry 24, 3107-3113.
15. Chejanovsky, N. and Loyter, A. (1985) J. Biol. Chem. 260, 7911-7918.
16. Struck, D.K., Hoekstra, D. and Pagano, R.E. (1981) Biochemistry 20, 4093-4099.
17. Harmsen, M., Wilschut, J., Scherphof, G., Hulstaert, C. and Hoekstra, D. (1985) Eur. J. Biochem. 149, 591-599.
18. Wiener, J.R., Pal. R., Barenholz, Y. and Wagner, R.R. (1985) Biochemistry 24, 7651-7658.
19. Hoekstra, D., Klappe, K., De Boer, T. and Wilschut, J. (1985) Biochemistry 23. 4739-4745.
20. Hoekstra, D. and Klappe, K. (1987). Bioscience Reports, in press.
21. Hoekstra, D. and Klappe, K., (1986) J. Virol. 58, 87-95.
22. Stegmann, T., Hoekstra, D. Scherphof, G. and Wilschut, J. (1986) J. Biol. Chem. 261, 10966-10969.
23. Nir, S., Klappe, K. and Hoekstra, D. (1986) Biochemistry 25, 2155-2161.
24. Lee, P.M., Cherry, R.J. and Bächi, T. (1983) Virology 128, 65-76
25. Junankar, P.R. and Cherry, R.J. (1986) Biochim. Biophys. Acta. 854, 198-206.
26. Hsu, M.C., Scheid, A. and Choppin, P.W. (1982) Proc. Natl. Acad, Sci. U.S.A. 79, 5862-5866.
27. Aroeti, B. and Henis, Y.I. (1986) Biochemistry 25, 4588-4596.
28. Klappe, K., Wilschut, J., Nir, S. and Hoekstra, D. (1986) Biochemistry 25, 8252-8260.
29. Eidelman, O. Schlegel, R. Tralka, T.S. and Blumenthal, R. (1984) J. Biol. Chem. 259, 4622-4628.
30. Haywood, P.M. and Boyer, B.P. (1984) Biochemistry 23, 4061-4066.
31. Doms, R.W., Helenius, A. and White, J. (1985) J. Biol. Chem. 260, 2973-2981.
32. Chejanovski, N., Zakai, N., Amselem, S., Barenholz, Y. and Loyter, A. (1986) Biochemistry 25, 4810-4817.
33. Parsegian, V.A. and Rand, R.P. (1983) Ann. N.Y. Acad. Sci. 416, 1-11.
34. Hoekstra, D. and Wilschut, J. (1987) in Water transport in biological membranes (Benga, G., Ed.) CRC Press Inc., Boca Raton, in press.
35. Nir, S. Klappe, K. and Hoekstra, D. (1986) Biochemistry 25, 8261-8266.
36. Hong, K. and Vacquier, V.D. (1986) Biochemistry 25, 543-549.

ACTIVE FUNCTION OF MEMBRANE RECEPTORS IN FUSION OF ENVELOPED VIRUSES WITH CELL PLASMA MEMBRANES

Abraham Loyter, Ofer Nussbaum, and Vitaly Citovsky

Department of Biological Chemistry
Institute of Life Sciences
The Hebrew University of Jerusalem
91904 Jerusalem, Israel

INTRODUCTION

Viral systems have provided many of the models for studying the organization and regulation of eukaryotic genes. However, very little is known regarding the first steps of virus infection, especially about the correlation between these steps and the promotion of infectious diseases (1,2). The recently acquired data on the function of individual components of viruses and their interaction in the ultimate production of infection (3), have enabled to commence research directed towards defining the molecular mechanism of virus-membrane interaction and for the elucidation of the role of membrane receptors in the process of virus penetration.

Two basic steps are involved in the penetration of viruses into animal cells: (i) adsorption of virus particles to cell plasma membranes, and (ii) penetration of the viral material and uncoating of the nucleic acid within the cell. Many studies (3,4) suggest that there are specific sites for attachment of the virus particles on the cell membrane. The specificity of the cell receptors for viruses seems to be the most important factor in determining species specificity, tissue tropism and selective vulnerability of selective cell populations. In our laboratory (5), we have used reconstituted envelopes of Sendai virus to transfer membrane fragments from SV_{40}-receptor-positive into SV_{40}-receptor-negative cells. Fusion-mediated implantation of SV_{40}-receptor into the rat cell line Fl'1-4 made these cells susceptible to infection by intact SV_{40} viruses (5).

Interaction of enveloped viruses with animal cells is also determined by membrane receptors, as was demonstrated by specific binding between viruses belonging to the paramyxo- and orthomyxovirus groups and sialic acid residues of membrane components (4). Discrete membrane receptors are also utilized by the Epstein Barr virus (EBV), an oncogenic Herpes virus of humans, which selectively infects human lymphocytes. Our method of fusion mediated implantation of membrane components was used to demonstrate the existence of membrane receptors for EBV (5). Fusion-mediated implantation of EBV receptors into membranes of receptor-negative cells resulted in the acquisition of virus binding activity by these cells (5). Also, susceptibility and infection of cells to the T-lymphotropic virus, Type III/lymphadenopathy-associated virus (HTLV-III/LAV) (which is involved in inducing AIDS, acquired immune deficiency syndrome) is dependent on the expression of a cell surface-specific antigen of the human T-cells (CD_4) (6).

Penetration of enveloped viruses into host cells is mediated, in addition to virus-membrane binding, also by a fusion process of the virus envelope with the cell plasma membrane, thus releasing the viral nucleocapsid directly into the cell cytoplasm (3-5). Enveloped animal viruses are composed of a membrane and an internal space containing the viral genome as well as accessory proteins.

Two pathways are known whereby enveloped viruses enter into living cells (3). Envelopes of viruses belonging to the paramyxovirus group such as Sendai virus or New Castle Disease virus (NDV), fuse with the cells' plasma membrane at pH 7.4, and consequently introduce their content — the viral nucleocapsid — directly into the cell cytoplasm. A different pathway of entry has been described for most other enveloped virions such as those belonging to the orthomyxovirus, toga or Herpes groups. Fusion of the viral envelope with the endosomal or lysosomal membrane is triggered by the intraorganelle, low pH environment (3). The low pH induces a change in the virus particle, which triggers the viral envelope fusion activity. The presence of envelope glycoproteins of these viruses on cell surfaces leads, in most cases, to induction of cell-cell fusion and formation of syncytia (3). Evidently, such cell-cell fusion can be induced, if the pH of the incubation medium is decreased to around pH 5.0 (3).

Until recently, virus-membrane fusion processes have mainly been studied through the use of phase and electron microscopic techniques (3) or by following processes such as virus-induced lysis of recipient cells (7). Infection of cultured cells by Sendai virus, for example, is accompanied by an increase in the cell membrane's permeability to both low and high molecular weight components (8). However, this assay does not determine the number of virions that actually fuse with recipient membranes and depends also on the condition of the virus, with damaged virus particles being more lytic than intact ones. Furthermore, recent experiments in our laboratory (unpublished results) have clearly shown that enveloped virions such as Semliki Forest virus (SFV) and *Herpes simplex* virus (HSV-1), although readily fusing with animal cell plasma membranes, do not increase their permeability even with molecules of low molecular weight such as Rb^+. It seems, therefore, that the correlation between the hemolytic and fusogenic activities of enveloped virions is as yet poorly understood.

Recently, we (9-11) as well as other groups (12), have used energy transfer and fluorescence dequenching methods to follow, on a quantitative basis, fusion processes occurring between enveloped virions and recipient living cell membranes. The fluorescent probe, octadecylrhodamine-B-chloride (R_{18}), was inserted, in our laboratory, into envelopes of Sendai, influenza, SFV and HSV-1, resulting in self-quenching (9,10,13). Our results showed that under such conditions, the infectivity of the virions is not affected, and the decrease in R_{18} surface density is proportional to the fluorescence dequenching. Incubation of fluorescently-labeled fusogenic virions with living cells, isolated membranes or phospholipid vesicles (liposomes) of the appropriate composition, resulted in fluorescence dequenching (13). All our results as well as those of others (12,13) support the view that the fluorescence dequenching observed is due to fusion between the viral envelope and the recipient cell membranes, and that this results from a dilution of the virus-associated probes in the recipient membranes.

FUSION OF SENDAI AND INFLUENZA VIRIONS WITH ERYTHROCYTES AND CULTURED CELLS REQUIRES VIRUS RECEPTORS AND IS ACCOMPANIED BY CELL LYSIS

A. Experiments with Sendai virions

Binding of Sendai virions to cell plasma membrane receptors, especially sialic acid residues of membrane glycolipids, is mediated by the viral hemagglutinin/neuraminidase (HN) glycoprotein, while virus-membrane fusion is promoted by the viral fusion (F) polypeptide (14). These two viral envelope glycoproteins are sufficient to promote virus-membrane fusion, as was inferred from experiments using reconstituted Sendai virus envelopes (RSVE) (5).

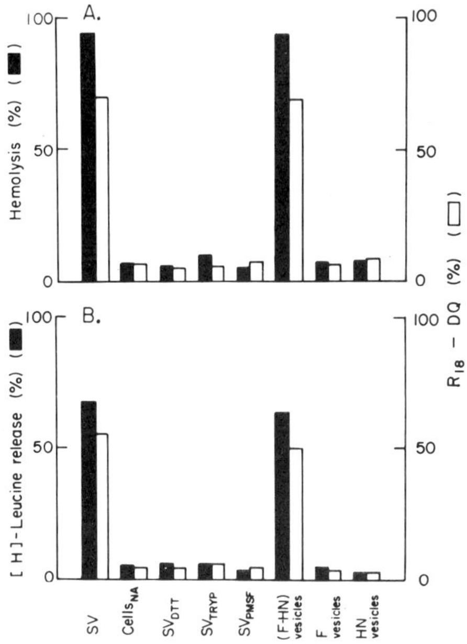

Fig. 1. Correlation between lytic and fusogenic activities of Sendai virions. A. Interaction with human erythrocytes and human erythrocyte ghosts (HEG): R_{18}-labeled Sendai virions (F, HN and F-HN vesicles) (5 µg) were prepared as described before (10,11) and incubated for 30 minutes at 37°C with either 250 µl of 3% red blood cells in PBS (induction of hemolysis) or with 200 µg of HEG in 200 µl of PBS (induction of fluorescence dequenching (R_{18}-DQ). After the incubation period, the degree of hemolysis and the extent of DQ were determined, as previously described (10). B. Interaction with hepatoma tissue culture cells (HTC): R_{18}-labeled (5 µg) viral preparations were incubated for 30 minutes at 37°C with 2 x 10^6 cells in 200 µl of PBS. After incubation, the degree of R_{18}-DQ was determined as described earlier (10,11). Virus-induced cell lysis was estimated by measuring the release of radioactivity from the cells labeled with ^{3}H-leucine, as described elsewhere (16). SV, Sendai virus; SV_{Tryp}, SV preincubated for 30 minutes at 37°C with trypsin (60 µg/mg of viral protein (17); SV_{DTT} and SV_{PMSF}, SV preincubated for 30 minutes at 37°C with 3 mM dithiothreitol (DTT) or 7 mM phenyl-methyl-sulfonyl-fluoride (PMSF), respectively (17,18); $Cells_{NA}$, SV incubated with neuraminidase-treated cells (30 mU/10^7 cell); F-vesicles, HN-vesicles, and (F-HN)-vesicles, reconstituted viral envelopes bearing purified F, HN, or a mixture (1:1) of both F and HN glycoproteins.

The results in Figure 1A confirm previous observations (7,15), showing that incubation of intact Sendai virus particles with human erythrocytes at pH 7.4, induced a high degree of hemolysis. It has been widely accepted that such virus-induced hemolysis reflects a process of virus-cell fusion (7).

The results in Figure 1B show that Sendai virions are also able to induce, at pH 7.4, lysis of living cultured cells, a process which has mainly been studied by Pasternak and his colleagues (16). In these experiments (Fig. 1B), lysis of the cultured cells was estimated by monitoring the amount of radiolabeled polypeptides released from the recipient cells upon incubation with Sendai virions at 37°C and at pH 7.4. The requirement for membrane receptors is evident from the results showing that no release of hemoglobin or of ^{3}H-labeled polypeptides was observed following incubation of Sendai virions with neuraminidase-treated erythrocytes or hepatoma tissue culture (HTC) cells, respectively (Fig. 1).

The view that virus-induced lysis indeed reflects a process of virus--cell fusion is strengthened by the experiments (Fig. 1) showing that very little, if any, release of intracellular material has been observed after incubation with non-infective, non-fusogenic virions. Sendai virions can be rendered non-fusogenic by treatment with trypsin, PMSF or DTT (17,18, Fig. 1). Trypsinization and treatment with PMSF significantly inhibited the viral fusion protein, whereas DTT irreversibly reduced the disulfide bonds of the viral HN glycoprotein (17,18). Trypsinized and PMSF-treated virions are able to bind to but not fuse, while DTT-treated viruses neither bind to nor fuse with cell membranes.

In addition, the results in Figure 1 show, as reported before (19), that virus-induced lysis was promoted only by reconstituted virions bearing the two viral envelope glycoproteins, *i.e.* the HN and F polypeptides. Very little, if any, lysis was observed following incubation of erythrocytes or HTC with membrane vesicles bearing either the viral HN or the F polypeptides. Furthermore, incubation of human erythrocytes or HTC with a suspension containing a mixture of HN and F vesicles, did not result in induction of cell lysis either (not shown).

As mentioned above, fusion between enveloped viruses and biological membranes can quantitatively be estimated by following intermixing of the viral envelope lipids with the lipids of the recipient membranes (9-12). We (10) and others (12) have inserted the fluorescent probe, R_{18}, into envelopes of Sendai virions as well as those of other animal viruses. Incubation of fluorescently-labeled Sendai virions and HEG or with HTC resulted in a relatively high degree of fluorescence dequenching (DQ) (10, Fig. 1). (HEG have been used as recipient membranes, since the hemoglobin present in intact erythrocytes hampers accurate fluorescence measurements.)

From the results shown in Figure 1, it can be inferred that the increase in the DQ occurs only when fluorescently-labeled lytic virions were incubated with HEG or HTC. Very little, if any, dequenching was observed following incubation of non-lytic, namely PMSF-, DTT- or trypsin-treated virions, with HEG or HTC or of lytic viruses with neuraminidase-treated cells (Fig. 1). Moreover, no significant increase in DQ was observed when fluorescently-labeled F or HN vesicles were used (Fig. 1). Increase in DQ was observed only after incubation with reconstituted vesicles bearing the two viral glycoproteins within the same membrane.

B. Experiments with influenza virions

Orthomyxoviruses, especially influenza virions, resemble paramyxoviruses in several aspects: their genome consists of negatively-stranded RNA, the virus particles are spherical, and their membranes contain two types of spiked glycoproteins which possess neuraminidase, hemagglutinin and fusion activities (20). The most important differences are: that in orthomyxoviruses (influenza), the genome is sequestered and that one of the spiked glycoproteins (NA) is a neuraminidase, while the other, the hemagglutinin (HA), contains both the capacity to bind to cell surface sialic acid residues of membrane glycolipids and to catalyse fusion (20). The HA glycoprotein is derived by proteolytic cleavage from a precursor glycoprotein designated as HA_O (21).

It has been clearly demonstrated that a mildly acidic pH is necessary for the expression of influenza's fusogenic activity (3). Various assays have been carried out to demonstrate the requirement for low pH values to activate these influenza virus' fusogenic abilities, among them induction of hemolysis (22) or of cell lysis (16).

Our experiments also showed that influenza virions at pH 5.0, indeed hemolyse human erythrocytes (Fig. 2A) and cause the release of intracellular polypeptides from HTC (Fig. 2B). No release of intracellular material was observed, following incubation of influenza virions with virus receptor-depleted cells, namely neuraminidase-treated human erythrocytes (Fig. 2A) or HTC (Fig. 2B). Treatment of influenza virions with glutaraldehyde (GA) (23) or NH_2OH (24) has been shown to inactivate the viral fusogenic activity. The results in Figure 2 clearly show that the lysis induced by treated virions was significantly lower than that observed with untreated virus particles. It is noteworthy that GA- or hydroxylamine-treated influenza virions are able to agglutinate and bind to cultured cells, although unable to induce cell-cell fusion (23,24).

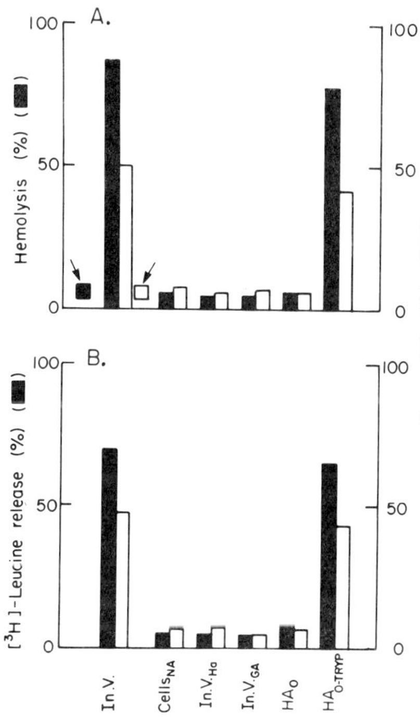

Fig. 2. Correlation between the lytic and fusogenic activities influenza virions. A. Interaction with human erythrocytes and HEG. B. Interaction with HTC. All experimental conditions were as described in Figure 1 for Sendai virions, except that the incubation was performed at pH 5.0, adjusted by the addition of 50 mM Na acetate, pH 5.0 (23). Arrows indicate the extent of the viral lytic and fusogenic activities at pH 7.4. InV, influenza virions; InV_{Ha}, InV preincubated for 60 minutes at 37°C with 1 M hydroxylamine; InV_{GA}, InV preincubated for 30 minutes at 37°C with 0.1% of glutaraldehyde; HA_O, InV bearing an uncleaved HA glycoprotein; $HA_{O_{Tryp}}$, HA_O pretreated for 20 minutes at 37°C with 15 μg/ml of trypsin (21).

The results in Figure 2 also show that incubation of HA_O-influenza virions with human erythrocytes or HTC promoted a very low degree of lysis. It has previously been shown (21) that trypsinization of HA_O-influenza virus activates its fusogenic polypeptide and, consequently, the virus is able to penetrate and infect cells. Indeed, a high degree of cell lysis was observed following incubation of human erythrocytes or HTC at pH 5.0 with trypsinized HA_0-influenza virions (Fig. 2).

Incubation of fluorescently-labeled influenza virions at pH 5.0 with HEG or HTC resulted in a relatively high degree of DQ (Fig. 2). As can be seen, the increase in the degree of fluorescence showed the same features as those of virus-induced lysis. No increase in DQ was observed after incubation with neuraminidase-treated cells or with GA- or NH_2OH-treated virions. Similarly, no increase in DQ was observed at pH 5.0 with HA_0-influenza virions, while a relatively high increase in DQ was obtained with trypsinized HA_0-virions (Fig. 2).

The experiments described in the present chapter and those reported before (10 - 12, 24) should lead to the following conclusions: (a) The increase observed in the degree of DQ with fluorescently-labeled (R_{18}-labeled) Sendai and influenza virions reflects a process of virus-membrane fusion. This conclusion is based principally on our observations showing that very little, if any, DQ was obtained with non-infective, non-fusogenic, inactivated virus particles. (b) Fusion of both virus particles, *i.e.* Sendai and influenza virions, with biological membranes or living cells is absolutely dependent on the presence of membrane receptors with which the virus particles interact. Practically no increase in the degree of fluorescence was observed upon incubation with membranes from which receptors for both viruses had been removed by treatment with neuraminidase. (c) Virus-cell fusion is always accompanied by virus-induced lysis of the recipient cells.

VIRUS-LIPOSOME FUSION: FUSION OF SENDAI AND INFLUENZA VIRIONS WITH PHOSPHATIDYLCHOLINE AND CHOLESTEROL LIPOSOMES LACKING VIRUS RECEPTORS IS A NON-LEAKY PROCESS

Fusion of Sendai and influenza virions with liposomes of various compositions has been extensively studied during the past few years by various groups. Haywood (25) was the first one to demonstrate fusion between Sendai virus and liposomes composed of neutral lipids and bearing sialoglycolipids. From this work, it was concluded that fusion of Sendai virus particles with liposomes displays the same features as fusion with biological membranes, that is, it was optimal at pH 7.0-8.0, at temperatures of 37-40°C, and required the presence of virus receptors. The use of R_{18}-labeled Sendai and influenza virions permitted the study of virus-liposome fusion in a quantitative manner, which does not rely on the viral lytic activity but monitors the intermixing of the virus and the liposome membranes. Indeed, our studies (10,26, and Table 1), as well as those of others (27), showed that incubation of Sendai or influenza virions with negatively-charged liposomes such as those composed of phosphatidylserine (PS) or cardiolipin (CL), resulted in both a high degree of DQ and virus release of the liposome content. These results unequivocally show that no virus receptors - sialoglycolipids - are required to allow fusion between these two enveloped virions and negatively-charged phospholipids. As opposed to incubation with PS, very little DQ was observed following incubation of either viruses with liposomes composed of phosphatidylcholine (PC) (10,13, and Table 1). However, insertion of cholesterol (chol) molecules into the PC liposomes renders them susceptible to the viral fusogenic activity (Table 1). Similarly to fusion with negatively-charged liposomes, also fusion with liposomes composed of PC/chol did not require the presence of virus receptors in the liposome bilayer (Table 1). However, fusion of both Sendai and influenza virions with liposomes composed of PC and chol, showed different characteristics from that with negatively-charged phospholipid vesicles. Very little DQ was observed upon incubation of inactivated, non-hemolytic Sendai or influenza virions with PC/chol liposomes (Table 1). As opposed to the latter, a high degree of DQ was observed when these non-hemolytic virions were incubated with negatively-charged liposomes. A high degree of DQ was also observed upon incubation of Sendai virus HN-vesicles with PS or liposomes containing dicetylphosphate (DCP) (11, and Table 1). As mentioned before (19) and shown above (Fig. 1), Sendai HN-vesicles neither induce lysis of cells nor are able to fuse with biological membranes. The results in Table 1 show that Sendai HN-vesicles also failed to fuse with PC/chol liposomes. A low degree of DQ was obtained only after incubation of trypsinized Sendai virions with negatively-charged liposomes, indicating a requirement for intact viral glycoproteins for such virus-liposome fusion.

Table 1. Fusion of Sendai and Influenza Virions with Neutral and Negatively-Charged Liposomes

Liposomes incubated with:	PC	PC/chol	PS	DCP/chol
		R_{18}-DQ (%)		
Exp. 1				
SV	2	45	65	50
SV_{DTT}	2	7	74	55
SV_{Tryp}	N.D.*	2	15	10
SV_{PMSF}	N.D.	2	68	55
RSVE	3	44	68	48
F-vesicles	2	2	5	7
HN-vesicles	N.D.	4	72	50
(F-HN)-vesicles	N.D.	45	65	50
Exp. 2				
InV	5	32	62	N.D.
InV_{GA}	4	7	20	N.D.
InV_{Ha}	N.D.	5	50	N.D.
InV_{HA_0}	4	8	30	N.D.
$InV_{HA_0\text{-}Tryp}$	4	30	45	N.D.

*N.D., not determined.

R_{18}-labeled viral preparations were obtained as previously described (10, 28, Figs. 1 and 2). Large unilamellar phospholipid vesicles, composed of phosphatidylcholine (PC), PC/cholesterol (PC/chol, 1:0.5 molar ratio), phosphatidylserine (PS) or dicetylphosphate/chol (DCP/chol, 1:0.5 molar ratio), were prepared as described before (10). Virus-liposome interaction was promoted by incubating the viral preparations (5 µg) for 30 minutes at 37°C with 50 µg of phospholipids, in a final volume of 200 µl of PBS (for Sendai virions) or PBS adjusted to pH 5.0, as described in Fig. 2 (for influenza virions). At the end of the incubation period, the R_{18}-DQ was determined as described in Fig. 1. Symbols of the various virus preparations are as described in Fig. 1. RSVE, reconstituted Sendai virus envelopes.

Virus-induced release of the self-quenched carboxyfluorescein (CF) from loaded liposomes was also used to study the correlation between the lytic and fusogenic activities of enveloped virions (29,30). Our results (29, and Table 2) showed that fusion of Sendai and influenza virions with loaded PC/chol liposomes did not result in release of the liposome content. Lysis of loaded PC/chol liposomes by Sendai or influenza virions was induced only when these liposomes carried virus receptors, *i.e.* sialoglycolipids (gangliosides, gang) or sialoglycolipids plus sialoglycoproteins (GP) (Table 2). As can be seen in Table 2, about 15% and 58% of the enclosed CF were released upon incubation of Sendai or influenza virions, respectively, with PC/chol liposomes bearing sialoglycolipids. Insertion of sialoglycoproteins into the PC/chol/gang liposomes increased their susceptibility to the Sendai virus' lytic activity (Table 2). However, insertion of only sialoglycoprotein molecules (the human erythrocyte glycophorin) into PC/chol liposomes, did not have any effect and they remained resistant to the viral lytic activity (Table 2). Virus-induced lysis of PC/chol/gang or PC/chol/gang/GP showed the same features as virus-induced hemolysis of erythrocytes or lysis of living cultured cells. Negligible amounts of the entrapped CF were released from the liposomes upon incubation with inactive, treated virions (Table 2). Similarly, no lysis was induced by the Sendai virus' HN or F vesicles or by the HA_0-influenza virions (Table 2).

Table 2. Virus-Liposome Interaction: Effect of Virus Receptors

Liposomes incubated with:	PC/chol		PC/chol/GP		PC/chol/gang		PC/chol/gang/GP	
	R_{18}-DQ	CF-release	R_{18}-DQ	CF-release	R_{18}-DQ	CF-release	R_{18}-DQ	CF-release
				(%)				
Exp. 1								
SV	45	1	44	1	45	15	45	44
SV_{DTT}	7	2	7	N.D.*	4	3	4	3
SV_{PMSF}	4	2	3	N.D.	4	5	4	0
F-vesicles	4	1	3	1	4	4	5	0
HN-vesicles	4	1	3	1	2	4	3	2
(F-HN)-vesicles	44	2	45	2	42	13	44	42
Exp. 2								
InV	30	1	31	2	52	58	50	57
InV_{Ha}	7	1	7	N.D.	9	3	7	2
InV_{HA_0}	8	2	7	N.D.	9	3	8	3
$InV_{HA_0\text{-Tryp}}$	28	1	27	N.D.	48	55	50	50

*N.D., not determined.

Empty (50 µg) or CF-containing liposomes (5 µg) (PC/chol, 1:0.5 molar ratio; PC/chol/GP, 1:0.5:0.001 molar ratio; PC/chol/gang, 1:0.5:0.3 molar ratio; and PC/chol/gang/GP, 1:0.5:0.3:0.001 molar ratio) were prepared as described elsewhere (29,30), and incubated for 30 minutes at 37°C with either 5 µg of R_{18}-labeled virions or with 10 µg of non-labeled virus preparations, respectively. The extent of R_{18}-DQ and CF-release were determined as described in Fig. 1 and by Citovsky et al. (10,29). All other experimental conditions were as in Table 1. For determination of CF fluorescence, all the samples were adjusted to pH 7.4 just before the measurements (31). Symbols of the various virus preparations are as in Fig. 1.

Based on the observations described in Tables 1 and 2, we suggested (10,13) that only fusion with liposomes composed of neutral lipids but not with those containing negatively-charged phospholipids, reflects the viral fusogenic activity needed for infection of living cultured cells. Our results also show that only fusion with liposomes composed of neutral lipids (PC/chol) and bearing virus receptors results in release of the liposome content, suggesting the formation of a leaky "fusion product". If, indeed, fusion of virus particles with such liposomes results in the formation of permanently leaky vesicles, it is expected that also the components entrapped within the virus envelopes will be released into the external medium during the fusion process. In our experiments (29, and Table 3), we have enclosed CF within reconstituted Sendai virus envelopes (RSVE). Fusion of such CF--loaded RSVE with HEG resulted, as expected, in the release of the RSVE content. This is due to the fact that the HEG used in these experiments were unsealed and therefore permeable to components of low molecular weight. On the other hand, a negligible percentage of the virus-entrapped CF leaked out from the viral envelopes upon incubation with PC/chol/gang/GP liposomes (Table 3). Under the same conditions, the RSVE fused efficiently with these liposomes. As can be seen (Table 3), up to 44% release was observed when empty RSVE were fused with CF-loaded liposomes.

Table 3. Interaction of CF-Containing RSVE with Liposomes Bearing Virus Receptors

System	R_{18}-DQ	CF-release
	(%)	
R_{18}-RSVE + PC/chol/gang/GP	45	N.D.*
RSVE + CF-PC/chol/gang/GP	N.D.	44
CF-RSVE + PC/chol/gang/GP	N.D.	2
CF-RSVE + HEG	N.D.	60

*N.D., not determined.

Empty (50 μg) PC/chol/gang/GP liposomes were incubated for 30 minutes at 37°C with 5 μg of either R_{18}-labeled or CF-containing RSVE which were obtained as described earlier (29,30). Alternatively, CF-containing RSVE (5 μg) were incubated with 100 μg of HEG, while non-labeled RSVE (10 μg) were incubated with 5 μg of CF-containing PC/chol/gang/GP liposomes. After the incubation period, the CF-release and R_{18}-DQ were determined as described in Table 2.

From these observations, it can be inferred: (a) Sendai virus envelopes are impermeable and able to retain their content, including components of low molecular weight such as CF. Hence, virus-induced lysis cannot be due to the release of cell components through the leaky virus envelopes, as has been suggested before (7). (b) Fusion of virus envelopes with liposomes composed of PC/chol — with or without virus receptors — does not lead to the formation of leaky vesicles. We suggest that the leakage observed upon the interaction of RSVE (or intact virions) with liposomes bearing sialic acid-containing components, precedes the fusion event and is not a consequence of it. We propose that, once the fusion process is completed, the leakage is stopped. This probably explains our observations (Tables 1-3) that virus-induced lysis of loaded liposomes never exceeds 40-60% of the liposome content.

OSMOTIC SWELLING IS REQUIRED TO ALLOW FUSION OF SENDAI AND INFLUENZA VIRIONS WITH BIOLOGICAL MEMBRANES LACKING VIRUS RECEPTORS

The above described experiments clearly show that virus receptors are required for fusion of Sendai and influenza virions with biological membranes (HEG) and with living cultured cells but not for fusion with liposomes. These results suggest that no specific receptors are needed for the fusion of membranes possessing exposed phospholipids such as liposomes.

However, the phospholipid molecules in biological membranes are masked, as was demonstrated by their unavailability to external ligands such as phospholipases (32). It is conceivable that the phospholipid bilayer of cell plasma membranes is also unsusceptible to interaction with viral envelope glycoproteins. Therefore, unmasking of membrane phospholipids may be required to allow their interaction with the viral envelope glycoproteins, which will subsequently results in virus-membrane fusion. It has been established that the interaction of Sendai and influenza virions with membrane receptors induce lysis. Such virus-induced increase in membrane permeability may result in cell swelling and the exposure of membrane phospholipids. Indeed, osmotic swelling of sealed human erythrocyte ghosts or vesicles have been shown to promote unmasking of their membrane phospholipids, as demonstrated by their availability to added phospholipases (32).

The results in Table 4 confirm previous observations (28,30) showing that incubation of Sendai virions with pronase-neuraminidase-treated, *i.e.* virus receptor-depleted, right-side-out erythrocyte vesicles (ROV) resulted in very little DQ. In contrast, a high degree (52%) of DQ was observed upon incubation of R_{18}-labeled Sendai virions with untreated ROV. The view that the DQ observed following incubation with ROV reflects a process of virus-membrane fusion, is evident from the results showing that very little DQ was obtained upon incubation of unfusogenic Sendai virions with untreated ROV (Table 4). When active, fusogenic virus particles were incubated with virus receptor-depleted ROV under hypotonic conditions (270 mOsm), a marked increase in DQ was observed. No significant DQ was observed at 270 mOsm, following incubation of unfusogenic virus particles with the pronase-neuraminidase-treated ROV.

Table 4. Fusion of Sendai Virions with Biological Membranes Lacking Virus Receptors: Effect of Osmolarity

System		ROV		NP-ROV		CGMV		PC/chol	
	mOsm:	330	270	330	270	600	240	330	270
					R_{18}-DQ (%)				
SV		52	56	2	48	3	36	45	44
SV_{DTT}		5	3	1	4	2	12	5	4
SV_{PMSF}		4	2	2	4	N.D.*	11	2	4
SV_{Tryp}		4	2	2	4	N.D.	11	2	4

*N.D., not determined.

Intact (ROV) or neuraminidase-pronase-treated right-side-out vesicles (NP--ROV) of human erythrocytes were prepared and sealed in 330 mOsm of NaCl, as previously described (28,30). Membrane vesicles of chromaffin granules (CGMV) were prepared and sealed in 600 mOsm of KCl, as described before (28). R_{18}-labeled viral preparations (5 μg), obtained as described in Fig. 1, were incubated for 45 minutes at 37°C with ROV, NP-ROV (200 μg of each), CGMV (400 μg) or PC/chol liposomes (50 μg of PC). The osmolarity of the incubation medium was adjusted to the indicated values, as described earlier (28, 30). At the end of the incubation period, the degree of R_{18}-DQ was determined as described in Fig. 1 and before (10,12). The symbols of the various virus preparations are as in Fig. 1.

Our previous results (28) unequivocally showed that the phenomenon observed, namely increase in DQ upon incubation with ROV under hypotonic conditions, is not due to a decrease in the ionic strength. The same degree of DQ was observed when sucrose was used instead of NaCl during the preparation of the ROV as well as to maintain the osmolarity of the incubation medium (not shown). Maximum dequenching was observed in a medium adjusted to 270 mOsm with NaCl or with sucrose (28).

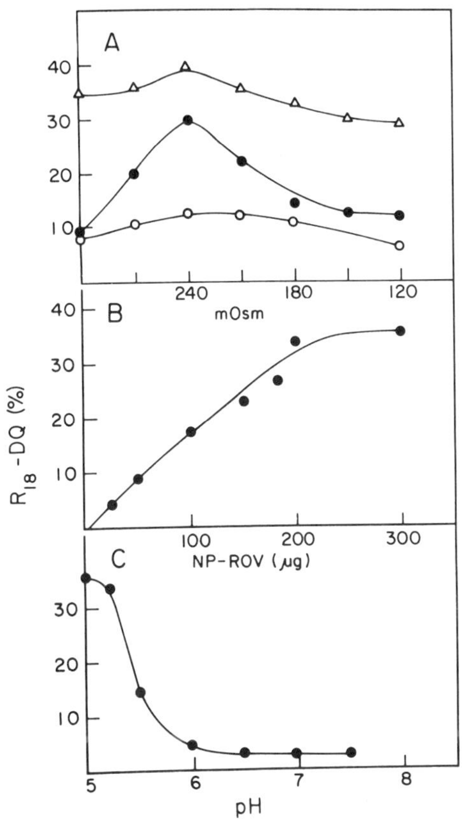

Fig. 3. Fusion of influenza virions with ROV lacking virus receptors: Effect of medium osmolarity. ROV were prepared and treated with neuraminidase and with neuraminidase and pronase, as described in Table 4 and before (28,30). A. Fluorescently labeled influenza virus particles (5 μg protein) were added to ROV (Δ), neuraminidase-treated ROV (o), or neuraminidase-pronase-treated ROV (●) (200 μg protein of each) suspended in 250 μl of PBS-Na-acetate, pH 5.0, and adjusted by addition of water to the indicated osmolarity. Following 30 minutes of incubation at 37°C, the extent of DQ was estimated. B. Fluorescently labeled influenza virions (5 μg viral protein) were incubated with increasing amounts of neuraminidase-pronase-treated ROV suspended in PBS-Na-acetate, pH 5.0, and adjusted by addition of water to 240 mOsm. After 30 minutes at 37°C, the extent of DQ was monitored. C. Fluorescently-labeled influenza virions (5 μg) were incubated with 200 μg of neuraminidase-pronase-treated ROV in PBS-Na-acetate, adjusted to 240 mOsm and the indicated pH values (28,30). NP-ROV, neuraminidase-pronase-treated ROV.

Essentially the same results were obtained upon incubation of influenza virions with ROV. A high degree of DQ was observed upon incubation of influenza virions with untreated ROV at pH 5.0 (Fig. 3). A much lower degree of DQ was observed when influenza virions were incubated with ROV from which virus receptors had been removed by treatment with neuraminidase and pronase (Fig. 3). An increase in the DQ was observed only under hypotonic conditions following incubation with the pronase-neuraminidase-treated ROV. It appears that in order to allow exposure of the membrane phospholipids, the erythrocyte glycoproteins must also be removed by treatment with pronase. As can be seen, alteration of the medium's osmolarity did not have any significant effect on the degree of DQ following incubation of influenza virions with ROV which were treated with neuraminidase only (Fig. 3A). The increase in DQ under hypotonic conditions was dependent on the amount of ROV used (Fig. 3B) as well as on the pH of the medium (Fig. 3C).

The view that no specific receptors are needed for fusion of hemolytic enveloped viruses with biological membranes whose phospholipid bilayer is exposed, is further strengthened by the results in Table 4. These results show that Sendai virions are able to fuse, under hypotonic conditions, with membrane vesicles obtained from chromaffin granules of bovine medulla (CGMV).

Incubation of R_{18}-labeled Sendai virions, under isotonic conditions (in this case, 600 mOsm), with CGMV resulted in a very low degree of DQ, while a high stimulation of DQ was observed in a medium containing low concentrations of salt (240 mOsm). A relatively low DQ was observed upon incubation of CGMV with non-fusogenic Sendai virions. Essentially the same results were obtained at pH 5.0 upon incubation with influenza virions (not shown). Fusion of these virions with CGMV cannot be mediated by virus receptors such as GD_{1a}, GT_{1b} and GQ_{1b} gangliosides (33) because chromaffin granule membranes are known to lack such sialoglycolipids (34). Furthermore, treatment with neuraminidase did not have any effect on the DQ observed under hypotonic conditions (28).

The results in Table 4 also show that no change in the extent of virus--liposome fusion (as determined by the degree of DQ) was observed, following incubation with liposomes (PC/chol) in a medium of low osmolarity. From these results, it should be inferred that the increase in DQ observed with virus receptor-depleted biological membranes can not result from activation of the viral fusion protein.

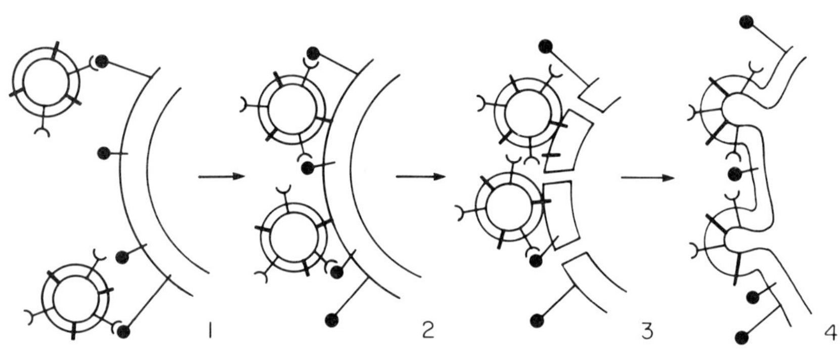

Fig. 4. Fusion of lytic enveloped viruses with cell plasma membranes: A suggested model. 1. Incubation of virus particles with cultured cells leads to binding of the virus particles to the recipient cell membranes. Binding is mediated through the attachment of the viral HN glycoprotein mainly to membrane sialoglycoproteins. Such attachment does not promote any change in the recipient cells. 2. Subsequently, the virions interact with membrane sialoglycolipids. 3. Virus-sialoglycolipids' interaction induces an increase in the recipient cell permeability (formation of defined pores?), osmotic swelling and exposure of membrane phospholipids. This allows the direct interaction of the viral envelope glycoprotein (F glycoprotein) and the membrane phospholipid bilayer. 4. Following virus--membrane fusion, the recipient cell plasma membrane is resealed and becomes unleaky once more. ●——, membrane sialoglycoproteins; ●– membrane sialoglycolipids; ≻ , the viral binding polypeptide; – , the viral fusion polypeptide.

CONCLUSIONS

The results described here may suggest that membrane receptors for enveloped lytic viruses, those belonging to the paramyxovirus or orthomyxovirus groups, do not only serve as passive binding sites which merely stimulate the contact between the virus particles and the cell plasma membranes, but it seems that membrane receptors for these viruses, the sialoglycolipids, also actively participate in the fusion event itself. Based on our observations (28-30), we suggest that the interaction between the viral envelope glycoproteins and membrane receptors induces a structural rearrangement of the phospholipid bilayer, which results in an increase in the membrane permeability to molecules of low molecular weight. The formation of specific pores of small diameter , following the interaction with NDV, has already been suggested before (35). From our work, it can be inferred that the increase in the membrane permeability occurs prior to the fusion step (Fig. 4). Thus, an increase in membrane permeability will permit water molecules to enter the living cells, resulting in osmotic swelling and a limited "stretching" of the cell plasma membrane. This process will lead to exposure of the membrane phospholipids to the viral glycoproteins. After fusion with the viral envelope, the cell plasma membrane is resealed (Fig. 4), as was demonstrated by experiments showing that virus induced leakage of cultured cells is a transient phenomenon (16). Hence, we propose a specific active role for membrane sialoglycolipids in the process of virus-induced osmotic swelling of recipient cells and exposure of membrane phospholipids, steps which may be a prerequisite to allow virus--membrane fusion.

ACKNOWLEDGMENTS

This work was supported by grants from the National Council for Research and Development, Israel, and from the Gesellschaft für Strahlenforschung (GSF), Munich (F.R.G.). We wish to thank Mrs. Edith Dicker for her assistance in the preparation of this manuscript.

REFERENCES

1. R.T. Johnson, "Viral Infections of the Nervous System," Raven Press, New York (1982).
2. M. Tardiew, R.L. Epstein and H.L. Weiner, Int. Rev. Citol. 80:27 (1982).
3. J. White, M. Kielian and A. Helenius, Quart Rev. Biophys. 16:191 (1983).
4. R. Rott and H.O. Klenk, in "Cell Surface Reviews," G. Poste and G.L. Nicolson, eds., North Holland Publishers, Amsterdam, Vol. 2, pp. pp. 77-87 (1977).
5. A. Loyter and D.J. Volsky, in "Membrane Reconstitution," G. Poste and G.L. Nicolson, eds., Elsevier/North Holland Biomedical Press, Amsterdam, pp. 225-266 (1982).
6. J.D. Lifson, G.R. Reeves, M.S. McGrath, M.S. Stein and E.G. Engleman, Science 232:1125 (1986).
7. M. Homma, K. Shimizu, Y.-K. Shimizu and N. Ishida, Virology 71:41 (1976).
8. C.L. Bashford, K.J. Micklem and C.A. Pasternak, Biochim. Biophys. Acta 814:247 (1985).
9. N. Chejanovsky and A. Loyter, J. Biol. Chem. 260:135 (1985).
10. V. Citovsky, R. Blumenthal and A. Loyter, FEBS Lett. 193:135 (1985).
11. N. Chejanovsky, S. Amselem, Y. Barenholz and A. Loyter, Biochemistry 25:4810 (1986).
12. D. Hoekstra, J. de Boer, K. Klappe and J. Wilschut, Biochemistry 23: 5675 (1984).
13. A. Loyter, V. Citovsky and R. Blumenthal, in "Methods of Biochemical Analysis," D. Glick, ed., John Wiley & Sons, New York, (1987) in press.
14. G. Poste and C.A. Pasternak, in "Cell Surface Reviews," G. Poste and G.L. Nicolson, eds., Elsevier/North Holland, Amsterdam, Vol. 2, pp. 305-367 (1978).
15. S. Knutton and T. Bächi, J. Cell Sci. 42:153 (1980).
16. K. Patel and C.A. Pasternak, J. Gen. Virol. 66:767 (1985).
17. S. Israel, D. Ginsberg, Y. Laster, N. Zakai, Y. Milner and A. Loyter, Biochim. Biophys Acta 732:337 (1983).

18. M. Ozawa, A. Asano and Y. Okada, J. Biochem. 86:1361 (1979).
19. V. Citovsky, P. Yanai and A. Loyter, J. Biol. Chem. 261:2235 (1986).
20. R.A. Lamb and P.W. Choppin, Annu. Rev. Biochem. 52:467 (1983).
21. H.D. Klenk, R.Rott, M. Orlich and J. Blodorn, Virology 68:426 (1975).
22. R.T. Huang, R. Rott and H.D. Klenk, Virology 110:243 (1981).
23. O. Nussbaum, M. Lapidot and A. Loyter, J. Virol., in press (1987).
24. M.F.G. Schmidt and B. Lambrecht, J. Gen. Virol. 66:2635 (1985).
26. S. Amselem, A. Loyter, D. Lichtenberg and Y. Barenholz, Biochim. Biophys. Acta 820:1 (1985).
27. T. Stegmann, D. Hoekstra, G. Scherphof and J. Wilschut, J. Biol. Chem. 261:10966 (1986).
28. V. Citovsky, Y. Laster, S. Schuldiner and A. Loyter, Biochemistry, in press (1987).
29. V. Citovsky, N. Zakai and A. Loyter, Exp. Cell Res. 166:279 (1986).
30. V. Citovsky and A. Loyter, J. Biol. Chem. 260:12072 (1985).
31. J. Barbet, P. Machy, A. Truneh and L.D. Leserman, Biochim. Biophys. Acta 772:347 (1984).
32. Y. Laster, E. Sabban and A. Loyter, FEBS Lett. 20:307 (1972).
33. M.A.K. Markwell, L. Svennerholm and J.C. Paulson, Proc. Natl. Acad. Sci. USA 78:5406 (1981).
34. M.T. Abbs and J.M. Phillips, Biochim. Biophys. Acta 595:200 (1980).
35. A. Szeigoleit, R. Fussle and R. Rott, Med. Microbiol.Immunol. 165:101 (1978).

'ENTRY' OF ENVELOPED VIRUSES INTO LIPOSOMES

Anne M. Haywood

Departments of Pediatrics, Medicine, and Microbiology
Box 689, University of Rochester Medical Center
Rochester, N.Y. 14642, U. S. A.

INTRODUCTION

'Enveloped' viruses, e. g., paramyxoviruses, myxoviruses, herpesviruses and human immunodeficiency virus, are viruses which have a membrane surrounding their nucleoprotein. The main function of the membrane of enveloped viruses is to provide a mechanism for entry and exit of the virion (the virus particle). Viral entry into cells comprises viral attachment to receptors, envelopment of the virus by the host, and fusion of viral and host membranes followed by viral disassembly (uncoating), activation of viral enzymes and messenger RNA synthesis. Membrane fusion is part of a continuum of membrane-related steps involved in viral entry and is probably only fully understood in the context of all the steps in viral entry.

As first shown in 1968, Sendai virus membranes can fuse with cellular plasma membranes.[1] Sendai virus is also taken into the cell by endocytosis. For many other viruses it has been difficult to get unequivocal micrographs of fusion of the viral and host plasma membranes. For this reason it has been assumed that these viruses enter by endocytosis followed by fusion of the viral membranes with the membranes of endocytic vesicles. In recent years this assumption has been further modified to state that the viral membranes must fuse with vesicular membranes at low pH. Fusion of viral membranes with a vesicular membrane, however, has never been demonstrated by microscopy. Direct evidence is lacking both for the assumption that the infecting Sendai virion enters by direct fusion with plasma membranes and for the assumption that for many other viruses the infecting virion enters by fusion with the membrane of an endocytic vesicle. The uncertainty about the route of entry that leads to initiation of infection arises from the fact that only a small proportion of the virions initiate infection and from the fact that cells ingest virions as they do other foreign particles. For electron microscope studies very high multiplicities (number of virions per cell) are used, so most of the particles visualized are not initiating infection. As will be discussed, further complicating the interpretation of the mode of entry is the fact the processes of endocytosis and fusion are interrelated.

Sendai virus has been extensively utilized to study membrane fusion both because of its marked ability to cause fusion and because Sendai virus membranes contain only three proteins. These proteins are the two glycoproteins, which form 120Å long spikes on the surface of the virus, and the M or matrix protein which is present on the inner side of the bilayer. One glycoprotein is the HN protein which binds to the viral receptor and in addition has neuraminidase (sialidase) activity. Binding of the HN protein to receptors on erythrocytes causes hemagglutination. There are indications that the HN protein participates in membrane fusion not only by causing binding but also by supplying other necessary functions.[2] The second glycoprotein is the F (fusion) protein. It is synthesized as an F_0 precursor and is cleaved to the active F protein, which is necessary for fusion of the viral and host (or liposomal) membranes and for infectivity.[3,4] The M protein is basic,[5] associated with actin,[6] and forms a matrix on the inner side of the bilayer.[7] The M protein is important in forming and maintaining the viral structure and is a major component in what has been referred to as the "viroskeleton".[8] The interior of the virus contains a single-stranded, minus-strand RNA which is associated with many nucleoprotein (NP) subunits to make a nucleocapsid that is a flexible rod. The interior of the virus also contains the L and P proteins which contribute to the transcriptase (RNA polymerase) activity[9] that is necessary to copy the minus strand virion RNA into plus strand messenger RNA's. Various enzyme activities have been found associated with the virion. In addition to the transcriptase these include methyltransferase,[10] polyadenylase,[11] protease,[12,13,14] protein kinase,[15,16] protein phosphatase,[15] and ATPase and ADPase[17] (ATP diphosphohydrolase) activities. It seems likely that these enzymatic activities participate in viral entry.

The use of liposomes in place of the host cell to study viral entry results in an experimental system that is simple enough to allow physical studies but permits the viral membrane to direct the interactions. When care is taken to use conditions that are comparable to those occurring *in vivo*, the characteristics of the viral membrane interactions are similar to those observed during virus entry, and several of the principles discovered by using liposomes have later been shown to apply also to cells.

In this paper the 'entry' of Sendai virus and of influenza virus into liposomes is discussed. The modulation of Sendai virus binding to receptors by carbohydrate polymers is described. A mechanism of fusion is proposed for Sendai virus membranes whereby the bending of liposomal or host membranes caused by envelopment of virus creates a very curved region of membrane and this curvature is necessary for fusion. The fusion of influenza virus membranes with liposomes at pH 7.5 is described, and the question is raised whether the mechanisms found for Sendai virus entry also apply to influenza virus entry. Preliminary work indicates that, if the virus structure is completely intact, the virus does not disassemble after fusion of the viral membranes with liposomes. This suggests that viral disassembly is an active process that either may involve viral enzymes (with substrates taken from the host) or may involve host factors.

VIRAL BINDING TO RECEPTORS IN LIPOSOMES

The receptors for both the myxovirus and the paramyxoviruses contain sialic acid, and both viruses contain a sialidase (neuraminidase) that is capable of digesting the viral receptors and releasing the virus. Since binding to receptors occurs at 0-4°C and the sialidase is only active at higher temperatures, binding of these viruses is usually studied at 0-4°C. Sialoglycoproteins have long been known to be able to bind these viruses.

Gangliosides when incorporated into liposomes also are capable of multivalent binding with the many HN proteins of Sendai virus and can serve as Sendai virus receptors.[18] When centrifuged at 8,800 x g for 10 min, viruses cosediment with large liposomes if they contain phosphatidylcholine (PC) and gangliosides but not if they contain PC and other negatively charged or zwitterionic lipids. Liposomes containing gangliosides cause inhibition of hemagglutination. Not all gangliosides can serve as Sendai virus receptors.[19,20] Ganglioside G_{D1a} is an example of a ganglioside with receptor activity for Sendai virus. Currently it is not known whether sialoglycoproteins, gangliosides, or both are receptors in the cell. The addition of specific gangliosides can make cells that have been treated with sialidase and lost their susceptibility to Sendai virus infection regain their susceptibility.[21] Therefore gangliosides must have a role in infection of cells.

Adhesion of virus to G_{D1a}-containing liposomes is weak enough that it can be released by high speed (300,000 x g) centrifugation.[22] The adhesion of virus to G_{D1a}-containing liposomes, however, is increased so it can withstand such centrifugation when polymers such as ficoll or dextran are present.[23] The increased adhesion of virus caused by these polymers occurs only with liposomes containing receptor gangliosides and not with liposomes containing gangliosides that are not receptors or other negatively charged lipids. This then represents a new kind of modulation of receptor binding. Similar modulation has been reported for the binding of lectins to receptor glycoproteins reconstituted in liposomes .[24] This is of technical importance, since biologic complexes are often isolated on ficoll or dextran gradients. Since there are polymers in serum and extracellular fluid and considerable carbohydrate in the glycocalyx, similar modulation of receptor binding may also occur *in vivo.*

Sendai virus membranes can fuse with liposomes that do not contain gangliosides, so binding to specific receptors is not an absolute requirement for membrane fusion.[25] However, it seems likely that binding to receptors does more than merely supply the specificity needed for viruses to become attached to their host cells. It is possible that virus receptors are also receptors for other ligands related to host cell functions and that binding stimulates the host function thereby preparing the host cell for infection. Alternatively binding could initiate changes in the virion structure which are required for one of the steps in entry. Knutton showed that major rearrangements in the surface of Sendai virus occur after viral binding to erythrocytes at 37°C.[26,27]

ENVELOPMENT ('ENDOCYTOSIS') OF VIRUS

Although binding of virus occurs at 0-4°C, both endocytosis of viruses by cells and viral membrane fusion occur only at higher temperatures. In the case of endocytosis by cells this requirement for higher temperatures is thought to be related to a requirement for energy production. Liposomes, however, also envelop viruses only at higher temperatures.[28] Envelopment of the virus by liposomes occurs when the liposomes contains gangliosides. Liposomes that contain only PC and no charged lipid, however, do not envelop the virus. Therefore it was proposed that the envelopment results from binding of many HN proteins over the surface of the virus to the receptors in the liposome with consequent deformation of the lipósome as diagrammed in Fig. 1. A similar mechanism involving binding of ligands over a particle surface with receptors on the cell surface has been described for endocytosis of particles by macrophages and has been dubbed the" zipper" hypothesis.[29]

Tsao and Huang[30] investigated the kinetics of binding and fusion of Sendai virus to ganglioside G_{D1a}-containing liposomes. When they used virus in great excess, they were able to define a step not previously described which they termed "binding stabilization". Binding stabilization does not occur at 0-4°C, but requires higher temperatures and is the rate-limiting step for Sendai virus membrane fusion when the temperature is above the transition temperature of the liposomal lipids. Both the binding stabilization and the fusion step require the F protein. It is possible that binding stabilization results from viral envelopment, since envelopment greatly increases the number of bonds between the virus HN proteins and receptors.

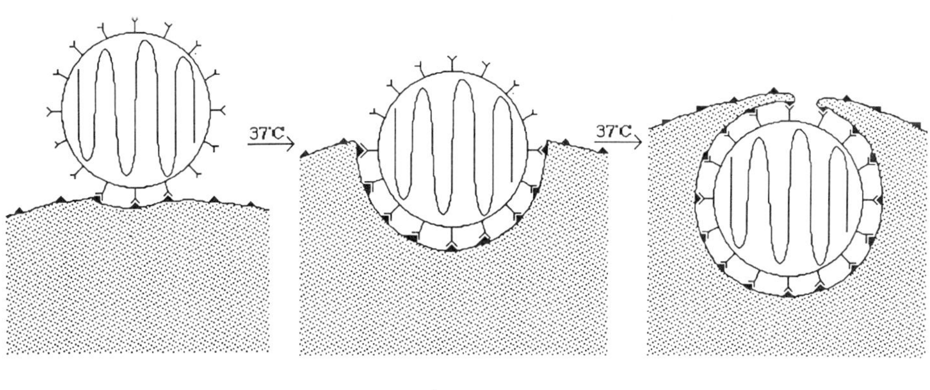

Fig. 1. Diagram of envelopment of a virus by a receptor-containing liposome. a) Binding of virus to a ganglioside-containing liposome at 0-4°C. b and c) Gradual envelopment of virus at 37°C. ▲ - Ganglioside receptor; Y - Sendai virus HN protein

Fusion of Sendai Virus Membranes with Liposomes

When Sendai viruses are incubated at 37°C with liposomes containing gangliosides, the virus membranes fuse with the liposomes. When the virus is able to disassemble, fusion of Sendai virus membranes with liposomes is easy to detect by electron microscopy, since the viral glycoprotein spikes move into the liposomal membrane and can be seen over the surface, often in patches. The viral ribonucleoprotein (RNP) can be seen inside the liposome and, in the case of multilamellar liposomes (MLV), is present between the outer lamella and the inner lamellae.[31] Fusion of Sendai virus can be quantitated by using ^{35}S-methionine-labeled virus and measuring the transfer of labeled protein to liposomes ("protein transfer fusion assay").[22] In the case of Sendai virus it is possible to separate viruses that are bound but not fused from liposomes by high speed centrifugation (300,000 x g) through sucrose. When the bound viruses are removed, the amount of ^{35}S-labeled protein associated with the liposomes indicates the amount of fusion. The ease with which bound Sendai viruses are removed from liposomes by centrifugation is not duplicated for all virus groups.

Role of Liposomal Lipid Composition in Membrane Fusion

The protein transfer fusion assay was used to determine if the lipid composition of the liposomes plays a major role in the fusion process.[25] Sendai virus membranes can fuse not only with liposomes containing PC and ganglioside G_{D1a} (which is a Sendai virus receptor) but also with liposomes containing PC and any of a range of other negatively charged lipids including phosphatidylserine (PS), phosphatidylinositol (PI), phosphatidic acid (PA), phosphatidylglycerol (PG), dicetyl phosphate, and ganglioside G_{M1} (which is not a Sendai virus receptor). However, the concentration of the receptor ganglioside G_{D1a} required for fusion is tenfold lower than the concentration of the other negatively charged lipids required for fusion. Cholesterol need not be present for fusion to occur. In fact when the liposomes contain PC, PE, and a suboptimal concentration of ganglioside G_{D1a}, cholesterol decreases the amount of fusion. The literature on the role of cholesterol in fusion of liposomes with Sendai virus membranes is very contradictory[32,33,34] and probably reflects the differing effects of cholesterol upon membranes of different compositions. Fusion occurs in the presence of EDTA.[22] The fact that Sendai virus membranes can fuse with liposomes containing PC and any of a wide range of negatively charged lipids makes it possible to eliminate most mechanisms for fusion of Sendai virus membranes with liposomes that depend upon properties of the membrane imparted by specific lipids. Thus it is possible to conclude that fusion can occur in the absence of gangliosides and that a lamellar to hexagonal phase transition, a PI to PA conversion, an anhydrous complex with Ca^{2+}, or a Ca^{2+}-induced lateral phase transition is not necessary. It is not possible, however, to eliminate the possibility that the binding of viral spike glycoproteins to gangliosides or other lipids produce a lateral phase transition and hence a packing defect. There are, however, no data indicating that this is so.

A possible difficulty in using liposomes with compositions that are completely unlike those in cellular membranes is that the virus may be induced to fuse by mechanisms that are different from those occurring *in vivo* . For instance Sendai virus membranes fuse with liposomes containing only PS[25,35] or cardiolipin.[35] However, fusion of viral membranes with liposomes with compositions such as pure cardiolipin did not strictly depend on viral protein conformation,[35] which implies that a different mechanism of fusion may be involved.

Close approach and destabilization of membrane bilayers

For two bilayers to fuse they must approach to within about 10Å and must be destabilized. When the virus HN proteins binds to receptor, the bilayers are still over 100Å apart. The barriers to close approach include the membrane proteins, the negative potential of the two membranes, and water.

Proteins. Knutton reported that after binding of the virion to erythrocytes at 37°C, the surface of the virion undergoes major changes which include rearrangements of the intramembranous particles and the formation of ridges and valleys on the surface of the liposome.[26,27] This suggests that the binding of the virus might lead to rearrangements of membrane proteins as well as of internal viral structures.

Water One of the major obstacles to close approach of bilayers is water.[36] Abidi and Yeagle have concluded from ^{31}P nuclear magnetic resonance (NMR) studies that the phospholipid headgroups in Sendai virus membranes are highly motionally restricted.[37] When the F protein on the virus is digested with trypsin, the ^{31}P-NMR spectrum becomes characteristic of a phospholipid bilayer. This indicates that the carbohydrate-bearing part of the F protein is involved in the motional restriction of the phospholipid headgroups of the intact virus and raises the possibility that the F protein might remove water from the viral and host bilayers.

Negative Potential on Virus and Host Membrane In PBS (phosphate buffered saline) Sendai virus has a ζ potential of -17 mV and ganglioside-containing liposomes and cells have a ζ potential in the same range.[18] The virus has to find a way to circumvent the considerable charge repulsion between the virus and the liposome or cell. The expression for the charge repulsion between two negatively charged particles is:[38]

$$V_R = \frac{D}{4} \frac{r_1 r_2}{r_1 + r_2} [(\psi_1 + \psi_2)^2 \ln(1 + e^{-kH}) + (\psi_1 + \psi_2)^2 \ln(1 - e^{-kH})]$$

where Ψ_1, Ψ_2- surface potentials of the particles; r_1, r_2 - radii of curvature of the particles; H is distance of closest approach of the particles; D - dielectric constant; k - Inverse Debye-Hückel length. Since Sendai virus undergoes membrane fusion external to the cell contents, it has no opportunity to change the ionic strength and thereby the Debye-Hückel length. There are no data to suggest that it can change its own surface potential. While the neuraminidase might change the potential of the liposome or cell by removing sialic acid from gangliosides, the action of the neuraminidase is relatively slow and the gangliosides should be free to diffuse in the liposome.

Fusion Occurs at the Curved Leading Edge of a Region of Membrane Enveloping Virus

Virus binding at 37°C does, however, cause a reduction in the radius of curvature, since the leading edge of the region of liposome enveloping the virus is very curved. Furthermore the ridges formed on the virus surface during binding[26,27] have a decreased radius of curvature. Electron micrographs taken of Sendai virus with liposomes (containing PC, PE, cholesterol and gangliosides) after only a few minutes of 37°C incubation made it clear that fusion occurs at the curved leading edge of the region of liposome that is enveloping virus (Figure 2).[39,40] The curvature on the leading edge of a region of liposome enveloping virus or of a developing cellular endocytic vacuole should not only greatly decrease the charge repulsion between the virus and liposome or plasma membrane in this region but it should also destabilize the bilayers by separating the phospholipids on the convex surface. Thus the formation of a very curved region in the liposome results in both close approach and destabilization of only a small region of the liposome or host cell membrane. This allows

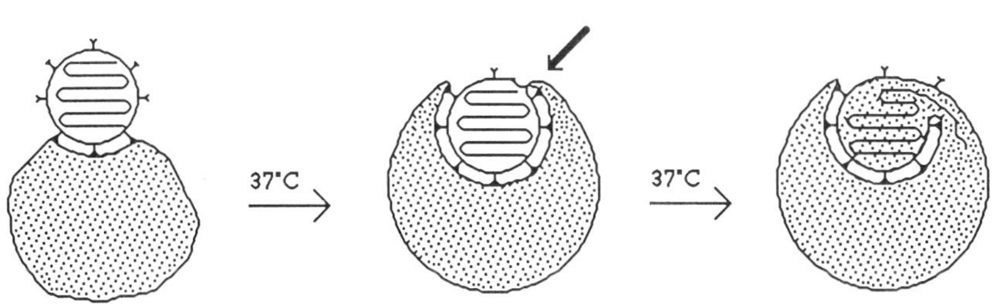

Fig. 2. Fusion of a virus and liposome occurs at the leading edge of a region of the liposome that is enveloping the virus. At this leading edge the curvature of the membrane facilitates close approach of the bilayers and destabilizes the bilayers by separating the phospholipids in the convex monolayer.

localization of the fusion process, so that the cell is not generally perturbed. Examples of fusion of virus with the leading edge of a developing endocytic vacuole are evident in earlier literature on fusion of Sendai virus membranes with cells including the initial paper on Sendai virus membrane fusion.[1]

According to the proposed mechanism deformation of a membrane with the formation of a very curved region is the important event, and this can occur without complete envelopment of the virus. Fig. 3 shows a liposome which has fused with a virus at the curved leading edge of a region of partial envelopment. For a cell or liposome to be deformable and to at least partially envelop a virus, it must have excess surface area, i.e., area in addition to that required to enclose its own volume.[41] By the proposed mechanism the number of viruses that can fuse with a cell or liposome should depend upon the excess surface area available and should be considerably less than the number that can bind. This has been shown to be true for erythrocyte ghosts.[42] If the binding stabilization step described by Tsao and Huang[30] does correspond to envelopment of the virus, the proposed mechanism of fusion fits with their description of the kinetics of fusion. Thus envelopment is likely to be the rate-limiting step. Since formation of a curved region also destabilizes the membrane, the fusion event that follows consists mainly of mixing of the viral and liposomal components. This is consistent with their suggestion that the activation of the fusion step is entropy controlled.

Ohki has stated that conditions that lead to membrane fusion all have in common events that cause an increased surface hydrophobicity or surface free energy.[43] The above mechanism for viral membrane fusion is an illustration of this more general hypothesis. It is also an example of how mechanical factors as well as biochemical factors can change the properties of the membrane.

The Sendai virus F protein is required for membrane fusion. The proposed mechanism does not involve a direct role for the F protein in the destabilization of the bilayer. Another postulated mechanism for Sendai virus membrane fusion is direct destabilization of the viral membrane by insertion of the hydrophobic NH_2-terminus of the viral F_1 protein.[44] The role of the F protein in fusion, however, does not have to be in the destabilization step. Several studies suggest there is a requirement for the F protein in the steps leading up to the destabilization. Thus Knutton[27] showed a requirement for the F protein for reorganization of the viral surface following binding at 37°C, Tsao and Huang[30] indicated a role for the F protein in the binding stabilization and Abidi and Yeagle's[37] work raised the question whether this glycoprotein can participate in dehydration of the viral and host membranes.

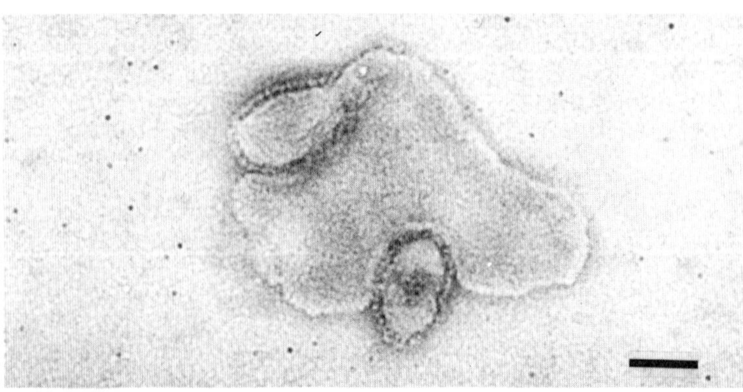

Fig. 3. Liposomes were made in PBS from 0.7 μmol PC, 0.3 μmol PE, 0.66 μmol cholesterol and brain gangliosides containing 0.11 μmol N-acetyl neuraminic acid. Sendai viruses (8,192 hemagglutinating units) were added at 0°C for 1 hr. The temperature was then raised to 37°C for 1 minute. The sample was stained with 2% ammonium molybdate at pH 7.8. Bar equals 100 nm.

Time, Temperature and pH Requirements for Sendai Virus Membrane Fusion

The fusion of Sendai virus membranes with liposomes at neutral pH like the entry of Sendai virus into cells is neither fast nor efficient.[22] Fusion occurs most rapidly during the first 15 minutes of incubation and then continues to increase steadily for over two hours. Even after 18 hrs at 37°C only 20-30% of the virus particles have fused and this number does not increase when the incubation is continued for a total of 24 hrs. When the amount of fusion occurring over two hours at different temperatures is measured, the amount of fusion increases with temperature between 15 and 40°C. Around 42°C the amount of fusion reaches a maximum and begins to decrease, which is consistent with the fact that hemagglutination has been found to be sensitive to temperature between 40 and 45°C.[45]

When the pH is varied, there is a broad maximum for Sendai virus membrane fusion between pH 7.5 and 9.0. Also at either extreme of pH the amount of fusion increases. At pH 11 the amount of fusion is increased by at least 50%. At pH 4.0 the amount of fusion is well over double that at neutral pH and the virus forms aggregates.[22] Not only the extent but the rate of fusion is greater at low pH.[35] Sendai virus which has had the F protein removed by trypsin treatment fuses with cardiolipin (CL) liposomes at pH 4.[35] When liposomes are reconstituted with the HN protein but not the F protein, they can undergo fusion with PS liposomes at pH 4.0.[46] Because fusion at low pH does not require the same participation of viral proteins as viral entry and because it is highly unlikely that there is a cellular compartment with a pH of 4.0 or lower, it seems probable that the efficient fusion of Sendai virus membranes at pH 4.0 or lower does not reflect an *in vivo* process. This again underscores the need to study viral interactions with liposomes under physiologic conditions.

FUSION OF INFLUENZA VIRUS MEMBRANES WITH LIPOSOMES

Influenza virus is structurally very similar to Sendai virus. Its membrane also contains two glycoprotein spikes, a hemagglutinin protein which binds receptors and appears to have a function similar to that of the F protein of Sendai virus and a neuraminidase protein which has neuraminidase activity. On the basis of its structure, influenza virus should be expected to enter cells by a mechanism similar to that of Sendai virus. Since the late 1960's there have been reports of fusion of influenza virus membranes at neutral pH.[47,48,49,50] The most convincing report is that of Fidgen and Tisdale[50] who incubated two strains of influenza virus with erythrocyte ghosts and showed electron micrographs where the viral glycoprotein spikes were clearly distributed over the surface of the ghosts.

Influenza virus membranes, as well as the membranes of very different viruses such as vesicular stomatitis virus and Semliki Forest virus, can be induced to fuse rapidly and efficiently by lowering the pH. A conformational change in the influenza virus hemagglutinin has been observed to occur at a pH that correlates with the pH at which fusion occurs.[51] The hemagglutinin protein contains a hydrophobic region that resembles the NH_2-terminus of the Sendai virus F protein. One postulated mechanism for influenza virus fusion states that, after a low pH-induced conformational change of the hemagglutinin, the hydrophobic region is inserted into the host membrane thereby destabilizing it.[51,52,53] Low pH also has other effects on viruses. It affects the binding of viruses and of their isolated spike glycoproteins.[53,54] For Semliki Forest virus it was suggested that at neutral pH the spike protein binds specifically and that at low pH it binds nonspecifically.[54] BHA (influenza virus hemagglutinin released from the virion by bromelain) can bind to only those liposomes that contain gangliosides at pH 7.0 but can bind to liposomes made of a variety of compositions at low pH.[53] Another effect of lowering pH is to change the net charge on the virus. For a few strains of influenza the isoelectric points have been published, and for these few strains the isoelectric points are close to the low pH at which fusion is reported.[22] Since the different influenza virus strains fuse at pH's above 5.0 and since endosomes have a pH between 5 and 6, it has been proposed that viral membrane fusion occurs in an endocytic vesicle at low pH. The hypothesis that a low pH-induced conformational change in the influenza virus hemagglutinin is required for membrane fusion and that influenza virus entry must be through endosomes was initially based on the assumption that fusion can not occur at neutral pH but only at low pH.

Because previous reports suggested fusion of influenza virus membranes with cell membranes occurs at neutral pH and because it seemed possible that low pH greatly facilitates binding and/or close approach rather than causes an essential change in viral components, experiments were undertaken to determine if there are experimental conditions where influenza virus membranes fuse with liposomes at neutral pH. If the effect of low pH is to affect binding and close approach, then fusion at neutral pH should have much slower kinetics than fusion at low pH and should require a composition that facilitates binding. Influenza virus X-31 fuses with liposomes containing 3 mol% G_{D1a} at pH 7.5.[55] The fusion can be detected by electron microscopy, since after fusion the influenza viral glycoproteins spikes are distributed in the outer lamella of the liposomes. Fusion can also be detected by transfer of ^{35}S-methionine labeled viral proteins to liposomes. Influenza virus binds more tightly to liposomes than does Sendai virus, so in order to be able to remove bound virus by centrifugation it is necessary first to add fetuin, a sialoglycoprotein that binds influenza. This assay showed that the amount of fusion of influenza virus membranes with liposomes containing PC and 5 mol% PS was less than with liposomes containing 3 mol% G_{D1a}. Thus influenza virus membranes like Sendai virus membranes fuse with liposomes containing negatively charged lipids but the presence of G_{D1a} in the liposomes facilitates fusion more than the presence of PS.

Liposomes with no net charge such as those containing only 1 μmol of PC or a combination of 0.33 μmol PC, 0.33 μmol PE, 0.33 μmol sphingomyelin and 0.5 μmol cholesterol do not fuse with influenza virus membranes at neutral pH. They do, however, fuse with influenza virus membranes at pH 5.2. Thus a requirement for low pH similar to that described by other investigators can be demonstrated for fusion of influenza virus membranes with liposomes containing no net charge. This suggests that one effect of low pH may be to cause nonspecific aggregation of influenza virus with surfaces that do not carry virus receptors.

Studies of the time course of fusion of influenza virus membranes with liposomes indicated that at 37°C only 6% of influenza viruses have fused by two minutes and 11% by 5 minutes.[56] After 5 minutes the rate of fusion slows down, although fusion continues up to at least one hour when the amount of fusion is 23%. This is considerable contrast to fusion at low pH where as much as 90% of the viruses have fused by 2-5 minutes.[57,58] This difference in rate and extent of fusion may in large part be responsible for the suggestion that fusion can not occur at neutral pH, since a comparison of the amount of fusion after an incubation of only a few minutes could easily give this impression. The rate and amount of fusion of influenza virus membranes with liposomes at neutral pH are consistent with the rate of viral entry into cells. The rate and amount of fusion are also consistent with the rate and amount of fusion of Sendai virus membrane fusion with liposomes where it is postulated the envelopment may be the rate-limiting step. The postulated entry of viruses into endosomes followed by fusion at low pH also has endocytosis as the rate-limiting step. Therefore the rate of influenza virus membrane fusion can not be used to distinguish whether entry *in vivo* is associated with fusion at low or neutral pH.

When the amount of fusion over the course of one hour was measured at different temperatures, the logarithm of the amount of fusion increased directly in proportion to the increase of temperature between 15°C and 50°C.[56] While Sendai virus hemagglutinin has been reported to be inactivated between 40°C and 45°C, the hemagglutinin of most influenza viruses is exceptionally heat resistant and is not inactivated at 50°C.[45,59] A change in the circular dichroism of influenza virus X-31 hemagglutinin, which is thought to correspond to denaturation, has been shown to occur at 62-63°C.[60]

Several technical details probably explain the fact that fusion of influenza virus at neutral pH is often not detected. The most important reason is the difference in kinetics at neutral and low pH. Because of the much slower kinetics at neutral pH it is necessary not only to use an adequate incubation time (e.g., one hour rather than five minutes) but also to use adequate concentrations of virus and liposomes. Because of the problem of light scattering very low concentrations of liposomes are often used in experiments using fluorescence assays[61] and such low concentrations of liposomes result in the rate of fusion being even slower. Also, as is evident from the above experiments, fusion of influenza virus membranes at neutral pH is dependent upon the liposome composition. In this context it must be recognized that different strains of influenza as well as different paramxyoviruses have different receptors.[62] Even if gangliosides are included in the liposome composition the viral receptor may not be present. Also not only will too low concentrations of gangliosides fail to promote binding but too high concentrations of gangliosides decrease the amount of fusion. Another reason for failing to observe fusion at neutral pH is that influenza viruses tend to aggregate and aggregated virus do not fuse at neutral pH. Lastly, disassembly makes fusion more obvious, and damage to the virus caused by low pH probably greatly assists disassembly. To facilitate study of fusion at neutral pH, it is possible, at least with fusion with liposomes, to take advantage of the fact that at 50°C the fusion occurs at a greatly increased rate.

The occurrence of fusion at neutral pH makes it possible to conclude that an acid dependent configurational change of the influenza hemagglutinin is not a requirement for fusion to occur. There is no information about whether the mechanism of fusion at neutral pH is different from that at low pH. Since fusion of influenza virus membranes can occur at both neutral and low pH, at present it is not possible to say whether viral entry occurs by fusion at low pH with the membrane of a vesicle or at neutral pH with the plasma membrane or with the membrane of a vesicle. The fact that the details of membrane fusion of influenza virus and Sendai virus membranes at both low and neutral pH are generally similar suggests, as one would also surmise from the similarities in structures of the viral membranes, that they may enter by the same mechanism. Since the fusion of Sendai virus membranes at low pH is at a pH that is less than that known for any cellular compartment and appears to proceed by a mechanism that is not physiologic, it is unlikely that Sendai virus enters the cell at low pH. Whether influenza virus enters by the same mechanism as Sendai virus deserves further investigation.

DISASSEMBLY (UNCOATING) OF ENVELOPED VIRUSES AFTER MEMBRANE FUSION

A critical part of viral entry is disassembly which includes the rearrangements of viral components that must occur for the viral components to be released from each other and distributed in the host cell, for viral enzymes to become active within the host cell and for RNA to be synthesized. Many of the interactions that were formed during assembly are probably reversed during disassembly. When virus is assembled, viral glycoproteins are inserted into the host membrane and then organized in the region of host membrane that contains M protein, the M protein interacts with the nucleocapsid, and then this region of host membrane which contains viral proteins and is associated with the nucleocapsid buds away from the host membrane and is released.[63] Most of what is known about the disassembly process comes from morphologic investigations, and extremely little is known about the biochemical aspects of disassembly. Disassembly is important for understanding membrane fusion, since morphologically fusion is very hard to identify in the absence of disassembly. If the virus fuses and does not disassemble, it may have the option of budding away from the cell again, so disassembly may make fusion irreversible.

For the HN proteins to be able to migrate into the liposomal membrane during disassembly, their binding to receptors must be released. Release of the HN protein from sialic acid-containing receptors during disassembly is one possible function of the viral sialidase. After release of the HN proteins from receptors, two patterns are seen.[39,64] In one pattern the viral glycoproteins move across the fusion bridge into the liposomal membrane and the RNP is released. Intermediate structures such as that diagrammed in Fig. 4a are seen. Finally the viral glycoprotein spikes are distributed over the surface of liposome and most of the RNP is released, although one section remains associated with the membrane (Fig. 4b). Other viruses are released from envelopment upon continued incubation but do not disassemble, so structures such as that diagrammed in Fig. 4c are seen. The presence of these structures is probably one reason why the fusion at the leading edge of a region of envelopment was not recognized earlier. During the first few minutes of incubation fused virus and liposome membranes are seen only at the leading edge of an enveloping region. The structures diagrammed in Fig. 4c are seen only subsequently after further incubation, but might be interpreted as the main fusion intermediates if micrographs were not also taken after shorter incubation times. If the fusion bridge in this structure is narrow, it may be hard to recognize by microscopy that the virus is fused rather than just bound and the contents of the virus and liposome may not be mixed, so that failure to disassemble may result in failure to identify fusion.

Differences in patterns of disassembly of viruses into erythrocytes also have been previously noted. These have been correlated with whether early harvest or late harvest viruses were used. Sendai virus is usually grown in embryonated eggs. If the viruses are harvested after 24 hrs incubation, they have undergone one round of replication and are termed "early harvest" viruses. If instead they are harvested after 48-72 hrs ("late harvest" viruses), some of the viruses formed during the first 24 hrs have initiated additional rounds of replication and others have been incubated in the egg an additional 24-48 hrs, which results in changes in the viruses. The structures of early and late harvest viruses are quite different, and it appears that the interactions between viral proteins, especially the M protein, are changed in late harvest viruses.[8] Early harvest viruses fuse with erythrocytes but do not cause hemolysis, but late harvest viruses cause hemolysis.[65,66] Diffusion of viral antigens (glycoproteins) over the erythrocyte surface, as demonstrated by immunofluorescence, occurs when late harvest viruses fuse with erythrocytes. When early harvest viruses fuse with erythrocytes, however, the diffusion of the viral glycoproteins is restricted unless the fused viruses and erythrocytes are exposed to low osmotic pressure and swell.[67] Preliminary data suggest that the differences between early harvest and late harvest Sendai viruses also determines the pattern of disassembly in liposomes. Initial studies indicate that, after fusion with liposomes, early harvest viruses exhibit the pattern shown in Fig. 4c.

The changes that occur in Sendai viruses with prolonged incubation seem to be a particular characteristic of Sendai virus. Other viruses can be altered by laboratory procedures such as freezing and thawing, sonication, osmotic changes, and low pH. Influenza viruses even after 48-72 hours incubation in embryonated eggs appear to be intact. In those few situtations where influenza viral glycoproteins have been seen to be distributed over the surface of cellular plasma membranes or liposomes, influenza viruses have undergone treatment which is likely to have damaged them. Thus in the studies of Fidgen and Tisdale,[50] the viruses were incubated with the erythrocyte ghosts on a grid and then washed with distilled water. This would have resulted in osmotic swelling of the fused viruses and erythrocytes and would have been analogous to the effects of osmotic swelling on the disassembly of early harvest Sendai viruses into erythrocytes.[66] Those viruses that were used for electron microscopic studies of fusion of influenza virus with liposomes had undergone a mild

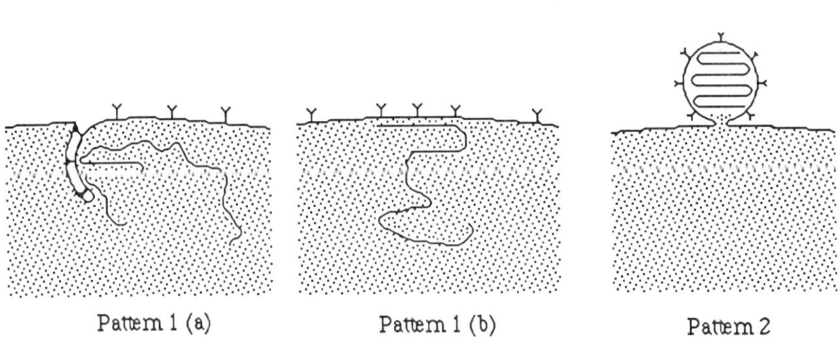

Fig. 4. Patterns of Sendai virus disassembly after fusion of viral and liposome membranes.
Pattern 1. Disassembly of the virion is concurrent with release of viral HN proteins from receptor binding. a) Partial disassembly. b) Complete disassembly.
Pattern 2. Viral HN proteins are released from receptor binding, but the virion stays mainly intact.

sonication, since during the initial studies the viruses were often aggregated.[55] (The viruses used in later work were not sonicated since it was found that aggregation could be avoided by freezing the virus with a low concentration of bovine serum albumin.[56]) Low pH changes the structure of influenza viruses[53,55,60,68], and, after fusion of influenza virus membranes at low pH, the viruses undergo disassembly. Preliminary data suggest that intact influenza viruses may not disassemble after fusion with liposomes at neutral pH.

If disassembly does not follow fusion of virus membranes unless laboratory procedures have caused some structural changes in the virus, several interesting possibilities arise. 1) The present concepts about which viruses fuse with the host plasma membrane may reflect whether the virus is damaged and therefore can disassemble into the plasma membrane. 2) If, when viruses fuse during the course of envelopment by the host plasma membrane, the viruses do not disassemble, endocytosis of the virus may be completed even though fusion has occurred. Release of the viral components from endocytic vesicles would then be a function of disassembly rather than of membrane fusion. 3) Since viral membrane fusion can occur both at neutral pH and at low pH, it is difficult to definitively establish the pH at which viral entry occurs *in vivo* . This question might be resolved by determining whether the viral enzymatic activities and RNA replication are normal after fusion at only one pH. 4) The virion enzymes are candidates for participation in disassembly, in which case disassembly could be an excellent target for antiviral agents.

As has been described in this paper, the steps of viral entry are interrelated. Thus a complete understanding of viral membrane fusion requires study of fusion in the context of both binding to viral receptors and viral disassembly.

ACKNOWLEDGEMENTS

Part of this work was supported by grants PCM 78-08931 and PCM 82-05896 from the National Science Foundation and by grant AI-15540 from the National Institutes of Health.

REFERENCES

[1]Morgan, C., and Howe, C., 1968, Structure and development of viruses as observed in the electron microscope. IX. Entry of Parainfluenza I (Sendai) virus, J. Virol., 2:1122-1132.

[2]Miura, N., Uchida, T., and Okada, Y., 1982, HVJ (Sendai virus)-induced envelope fusion and cell fusion are blocked by monoclonal anti-HN protein antibody that does not inhibit hemagglutination activity of HVJ, Exp. Cell Res., 141:409-420.

[3]Homma, M., and Ohuchi, M., 1973, Trypsin action on the growth of Sendai virus in tissue culture cells. III. Structural differences of Sendai viruses grown in eggs and tissue culture cells, J. Virol., 12.1457-1465.

[4]Scheid, A., and Choppin, P. W., 1974, Identification of biological activities of paramyxovirus glycoproteins. Activation of cell fusion, hemolysis and infectivity by proteolytic cleavage of an inactive precursor protein of Sendai virus, Virology, 57:475-490.

[5]Blumberg, B. M., Rose, K., Simona, M. G., Roux, L., Giorgi, C., and Kolakofsky, D., 1984, Analysis of the Sendai virus M gene and protein, J. Virol., 52:656-663.

[6]Giuffre, R. M., Tovell, D. R., Kay, C. M., and Tyrrell, D. L. J., 1982, Evidence for an interaction between the membrane protein of a paramyxovirus and actin, J. Virol., 42:963-968.

[7]Shimizu, K., and Ishida, N., 1975, The smallest protein of Sendai virus: Its candidate function of binding nucleocapsid to envelope, Virology, 67:427-437.

[8]Kim, J., Hama, K., Miyake, Y., and Okada, Y., 1979, Transformation of intramembrane particles of HVJ (Sendai virus) envelopes from an invisible to visible form on aging of virions, Virology, 95:523-535.

[9]Hamaguchi, M., Yoshida, T., Nishikawa, K., Naruse, H., and Nagai, Y., 1983, Transcriptive complex of Newcastle disease virus. I. Both L and P proteins are required to constitute an active complex, Virology, 128:105-117.

[10]Colonno, R. J., and Stone, H. O., 1976, Isolation of a transcriptive complex from NDV virions, J. Virol., 19:1080-189.

[11]Weiss, S. R., and Bratt, M. A., 1974, Polyadenylate sequences on NDV mRNA synthesized in vivo and in vitro, J. Virol., 13:1220-1230.

[12]Zhirnov, O. P., and Bukrinskaya, A. G., 1977, Study of Sendai virus proteins: Proteolytic activity in virus particles, Voprosy Virusologii, 5:571-577.

[13]Israel, S., Ginsberg, D., Laster, Y., Zakai, N., Milner, Y., and Loyter, A., 1983, A possible involvement of virus-associated protease in the fusion of Sendai virus envelopes with human erythrocytes, Biochim. Biophys. Acta, 732:337-346.

[14]Sugawara, K.-E., Tashiro, M., and Homma, M., 1982, Intermolecular association of HANA glycoprotein of Sendai virus in relation to the expression of biological activities, Virology, 117:444-455.

[15]Lamb, R. A., 1975, The phosphorylation of Sendai virus proteins by a virus particle-associated protein kinase, J. Gen. Virol., 26:249-263.

[16]Roux, L., and Kolakofsky, D., 1974, Protein kinase associated with Sendai virions, J. Virol., 13:545-547.

[17]Neurath, A. R., 1965, Study on the adenosine diphosphatase (adenosine triphosphatase) associated with Sendai virus, Acta Virol., 9:313-322.

[18]Haywood, A. M., 1974, Characteristics of Sendai virus receptors in a model membrane, J. Mol. Biol., 83:427-436.

[19]Haywood, A. M., 1975, Model membranes and Sendai virus: Surface-surface interactions, in:"Negative Strand Viruses," Barry, R. D., and Mahy, B. W. J., eds., Vol.2, pp. 923-928, Academic Press, London.

[20]Holmgren, J., Svennerholm, L., Elwing, H., Fredman, P., and Strannegärd, Ö., 1980, Sendai virus receptor: Proposed recognition structure based on binding to plastic-adsorbed gangliosides., Proc. Natl. Acad. Sci., USA, 77:1947-1950.

[21]Markwell, M. A. K., Svennerholm, L., and Paulson, J. C., 1981, Specific gangliosides function as host cell receptors for Sendai virus, Proc. Natl.Acad. Sci. USA, 78:5406-5410.

[22]Haywood, A. M., and Boyer, B. P., 1982, Sendai virus membrane fusion: Time course and effect of temperature, pH, calcium, and receptor concentration, Biochemistry, 21:6041-6046.

[23]Haywood, A. M., and Boyer, B. P., 1986, Ficoll and dextran enhance adhesion of Sendai virus to liposomes containing receptor (ganglioside G_{D1a}), Biochemistry, 25:3925-3929.

[24]Grant, C. W. M., and Peters, M. W., 1984, Lectin-membrane interactions. Information from model systems, Biochim. Biophys. Acta, 779:403-422.

[25]Haywood, A. M., and Boyer, B. P., 1984, Effect of lipid composition upon fusion of liposomes with Sendai virus membranes, Biochemistry, 23:4161-4166.

[26]Knutton, S., 1976, Changes in viral envelope structure preceding infection, Nature, 264:672-673.

[27]Knutton, S., 1978, The mechanism of virus-induced cell fusion, Micron, 9:133-154.

[28]Haywood, A. M., 1975, 'Phagocytosis' of Sendai virus by model membranes, J. Gen. Virol., 29:63-68.

[29]Griffin, F. M. Jr., Griffin, J. A., Leider, J. E., and Silverstein, S. C., 1975, Studies on the mechanism of phagocytosis. I. Requirements for circumferential attachment of particle-bound ligands to specific receptors on the macrophage plasma membrane, J. Exp. Med., 142:1263-1282.

[30]Tsao, Y-s., and Huang, L., 1986, Kinetic studies of Sendai virus-target membrane interactions: Independent analysis of binding and fusion., Biochemistry, 25:3971-3976.

[31]Haywood, A. M., 1974, Fusion of Sendai viruses with model membranes, J. Mol. Biol., 87:625-628.

[32]Kundrot, C. E., Spangler, E. A., Kendall, D. A., MacDonald, R .C., and MacDonald, R. I., 1983, Sendai virus-mediated lysis of liposomes requires cholesterol, Proc. Natl. Acad. Sci., USA, 80:1608-1612.

[33]Tsao, Y.-s., and Huang, L., 1985, Sendai virus induced leakage of liposomes containing gangliosides, Biochemistry, 24:1092-1098.

[34]Umeda, M., Nojima, S., and Inoue, K., 1985, Effect of lipid composition on HVJ-mediated fusion of glycophorin liposomes to erythrocytes, J. Biochem., 97:1301-1310.

[35]Klappe, K., Wilschut, J., Nir, S., and Hoekstra, D., 1986, Parameters affecting fusion between Sendai virus and liposomes. Role of viral proteins, liposome composition, and pH, Biochemistry, 25:8252-8260.

[36]Rand, R. P., 1981, Interacting phospholipid bilayers: measured forces and induced structural changes, Ann. Rev. Biophys. Bioeng., 10:277-314.

[37]Abidi, T. F., and Yeagle, P. L., 1984, Surface properties of Sendai virus envelope, Biochim. Biophys. Acta, 775:419-425.

[38]Weiss, L., and Harlos, J. P., 1972, Short-term interactions between cell surfaces, in: "Progress in Surface Science" 1 pp. 355-405, Pergamon Press, New York.

[39]Haywood, A. M., and Boyer, B. P., 1981, Initiation of fusion and disassembly of Sendai virus membranes into liposomes, Biochim. Biophys. Acta, 646:31-35.

[40]Haywood, A.M., 1983, Virus infection of liposomes (or virology for the microveterinarian), in:"Liposome Letters," Bangham, A.D., ed., pp. 277-287, Academic Press, London.

[41]Evans, E. A., and Parsegian, V. A., 1983, Energetics of membrane deformation and adhesion in cell and vesicles aggregation, Ann. N.Y. Acad. Sci., 416:13-33.

[42]Nir, S., Klappe, K., and Hoekstra, D., 1986, Kinetics and extent of fusion between Sendai virus and erythrocyte ghosts: Application of a mass action kinetic model, Biochemistry, 25:2155-2161.

[43]Ohki, S., 1984, Effects of divalent cations, temperature, osmotic pressure gradient, and vesicle curvature on phosphatidylserine vesicle fusion, J. Membrane Biol., 77:265-275.

[44]Richardson, C. D., Scheid, A., and Choppin, P.W., 1980, Specific inhibition of paramyxovirus and myxovirus replication by oligopeptides with amino acid sequences similar to those at the N-termini of the F_1 or HA_2 viral polypeptides, Virology, 105:205-222.

[45]Fukai, K., and Suzuki, T., 1955, On the characteristics of a newly-isolated hemagglutinating virus from mice, Med. J. Osaka Univ., 6:1-15.

[46]Chejanovsky, N., Zakai, N., Amselem, S., Barenholz, Y., and Loyter, A., 1986, Membrane vesicles containing the Sendai virus binding glycoprotein, but not the viral fusion protein, fuse with phosphatidylserine liposomes at low pH, Biochemistry, 25:4810-4817.

[47]Čiampor, F., and Križanová, O., 1971, Interaction of plasma membranes with influenza virus. III. Electron microscopic study of interactions, Acta Virol., 15:361-366.

[48]Blaskovic, P., Rhodes, A. J., Doane, F. W., and Labzoffsky, N. A., 1972, Infection of chick embryo tracheal organ cultures with influenza A2 (Hong Kong) virus, Archiv. Gesamte Virusforsch., 38:250-266.

[49]Morgan, C., and Rose, H. M., 1968, Stucture and development of viruses as observed in the electron microscope. VIII. Entry of influenza virus, J. Virol., 2:925-936.

[50]Fidgen, K. J., and Tisdale, M., 1981, An 'on grid' electron microscopic method for studying the interaction and fusion of influenza A virus with human erythrocyte membranes, J. Virol. Methods, 3:271-276.

[51]Skehel, J. J., Bayley, P. M., Brown, E. B., Martin, S. R., Waterfield, M D., White, J. M., Wilson, I. A., and Wiley, D. C., 1982, Changes in the conformation of influenza virus hemagglutinin at the pH optimum of virus-mediated membrane fusion, Proc. Natl. Acad. Sci., USA, 79:968-972.

[52]Daniels, R. S., Downie, J. C., Hay, A. J., Knossow, M., Skehel, J. J., Wang, M. L., and Wiley, D. C., 1985, Fusion mutants of the influenza virus hemagglutinin glycoprotein, Cell, 40:431-439.

[53]Doms, R. W., Helenius, A., and White, J., 1985, Membrane fusion activity of the influenza virus hemagglutinin. The low pH-induced conformational change, J. Biol. Chem., 260:2973-2981.

[54]Fries, E., and Helenius, A., 1979, Binding of Semliki Forest virus and its spike glycoproteins to cells, Eur. J. Biochem., 97:213-220.

[55]Haywood, A. M., and Boyer, B. P., 1985, Fusion of influenza virus membranes with liposomes at pH 7.5, Proc. Natl. Acad. Sci., USA, 82:4611-4615.

[56]Haywood, A. M., and Boyer B. P., 1986, Time and temperature dependence of influenza virus membrane fusion at neutral pH, J. Gen. Virol., 67:2813-2817.

[57]Yoshimura, A., Kuroda, K., Kawasaki, K., Yamashina, S. Maeda, T., and Ohnishi, S.-I., 1982, Infectious cell entry mechanism of influenza virus, J. Virol., 43:284-293.

[58]White, J., Helenius, A., and Kartenbeck, J., 1982, Membrane fusion activity of influenza virus, EMBO J., 1:217-222.

[59]Howe, C., Lee, L. T., and Rose, H. M., 1961, *Collocalia* mucoid: A substrate for myxovirus neuraminidase, Arch. Biochem. Biophys., 95:512-520.

[60]Ruigrok, R. W. H., Martin, S. R., Wharton, S. A., Skehel, J. J., Bayley, P.M., and Wiley, D. C., 1986, Conformational changes in the hemagglutinin of influenza virus which accompany heat-induced fusion of virus with liposomes, Virology, 155: 484-497.

[61]Stegmann, T., Hoekstra, D., Scherphof, G., and Wilschut, J., 1986, Fusion activity of influenza virus. A comparison between biological and artificial target membranes, J. Biol. Chem., 261:10966-10969.

[62]Rogers, G. N., and Paulson, J. C., 1983, Receptor determinants of human and animal influenza virus isolates: Differences in receptor specificity of the H3 hemagglutinin based on species of origin, Virology, 127:361-373.

[63]Dubois-Dalcq, M., Holmes, K. V., and Rentier, B., 1984, Assembly of paramyxoviridae, in:"Assembly of Enveloped RNA Viruses," pp. 44-65, Springer Verlag, New York.

[64]Haywood, A. M., and Boyer, B. P., 1982, Fusion and disassembly of Sendai virus membranes with liposomes, Biophys. J., 37:128-129.

[65]Homma, M., Shimizu, K., Shimizu, Y., Shimizu, Y. K., and Ishida, N., 1976, On the study of Sendai virus hemolysis. I. Complete Sendai virus lacking in hemolytic activity, Virology, 71:41-47.

[66]Shimizu, Y.K., Shimizu, K., Ishida, N., and Homma, M., 1976, On the study of Sendai virus hemolysis. II. Morphological study of envelope fusion and hemolysis, Virology, 71:48-60.

[67]Knutton, S., and Bächi, T., 1980, The role of cell swelling and haemolysis in Sendai virus-induced cell fusion and in the diffusion of incorporated viral antigens, J. Cell Sci., 42:153-167.

[68]Maeda, T., Kawasaki, K., and Ohnishi, S.-I., 1981, Interaction of influenza virus hemagglutinin with target membrane lipids is a key step in virus-induced hemolysis and fusion at pH 5.2, Proc. Natl. Acad. Sci. USA, 78:4133-4137.

FUSION ACTIVITY OF INFLUENZA VIRUS AND RECONSTITUTED VIRAL ENVELOPES: DIRECT EVIDENCE FOR FUSION IN AN INTRACELLULAR ACIDIC COMPARTMENT

Jan Wilschut and Toon Stegmann

Laboratory of Physiological Chemistry, University of Groningen, Bloemsingel 10, 9712 KZ Groningen, The Netherlands

INTRODUCTION

Many viruses infect their host cells by transfer of the genetic material contained within the viral nucleocapsid to the cytoplasm of the cell. For enveloped viruses, fusion of the viral membrane with a cellular target membrane is a key step in the transfer mechanism (White et al., 1983). Certain viruses, such as Sendai virus, fuse directly with the cellular plasma membrane and, consequently, fusion occurs at the neutral pH of the extracellular medium. Other viruses are first internalized via a process of receptor-mediated endocytosis, routing the viral particles into the endosomal compartment of the cell. The mildly acidic pH in this compartment then triggers a fusion reaction between the viral membrane and the limiting membrane of the endosomes, which also results in release of the viral nucleocapsid into the cytoplasm of the cell.

There is no complete consensus as to the mechanism of host-cell infection by influenza virus. Most of the available evidence points to a mechanism involving entry of the virus particles via the endocytic pathway. First, in model systems the virus fuses at low pH with either artificial or biological target membranes (Maeda et al., 1981; Matlin et al., 1981; White et al., 1982a; Stegmann et al., 1985, 1986). Second, at low pH a conformational change occurs in one of the viral spike glycoproteins, the hemagglutinin, HA (Skehel et al., 1982). It has been unequivocally demonstrated that the fusion activity of influenza virus resides exclusively on the HA molecule (White et al., 1982b). The major structural and functional features of the HA are described in the contribution by Doms and Helenius, elsewhere in this volume. The low-pH-induced conformational change in the molecule leads to exposure of the previously buried N-terminus of its HA_2 subunit (Skehel et al., 1982; Doms et al., 1985). This N-terminus is hydrophobic and has been shown to be crucial for the expression of the fusion activity of the virus (Gething et al., 1986). The exposure of a hydrophobic segment of the HA molecule at the pH of fusion is suggestive of a fusion mechanism involving penetration of this segment into the target membrane. Third, Yoshimura and Ohnishi (1984) have demonstrated that fluorescently labelled influenza virus particles appear in intracellular acidic vacuoles on a time scale corresponding to that of infection, while recently it has also been shown that influenza virus particles enter the cell as intact entities that do not leave their spike proteins behind on the cellular plasma membrane (Richman et al., 1986).

Finally, infection of cells by influenza virus is completely blocked by NH_4Cl (Matlin et al., 1981), which is known to raise the internal pH of intracellular acidic vacuoles (Mellman et al., 1986).

In contrast with the above view on the infectious entry of influenza virus into its host cell, Haywood and Boyer (1985) have suggested that infection of cells occurs via direct fusion at neutral pH with the cellular plasma membrane. This proposal was made on the basis of the observation that the virus fuses at neutral pH with large multilamellar zwitterionic liposomes containing the ganglioside G_{D1a}. Electron micrographs of negatively stained preparations demonstrated engulfment of the virus particles by the liposomes and the appearance of viral spikes on the liposomal surface. In parallel a fraction of the virus particles was shown to become tightly associated with the liposomes. This association was taken as a measure of fusion of the virus with the liposomes, as it could not be reversed by fetuin, which competes with the viral HA for the ganglioside receptor on the liposomes. It should be noted that also in the system of Haywood and Boyer (1985) the extent of fusion was higher at pH 5.0 than at pH 7.4.

The chapters by Hoekstra et al. and Nir et al. in this volume describe some of our results on fusion of influenza virus with erythrocyte ghosts or liposomes of various compositions. In these studies fluorescence assays were employed to monitor the kinetics of the fusion process. For a detailed description of these assays we refer to the chapter by Hoekstra et al. in this volume. In the following section we will show that one of the assays, involving labelling of the virus particles with octadecyl rhodamine B (R_{18}) (Hoekstra et al., 1984), can be utilized to directly measure fusion of intact influenza virus in an acidic compartment of cultured cells (Stegmann et al., 1987a).

Additional evidence for influenza virus entry into cells via the endocytic pathway would be provided by the demonstration that functionally reconstituted influenza "virosomes" have the capacity to deliver compounds encapsulated in their aqueous interior to the cellular cytoplasm via endocytosis and subsequent fusion of the virosomal with the endosomal membrane. Recently, we have achieved the functional reconstitution of influenza virus envelopes (Stegmann et al., 1987b). The third section of this chapter presents a brief description of the reconstitution procedure and the structural and functional characteristics of the virosomes.

FUSION OF INFLUENZA VIRUS WITH CULTURED CELLS

When R_{18}-labelled influenza virus is added to subconfluent monolayers of LLC-MK_2D or BHK-21 cells in the cold, a fraction of the virus particles becomes cell-associated (Fig. 1). The binding does not saturate over a period of 60 min. At that time under the conditions of the experiment approximately 4.5% of the added virus is associated with the LLC-MK_2D cells and 10.5% with the BHK-21 cells. This corresponds to about 2100 virus particles per LLC-MK_2D cell and about 4200 particles per BHK-21 cell.

Table 1 shows that R_{18}-labelled influenza virus, bound to LLC-MK_2D cells at 4°C, has the capacity to fuse with the cellular plasma membrane upon a brief exposure to pH 5.0 at 37°C. Fusion is revealed by an increase of R_{18} fluorescence intensity, resulting from dilution of the probe from the viral into the cell membrane and the concomitant relief of its fluorescence selfquenching. The observed increase of fluorescence intensity of 48% indicates that approximately half of the virus particles that are still associated with the cells after the incubation at pH 5.0, 37°C, have

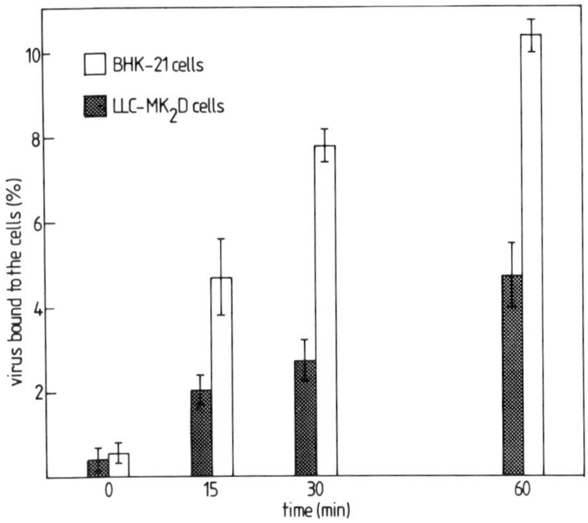

Fig. 1. Binding of influenza virus to BHK-21 or LLC-MK$_2$D cells. 1.5-2 x 10^6 BHK-21 or 1-1.5 x 10^6 LLC-MK$_2$D cells on 35-mm plastic petri dishes were incubated with R$_{18}$-labelled influenza virus X-47 (10 nmol of membrane phospholipid) at 4°C in 0.5 ml Hanks medium, buffered with 10 mM HEPES to pH 7.4, for the indicated periods of time. Cell-associated fluorescence was determined in the presence of Triton X-100.

actually fused. In this respect it should be noted that a fraction of the virus particles that become cell-associated during the preincubation for 60 min at 4°C is released into the medium during the warming to 37°C (results not shown).

When, after binding of the virus at 4°C, the cells are washed and warmed to 37°C at pH 7.4, neither with the LLC-MK$_2$D cells nor with the BHK-21 cells any significant increase in fluorescence intensity is observed during the first 10 min of incubation (Table 1). This result is in sharp contrast with the increase in fluorescence seen after 5 min at pH 5.0 and clearly indicates that at neutral pH there is essentially no fusion of the virus with the cellular plasma membrane.

Continued incubation at neutral pH does produce an increase in fluorescence intensity of the R$_{18}$ probe (Table 1). With the BHK-21 cells the increase starts after about 10 min, for the LLC-MK$_2$D cells the lag phase appears to be slightly longer. With either cell type after 60 min the fluorescence increase is around 30%, representing approximately 500 and 300 virus particles fused per BHK-21 and LLC-MK$_2$D cell, respectively.

The lack of fluorescence increase during the first 10 min of incubation at neutral pH (Table 1), suggests that the subsequent increase is due to a fusion reaction inside the cell, the lag phase representing the time required for the virus particles to reach the intracellular compartment. This explanation for the lag phase is supported by the effects of NH_4Cl on the fusion reaction. NH_4Cl is known to raise the internal pH of acidic vacuoles inside the cell (Mellman et al., 1986). The results in Table 2 show that in the presence of NH_4Cl the increase in fluorescence is almost completely blocked. Fluorescence dequenching is also blocked by monensin (results not shown), another known inhibitor of vacuolar acidification (Mellman et al., 1986). These results indicate that, in the absence of inhibitors, the observed fusion reaction occurs in an intracellular acidic compartment.

Table 1. Fusion with the cellular plasma membrane and intracellular fusion of R_{18}-labelled influenza virus[a]

Temp. (°C)	Time (min)	Fluorescence increase (%)		
		BHK-21	$LLC\text{-}MK_2D$	
		pH 7.4	pH 7.4	pH 5.0
37	5	0 ± 0.2	0 ± 3.0	48 ± 9
	10	0 ± 0.2	n.d.	
	15	6.6 ± 0.3	0.1 ± 3.6	
	30	18.3 ± 1.3	16.0 ± 5.7	
	60	28.3 ± 3.7	33.1 ± 0.9	
20	60	5.8 ± 0.8		
	120	11.1 ± 1.3		

[a]Virus was allowed to bind to the cells at 4°C (see Fig. 1) for 30 min. Then the cells were washed at 4°C and 2 ml of prewarmed Hanks/HEPES medium at the indicated pH was added and the incubation continued at either 37°C or 20°C for the periods of time indicated. The cells were washed, scraped off and the fluorescence was measured at 4°C before and after addition of Triton X-100. Fluorescence of the virus bound at 4°C was set to zero, while the value in the presence of Triton, representing infinite probe dilution, was taken as 100%. Average values (± SEM) of 4 determinations are presented.

Table 2. Effects of NH_4Cl on the intracellular fusion of R_{18}-labelled influenza virus[a]

Time (min)	Condition	Fluorescence increase (%)
60	control	30.0 ± 1.1
60	NH_4Cl	0.6 ± 3.1
10	control	0 ± 0.4
10	after preincubation with NH_4Cl	14.9 ± 0.8

[a]Intracellular fusion of the virus in BHK-21 cells was measured at 37°C, as described in the footnote to Table 1, in the absence or presence of 20 mM NH_4Cl. Alternatively, the cells were allowed to internalize the virus at 37°C for 60 min in the presence of 20 mM NH_4Cl, after which the cells were washed to remove the NH_4Cl and the incubation continued for 10 min.

In order to distinguish between fusion of the virus at the level of the endosomes or at the level of the subsequent acidic compartment on the endocytic pathway, i.e. the lysosomal compartment, we investigated the intracellular fusion reaction at 20°C. At this temperature transfer from endosomes to lysosomes is blocked (Dunn et al., 1980). Yet, a significant degree of fluorescence dequenching occurs under these conditions (Table 1), indicative of fusion at the level of the endosomes. The extent of fusion is less at 20°C than at 37°C (Table 1). This must, at least in part, be due to the effect of temperature on the rates of virus internalization and fusion. However, the possibility can not be excluded that at 37°C the observed dilution of the fluorophore occurs also to some extent at the level of the lysosomes, in addition to the fusion occurring in the endosomal compartment of the cell.

The kinetics of fusion that can be deduced from the data in Table 1 are a composite of the rates of virus internalization, vacuolar acidification and fusion. After preloading of the cells at 37°C with fluorescently labelled virus in the presence of NH_4Cl, such that fusion does not occur, one can obtain an estimate for the rate of the actual fusion event by rapidly removing the NH_4Cl block. It has been demonstrated that removal of NH_4Cl results in a very fast reacidification of intracellular vacuoles (Ohkuma and Poole, 1978). Under these conditions there is no lag phase; rather, an immediate fusion reaction of the internalized virus is observed, resulting in a fluorescence increase of approximately 15% after 10 min (Table 2). This value represents a minimum of 400 virus particles fused per BHK-21 cell.

The above results give a direct demonstration of intracellular fusion of influenza virus at low pH after internalization of the virus particles by endocytosis. The reaction occurs to a significant extent in the endosomal compartment of the cell. Importantly, essentially no fusion is observed at the level of the cellular plasma membrane. This contrasts with the view of Haywood and Boyer (1985), who have suggested that influenza virus may infect cells via direct fusion with the plasma membrane at neutral pH.

RECONSTITUTION OF INFLUENZA VIRUS ENVELOPES

A detailed study of the molecular events involved in the intracellular fusion of influenza virus and a further quantitation of the process require the ability to manipulate the components of the viral envelope. Functional reconstitution of the viral spike glycoproteins into "virosomes" provides the possibility for such a manipulation. Despite various efforts (Huang et al., 1979; Kawasaki et al., 1983; Hosaka et al., 1983) reconstitution of influenza virus envelopes, displaying biological fusion activity, has not been reported as yet.

In several studies on the reconstitution of viral envelopes the non-ionic detergent octylglucoside has been utilized to solubilize the viral membrane lipids and proteins (Helenius et al., 1977, 1981; Petri and Wagner, 1979; Huang et al., 1979; Eidelman et al., 1984; Harmsen et al., 1985). Due to its relatively high critical micelle concentration (cmc), octylglucoside offers the advantage over other detergents that it can be easily removed by dialysis. However, there is evidence to indicate that the use of octyl-glucoside may result in a poor and/or non-functional incorporation of viral spike proteins in the reconstitution products (Helenius et al., 1977, 1981; Eidelman et al., 1984; Metsikkö et al., 1986). In our initial attempts to produce functional influenza virosomes employing octylglucoside, we also obtained a poor protein incorporation in the reconstituted vesicles (Fig. 2A) and after sucrose-gradient centrifugation the virosomes had essentially no fusion activity toward erythrocyte ghosts as target membranes (results

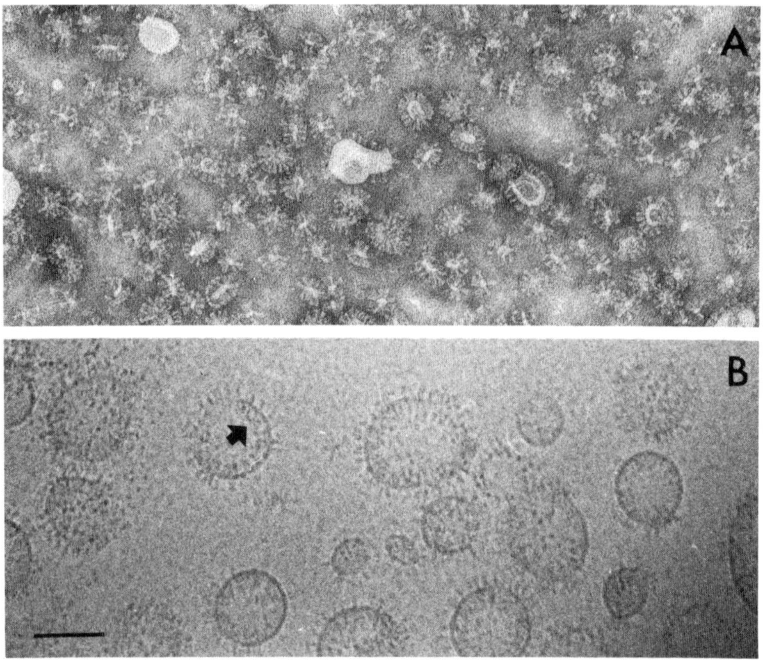

Fig. 2. Electron micrographs of reconstituted influenza virus envelopes. A, reconstitution products, obtained after octylglucoside dialysis, negatively stained with phosphotungstate; B, an unstained, frozen sample of virosomes, obtained after $C_{12}E_8$ reconstitution (arrow indicates the symmetrical distribution of spikes across the membrane). Bar represents 100 nm.

not shown). Presumably, during the course of octylglucoside dialysis, at the stage of vesicle formation, the viral spike proteins are stabilized in oligomeric protein/lipid/detergent complexes, precluding their insertion into the lipid bilayer (Helenius et al., 1977, 1981). After eventual complete removal of the detergent most of the protein is recovered in the form of rosettes or very small vesicles (Fig. 2A).

Recently, Metsikkö et al. (1986) reported the functional reconstitution of the G protein of vesicular stomatitis virus (VSV) employing the Triton-like detergent octaethyleneglycol mono(n-dodecyl)ether, $C_{12}E_8$. Previously, Triton X-100 had been used successfully in the reconstitution of Sendai virus envelopes (Volsky and Loyter, 1978; Vainstein et al., 1984). Because of its low cmc Triton X-100 can not readily be dialyzed. Therefore, Vainstein et al. (1984) used Bio-Beads SM-2, added directly to the solubilized Sendai virus membrane, to remove the detergent. Likewise, Metsikkö et al. (1986) employed Bio-Beads SM-2 for removal of the $C_{12}E_8$.

Following the protocol of Metsikkö et al. (1986) we achieved functional reconstitution of influenza virus envelopes (Stegmann et al., 1987b). In this procedure the viral membrane is solublized in 100 mM $C_{12}E_8$. After removal of the nucleocapsid by ultracentrifugation, approximately 90% of the viral lipid and 36% of the protein (representing essentially all of the envelope protein) are recovered in the supernatant. Subsequently, the bulk of the detergent is removed in two steps by shaking the supernatant with Bio-Beads SM-2. Further detergent removal is achieved by centrifugation of the preparation on a discontinuous sucrose gradient.

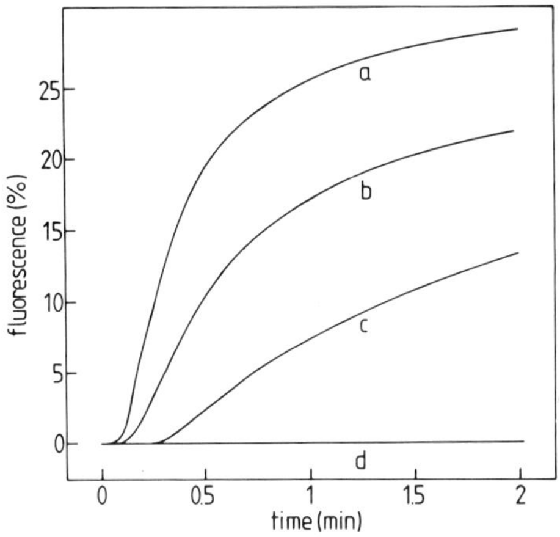

Fig. 3. Fusion between $C_{12}E_8$ influenza virosomes, labelled with N-NBD-PE and N-Rh-PE, or intact virus, labelled with R_{18}, and erythrocyte ghosts at 37°C. Curve a, fusion of the virus at pH 5.1; curves b, c and d, fusion of the virosomes at pH 5.1, 5.3 and 5.6, respectively. Virus or virosomes were present at 2.5 μM membrane phospholipid and ghosts at 50 μg of protein per ml. The residual fluorescence of virus or virosomes was set to zero and the fluorescence in the presence of Triton X-100, after appropriate correction, representing infinite probe dilution, was taken as 100%.

The reconstitution yields a single population of virosomes with a uniform buoyant density (1.12 g/ml) and a protein to phospholipid ratio identical to that in the solubilized viral supernatant. The recovery of virosomes is low (about 25%), since the Bio-Beads bind significant amounts of viral lipid and protein. Based on experiments employing ^{14}C-labelled $C_{12}E_8$ we estimate that the amount of residual detergent in the virosomes is about 9 mol % relative to the total virosomal lipid (Stegmann et al., 1987b). Fig. 2B shows an unstained frozen preparation of the virosomes. They appear unilamellar and are relatively homogeneous in size with a mean diameter of about 100 nm. There is a high density of spikes on the virosomes and spikes are present on both sides of the membrane (Fig. 2B, arrow). This latter observation suggests that, during detergent removal, the spike proteins become membrane-associated at the point of lipid bilayer formation. It may well be that this is a prerequisite not only for efficient protein incorporation, but perhaps also for functional reconstitution.

To examine the fusion activity of the virosomes the fluorescent phospholipids, N-NBD-PE and N-Rh-PE were added to the viral supernatant to a concentration of 0.6 mol % each relative to the total lipid, before detergent removal. Thus, fusion could be monitored continuously as an increase of N-NBD-PE fluorescence intensity due to dilution of the probes into the target membrane and the concomitant relief of resonance energy transfer between the donor, the N-NBD-PE, and the acceptor, the N-Rh-PE (Struck et al., 1981). Fig. 3 (curve b) shows the fluorescence increase due to fusion of the virosomes with human erythrocyte ghosts at pH 5.1, 37°C. For comparison, fusion between the intact virus, labelled with R_{18}, and erythrocyte ghosts is also presented. Clearly, the fusion activity of the virosomes is similar to that of the intact virus under comparable conditions. The somewhat lower initial rate of fusion of the virosomes may be due to the symmetrical distribution of spikes, resulting in a reduced spike

Table 3. Fusion of reconstituted influenza virosomes with BHK-21 cells[a]

Condition		Fluorescence increase (%)
virosomes, control	pH 5.0	26.7 ± 1.5
virosomes, preincubated at pH 5.0	pH 5.0	0 ± 0.7
virosomes, control	pH 7.4	25.0 ± 1.8
virosomes, 20 mM NH_4Cl	pH 7.4	8.8 ± 5.2

[a]Virosomes, labelled with *N*-NBD-PE and *N*-Rh-PE, were allowed to bind to BHK-21 cells for 1 *h* at 4°C. Then the cells were washed and incubated at pH 5.0 for 5 min or pH 7.4 for 1 h (in the absence or presence of 20 mM NH_4Cl), and the *N*-NBD-PE fluorescence was measured before and after addition of Triton X-100 (see footnotes to Tables 1 and 2 and legend to Fig. 3).

density on the external surface relative to the spike density on the intact virus. The pH dependence of fusion of the virosomes with erythrocyte ghosts is identical to that of fusion of the virus with ghosts. In either case fusion is negligible above pH 5.6 (Fig. 3, curve d) and half-maximal at pH 5.3 (Fig. 3, curve c). The virosomes also display full hemolytic activity with essentially the same pH dependence (Stegmann et al., 1987b). Previously, we have shown that the detailed pH dependence profile in model systems is an important characteristic of the biological fusion activity of the virus (Stegmann et al., 1986). The virosomes meet this criterion for biological activity.

Preincubation of the virosomes alone at low pH produces a very fast irreversible loss of fusion activity toward erythrocyte ghosts, the half-time of inactivation being less than 15 s (results not shown). Also in this respect the virosomes closely resemble the intact virus (Stegmann et al., 1986). The inactivation of fusion capacity during low-pH incubation of virus or virosomes in the absence of target membranes represents another criterion for the biological nature of the fusion activity (Stegmann et al., 1986, 1987b).

The reconstituted virosomes fuse efficiently with the plasma membrane of cultured cells upon a brief exposure to low pH (Table 3). The reaction is completely inhibited by a pretreatment of the virosomes alone at pH 5.0, excluding the possibility that with the untreated virosomes the observed dilution of the fluorophores is due to a non-fusion process, such as exchange of individual molecules. Continued incubation at neutral pH, 37°C, of cells, to which fluorescently labelled virosomes were allowed to bind in the cold, also produces a considerable increase in fluorescence intensity, reaching a value of 25% after 1 h (Table 3). This fluorescence increase only occurs after a lag phase of about 10-15 min and is inhibited by NH_4Cl. This latter result (Table 3) indicates that the virosomes, after internalization through endocytosis, fuse with the membrane of an intracellular acidic compartment, presumably the endosome. The extent of intracellular fusion after 1 h is approximately the same as the extent of fusion at the level of the plasma membrane after brief exposure of the cells to low pH. Both represent roughly 1000 virosomes fused per cell.

Functionally active influenza virosomes represent a very promising carrier system for efficient delivery of biologically active substances into cells either via the endocytic pathway or via direct fusion of the virosomes with the cellular plasma membrane. The size of the virosomes and the method of their preparation in principle allow the encapsulation of significant amounts of water-soluble compounds, including compounds of low molecular weight, since the reconstitution procedure does not involve dialysis. Moreover, virosomes provide the potential possibility to introduce membrane-associated compounds into cells, either directly into the plasma membrane or into the membrane of the endosomal/lysosomal compartment.

ACKNOWLEDGEMENTS

We wish to thank Kalervo Metsikkö, Gerrit van Meer and Kai Simons for many helpful suggestions on the reconstitution procedure, Frank Booy and Jan van Breemen for carrying out the electronmicroscopy of the virosomes, Henriëtte Morselt and Janny Scholma for skillful technical assistance, Gerrit Scherphof for his interest in the project and Rinske Kuperus for expert preparation of the manuscript. This investigation has been carried out under the auspices of the Netherlands Foundation for Chemical Research (SON) with financial support from the Netherlands Organization for the Advancement of Pure Research (ZWO).

REFERENCES

Doms, R.W., Helenius, A. and White, J. (1985) Membrane fusion activity of the influenza virus hemagglutinin. The low pH-induced conformational change. J. Biol. Chem., 260:2973-2981.

Dunn, W.A., Hubbard, A.L. and Aronson, N.N. (1980) Low-temperature selectivity inhibits fusion between pinocytic vesicles and lysosomes during heterophagy of ^{125}I-asialofetuin by the perfused rat liver. J. Biol. Chem., 255:5971-5978.

Eidelman, O., Schlegel, R., Tralka, T.S. and Blumenthal, R. (1984) pH-dependent fusion induced by vesicular stomatitis virus glycoprotein reconstituted into phospholipid vesicles. J. Biol. Chem., 259:4622-4628.

Gething, M-J., Doms, R.W., York, D. and White, J. (1986) Studies on the mechanism of membrane fusion. Site-specific mutagenesis of the hemagglutinin of influenza virus. J. Cell Biol., 102:11-23.

Harmsen, M.C., Wilschut, J., Scherphof, G., Hulstaert, C. and Hoekstra, D. (1985) Reconstitution and fusogenic properties of Sendai virus envelopes. Eur. J. Biochem., 149:591-599.

Haywood, A.M. and Boyer, B.P. (1985) Fusion of influenza virus membranes with liposomes at pH 7.5. Proc. Natl. Acad. Sci. USA, 82:4611-4615.

Helenius, A., Fries, E. and Kartenbeck, J. (1977) Reconstitution of Semliki Forest virus membrane. J. Cell Biol., 75:866-880.

Helenius, A., Sarvas, M. and Simons, K. (1981) Asymmetric and symmetric membrane reconstitution by detergent elimination. Studies with Semliki-Forest virus spike glycoprotein and penicillinase from the membrane of B. licheniformis. Eur. J. Biochem., 116:27-35.

Hoekstra, D., De Boer, T., Klappe, K. and Wilschut, J. (1984) Fluorescence method for measuring the kinetics of fusion between biological membranes. Biochemistry, 23:5675-5681.

Hosaka, Y., Yasuda, Y. and Fukai, K. (1983) Hemolysis by liposomes containing influenza virus hemagglutinins. J. Virol., 46:1014-1017.

Huang, R.T.C., Wahn, K., Klenk, H.D. and Rott, R. (1979) Association of the envelope glycoproteins of influenza virus with liposomes. A model study on viral envelope assembly. Virology, 92:212-217.

Kawasaki, K., Sato, S. and Ohnishi, S-I. (1983) Membrane fusion activity of reconstituted vesicles of influenza virus hemagglutinin glycoproteins. Biochim. Biophys. Acta, 733:286-290.

Maeda, T., Kawasaki, K. and Ohnishi, S-I. (1981) Interaction of influenza virus hemagglutinin with target membrane lipids is a key step in virus-induced hemolysis and fusion at pH 5.3. Proc. Natl. Acad. Sci. USA, 78:4133-4137.

Matlin, K.S., Reggio,H., Helenius, A. and Simons, K. (1981) Infectious entry pathway of influenza virus in a canine kidney cell line. J. Cell Biol., 91:601-613.

Mellman, I., Fuchs, R. and Helenius, A. (1986) Acidification of the endocytic and exocytic pathways. Ann. Rev. Biochem., 55:663-700.

Metsikkö, K., Van Meer, G. and Simons, K. (1986) Reconstitution of the fusogenic activity of vesicular stomatitis virus. EMBO J., 5:3429-3435.

Ohkuma, S. and Poole, B. (1978) Fluorescence probe measurement of the intralysosomal pH in living cells and the perturbation of pH by various agents. Proc. Natl. Acad. Sci. USA, 75:3327-3331.

Petri, W.A. and Wagner, R.R. (1979) Reconstitution into liposomes of the glycoprotein of vesicular stomatitis virus by detergent dialysis. J. Biol. Chem., 254:4313-4316.

Richman, D.D., Hostetler, K.Y., Yazaki, P.J. and Clark, S. (1985) Fate of influenza A virion proteins after entry into subcellular fractions of LLC cells and the effect of amantidine. Virology, 151:200-210.

Skehel, J.J., Bayley, P.M., Brown, E.B., Martin, S.R., Waterfield, M.D., White, J.M., Wilson, I.A. and Wiley, D.C. (1982) Changes in the conformation of influenza virus hemagglutinin at the pH optimum of virus-mediated membrane fusion. Proc. Natl. Acad. Sci. USA, 79:968-972.

Stegmann, T., Hoekstra, D., Scherphof, G. and Wilschut, J. (1985) Kinetics of pH-dependent fusion between influenza virus and liposomes. Biochemistry, 24:3107-3113.

Stegmann, T., Hoekstra, D., Scherphof, G. and Wilschut, J. (1986) Fusion activity of influenza virus. A comparison between biological and artificial target membrane vesicles. J. Biol. Chem., 261:10966-10969.

Stegmann, T., Morselt, H.W.M., Scholma, J. and Wilschut, J. (1987a) Direct measurement of the fusion of influenza virus in an intracellular acidic compartment. Submitted for publication.

Stegmann, T., Morselt, H.W.M., Booy, F.P., Van Breemen, J.F.L., Scherphof, G.L. and Wilschut, J. (1987b) Functional reconstitution of influenza virus envelopes. EMBO J., in press.

Struck, D.K., Hoekstra, D. and Pagano, R.E. (1981) Use of resonance energy transfer to monitor membrane fusion. Biochemistry, 20:4093-4099.

Vainstein, A., Hershkowitz, M., Israel, S., Rabin, S. and Loyter, A. (1984) A new method for the reconstitution of highly fusogenic Sendai virus envelopes. Biochim. Biophys. Acta, 733:181-188.

Volsky, D.J. and Loyter, A. (1978) An efficient method for reassembly of fusogenic Sendai virus envelopes after solubilization of intact virions with Triton X-100. FEBS Lett., 92:190-194.

White, J., Helenius, A. and Gething, M-J. (1982b) Hemagglutinin of influenza virus expressed from a cloned gene promotes membrane fusion. Nature, 300:658-659.

White, J., Kartenbeck, J. and Helenius, A. (1982a) Membrane fusion activity of influenza virus. EMBO J., 1:217-222.

Yoshimura, A. and Ohnishi, S-I. (1984) Uncoating of influenza virus in endosomes. J. Virol., 51:497-504.

KINETICS AND EXTENT OF FUSION OF INFLUENZA VIRUS AND SENDAI VIRUS WITH LIPOSOMES

Shlomo Nir[a], Toon Stegmann[b], Dick Hoekstra[b] and Jan Wilschut[b]

a. Seagram Center for Soil & Water Sciences, Faculty of Agriculture, The Hebrew University of Jerusalem, Rehovot 76100, Israel. b. Laboratory of Physiological Chemistry, University of Groningen, Bloemsingel 10, 9712 KZ Groningen, The Netherlands

ABSTRACT

The kinetics and extent of low-pH induced fusion of influenza virus and Sendai virus with large unilamellar liposomes were investigated with fluorescence assays for lipid mixing. The model views the overall fusion reaction as a sequence of a second-order process of virus-liposome aggregation, which is partially reversible, followed by the first-order fusion reaction itself. The calculations gave good simulations and predictions for the kinetics and extent of fusion at different virus/liposome concentrations and ratios, and provided a pattern of the dependence of viral fusion activity on liposome composition, i.e., acidic and neutral phospholipids, cholesterol and gangliosides. The fusion products consist of a single virus particle and several liposomes, indicating negative cooperativity between viral glycoproteins and those present in the fusion products. At pH 5.0 and 37°C very high rate constants for aggregation and fusion were obtained and essentially all of the virus particles were capable of fusing with cardiolipin liposomes. At higher pH values the rate constants of fusion were considerably smaller for all liposome compositions and the percent of virus particles capable of fusing was lowered, whereas the aggregation rate constants were only moderately affected. The fusion activity of influenza virus was reduced by its exposure to low pH. The inactivation reflected intramembrane events and was independent of virus concentrations. A mode of virus inactivation due to formation of clusters of glycoproteins in its membrane at low pH is consistent with our results.

Abbreviations: CL, cardiolipin (bovine heart); C_{11}=C=forward dimerization rate constant ($M^{-1}s^{-1}$); D_{11}=D=dissociation rate constant from an aggregated dimer to monomer (s^{-1}); DOPC, dioleoylphosphatidylcholine; DOPE, dioleoylphosphatidylethanolamine; Chol, cholesterol; f_{11}=f=rate of fusion of an aggregated dimer; PE, phosphatidylethanolamine; PEG, poly(ethyleneglycol); PC, phosphatidylcholine (lecithin); PG, phosphatidylglycerol; PS, phosphatidylserine; HA, hemagglutinin; LUV, large unilamellar vesicles; N-NBD-PE, N-(7-nitro-2,1,3-benzoxadiazol-4-yl)phosphatidylethanolamine; BHA, bromelain-solubilized hemagglutinin; R_{18}, octadecyl rhodamine B chloride.

Introduction

The fusion process is viewed as a sequence of two steps: (1) adhesion of the virus particles to liposomes; (2) the fusion process itself, involving destabilization and merging of the membranes (Nir et al.1980; Bentz et al.1983). Recent development of fluorescent assays (Wilschut et al.1980; Struck et al.1981; Hoekstra 1982; Hoekstra, et al.1984; Ellens et al.1985) allows continuous and sensitive monitoring of membrane fusion. Studies of virus-cell or virus-liposome fusion have mostly employed fusion assays that monitor membrane mixing. Hoekstra et al.(1984,1985), Stegmann et al.(1985), and Klappe et al.(1986) supplemented the experimental procedures by tests to discriminate between lipid exchange and real fusion. The studies described here on the fusion of influenza virus and Sendai virus with liposomes will indicate additional proofs for the occurrence of fusion rather than exchange.

It is evident that the mechanisms of virus-vesicle fusion, where Ca and other cations have little effect (White et al.1982; Haywood and Boyer,1984; Stegmann et al.1985), and where viral glycoproteins play a major role, are quite distinct from those acting in cation-induced fusion of vesicles (Papahadjopoulos et al.1977; Portis et al.1979; Nir et al.1980,1983; Wilschut et al.1980,1985; Ohki and Ohshima, 1985; Duzgunes, 1985). However, the general stages of fusion are the same, i.e., membrane contact, destabilization and merging.

This study employs a theoretical analysis which converts the increase in fluorescence intensity to the percent of fused particles, by using analytical and numerical solutions of the equations describing adhesion and fusion processes, and gives simulation and prediction of experimental results. The analysis yields separately the rate constants of fusion, adhesion, and dissociation, and thus gives details of the functional involvement of viral glycoproteins in the fusion process, and its dependence on factors such as pH, liposome composition and agents such as PEG. The coupling of these studies with other studies (e.g., Wilson et al.1981; Skehel et al.1982; Gething et al.1986) which focus on molecular details would enable to correlate these details with functional stages of the fusion process, e.g., adhesion or membrane destabilization.

Mass Action Kinetics of Vesicle-Virus Fusion

The equations and solutions for the virus-vesicle system are given in Nir et al.(1986a). Here we illustrate only a few of the initially occurring reactions:

$$L + V \underset{D_{11}}{\overset{C_{11}}{\rightleftharpoons}} A(1,1) \qquad (1)$$

$$A(1,1) + L \underset{D_{21}}{\overset{C_{21}}{\rightleftharpoons}} A(2,1) \qquad (1\ )$$

$$A(1,1) \xrightarrow{f_{11}} F(1,1) \qquad (2)$$

in which A(I,J) and F(I,J) denote molar concentrations of aggregates or fusion products consisting of I liposomes and J virions, respectively. With this notation L = A(1,0) and V = A(0,1) are molar concentrations of liposomes and virions, respectively. The aggregation reaction, Eq (1), is of second order, depending on the product of the concentrations L and V, whereas the fusion reaction, Eq (2), is of first order. The aggregation rate constants, C_{ij}, have the units $M^{-1}.S^{-1}$, whereas the deaggregation rate constants, D_{ij}, and the fusion rate constants, f_{ij}, have the unit S^{-1}.

The number of non-linear differential equations describing the overall fusion reaction with aggregation fusion products consisting of up to 8 particles amounts to several hundred. In addition to numerical solutions, analytical solutions were found (Nir et al.1986a) for certain limiting cases.

In all systems studied, vesicle-vesicle or virus-virus aggregation did not occur and the fusion products consisted of a single virus and any number of liposomes. In attempts to simulate the experimental data we employed only three parameters, C, f and D.

In the application of fusion assays based on membrane mixing we have worked in the linear regime, i.e., at sufficiently dilute surface concentrations of the fluorescent probes. The percent of fluorescence intensity relative to the maximal at infinite dilution is denoted by I. Per molecule of N-NBD-PE (Struck et al.1981) or R_{18} (Hoekstra et al.1984),

$$I = 100\ (1 - x), \qquad (3)$$

in which x is probe concentration relative to the initial value. The program calculates I values for the evolved distributions of fusion products. The program can be easily adjusted for any fusion assays.

The program accounts explicitly for the percent of inactive virions that can adhere to liposomes but cannot fuse. In the current study the program has been extended to account for inactivation of the virus with time, which is expressed by a reduction in the rate constants.

Final Extents of Fusion

Final extents of fusion were deduced from the final levels of fluorescence intensity increase, which were reached after several minutes to several hours of incubation of virus particles with liposomes. These studies enabled to deduce the percent of active virus particles and provided answers to the question whether the viral glycoproteins could be recycled, i.e., whether a fusion product could fuse again. Nir et al. (1986a,c) derived analytical calculations for three limiting cases, and showed that the final fluorescence levels could only be explained in the framework of a model that assumes that the fusion products consist of any number of liposomes fused with a single virus particle, but prohibits the fusion of an additional virus particle with the fusion products for at least up to a 10/1 ratio between liposomal to viral lipid. Tables 1A and 1B illustrate the agreement between experimental results and the predictions of the above model. The calculations have assumed that in the system influenza virus or Sendai virus and Cardiolipin liposomes, all the virus particles are active at pH 5 (and 37°C). In Table 1A (influenza) the liposomes were labeled, whereas in Table 1B the virus particles were labeled. In these experiments the ratio between liposomes to virus particles was varied by one to two orders of magnitude.

Table 1A. Final Levels of N-NBD-PE Fluorescence (%) at pH 5.0 and 37°C for >1 h incubation times.

Ratio of liposomes:virus (lipid phosphorus)	Calculated for fusion products consisting of a single virus particle and any number of liposomes	Experimental values (average of two batches)
16:1	5.9	4.8
8:1	11.1	11.1
4:1	19.7	20.5
2:1	30.2	30.5
1:1	38.7	40.5
1:2	44.0	43.5
1:4	46.9	47.0
1:8	48.4	48.0
1:16	49.2	49.0

Large unilamellar CL liposomes were labeled by N-NBD-PE and N-Rh-PE at 0.6 mol% each on the basis of lipid phosphorus. A small volume of virus suspension was injected into the liposome-containing medium in the cuvette, and the increase of the N-NBD-PE fluorescence, due to dilution of the fluorophores into the viral membrane, was recorded continuously. More details are given in Nir et al.(1986a) where calculated values according to other models are also given. The (I) values in the Table are % of maximal fluorescence intensity increase at infinite dilution (See Eq(3)).

Table 1B. Experimental and Calculated Values of Final Levels of Fluorescence Intensity[a].

Liposome composition	Liposomal lipid (μM)	Medium	Experimental % fluorescence	Calculated % fluorescence
CL	2.5	pH 5	50.6	(100)[b] 44
	10		85.8	78
	25		84.7	90
	50		99	95
CL	2.5	pH 7.4	27.6	(60) 26.3
	10		44	46.9
	25		56.1	53.9
	50		62.7	56.8
CL/DOPC (1:1)	2.5	pH 7.4	13.0	(30) 13.1
	10		21.8	23.5
	25		25	27
	50		29.6	28.4
CL	2.5	pH 7.4 4% PEG	37.6	(80) 35.1
	10		56.7	62.5
	25		69.7	71.9
	50		77.5	75.8

[a] R_{18}-labeled Sendai virus particles were incubated with different liposomes for 3-24 h at 37°C. The virus concentration was 20μg of protein/2 mL, or 2.8 μM with respect to phospholipid.

[b] When all virus particles are active, Eq(7) reduces to Eq (4). Numbers in parentheses indicate percent active virus. See Nir et al.(1986c) for more cases.

The final value of V_f, molar concentration of virions fused, is given by

$$V_f = V_o \quad (1 - \exp - L_o/V_o), \tag{4}$$

in which L_o and V_o denote the initial molar concentrations of liposomes and virus particles, respectively. All liposomes are assumed to fuse after long incubation times, but all of the virions will have fused only if the ratio L_o/V_o is large, e.g., $\geqslant 4$. For equal concentrations of active virions and liposomes about 1/3 of the virions do not fuse, because of the fact that certain fusion products include more than a single liposome, but cannot fuse with the remaining virions. Direct tests for this consequence of Eq (4) were carried out for several cases. In Fig. 1 we demonstrate one example. Influenza virus and labeled CL liposomes were incubated at a 1:1 ratio. When the final level of fluorescence was attained, additional nonlabeled liposomes or virus particles were added to yield final virus/liposome ratios of 1:4 and 4:1, respectively. Addition of liposomes resulted in an increase in fluorescence, indicating fusion between the liposomes and the initially formed fusion products. By contrast, additional virus did not produce any further increase in fluorescence, demonstrating that the virus did not fuse with the fusion products. Amselem et al. (1986) and Klappe et al. (1986) provided a similar demonstration for Sendai virus. In the latter case the process described in Fig. 1 was iterated for three times.

It is remarkable that the predictions according to Eq (4) do not involve any parameters. The fact that such predictions can be made, and that the addition of blank virions to fusion products does not result in fluorescence increase, constitutes an additional proof that the experiments have indeed monitored fusion rather than lipid exchange. The lack of fusion between virions and fusion products can be interpreted to indicate a negative cooperativity, or repulsion, between glycoproteins in intact virus particles and those in the fusion products. In this context we note that the binding of the HA glycoprotein from influenza to liposomes is limited (Doms et al.1985), e.g., 100% binding of the BHA (at pH 5) required 5000 lipid molecules per BHA, whereas at a ratio of 2000:1 65% of BHA bound. We estimate that when the binding ratio of HA to lipid is 1:2000 it corresponds to less than the ratio of HA to lipid in fusion products consisting of a single virus and six liposomes.

At higher pH values the final extents of fusion are reduced for both influenza and Sendai virus, even in the presence of large excess of liposomes. The final extents of fusion were also reduced at pH 5 for a variety of liposome compositions. Accordingly, the model was extended by considering a fraction α of inactive virions. Two extreme cases have been considered (Nir et al., 1986c). If the binding of liposomes to inactive virions is reversible, then all the liposomes will eventually fuse, whereas V_f is given by

$$V_f = V_o \ (1 - \alpha) \ \{1 - \exp(- L_o/ (V_o(1 - \alpha)))\} \tag{5}$$

If the binding of liposomes to inactive virions is irreversible, then the concentration of fused liposomes, L_f, will be

$$L_f = L_o(1 - \alpha) \tag{6}$$

and

$$V_f = V_o \ (1 - \alpha) \ \ (1 - \exp \ (- L_o / V_o)) \tag{7}$$

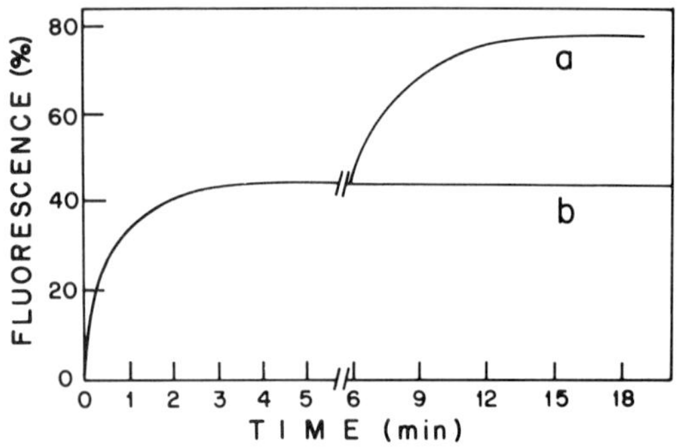

FIGURE 1. Continued fusion of virus-liposome fusion products with liposomes, but not with virus. Influenza virus and fluorescently labeled CL liposomes were fused at pH 5.0 and 37°C at concentrations of 25 μM lipid phosphorus each. After 6 min, additional nonlabeled liposomes (curve a) or virus particles (curve b) were added to yield final virus/liposome ratios of 1:4 and 4:1, respectively. Printed with permission from Biochemistry 1986, 25, p.257, copyright (1987) American Chemical Society.

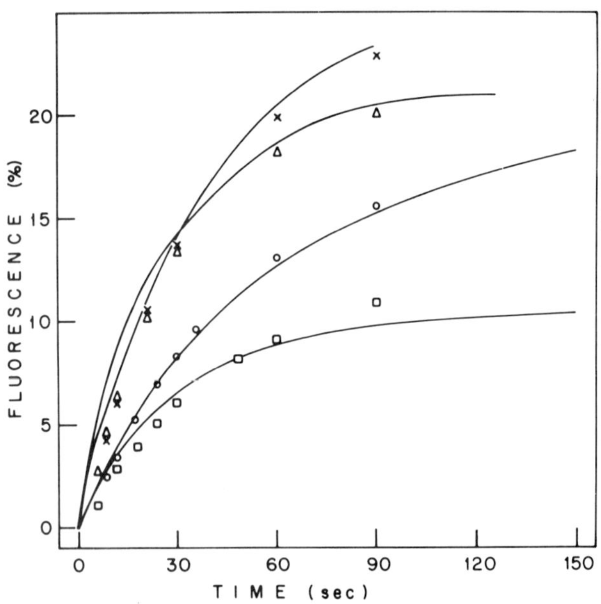

FIGURE 2. Fusion between large unilamellar vesicles of several compositions and influenza virus at pH 5.0 and 37°C. Drawn lines: experimental data. Symbols: calculated values using the parameters in Table 2A. Fusion at a 1:1 ratio of virus to liposomes, 5 μM of lipid phosphorous each. Liposome compositions: pure PS (X); 40% CL, 60% PC/PE (2:1) (Δ); 40% PS, 60% PC/PE (2:1) (O); 5% GD1a 39% Chol 56% PC/PE (2:1) (□).

Table 2A. Analysis of the Kinetics and Extent of Influenza Virus Fusion with Large Unilamellar Vesicles at 37°C.[a]

Liposome composition	% of active virus	Fusion rate const. f(s^{-1})	Aggr. rate const. C($M^{-1}s^{-1}$)	Diss. rate const. D(s^{-1})	Decay[d] rate const. γ(s^{-1})
pH 5					
pure CL	100	0.9	7.5×10^8	0.02	0.0-0.003
40%CL,60%PC/PE[b]	80	0.25	5×10^8	0.01	0.05
20%CL,80%PC/PE	70	0.20	4×10^8	0.01	0.05
pure PS	100	0.15	4.8×10^8	0.01	0.04
80%PS,20%PC/PE	90	0.14	4.5×10^8	0.01	0.045
40%PS,60%PC/PE	80	0.1	4.4×10^8	0.01	0.045
5%GD1a,95%PC/PE	20-40[c]	0.2	3.5×10^8	0.01	0.048
5%GD1a,39%Chol,56%PC/PE	20-45[c]	0.5	2.5×10^8	0.015	0.045
PC/PE	20-40	0.25	1.2×10^8	0.01	0.048
pH 6					
pure CL	25-40	0.25	4.3×10^8	0.025	N.D.

[a]The estimated uncertainties are 30, 20, 50 and 20% for f, C, D and γ, respectively. See Table 1A for experimental details. Fusion was analyzed as described in the text. [b]PC/PE ratio 2:1. [c]The higher value was used in the simulation of the kinetics. [d]See Eq(8).

Table 2B. Percent of Sendai Virus Fusion Activity and Rate Constants of Fusion (f), Aggregation (C), and Dissociation (D) as Functions of Liposomal Membrane Composition, pH, and PEG Concentration at 37°C.[a]

Liposome composition, pH	% PEG	% active virus[a]	f(s^{-1})	C($M^{-1}\cdot s^{-1}$)	D(s^{-1})
CL, 5		100	1	1.8×10^8	0.001
CL, 7.4	8	86	0.05	1.2×10^8	0.01
CL, 7.4	4	80	0.03	1.2×10^8	0.025
CL, 7.4		55	0.018	9×10^7	0.02
CL/DOPC, 5		80	0.2	1.2×10^8	0.01
CL/DOPC, 7.4	8	70	0.013	8×10^7	0.02
CL/DOPC, 7.4	4	50	0.01	8×10^7	0.02
CL/DOPC, 7.4		30	0.009	8×10^7	0.02
PS, 5		30[b]	0.03	5×10^7	0.04
PS, 7.4	8	30	0.013	7×10^7	0.04
PS, 7.4	4	30	0.006	7×10^7	0.04
PS, 7.4		25	0.005	7×10^7	0.04

[a]Each case in this Table represents a summary of a comparison between calculated and experimental levels of final fluorescence for four ratios of liposomes to viruses. [b]The actual values in this case may vary between 20% and 40%, but the results are better described by assuming irreversible binding of liposomes to inactive virus particles.

The final extents of fluorescence increase were adequately simulated and predicted for a large range of ratios between virions and liposomes. The results were best explained by assuming reversible binding in the case of influenza whereas with Sendai virus the results were better explained by an essentially irreversible binding of liposomes to inactive virions. The difference between V_f in Eqs (5) and (7) becomes small for large L_o/V_o.

Table 2 provides a summary of percents of virions capable of fusing as a function of liposome composition, pH and PEG. At pH 5 the activity of influenza virus is 100% when the target membranes are negatively charged liposomes, such as CL or PS. In the case of Sendai virus 100% activity ($\alpha = 0$) is found with CL, but with PS it is 30%, i.e., less than with CL/DOPC (1:1). Studies of Amselem et al. (1986) indicated a similar low fusion activity of Sendai virus with small unilamellar vesicles of PS or PG. Hence, it appears that CL is a unique phospholipid which promotes the activity of both influenza and Sendai virus.

Elevation of pH from 5 to 7.4 results in significant loss of activity of Sendai virus, with CL and CL/DOPC (and other compositions given in Klappe et al.1986), whereas the effect is less dramatic with PS. However, further reduction of pH to 4 enhances the fusion activity towards PS liposomes significantly (Amselem et al.1986; Klappe et al.1986). Haywood and Boyer (1985) showed a certain degree of fusion of influenza virus with negatively charged lipsomes at pH 7.5. Stegmann et al. (1985) demonstrated that elevation of pH resulted in significant loss of fusion activity of influenza virus. The results in Table 2A indicate a reduction in the percent of active virions to 25-40 at pH 6. We note in this context that Doms et al. (1985) found almost 100% binding of BHA to liposomes of several compositions at pH 5, whereas in most cases they found less than 5% binding at neutral pH. While the addition of the ganglioside GD1a to PC/PE liposomes promoted virus binding, it did not affect the final extents of fusion.

The addition of the dehydrating agent PEG results in a significant increase in the percent of Sendai virus capable of fusing with CL or CL/DOPC liposomes at neutral pH. The interpretation of the effect of pH, PEG, and the phenomenon of inactive virions will be presented in the last section.

Kinetics of Fusion

We have varied the lipid concentration from 1 to 50 μM, and in certain cases the ratios of virions to liposomes were varied by more than an order of magnitude. The results obtained with suspensions of low lipid concentrations where aggregation is rate limiting, are insensitive to f values, as long as the condition $L_oC \ll f$ holds (say, in 1/1 suspensions), and thus can be used to fix the value of C, the rate constant of aggregation. Then we determined the value of f, the rate constant of fusion, by simulating the results of fusion kinetics in more concentrated suspensions. We note that with labeled liposomes and a 1/1 suspension the initial I values ($\lesssim 10$) multiplied by 2 yield good estimates for the percent of fused liposomes or virions. At later stages the percents of fused liposomes and virions are above and below 2I, respectively, since the fusion products consist of a single virus and several liposomes.

The third parameter, D, the rate constant of deaggregation, which was determined from the results at the later stages of fusion, had a minor effect on the results at the early stages of fusion.

When the activity of virions is not complete, it is necessary to determine first the fraction of inactive virions, α, from the final extents, and then the determination of the rate constants proceeds as described. When $\alpha \neq 0$ deaggregation can promote the fusion kinetics and extent at later stages ($I \geq 15 - 30$) due to detachment of liposomes bound to inactive virions. Thus the released liposomes can adhere to and fuse with active virions and fusion products. This behavior was not found in vesicle-vesicle (Bentz et al.1983,1985; Nir et al.1983) or virus-cell fusion (Nir et al.1986b).

Several studies (Nir et al.1986a,c; Amselem et al.1986) demonstrated that the kinetics of virus-liposome fusion can be well simulated and predicted for a wide range of incubation times and concentrations. Fig. 2 provides a few recent results on the kinetics of fusion of influenza virus with liposomes. Tables 2A and 2B give a summary of the rate constants for several representative cases.

The largest variability is exhibited by the rate constant of fusion, f, the lowest value of which is 0.005 S^{-1} for Sendai virus fusing with PS liposomes (at pH 7.4) and the highest value is about $1S^{-1}$ for both Sendai and influenza virions fusing with CL liposomes at pH 5. Klappe et al.(1986) who fused Sendai virus with liposomes consisting of PC, PE, cholesterol and the viral receptor ganglioside GD1a observed significantly lower initial rates of fusion in this case than with PS liposomes, so that $f \ll 0.005S^{-1}$ in the former case.

The value $f=1S^{-1}$ is similar to that found for CL-containing liposomes induced to fuse among themselves by Ca (Wilschut et al.1985). When $f=1S^{-1}$ 10% of the virions attached to target membranes will fuse within 100 ms upon the induction of a signal such as the lowering of pH. For $f=0.001S^{-1}$ the corresponding time will be 100S. This difference in the rate constant of fusion can affect the fate of the virus within the endosome, which may be routed into other compartments before it has fused (Matlin et al.1981; Yoshimura and Ohnishi 1984).

The range of C values is $5x10^7$ to $1.8x10^8$ $M^{-1}S^{-1}$ for Sendai virus and $2.5x10^8$ to $10^9M^{-1}S^{-1}$ for influenza virus. Remarkably there is little change in C values in the system influenza virus-CL when the pH is raised from 5 to 6, which results in four-fold reduction in f and in a similar reduction in the percent of active virions. The rates of aggregation are just one to two orders of magnitude below the value $C=5x10^9$ M^{-1} S^{-1} in diffusion controlled aggregation (Smoluchowski, 1917). The implication is that both viruses encounter small potential barriers for adhesion to liposomes under the conditions reported in Table 2.

Table 2A (influenza) indicates that the incorporation of gangliosides GD1a into neutral liposomes, such as PC/PE, results in three-fold increase in C values. On the other hand the gangliosides had no effect on the fusion rate constants, or on the percent of active virions. In the case of Sendai virus the incorporation of gangliosides was essential (Haywood and Boyer, 1982) to enable its attachment to neutral liposomes (Hoekstra et al.1985).

Inactivation of Fusion Capacity

Preincubuation of influenza virus alone under conditions otherwise optimal for fusion, i.e., pH 5 and 37 °C, results in a rapid and irreversible loss of viral capacity to fuse or to cause hemolysis (Sato et al.1983; Junankar and Cherry,1986; Stegmann et al.1986). It has been demonstrated that HA undergoes a conformational change (Skehel et al.1982) which leads to an exposure of a hydrophobic segment of the HA2 subunit to the aqueous medium. This conformational change occurs at or near the optimal pH for fusion (Daniels et al.1985; Doms et al.1986). It was pointed out (White et al.1983) that the above hydrophobic segment bears close resemblance to a hydrophobic segment of Sendai virus (Richardson et al.1980; Blumberg et al.1985) except for the presence (in influenza) of three acidic amino acids (Min Jou et al.1980). An inactivation of fusion capacity was also noted upon preincubation of Sendai virus at low pH (Amselem et al.1986), but detailed studies are only available for influenza virus. Our first aim was to elucidate the effect of preincubation of the virus at low pH on the final extents of fusion and on the rate constants of fusion and aggregation.

The preincubation of influenza virus at pH 5 ($37^{o}C$) had a relatively small effect on the kinetics of its fusion with CL liposomes. With CL, CL/PC/PE and pure PS the final extents of fluorescence increase were almost unaffected by such preincubation of the virus up to 30 min. With PC/PE/cholesterol/GD1a liposomes the final fluorescence increase remained unchanged following 1 min of preincubation, but it dropped from 19 to 12% following 30 min of preincubation. Within experimental uncertainty the same trend was observed with PC/PE/GD1a liposomes. On the other hand the initial rates of fluorescence increase exhibited in all cases (excluding CL) rapid and dramatic reduction as a function of preincubation period. For instance, for PC/PE/GD1a the initial rates dropped from 17.5 (% fluorescence increase/min.) to ca. 6.5, 2.6 and≈0 following 30 sec., 1 min., and 30 min. of preincubation, respectively. We have established that the inactivation reflected intramembrane processes, rather than viral aggregation. No changes in light scattering or absorbance were found during preincubation, and viral inactivation was independent of its concentration during preincubation. We have extended the analysis by recording I values for several virus and liposome concentrations following 0 or 30 sec. of preincubation. The rate of inactivation was significantly lower in more dilute suspensions, where aggregation is the rate limiting step. Consequently we concluded that the aggregation rate constant was relatively little affected by the intramembrane changes of the HA. Additional support for this conclusion was provided by binding experiments. Except for the case of PC/PE (2:1) liposomes (without gangliosides) binding activity of the virus remained unchanged after 5 min. of preincubation. Accordingly, we kept the parameters C and D constant and assumed a decay of the fusion rate constants according to

$$f(t) = f(0)(\exp - \gamma t) \tag{8}$$

or

$$f(t) = f(0)/(1+\gamma t)^2 \tag{9}$$

The results in Fig. 2 and Table 2A are based on the use of Eq (8), but the use of Eq (9) would yield comparable simulations. As is illustrated by the representative cases in Fig. 2, good fit of the calculated to experimental I values was obtained for all types of liposomes. The simulation yielded similar values of γ with or without preincubation of the virus at low pH, and lower values of f(0) were found with preincubated virus, in accord with Eq (8). The range of γ values varied little with liposome composition, being 0.04 to 0.05 S^{-1}. It should be realized that Eq (8) has been applied for simulating the first two minutes of the process of fusion and inactivation. We present a simple model that can yield both the initial exponential decay and the residual fusion activity that must be retained. It may be recalled that even after 30 min. of preincubation the final I levels are reduced by 50% at most, despite the apparent vanishing initial rates of fluorescence increase.

Let us assume that the rate constant of fusion, f, is proportional to Na, the number of HA molecules in the optimal conformation for fusion. For simplicity we consider just two states and set

$$N = Na + Nb, \tag{10}$$

in which N denotes the total number of HA molecules. The b conformation(s) can result when the HA molecule in conformation a (at low pH), which is characterized by an exposed hydrophobic segment, does not encounter the target membrane and instead will associate with the viral membrane in an essentially irreversible manner. Mathematically this can be described by

$$dNa/dt = -\gamma_1 Na + \gamma_2 Nb \tag{11}$$

where initially Na=N and $\gamma_2 \ll \gamma_1$. The solution of Eq (11) is

$$Na(t) = N \exp - \gamma t + N \gamma_2 (1 - \exp(-\gamma t)/\gamma) \tag{12}$$

in which $\gamma = \gamma_1 + \gamma_2$

Since $\gamma_2 \ll \gamma_1$ it follows that initially Na decays exponentially according to

$$Na(t) = N \exp(-\gamma t) \tag{13}$$

as given by Eq (8), whereas after prolonged inactivation periods

$$Na = N \gamma_2 / \gamma \tag{14}$$

A decay according to Eq (9) would result if the number of HA monomers is reduced due to their clustering in the viral membrane as proposed by Junankar and Cherry (1986). The form of Eq (9) is identical with the Smoluchowski (1917) equation which gives the reduction in the number of monomers upon initiation of aggregation. Again, residual fusion activity can be retained if the equilibrium distribution of (HA) particles is taken into account (Nir et al.1980b, 1983; Bentz and Nir,1981). Studies are underway to determine more details of the mode of HA inactivation.

Viral Inactivation and Percent of Inactive Virus

Our interpretation is that inactivation due to preincubation of the virus at low pH, which amounts mostly to a reduction in f values, reflects the loss in the number of HA molecules whose hydrophobic segment is exposed. While adhesion can apparently proceed by means of inactivated HA or the NA glycoproteins (except for the case of PE/PC liposomes without gangliosides), the process of fusion may require the penetration of the hydrophobic segment of HA into the target membrane. It is not resolved whether the penetration of the hydrophobic segment is needed for overcoming the potential barrier due to "hydration forces" (Rand, 1981; Lis et al.1982). Alternatively, this penetration destabilizes both target and viral membranes. The fact that the value of f is much larger and the inactivation is rather small with CL liposomes may imply that a different mechanism of fusion operates in this case, despite the fact that viral activity is significantly lower at higher pH. Stegmann et al. (1986) found that the fusion of influenza virus with PC/PE/GD1a liposomes bears close similarity to its fusion with biological membranes such as those of eryhrocytes. Our current studies indicate that PS is an intermediate case between CL and PC/PE/GD1a, being charged and yielding 100% viral fusion activity, but exhibiting a reduction of fusion rates upon preincubation of the virus at low pH. In studies on Sendai virus whose fusion activity towards biological membranes is optimal at neutral pH (and 37°C), it was proposed that its fusion with liposomes at low pH relies mainly on the HN rather than on the F(fusion) glycoprotein (Citovsky et al.1985; Chejanovsky et al.1986; Amselem et al.1986; Klappe et al.1986).

The values of f and percent of virions fused may reflect different molecular characteristics of fusing systems. For example, with PS liposomes f values are relatively small whereas all virions can fuse. We also distinguish between virus inactivation due to its preincubation at low pH and the existence of virions incapable of fusing with certain target membranes. For instance, extensive inactivation of influenza virus is found at pH 5 when PS liposomes are the target membranes. On the other hand, inactivation is rather slow at higher pH values, while the percent of virions capable of fusing is reduced under such conditions.

Our formulation for the finding of reduced fractions of virions capable of fusing (Eqs (5) or (7)) implies that the virus preparation is heterogeneous in the sense that a certain fraction of virions can not form sufficiently close contacts with given liposomal membranes, although binding can occur. Thus, in the case of Sendai virus, addition of the dehydrating agent PEG can promote membrane contact and increase the percent of fused particles (see Table 2B). A change in pH affects the viral glycoproteins and thus the potential of the virus to fuse with given liposomes, whose affinities for the proteins are also affected by pH (Klappe et al.1986).

Alternatively, instead of having active and inactive virions the membrane of each virion can include a spectrum of HA molecules. If binding to a liposome does not occur at a region of the "active" HA, fusion will not occur. If this binding is reversible then eventually all virions will be able to fuse with the available liposomes. This possibility is ruled out if indeed the limited I values measured after 24 h (see Table 2) would not increase further, but this has not been persued as yet. If the binding of inactive HA to liposomes is irreversible, then Eq (7) holds albeit in a different sense.

Acknowledgements

Ms Sue Salomon is acknowledged for the expert typing of the manuscript. This study was supported in part by the National Institutes of Health Grant GM-31506 (to J. Bentz and S.N.) and by the Netherlands Organization for the Advancement of Pure Research (ZWO). Prof. Dov Lichtenberg is acknowledged for reading a part of the manuscript.

References

Amselem, S., Barenholz, Y., Loyter, A., Nir, S. and Lichtenberg, D. 1986. Fusion of Sendai virus with negatively charged liposomes as studied by pyrene-labelled phospholipid liposomes. Biochim. Biophys. Acta 860: 301-313.

Bentz, J. and Nir, S. 1981. Kinetic and equilibrium aspects of reversible aggregation. J. Chem. Soc. Faraday I, 77: 1249-1275.

Bentz, J., Nir, S. and Wilschut, J. 1983. Mass action kinetics of vesicle aggregation and fusion. Colloids and Surfaces 6: 333-363.

Bentz, J., Duzgunes, N. and Nir, S. 1985. Temperature dependence of divalent cation induced fusion of phosphatidylserine liposomes: Evaluation of the kinetic rate constants. Biochemistry 24: 1064-1072.

Blumberg, B.M., Giorgi, C., Rose, K. and Kolakofsky, D. 1985. Sequence determination of the Sendai virus fusion protein gene. J. Gen. Virol. 66: 317-331.

Chejanovsky, N., Zakai, N., Amselem, S., Barenholz, Y. and Loyter, A. 1986. Membrane vesicles containing the Sendai virus binding glycoprotein, but not the viral fusion protein, fuse with phosphatidylserine liposomes at low pH. Biochemistry 25: 4810-4817.

Citovsky, V., Blumenthal, R. and Loyter, A. 1985. Fusion of Sendai virions with phosphatidylcholine-cholesterol liposomes reflects the viral activity required for fusion with biological membranes. FEBS Lett. 193, 135-140.

Daniels, R.S., Downie, J.C., Hay, A.J., Knossow, M., Skehel, J.J., Wang, M.L. and Wiley, D.C. 1985. Fusion mutants of the influenza virus hemagglutinin glycoprotein. Cell 40: 431-439.

Doms, R., Helenius, A. and White, J. 1985. Membrane fusion activity of the influenza virus hemagglutinin. The low pH-induced conformational change. J. Biol. Chem. 260: 2973-2981.

Doms, R.W., Gething, M.-J., Henneberry, J., White, J. and Helenius, A. 1986. Variant influenza virus hemagglutinin that induces fusion at elevated pH. J. Virology 57: 603-613.

Düzgüneş, N. 1985. Membrane fusion. Subcell. Biochemistry 11: 195-286.

Ellens, H., Bentz, J. and Szoka, F.C. 1985. H^+- and Ca^{2+}-induced fusion and destabilization of liposomes. Biochemistry 24: 3099-3106.

Gething, M.-J., Doms, R.W., York, D. and White, J. 1986. Studies on the mechanism of membrane fusion: Site specific mutagenesis of the hemagglutinin of influenza virus. J. Cell Biol. 102: 11-23.

Haywood, A.M. and Boyer, B.P. 1982. Sendai virus membrane fusion: Time course and effect of temperature, pH, calcium, and receptor concentration. Biochemistry 21: 6041-6046.

Haywood, A.M. and Boyer, B.P. 1984. Effect of lipid composition upon fusion of liposomes with Sendai virus membranes. Biochemistry 23: 4161-4166.

Haywood, A.M. and Boyer, B.P. 1985. Fusion of influenza virus membranes with liposomes at pH 7.5. Proc. Natl. Acad. Sci. U.S.A. 82: 4611-4615.

Hoekstra, D. 1982. Role of lipid phase separations and membrane hydration in phospholipid vesicle fusion. Biochemistry 21: 2833-2840.

Hoekstra, D., De Boer, T., Klappe, K. and Wilschut, J. 1984. Fluorescence method for measuring the kinetics of fusion between biological membranes. Biochemistry 23: 5675-5681.

Hoekstra, D. and Klappe, K. 1986. Sendai virus-erythrocyte membrane interaction: Quantitative and kinetic analysis of viral binding, dissociation and fusion. J. Virology 58: 87-95.

Hoekstra, D., Klappe, K., De Boer, T. and Wilschut, J. 1985. Characterization of the fusogenic properties of Sendai virus: Kinetics of fusion with erythrocyte membranes. Biochemistry 24: 4739-4745.

Junankar, P.R. and Cherry, R.J. 1986. Temperature and pH dependence of the haemolytic activity of influenza virus and of the rotational mobility of the spike glycoproteins. Biochim. Biophys. Acta 854: 198-206.

Klappe, K., Wilschut, J., Nir, S. and Hoekstra, D. 1986. Parameters affecting fusion between Sendai virus and liposomes. Role of viral proteins, liposome composition, and pH. Biochemistry 25: 8252-8260.

Lis, L.J., McAlister, M., Fuller, N., Rand, R.P. and Parsegian, V.A. 1982. Interactions between neutral phosphilipid bilayer membranes. Biophys. J. 37: 657-666.

Matlin, K., Reggio, H., Helenius, A. and Simons, K. 1981. The infective entry of influenza virus into MDCK cells. J. Cell Biol. 91: 601-613.

Min Jou, M., Verhoeyen, M., Devos, R., Saman, E., Fang, R., Hylebroeck, D., Friers, W., Threlfall, G., Barber, C., Carey, N. and Emtage, S. 1980. Complete structure of the Hemagglutinin gene from the human influenza A/Victoria/3/75 (H_3N_2) strain as determined from cloned DNA. Cell 19: 683-696.

Nir, S., Bentz, J. and Wilschut, J. 1980a. Mass action kinetics of phosphatidylserine vesicle fusion as monitored by coalescence of internal vesicle volumes. Biochemistry 19: 6030-6036.

Nir, S., Bentz, J. and Portis, A.R.Jr. 1980b. Effect of cation concentrations and temperature on the rates of aggregation of acidic phospholipid vesicles: Application to fusion. Adv. Chem. Ser. 188: 75-106.

Nir, S., Bentz, J., Wilschut, J. and Düzgüneş, N. 1983. Aggregation and fusion of phospholipid vesicles. Prog. Surface Sci. 13: 1-124.

Nir, S., Stegmann, T. and Wilschut, J. 1986a. Fusion of influenza virus and cardiolipin liposomes at low pH: Mass action analysis of kinetics and extent. Biochemistry 25: 257-266.

Nir, S., Klappe, K. and Hoekstra, D. 1986b. Kinetics and extent of fusion between Sendai virus and erythrocyte ghosts: Application of a mass action kinetic model. Biochemistry 25: 2155-2161.

Nir, S., Klappe, K. and Hoekstra, D. 1986c. Mass action analysis of kinetics and extent of fusion between Sendai virus and phospholipid vesicles. Biochemistry 25: 8261-8266.

Ohki, S. and Ohshima, H. 1985. Divalent cation-induced phosphatidic acid membrane fusion: Effect of ion binding and membrane surface tension. Biochim. Biophys. Acta 812: 147-154.

Papahadjopoulos, D., Vail, W.J., Newton, C., Nir, S., Jacobson, K., Poste, G. and Lazo, R. 1977. Studies on membrane fusion. III. The role of calcium-induced phase changes. Biochim. Biophys. Acta 465: 579-598.

Portis, A., Newton, C., Pangborn, W. and Papahadjopoulos, D. 1979. Studies on the mechanism of membrane fusion: Evidence for an intermembrane Ca^{2+} -phospholipid complex, synergism with Mg^{2+}, and inhibition by spectrin. Biochemistry 18: 780-790.

Rand, R. 1981. Interacting phospholipid bilayers: measured forces and induced structural changes. A. Rev. Biophys. Bioengng. 10: 277-314.

Richardson, C., Scheid, A. and Choppin, P. 1980. Specific inhibition of paramyxo-virus and myxovirus replication by oligopeptides with amino acid sequences similar to those at the N-termini of the F_1 or HA_2 viral polypeptides. Virology 105: 205-222.

Sato, S.B., Kawasaki, K. and Ohnishi, S. 1983. Hemolytic activity of influenza virus hemagglutinin glycoproteins activated in mildly acidic environments. Proc. Natl. Acad. Sci. U.S.A. 80: 3153-3157.

Skehel, J.J., Bayley, P.M., Brown, E.B., Martin, S.R., Waterfield, M.D., White, J.M., Wilson, I.A. and Wiley, D.C. 1982. Changes in the conformation of influenza hemagglutinin at the pH optimum of virus mediated membrane fusion. Proc. Natl. Acad. Sci. U.S.A. 79: 968-972.

Smoluchowski, M. 1917. Investigation into a mathematical theory of the kinetics of coagulation of colloidal solutions. Z. Phys. Chem. Abt. A92: 129-168.

Stegmann, T., Hoekstra, D., Scherphof, G. and Wilschut, J. 1985. Kinetics of pH - Dependent fusion between influenza virus and liposomes. Biochemistry 24: 3107-3113.
Stegmann, T., Hoekstra, D., Scherphof, G. and Wilschut, J. 1986. Fusion activity of influenza virus. A comparison between biological and artificial target membrane vesicles. J. Biol. Chem. 261: 10966-10969.
Struck, D.K., Hoekstra, D. and Pagano, R.E. 1981. Use of resonance energy transfer to monitor membrane fusion. Biochemistry 20: 4093-4099.
White, J., Helenius, A. and Kartenbeck, J. 1982. Membrane fusion activity of influenza virus. EMBO J. 1: 217-222.
White, J., Kielian, M. and Helenius, A. 1983. Membrane fusion proteins of enveloped animal viruses. Q. Rev. Biophys. 16: 151-195.
Wilschut, J., Duzgunes, N., Fraley, R. and Papahadjopoulos, D. 1980. Studies on the mechanism of membrane fusion: Kinetics of calcium ion induced fusion of phosphatidylserine vesicles followed by a new assay for mixing of aqueous vesicle contents. Biochemistry 19: 6011-6021.
Wilschut, J., Nir, S., Scholma, J. and Hoekstra, D. 1985. Kinetics of Ca-induced fusion of cardiolipin/phosphatidylcholine vesicles: Correlation between vesicle aggregation, bilayer destabilization and fusion. Biochemistry 24: 4630-4636.
Wilson, I.A., Skehel, J.J. and Wiley, D.C. 1981. The haemagglutinin membrane glycoprotein of influenza virus: structure at 3A resolution. Nature (London.) 289: 366-373.
Yoshimura, A. and Ohnishi, S.I. 1984. Uncoating of influenza virus in endosomes. J. Virology 51: 497-504.

STRATEGIES FOR THE INVESTIGATION OF EXOCYTOTIC MEMBRANE FUSION

Manfred Gratzl

Abteilung Anatomie und Zellbiologie
Universität Ulm, Bundesrepublik Deutschland

INTRODUCTION

The transfer of hormones from the cell into the extracellular space during secretion by exocytosis is accomplished by the fusion of the secretory vesicle membrane with the plasma membrane. As opposed to the study of artificial membranes or isolated natural membranes as models, suitable modified secretory cells have been introduced in the past few years to investigate exocytotic membrane fusion. The groups working in the latter field are attempting to learn as much as possible from nature with the aim of characterizing the properties of membrane fusion as it occurs in the intact cell.

The inherent problem in such investigations is that the interacting membranes are not accessible from outside the cells. Manipulation of the cytoplasm in a rather limited way can be carried out using ionophores. Direct access to the cytoplasm and its complete control requires the breakdown of the plasma membrane as a permeability barrier. With single cells this can be done using cells attached to pipettes (c.f. Fernandez et al., 1984). Cell suspensions have been permeabilized with different techniques without impairing the process of exocytosis. With high-voltage discharges and detergents, it has been possible to permanently permeabilize cells for secretion studies. In both ways, if carefully controlled, the cells seem not to be altered greatly in so far as the release process is concerned (c.f. Knight and Baker, 1982; Wilson and Kirshner, 1983; Dunn and Holz, 1983). The lesions evoked in the cell membrane with these methods differ in size. With detergents, cellular macromolecules (like cytoplasmic lactate dehydrogenase) are lost, while with high voltage discharges few such molecules escape. In contrast to the techniques mentioned above, defined pores can be inserted into the cell membrane of secretory cells using staphylococcal alpha-toxin. This allows small solutes to be exchanged and keeps macromolecules necessary for secretion within the cells. The application of alpha-toxin for the analysis of exocytotic membrane fusion (Ahnert-Hilger et al., 1985a,b, 1987;

Bader et al., 1986; Lind et al., 1987; Ahnert-Hilger and Gratzl, 1987) will be described in this chapter. The characteristics of hormone release from permeabilized cells will be compared with secretion by intact cells and fusion of isolated secretory vesicle membranes and of liposomes.

PERMEABILIZATION OF SECRETORY CELLS WITH ALPHA-TOXIN

Alpha-toxin is a water soluble protein produced by most strains of Staphylococcus aureus. The mechanism of the toxins attack of mammalian cells has been established in a series of investigations (Füssle et al., 1981; Bhakdi et al., 1981, 1984; for review see Bhakdi and Tranum-Jensen, 1987). First the toxin (molecular weight 34 KD) is bound to the cells, then it hexamerizes in the plasma membrane and forms stable transmembrane pores, which do not permit the passage of macromolecules with a molecular weight of 4 KD and larger but allows for the release of small molecules from erythrocytes.

Secretory cells can also be permeabilized by alpha-toxin. Bovine chromaffin cells, rat pheochromocytoma or insulinoma cells in culture, upon addition of alpha-toxin, release small solutes like monovalent cations or ATP and take up trypan blue while large proteins such as cytoplasmic lactate dehydrogenase are retained by the cells (Ahnert-Hilger et al., 1985; Bader et al., 1986; Lind et al., 1987). In other words the size of the alpha-toxin pore allows for the removal of all small molecules from the cytoplasm and their replacement by other ones. Therefore this technique can serve to add buffer substances for the control of pH and divalent cations. In addition, drugs known to interfere with defined intracellular regulator sites, but ineffective in intact cells due to limited access, can easily be introduced into secretory cells. Finally, the remarkable insensitivity of secretory vesicle fusion with the plasma membrane to the incorporation of alpha-toxin allows sufficient time for the experimental manipulation of the permeabilized cells.

As a consequence of the small size of the alpha-toxin pore, the toxin monomer itself is excluded from the cytoplasm. Thus the toxin's attack is restricted to the plasma membrane which is an advantage when compared to the action of detergents used for cell permeabilization. Also detergents like digitonin harbor the hazards of modifying the composition of the interacting membrane and thus distorting the data obtained.

As evidence for exocytosis in alpha-toxin permeabilized cell preparations, the same criteria can be used which were applied to intact cells: E. g. in order to exclude simple leakage of hormones from secretory vesicles the parallel release of small and large molecular weight secretory products present within the vesicles can be measured. This co-release, in conjunction with the observation that large molecules residing in the cytoplasm remain within the cells clearly indicates that hormone release occurs via exocytosis (Ahnert-Hilger et al., 1985b; Bader et al., 1986). Another approach to demonstrate exocytosis makes use of the compartmentation of enzymes metabolizing catecholamines (Ahnert-Hilger et al., 1987). Catecholamines are released unchanged from intact pheochromocytoma cells because the cytoplasm containing the oxidizing enzymes is circumvented if secretion occurs by exocytosis. When transferred into the cytoplasm (by reversing the driving force for vesicular uptake of catecholamines) oxidized catecholamines are formed. Since the same situation is seen in alpha-toxin permeabilized cells, it is safe to conclude that the secretory vesicle contents are directly transported by exocytosis into the extracellular space.

Recently, in addition to the other documented procedures (c.f. references Bhakdi and Tranum-Jensen, 1987) a rapid and effective purification procedure for alpha-toxin from the culture medium of Staphylococcus aureus (strain Wood 46) has been described (Lind et al., 1987). This technique facilitates again the study of membrane fusion in permeabilized secretory cells.

CHARACTERIZATION OF EXOCYTOSIS IN ALPHA-TOXIN PERMEABILIZED CELLS

In the first series of experiments the minimal (cytoplasmic) requirements for secretory vesicle/plasma membrane fusion have been established in alpha-toxin permeabilized PC12 cells, a rat pheochromocytoma cell line. It was demonstrated that the release of dopamine from these cells upon addition of alpha-toxin was reduced from "extracellular" (mM) to intracellular (μM) calcium concentrations (Ahnert-Hilger et al., 1985a). In media containing defined free concentrations of calcium (adjusted with chelators and controlled with a calcium-specific electrode) a characteristic biphasic response as a function of the free calcium concentration could be observed (Fig. 1 and Ahnert-Hilger et al., 1985a,b; Ahnert-Hilger and Gratzl, 1987). One phase already activated below 1 μM free calcium, plateaued at about 1-5 μM free calcium and a second occurred in the presence of larger amounts of free calcium (10-100 μM). Calcium-induced dopamine release could be observed at pH 6.6 as well as at pH 7.2, it was insensitive to changes in the nature of monovalent cations or anions. Magnesium was unable to replace calcium. However, when present simultaneously with calcium in 1-5 μM concentrations it augmented hormone release (Fig. 2 and Ahnert-Hilger and Gratzl, 1987). It is interesting to note that the effect of magnesium was only seen at the low calcium concentrations which are likely to occur in stimulated cells (see next chapter). Further attempts to characterize or identify the targets (receptors) of calcium within the pheochromocytoma cells using group-specific reagents or drugs interfering with calmodulin (trifluoperazine) have not yet been successful.

One reason for that may be that additional modulatory reactions involved in the regulation of exocytosis are working in alpha-toxin permeabilized cells. From the inhibitory action of GTP-gamma-S on calcium-induced exocytosis, a modulatory role of GTP binding proteins has been inferred as for the activation by TPA an involvement of the proteinkinase C cascade (Ahnert-Hilger et al., 1987). The latter effect requires the presence of ATP while the calcium-induced exocytosis in the pheochromocytoma cells used is entirely independent on this nucleotide (Ahnert-Hilger et al., 1985a,b; Ahnert-Hilger and Gratzl, 1987).

Interestingly, chromaffin cells in primary culture also in the absence of TPA require additional ATP for calcium-induced exocytosis (Bader et al., 1986). Other nucleotides could not replace ATP suggesting a very specific role in these cells. The ATP requirement of secretion in intact cells will be discussed in detail in the next section of this article.

COMPARISON OF EXOCYTOSIS IN ALPHA-TOXIN PERMEABILIZED CELLS WITH MEMBRANE FUSION OF BIOLOGICAL MEMBRANES AND OF LIPOSOMES

The importance of extracellular calcium in stimulus-secretion coupling was realized more than 25 years ago (Douglas and Rubin, 1961). Injection studies with mast cells and the giant synapse indicated that intracellular calcium may trigger secretion (Kanno et al., 1973; Miledi, 1973). After it was observed that transmitter release is actually paralleled by an increase in intracellular free calcium (Llinàs and Nicholson) this issue was further investigated in a variety of secretory cells. These studies soon became experimentally easier to be carried out because new techniques became available. Resting pheochromocytoma cells or adrenal medullary chromaffin cells in primary culture, i. e. the cells used up to now for permeabilization with alpha-toxin, contain about 0.1 µM free calcium.

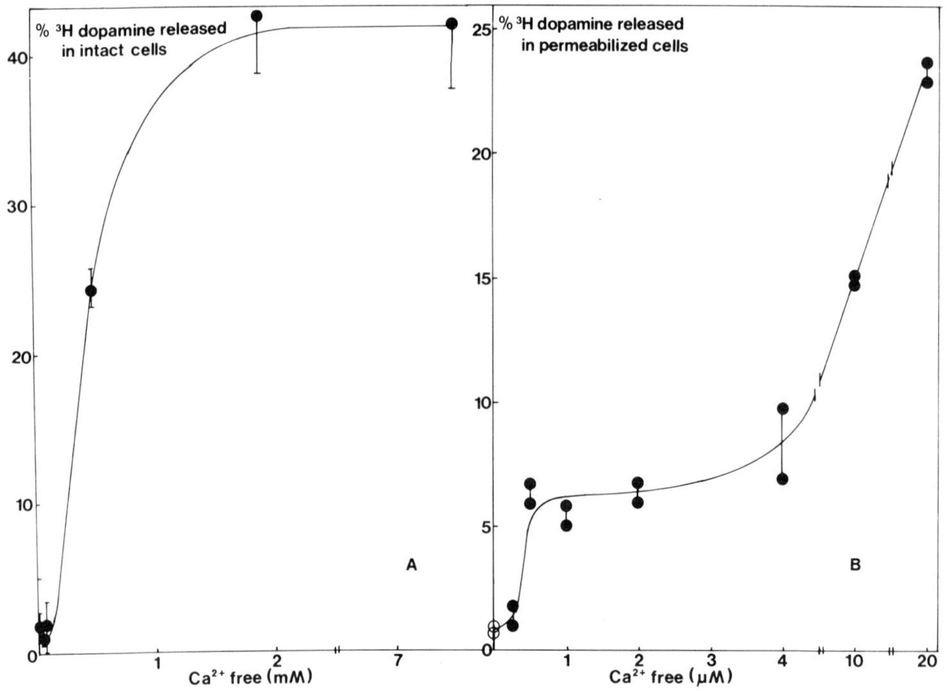

Fig. 1. Calcium requirement of dopamine release by intact and by alpha-toxin permeabilized pheochromocytoma cells (from Ahnert-Hilger et al., 1985a, by permission).

When stimulated, the free calcium increases to about 0.5 μM (Knight and Kesteven, 1983; Meldolesi et al., 1984). It is worth mentioning, however, that an average intracellular free calcium concentration has been measured in these studies and the true concentrations just beneath the plasma membrane may be higher as a consequence of the calcium influx during stimulation. In another study (Kao and Schneider, 1986) values of about 10 μM free calcium have been obtained in chromaffin cells stimulated with nicotine.

In any case, the calcium dependence of catecholamine secretion by intact cells is in accordance with the situation seen in alpha-toxin permeabilized pheochromocytoma cells and adrenal medullary chromaffin cells in primary culture (see previous chapter). Interestingly enough, also the dependence on ATP in alpha-toxin permeabilized chromaffin cells (Bader et al., 1986) is paralleled by a similar observation in intact cells (Rubin, 1969; Kirshner and Smith, 1965). Intact pheochromocytoma cells, on the other hand, do not depend on cytoplasmic ATP during secretion (Reynolds et al., 1982), a fact which is supported by the insensitivity of calcium-induced dopamine release to ATP in the same cells permeabilized with alpha-toxin (Ahnert-Hilger et al., 1985a,b; Ahnert-Hilger and Gratzl, 1987).

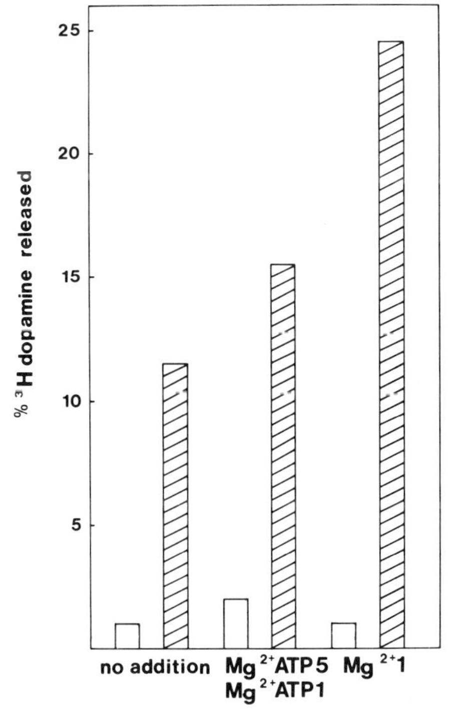

Fig. 2. Effect of magnesium with or without ATP on calcium-induced dopamine release by permeabilized pheochromocytoma cells. The open bars represent controls, the hatched bars alpha-toxin treated cells. The medium contained 10 μM free calcium and the indicated amounts of free magnesium (from Ahnert-Hilger and Gratzl, 1987, by permission).

The exact reason for the necessity of ATP in permeabilized chromaffin cells in primary culture is not known. One possibility is that regulatory sequences (like the proteinkinase C system) have all components activated but one key protein is dephosphorylated and requires ATP for rephosphorylation. In order to identify such sites phosphorylation studies have been carried out using intact cells, permeabilized cells, or subcellular fractions (Schulman and Greengard, 1978; Amy and Kirshner, 1981; Treiman et al., 1983; Niggli et al., 1984). However, also the topography of chromaffin vesicle distribution in the different cells investigated must be considered. In cultured pheochromocytoma cells secretory vesicles are concentrated near the plasma membrane (Watanabe et al., 1983) and thus are close to the site of exocytosis. On the other hand within adrenal medullary chromaffin cells secretory vesicles are more randomly distributed (c.f. Bader et al., 1981) and ATP dependent vesicle movement (Allen et al., 1985; Vale et al., 1985) could be necessary.

The step requiring ATP is not the process of membrane fusion itself but a preceding one, since isolated secretory vesicles isolated from chromaffin and other secretory cells fuse without ATP added. These biological membranes fuse half maximal with about 1-10 μM free calcium, i. e. the same concentration observed in stimulated intact cells or the amount of calcium effective in alpha-toxin permeabilized cells (Dahl and Gratzl, 1976; Ekerdt et al., 1981; Gratzl and Dahl, 1976, 1978; Gratzl et al., 1977, 1980). When the membranes investigated were further simplified by the preparation of liposomes from the membrane lipids of chromaffin vesicles, characteristic properties like the susceptibility to low concentrations of calcium for the fusion process were lost (Ekerdt et al., 1981). In conjunction with the observation that treatment with chemicals or enzymes of the secretory vesicle membranes led to the same changes in membrane fusion properties, it was concluded that membrane proteins are responsible for these properties (Ekerdt et al., 1981). Actually liposomes containing acidic phospholipids exhibit very similar fusion properties as do the liposomes prepared from the membrane lipids of isolated secretory vesicles (Ekerdt et al., 1981; c.f. for ref. Düzgünes, 1985). In attempts to reduce the calcium requirements of these membranes from mM to more physiological (μM) concentrations the effect of synexin, a cytoplasmic protein present in a variety of secretory cells has been investigated (Hong et al., 1982a,b). It turned out that with this protein, in the presence of mM magnesium, the threshold of the liposomes to undergo fusion could be reduced to about 10 μM calcium. This indicates that fusion of liposomes can be modulated in a way to approach the milieu existing in resting or stimulated cells.

CONCLUSION

Permeabilization of secretory cells by staphylococcal alpha-toxin indeed is a valuable technique to investigate membrane fusion during exocytosis because it allows the permanent control of the composition of the cytoplasm with respect to small molecules (MW 1 KD) without perturbing the fusion process itself. The secretion observed in such permeabilized cells is compatible with secretion by intact cells and closely resembles membrane fusion studied with isolated secretory vesicle membranes.

After extensively studying the effects of small solutes, inhibitors or activators, the concept of using natural occurring pore forming substances for cell permeabilization opens further possibilities. Permanent pores which are big enough to release even large molecular weight substances from the cells or to introduce antibodies directed against membrane components still residing in permeabilized cells can be used to identify the relevant membrane proteins involved in exocytotic membrane fusion. Then they can be isolated and incorporated into liposomes with the aim of modifying their properties in a way that the fusion process in this simplified system is compatible with the situation seen in intact secretory cells. This would then elucidate the molecular mechanism of membrane fusion during exocytosis.

ACKNOWLEDGEMENTS

The work of the author referred to in this article was supported by the Deutsche Forschungsgemeinschaft (Gr 681) and the Forschungsschwerpunkt No. 24 of the State of Baden-Württemberg. The author thanks Mrs. M. Henning for her helpful advice on how to word this manuscript and Mrs. B. Mader for typing it.

REFERENCE LIST

Ahnert-Hilger, G., and Gratzl, M., 1987, Further characterization of dopamine release by permeabilized PC12 cells, J. Neurochem., in press.

Ahnert-Hilger, G., Bhakdi, S., and Gratzl, M., 1985a, Alpha-toxin permeabilized cells - A new approach to investigate stimulus secretion coupling, Neurosci. Lett., 58:107.

Ahnert-Hilger, G., Bhakdi, S., and Gratzl, M., 1985b, Minimal requirements for exocytosis - a study using PC 12 cells permeabilized with staphylococcal alpha-toxin, J. Biol. Chem., 260:12730.

Ahnert-Hilger, G., Bräutigam, M., and Gratzl, M., 1987, Calcium-stimulated catecholamine release from alpha-toxin permeabilized PC12 cells: Biochemical evidence for exocytosis and its modulation by TPA and GTP-gamma-S, submitted

Allen, R. D., Weiss, D. G., Hayden, J. H., Brown, D. T., Fujiwake, H., and Simpson, M., 1985, Gliding movement of and bidirectional transport along single native microtubules from squid axoplasm: Evidence for an active role of microtubules in cytoplasmic transport, J. Cell Biol., 100:1736.

Amy, C. M., and Kirshner, N., 1981, Phosphorylation of adrenal medulla cell proteins in conjunction with stimulation of catecholamine secretion, J. Neurochemistry, 36:847.

Bader, M. F., Ciesielski-Treska, J., Thiersé, D., Hesketh, J. E., and Aunis, D., 1981, Immunocytochemical study of microtubules in chromaffin cells in culture and evidence that tubulin is not an integral protein of the chromaffin granule membrane, J. Neurochem., 37:917.

Bader, M. F., Thiersé, D., Aunis, D., Ahnert-Hilger, G., and Gratzl, M., 1986, Characterization of hormone and protein release from alpha-toxin permeabilized chromaffin cells in primary culture, J. Biol. Chem., 261:5777.

Bhakdi, S., Muhly, M., and Füssle, R., 1984, Correlation between toxin binding and hemolytic activity in membrane damage by staphylococcal alpha-toxin, Infect. Immun., 46:318.

Bhakdi, S., Füssle, R., and Tranum-Jensen, J., 1981, Staphylococcal alpha-toxin: Oligomerization of hydrophilic monomers to form amphiphilic hexamers induced through contact with deoxycholate detergent micelles, Proc. Natl. Acad. Sci. USA, 78:5475.

Bhakdi, S., and Tranum-Jensen, J., 1987, Damage to mammalian cells by proteins that form transmembrane pores, Rev. Physiol. Biochem. Pharmacol., 107:147.

Dahl, G., and Gratzl, M., 1976, Calcium-induced fusion of isolated secretory vesicles from the islet of Langerhans, Cytobiologie, 12:344.

Douglas, W. W., and Rubin, R. P., 1961, The role of calcium in the secretory response of the adrenal medulla to acetylcholine, J. Physiol. (London), 159:40

Dunn, L. A., and Holz, R. W., 1983, Catecholamine secretion from digitonin-treated adrenal medullary chromaffin cells, J. Biol. Chem., 258:4989.

Düzgünes, N., 1985, Chapter 5: Membrane Fusion, in: "Subcellular Biochemistry," Vol. 11, D. B. Roodyn, ed., Plenum Press, New York.

Ekerdt, R., Dahl, G., and Gratzl, M., 1981, Membrane fusion of secretory vesicles and liposomes: two different types of fusion, Biochim. Biophys. Acta, 646:10.

Fernandez, J. M., Neher, E., and Gomperts, B. D., 1984, Capacitance measurements reveal stepwise fusion events in degranulating mast cells, Nature, 312:453.

Füssle, R., Bhakdi, S., Sziegoleit, A., Tranum-Jensen, J., Kranz, T., and Wellensiek, H.J., 1981, On the mechanism of membrane damage by Staphylococcus aureus alpha-toxin, J. Cell Biol., 91:83.

Gratzl, M., and Dahl, G., 1976, Calcium-induced fusion of Golgi derived secretory vesicles isolated from rat liver, FEBS Letters, 62:142.

Gratzl, M., and Dahl, G., 1978, Fusion of secretory vesicles isolated from rat liver, J. Membrane Biology, 40:343.

Gratzl, M., Dahl, G., Russell, J. T., and Thorn, N. A., 1977, Fusion of neurohypophyseal membranes in vitro, Biochim. Biophys. Acta, 470:45.

Gratzl, M., Schudt, Ch., Ekerdt, R., and Dahl, G., 1980, Fusion of isolated biological membranes. A tool to investigate basic processes of exocytosis and cell-cell fusion. in: "Membrane structure and function," E. E. Bittar, ed., John Wiley, New York, p 59.

Hong, K., Düzgünes, N., Ekerdt, R., and Papahadjopoulos, D., 1982, Synexin facilitates fusion of specific

phospholipid vesicles at divalent cation concentrations found intracellularly, Proc. Natl. Acad. Sci. USA, 79:4942.
Hong, K., Düzgünes, N., and Papahadjopoulos, D., 1982, Modulation of membrane fusion by calcium-binding proteins, Biophys. J., 37:297.
Kanno, T., Chochrane, D. E., and Douglas, W. W., 1973, Exocytosis (secretory granule extrusion) induced by injection of calcium into mast cells, Can. J. Physiol. Pharmacol., 51:1001.
Kao, L. S., and Schneider, A. S., 1986, Calcium mobilization and catecholamine secretion in adrenal chromaffin cells, J. Biol. Chem. 261:4881.
Knight, D. E., and Baker, P. F., 1982, Calcium-dependence of catecholamine release from bovine adrenal medullary cells after exposure to intense electric fields, J. Membrane Biol., 68:107.
Knight, D. E., and Kesteven, N. T., 1983, Evoked transient intracellular free calcium changes and secretion in isolated adrenal medullary cells, Proc. R. Soc. Lond. B, 218:177.
Lind, I., Ahnert-Hilger, G., Fuchs, G., and Gratzl, M., 1987, Purification and application for cell permeabilization of alpha-toxin from staphylococcus aureus, Anal. Biochem., in press.
Llinás, R., and Nicholson, C., 1975, Calcium role in depolarization-secretion coupling: An aequorin study in squid giant synapse, Proc. Natl. Acad. Sci. USA, 72:187.
Meldolesi, J., Huttner, W. B., Tsien, R. Y., and Pozzan, T., 1984, Free cytoplasmic Calcium and neurotransmitter release: studies on PC12 cells and synaptosomes exposed to alpha-latrotoxin, Proc. Natl. Acad. Sci. USA, 81:620.
Miledi, R., 1973, Transmitter release induced by injection of calcium ions into nerve terminals, Proc. R. Soc. Ser. B, 183:421.
Niggli, V., Knight, D. E., Baker, P., F., Vigny, A., and Henry, J.-P., 1984, Tyrosine hydroxylase in "leaky" adrenal medullary cells: Evidence for in situ phosphorylation by separate calcium and cyclic AMP-dependent systems, J. Neurochem., 43:646.
Schulman, H., and Greengard, P., 1978, Calcium-dependent protein phosphorylation system in membranes from various tissues, and its activation by "calcium-dependent-regulator", Proc. Natl. Acad. Sci. USA, 75:5432.
Treiman, M., Weber, W., and Gratzl, M., 1983, 3', 5'-cyclic adenosine monophosphate- and Calcium-calmodulin-dependent endogenous protein phosphorylation activity in membranes of the bovine chromaffin secretory vesicles: Identification of two phosphorylated components as tyrosine hydroxylase and protein kinase regulatory subunit type II, J. Neurochem., 40:661.
Vale, R. D., Schnapp, B. J., Reese, T. S., and Sheetz, N. P., 1985, Organelle, bead, and mikrotubule translocations promoted by soluble factors from the squid giant axon, Cell, 40:559.

Watanabe, D., Torda, M., and Meldolesi, J., 1983, The effect of alpha-latrotoxin on the neurosecretory PC12 cell line: Electron microscopy and cytotoxicity studies, Neuroscience, 10:1011.

Wilson, S. P., and Kirshner, N., 1983, Calcium-evoked secretion from digitonin-permeabilized adrenal medullary chromaffin cells, J. Biol. Chem., 258:4994.

ENZYMATIC REGULATION OF MEMBRANE FUSION DURING SYNCHRONOUS EXOCYTOSIS IN PARAMECIUM CELLS

Helmut Plattner, Christine J. Lumpert, Ute Gras, Jeannine Vilmart-Seuwen, Brigitte Stecher, Barbara Höhne, Massoud Momayezi, Regina Pape, and Helmut Kersken

Faculty of Biology, University of Konstanz, POB 5560
D-7750 Konstanz, Federal Republic of Germany

ABSTRACT

We have used Paramecium tetraurelia cells to analyze parameters relevant for membrane fusion regulation during exocytosis for different reasons: This cell type secretes trichocysts synchronously (1 sec); there exist secretory mutants; finally, the cell cortex can be isolated and used to analyze exocytosis in vitro. We have found with SDS-FAGE-ARG from ^{32}P-prelabeled cells that the selective dephosphorylation of a 65 kD PP always parallels exocytosis. Its extent depends on the amount of trichocysts actually released (dose-dependence in "normal" cells; none in non-discharge mutations; variable in cells during replenishment of the trichocyst pool after a preceding trigger). SDS-PAGE-ARG or immunolabeling reveal 65 kD PP in the cell cortex, which also contains CaM (localized by affinity- or immuno-labeling specifically on the preformed exocytosis sites) and a CaN-like protein (shown on nitrocellulose blots). Ca+CaM+CaN stimulates exocytosis in vitro, as it does, after microinjection, in vivo. (Alkaline phosphatase can mimic this effect). Concomitantly, αCaM- or αCaN-AB suppress exocytosis. We could also show that Ca+CaM+CaN causes 65 kD PP dephosphorylation during exocytosis. The Ca^{2+}-CaM triggered dephosphorylation of the 65 kD PP, mediated by a CaN-type phosphatase, is therefore considered as the/a primary step in fusion regulation during exocytosis. ATP possibly reduces the fusion capacity, at least in part, via protein rephosphorylation. In our system we could also ascertain that ATP consumption as well as protein phosphorylation occur only ≥5 seconds after synchronous exocytosis and protein dephosphorylation. The synchronicity of our system allows to avoid any possible overlaps of these phenomena.

ABBREVIATIONS: α...-AB = anti...antibodies, AED = aminoethyldextran, AIDP = adenylylimidodiphosphate, ATP = adenosintriphosphate, $[Ca^{2+}]_i$ = intracellular Ca-concentration, CaM = calmodulin, CaN = calcineurin, EGTA = ethyleneglycol bis(β-aminoethyl ether)-N,N'-tetraacetate, EM = electron microscope (-ic), nd = "non-discharge" mutation, PAGE = polyacrylamide gel electrophoresis, P_i = inorganic phosphate, PP = phosphoprotein, $pNO_2ØP$ = p-nitrophenyl phosphate, SDS = sodiumdodecyl sulphate, tl= trichless mutant, wt = wild type.

INTRODUCTION

Many different mechanisms have been ascertained to be involved in the release of secretory materials by exocytosis from a variety of cell types (Campbell, 1983; Rubin et al., 1985; Hutton, 1986). Yet it has proved difficult to pinpoint the mechanism(s) that actually cause(s) the two membranes involved in exocytosis to undergo fusion. Most of the secretory mechanisms described depend on an increase of $[Ca^{2+}]_i$ (Poste & Allison, 1973; Douglas, 1978; Baker & Knight, 1981;1984; Campbell, 1983; Rubin et al., 1985; Plattner et al., 1985; DeLisle & Williams, 1986; Hutton, 1986; Cobbold et al., 1987; Plattner, 1987b). This, however, could encompass preparatory steps (including, e.g., transcellular transport phenomena, that also involve Ca^{2+}-regulated cytoskeletal elements etc.), fusion proper, or phenomena occurring after fusion (including membrane retrieval and the reduction of $[Ca^{2+}]$ levels). Ca^{2+} also promotes membrane fusion *in vitro* (Düzgünes, 1985), either with organelles (Dahl et al., 1979) or with liposomes (Papahadjopoulos et al., 1978; DeKruijff et al., 1985; Rand & Parsegian, 1985), though yields are very low or $[Ca^{2+}]$ to be used is very high.

We consider the *Paramecium tetraurelia* cell, which we introduced as a model system for exocytosis research some time ago (Plattner et al., 1973), a particularly suitable system for analyzing the final steps of exocytosis for the following reasons. (a) Cells can be cultivated in sterile mass cultures; (b) mutations on different levels of secretory activity are available; (c) cells can be easily microinjected. (d) The majority of a large number (~10^3) of secretory organelles ("trichocysts") can be instantaneously (1 sec) released by exocytosis, e.g., in response to aminoethyldextran (AED), (Plattner et al., 1984, 1985; Plattner, 1987a,b), since they are docked to the cell membrane "ready to fire". (e) Finally, we have recently established an *in vitro* exocytosis system, using isolated cell "cortices" (Vilmart-Seuwen et al., 1986, 1987). In this system concentrations of ions or nucleotides can be deliberately manipulated; certain enzymes, antigens or specific antibodies can be tested and the results quantified by light microscopy.

All this should be helpful (a) to pinpoint aspects pertinent to membrane fusion proper, i.e., with a minimal interference of other processes (like intracellular transport) and (b) to determine threshold concentrations for relevant compounds. Even function-repair experiments have been possible by organelle and/or cytoplasm transfer (Beisson et al., 1980; Lefort-Tran et al. 1981). Here we try to demonstrate as a crucial step in fusion regulation the de-phosphorylation of a 65 kD PP via a Ca^{2+}-CaM-stimulated

CaN-type PP-phosphatase (Momayezi et al., 1987 a, b).

MATERIALS AND METHODS

Paramecium tetraurelia cells were cultivated as indicated by Plattner et al. (1984) and Zieseniss & Plattner (1985), i.e., either monoxenically (with _Enterobacter aerogenes_ as a food supply) or axenically. Whereas most experiments were done (unless indicated otherwise) with wild type (_wt_ = _7S_) or closely related (_K401_) cells we occasionally also used mutants [summarized by Cohen & Beisson (1980), Pouphile et al. (1986)]. Among them are mutants devoid of trichocysts ("trichless":_tl_) and "non-discharge" mutants [_nd6_, _nd7_ (Lefort-Tran et al., 1981) without exocytotic membrane fusion capacity; similarly _nd9_ cultivated at 28°C, in contrast to permissively cultivated _nd9-18°C_ cells (Beisson et al., 1976) showing _wt_ behavior.] Eventually cells were thoroughly labeled (4 hrs, 25°C) with ^{32}P, eventually triggered and rapidly (0.2 sec) inactivated and then subjected to SDS-PAGE as in (Zieseniss & Plattner, 1985). The degree of 65 kD PP phosphorylation was determined either by liquid scintillation counting or by scanning SDS-PAGE-ARG overlays as in (Zieseniss & Plattner, 1985).

Microinjections were done as before (Momayezi et al., 1986; Vilmart-Seuwen et al., 1986), labeling and EM or LM analyses as in (Momayezi et al., 1986). To induce synchronous exocytosis cells were exposed to AED, MW = 40 kD, (Plattner et al., 1984, 1985) in the presence of 10^{-4}M Ca^{2+}. For exocytosis _in vitro_ isolated cell cortex fragments (cell surface membranes with trichocysts attached) were prepared as described before (Vilmart-Seuwen et al., 1986). They were usually triggered by reducing [Mg^{2+}] from 10 to 0.5 mM in the presence of 10^{-5}M free [Ca^{2+}]$_i$; free [Ca^{2+}] was calculated and eventually, after measurement in the presence of all components (pH 7.0), slightly adjusted to 10^{-5}M, unless indicated otherwise. The amount of exocytosis _in vivo_ or _in vitro_ was determined by light microscopy. Exocytosis _in vivo_ is terminated within 1 sec (Plattner & Deitmer, 1985; Plattner, 1987a), whereas _in vitro_ ~20-30 sec are required (Vilmart-Seuwen et al., 1986). We applied, either _in vivo_ (by microinjection, when required) or _in vitro_, different nucleotides (Vilmart-Seuwen et al., 1986, 1987) or drugs (Lumpert et al., 1987) as well as Ca^{2+}, CaM, CaN or αCaM-AB and αCaN-AB (Momayezi et al., 1987a, b), in order to test inhibitory (when combined with the above trigger procedures) or stimulative effects (using only 5 mM [Mg^{2+}] in that case for _in vitro_ studies). The AB used were characterized by Momayezi et al. (1986) for CaM, by Momayezi et al. (1987b) for CaN and by Stecher et al. (1987) for 65 kD PP. Sometimes we applied a "protease inhibitor cocktail" made up of phenylmethylsulfonyl fluoride,

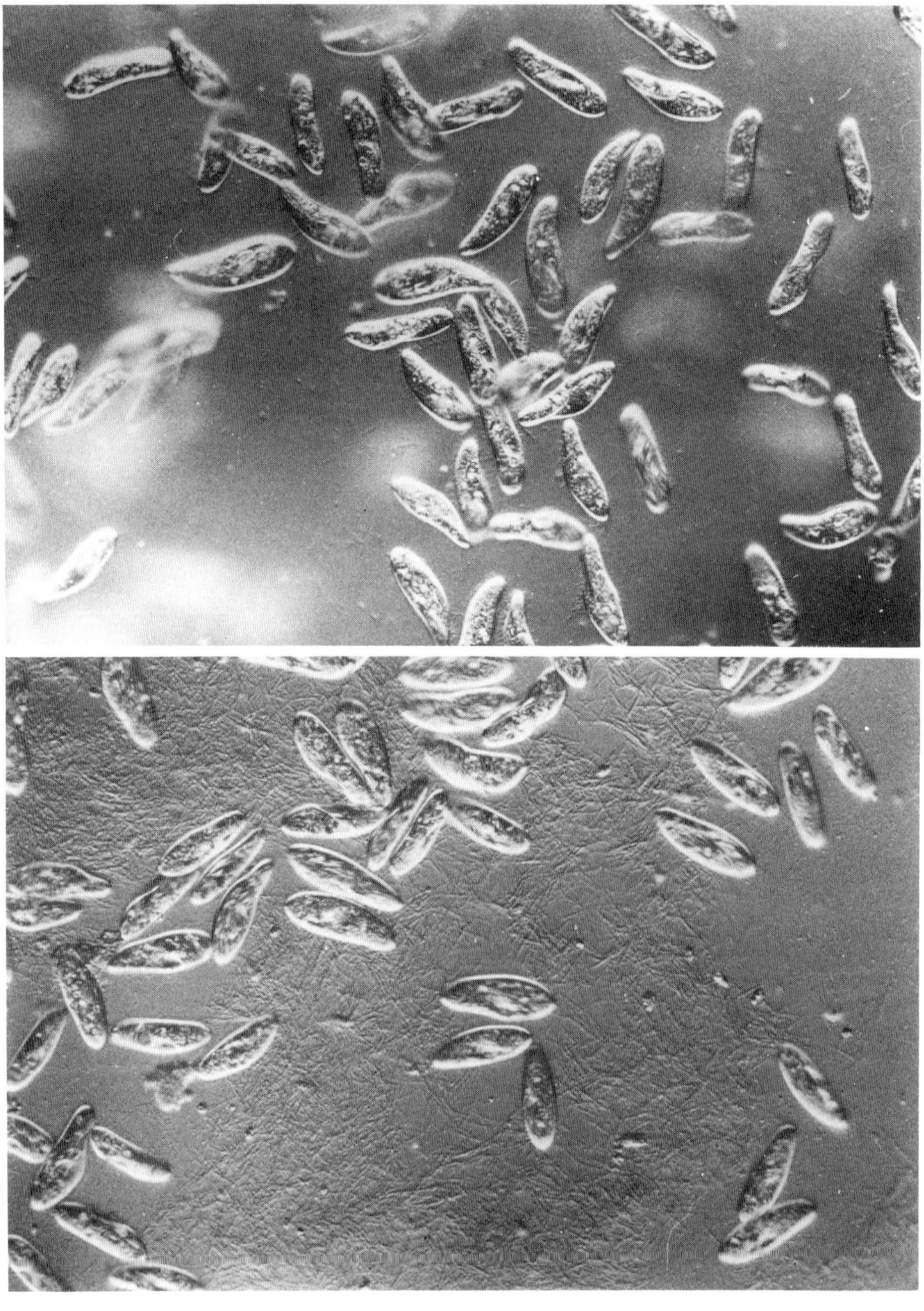

Fig. 1. P. tetraurelia culture before (top) and immediately after (bottom) AED triggered exocytosis of trichocysts which, by a 4-fold expansion of their contents, become visible as long rods. Magnification 130 x. Reproduced with permission from Academic Press from Plattner et al. (1984).

p-tosyl-L-arginin-methylester, aprotinin and leupeptin as in Momayezi et al., 1987b).

RESULTS AND DISCUSSION

Exocytosis In Vivo and In Vitro

Exocytosis in vivo: When $1.5 \cdot 10^{-6}$M AED is added to cells (except for tl or nd mutants), in the presence of 10^{-4}M Ca^{2+}, they secrete the majority of their trichocysts (of a total of ~1000), (Plattner et al., 1984, 1985), within ~1 sec (Fig. 1). 1 sec is, thus, the degree of synchronicity, as determined by morphometry after fixation and by electrophysiology (Plattner & Deitmer, 1985; Plattner, 1987a, b). This also corresponds to direct light microscopic observation. This massive exocytotic response does not entail any loss of vitality (Plattner et al., 1985). Cells replenish a complete set of trichocysts within ~10 hrs (Pape & Plattner, 1985). Trichocysts, once docked to the cell membrane, become releasable with only a small time delay of ~5 min; This might correspond to the time required to assemble the structural components which are found on a normally functioning membrane fusion site (Pape & Plattner, 1985). The main structural events occurring during exocytosis in vivo and in vitro are summarized in Figure 2.

Exocytosis in vitro: Gentle mechanical disruption of cells in the presence of high [Mg^{2+}] and low [Ca^{2+}] (see "Materials and Methods") allows for the preparation of cell "cortices". These are made up of the cell surface membrane complex [cell membrane with alveolar membranes (of unknown function) still attached] and of numerous trichocysts with their intact membrane (Vilmart-Seuwen et al., 1986, 1987). Still attached are some microfilaments and some mitochondria; in the absence of metabolic substrates [ATP] is $<10^{-6}$M (Vilmart-Seuwen et al., 1986). Reversing the [Mg^{2+}]/[Ca^{2+}] ratio, as indicated in "Materials and Methods", causes membrane fusion and stretching ("decondensation") of the contents that remain stuck in the exocytotic opening (Fig. 2). This system, therefore, involves only exocytosis but no endocytosis coupling. Also in contrast to the in vivo system, exocytosis goes for ~20-30 sec.

We used cortices not only (a) to localize endogenous components possibly relevant for exocytosis, but also (b) to test the effect of exogenous components (antigens or drugs) and (c) of antibodies to endogenous antigens (see below). Cortex preparations also allowed us (d) to establish the possible effects of nucleotides (see below for ATP) and "minimum conditions" for exocytosis. For instance, we observed that the usual [Ca^{2+}] of 10^{-6}M is required for inducing membrane fusion (Vilmart-Seuwen et al., 1986,

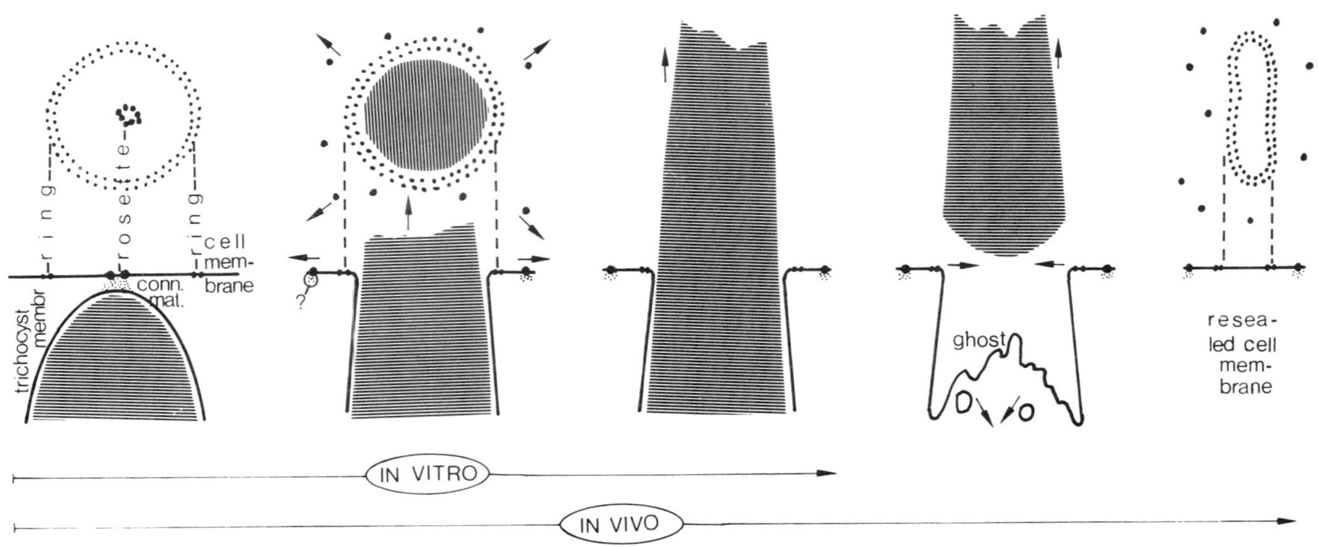

Fig. 2. Ultrastructural organization of trichocyst-exocytosis-competent strains of P. tetraurelia. Top row: En face view seen after freeze fracturing. Bottom row: View derived from median sections. Arrows indicate directions of movements, e.g., of rosette particles, of trichocyst contents (hatched area), of the cell membrane during resealing and of ghost vesicles during membrane retrieval. The association of connecting material (stippled areas) with rosette particles is ascertained only in the resting state, but hypothetically possible during exocytosis. The scheme differentiates between the steps occuring during exocytosis in vivo and in vitro. Data combined from (Plattner et al., 1973; Plattner, 1987a, b).

1987), as in most other systems (Baker & Knight, 1981, 1984; Douglas, 1978; Campbell, 1983; Rubin et al., 1985; Cobbold et al., 1987). It has to be born in mind, of course, that cortices still retain all relevant ultrastructural and functional components relevant for exocytotic membrane fusion.

Ultrastructural Aspects Pertinent to Membrane Fusion

These aspects (Fig. 2) have recently been summarized in refs. (Plattner 1987a, b). Briefly, the "resting" exocytosis sites in normally secreting strains display a "rosette" of 9 membrane intercalated particles (Plattner et al., 1973; Beisson et al.,1976). The trichocyst membrane, devoid of particles in this area, is closely approaching this cell membrane region and a diffuse mass of "connecting material" is found between the two membranes (Plattner et al., 1980; Westphal & Plattner, 1981; Plattner et al., 1982). This material as well as rosettes are missing in nd6, nd7, and nd9-28°C (Beisson et al., 1980; Plattner et al., 1980; Pouphile et al., 1986). These features are, thus, characteristic of normally secreting cells (7S, K401, nd9-18°C). A "ring", 300 nm in diameter, also made up of membrane particles, delineates the fusogenic zone proper and has no implications on fusogenicity. During exocytosis rosette particles and connecting materials are spread (beyond the particle rings) within the cell membrane (Plattner 1987a, b). When exocytosis is induced by AED *in vivo*, exocytosis sites are rapidly resealed (Plattner & Deitmer, 1985; Plattner 1987a, b). During *in vitro* triggering, exocytosis is incomplete, since trichocyst contents remain stuck in the exocytotic opening and, thus, impede resealing (Fig. 2), (Vilmart-Seuwen et al., 1986).

Screening of Enzymatic Mechanisms Possibly Involved

This was done either with the use of different enzymes or of different drugs (activators or inhibitors), either *in vivo* by microinjection, or *in vitro* (Lumpert et al., 1987). The materials used were selected according to current discussions on fusogenic mechanisms [see (Lucy, 1984; Lumpert et al. 1987)]. Control experiments were conducted with inert proteins (bovine serum albumin) or zymogens (for proteases) as well as with the mere solvent (for drugs).

Among the drugs we tested with *P. tetraurelia* cells there were inhibitors of protein kinase A or C and of different proteases, including metalloprotease, as well as a protein kinase C activating phorbol ester (Lumpert et al., 1987). None produced an exocytotic response; none inhibited exocytosis in response to AED. Only inhibitors of Ca^{2+}-CaM-dependent protein kinases, like alloxan or diazepam slightly (10 to 25%) stimulated exocytosis within the usual concentration range. This corresponds to our hypothe-

TABLE 1

Effect of CaM, CaN, αCaM-AB, αCaN-AB or of a protease inhibitor cocktail on exocytosis in *P. tetraurelia*. Data from (Momayezi et al., 1987b).

(A) *In vivo* (microinjection) experiments	exocytotic response achieved*)
Ca^{2+}+CaM+CaN (10^{-5}M; 0.1 and 0.08 mg/ml)	+++
Ca^{2+} (10^{-5}M)	-
αCaM-AB (0.1 mg/ml) → AED (extracellular)	+++
αCaN-AB (0.1 mg/ml) → AED (extracellular)	-
protease inhibitor cocktail → AED (extracellular)	+++
(B) *In vitro* experiments with isolated cortices	
Ca^{2+} (10^{-5}M)+ 5 mM Mg^{2+} ("basic buffer")	-
"basic buffer"+CaM+CaN(27 or 130 μg/ml, respectively)	+ to ++
like before, but with protease inhibitor cocktail	+ to ++
10^{-5}M Ca^{2+}+ 0.5 mM Mg^{2+} ("stimulation buffer")	++
"stimulation buffer"+αCaM-AB (0.1 mg/ml)	-
like before + αCaN (0.08 mg/ml)	- to (+)
"stimulation buffer" + αCaM+αCaN;+CaM+CaN in excess**)	++

*) -, (+), +, ++, +++ indicate no, occasional, weak, moderate, strong exocytosis.

**) concentrations as before.

sis, that protein dephosphorylation, rather than phosphorylation, would be involved in the fusogenic mechanism(s), (see below).

In vitro exocytosis could be slightly triggered by trypsin or chymotrypsin (yet not be the corresponding zymogens). Microinjected trypsin caused a massive exocytosis (Lumpert et al., 1987). How can this be reconciled with our hypothesis (Momayezi et al. 1987a, b) that PP dephosphorylation, rather than partial proteolysis [as proposed by Lucy (1984)] accounts for fusion induction in paramecia? We have the following arguments: (a) After AED triggered exocytosis we could not recognize any protein splitting, at least not on 1D SDS-PAGE (Zieseniss & Plattner, 1985). (b) Previous addition of a protease inhibitor cocktail (*in vitro* or *in vivo*, by microinjection) did not reduce the exocytotic response (Tab. 1). (c) Some proteases, like papain or pronase, remained without any effect. (d) The results achieved with (chymo-)trypsin could thus be due to a circumstantial dephosphorylation effect (Lumpert et al., 1987). Yet more work on protease effects appears advisable.

In conclusion, none of the data achieved with different drugs or exogenous enzymes contradicts our hypothesis summarized at the end of this chapter. The data obtained even provide additional support. Positive evidence for our hypothesis comes also from experiments with specific phosphatases and antibodies (see below). This again is in line with the fact that exocytosis can be inhibited by different polyanions (Plattner et al., 1985; Lumpert et al., 1987) known to block phosphatases.

Dephosphorylation of a 65 kD PP In Vivo and In Vitro

$^{32}P_i$ added during 4 hrs to cells (sterilly cultured and starved by transfer into a buffer allowing for AED triggering) allowed for the phosphorylation of many proteins, according to SDS-PAGE-ARG (Gilligan & Satir, 1982; Zieseniss & Plattner, 1985). (Secretory contents are not phosphorylated.) Yet when such cells were fixed and simultaneously triggerd by picric acid (Gilligan & Satir, 1982) or synchronously triggered by AED (Zieseniss & Plattner, 1985) and rapidly inactivated, one band of a M_r = 65 kD

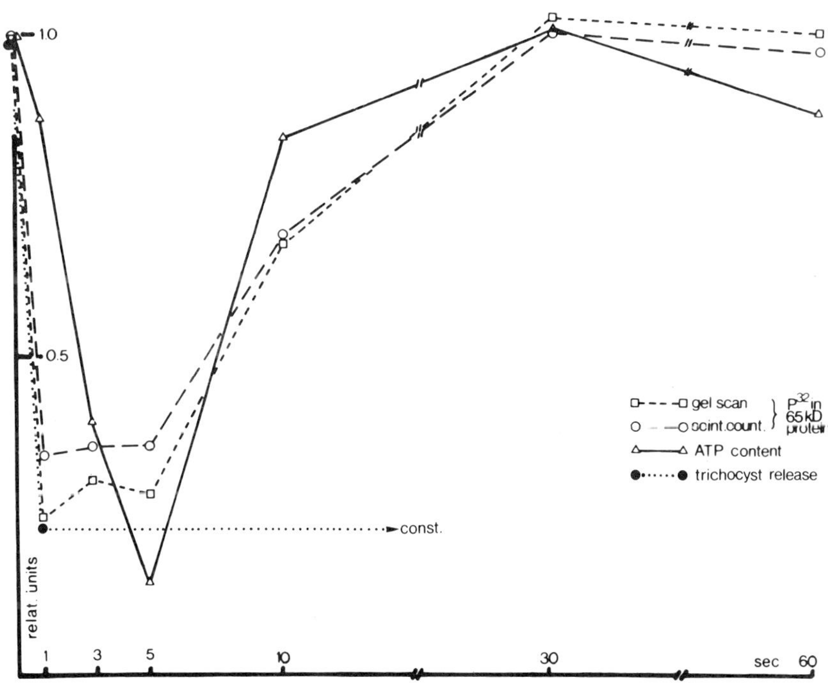

Fig. 3. Time course of trichocyst release, of 65 kD PP dephosphorylation and of ATP consumption in the course of synchronous trichocyst exocytosis (wt cells) in response to AED added at time 0 (to which all data are normalized). Within the time resolution of these experiments (1 sec) trichocyst release is paralleled by 65 kD PP dephosphorylation, but clearly not by ATP consumption (possibly required to re-establish resting $[Ca^{2+}]_i$ levels). Data compiled from (Plattner 1987a; Zieseniss & Plattner, 1985; Vilmart-Seuwen et al.,1986).

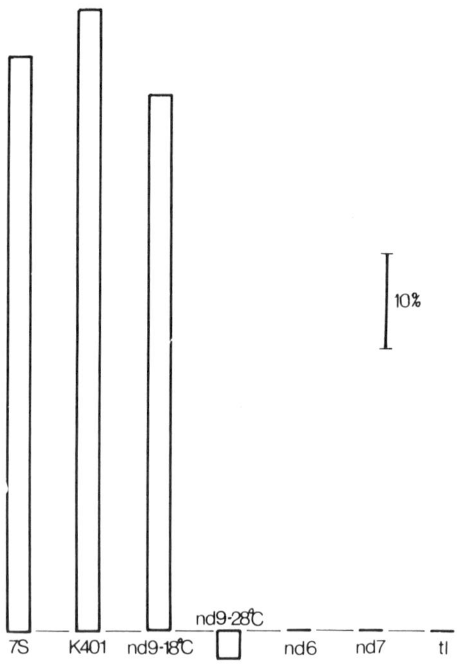

Fig. 4. Extent of 65 kD dephosphorylation in *P. tetraurelia* strains with normal (left) and absent (right) secretory response to AED. Redrawn after Zieseniss & Plattner (1985).

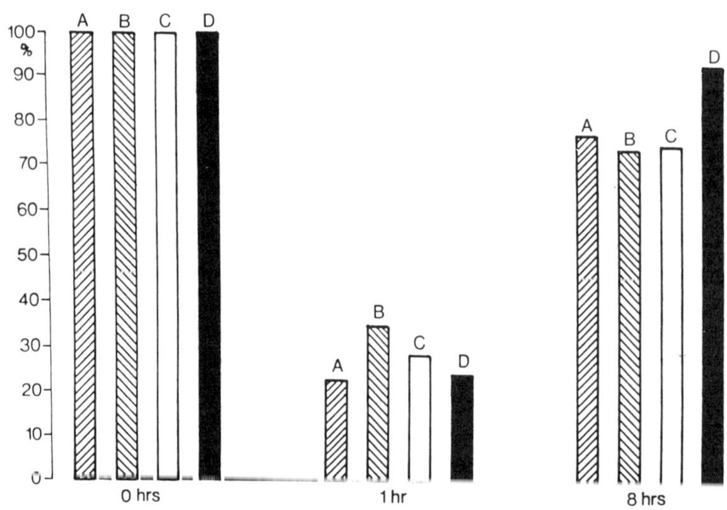

Fig. 5. Coincidence of (A) trichocysts present on potential docking sites (EM), (B) presence of rosettes (EM), (C) light microscopically determined trichocysts releasable in response to AED and (D) extent of 65 kD dephosphorylation in response to AED. Strain: 7S. Data are normalized to 0 hr values. A second AED trigger was applied later on when little (1 hr) or most (8 hrs) of the set of releasable trichocysts had been re-installed on the cell membrane. Data combined from Pape & Plattner (1985) and Zieseniss & Plattner (1985).

was selectively dephosphorylated. Coomassie blue staining revealed no change of protein patterns. Dephosphorylation was rapidly reversed, i.e., within 10-30 sec (Fig. 3). Dephosphorylation occurred only with strains capable of exocytosis (Fig. 4), although other strains had similar, some even higher amounts of the 65 kD PP. We know that defective strains are unable to assemble the structural components ["rosette" particles in the cell membrane (Beisson et al., 1976; Pouphile et al., 1986), "connecting material" between trichocyst and cell membrane (Beisson et al., 1980; Plattner et al., 1980, 1982; Pouphile et al., 1986)]. These components are also dispersed during exocytosis (Fig. 2) and reassembled with only a small lag time, when new trichocysts are docked to the cell membrane (Pape & Plattner, 1985). These phenomena are strictly correlated not only with the recovery of trichocyst release but also with the amount of 65 kD PP dephosphorylation in response to a second AED trigger (Zieseniss & Plattner, 1985), as shown in figure 5.

Since we could induce exocytosis in vitro either by a Ca/Mg shift [Vilmart-Seuwen et al. (1986); see "Materials and Methods"] or by adding an excess of CaM+CaN (Momayezi et al., 1986a, b; Tab. 1) we subjected

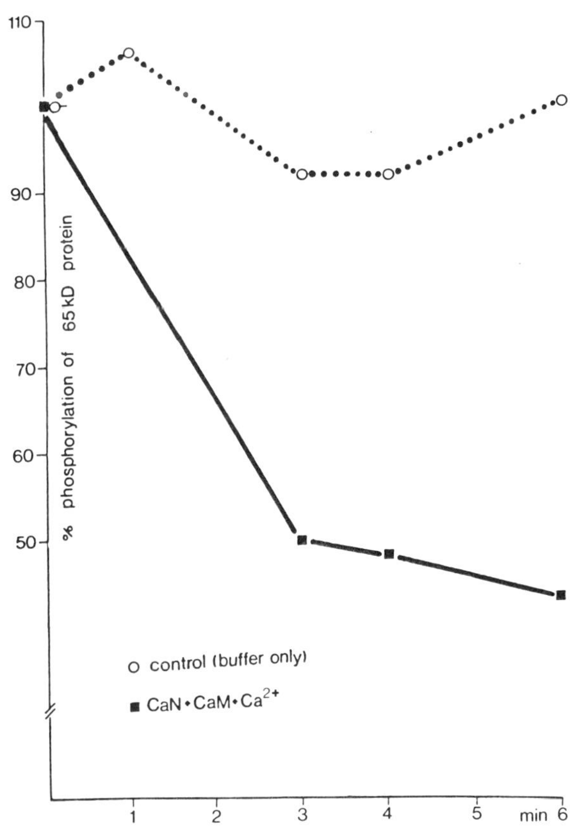

Fig. 6. Dephosphorylation of the 65 kD PP in isolated cortex preparations during in vitro exocytosis triggered by adding Ca^{2+}+CaM+CaN. Dotted line represents the control. According to (Momayezi et al., 1987b).

cortices from ^{32}P prelabeled cells to these trigger procedures, before they were analyzed on SDS-PAGE gels. Again we found the dephosphorylation of the 65 kD PP (Momayezi et al., 1987b), as shown in Fig. (6).

We conclude, (a) that the dephosphorylation of the 65 kD PP is functionally connected with the membrane fusion capacity and (b) that this depends on the proper assembly of the ultrastructural components mentioned above. However, we do not yet know precisely their biochemical identity (see also "Localization of antigens...", below).

Effects of CaM, CaN, αCaM-AB and αCaN-AB on Exocytosis

Given (a) the fact that protein dephosphorylation parallels exocytosis (Gilligan & Satir, 1982; Zieseniss & Plattner, 1985) and (b) the occurrence of CaM precisely at the fusogenic sites (Momayezi et al., 1986) we had envisaged the possibility that a Ca^{2+}-CaM-dependent PP phosphatase of a CaN-like type [see (Manalan & Klee, 1985)] could represent a key enzyme in the regulation of membrane fusion during exocytosis in *Paramecium* cells (Momayezi et al., 1987a, b). Evidence for this hypothesis could be obtained *in vivo* (microinjections) and *in vitro* (cortex preparations), either by adding an excess of CaN (in an activated form as a Ca^{2+}+CaM+CaN complex) or by introducing specific antibodies (IgG),(Momayezi et al., 1987a,b); see Tab. 1.

In cooperation with M.H. Krinks and C.B. Klee (NIH, Bethesda, USA) we have proved the occurrence of a CaN-like protein in *P. tetraurelia* (Momayezi et al., 1987b). Microinjected αCaN-AB fully inhibited AED-triggered exocytosis. The high percentage of CaM in these cells (Momayezi et al., 1986) explains why microinjected αCaM-AB had no effect, since they could bind only a small antigen fraction. *In vitro* αCaM-AB as well as αCaN-AB strictly inhibit exocytosis. Also noteworthy are the following observations (Tab. 1): (a) Microinjection of 10^{-5} M Ca^{2+} (estimated final concentration in the cell, like all concentrations indicated) does not provoke exocytosis (Vilmart-Seuwen et al., 1986), (probably due to rapid sequestration). (b) Protease inhibitors exert no effect on exocytosis elicited by AED (Lumpert et al., 1987). This invalidates the relevance of proteases [c.f. Lucy (1984)] for fusion induction in our system.

Localization of Antigens and Enzymes Relevant for Membrane Fusion in the Cell Cortex

Since exocytosis in *P. tetraurelia* cells involves the fusion of the membrane of firmly attached trichocysts with the cell membrane, one should be able to localize antigens or enzymes relevant for membrane fusion either (a) *in situ* by affinity labeling, by immunocytochemistry or by enzyme cyto-

TABLE 2

Localization of components relevant to exocytosis regulation in the cortex of P. tetraurelia cells and methods used for localization studies*).

Component	presence in cortex	presence on fusion site	methods used	references
65 kD PP	+	n.y.a.	fractionation**)	(Momayezi et al.,1986, 1987b;Gras et al.,1987)
	+	n.y.a.	^{32}P-SDS-PAGE-ARG	
	+	n.y.a.	immunofluorescence	(Stecher et al.,1987)
CaM	+	+	EM affinity labeling	(Plattner, 1987)
	+	+	LM affinity labeling	(Momayezi et al.,1986)
	+	+	EM immuno-labeling	(Momayezi et al.,1986)
CaN	+	n.y.a.	fractionation**) → immuno-blotting	(Momayezi et al.,1987b)

*) The localization indicated does not preclude that components (e.g., CaM) might occur also outside the cortex. + = presence of a component; n.y.a. = not yet analyzed.

**) Isolated cell cortices.

chemistry as well as (b), after cell fractionation, by immunoblotting or by spectralphotometric enzyme tests. As to enzyme measurements and localizations we refer to the following subchapter on "Effects of nucleotides". Briefly, there is some evidence for the possible occurrence of a Ca^{2+}-ATPase based on EM enzyme cytochemistry (Plattner et al., 1977; 1980) and on enzyme measurements with isolated cell surface membrane fragments (Bilinski et al., 1981; Tiggemann & Plattner, 1982). However, some of this activity could possibly be due to an overlap with a protein kinase/phosphatase system. As far as the P. tetraurelia cells are concerned we have no information on the kinase(s) involved.

Data on the cortical localization are available for the 65 kD PP, for CaM and for CaN (Tab.2). Again only for CaM we have positive evidence for its occurrence on the fusogenic sites (Momayezi et al., 1986). We have found evidence that not only CaM (Momayezi et al.,1986) and CaN (Momayezi et al., 1987a, b) but also the 65 kD PP is bound to the cell surface complex (Gras et al., 1987; Stecher et al., 1987); see Fig. 7. All three might form part of the "connecting material" found between the trichocyst and the cell membrane (see above). As mentioned before, this material is seen in the EM only in exocytosis competent strains (Plattner et al.,1980, 1982; Pouphile et al., 1986). The same has been inferred from functional data obtained by microinjection experiments (Beisson et al., 1980). We now try to analyze the more precise localization of the 65 kD PP and of CaN on the EM level. Studies of this kind are still greatly hampered by possible rapid redistribution caused by activation-deactivation cycles or simply by EM preparation procedures. However, the cortical localization of all three components suggests a causal connection during exocytotic membrane fusion. Exocytosis can be inhibited by αCaM-AB also in other systems, like sea urchin eggs (Steinhardt & Alderton, 1982) or chromaffin cells (Kenigsberg & Trifaró, 1985). Although in the literature a mechanism of membrane fusion induction via PP dephosphorylation had never been taken into consideration before, the data available would appear fully compatible with ours, if one assumes that our system is arrested in a primed late pre-fusion stage.

Effects of Nucleotides

If protein dephosphorylation rather than phosphorylation were a fusogenic mechanism, ATP as a P_i donor should inhibit exocytosis. Indeed ATP actually inhibits exocytosis *in vitro* or *in vivo* (after microinjection). Alternatively exocytosis can be provoked when cells are exposed to metabolic inhibitors or microinjected with apyrase, an ATP diphosphohydrolase (Vilmart-Seuwen et al., 1986).

However, there might also be another factor involved in ATP depression of exocytotic membrane fusion. Some time ago we had shown, using methods of EM enzyme cytochemistry, that preformed exocytosis sites in *P. tetraurelia* contain a set of enzymes with (a) Ca^{2+}-dependent ATP splitting activity and (b) with the capacity to hydrolyze $pNO_2ØP$ (Plattner et al., 1977, 1980). This finding was considered relevant for membrane fusion since *nd* or *tl* mutations were devoid of these reactions (Plattner et al.,1980). At that time we had assumed the occurrence of a Ca^{2+}-ATPase with a (de-)phosphorylation step involved. It would now appear possible that an overlapping activity of a Ca^{2+}-dependent protein kinase/phosphatase system could have mimicked an ATPase activity. Yet a Ca^{2+}-dependent (Plattner et al., 1977, 1980; Bilinski et al., 1981) or, more precisely, a Ca^{2+}-CaM-dependent ATPase activity still could be assumed to be part of a multifactorial membrane fusion mechanism, since we could recently ascertain that the non-hydrolyzable ATP analogue AIDP also inhibits exocytosis *in vivo* and *in vitro* (Vilmart-Seuwen et al., 1986). The amount of ATP possibly hydrolyzed during membrane fusion must be unmeasurably small, since a massive decay of the ATP pool occurs only *after* the actual trichocyst release (Fig. 3), (Vilmart-Seuwen et al., 1986). Therefore, the last hypothesis is difficult to analyze at this time. A Ca^{2+}-ATPase could also deal with local $[Ca^{2+}]$ regulation.

The inhibitory effect achieved with $pNO_2ØP$ substrate (Vilmart-Seuwen et al., 1987) could be assigned to substrate inhibition either of an ATP

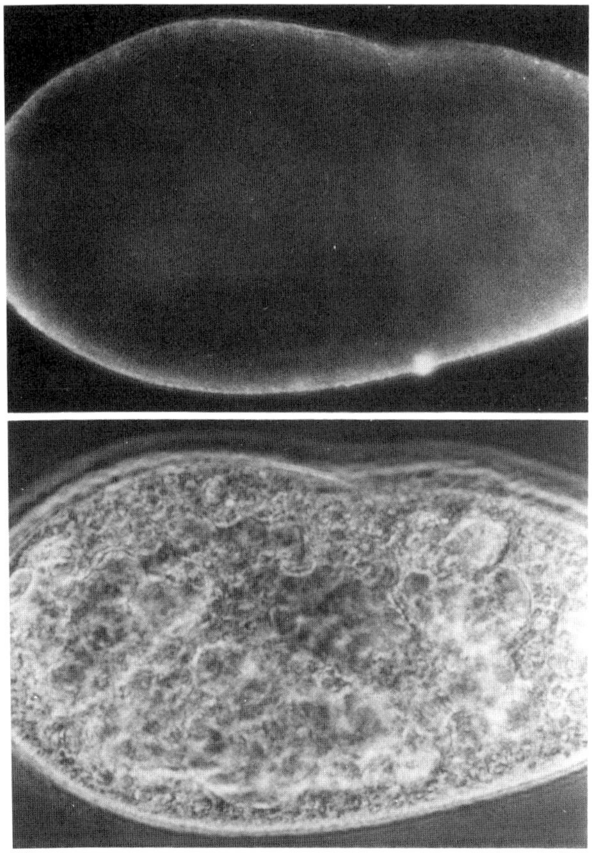

Fig. 7. Left: Localization of the 65 kD PP in a permeabilized P. tetraurelia cell by immunofluorescence using α-mouse second AB-FITC; right: corresponding phase contrast picture. Antisera against 65 kD PP (controlled by Western blots) have been prepared in mice by injecting electrophoretically purified 65 kD PP (from sterilly cultured P. tetraurelia cells). Controls with preimmune sera or unspecific AB gave no significant labeling. Under identical conditions, sera or AB against cytosolic antigens did not yield the cortical labeling pattern (cilia being negative) that is typical of the 65 kD PP (left). Non-permeabilized cells were also not stained. Magnification 1000 x. Data from Stecher et al. (1987).

splitting enzyme activity or via the CaN-type PP phosphatase, since CaN also hydrolyzes $pNO_2ØP$ (Pallen & Wang, 1983). As stated, both, an ATPase (Tiggemann & Plattner, 1982) as well as the PP phosphatase (Manalan & Klee, 1985; Momayezi et al., 1987a,b) would be stimulated by Ca^{2+}-CaM.

CONCLUSIONS AND OUTLOOKS

From all the facts presented above we conclude that membrane fusion during exocytosis may follow the scheme in Fig. 8. This scheme accounts for the following facts: (a) An increased free $[Ca^{2+}]_i$ is required for trichocyst exocytosis (Plattner et al., 1984, 1985). (b) The dephosphorylation of a 65 kD PP always accompanies exocytosis (Zieseniss & Plattner,

1985). (c) Dephosphorylation of the 65 kD PP occurs also when Ca+CaM+CaN or alkaline phosphatase is added to cortices which then release their trichocysts (Momayezi et al., 1987a,b). (d) These enzymes, when microinjected, also cause exocytosis (Momayezi et al., 1987a,b). (e) The 65 kD PP (Stecher et al., 1987), CaM (Momayezi et al., 1986) and CaN (Momayezi et al., 1987a,b) are present in the cell cortex. (f) αCaM-AB or αCaN-AB (Momayezi et al., 1987a,b) as well as α65 kD-AB (Stecher et al.,1987) inhibit exocytosis.

Therefore, it appears very likely that a Ca^{2+}-CaM-dependent dephosphorylation of the 65 kD PP via a CaN-like PP-phosphatase is involved in the regulation of exocytotic membrane fusion (Fig. 8). It remains to be seen whether an increased hydrophobicity of the dephosphorylated 65 kD protein species would - in analogy to some studies with liposomes (Batenburg et al., 1985) - directly act as a fusogen, e.g., by a local perturbation of the membrane lipids. Previous work on exocytosis regulation had focussed the attention on protein phosphorylation, rather than dephosphorylation. yet its relevance has largely remained questionable (Burgoyne, 1984).

Multiple fusion control is also possible, e.g., via direct Ca^{2+}

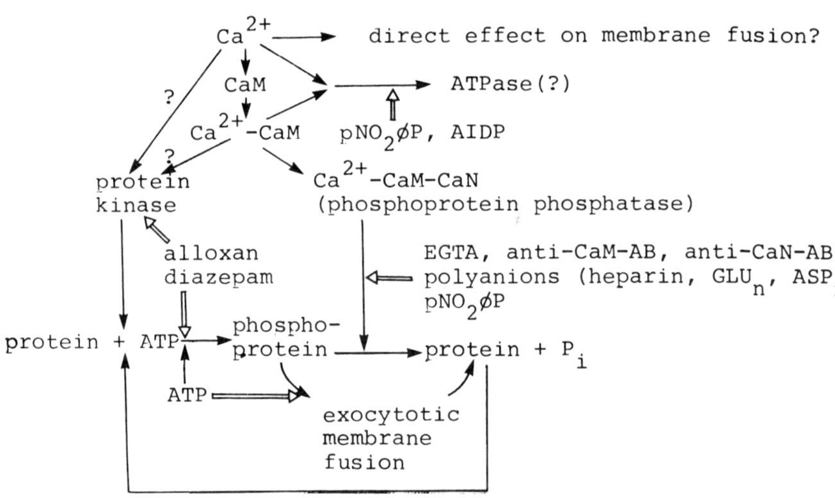

Fig. 8. Scheme proposed for mechanisms involved in the regulation of membrane fusion during exocytosis in P. tetraurelia cells, as derived from the arguments presented throughout the text. Big arrows mark sites of action of inhibitory compounds. Question marks indicate some of the uncertainties of this scheme. Data combined from Plattner et al.(1977, 1980), Zieseniss & Plattner (1985), Vilmart-Seuwen et al. (1986), Momayezi et al. (1986, 1987b) and Lumpert et al. (1987).

effects (for reviews, see Papahadjopoulos et al., 1978; Dahl et al., 1979; Düzgünes, 1985; Rand & Parsegian, 1986) or via local ATP hydrolysis (Poste & Allison, 1973; Vilmart-Seuwen et al., 1986). Yet so far we can not prove any direct role of proteases - a mechanism considered relevant for cell-cell fusion (Lucy, 1984).

An important aspect is the requirement of a proper assembly of fusogenic components, notably of "rosette" freeze-fracture particles and of "connecting material" which had been shown to represent membrane integrated (Vilmart & Plattner, 1983) and peripheral proteins (Westphal & Plattner, 1981), at the fusion sites for fusion to occur. This is in line with the requirements of a "focal fusion concept" proposed some time ago (Plattner, 1981). Future studies will show whether all the above mentioned molecular species reside precisely at the fusion sites, how they are anchored there, what the structure-function equivalents are and whether our scheme derived from P. tetraurelia can be extended to other cell types. Possibly in this cell type fusogenic sites are kept in a primed state by the presence of the 65 kD PP and by ATP. It then could immediately result in membrane fusion. It cannot be excluded that other systems would require additional steps before the actual fusion process can take place.

ACKNOWLEDGEMENTS

We thank Dr. Janine Beisson (CNRS, Gif-sur-Yvette) for her cooperation with the mutants. We also gratefully acknowledge the secretarial and photographic assistance provided by Ms. E. Rehn and Ms. D. Bliestle and the technical assistance by Mss. C. Braun, R. Hildebrand, R. Hohenberger, R. Stürzl, and C. Wolf. - This work was supported by SFB 156 from the Deutsche Forschungsgemeinschaft.

REFERENCES

Baker, P.F. & Knight, D.E., 1981, Phil. Trans. R. Soc. Lond. B Biol. Sci. 296:83.
Baker, P.F. & Knight, D.E., 1984, Trends Neurosci. 7:120.
Batenburg, A.M. Bougis, P.E., Rochat, H., Verkleij, A.J. & DeKruijff, B., 1985, Biochemistry 24:7101.
Beisson, J., Cohen, J., Lefort-Tran, M., Pouphile, M. & Rossignol, M., 1980, J. Cell Biol. 85:213.
Beisson, J., Lefort-Tran, M., Pouphile, M., Rossignol, M. & Satir, B., 1976, J. Cell Biol. 69:126.
Bilinski, M., Plattner, H. & Tiggemann, R., 1981, Eur. J. Cell Biol. 24:108.
Burgoyne, R.D., 1984, Biochim. Biophys. Acta 779:201.
Campell, A.K., 1983, Intracellular Calcium. Its Universal Role as a Regulator, John Wiley & Sons, Chichester, New York etc.
Cobbold, P.H., Cheek, T.R., Cuthbertsons, K.S.R. & Burgoyne, R.D., 1987, FEBS Lett. 211:44.
Cohen, J. & Beisson, J., 1980, Genetics 95:797.
Dahl, G., Ekerdt, R. & Gratzl, M., 1979, in "Secretory Mechanisms" eds. Hopkins, C.R. & Duncan, C.J., Cambridge University Press, Cambridge (GB), p. 349.
DeKruijff, B., Cullis, P.R., Verkleij, A.J., Hope, M.J., VanEchteld, C.J.A. & Taraschi, T.F., 1985, in "The Enzymes of Biological Membranes", Vol.

1, Membrane Structure and Dynamics, ed. Martonosi, A.N., Plenum Press, New York, London, p. 131.
DeLisle, R.C. & Williams, J.A., 1986, Ann.Rev. Physiol. 48:225.
Douglas, W.W., 1978, Ciba Found. Symp. 54:61.
Düzgünes, N., 1985, in "Subcellular Biochemistry", ed. Roodyn, D.B., vol. II. Plenum Press, New York, p. 195.
Gilligan, D.M. & Satir, B.H., 1982, J. Biol. Chem. 257:13903.
Gras, U., Stecher, B., Höhne, B., Momayezi, M. & Plattner, H., 1987, Eur. J. Cell Biol. Suppl. 17:20.
Hutton, J.C., 1986, Cell Calcium 7:339.
Kenigsberg, R.L. & Trifaró, J.M., 1985 Neurosci. 14:335.
Lefort-Tran, M., Aufderheide, K., Pouphile, M., Rossignol, M. & Beisson, J., 1981, J. Cell Biol. 88:301.
Lucy, J.A., 1984, FEBS Lett. 116:223.
Lumpert, C.J., Kersken, H., Gras, U. & Plattner, H., 1987, Cell Biol. Int. Rep. (in press).
Manalan, A.S. & Klee, C.B., 1985, in "Calcium in Biological Systems" eds. Rubin, R.P., Weiss, G.B. & Putney, J.W., Plenum Press, New York, London, p. 307.
Momayezi, M., Kersken, H., Gras, U., Vilmart-Seuwen, J. & Plattner, H., 1986, J. Histochem. Cytochem. 34:1621.
Momayezi, M., Lumpert, C., Kersken, H., Gras, U. & Plattner, H., 1987a, Eur. J. Cell Biol. Suppl. 17:38.
Momayezi, M., Lumpert, C.J., Kersken, H., Plattner, H., Krinks. M.H. & Klee, C.B., 1987b, J. Cell Biol. (in press).
Pallen, C.J. & Wang, J.H., 1983, J. Biol. Chem. 258:8550.
Papahadjopoulos, D., Portis, A., Pangborn, W. & Newton, C., 1978, Life Sci. Res. Rep. 11:413.
Pape, R. & Plattner, H., 1985, Eur. J. Cell Biol. 36:38.
Plattner, H., 1981, Cell Biol. Int. Rep. 5:435.
Plattner.H., 1987a, in "Cell Fusion", ed. Sowers, A.E., Plenum Press, New York, p. 69.
Plattner, H., 1987b, in "Cellular Membrane Fusion: Fundamental Mechanisms and Applications of Membrane Fusion Techniques",eds. Wilschut, J. & Hoekstra, D., Marcell Dekker Inc., New York, in press.
Plattner, H. & Deitmer, J., 1985, Eur. J. Cell Biol. Suppl. 7:51.
Plattner, H., Matt, H., Kersken, H., Haacke, B. & Stürzl, R., 1984, Exp. Cell Res. 151:6.
Plattner, H., Miller, F. & Bachmann, L., 1973, J. Cell Sci. 13:687.
Plattner, H., Reichel, K. & Matt, H., 1977, Nature 267:702.
Plattner, H., Reichel, K., Matt, H., Beisson, J., Lefort-Tran, M. & Pouphile, M., 1980, J. Cell Sci. 46:17.
Plattner, H., Stürzl, R. & Matt, H., 1985, Eur. J. Cell Biol. 36:32.
Plattner, H., Westphal, C. & Tiggemann, R., 1982, J. Cell Biol. 92:368.
Poste, G. & Allison, A.C., 1973, Biochim. Biophys. Acta 300:421.
Pouphile, M., Lefort-Tran, M., Plattner, H., Rossignol, M. & Beisson, J., 1986, Biol. Cell 56:151.
Rand, R.P. & Parsegian, V.A., 1986, Ann. Rev. Physiol. 48:201.
Rubin, R.P., Weiss, G.B. & Putney, J.W. (eds.), 1985, "Calcium in Biological Systems", Plenum Press, New York, London.
Stecher, B., Höhne, B., Gras, U., Momayezi, M., Glas-Albrecht, R. & Plattner, H., 1987, (submitted).
Steinhardt, R.A. & Alderton, J.M., 1982, Nature 295:154.
Tiggemann, R. & Plattner, H., 1982, FEBS Lett. 148:226.
Vilmart, J. & Plattner, H., 1983, J. Histochem. Cytochem. 31:626.
Vilmart-Seuwen, J., Kersken, H., Stürzl, R. & Plattner, H., 1986, J. Cell Biol. 103:1279.
Vilmart-Seuwen, J., Lumpert, C.J., Kersken, H. & Plattner, H., 1987, (submitted).
Westphal, C. & Plattner, H., 1981, Biol. Cell 42:125.
Zieseniss, E. & Plattner, H., 1985, J. Cell Biol. 101:2028.

STUDIES ON THE STRUCTURE AND FUNCTION OF THE ASIALOGLYCOPROTEIN RECEPTOR IN THE CELL, IN VITRO, AND IN RECONSTITUTED MEMBRANES

Darrell Doyle, James Petell, and James Sawyer

State University of New York at Buffalo
Department of Biological Sciences
Buffalo, New York 14260

The hepatocyte of mammalian liver is a highly differentiated polarized cell. The plasma membrane of the hepatocyte is also complex consisting of the following differentiated regions or domains: the perisinusoidal domain facing the liver sinusoid which is carrying blood from the gastrointestinal system of the animal via the hepatic portal vein; lateral domains between two adjacent hepatocytes; the bile canaliculus formed by the opposed membranes of two adjacent hepatocytes; desmosomes, tight and gap junctions; and the basal domain attaching the hepatocyte to its connective tissue substratum (Figure 1). Each of these domains is comprised of different sets of proteins and possibly lipids which endow the domain with its characteristic function and in some cases a recognizably characteristic morphology at the electron microscope level. The perisinusoidal domain is involved in secretion of serum proteins and the uptake of small and large molecular weight material from the blood. This domain would contain receptors involved in the cellular uptake of hormones like insulin[1], other constituents in the blood including low density lipoproteins[2], and glycoproteins terminated in penultimate galactose residues rather than the normal sialic acid residues (asialoglycoproteins[3-5]). The perisinusoidal domain in addition would contain factors involved in fusion of secretory vesicles arising from the trans region of the Golgi apparatus with this area of the plasma membrane. The bile canalicular domain of the plasma membrane would be expected to contain proteins and lipids that are involved in the transport of bile salts, either newly synthesized in the hepatocyte or recaptured from the serum at the sinusoidal domain, across the plasma membrane and eventually to the lumen of the bile duct. Similarly the gap junction contains proteins that are involved in the transport of small molecules[6] between cells and the basal domain contains proteins and lipids that are involved in the secretion of a basal lamina and the attachment of the hepatocyte to the basal lamina. How the cells establish and maintain the protein and the lipid composition of these different domains of its plasma membrane is an important area of investigation and has been the subject of research in our laboratory for several years.

We have chosen to study in some detail two domains of the plasma membrane of the hepatocyte, the perisinusoidal domain and the bile canalicular domain. The receptor for asialoglycoproteins (galactose terminated glycoproteins) is present primarily in coated pits of the perisinusoidal domain[7-9]. This receptor system is composed of three related glycoproteins having M_rs of 42-43000, 50-52000, and 59-60000 (Figure 2). This system

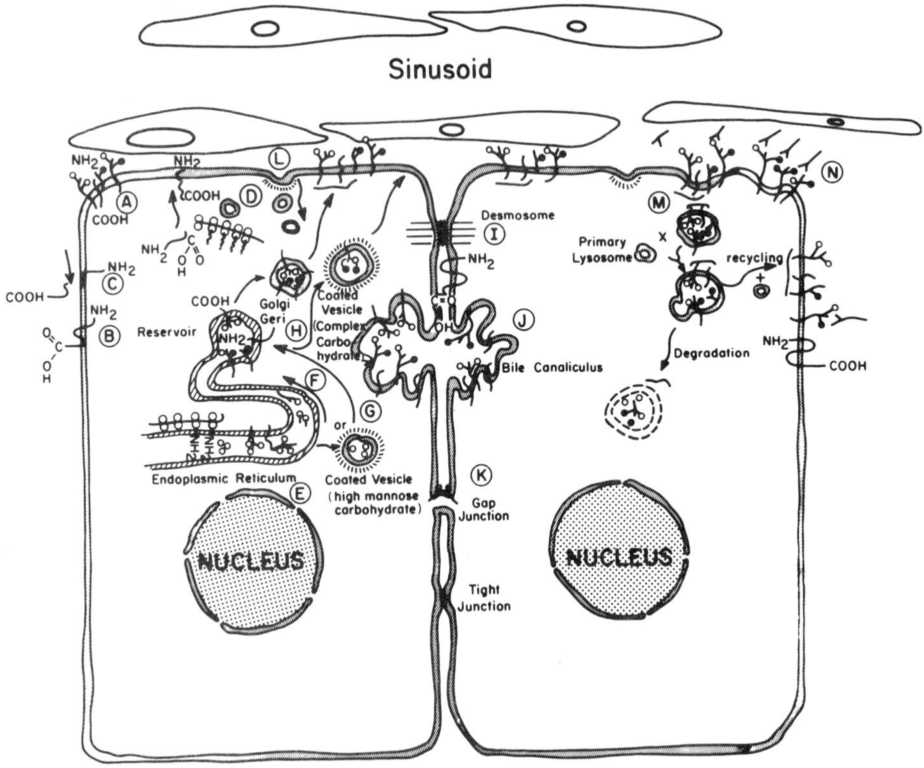

Figure 1. Diagrammatic representation of the polarized structure of hepatocytes. The perisinusoidal domain of the plasma membrane facing the sinusoid is involved both in the secretion of proteins synthesized on rough endoplasmic reticulum and transported through the Golgi system and in the uptake of small and large molecular weight material in the serum. The bile canalicular domain is formed by the opposed plasma membrane of two adjacent hepatocytes. It is involved in the transport of bile salts and is separated from other domains of the plasma membrane including the perisinusoidal domain and the gap junction by tight junctions. The basal domain is involved in the secretion of a basal lamina and in the attachment of the hepatocyte to the basal lamina. A: transmembrane plasma membrane glycoprotein with amino-terminal end facing the outside of the cell. B,C: transmembrane proteins with carboxyterminal ends facing the outside of the cell, one of which has had its carboxyterminal end cleaved by an extracellular protease. The receptor for asialoglycoproteins has its carboxyterminal end facing the outside of the cell and is localized in a coated pit (L) in the perisinusoidal domain. The 110000 M_r glycoprotein is localized in the bile canaliculus with its amino-terminal end facing the lumen of this structure. E-H: Intracellular pathway for the biogenesis of a secretory or plasma membrane bound glycoprotein. M-N: interiorization of membrane provoked by ligand-binding and recycling of interiorized units of plasma membrane.

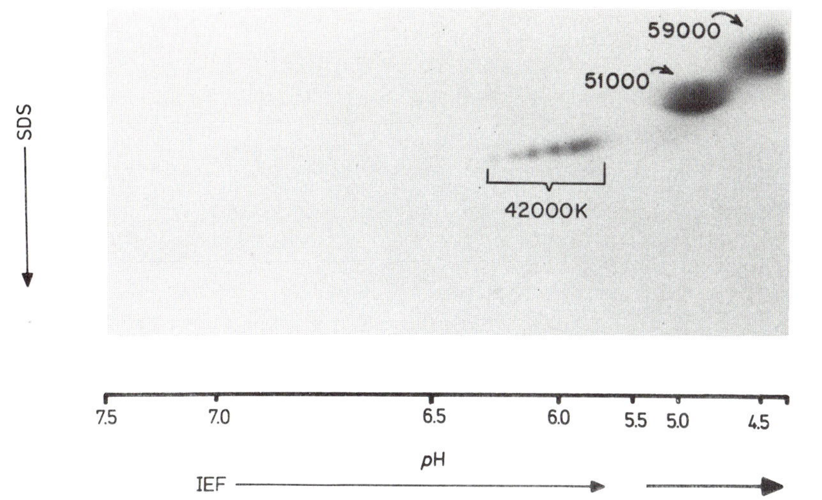

Figure 2. Two dimensional polyacrylamide gel electrophoretic analysis of the asialoglycoprotein receptor (galactose terminated glycoproteins) from rat hepatocytes (see Warren and Doyle[5] for details of isolation of the receptor polypeptides from in situ iodinated rat liver hepatocytes). Separation is by M_r in the sodium dodecyl sulfate dimension (SDS) and by charge in the isoelectric focusing dimension (IEF).

endows the hepatocyte with the ability to efficiently clear asialoglycoproteins from the circulation or from the culture medium if the cells are maintained in primary culture (Figure 3). The asialoglycoproteins bound to the receptor are internalized and delivered to lysosomes where the asialoglycoprotein is degraded to amino acids; the receptor however is segregated at an early step in endocytosis and is recycled to the surface where it is available to bind more ligand[4,5].

It is possible by cell fractionation methods to isolate a fraction, denoted P2, from rat liver enriched in the perisinuoidal plasma membrane domain (Figure 4)[9,10]. It is also possible to isolate from the same homogenate of rat liver a fraction, denoted N2 upper, which is enriched in the bile canalicular domain of the plasma membrane (Figure 4). From this latter fraction we have purified to homogeneity a transmembrane glycoprotein of M_r 110000 (Figure 5)[11]. Antibodies specific for both the 110000 M_r bile canalicular glycoprotein and the perisinusoidal receptor for asialoglycoproteins were used to confirm the localization of these proteins in rat liver (Figure 6).

One goal of research in our laboratory is to reconstitute these domain-specific membrane proteins in lipid bilayers and to implant them in a functional way in cells that lack the functions specified by these proteins. Using this approach we hope to be able to eventually identify other proteins, and perhaps lipids, that are involved in directing to and restricting the domain-specific protein in its plasma membrane domain of residence.

In fact several years ago we fused the P2 perisinusoidal plasma membrane fraction of liver to mouse L cells utilizing polyethylene glycol as fusogen (Figure 7)[12]. We were in fact able to transfer the asialoglycoprotein receptor activity to the L cell and the receptor did impart to the L cell the ability to bind, interiorize, and degrade asialoglycoproteins (Figure 8), properties that the L cell does not normally possess. We carried these fusion experiments one step further by reconstituting purified receptor in a

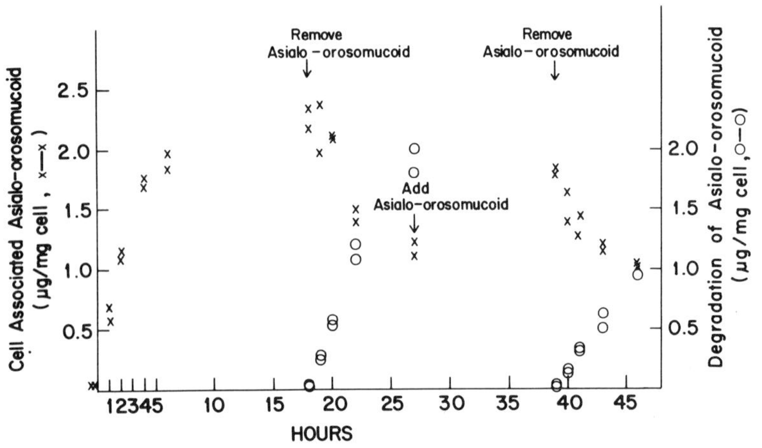

Figure 3. Metabolism of iodinated asialo-orosomucoid by primary cultures of rat hepatocytes. Primary cultures of rat hepatocytes prepared as described elsewhere (Warren and Doyle[5]) were exposed to ^{125}-I labeled asialo-orosmucoid in the medium. Cell-associated phosphotungstic acid insoluble radioactivity (X-X) was determined at the times indicated after washing the cells three times with phosphate buffered saline. At the times indicated the medium containing the labeled asialo-orosomucoid was removed and replaced with fresh medium not containing asialo-orosomucoid and cell associated acid insoluble as well as acid soluble radioactivity in the medium was determined.

mixture of phospholipids purified from liver in a right-side out configuration (Figure 9)[13]. These closed unilamellar membrane vesicles (Figure 10) were fused with L cells by the method outlined in figure 7 and again were able to impart to the L cell the ability to bind, interiorize, and degrade asialoglycoproteins present in the medium (Figure 11). However, most of the receptor activity that was transferred to the L cell by fusion with the reconstituted vesicles was inactive in mediating endocytosis of asialoglycoproteins. It appeared that rather than being authentically fused much of the reconstituted material was tightly bound to the surface of the L cell after polyethylene glycol mediated fusion. In fact, we could show by

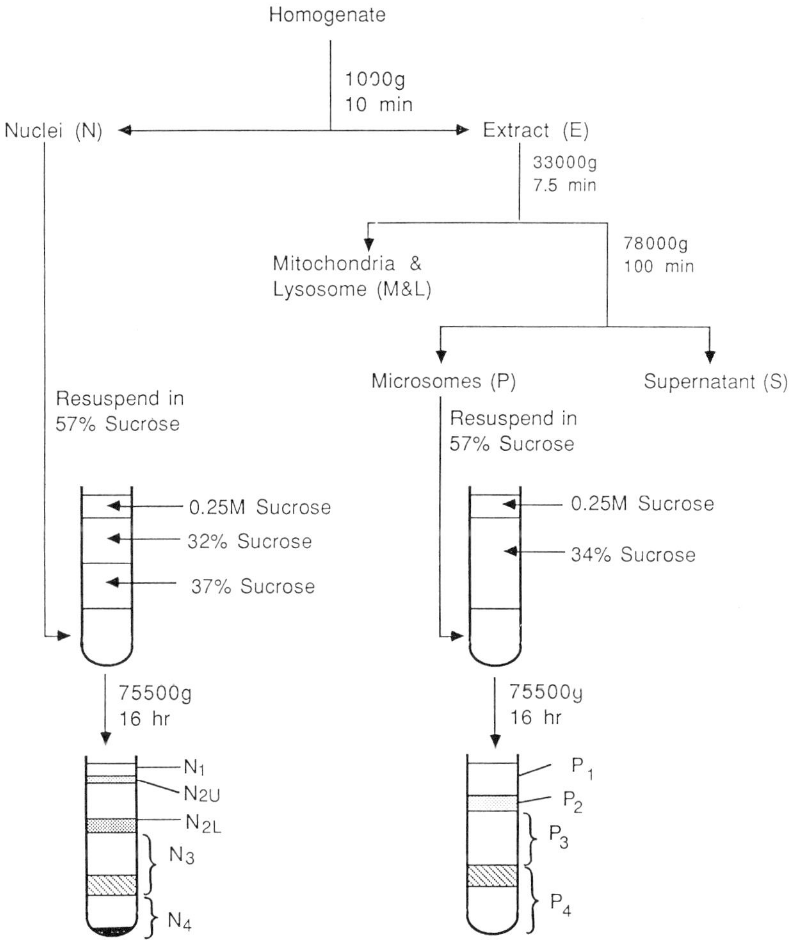

Figure 4. A fractionation protocol for the preparation of plasma membrane fractions from rat liver (for details see Cook, et al.[9]).

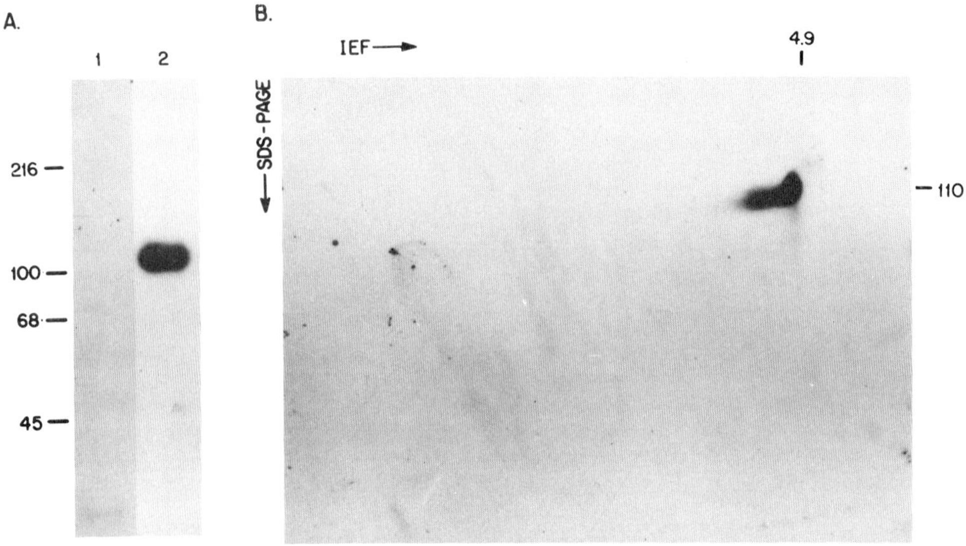

Figure 5. Two dimensional polyacrylamide gel electrophoretic analysis of the 110000 M_r glycoprotein of rat liver bile canaliculus. Hepatocytes were surface labeled by lactoperoxidase catalyzed iodination at 4°C. Radioactive GP 110 was immunoprecipitated from Triton X-100 cell extracts and subjected to PAGE and autoradiography. A) SDS PAGE: 1, preimmune sera; 2, immune sera. B) Two dimensional PAGE: first dimension, isoelectric focusing (IEF); second dimension, SDS PAGE. The 110,000 M_r glycoprotein has an isoelectric point of 4.9.

cytochemical methods that many of the reconstituted vesicles were tightly bound to the surface of the L cell and remained bound for relatively long intervals of time. The receptor in these attached vesicles was able to bind asialoglycoproteins but the bound material was not interiorized to be degraded in lysosomes.

In order to increase the efficiency of fusion and to decrease the background produced by tight "sticking" of vesicles in the fusion reaction we decided to (1) learn more about the mechanism by which the asialoglycoprotein receptor performs its function in the hepatocyte and (2) learn more about the composition of the domain in which each of these membrane proteins resides. Our hope being that such knowledge when in hand would enable us to better design a reconstituted vesicle which could be efficiently fused with mouse L cells.

The first question that we attempted to answer was whether all three M_r species of the asialoglycoprotein receptor existed as a multisubunit complex, and if so, to determine the relative stoichiometry of the different subunits in the complex. In order to answer this question we required

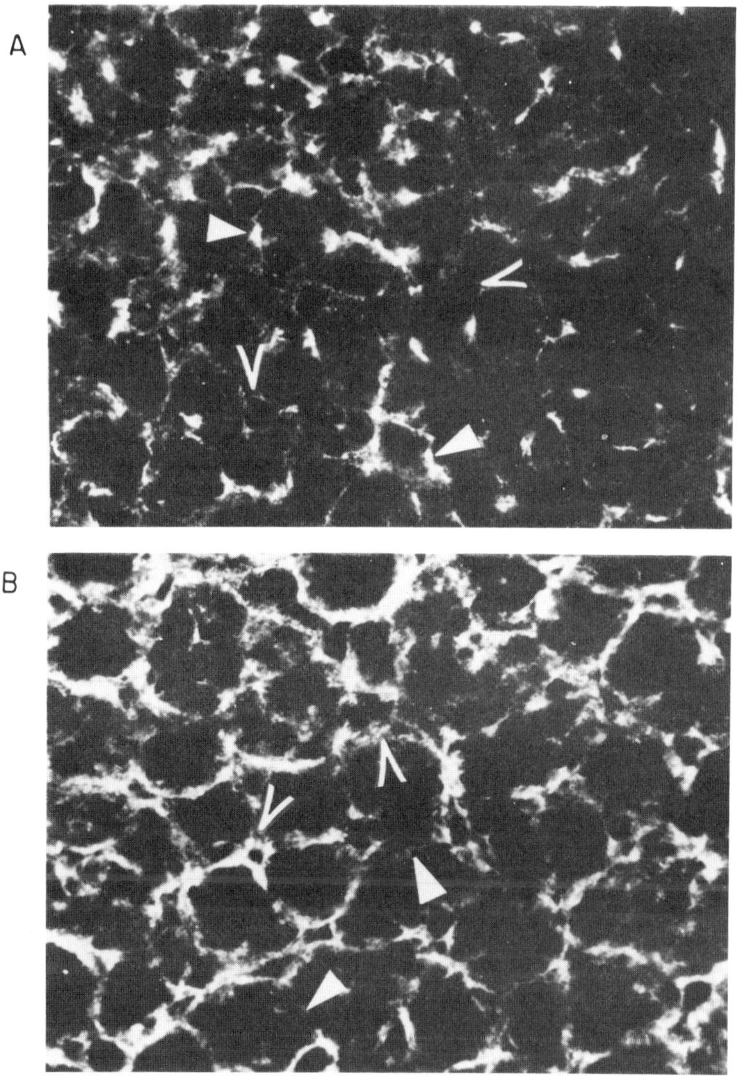

Figure 6. Immunolocalization of the asialoglycoprotein receptor and the 110000 M_r glycoprotein to the perisinusoidal and bile canalicular domains respectively. Frozen liver sections were immunofluorescently stained using rabbit antibodies directed against the 110000 M_r glycoprotein (A) or the asialoglycoprotein receptor (B). White solid arrows indicate bile canaliculi. White open arrows indicate sinusoidal domains.

antibodies that could distinguish between each of the different M_r species of the asialoglycoprotein receptor. Knowledge about the primary amino acid sequence of the 43000 and 50000/59000 M_r forms of the receptor published by other investigators revealed a region of diversity in the

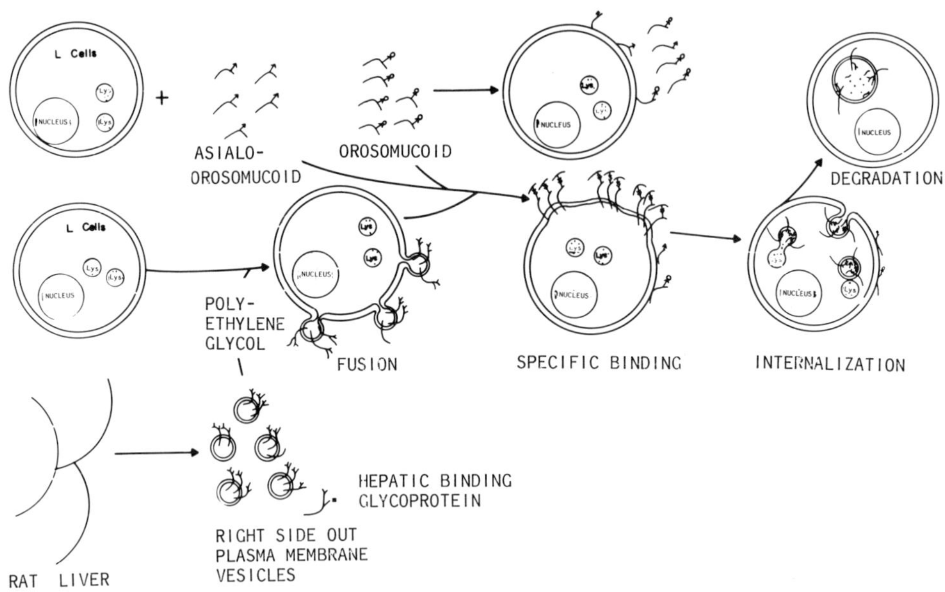

Figure 7. Schematic representation of the method for fusion of membrane vesicle with mouse L cells.

carboxy terminal domain of these polypeptide species (Figure 12)[14]. Twelve of sixteen amino acid residues in this region differed between the 43000 and 50000 M_r polypetides. The two divergent regions of sixteen amino acids were chemically synthesized and used as immunogens in rabbits. Each of the two rabbits responded producing antibodies that reacted with each of the divergent peptides; there was little if any cross reaction of the antibodies between the divergent polypeptides (Figure 13). These antibodies could also recognize and distinguish between the 43000 and the 50000/59000 M_r species of asialoglycoprotein receptor (Figure 14). We next used these antibodies to determine if the three M_r species of the asialoglycoprotein receptor existed as a complex or as individual entities in the perisinusoidal domain of the plasma membrane of the hepatocyte. To do this experiment externally oriented proteins of the plasma membrane including the three different M_r species of the asialoglycoprotein receptor were radioactively tagged using the method of lactoperoxidase catalysed iodination[15,16]. Detergent extracts of the membrane were then prepared and reacted with each of the antibodies which can distinguish between the 43000 M_r and the 50000/59000 M_r species of receptor. If the three different M_r species of receptor were present as a complex in the plasma membrane of the hepatocyte then each of the two different antibodies should precipitate in all three M_r species of receptor, which was exactly the result obtained (Figure 15). Hence in future reconstitution experiments we will control the composition of the purified different M_r species of the asialoglycoprotein receptor prior to insertion into the lipid bilayer. In this way we will be able to elucidate the function of individual receptor components in the metabolism of asialoglycoproteins.

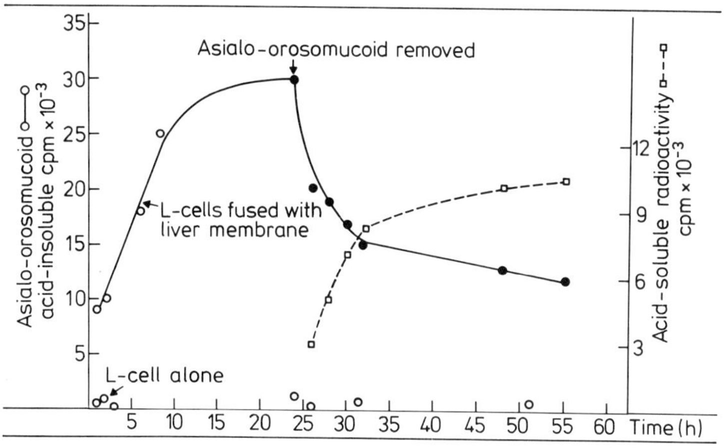

Figure 8. Metabolism of asialo-orosomucoid by L-cells fused with rat liver plasma membrane. L-929 cells were fused with the P_2 fraction of rat liver. The cells were distributed among 40 plates (35 x 10 mm) which were incubated at 37°C overnight. The next day, the attached cells were washed three times with phosphate-buffered saline. Medium (1 ml), containing ^{125}I-asialo-orosomucoid (4 x 10^6 cpm, 10 ng) and 5 µg of unlabeled asialo-orosomucoid was added to each culture. Cell-associated phosphotungstic acid-insoluble radioactivity was determined at the times indicated. Cells were washed three times with phosphate-buffered saline and material insoluble in phosphotungstic acid was counted in the Biogamma spectrometer. After 24 h in the presence of asialo-orosomuoid, the medium was removed from the cells and the cells were washed twice with medium not containing asialo-orosomucoid. Phosphotungstic acid-soluble radioactivity in the medium and acid-insoluble radioactivity associated with the cells was determined at the times indicated. L-cells not fused with the liver membranes will bind 500 cpm or less of asialo-orosomucoid when added at the same concentration and under the same conditions as specified above for the fused cells.

Is the composition of the lipid bilayer different in the different domains of the plasma membrane of the hepatocyte and is lipid composition important in producing reconstituted vesicles that can efficiently transfer membrane proteins between cells and still maintain the function specified by these proteins? To help answer these questions we are attempting to isolate in as homogeneous a form as possible the different domains of the plasma membrane of the hepatocyte[17]. Our approach for the isolation of a homogeneous domain of the plasma membrane is to first fracture the plasma membrane by controlled sonication into closed right-side out vesicles having a size smaller than the domain of origin (Figure 16). Next an antibody specific for a protein in the plasma membrane domain of interest is coupled directly or indirectly to nitrocellulose. We have coupled either a polyclonal monospecific antibody recognizing shared epitopes among the three different M_r species of the asialoglycoprotein receptor directly to nitrocellulose or indirectly via a immunoblotting protocol as a specific immunoreagent for the isolation of the perisinusoidal domain from the mixed

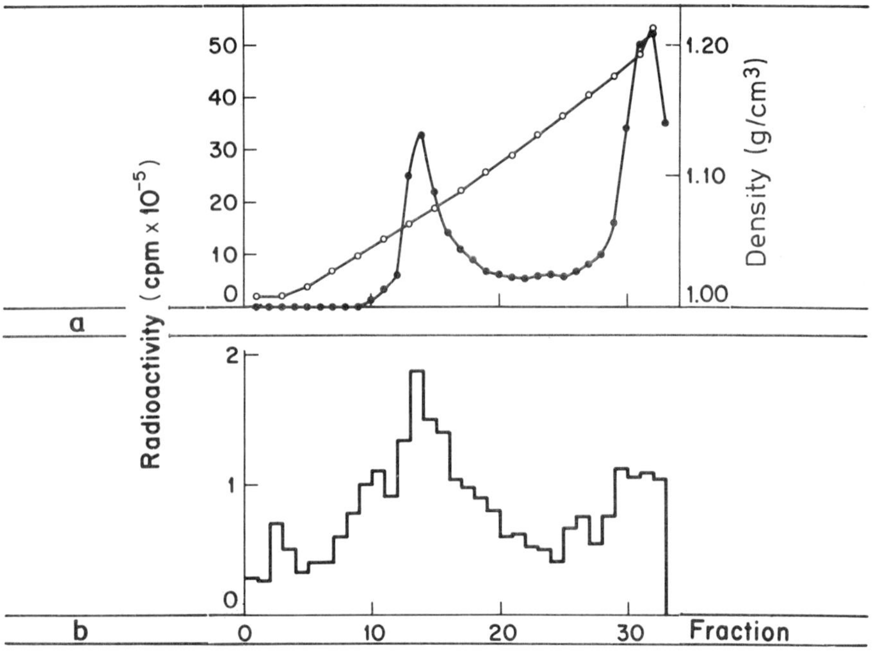

Figure 9. Incorporation of the hepatocyte asialoglycoprotein receptor into reconstituted membrane vesicles. A: Iodinated asialoglycoprotein receptor was mixed with total liver phospholipids and dialysed for three days against phosphate buffered saline. B: Non-labeled asialoglycoprotein receptor was reconstituted similar to A. Both preparations were subjected to isopycnic centrifugation as described in Baumann, et al.[13]. In A fractions were assayed for ^{125}I radioactivity; in B the binding of iodinated asialo-orosomucoid ligand by the reconstituted vesicles was monitored.

closed vesicle fraction of the plasma membrane of the hepatocyte. Using similar techniques we have coupled a polyclonal monospecific antibody to the 110000 M_r glycoprotein of the bile canaliculus to isolate this domain from the mixed closed vesicle population of the plasma membrane fraction of liver. These immunochips as we call them can be used to specifically isolate both the perisinusoidal domain and the bile canalicular domain from the same closed vesicle population of the sonicated plasma membrane fraction of rat liver (Figure 17). By making certain assumptions about the relative amount of the asialoglycoprotein receptor system in the perisinusoidal domain and the 110000 M_r glycoprotein in the bile canalicular domain and the relative amounts of each of these domains in the plasma membrane of the hepatocyte in the rat liver, we can estimate a ratio of asialoglycoprotein:110000 M_r glycoprotein in the immunoaffinity purified vesicles of 30-40 when each vesicle is 90% enriched relative to the other. However during the preparation of these vesicles from isolated rat hepatocytes there is some mixing of domain constituents. By forming even a small right-side out closed vesicle from the primary cultures of hepatocytes we have obtained a relative differential enrichment of the two domain markers of over 50 fold, indicating an even greater than 90% relative enrichment of the two immunoaffinity purified membrane domains.

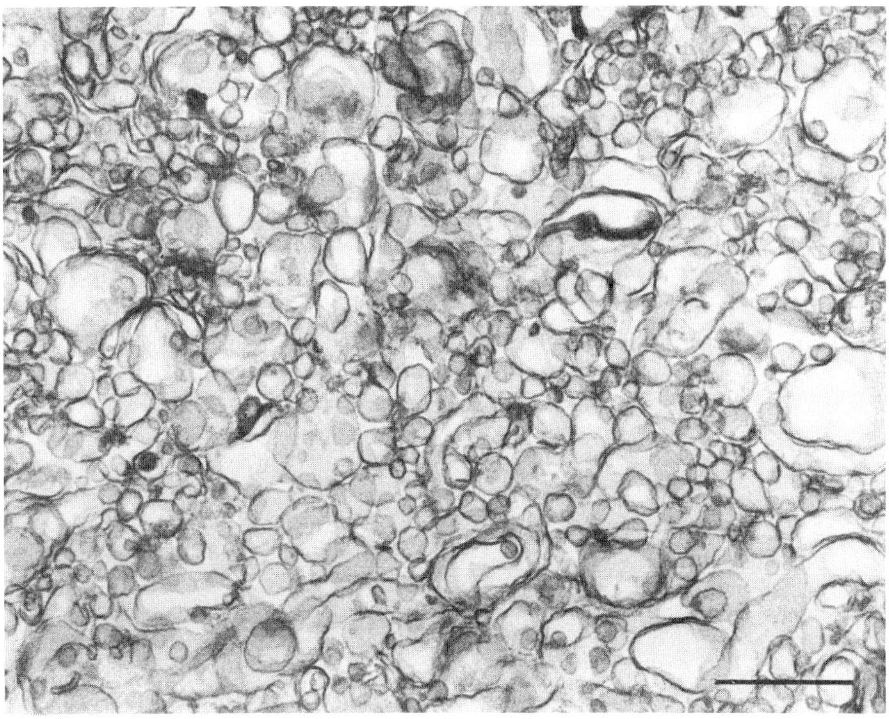

Figure 10. Electron micrograph of reconstituted vesicles (for details see Baumann, et al.[13]).

At present we are modifying somewhat the procedures for the large scale isolation of the perisinusoidal domain, the bile canalicular domain, and some of the other domains of the hepatocyte using the immunoblot methodology. Our goals in the future will be to assess both the protein and the lipid composition of individual membrane domains, to use isolated domains as the donor source for transferring specific membrane functions of the hepatocyte to other types of cells; and finally to reconstitute in a more physiological way domain specific membrane proteins in a lipid bilayer which more closely resembles the lipid bilayer in which they normally reside. Using this newer generation of reconstituted membrane systems we hope to learn more about specific relationships and interactions among the proteins and lipids forming the membrane domain together with other proteins and lipids in different cell types that are recipients for transfer of the reconstituted membrane.

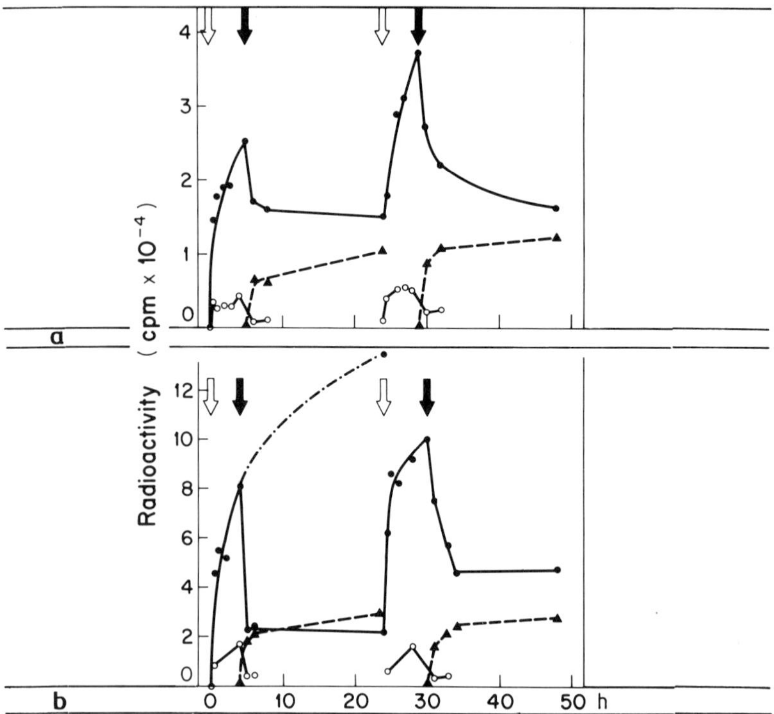

Figure 11. Binding and metabolism of asialo-orosomucoid by mouse L cells containing transferred rat hepatocyte receptor for asialoglycoproteins. Reconstituted phospholipid vesicles containing either rat liver membrane proteins from the P2 membrane fraction or the asialoglycoprotein receptor were used for the fusions. In A, vesicles containing P2 proteins were fused with L cells. The same preparation of vesicles was used for treatment of control cells after these cells had been treated with polyethylene glycol and washed free of fusogen. In B, vesicles containing the asialoglycoprotein receptor were fused with L cells. As a control L cells pretreated and washed free of polyethylene glycol were also treated with vesicles containing the asialoglycoprotein receptor. After 16 h, the adherent cells were washed with culture medium and exposed to medium containing [^{125}I]-asialo-orosomucoid (0 time). After incubation for 4-5 h at 37°C the labeled medium was removed and the cells washed. The cells were then maintained at 37°C. Two culture dishes in B were left with labeled medium for 24 h (------). After 24 h the cells were again exposed to medium containing [^{125}I]-asialo-orisomucoid. At the times indicated, duplicate culture dishes were washed and the acid insoluble radioactivity associated with the experimental cells (closed circles) or control cells (open circles) as well as the acid soluble radioactivity released into the medium by the former cells (closed triangles) were determined (for details see Baumann, et al.[13]).

P1:ARG-ARG-PRO-TYR-ARG-TRP-VAL-CYS-GLU-THR-GLU-LEU-GLY-LYS-ALA-ASN

P2:GLN-GLN-VAL-ASN-ARG-TRP-ALA-CYS-GLU-ARG-LYS-ARG-ASP-ILE-THR-TYR

Figure 12. Peptides used for production of specific antibodies to the asialoglycoprotein receptor. Peptides corresponding to the sixteen carboxy terminal amino acids (Drickamer et al.[14]) of the 43000 (P1) and 50000/60000 (P2) forms of the receptor were prepared by automated solid phase synthesis. The peptides were coupled to keyhole limpet hemocyanin and injected into rabbits.

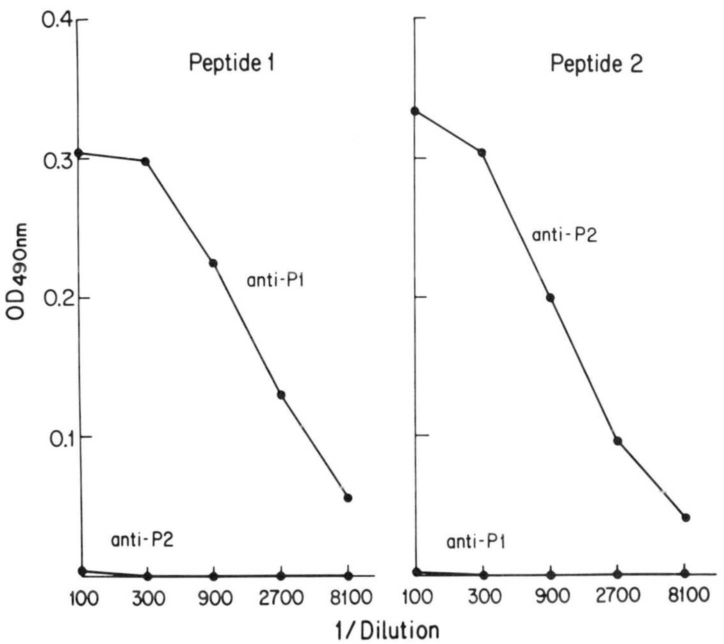

Figure 13. ELISA analysis of the antisera to the asialoglycoprotein receptor derived peptides. Microtiter plate wells were coated with either peptide, washed, and then incubated successively with 1% BSA, peptide antiserum at the indicated dilution, and protein A conjugated with horseradish peroxidase. Peroxidase was detected with O-phenylenediamine and the OD490 nM of a 1/20 dilution of the product was determined.

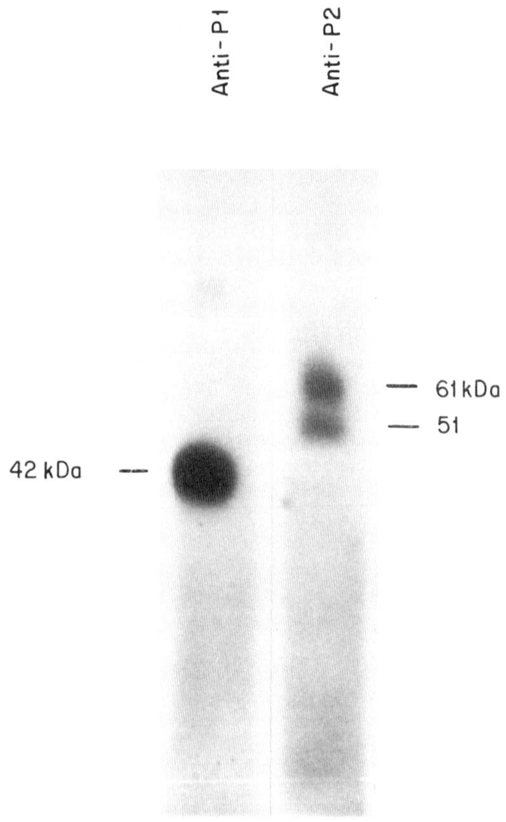

Figure 14. Western blot analysis of antisera to the receptor derived peptides. Purified asialoglycoprotein receptor was subjected to SDS-polyacrylamide gel electrophoresis and then transferred to nitrocellulose paper. After blocking with carrier protein the nitrocellulose was incubated with anti-P1 serum (1:500) or anti-P2 serum (1:125) followed by ^{125}I-labelled protein A.

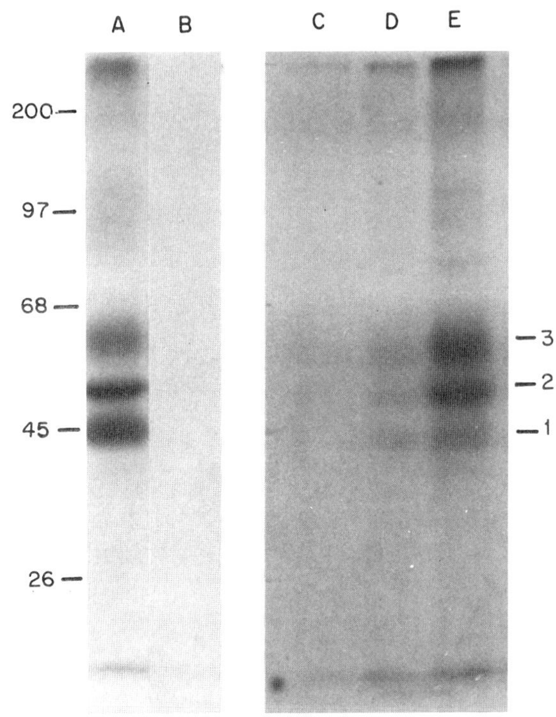

Figure 15. Immunoprecipitation of the asialoglycoprotein receptor from hepatocyte extracts. Hepatocytes isolated by collagenase perfusion were cultured overnight and labelled with ^{125}I (see Warren and Doyle[5] for details). A cell pellet was extracted in detergent containing buffer and aliquots of the supernatant incubated with antiserum to the purified receptor (A), control antiserum (B), 15 μg of purified preimmune IgG (C), 15 μg affinity purified antibody to P1 (D), 15 μg affinity purified antibody to P2 (E). Immune complexes were collected by incubation with protein A coupled to Sepharose and analyzed by SDS-polyacrylamide gel electrophoresis.

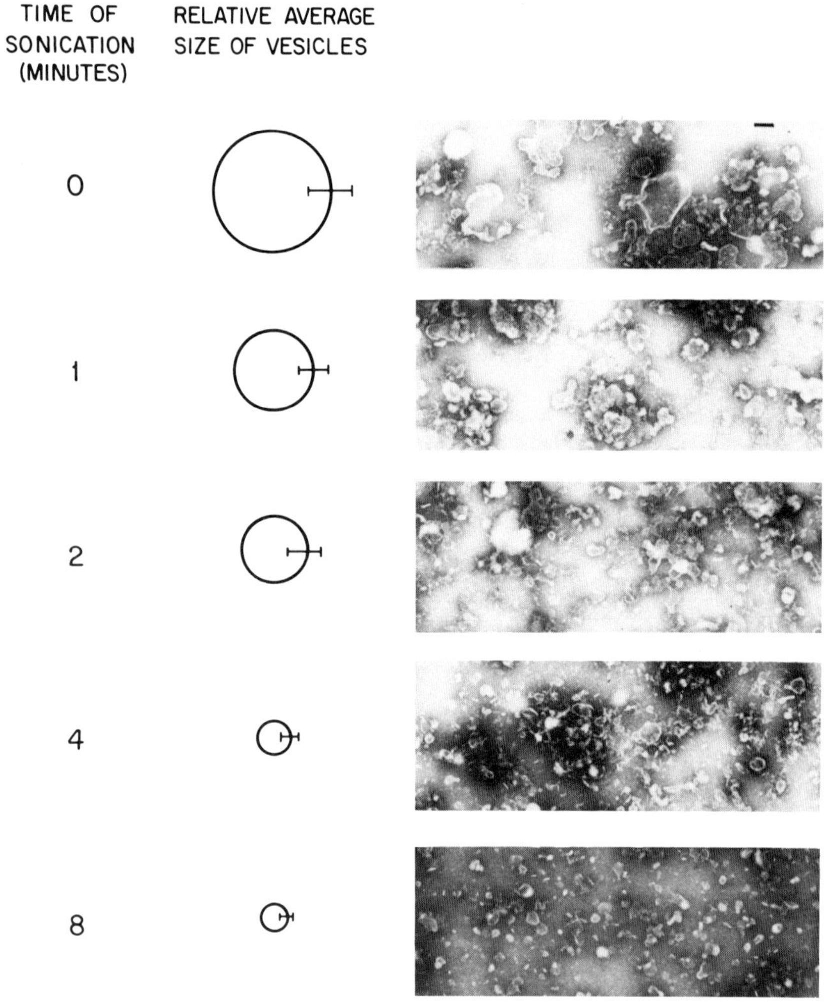

Figure 16. Electron micrograph of membrane vesicles obtained by sonication. Membrane preparations were sonicated for 1, 2, 4 or 8 minutes. Vesicle preparations were subjected to negatively staining followed by electron microscopy. Approximate sizes of sonicated vesicles were 62±2.3, 48±2.9, 26±1.4 and 18±1.0 μm, respectively. A depection of the mean vesicle size is shown. Bar, 0.1 μm; x 45000.

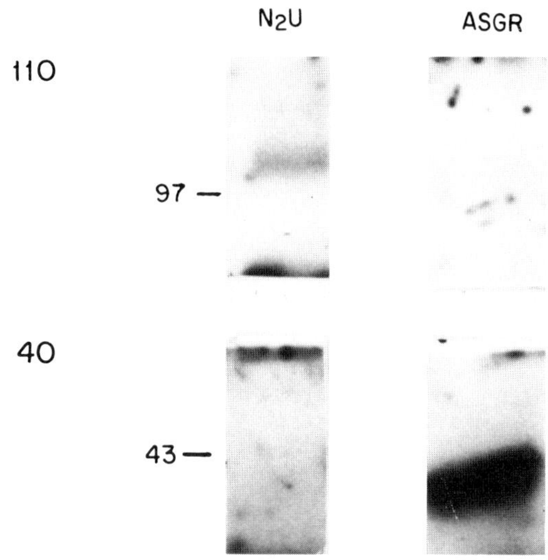

Figure 17. Analyses of iodinated vesicles bound to affinity nitrocellulose immunoabsorbents. A. Iodinated vesicles obtained by sonication for 8 minutes, initial (I), were eluted from the 110000 M_r region (110) of N_{2u} or the 40000 M_r region of ASGR affinity nitrocellulose strips and subjected to SDS PAGE followed by autoradiography. B, background or non-specific region of immunoabsorbent. B. Autoradiograph of SDS PAGE of iodinated vesicles sonicated for 1, 2, or 4 minutes bound to N_{2u} (N) or ASGR (A) affinity nitrocellulose strips.

REFERENCES

1. Czech, M.P., "Molecular Basis of Insulin Action," Plenum, New York (1985).
2. Brown, M.S. and Goldstein, J.L., Lipoprotein Metabolism in the Macrophage: Implications for Cholesterol Deposition in Atherosclerosis, Ann. Rev. Biochem. 52:223-261 (1983).
3. Ashwell, G. and Morell, A.G., Role of Surface Carbohydrates in the Hepatic Recognition and Transport of Circulating Glycoproteins, Adv. Enzymol. Relat. Areas Mol. Biol. 41:99-128 (1974).
4. Ashwell, G., and Harford, J., Carbohydrate-Specific Receptors of the Liver, Ann. Rev. Biochem. 51:531-554 (1982).
5. Warren, R. and Doyle, D., Turnover of the Surface Proteins and the Receptor for Serum Asialoglycoproteins in Primary Cultures of Rat Hepatoytes, J. Biol. Chem. 256:1346-1355 (1981).
6. Paul, D.L., Molecular Cloning of cDNA for Rat Liver Gap Junction Protein, J. Cell Biol. 103:123-134 (1986).
7. Wall, D.A., Wilson, G. and Hubbard, A.L., The Galactose-Specific Recognition System of Mammalian Liver: The Route of Ligand Internalization in Rat Hepatocytes, Cell 21:79-93 (1980).
8. Geuze, H.J., Slot, J.W., Strous, G.J.A.M., Lodish, H.F. and Schwartz, A.L., Immunocytochemical Localization of the Receptor for Asialoglycoprotein in Rat Liver Cells, J. Cell Biol. 92:865-870 (1982).
9. Cook, J., Hou, E., Hou, Y., Cairo, A. and Doyle, D., Establishment of Plasma Membrane Domains in Hepatocytes. I. Characterization and Localization to the Bile Canaliculus of Three Antigens Externally Oriented in the Plasma Membrane, J. Cell Biol. 97:1823-1833 (1983).
10. Aronson, N.N. and Touster, O., Isolation of Rat Liver Plasma Membrane Fragments in Isotonic Sucrose, Methods Enzymol. 31:90-102 (1978).
11. Petell, J.K., Diamond, M., Hong, W.J., Bujanover, Y., Amarri, S., Pittschieler, K. and Doyle, D., Isolation and Characterization of a 110,000 Molecular Weight Glycoprotein Localized to the Hepatocyte Bile Canaliculus, J. Biol. Chem. in press (1987).
12. Doyle, D., Hou, E. and Warren, R., Transfer of the Hepatocyte Receptor for Serum Asialoglycoproteins to the Plasma Membrane of a Fibroblast: Acquisition of the Hepatocyte Receptor Functions by Mouse L Cells, J. Biol. Chem. 254:6853-6856 (1979).
13. Baumann, H., Hou, E. and Doyle, D., Insertion of Biologically-Active Membrane Proteins from Rat Liver into the Plasma Membrane of Mouse Fibroblasts, J. Biol. Chem. 255:10001-10012 (1980).
14. Drickamer, K., Mamon, J.F., Binns, G. and Leung, J.O., Primary Structure of the Rat Liver Asialoglycoprotein Receptor, J. Biol. Chem. 259:770-778 (1984).
15. Hubbard, A.L. and Cohn, Z.A., The Enzymatic Iodination of the Red Cell Membrane, J. Cell Biol. 55:390-405 (1972).
16. Tweto, J., Friedman, E. and Doyle, D., Proteins of the Hepatoma Tissue Culture Cell Plasma Membrane, J. Supramolecular Structure 4:141-159 (1976).
17. Petell, J.K., Bujanover, Y., Gocayne, J., Amarri, S. and Doyle, D., Isolation of Domains of the Plasma Membrane of Hepatocytes, Exp. Cell Res. in press (1987).

PHOSPHORYLATION EVENTS IN REGULATION OF EXOCYTOSIS

Birgit H. Satir and Timothy J. Murtaugh

Department of Anatomy
Albert Einstein College of Medicine
Bronx, New York 10461

BACKGROUND

The ciliated protozoan *Paramecium tetraurelia* is an especially well suited organism with which to study the process of signal transduction associated with stimulus exocytosis coupling. Wild type (wt) *Paramecium* contains thousands of membrane-bounded secretory organelles, called trichocysts. These organelles are docked at specific sites below the plasma membrane. The site of a docked trichocyst is marked in freeze-fracture electron microscopy by a specific membrane microdomain consisting of an outer ring of intramembrane particles (IMPs) with a central rosette composed of 9-11 IMPs (Janish, 1972, Satir 1974). Upon stimulation with secretagogue, such as trinitrophenol (TNP), in the presence of Ca^{2+}, the trichocyst membrane and the plasma membrane fuse and the trichocyst content, called trichocyst matrix (tmx), expands lengthwise, leaving the cell as a paracrystalline needle-like structure 40μm long. Both membrane fusion and matrix expansion are Ca^{2+}-dependent irreversible events. The released paracrystalline tmx is easily observed in the light microscope. This has been used to assay the isolation of several secretory mutants of this cell (Beisson et al., 1976; Pollack, S., 1974; Beisson and Rossignol, 1975). In addition to the two Ca^{2+}-dependent events mentioned above, a third event has been described which involves a covalent modification of a protein. Gilligan and Satir (1982) showed that wt cells dephosphorylate a M_r 63kDa phosphoprotein (63kDa PP) when stimulated to release with TNP in the presence of external Ca^{2+}. Both dephosphorylation and exocytosis are blocked when external Mg^{2+} is substituted for Ca^{2+}. The potential role of this 63kDa PP in exocytosis was investigated using the secretory mutants to dephosphorylate the 63kDa PP in response to stimulation. Specifically a temperature sensitive signal transduction mutant (nd9) has been studied in detail. The exocytic membrane microdomain (rosette) is present and trichocyst release occurs normally when nd9 cells are grown at 18°C (permissive temperature), but when cells are grown at 27°C (non-permissive temperature) the microdomain is aberrant and the cells do not secrete. In the presence of external Ca^{2+}, addition of the divalent cation ionophore A23187 to the nd9 cells at the non-permissive temperature (27°C), results in expansion of the tmx without concomitant membrane fusion. Therefore, tmx expansion is not inherently defective in this mutant and the lack of exocytosis may be directly related to membrane events, particularly to the lack of a correctly assembled exocytic membrane microdomain the components of which evidently plays an important role in signal transduction in membrane fusion. In addition, nd9 cells dephosphorylate the 63kDa PP when grown at the permissive temperature (18°C), but do not dephosphorylate the protein when grown at the non-permissive temperature (27°C) (Gilligan and Satir, 1982). Dephosphorylation of this protein is also absent in other exocytosis mutants that show a defective exocytic membrane microdomain. Rephosphorylation normally takes place within 10 seconds (Zieseniss and Plattner, 1985).

Previous results from cell fractionation experiments in Paramecium has shown that when cells are labeled in vivo with $^{32}P_i$ the 63kDa PP is the most highly phosphorylated protein and that most of this labeled protein is found in the high speed supernatant (S2, 100,000 X g). This fraction has recently been used in isolating the 63kDa protein and in the production of an affinity purified antibody against the resulting purified protein (Murtaugh et al. 1987). This should be useful in a more comprehensive study of the role of protein phosphorylation/dephosphorylation in Paramecium exocytosis. Evidence is accumulating that protein phosphorylation and/or dephosphorylation may be a regulatory step in the signal transduction pathway during regulative exocytosis in many secretory systems.

Aspects of Characterization of the 63kDa Phosphoprotein

As part of the isolation procedure, preliminary investigation of the native size of the 63kDa PP was done using gel filtration chromatography and non-reducing SDS-PAGE. ^{32}P-labeled 63kDa PP in a fresh S2 fraction elutes from a gel filtration column with an apparent native size of 60-65kDa (Fig. 1).

The presence or absence of a disulfide reducing agent (such as 2-mercaptoethanol) has no effect on the elution from the column or on the migration of the 63kDa protein in SDS-PAGE. This suggests that the labeled protein probably exists as a monomeric polypeptide in the native state. The labeled phosphoamino acid in the 63kDa protein is phosphoserine (Fig. 2).

It is evident from Fig. 1 that the 63kDa PP is only a minor constituent of the S2 fraction. Two dimensional PAGE shows that the labeled 63kDa PP appears as two spots with pI's of approximately 5.8 and 6.2. Using the affinity purified antibody made against the purified protein we identified a third form with a pI of 6.3 that was not labeled (Murtaugh et al., 1987). The presence of multiple slightly acidic labeled isoelectric forms (pI 5.8-6.3) could be due to multiple phosphorylation sites on the 63kDa PP. The unlabeled species might also be phosphorylated at sites that are inaccessible to labeling under our conditions. Questions as to whether one particular labeled isoelectric form specifically becomes dephosphorylated upon stimulation to exocytosis or whether unlabeled polypeptides identified by the antibody in two dimensional gels are the important dephosphorylated species associated with exocytosis are important to pursue further in the future.

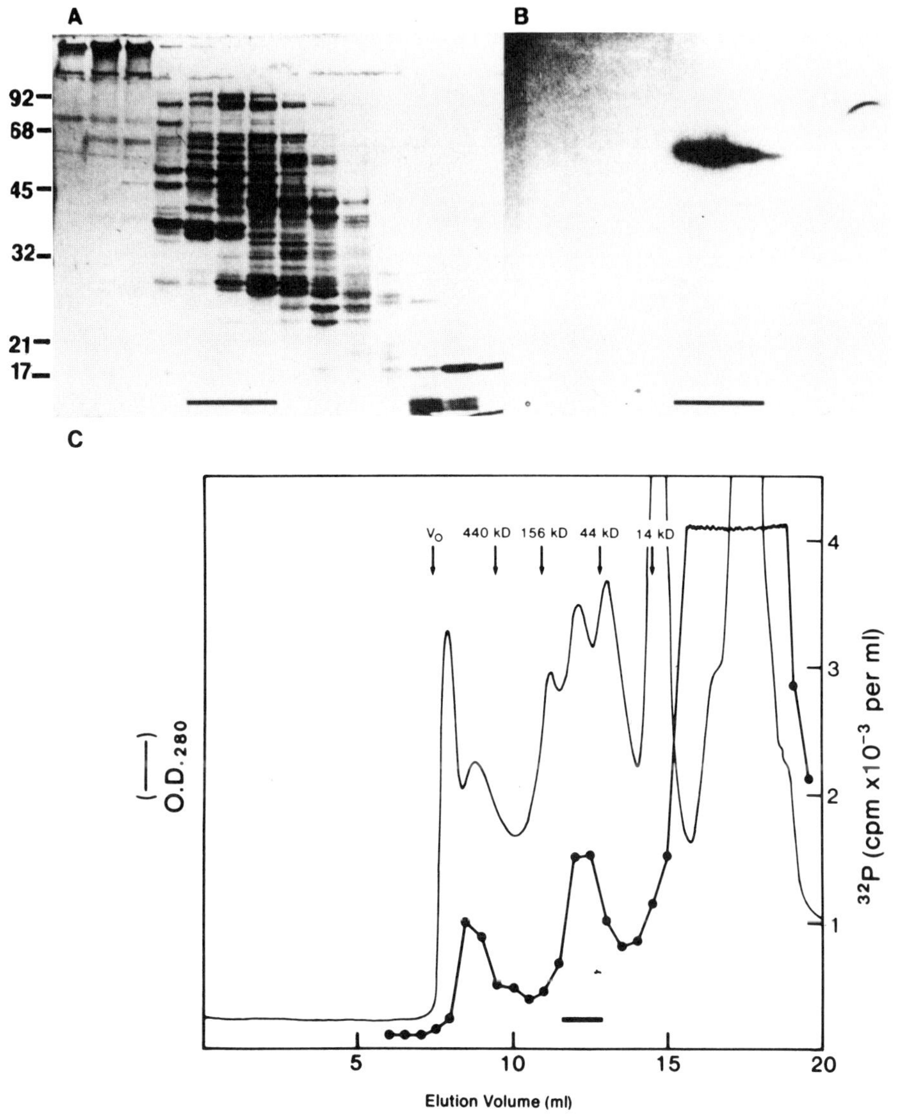

Fig. 1. Gel filtration chromatography of ^{32}P-labeled supernatant fraction. 0.5 ml of a 100,000xg supernatant was applied to a Superose 12 column, eluted with 0.1 M ammonium bicarbonate and collected in 0.5 ml fractions. The bars at the bottom of each panel indicate the fractions which contained peak ^{32}P-labeled 63kDa protein. Panel A: SDS-PAGE of column fractions. Panel B: Autoradiogram of gel in A. Pancel C: Chromatogram showing the elution of protein (———) and ^{32}P (●——●). Calibration standards include: Vo, blue dextran; 440kDa, ferritin; 156kDa, gamma globulin; 44kDa, ovalbumin; 14kDA, cytochrome c.

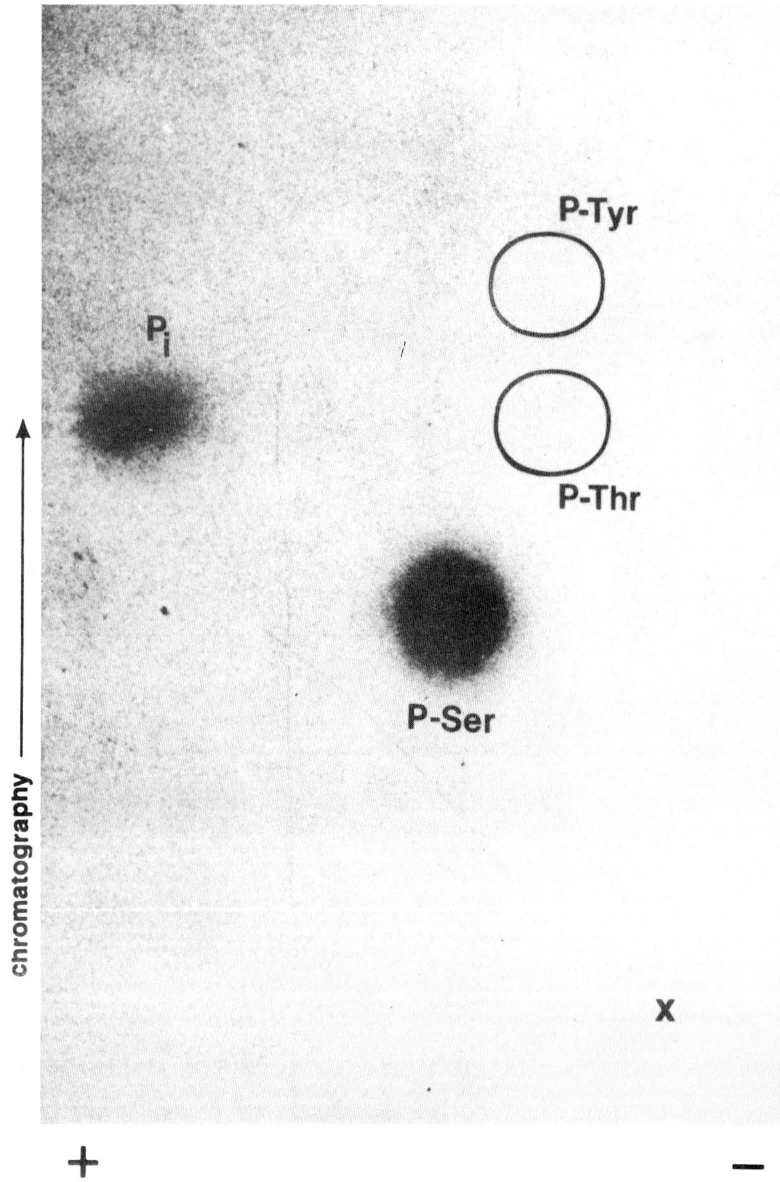

Fig. 2. ^{32}P-autoradiogram of two-dimensional thin layer electrophoresis and chromatography of phosphorylated amino acids from hydrolyzed ^{32}P-labeled 63kDa protein. Locations of ninhydrin-stained standards are indicated: P-ser, phosphoserine; P-Thr, phosphothreionine; P-Tyr, phosphotyrosine. "X" marks the origin.

Localization of the 63kDa PP

We have indicated that the labeled 63kDa PP is found in the high speed, supernatant fraction (S2). This is shown definitively in Fig. 3 (left panel).

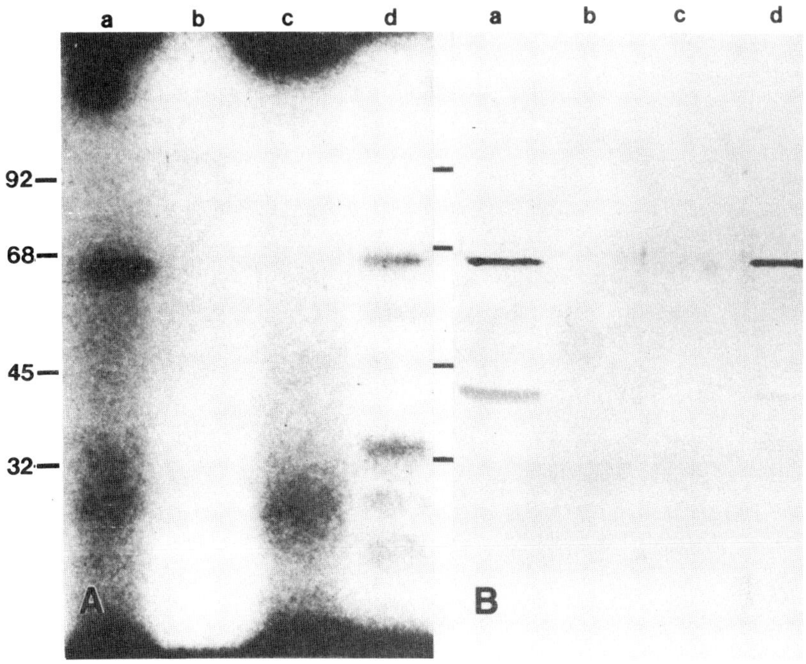

Fig. 3. Panel shows autoradiogram (panel A) and immunoblot (panel B) of SDS-PAGE separated Paramecium samples. Cells were labeled with $^{32}P_i$, homogenized and fractionated in the presence of EDTA. Lane a, homogenate; lane b, low-speed pellet (P1); lane c, high-speed pellet (P2); lane d, high-speed supernatant (S2).

The labeled 63kDa PP of whole cell homogenates, found almost exclusively in the S2, is recognized both in the homogenate and in the S2 in immunoblots with our affinity purified antibody (Fig. 3; right panel). No other labeled species is so recognized. The antibody also recognizes an unlabeled polypeptide at ca. 40kDa in these fractions, which may be a breakdown product of the 63kDa protein.

The dephosphorylation of the 63kDa protein that accompanies membrane fusion in trichocyst exocytosis _in vivo_ requires the presence of Ca^{2+}. We repeated the fractionation experiment subsituting 5mM Ca^{2+} for EDTA in the homogenization buffer, effectively lowering pCa by several orders of magnitude. The labeled 63kDa PP of the homogenate still is localized specifically to the S2 fraction (Fig. 4; left panel). However, the

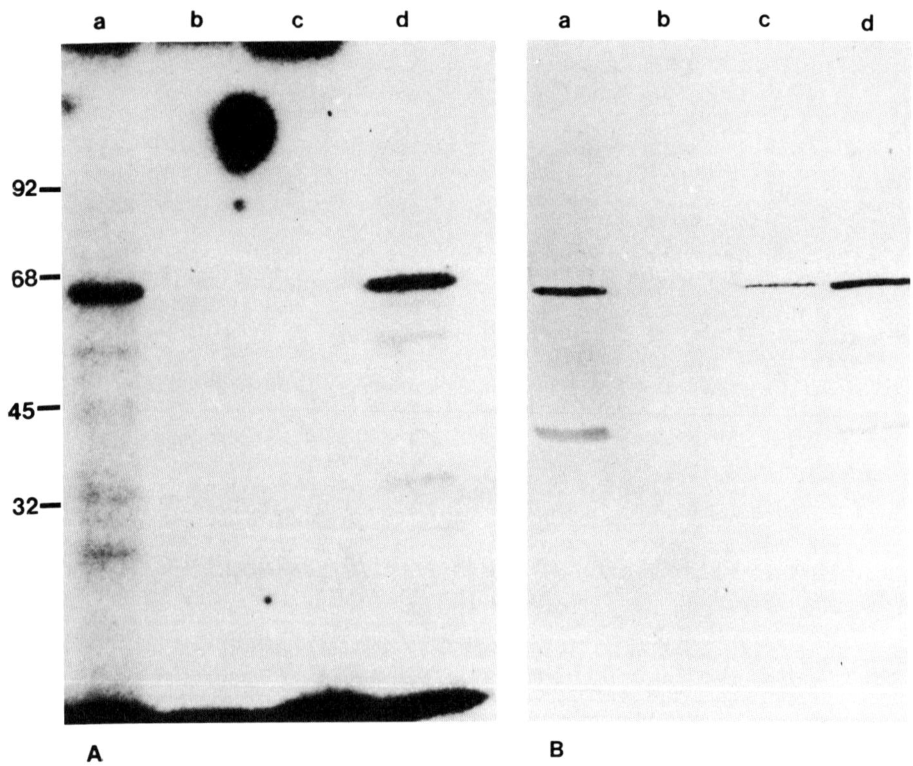

Fig. 4. Panel shows autoradiogram (panel A) and immunoblot (panel B) of SDS-PAGE separated _Paramecium_ samples. Cells were labeled with $^{32}P_i$, homogenized and fractionated in the presence of 5 mM Ca^{2+} and no EDTA. Lane a, homogenate; lane b, low-speed pellet (P1); lane c, high-speed pellet (P2); lane d, high-speed supernatant (S2).

immunoblot of this preparation (Fig. 4; right panel) now shows that an unlabeled, and presumably dephosphorylated form of the 63kDa PP is now present in the high speed pellet (P2).

This may represent 63kDa PP associated with the microsomal fraction. In addition, it has been shown that upon deciliation of Paramecium the ^{32}P labeled 63kDa PP is released from the cells; however, using Western immunoblot analysis some unlabeled 63kDa PP remains associated with the deciliated cell bodies (Hamasaki et al., 1987).

The 63kDa PP may be a cytoplasmic protein that lies between the secretory vesicle (trichocyst) and the plasma membrane. As pCa in this region is lowered, when stimulus sensitive Ca^{2+} channels in the plasma membrane open (Vuoso and Satir, 1987), this protein becomes dephosphorylated. The dephosphorylated protein now may bind to specific sites on the cell or vesicle membrane, possibly to facilitate the subsequent membrane fusion. A similar translocation has been shown for synapsin I, which is soluble when phosphorylated, but can bind to synaptic vesicles in its dephosphorylated form (Huttner et al., 1983).

Presence of the 63kDa protein in other species

Tetrahymena thermophila has a secretory system comparable to that of *Paramecium*, except that the secretory organelles release a mucus-like product, capable of forming a capsule around the organism (Tiedtke, 1976; Orias et al., 1983). When *Tetrahymena* are labeled *in vivo* with $^{32}P_i$, a phosphoprotein at ca. 62kDa prominently incorporates label. Fig. 5 shows that this band is dephosphorylated in a Ca^{2+} dependent manner when mucocyst secretion is induced by TNP.

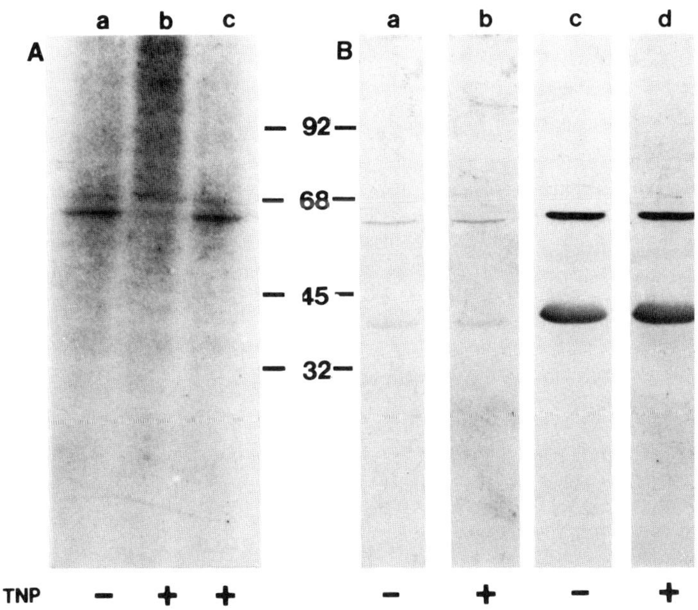

Fig. 5. Presence of 63kDa-like protein in *Tetrahymena*. Panel A: Autoradiogram of an SDS-PAGE of whole *Tetrahymena* cells, labeled with $^{32}P_i$. Lane a, unstimulated cells; lane b, cells stimulated with trinitrophenol (TNP) in the presence of extracellular Ca^{2+} (5mM; no Mg^{2+}); lane c, cells stimulated with TNP in the presence of extracellular Mg^{2+} (20mM; no Ca^{2+}). Panel B: Immunoblots of *Tetrahymena* cells (a,b) and *Paramecium* cells (c,d), either stimulated or unstimulated by TNP, all in the presence of extracellular Ca^{2+}.

The presence of high external concentrations of Mg^{2+} at the time of stimulation blocks both secretion and dephosphorylation. The labelled 62kDa phosphoprotein of *Tetrahymena* reacts with the affinity-purified antibody to the *Paramecium* 63kDa PP on immunoblots. In addition, the antibody binds to an unlabeled 38kDa band, that corresponds to the similar 40kDa band in *Paramecium*.

In both *Paramecium* and *Tetrahymena* Fig. 5 also demonstrates that despite the impressive loss of label that accompanies stimulation, the antibody still reacts with the 62-63kDa band without obvious change in the quantity of the polypeptide. Similarly, the binding to the 38-40kDa band is unchanged. Therefore, the secretory stimulus produces dephosphorylation without significant artifactual loss or degradation of the 63kDa phosphoprotein.

The presence of a 62kDa phosphoprotein in *Tetrahymena* that appears to be both antigenically and functionally related to the stimulus-sensitive 63kDa phosphoprotein in *Paramecium* suggests that a homologous class of proteins may be conserved during evolution. Preliminary evidence indicates that the antibody to the 63kDa PP identifies members of this class in a variety of metazoan organisms, including human tissue (Satir et al., 1987). Further work will be necessary to determine the degree to which these proteins are stucturally and functionally related.

References

Beisson J. and Rossignol, M., 1975, *in*: "Molecular Biology of Nucleocytoplasmic Relationships," Elsevier-North Holland Publishing Co., Amsterdam.
Beisson, J., Lefort-Tran, M. Pouphile, M., Rossignol, M. and Satir, B., 1976, J. Cell Biol. 69:126
Gilligan, D.M. and Satir, B.H., 1982, J. Biol. Chem. 257:13903
Hamasaki, T., Murtaugh, T.J., Satir, P. and Satir, B.H. 1987, (in preparation)
Huttner, W.B., Schiebler, W., Greengaard, P. and De Camilli, P. 1983, J. Cell Biol., 96:1374
Janish, R., 1972, J. Protozool., 19:470
Murtaugh, T.J., Gilligan, D.M. and Satir, B.H., 1987, J. Biol. Chem., (in press)
Orias, E., Flacks, M. and Satir, B.H., 1983, J. Cell. Sci., 64:49
Pollack, S., J. Protozool, 1974, 21(2):352
Satir, B., Schooley, C. and Satir, P., 1972, Nature London 235:53
Satir, B., 1974, Soc. Exp. Biol. Symp. XXVIII, 399:418
Satir, B. H., Busch, G., Vuoso, A., and Murtaugh, T.J., 1987, J. Cell. Biochem. (in press).
Tiedtke, A., 1976, Naturwissenschaffen, 63:93
Vuoso, A. and Satir, B.H., 1986, J. Cell Biol. 103(5):1714
Zieseniss, E., Plattner, H., 1985, J. Cell Biol. 101:2028

Acknowledgements

This work was supported by USPHS GMS 32767 to B.H. Satir. The authors thank Dr. P. Satir for valuable discussions and Marie Frees and Edna Horowitz for valuable secretarial assistance.

GEOMETRIC TOPOLOGY OF MEMBRANE FUSION: FROM SECRETION TO INTERCELLULAR JUNCTIONS

Pedro Pinto da Silva

Membrane Biology Section, Laboratory of Mathematical Biology, Bldg. 538, Room 104, National Cancer Institute Frederick Cancer Research Facility, Frederick, MD 21701

A. Topological invariance in an Eucaryotic Cell

The geometric topology of plasma membranes and of intracellular membranes defines the division of spaces that characterize the eucaryotic cell. This topology is invariant: for instance, mitochondria are composed of two topologically concentric membranes. So are nuclei, with the nuclear pores as an important added feature. Topological invariants are such constant features that we cannot, for instance, imagine a nuclear envelope positioned across a plasma membrane.

Many cellular processes involve non-elastic transformations and must, therefore, be mediated by topological changes. Often, these processes involve local changes in geometric topology that do not affect our general account of the system. For instance, the budding of vesicles from the transmost cisternae of the Golgi apparatus involves a topological change which does not lead to an altered description of the cell. Other processes must, however, be mediated by changes that involve radical -- even if transient -- changes of membrane topology. Secretion is one of these.

Any transition to a distinct topology must involve transformation that cannot be achieved by elastic means, i.e, without breaking the prevailing structural continuity[1,2]. Therefore, cellular processes must exist which effect the required topological transitions.

B. Secretion

Secretion is the transfer of cellular products to the exterior of the cell. These products may be part of the cytoplasm or may be stored in membrane delimited exoplasmic spaces (secretory vesicles). The first process can involve the separation and "pinch-off" of portions of the cytoplasm surrounded by plasma membrane ("apocrine secretion"). It happens, for instance, during secretion of milk fat globules and the budding of viruses. The process is relatively simple and does not result in the mixing of cytoplasmic and exoplasmic components. The "pinched off" membranes may remain sealed (at least for some time) and their interior can be considered as the equivalent of a "vacuole" of cytoplasm floating in an exoplasmic space.

The second process in trickier: it involves the establishment of a continuity of space between an exoplasmic space "inside" the cell, and the external environment. To be effective, this transfer must be achieved without the establishment of structural continuity with the endo- (or cyto-) plasm. This topological transition can only be mediated by catastrophic events[3]. We will see that these "catastrophes" are in fact well controlled and do indeed avoid the real catastrophe that would be plasma membrane rupture and cell lysis.

I will describe and analyze here a sequence of membrane events which Nogueira and I proposed in 1977[4,5] based on our observations of freeze-fracture and thin-section images of the massive release of secretory products which occurs during transformation of a fungal zoospore (Phytophthora palmivora) into a cyst. In these cells, for a very short period, numerous vacuoles can be induced to fuse with the plasma membrane and release the precursors of material needed to quickly fabricate a thick cell wall. The entire transformation of zoospore into a thick walled cyst takes no more than two minutes. I will not re-analyze here the complex arguments of morphological interpretation of freeze-fracture images. We proposed the following topological sequence for membrane fusion (Figs. 1 and 2):

1. <u>Approximation and structural apposition of membrane areas of the plasma membrane and secretory vesicle membrane</u>.
This is an obvious step (a minimum of contact is needed for kissing). This step involves no topological changes but makes them possible. From a non-topological point of view this stage is of enormous significance: it is contemporaneous with biochemical mechanisms that lead to fusion itself. This is the time for action of calcium-binding proteins, and for the initial interaction of components of the two membranes. This stage may also involve lateral displacement and reorganization of components in both membranes: for instance, some investigators have observed a clearing of intramembrane particles (integral membrane proteins, in general) from the area of the plasma membrane where fusion is due to occur (see, for instance ref 6).

2. <u>The first catastrophe: intermixing and restructuring of lipid and/or protein components at the site of fusion</u>.
The apposition of components from both membranes (together with the migration of cytoplasmic components to the region of fusion) makes possible the first topological change in the system. The mechanisms that elicit this change are probably varied and appear to involve lipids as well as proteins, as discussed in many of the contributions to this volume. From a topological point of view what is significant is that two separate, concentric membranes become a single system. We proposed that in secretory events this will lead to the establishment of a continuity between the cytoplasmic leaflet of the plasma membrane and that of the secretory vesicle membrane. The molecules at the site of fusion -- which played the central role in the catastrophic reorganization of the membrane system -- may now reorganize into an inverted membrane micelle that, if stable, may remain sequestered (but potentially mobile) within the apolar matrix. This micelle(s) can include both lipids and proteins. It is important to realize that the micelle itself was proposed not as an "engine" of the fusion process but as a consequence (or "fossil") of a catastrophe. Freeze-fracture observations showed spherical

cavities at the sites of fusion. They were interpreted to correspond to the emplacement of inverted micelles.

3. Formation of a bilayer diaphragm

The next event is the formation of a bilayered membrane diaphragm through the apposition and flow of membrane lipids (peripheral membrane proteins cannot be excluded) from the exoplasmic halves of both membranes. This structure abutts the first topological

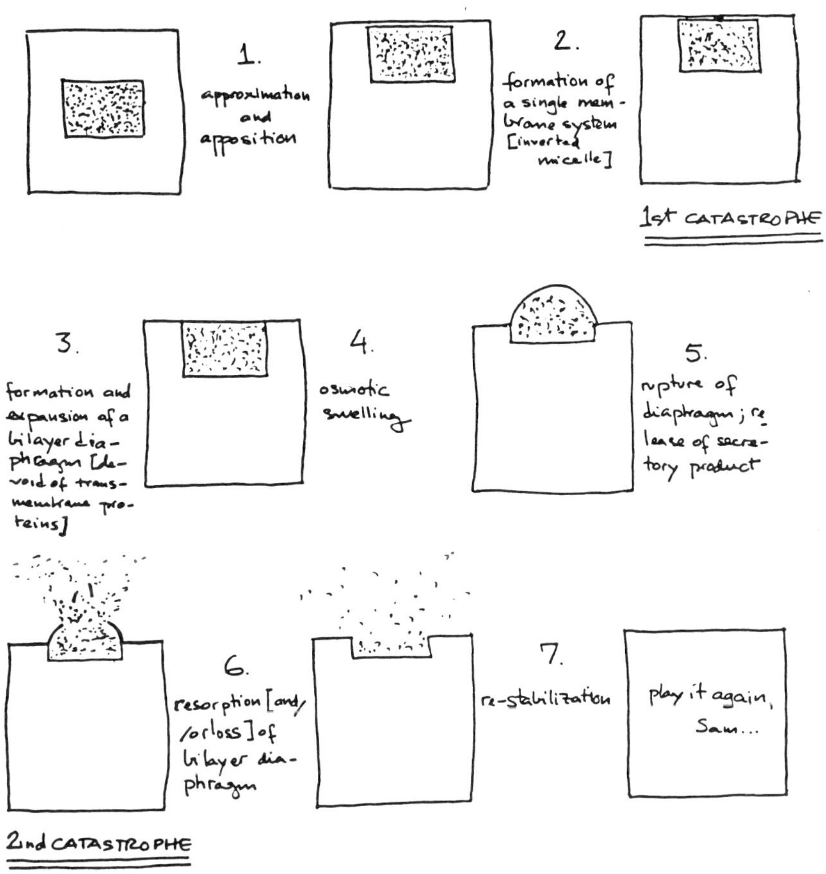

Figure 1. Topological sequence proposed for membrane fusion during secretion. This schematic line drawing is a topological diagram and represents the geometric surface formed by the apolar juncture of the bilayer continuum in membranes. Topological changes occur only in steps 2 and 5. Structurally (but not topologically) many phenotypically diverse aspects can result within this sequence. For instance: a) if secretory vesicle is small, the diaphragm (even if stable) will be small because its diameter is closely related to that of the vesicle; b) if the area of apposition is large and osmotic instability is manifested slowly, several fusion events might occur leading to multiple bilayer diaphragms and a tubular network of membranes (see Pinto da Silva and Nogueira, 1977); c) if osmotic instability is sudden, the bilayer diaphragm will rupture soon after its formation, so the diaphragm will never be large and will be rarely observed.

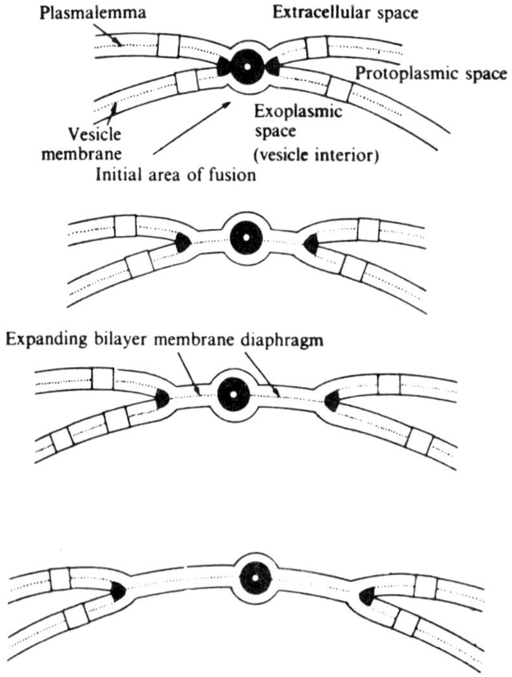

Figure 2. Diagram represents steps during membrane fusion leading to formation of bilayer membrane diaphragm. Dotted line follows hydrophobic juncture which is split during freeze-fracture. Heavily dotted areas represent postulated inverted membrane micelle (circular) and toroid hemi-micelle (wedges) (from Pinto da Silva and Nogueira, 1977). Lipids as well as peripheral membrane proteins (and also, putative integral membrane proteins associated with one membrane half) may flow from either the outer half of the plasma membrane or the inner (exoplasmic) half of the vesicle membrane into the bilayer membrane diaphragm. Membrane particles (squares) are thought to represent protein-containing intercalations, and cannot flow into the bilayer membrane diaphragm.

reorganization. It becomes clear that the contents of the secretory vesicle are now bound by an hybrid membrane: topologically, it is a lens facing both the cytoplasm and the external environment. Inspection of such a structure shows that transmembrane proteins from either the plasma membrane or the secretory vesicle membrane cannot migrate into the bilayer diaphragm: to do so, their polar regions at either surface would have to cross the apolar matrix of the membrane. In freeze-fracture specimens we have observed that this bilayer diaphragm is singularly devoid of particles, as predicted (Fig. 4). Thin sections show its structure as the trilaminar ("unit membrane") profile typical of biomembranes (Fig. 3). This structure appears to be in most cases short lived mainly, I believe, because it immediately leads to a series of events which abuts with the second catastrophe.

4. The second catastrophe: rupture of the bilayer diaphragm; release of secretory product.

The membrane system (a lens) created by the formation of the diaphragm is without precedent in the eucaryotic cell where topological concentricity is the norm. The ionic equilibrium within the lens must respond to two distinct ionic environments: the outside, from which it is separated by the bilayered diaphragm; and the cytoplasm, separated by the secretory vesicle membrane. Physiologically, it is plausible that the bilayer diaphragm is comparatively unresponding [it lacks transmembrane proteins]. Structurally, it is weaker [as it is originated from two exoplasmic membrane "halves" it lacks a membrane skeleton][4]. Thus, any osmotic swelling that may occur (see other contributions in this volume)

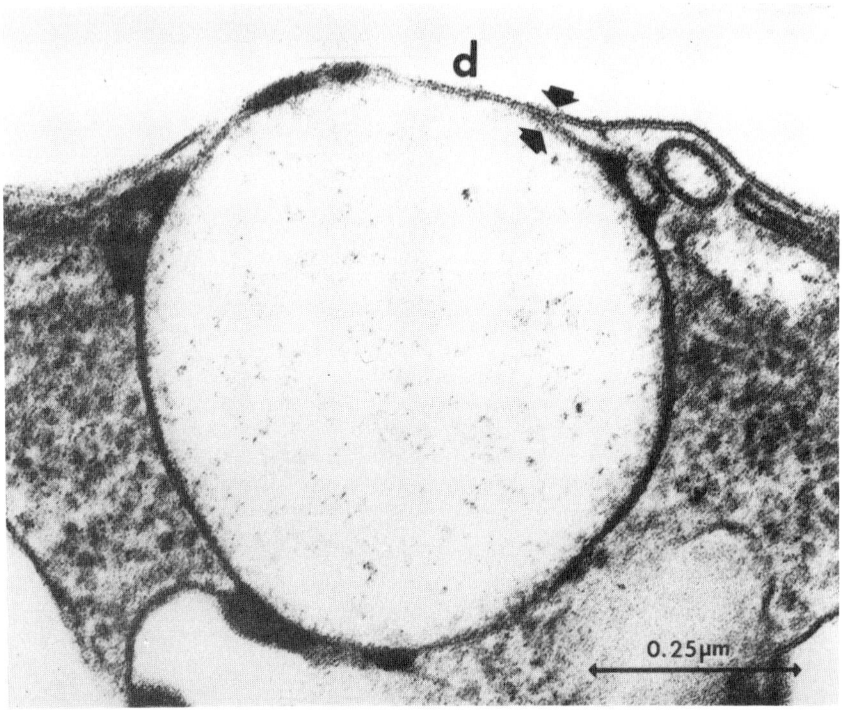

Figure 3. Peripheral vesicle fused with plasmalemma in a P. palmivora zoospore during secretion. Trilaminar profile of the bilayer membrane diaphragm is clear. Double arrows indicate region where the outer dense line of diaphragm appears continuous with the inner dense line of the peripheral vesicle membrane. Continuity between the inner dense line of the plasma membrane and the outer dense line of the peripheral vesicle membrane is suggested x 140,000 (from Pinto da Silva and Nogueira, 1977).

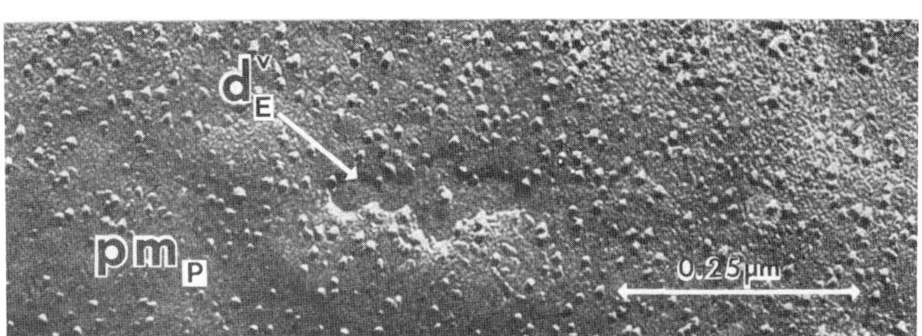

Figure 4. Fracture face P (pm_P) and a bilayer membrane diaphragm ($d_E{}^v$). Note presence of single particle in center of otherwise particle-free diaphragm x 135,000 (from Pinto da Silva and Nogueira, 1977).

will lead to tension, rupture, and partial resorption of the bilayer diaphragm. The secretory product is now released. This completes the transfer of a secretory product from an internal exoplasmic space to the outside without intermixing with the cytoplasm.

C. Structural variations within a single topological sequence

The sequence of events outlined above can manifest itself in many structurally distinct phenotypes. Several parameters influence this. I will discuss just a few.

1. Diameter of the secretory vesicle
The smaller the vesicle, the fewer events of fusion can take place. The area of close contact is clearly dependent on the size and also on the much sharper angle of curvature of smaller vesicles. A larger secretory vesicle may establish a large area of contact.

2. Surface of membrane apposition
This surface can be considered as a time dependent field with the probability of fusion dependent on local -- changing -- molecular conditions. If the osmotic disequilibrium introduced by the first fusion event (and first bilayer diaphragm) is sudden and violent rupture ensues, further fusion events become superfluous.

3. Bilayer diaphragm
This maybe (in fact, it appears to be) a short lived structure that bursts soon after the onset of the osmotic disequilibrium that it provokes. During encystment of the zoospores of *P. palmivora* its life may be longer, leading to the occurrence of multiple fusion sites and multiple diaphragms in a single vesicle[4]. In most other cases I believe that the diaphragm is a very short lived structure.

D. Intercellular junctions as metastable intermediates of membrane fusion

Clearly, fusion occurs through many and diverse mechanisms where proteins in some instances and, in others, lipids may play crucial roles in the provocation and explicitation of the first catastrophe. In addition, variations of some of the parameters outlined in the previous section may lead to phenotypically diverse structural sequences.

This sequence of fusion has also lead to the formulation of an alternative solution[7] for the organization of lipids and proteins in the *zonula occludens* (or "tight junction"), an intercellular specialization that selectively seals epithelia and is responsible for the definition of "outside" and "inside" environments in animal tissues. In my view the *zonula occludens* may represent a case of linear fusion between the apical regions of plasma membranes in epithelial cells. Freeze-fracture shows the tight junction as an anastomosed array of strands (Fig. 5). These strands were previously interpreted as the product of lateral cross-linkage (during glutaraldehyde fixation) of linear arrays of intercellular, transmembrane protein rivets. I proposed that, instead, the strands represent inverted cylindrical membrane micelles composed of lipids and proteins (Figs. 6 and 7). Basically, the situation is similar to the scheme of membrane fusion that I outlined above but in this case, linearly prolonged. The tight junction would, therefore, represent a membrane fusion intermediate stabilized at the end of the first catastrophe. The second catastrophe has no reason to occur because of the topological symmetry of the system. At

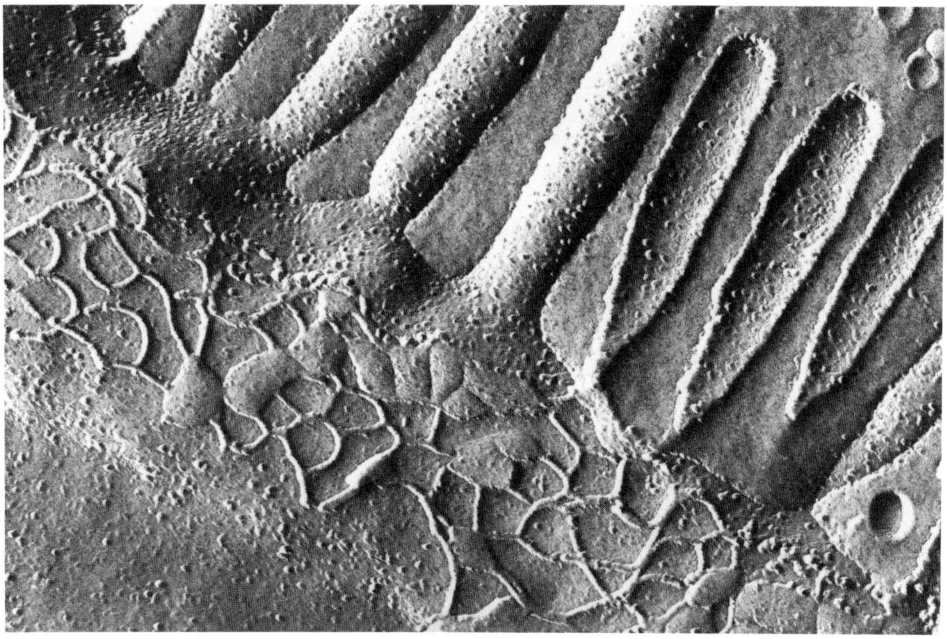

Figure 5. Tight junctions seal epithelia. In freeze-fractured cells the junctional elements appear as anastomosed strands at the apical portion of lateral membranes (epithelial cell from large intestine, courtesy of Dr. Daniel S. Friend) x 105,000.

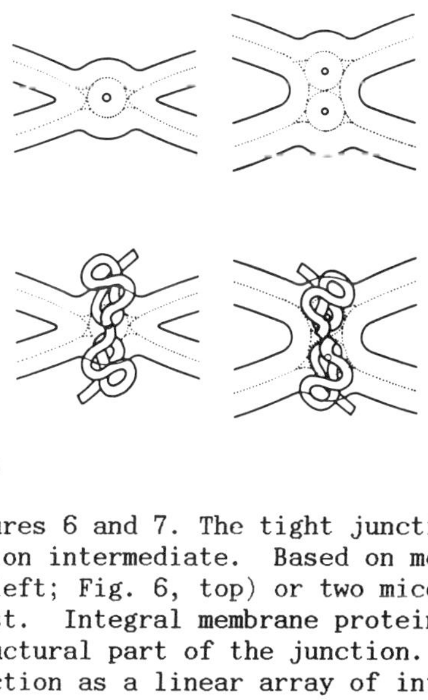

6

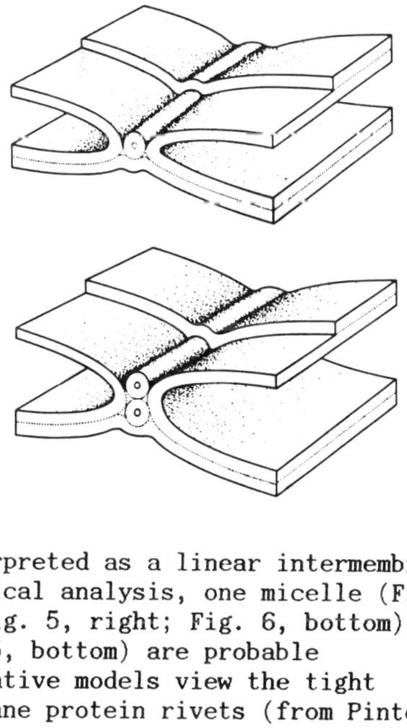

7

Figures 6 and 7. The tight junction interpreted as a linear intermembrane fusion intermediate. Based on morphological analysis, one micelle (Fig. 5, left; Fig. 6, top) or two micelles (Fig. 5, right; Fig. 6, bottom) may exist. Integral membrane protein (Fig. 5, bottom) are probable structural part of the junction. Alternative models view the tight junction as a linear array of intermembrane protein rivets (from Pinto da Silva and Kadar, 1982).

present, there is no consensus as to the structure of the tight junction. Both membrane proteins and lipids are expected to have roles (Fig. 5, bottom). The structure is physiologically dynamic, time dependent and is modulated in an almost endless variety of appearances in different epithelia. The complexity of the network of linear fusion appears related to permeability properties but the relationship is far from straightforward. It is likely that the junctional strand itself has varied structure and even topology: The inverted cylindrical micelle as well as the continuity of exoplasmic leaflets are not a physiological "must" and on the contrary, may represent an unfavorable topology in junctions that can, in many instances, function as a filter rather than as an occluding barrier to the passage of ions and small molecules.

The gap junction is another type of intercellular contact that may be considered under the light of membrane fusion. Here, the protein component is well characterized. The "connexon" comprises six protein subunits that form an intercellular channel[8]. The regulation of the intercellular traffic of ions and metabolites (and, even, small macromolecules) across this channel is thought to be mediated by allosteric changes of the protein. It is clear, however, that to be effective, the channel must have a competent - i.e., high resistance - lining over the portion that spans the intercellular space. While such high resistance properties might be secured by proteins alone, it is plausible that lipids play a significant, even crucial role. At one extreme of this view, the exoplasmic leaflets of both plasma membranes would be fused into a continuum around each connexon. Future research will clarify any role that membrane lipids may have in the biogenesis, structural maintenance and physiology of gap and of tight junctions.

E. Topology and structure: some comments on experimental morphology

The topological sequence outline here was published in 1977 and, while much adopted by others (in particular the structural intermediates of membrane fusion, such as the inverted membrane micelle), it has generally been misunderstood. It is a topological sequence, not a structural description. The sequence is based on my long held belief that the basic problems of membrane organization must be first treated topologically and only after can the search for structural solutions begin[9]. This led me to propose the distribution of polar/apolar regions in intramembrane particles (apolar residues distributed in a "torus" in register with the apolar matrix of the bilayer)[10]. Recent structural solutions for transmembrane proteins show that the topology proposed is correct. In a way that I fail to understand, topology and thermodynamics appear to go hand in hand. Attention to their guidelines could have avoided the acceptance of absurd models of membrane transport (tumbling carriers) and also of membrane fusion[11].

Over the past thirty years the electron microscope provided images that, we felt, brought us close to the visual reality of the microcosmos. Mere observation has resulted in efforts to perturb the biological systems as least as we know how. Yet, to study many of the dynamic cellular processes that involve biological membranes I believe that it is necessary to perturb and to provoke. In many cases, in particular during secretion, events are so asynchronous, infrequent, and short-lived that, even if they produce distinct morphological phenotypes they cannot be discerned against the background noise of the ultrastructural image (in particular, the freeze-fracture image). Visual images, even the most crisp and beautiful, contain misinformation and random noise. It becomes, therefore, necessary to experimentally perturb the system in order

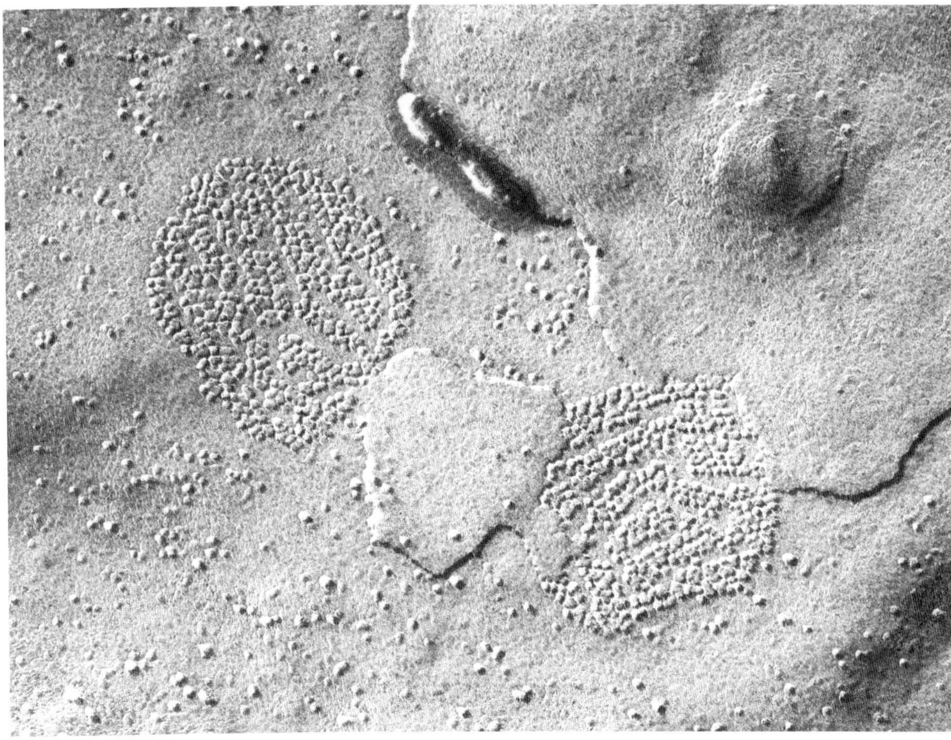

Figure 8. Three gap junctions stand out against a phenotypically undifferentiated region of plasma membrane of a fibroblast grown in tissue culture. Each subunit of a gap junction (a "connexon") contains a low-resistance intercellular channel. x 160,000. (From Pinto da Silva and Gilula, Exptl. Cell Res., Vol. 71, pp. 393 (1972).

to be able to resolve new signals in an intelligible pattern[12]. Membrane fusion is a dynamic, vectorial process, thus, inherently difficult to analyze though the observation of micrographs. Freeze-fracture electron microscopy cannot overcome these difficulties. However, in contrast to other visual and spectroscopic and biochemical approaches provides non-averaged views of the membranes during fusion. This, and the power to closely inspect a variety of organisms and physiologic circumstances makes it an essential partner in the multidisciplinary approach that has characterized the study of fusion of liposomes and of biomembranes.

Acknowledgements

I wish to thank Mr. Steven Samuelsson and the staff of the NCI-FCRF Central Clerical Pool for expert editorial assistance. I am also grateful to Dr. Daniel S. Friend, The Rockefeller University Press, the MIT Press and Academic Press for permission to reproduce work.

References

1. W. Lietzmann, "Visual Topology," Chatto & Windus Ltd., London (1962).
2. B. H. Arnold, "Intuitive concepts in elementary topology," Prentice-Hall Inc., Englewood Cliffs, New Jersey (1962).

3. R. Thom, "Stabilité structurelle et morphogénese. Essai d'une théorie générale des modéles," W. A. Benjamin, Inc., Reading, Massachussets (1972).
4. P. Pinto da Silva and M. L. Nogueira, Membrane fusion during secretion; a hypothesis based on electron microscopic observation of *Phytophthora palmivora* zoospores during encystment, *J. Cell. Biology* 73:161 (1977).
5. D. Shotton, Freeze-fracture view of membrane fusion, *Nature* 272:16 (1978).
6. E. Chi, D. Langunoff, and J. K. Koehler, Freeze-fracture study of mast cell secretion, *Proc. Natl. Acad. Sci. U.S.A.* 73:2823 (1976).
7. P. Pinto da Silva and B. Kachar, On tight junction structure, *Cell* 28:441 (1982).
8. P. T. N. Unwin and G. Zampighi, Structure of the junction between communicating cells, *Nature* 283:545 (1975).
9. P. Pinto da Silva, Topology, dynamics and molecular cytochemistry of integral membrane proteins: a freeze-fracture view, *in*: "Electron Microscopy of Proteins," J. R. Harris, ed., Vol. 6, pp. 1-38, Academic Press, Inc., London (1987).
10. P. Pinto da Silva, Translational mobility of the membrane intercalated particles of human erythrocyte ghosts: pH-dependent, reversible aggregation, *J. Cell Biol.* 53:777 (1972).
11. B. Satir, C. Schooley, and P. Satir, Membrane re-organization during secretion in *Tetrahymena*, *Nature (London)* 235:53 (1972).
12. P. Pinto da Silva and B. Kadar, Quick-freezing vs. chemical fixation: capture and identification of membrane fusion intermediates, *Cell Biol. Int. Rep.* 4:625 (1980).

'SLOW ARTIFACTS' IN ASSAYS OF LIPID MIXING BETWEEN MEMBRANES

John R. Silvius, Rania Leventis and Pamela M. Brown

Department of Biochemistry
McGill University
Montreal, Quebec, Canada H3G 1Y6

ABSTRACT

Many of the fluorescence artifacts that are encountered in assays of lipid mixing based on energy transfer develop very rapidly upon addition of the fusion-inducing agent(s). These 'rapid artifacts' often arise when the lipid probes employed are directly sensitive to the physical state of the lipid environment. In such cases, it is often possible to correct for the artifact by recalibrating the fluorescence levels corresponding to 0% and 100% lipid mixing, or by choosing an alternative set of lipid probes with lower environmental sensitivity. However, one can also encounter 'slow artifacts' that arise directly from the processes of vesicle aggregation and coalescence that are required to promote lipid mixing between vesicles. Using novel lipid probes with minimal environmental sensitivity, we have observed several types of 'slow artifacts' in different lipid vesicle systems. The mechanisms of these artifacts appear to include the following: (1) development of energy transfer between probes in apposed but unfused membranes; (2) enhancement of energy transfer between probes when labeled lipid vesicles collapse; (3) slow changes in probe fluorescence as perturbing agents in the aqueous phase enter lipid vesicles; (4) lateral segregation of lipid probes; (5) 'hemifusion' between membranes without true coalescence; and (6) differential rates of transfer of different probes between interacting bilayers. We describe examples of systems where these 'slow artifacts' can arise, how they can be detected and some possible strategies to eliminate or to correct for them.

Fluorescence assays of lipid probe mixing between membranes offer an attractive and relatively simple means to monitor the interactions between membranes that lead to intermixing of membrane lipids, including membrane fusion. The first such assays to be developed (1-3) were used mainly in systems in which at least one of the participating membranes was an artificial one, since the lipid probes employed could be incorporated only with difficulty into preformed, intact membranes. More recently, the range of these assays has been extended to include fusion events between two biological membranes by the use of fluorescent labels that can be incorporated into intact membranes (4). The relative simplicity and ease of application of these assays has made them very popular for use in a wide variety of systems (for a review, see [5]).

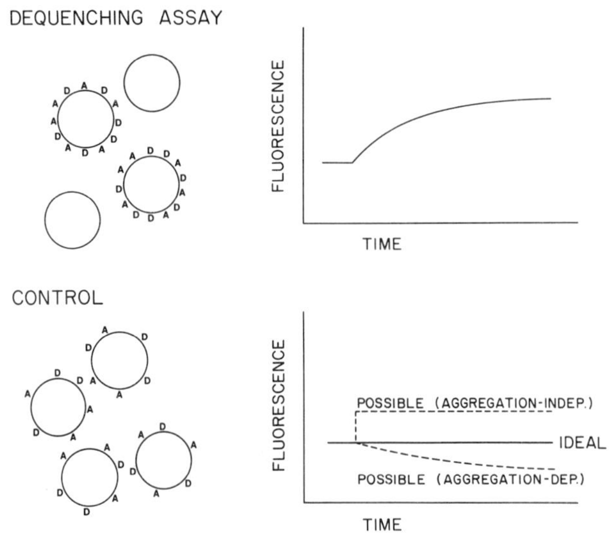

Figure 1. Schematic illustration of the design of a lipid mixing/dequenching (LMD) assay. In the actual assay, vesicles co-labeled with a fluorescent donor lipid (D) and an acceptor species (A) are incubated with unlabeled vesicles, and the time course of lipid mixing is monitored by fluorescence dequenching as shown at the upper right. In a parallel control experiment, labeled vesicles alone are incubated under the normal assay conditions but in the absence of unlabeled vesicles. The ideal behavior of such a control sample, and possible 'fast' (aggregation-dependent) and 'slow' (aggregation-independent) artifacts are illustrated at the lower right.

As assays of lipid probe mixing between membranes have been utilized in an increasing number of systems, it has become clear that these assays can encounter a variety of artifacts in some applications. In this article, we will discuss some of the artifacts that can arise in these assays, some guidelines to identify the types of systems in which particular artifacts are likely to arise, and some strategies to detect and to avoid such artifacts. We will concentrate our discussion on lipid-mixing assays that monitor resonance energy transfer (RET) between fluorescent lipid probes. Alternative types of assays have been proposed that measure probe fluorescence lifetimes or excimer formation to monitor lipid probe mixing; examples of such assays are discussed in other chapters in this volume.

RET-based assays of lipid mixing between membranes fall into two main classes. 'Lipid mixing/quenching' (LMQ) assays monitor the interactions between two populations of vesicles that are labeled separately with fluorescent donor and acceptor species (for an example, see [6]). 'Lipid mixing/dequenching' (LMD) assays monitor the interactions between two populations of membranes, one of which is co-labeled with donor and acceptor species and the other of which is unlabeled. In LMQ assays, lipid mixing leads to a quenching of the donor fluorescence, while in LMD assays, the fluorescence of the donor becomes enhanced (due to a reduced efficiency of energy transfer to the acceptor) during mixing. Because many of the common artifacts that can arise in these assays lead to a quenching of the donor fluorescence, LMD assays are often prefered in practice because they may be less prone to yield 'false positive' signals (7). In this article we

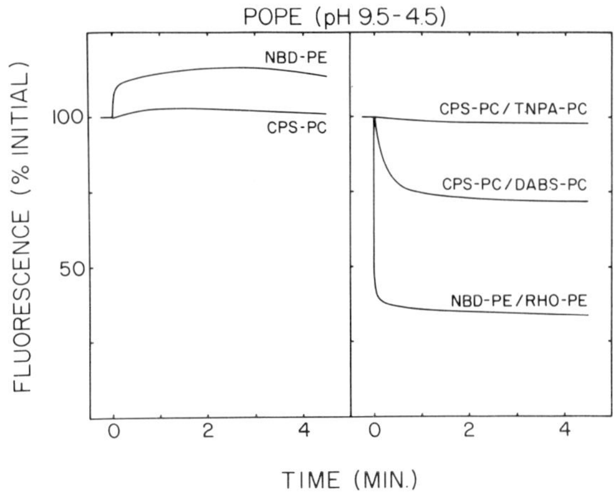

Figure 2. (A) Fluorescence changes observed when POPE vesicles, bath-sonicated at pH 9.5 and containing 0.5 mole % of either NBD-PE or (12-CPS)-18 PC, were abruptly exposed to pH 4.5 at time zero (temperature = 37 ^{o}C, lipid concentration = 30 μM).
(B) As in (A), but the vesicles were co-labeled with 0.4 mole % each of NBD-PE and Rho-PE, 0.5 mole % each of (12-CPS)-18 PC and (12-DABS)-18 PC, or 0.5 mole % of (12-CPS)-18 PC plus 1 mole % of (16-TNPA)-16 PC, as indicated.

will discuss primarily fluorescence artifacts that can influence the results obtained in assays of the LMD type.

The design of a typical LMD assay is summarized in Figure 1. A key control in this assay examines the effects of the fusogenic agent or stimulus on the fluorescence of the labeled membranes in the absence of any unlabeled membranes. Ideally, such controls should demonstrate that the fusogenic stimulus *per se* has no effect on the fluorescence of the labeled membranes. However, in practice it is not uncommon to observe that the fluorescence of these control samples changes significantly under the conditions that promote lipid mixing between membranes. As indicated schematically in Figure 1, these effects can be loosely grouped into two classes: very fast perturbations of the sample fluorescence, which can develop more quickly after application of the fusogenic stimulus than do membrane-membrane interactions; and more gradual changes in fluorescence that develop through membrane-membrane interactions but nonetheless do not reflect true lipid mixing between membranes. These effects exemplify what we will refer to as 'fast artifacts' and 'slow artifacts,' respectively, in lipid-mixing assays.

'Fast artifacts' often indicate that the fluorescence of the lipid probes is directly sensitive either to the fusogenic agent (*e.g.*, through direct binding of a fusogenic cation to a fluorescent phospholipid probe) or to a change in the physical properties of the probe environment in the presence of the fusogenic agent. A number of examples of such behavior have been reported (8 11), and one such example is illustrated in Figure 2A. NBD-PE, when incorporated into vesicles of 1-palmitoyl-2-oleoyl phosphatidylethanolamine (POPE) that are prepared at pH 9.5, shows fluorescence that is rapidly and substantially enhanced when the pH is decreased to 4.5 or 6.5. When similar vesicles are co-labeled with NBD-PE and Rho-PE, a reduction of the pH causes a strong and rapid *quenching* of the NBD-

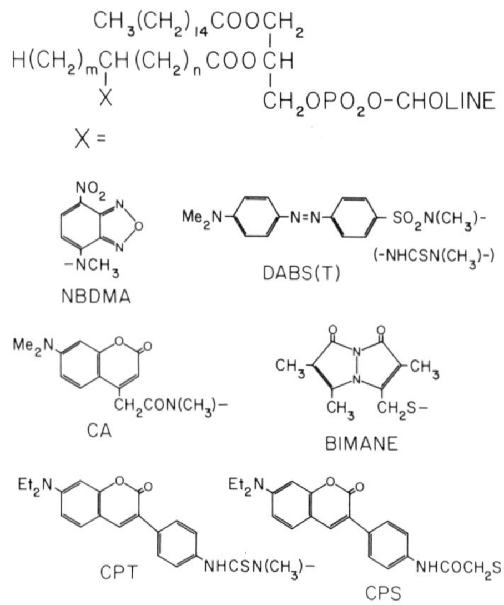

Figure 3. General structures of the novel fluorescent phosphatidylcholines employed in this study. For probes designated as (12-X)-18 PC, m = 6 and n = 10. Probes with headgroups other than phosphorylcholine are named as (12-X)-18 PE, PA, etc.

PE fluorescence (Figure 2B). In both cases, the fluorescence changes induced by the reduction in pH are faster than the rate of vesicle aggregation, a point that is demonstrated most clearly by the fact that the rates of the fluorescence changes remain unchanged even at very low vesicle concentrations, where aggregation rates are relatively slow.

In some cases, 'fast artifacts,' when they occur in lipid mixing assays, can be clearly resolved from slower fluorescence changes that arise from genuine lipid mixing. It is then possible to obtain useful information about rates of lipid mixing simply by recalibrating the fluorescence signals corresponding to 0% and 100% mixing in the assay system. Another approach to the problem of 'fast artifacts,' which can be the only feasible approach for some systems, is to employ an alternative set of lipid probes whose fluorescence shows less direct sensitivity to the physical state of the membrane environment.

In an effort to obtain lipid probes whose fluorescence shows minimal sensitivity to the properties of the local environment, we have synthesized and characterized a number of labeled phospholipid analogues, whose structures are shown in Figure 3. For the experiments that we describe in this paper, we have used labeled phospholipids whose fluorescent or chromophoric groups are attached to the 12-position of an eighteen-carbon (stearoyl) chain, which is linked to the 2-position of the glycerol backbone. These probes are abbreviated as (12-X)-18 PC (or PE, etc., depending on the probe headgroup), where X is the abbreviation for the fluorescent or chromophoric group. The absorption and emission maxima for these probes are summarized in Table I. The coumarin- and bimane-labeled probes can serve as energy-transfer donors in combination with NBDMA-, DABS- or DABT-labeled probes as acceptors. The efficiencies of energy transfer between (12-CPS)-18 PC and various acceptor species are summarized in Table II, which lists the concentrations of acceptors that quench the donor fluorescence by 50% when both species are incorporated into POPC vesicles. The same acceptor species quench the fluorescence of (12-Bimane)-18 PC with comparable efficiencies. The efficiencies of energy transfer between these

TABLE I

Excitation and Emission Maxima for Various Acyl Chain-labeled Phosphatidylcholines in Phosphatidylcholine Vesicles

Probe	Excitation Maximum (nm)	Emission Maximum (nm)
(12-CPS)-18 PC	390	473
(12-Bimane)-18 PC	394	471
(12-NBDMA)-18 PC	481	538
(12-DABS)-18 PC	447	-
(12-DABT)-18 PC	433	-

probes are comparable to or just slightly lower than those observed under similar conditions using the NBD-PE/Rho-PE donor-acceptor pair. As a result, the acyl chain-labeled probes can be usefully employed in lipid mixing assays using very low bilayer concentrations of the donor and acceptor species (<0.5 mole %).

As we have already noted, our prime reason for constructing the novel fluorescent phospholipids shown in Figure 3 was to identify probes whose fluorescence is not directly sensitive to changes in the physical properties of the membrane environment. Among the fluorescent donor species that we have prepared, the CPS- and CPT-labeled species show the least environmental sensitivity of their fluorescence. For example, as shown in Figure 2A, the fluorescence of (12-CPS)-18 PC, when the probe is incorporated into DOPE vesicles that are generated at pH 9.5, remains almost unchanged when the pH is abruptly reduced to 4.5. The fluorescence of this probe is also very stable to a number of other perturbations of its membrane environment, including gel-to-liquid-crystalline phase transitions and the disruption of the lipid bilayer by addition of an excess of Triton X-100 (not shown). The bimane-labeled phospholipids shown in Figure 3 also give a fairly constant fluorescence in lipid dispersions with varying physical properties, while the fluorescence intensity for the NBDMA-labeled lipids can vary by as much as twofold in different lipid systems.

Of the fluorescent phospholipids shown in Figure 3, the coumarin-labeled species have the best potential to minimize 'fast artifacts' when used as fluorescence donors in lipid-mixing assays, since the fluorescence of these species shows minimal sensitivity to abrupt changes in the physical properties of the probe environment. We have confirmed this expectation, using (12-CPS)-18 PC as an energy-transfer donor together with various acceptor probes, in several systems where the use of NBD-PE as an energy-transfer donor leads to substantial 'fast artifacts,' such as the system described in Figure 2. However, in some systems the removal of 'fast artifacts' may unmask another set of artifactual fluorescence changes that result from interactions between vesicles but which do not reflect true lipid mixing. An example of this phenomenon is presented in Figure 2B. As was already noted above, POPE vesicles that are labeled with NBD-PE and Rho-PE show a dramatic and very rapid decrease in fluorescence when the pH of the medium is reduced from 9.5 to 4.5, and this decrease is not correlated with vesicle-vesicle aggregation. When the experiment is repeated with vesicles labeled with (12-CPS)-18 PC and (12-DABS)-18 PC, a reduction in pH causes no immediate decrease in sample fluorescence, but a more gradual fluorescence decrease is observed instead, whose rate is limited by the rate of vesicle aggregation.

To explain (and thus hopefully to eliminate) 'slow artifacts' of the type just described for POPE vesicles labeled with (12-CPS)-18 PC and (12-

TABLE II

Bilayer Concentrations of Energy-Transfer Acceptors Giving 50% Quenching of (12-CPS)-18 PC or NBD-PE Fluorescence in Phosphatidylcholine Vesicles

Donor/Acceptor	Mole % Acceptor Giving 50% Quenching
(12-CPS)-18 PC/(12-DABS)-18 PC	0.34
(12-CPS)-18 PC/(12-NBDMA)-18 PC	0.31
(12-CPS)-18 PC/(16-TNPA)-16 PC	1.0
NBD-PE/Rho-PE	0.33

DABS)-18 PC, we must consider the possible mechanisms by which such artifacts could arise. In the case just described, we can use appropriate control experiments, including that shown in Figure 2A, to rule out trivial explanations such as a major shift in the spectral properties of either probe at the lower pH, a rapid precipitation of the vesicles out of the fluorimeter beam upon reduction of the pH (although this does occur at much longer times), etc. The most reasonable explanations remaining are that the reduction in pH causes either a lateral redistribution of the probes within individual vesicles or a large-scale aggregation of vesicles, allowing energy transfer to develop between probes in distinct but apposed bilayers. To decide between these possibilities, we carried out experiments similar to those described above, using POPE vesicles labeled with (12-CPS)-18 PC plus each of a variety of alternative energy-transfer acceptors. When (12-DABS)-18 PC was replaced by either (12-NBDMA)-18 PC, (12-DABT)-18 PC or (12-DABS)-18 PE in these experiments, the fluorescence response to a reduction in pH was almost identical to that observed for vesicles labeled with (12-DABS)-18 PC as the acceptor probe (not shown). These different acceptors, while they are structurally rather different, all quench the fluorescence of (12-CPS)-18 PC with very similar efficiencies at equal concentrations of acceptors in POPC or POPE vesicles. These results suggest that vesicle aggregation and bilayer apposition, rather than lateral redistribution of lipid probes, are responsible for the 'slow' fluorescence artifact seen for POPE vesicles labeled with (12-CPS)-18 PC and (12-DABS)-18 PC.

To test further the suggestion just presented, we prepared an alternative quencher of (12-CPS)-18 PC fluorescence, 1-palmitoyl-2-(16'-trinitrophenylaminopalmitoyl)-phosphatidylcholine ([16-TNPA]-16 PC). This probe is a substantially less efficient energy-transfer acceptor for (12-CPS)-18 PC than are the acceptors discussed above (see Table II). When POPE vesicles labeled with (12-CPS)-18 PC plus (16-TNPA)-16 PC are subjected to a reduction in pH from 9.5 to 4.5, almost no change in fluorescence is seen, even when the level of the acceptor species in the vesicles is adjusted so that the quenching of the donor fluorescence at pH 9.5 is comparable to that observed with the other fluorescence acceptors discussed above (Figure 2B). Our explanation for this result is as follows. The (12-CPS)-18 PC-/(16-TNPA)-16 PC energy-transfer pair has a significantly shorter characteristic energy-transfer length (R_0) than do the other energy-transfer pairs discussed above. As illustrated in Figure 4, when POPE bilayers become apposed at low pH, we thus expect that vesicles incorporating the TNPA-labeled acceptor will exhibit significantly less energy transfer _between_ bilayers than will vesicles incorporating the DABS-labeled acceptor, even when the acceptor levels in the two types of vesicles are adjusted so that the efficiencies of energy transfer _within_ individual bilayers are comparable (12,13).

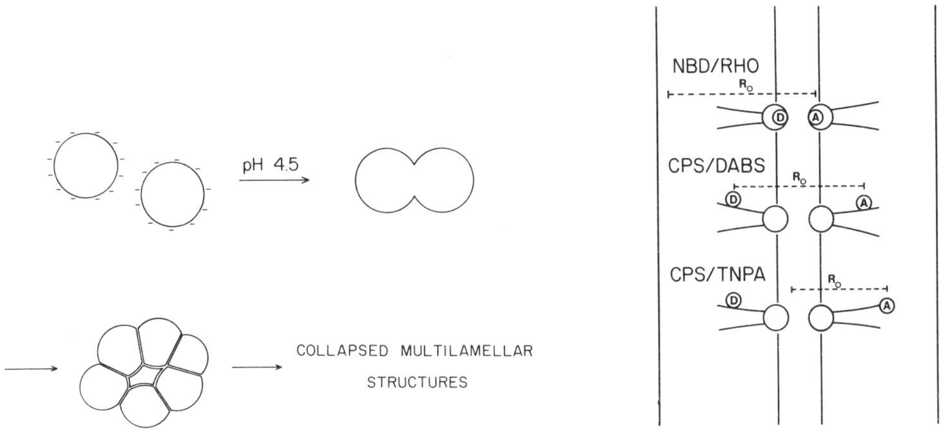

Figure 4. Left side: Schematic illustration of the process of vesicle coalescence and ultimate collapse when PE vesicles, prepared at alkaline pH, are exposed to pH 4.5.
Right side: Illustration of the possible separations between probe groups, and the Forster energy-transfer lengths, for three combinations of lipid probes (NBD-PE/Rho-PE, [12-CPS]-18 PC /[12-DABS]-18 PC and [12-CPS]-18 PC/[16-TNPA]-16 PC) in apposed but distinct PE bilayers at low pH.

The example just presented illustrates one potential source of 'slow artifacts' that can arise in lipid-mixing assays, namely the development of energy transfer between apposed but unfused bilayers. This artifact was among the first considered in lipid-mixing assays and was a prime motivation for the adoption of the lipid mixing/dequenching design for many such assays. From the example presented above, we can note that the presence of this artifact should be readily detectable, when studying interactions between membranes with very similar compositions, simply by noting the results of control experiments that examine the fluorescence behavior of labeled membranes alone under the conditions that promote membrane-membrane interactions. When such artifacts are observed and can be attributed to energy transfer between apposed membranes, two strategies are available to minimize them: either a high ratio of unlabeled to labeled vesicles can be used in an assay of the LMD type, or a donor-acceptor probe combination with a shorter energy-transfer length can be employed. The former strategy may not be successful in systems where vesicles can collapse, permitting different portions of the inner surfaces of individual vesicles to become apposed, on the time scale of the lipid-mixing assay.

A second type of 'slow artifact' that can complicate lipid-mixing assays is the possible lateral redistribution of lipid probes under the conditions used to promote interactions between membranes. We and others have shown previously that such redistributions of probes can occur on time scales comparable to those for lipid-mixing assays and can in fact be used to monitor these processes using appropriately chosen lipid probes (6,15). An example of probe lateral redistributions that can interfere with a lipid-mixing assay is given in Figure 5. As shown in Figure 5A, the fluorescence of (12-CPS)-18 PC incorporated into vesicles of bovine brain phosphatidylserine (PS) is only very modestly perturbed when the vesicles are exposed to 5 mM calcium, which induces a large-scale aggregation and fusion of vesicles, leading ultimately to formation of a highly ordered

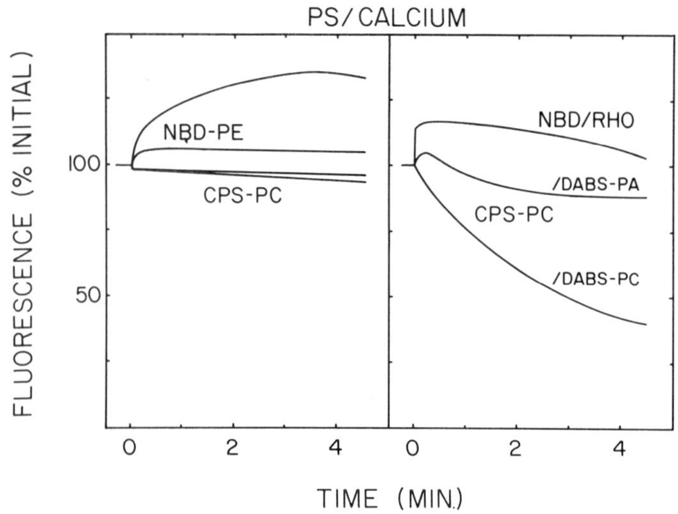

Figure 5. Left panel: Time courses of fluorescence observed when large unilamellar PS vesicles labeled with 0.5 mole % of either NBD-PE (upper two curves) or (12-CPS)-18 PC (lower two curves) were exposed to 2.5 or 5 mM calcium.
Right panel: 5 mM calcium was added to similar vesicles labeled with 0.5 mole % each of NBD-PE and Rho-PE, (12-CPS)-PC and (12-DABS)-18 PC, or (12-CPS)-18 PC and (12-DABS)-18 PA.

'cochleate' phase of the lipid (16,17). When (12-CPS)-18 PC is incorporated into the vesicles together with (12-DABS)-18 PC as an energy-transfer acceptor, the addition of calcium causes a large but gradual decrease in fluorescence. When the experiment is repeated using PS vesicles labeled with (12-CPS)-18 PC together with (12-DABS)-18 PA, which carries a phosphatidic acid headgroup, a much smaller quenching of the donor fluorescence is seen upon addition of calcium. This behavior appears to reflect the fact that the neutral phosphatidylcholine and the anionic phosphatidic acid probes adopt different lateral distributions in PS vesicles after the addition of calcium. Florine and Feigenson (17) have recently documented the differential solubilities of phospholipid probes with different headgroups in the highly ordered 'cochleate' phase formed by PS in the presence of calcium. Other workers have also observed the presence of significant fluorescence artifacts when lipid-mixing assays are used to examine the calcium-promoted coalescence of PS vesicles, and these artifacts have often been attributed at least in part to calcium-induced lateral redistributions of the probes employed (6,9,18). We have identified similar artifacts using fluorescent phosphatidylcholines as probes to examine the calcium-promoted interactions of phosphatidic acid - phosphatidylcholine vesicles (15,19 and unpublished results). Artifacts arising from this source may be reduced, if not eliminated, by using lipid probes that show more uniform partitioning among the different phases present after phase separation, or by reversing the phase separation (e.g., by chelation of calcium in the above examples) before attempting to quantify the extent of lipid mixing. The latter strategy requires the assumption, of course, that the treatment by which the phase separation is reversed does not itself alter the extent of lipid intermixing between membranes.

Another type of 'slow artifact' can arise in some systems when the fluorescence of the lipid probes employed is sensitive to an agent that only slowly gains access to some parts of the system during the course of the probe-mixing assay. One prime example of this type of 'slow artifact'

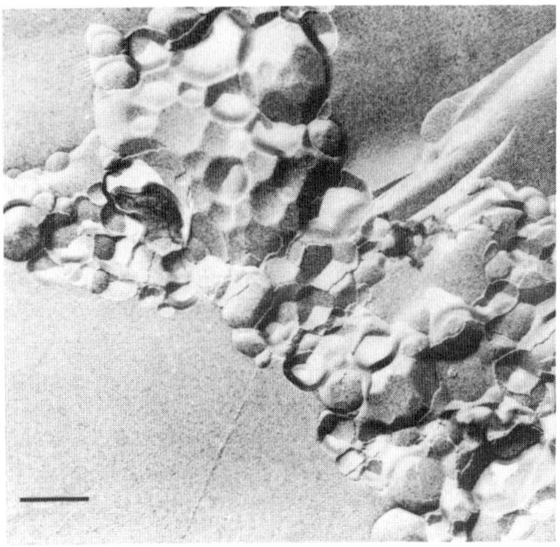

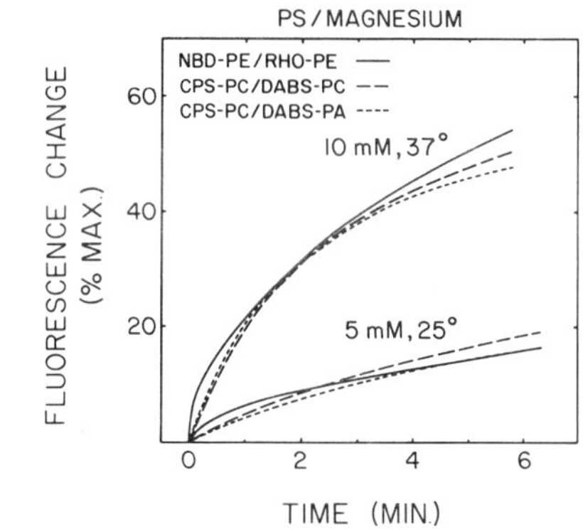

Figure 6. Freeze-fracture electron micrograph of vesicles composed of equimolar dioleoyl PC and dioleoyl PA in the presence of 10 mM magnesium. Vesicles were incubated for one minute at 25 °C in the presence of magnesium before spray-freezing. Bar = 200 nm.

Figure 7. Time courses of lipid probe mixing observed between PS vesicles at the indicated temperatures and magnesium concentrations. The time courses shown were obtained using three different combinations of lipid probes, as indicated: NBD-PE/Rho-PE, (12-CPS)-18 PC/(12-DABS)-18 PC, and (12-CPS)-18 PC/(12-DABS)-18 PA.

arises when an ion used to trigger a fusion process (Ca^{2+}, H^+, etc.) perturbs the fluorescence of the probes used and cannot permeate very rapidly through the labeled membranes. In such cases, the addition of the fusion-triggering ion to the sample may rapidly perturb the fluorescence of the lipid probes present at the outer faces of the labeled membranes, but the fluorescence of the probes located on the inner surface may change more slowly, as the ion gradually enters into the vesicle interiors. Since in many systems interactions between membranes can lead to a transitory increase in permeability as well as to fusion (5,10,20,21), this type of artifact can be seen even in systems where the fluorescence-perturbing agent does not normally cross the membranes at significant rates. We have observed such artifacts in studies of fluorescent lipid probes in PA/PC vesicles (15), and we have some evidence that similar artifacts can arise when using NBD-PE and Rho-PE to assay proton-induced fusion of PE vesicles in the absence of agents, such as acetate, that allow rapid permeation of protons to the vesicle interiors (unpublished observations).

The types of 'slow artifacts' described thus far are most likely to be significant in systems where lipid mixing is accompanied by a massive reorganization of the vesicle structure, including processes such as vesicle collapse or lipid phase transitions. These types of artifacts, when present, can usually be detected by control experiments using only labeled membranes as already described. A second class of potential 'slow artifacts' is more likely to complicate lipid-mixing assays in systems where interactions between membranes do not culminate in such dramatic changes in the entire state of the system. These latter artifacts produce 'lipid mixing' signals that represent genuine mixing of lipid probes between membranes but that nonetheless do not faithfully reflect the true rate of membrane coalescence. Within this category we can distinguish two types of artifacts: 'hemifusion,' in which membranes intermix lipids only between their outer monolayers; and the possible selective transfer of probes between interacting bilayers at rates that do not generally reflect the overall rate of lipid mixing between membranes.

'Hemifusion' between membranes was first demonstrated by Ellens et.al. (21) in a study of divalent cation-mediated interactions between vesicles composed of phosphatidylethanolamine and cholesterol hemisuccinate. It was observed that under certain conditions the vesicles aggregated and showed substantial intermixing of lipid probes yet failed to show either mixing or leakage of aqueous contents, as would be expected during true fusion of the vesicles. Studies with several other systems have since produced similar findings (19,22,23). As one example, in the presence of magnesium, vesicles composed of equimolar phosphatidic acid and phosphatidylcholine aggregate massively and show roughly 50% intermixing of lipid probes between vesicles (19). However, virtually no intermixing or leakage of vesicle contents can be demonstrated under the same conditions. Electron-microscopic examination of the vesicles under these conditions (Figure 6) reveals that the vesicles are highly aggregated in the presence of magnesium, with very extensive apposition of the walls of neighboring vesicles, but do not appear to have lost or intermixed any significant fraction of their internal contents. Examples such as these demonstrate that positive results in lipid-mixing assays must not be interpreted as evidence of full coalescence of interacting vesicles in the absence of corroborative evidence from other techniques (e.g., contents-mixing assays).

A second 'hidden' artifact that can complicate the interpretation of results obtained from lipid-mixing assays is the possibility that the probe(s) employed may be able to transfer between apposed membranes at a rate that is not representative of the overall rate of mixing of lipids between the two membranes. It is usually assumed in lipid-mixing assays that the rate of transfer of lipid probes between membranes is governed by the rate of bulk transfer of lipids at points of continuity that are established between membrane surfaces. While it is common to demonstrate that the lipid probes used in lipid-mixing assays do not exchange spontaneously between aggregated but unfused membranes (2-4,6), few studies have compared how different fluorescent probes report the rate of lipid mixing in a system where probe transfer between membranes does occur. We have examined this question in two systems where transfer of lipid probes between vesicles takes place but is much slower than the rate of vesicle aggregation, a situation in which we expect to be able to detect differences in the rates of transfer of different probes between membranes most readily.

Our first experiments of the type just described examined the mixing of lipid probes between PS vesicles in the presence of magnesium. As shown in Figure 7, LMD assays of this process under two different sets of conditions report very similar lipid-mixing rates when either the NBD-PE/Rho-PE, the (12-CPS)-18 PC/(12-DABS)-18 PC or the (12-CPS)-18 PC/(12-DABS)-18 PA energy-transfer pair are used as probes. This result strongly suggests that in this system,all three sets of lipid probes are transferred between labeled and unlabeled vesicles by a 'bulk' mechanism that transfers lipids between the vesicles without any notable discrimination among the different lipids present. In such cases, it is clearly justifiable to use measurements of rates of lipid probe transfer to estimate the rates at which lipids generally are transferred between interacting membranes.

We have also compared the rates of transfer of different lipid probes between lipid vesicles in a second, more unusual system. Large unilamellar vesicles with opposite surface charges can be prepared using mixtures of PE and PC together with low mole fractions (10-20 mole %) of either PS or a cationic lipid analogue, 1,2-dioleoyl-3-(trimethylammonium)-1,2-propanediol (DOTAP). In buffers of physiological ionic strength, we observe a gradual transfer of lipid probes from labeled vesicles with one type of surface charge to unlabeled vesicles of the opposite surface charge. Our findings to date suggest that the anionic probes NBD-PE and Rho-PE tend to transfer more rapidly from negatively to positively charged vesicles, and more slowly in the opposite direction, than do the zwitterionic probes (12-CPS)-18 PC and (12-DABS)-18 PC under the same conditions. These findings raise some question as to whether the rates of mixing of particular lipid probes between two populations of membranes can always be assumed to represent faithfully the rates of 'bulk' lipid mixing between membranes. Further comparisons of the rates of transfer of different lipid probes between membranes in a variety of systems will be useful to clarify this point.

A recitation of the potential artifacts that can arise in lipid-mixing assays may tempt an experimenter either to avoid the use of such assays or simply to ignore the possibility of such artifacts on the grounds that the results of a lipid-mixing assay will usually be 'roughly correct.' In fact, however, lipid-mixing assays based on measurements of resonance energy transfer can be used to give quantitative information about the extent and the kinetics of membrane coalescence in a wide variety of systems. Potential artifacts can often be detected by standard control experiments, particularly in assays of the LMD type, where measurements of the fluorescence of labeled vesicles alone under fusion-promoting conditions can signal the presence or absence of many common types of artifacts. When such artifacts are detected, the application of suitable corrections may require a knowledge of the nature of the process(es) responsible for the artifacts. To gain such information, and in favorable cases to reduce or to eliminate the artifact, it may be useful to employ an alternative, structurally different set of lipid probes to examine the same lipid-mixing process. Comparisons of results obtained using different lipid probes in the same system can also be useful to ensure that a particular probe does not undergo intermembrane transfer at an anomalously slow or rapid rate in the system of interest, a potential source of artifacts that has been frequently suggested but less often tested in lipid-mixing assays to date. The effort that these control experiments may require in some cases can often be rewarded not only by an increased confidence in the results obtained in a lipid-mixing assay, but also by a deeper understanding of the process of membrane coalescence in the system under study.

Acknowlegements

This research was supported by grants to J.R.S. from the Medical Research Council of Canada and les Fonds F.C.A.R. du Quebec.

References

(1) Vanderwerf, P. and Ullman,E.F. (1980) B.B.A. 596, 302.
(2) Uster, P.S. and Deamer, D.W. (1981) Arch.Biochem.Biophys. 209, 385.
(3) Struck, D.K. et.al. (1981) Biochemistry 20, 4093.
(4) Hoekstra, D.et.al. (1984) Biochemistry 23, 5675.
(5) Duzgunes, N. and Bentz, J. (1987) in Spectroscopic Membrane Probes (L.M. Loew. ed.), CRC Press, in press.
(6) Hoekstra, D. (1982) Biochemistry 21, 2833.
(7) Allen, T.M. and Duzgunes, N. (1985) Biophys.J. 47, 169a.
(8) Morris, S.J. et.al. (1985) J.Biol.Chem. 260, 4122.
(9) Parente, R.A. and Lentz, B.R. (1986) Biochemistry 25, 1021.
(10) Ellens, H. et.al. (1986) Biochemistry 25, 258.
(11) Silvius, J.R. et.al. (1987) Biochemistry 26, in press.
(12) Gibson, G.A. and Loew, L.M. (1979) B.B.R.C. 88, 135.
(13) Gibson, G.A. and Loew, L.M. (1979) B.B.R.C. 88, 141.

(14) Wolber, P.K. and Hudson, B.S. (1979) Biophys.J. 28, 197.
(15) Graham, I. et.al. (1985) Biochemistry 24, 7123.
(16) Portis, A. et.al. (1979) Biochemistry 18, 780.
(17) Florine, K.I. and Feigenson, G.W. (1987) Biochemistry 26, 1757.
(18) Hoekstra, D. (1982) Biochemistry 21, 1055.
(19) Leventis, R. et.al. (1986) Biochemistry 25, 6978.
(20) Wilschut, J. et.al. (1980) Biochemistry 19, 6011.
(21) Ellens, H. et.al. (1985) Biochemistry 24, 3099.
(22) Pryor, C. et.al. (1985) Biochemistry 24, 2203.
(23) Rosenberg, J. et.al. (1983) B.B.A. 735, 173.

WHY FUSION ASSAYS DISAGREE

Nejat Düzgüneş[a,b], Theresa M. Allen[d], Jane Fedor[a]
and Demetrios Papahadjopoulos[a,c]

Cancer Research Institute[a] and Departments of Pharmaceutical Chemistry[b] and Pharmacology[c], University of California, San Francisco, CA 94143-0128, U.S.A., and Department of Pharmacology[d], University of Alberta, Edmonton, Alberta, Canada T6G 2H7

INTRODUCTION

Membrane fusion can be viewed as a process involving several stages (Nir et al., 1983; Düzgüneş and Bentz, 1987):

(i) Adhesion or aggregation of the membranes;
(ii) Molecular contact between the membranes;
(iii) Membrane destabilization in the area of contact;
(iv) Intermixing of membrane components;
(v) Coalescence of the internal aqueous contents of the compartments bounded by the membranes.

A number of fluorescence assays have been developed that monitor these processes. It is important to ascertain which stages the assays indeed monitor. Membrane fusion, in our operational definition, necessarily includes step (v), and thus the most reliable assays of membrane fusion are those monitoring the intermixing of aqueous contents. If the aqueous markers used for the fluorescence assay leak into the medium before actual membrane fusion, the lack of a fluorescence signal may be misleading. In such instances membrane fusion could be monitored by the use of lipid mixing assays, or, ideally, assays could be developed that utilized macromolecules too large to readily leak into the medium. In certain cases, destabilization may result in the formation of a trilaminar (i.e. a single bilayer) diaphragm between the two compartments, causing intermixing of membrane components without the intermixing of contents (Palade, 1975; Neher, 1974; Pinto da Silva and Nogueira, 1977; Melikyan et al., 1983), in a process that has been called fusion (Palade, 1975). We have considered this process as semi-fusion (Düzgüneş, 1985; Düzgüneş and Bentz, 1987). Destabilization may also lead to the formation of the hexagonal phase, with extensive leakage of contents and lipid mixing, but without the coalescence of internal contents (Ellens et al., 1986).

FLUORESCENCE ASSAYS FOR MEMBRANE FUSION

Fluorescence assays monitoring the intermixing of aqueous contents include the Tb/DPA (terbium/dipicolinic acid) assay (Wilschut et al., 1980), and the ANTS/DPX (aminonaphthalene-trisulfonate/xylylene-bis-pyridinium bromide) assay (Ellens et al., 1985). The most widely used assay for the intermixing of lipids is based on resonance energy transfer (RET) between nitrobenzoxadiazol (NBD) and rhodamine (Rh) covalently attached to the head-group of phosphatidylethanolamine (PE). (Struck et al., 1981). Since RET depends on the proximity of the energy donor (NBD) and energy receptor (Rh), the change in surface density of the probes during membrane fusion can be monitored as a change in fluorescence intensity.

There are two versions of the lipid mixing assay:

(i) Probe dilution: Here both probes are incorporated in one population of vesicles ("labeled" vesicles), and their dilution into "unlabeled" vesicles is monitored as the increase in NBD fluorescence, resulting from decreased RET from the donor to the acceptor (Struck et al., 1981; Rosenberg et al., 1983; Eidelman et al., 1984: Ababei and Hildenbrand, 1984; Düzgüneş et al., 1985; Ellens et al., 1985; Wilschut et al., 1985a,b). Fluorophores such as NBD-PE or octadecyl-Rh can also be incorporated individually in one population of vesicles at a self-quenching surface concentration, and fusion can be monitored by the relief of self quenching as the probe is diluted into unlabeled vesicles (Hoekstra et al., 1984; Citovsky and Loyter, 1985). The dilution of pyrene-labeled phospholipids into unlabeled vesicles, monitored as the reduction of the excimer/monomer emission intensity, has also been used as a fusion assay (Amselem et al., 1986; Morris et al., 1987).

(ii) Probe mixing: In this version of the assay each probe is placed in a separate population of vesicles, and the quenching of donor fluorescence (or increase of acceptor fluorescence) is monitored (Hoekstra, 1982a,b; Wilschut et al., 1983; Silvius and Gagne, 1984a,b; Morris et al., 1985): RET pairs other than NBD and Rh have also been used for probe-mixing assays (Gibson and Loew, 1979a; Vanderwerf and Ullman, 1980; Uster and Deamer, 1981).

THE IDEAL FUSION ASSAY

We consider the following criteria to be essential for the reliability of fusion assays monitoring the intermixing of aqueous contents (Düzgüneş and Bentz, 1987):

(a) The fluorophores and their adjuvants (i.e. the component of the assay which interacts with the fluorophore to either generate or quench fluorescence) should not bind to the membranes.

(b) The fluorescence reactions must be faster than the fusion reaction.

(c) The encapsulated aqueous space markers must not leak before the fusion reaction; leakage of the markers during fusion must reflect the leakage of the encapsulated volume.

(d) The fluorophore and adjuvant should not interact outside the membrane vesicles after leakage into the medium; i.e. a quencher must be present in the medium.

Lipid mixing assays should meet the following criteria:

(a) The fluorescent probes should be inserted stably into the membrane and should not transfer between membranes through the aqueous phase.

(b) The probes should not affect the kinetics of aggregation and fusion, and thus should be present at sufficiently low surface density.

(c) The probes should reflect the flow of the bulk phase lipid; they should not transfer between membranes during intermembrane contact if there is no mixing of the bulk phase lipid.

(d) The fluorescence quantum efficiency of the probes should not be altered by cation binding to the probes or the membrane, or by membrane fusion.

KINETICS OF PROBE MIXING AND PROBE DILUTION DURING MEMBRANE FUSION

We have recently investigated the relative kinetics of probe mixing and probe dilution in a number of phospholipid vesicle systems whose fusion characteristics have been described previously using the contents mixing assays (Düzgüneş et al., 1987). Here we present a summary of our observations.

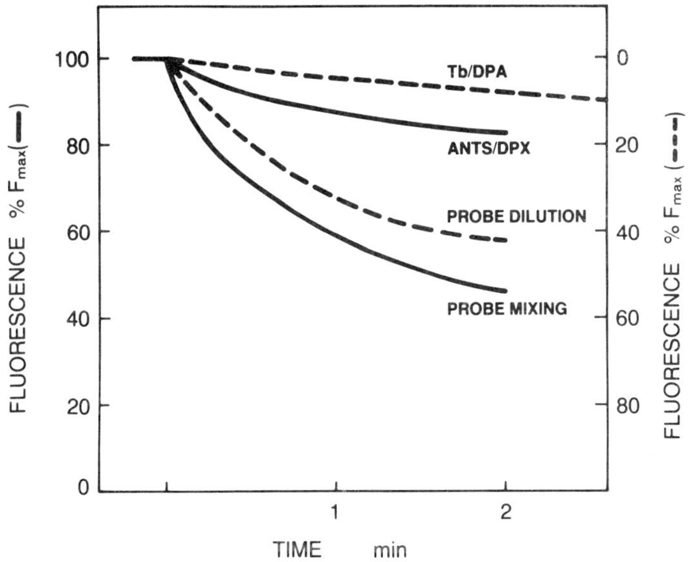

Figure 1. The time course of intermixing of contents and lipid mixing during fusion of LUV(PS) induced by 2 mM Ca^{2+}. Contents mixing was assayed by either the Tb/DPA or the ANTS/DPX assay. Lipid mixing was monitored by either the probe dilution or the probe mixing assay. The fluorescence scale for the solid lines is on the left and that for the dashed lines is on the right. Fusion results in an increase of the fluorescence in the Tb/DPA and probe dilution assays, and a decrease in the ANTS/DPX and probe mixing assays. From Düzgüneş et al. (1987).

Large unilamellar vesicles (LUV) used in these studies were prepared by reverse phase evaporation, followed by extrusion through polycarbonate membranes (Szoka et al., 1980; Düzgüneş et al., 1981, 1983). For the probe dilution assay, fluorescence-labeled vesicles were prepared with 0.6 mole % each of NBD-PE and Rh-PE in the membrane. (Düzgüneş et al., 1987). For determination of the kinetics of lipid mixing, the labeled vesicles were mixed with the unlabeled vesicles at a ratio of 1:1 and a total lipid concentration of 0.05 µmol/mL. The medium consisted of 100 mM NaCl, 5 mM TES, 0.1 mM EDTA, pH 7.4. Complete intermixing of all the bilayers upon fusion would be expected to result in a membrane containing 0.3 mole % each of the two fluorescent phospholipids. The fluorescence of such vesicles was set to 100%. The residual fluorescence of the labeled vesicles was taken as the zero level of fluorescence.

For the probe mixing method, NBD-PE was incorporated in one population of vesicles and Rh-PE in a second vesicle population, at 2 mole % each, and the vesicles were mixed at a 1:1 ratio (Hoekstra, 1982; Düzgüneş et al., 1987). Intermixing of the lipids of the two vesicle populations leads to RET between NBD and Rh and a decrease in the NBD fluorescence signal. NBD fluorescence measurements were made in an SLM-4000 fluorometer, using an excitation wavelength of 450 nm and an emission wavelength of 520 nm. Aggregation of the vesicles was monitored simultaneously in the second emission channel of the fluorometer, by using a Melles-Griot 450 nm band-pass filter.

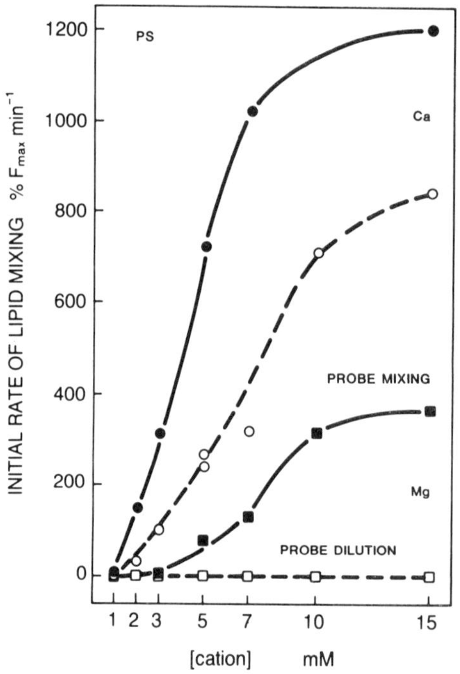

Figure 2. The initial rate of lipid mixing in LUV(PS) as a function of divalent cation concentration. Circles, Ca^{2+}; squares, Mg^{2+}. Filled symbols, probe mixing assay; open symbols, probe dilution assay. Temperature: 25° C. Aqueous medium: 100 mM NaCl, 5 mM TES, pH 7.4, 0.1 mM EDTA. Lipid concentration: 0.05 µmole/mL. From Düzgüneş et al. (1987).

The fusion of phosphatidylserine (PS) vesicles has been characterized well (Papahadjopoulos et al., 1979; Nir et al., 1983; Düzgüneş and Papahadjopoulos, 1983). This system was therefore used initially to compare the probe mixing and probe dilution assays. As shown in Figure 1, the fluorescence intensity changes more rapidly in the case of probe mixing than probe dilution when Ca^{2+} is added to PS vesicles. The initial rates of lipid mixing monitored by the two assays are plotted as a function of the Ca^{2+} or Mg^{2+} concentration in Figure 2. The rates obtained with the probe mixing assay in the presence of Ca^{2+} (filled circles) are higher than those obtained with the probe dilution assay (open circles). In the presence of Mg^{2+}, no lipid mixing is observed with the probe dilution assay (open squares). However, the probe mixing assay indicates rapid and extensive lipid mixing (filled squares). Earlier studies with the Tb/DPA assay and electron microscopy had revealed that Mg^{2+} induces the aggregation but not the fusion of LUV(PS) (Wilschut et al, 1981). The approach of two membranes closer than about 100 Å could result

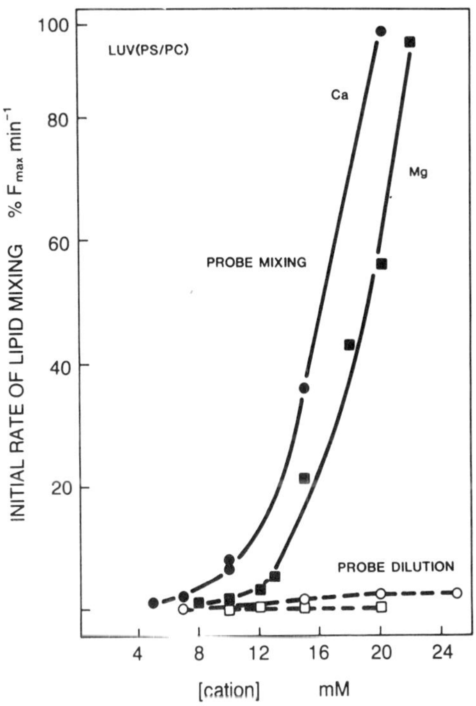

Figure 3. The initial rate of lipid mixing during aggregation of LUV(PS/PC; 1:1) induced by Ca^{2+} or Mg^{2+}. The description of the symbols is the same as in Figure 2. From Düzgüneş et al. (1987).

in RET between the donor and acceptor fluorophores (Gibson and Loew, 1979b). Thus, the probe mixing assay appears to monitor not only membrane fusion, as in the case of Ca^{2+}, but also membrane aggregation or adhesion. In the probe dilution assay, aggregation would not be expected to cause RET, since both probes are in the same membrane, unless probe transfer occurs during the close approach of the membranes. The absence of a change in fluorescence during Mg^{2+}-induced aggregation, indicates that probe transfer is not significant.

In the probe mixing assay, NBD fluorescence decreases as a result of RET to Rh-PE, when the fluorophores are in close proximity. If the proximity arises from interbilayer apposition, the fluorescence should return to the original (100%) level when the vesicles are de-aggregated. If RET occurs because of membrane mixing or probe transfer between the membranes, de-aggregation should not alter the level of fluorescence. In the case of Ca^{2+}, addition of EDTA arrests the fluorescence at the level obtained during fusion (Düzgüneş et al., 1987). Thus, the fluorescence intensity reflects the extent of probe intermixing during Ca^{2+} induced fusion of LUV(PS). In the case of Mg^{2+}, however, the fluorescence level is partially reversible upon EDTA addition. This observation suggests that the decrease in fluorescence signal in the presence of Mg^{2+} arises in part from RET between probes in apposed membranes, and in part from some lipid mixing or probe exchange between vesicles in the aggregated state.

The difference between the probe-mixing and probe-dilution methods is also apparent in the case of LUV composed of PS and phosphatidylcholine (PC), which aggregate but do not fuse in the presence of Ca^{2+} or Mg^{2+} (Düzgüneş et al, 1981; Uster and Deamer, 1981). Figure 3 shows that the probe mixing method is very sensitive to the aggregation of these vesicles, whereas the probe dilution method indicates a negligible rate of lipid mixing at high Ca^{2+} concentrations.

The Na^{+}-induced aggregation of small unilamellar vesicles (SUV) composed of PS is readily detected by the probe mixing assay, but not by the probe dilution assay (Düzgüneş et al., 1987). SUV are thought to aggregate in the primary minimum of the interaction energy vs. intermembrane distance curve, since the secondary minimum is not deep enough to sustain stable aggregation, while LUV may aggregate in the secondary minimum (Nir et al, 1981, 1983). The deduction is that probe molecules do not transfer from labeled to unlabeled vesicles even when the vesicles aggregate in a primary minimum.

EFFECTS OF ION-BINDING AND AGGREGATION ON FLUORESCENCE INTENSITY

The degree to which divalent cations affect the fluorescence of the probes used for the fusion assays, is important to determine for a correct interpretation of changes in fluorescence intensity as lipid mixing. When NBD-PE and Rh-PE are incorporated in separate populations of PC vesicles at 1 mole %, the addition of Ca^{2+} does not have any effect on the fluorescence (Düzgüneş et al., 1987). These data confirm earlier results obtained by Hoekstra (1982a), and indicate that changes in NBD fluorescence in the presence of Ca^{2+} are not due to the interaction of Ca^{2+} directly with the probe (cf. Morris et al., 1985; Parente and Lentz, 1986). The decrease in NBD fluorescence in the presence of Ca^{2+} reported by Parente and Lentz (1986) may be ascribed to the use of higher probe concentrations and of a lipid with saturated acyl chains.

When the probes are in PS membranes, addition of Ca^{2+} affects the fluorescence differently, depending on the concentration of probes. With vesicles containing 0.1 mole % of each probe, the fluorescence intensity increases considerably, whereas with vesicles containing 0.25 and 0.5 mole % probe the intensity decreases with time (Düzgüneş et al., 1987). Our interpretation of these observations is that Ca^{2+}-binding to PS increases NBD-PE fluorescence, but the presence of sufficiently high concentrations of Rh-PE in close proximity to NBD-PE reduces the fluorescence due to RET.

One of the factors which affect the fluorescence signal is vesicle aggregation. In the probe dilution assay, addition of Ca^{2+} or a mixture of Ca^{2+} and Mg^{2+} to LUV(PS) results first in an increase then in a decrease in fluorescence intensity (Rosenberg et al., 1983; Düzgüneş et al., 1987). When EDTA is added to stop the aggregation and fusion reactions, the fluorescence increases, suggesting that the decline in fluorescence was the result of massive aggregation and the reduction in the number of vesicles on which the light beam is incident.

KINETICS OF CONTENTS MIXING COMPARED TO LIPID MIXING

The relative kinetics of the coalescence of aqueous contents and intermixing of lipids during membrane fusion provides insights into the molecular mechanisms of the process. We have used both the Tb/DPA and the ANTS/DPX assays for these experiments.

LUV were prepared with the following encapsulated solutions: (i) 2.5 mM Tb chloride, 50 mM sodium citrate; (ii) 50 mM sodium DPA, 20 mM NaCl; (iii) 25 mM ANTS, with sufficient NaCl to adjust the osmolality to 190 mosM; (iv) 90 mM DPX. All solutions contained 5 mM TES and were adjusted to pH 7.4.

For the ANTS/DPX assay, an ANTS solution is entrapped in one vesicle population and a DPX solution in the other (Ellens et al., 1985; Düzgüneş et al., 1985). The quenching of ANTS fluorescence by DPX initially encapsulated in separate populations of vesicles, is a measure of the intermixing of aqueous contents. Fusion events between the two vesicle populations lead to a decrease in fluorescence. Release of contents from the liposomes and their dilution into the medium do not result in the decrease of fluorescence, since quenching by DPX is highly concentration-dependent. Leakage of quenched ANTS/DPX from liposomes leads to a fluorescence increase (Ellens et al., 1984). ANTS fluorescence (above 530 nm) was followed by means of a Corning 3-68 cut-off filter, with the excitation wavelength at 360 nm. Vesicles containing either ANTS or DPX were mixed at a 1:1 ratio, at a final lipid concentration of 0.05 µmol/mL in 100 mM NaCl, 5 mM TES, 0.1 mM EDTA. The initial fluorescence of the suspension was set at 100% maximum fluorescence, and the fluorescence of vesicles containing a 1:1 mixture of ANTS and DPX was taken as 0%.

In the Tb/DPA method, a solution of terbium citrate is entrapped in one population of vesicles and the Na salt of DPA is entrapped in a second vesicle population (Wilschut et al., 1980; Düzgüneş et al., 1981). Fusion events leading to intermixing of aqueous contents entrapped within the two vesicle populations result in an increase in fluorescence signal due to formation of a fluorescent Tb/DPA complex. Leakage of the Tb/DPA complex from liposomes (or entry of the medium into the internal aqueous space) leads to a decrease in fluorescence signal as a result of its interaction with Ca^{2+} and EDTA (Bentz et al., 1983; Bentz and Düzgüneş, 1985). For the Tb/DPA assay, Tb-containing and DPA-containing vesicles were mixed at a 1:1 ratio. The excitation wavelength was set at 276 nm and the emission at 545 nm was followed by using the Corning cut-off filter transmitting wavelengths above 530 nm. 100% Tb fluorescence was determined in the absence of EDTA after lysing 25 µM Tb-vesicles with 0.5% (w/v) sodium cholate or 0.8 mM octaethyleneglycoldodecyl ether in the presence of excess free DPA (20 µM).

The time-course of lipid mixing and of aqueous contents mixing during fusion of LUV(PS) induced by 2 mM Ca^{2+} is shown in Figure 1. The kinetics of both probe dilution and probe mixing are faster than that of contents mixing. The ANTS/DPX assay reveals consistently faster fusion rates than the Tb/DPA assay for PS vesicles. This difference could theoretically be due to several factors including the following: (i) different size distributions of the vesicle populations; (ii) different rates of reaction of the two components of the assay; and (iii) the different interaction of the encapsulated compounds with the vesicle membrane. Dynamic light scattering measurements indicate that the size distribution of the vesicles are very similar. Although the reaction mechanisms are different (chelation for the Tb/DPA reaction, and collisional quenching for the ANTS/DPX reaction), the rates of these reactions are very rapid, especially at the concentrations found in the vesicle interior, and thus would not be expected to contribute to the different rates of contents-mixing (Düzgüneş and Bentz, 1987). Although more ANTS/DPX binds to pre-formed SUV(PS) than Tb/DPA, binding of either complex to LUV(PS) is negligible (J. Bentz, D. Alford and N. Düzgüneş, in preparation). To test the possibility that the encapsulated compounds may interact with the membrane and thus affect the rate of destabilization and fusion, the probe-dilution assay was performed with vesicles encapsulating either Tb or ANTS. These compounds do not affect NBD fluorescence. Tb-containing vesicles exhibit slightly faster lipid-mixing rates than the ANTS-containing vesicles at 2 mM Ca^{2+}, but at higher concentrations the time-courses are indistinguishable (Düzgüneş et al., 1987). Thus, Tb and ANTS do not appear to interact with the membrane to an appreciable extent.

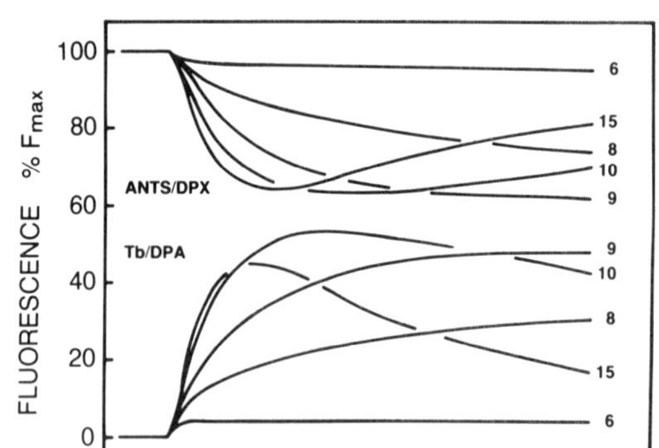

Figure 4. The time course of fusion of cardiolipin/dioleoyl PC (1:1) vesicles in the presence of various concentrations of Ca^{2+}, monitored by the ANTS/DPX or the Tb/DPA assay. Temperature: 25° C. From Düzgüneş et al. (1987).

With another vesicle fusion system, cardiolipin/dioleoyl PC LUV in the presence of Ca^{2+}, the kinetics of fusion monitored by the Tb/DPA assay are faster than that observed with the ANTS/DPX assay (Figure 4). The probe dilution and the Tb/DPA assays yield identical aggregation and fusion rate constants for this system at 25° C (Wilschut et al., 1985a). At 37° C where lipid mixing is faster than contents mixing, the difference between the two contents mixing assays is still observed. The difference is consistent over a range of Ca^{2+} concentrations, and also with 10 mM Mg^{2+}. The rate and time-course of leakage (or dissociation) of Tb/DPA or of ANTS/DPX are similar, indicating that the difference between the fusion assays could not be attributed to differences in leakage rates. It is possible that Tb citrate or DPX interact to a limited extent, but differentially, with the two kinds of negatively charged lipid, affecting their mobility or response to divalent cations, and thus their participation in the fusion reaction. These possibilities are being investigated currently.

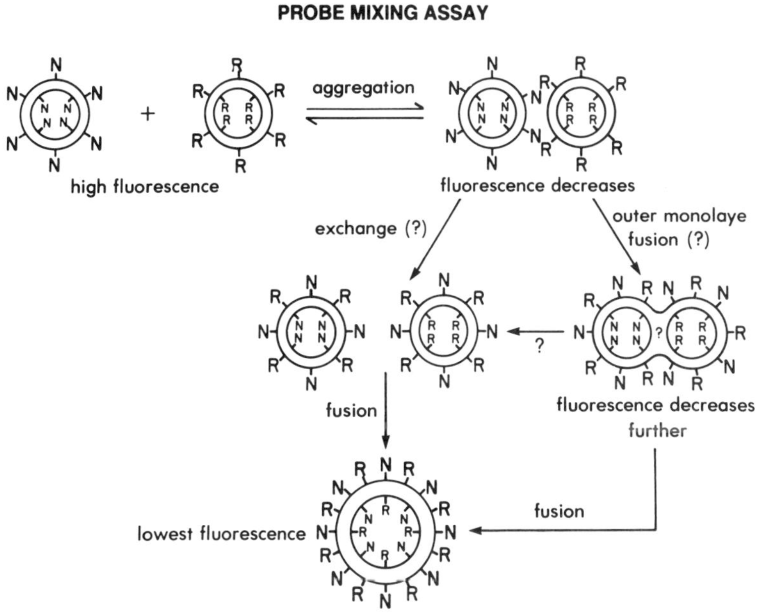

Figure 5. The behavior of lipid probes and of the corresponding fluorescence in the probe mixing assay. From Düzgüneş et al. (1987).

DISCUSSION

The observation that during Ca^{2+}-induced fusion of LUV(PS) the rate of probe dilution is higher than that of aqueous contents mixing, suggests that the outer monolayers of the vesicles begin to intermix before the bilayer destabilization step that leads to membrane fusion (in the sense of complete bilayer intermixing). Thus, the destabilization leading to lipid mixing between two membranes is a different process than the destabilization leading to communication between the two aqueous compartments. The rate constants for aggregation obtained with the probe dilution assay are higher than those obtained with the Tb/DPA assay. Thus, Wilschut et al. (1985b) have suggested that lipid mixing can take place without contents mixing during the reversible aggregation of the vesicles (the left arrows in the middle rows of Figures 5 and 6). They have also proposed that the formation of a fusion intermediate does not have to result in membrane fusion. This intermediate could be manifested as a lipid exchange process during close intermembrane contact, and does not have to impair bilayer integrity. However, the intermediate shown in Figures 5 and 6 is likely to be irreversible. Not in all liposome systems is lipid mixing faster than contents mixing. For example, SUV(PS) or LUV(cardiolipin/dioleoyl PC) fusing in the presence of Ca^{2+} at 25° C exhibit very similar rates of lipid mixing and aqueous contents mixing (Hoekstra, 1982b; Wilschut et al., 1983, 1985a). In these cases, the destabilization processes leading to lipid mixing and contents mixing apparently proceed at the same rate.

The probe mixing assay is extremely sensitive to vesicle aggregation and intermembrane contact in the absence of intermixing of aqueous contents (Figure 5). Aggregation may be accompanied by probe exchange, or may lead to the mixing of lipids located only in the outer monolayers of the

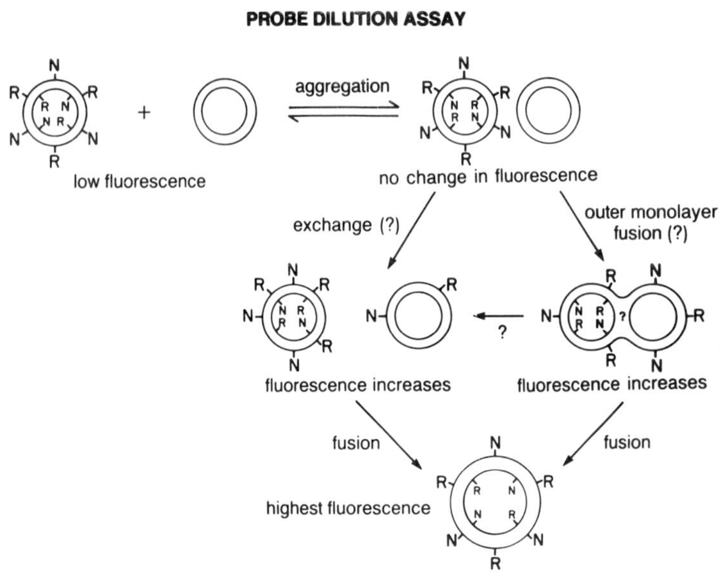

Figure 6. The behavior of lipid probes and of the corresponding fluorescence in the probe dilution assay. From Düzgüneş et al. (1987).

membranes (semi-fusion). When contents mixing occurs, its rate is equal to or slower than the rate of lipid mixing (Wilschut et al., 1985a), and the probe mixing assay indicates a higher rate of lipid mixing than the probe dilution assay (Figure 1). Thus, for LUV(PS) under the conditions used in our experiments, aggregation precedes lipid mixing, which in turn precedes the destabilization process leading to fusion. The probes do not appear to affect the aggregation and fusion characteristics of the membranes, since the rate of probe mixing does not change with the probe concentration in the membrane. Nevertheless, we are now investigating whether the interaction of the probe molecules in apposed membranes induces or enhances fusion (detected by the contents mixing assays) following aggregation mediated by divalent cations.

Probe dilution in the absence of any intermixing or leakage of contents has been observed with LUV(phosphatidylglycerol) in the presence of Mg^{2+} (Rosenberg et al., 1983), and LUV(PE/cholesteryl hemisuccinate) in the presence of low concentrations of Ca^{2+} or Mg^{2+} (Ellens et al., 1985). Since the assay is not sensitive to the mere aggregation of vesicles, lipid mixing observed in these cases probably reflects semi-fusion.

Comparison of the probe mixing assay with the probe dilution assay indicates that experiments utilizing the former assay have to be interpreted cautiously, since aggregated vesicles which have not fused or intermixed their outer monolayers would still cause a decrease in the fluorescence due to RET across the interbilayer space (cf. Silvius and Gagne, 1984a,b; Bondeson and Sundler, 1985). Furthermore, even in systems which do undergo fusion, rate constants of fusion could be overestimated (cf. Morris et al., 1985).

ACKNOWLEDGEMENTS

This work was supported by grants from the National Institutes of Health (GM28117; DP and ND), the American Heart Association (ND), the UCSF Academic Senate (ND), the Alberta Heritage Foundation for Medical Research and the Medical Research Council of Canada (MA-6487; TMA). We thank Dr. S. Nir for helpful discussions.

REFERENCES

Ababei, L. and Hildenbrand, K., 1985, Kinetics of calcium-induced mixing of lipids and aqueous contents of large unilamellar phosphatidylserine vesicles, Chem. Phys. Lipids, 35:39.

Amselem, S., Barenholz, Y., Loyter, A., Nir, S. and Lichtenberg, D., 1986, Fusion of Sendai virus with negatively charged liposomes as studied by pyrene-labelled phospholipid liposomes, Biochim. Biophys. Acta, 860:301.

Bentz, J. and Düzgüneş, N., 1985, Fusogenic capacities of divalent cations and the effect of liposome size, Biochemistry, 24:5436.

Bentz, J., Düzgüneş, N. and Nir, S., 1983, Kinetics of divalent cation induced fusion of phosphatidylserine vesicles: Correlation between fusogenic capacities and binding affinities, Biochemistry, 22:3320.

Bondeson, J. and Sundler, R., 1985, Lysine peptides induce lipid intermixing but not fusion between phosphatidic acid-containing vesicles, FEBS Lett., 190:283.

Citovsky, V. and Loyter, A., 1985, Fusion of Sendai virions or reconstituted Sendai virus envelopes with liposomes or erythrocyte membranes lacking virus receptors, J. Biol. Chem., 260:12072.

Düzgüneş,N., 1985, Membrane fusion, in: "Subcellular Biochemistry," Vol. 11, D.B. Roodyn, ed., pp. 195-286, Plenum Press, New York.

Düzgüneş, N. and Papahadjopoulos, D., 1983, Ionotropic effects on phospholipid membranes: Calcium-magnesium specificity in binding, fluidity, and fusion, in: "Membrane Fluidity in Biology," Vol. 2, R.C. Aloia, ed., pp. 187-213, Academic Press, New York.

Düzgüneş N. and Bentz, J., 1987, Fluorescence assays for membrane fusion, in: "Spectroscopic Membrane Probes," L.M. Loew, ed., CRC Press, Boca Raton, Florida (in press).

Düzgüneş, N., Wilschut, J., Fraley, R., and Papahadjopoulos, D., 1981, Studies on the mechanism of membrane fusion: Role of head-group composition in calcium- and magnesium-induced fusion of mixed phospholipid vesicles, Biochim. Biophys. Acta, 642:182.

Düzgüneş, N., Wilschut, J., Hong, K., Fraley, R., Perry, C., Friend, D. S., James, T. L., and Papahadjopoulos, D., 1983, Physicochemical characterization of large unilamellar vesicles prepared by reverse-phase evaporation, Biochim. Biophys. Acta, 732:289.

Düzgüneş, N., Straubinger, R. M., Baldwin, P. A., Friend, D. S., and Papahadjopoulos, D., 1985, Proton-induced fusion of oleic acid-phosphatidylethanolamine liposomes, Biochemistry, 24:3091.

Düzgüneş, N., Allen, T.M., Fedor, J. and Papahadjopoulos, D., 1987, Lipid mixing during membrane aggregation and fusion. Why fusion assays disagree, Biochemistry (submitted).

Eidelman, O., Schlegel, R., Tralka, T.S. and Blumenthal, R., 1984, pH-dependent fusion induced by vesicular stomatitis virus glycoprotein reconstituted into phospholipid vesicles, J. Biol. Chem., 259:4622.

Ellens, H., Bentz, J., and Szoka, F. C., 1984, pH-induced destabilization of phosphatidylethanolamine-containing liposomes: Role of bilayer contact, Biochemistry, 23:1532.

Ellens, H., Bentz, J. and Szoka, F. C., 1985, H^+- and Ca^{2+} -induced fusion and destabilization of liposomes, Biochemistry, 24:3099.

Ellens, H., Bentz, J. and Szoka, F.C., 1986, Destabilization of phosphatidylethanolamine liposomes at the hexagonal phase transition temperature, Biochemistry, 25:285.

Gibson, G.A. and Loew, L.M., 1979a, Phospholipid vesicle fusion monitored by fluorescence energy transfer, Biochem. Biophys. Res. Commun., 88:135.

Gibson, G.A. and Loew, L.M., 1979b, Application of Förster resonance energy transfer to interactions between cell and lipid vesicle surfaces, Biochem. Biophys. Res. Commun., 88:141.

Hoekstra, D., 1982a, Role of lipid phase separations and membrane hydration in phospholipid vesicle fusion, Biochemistry, 21:2833.

Hoekstra, D., 1982b, Kinetics of intermixing of lipids and mixing of aqueous contents during vesicle fusion, Biochim. Biophys. Acta, 692:171.

Hoekstra, D., de Boer, T., Klappe, K. and Wilschut, J., 1984, Fluorescence method for measuring the kinetics of fusion between biological membranes, Biochemistry, 23:5675.

Melikyan, G.B., Abidor, I.G., Chernomordik, L.V. and Chailakhyan, L.M., 1983, Electrostimulated fusion and fission of bilayer lipid membranes, Biochim. Biophys. Acta, 730:395.

Morris, S.J., Gibson C.C., Smith, P.D., Greif, P.C., Stirk, C.W., Bradley, D., Haynes, D.H. and Blumenthal, R., 1985, Rapid kinetics of Ca^{2+}-induced fusion of phosphatidylserine/ phosphatidylethanolamine vesicles: The effect of bilayer curvature on leakage, J. Biol. Chem., 260:4122.

Morris, S.J., Bradley, D., Gibson, C.C., Smith, P.D. and Blumenthal, R., 1987, Use of membrane-associated probes to monitor membrane fusion, in: "Spectroscopic Membrane Probes," L.M. Loew, ed., CRC Press, Boca Raton, Florida (in press).

Neher, E., 1974, Asymmetric membranes resulting from the fusion of two black lipid bilayers, Biochim. Biophys. Acta, 373:327.

Nir, S., Bentz, J. and Düzgüneş, N., 1981, Two modes of reversible vesicle aggregation: Particle size and the DLVO theory, J. Coll. Interface Sci., 84:266.

Nir, S., Bentz, J., Wilschut, J. and Düzgüneş, N., 1983, Aggregation and fusion of phospholipid vesicles, Prog. Surface Sci., 13:1.

Palade, G., 1975, Intracellular aspects of the process of protein synthesis, Science, 189:347.

Papahadjopoulos, D., Poste, G., and Vail, W.J., 1979, Studies on membrane fusion with natural and model membranes, Methods Memb. Biol., 10:1.

Parente, R.A. and Lentz, B.R., 1986, Fusion and phase separation monitored by lifetime changes of a fluorescent phospholipid probe, Biochemistry, 25:1021.

Pinto da Silva, P. and Nogueira, M. L., 1977, Membrane fusion during secretion. A hypothesis based on electron microscope observation of Phytophthora palmivora zoospores during encystment J. Cell Biol., 73:161.

Rosenberg, J., Düzgüneş, N. and Kayalar, C., 1983, Comparison of two liposome fusion assays monitoring the intermixing of aqueous contents and of membrane components, Biochim. Biophys. Acta, 735:173.

Silvius, J.R. and Gagne, J., 1984a, Calcium-induced fusion and lateral phase separations in phosphatidylcholine-phosphatidylserine vesicles. Correlation by calorimetric and fusion measurements, Biochemistry, 23:3241.

Silvius, J.R. and Gagne, J., 1984b, Lipid phase behavior and calcium-induced fusion of phosphatidylethanolamine-phosphatidylserine vesicles. Calorimetric and fusion studies, Biochemistry, 23:3232.

Struck, D. K., Hoekstra, D., and Pagano, R. E., 1981, Use of resonance energy transfer to monitor membrane fusion, Biochemistry, 20:4093.

Szoka, F., Olson, F., Heath, T., Vail, W., Mayhew, E. and Papahadjopoulos, D., 1980, Preparation of unilamellar liposomes of intermediate size (0.1-0.2 µm) by a combination of reverse phase evaporation and extrusion through polycarbonate membranes, Biochim. Biophys. Acta, 601:559.

Uster, P.S. and Deamer, D.W., 1981, Fusion competence of phosphatidylserine-containing liposomes quantitatively measured by a fluorescence resonance energy transfer assay, Arch. Biochem. Biophys., 209:385.

Vanderwerf, P. and Ullman, E.F., 1980, Monitoring of phospholipid vesicle fusion by fluorescence energy transfer between membrane-bound dye labels, Biochim. Biophys. Acta, 596:302.

Wilschut, J., Düzgüneş, N., Fraley, R., and Papahadjopoulos, D., 1980, Studies on the mechanism of membrane fusion: Kinetics of Ca^{2+}-induced fusion of phosphatidylserine vesicles followed by a new assay for mixing of aqueous vesicle contents, Biochemistry, 19:6011.

Wilschut, J., Düzgüneş, N. and Papahadjopoulos, D., 1981, Calcium/magnesium specificity in membrane fusion: Kinetics of aggregation and fusion of phosphatidylserine vesicles and the role of bilayer curvature, Biochemistry, 20:3126.

Wilschut, J., Düzgüneş, N., Hong, K., Hoekstra, D., and Papahadjopoulos, D., 1983, Retention of aqueous contents during divalent cation-induced fusion of phospholipid vesicles, Biochim. Biophys. Acta, 734:309.

Wilschut, J., Nir, S., Scholma, J. and Hoekstra, D., 1985a, Kinetics of Ca^{2+}-induced fusion of cardiolipin-phosphatidylcholine vesicles: Correlation between vesicle aggregation, bilayer destabilization, and fusion, Biochemistry, 24:4630.

Wilschut, J., Scholma, J., Bental, M., Hoekstra, D. and Nir, S., 1985b, Ca^{2+}-induced fusion of phosphatidylserine vesicles: Mass action kinetic analysis of membrane lipid mixing and aqueous contents mixing, Biochim. Biophys. Acta, 821:45.

CONCENTRATION DEPENDENCE OF DPHpPC FLUORESCENCE LIFETIME:

PHOTOPHYSICS AND UTILITY FOR MONITORING MEMBRANE FUSION

Barry R. Lentz+, Stephen W. Burgess+, and Enrico Gratton†

+Department of Biochemistry, 231H
University of North Carolina at Chapel Hill
Chapel Hill, NC 27514; and
†Department of Physics
University of Illinois
Urbana, IL 61801

INTRODUCTION

Lipid mixing due to membrane fusion has been measured by monitoring the concentration-dependent properties of membrane-associated probes, usually a fluorescent probe (Struck *et al.*, 1981; Hoekstra *et al.*, 1984). Aside from displaying concentration-dependent fluorescence, probes to be used for fusion assays should 1) not interchange rapidly between membranes, 2) resemble as closely as possible a phospholipid molecule, 3) not interfere with the natural packing and local structure of the bilayer, 4) have fluorescence properties that are insensitive to the presence of added fusogenic agents, and 5) have a high extinction coefficient and quantum yield over broad excitation and emission limits. While the combination of head-group-labeled N-(4-nitro-2,1,3-benzoxadiazol (NBD)- and rhodamine (Rh)-labeled phosphatidylethanolamines (PE) (Struck, *et al.*, 1983) meets many of these requirements, we (Parente and Lentz, 1986a; 1986b) and others (Morris *et al.*, 1985) have shown that the fluorescence of these probes is quenched by addition of fusogenic agents such as Ca^{2+} (Figure 1) and poly(ethylene glycol) (PEG). In addition, octadecyl Rhodamine B chloride (R_{18}), while having certain advantages for the study of biological membrane fusion, has the disadvantages of not resembling a phospholipid and of slowly interchanging between membranes (Hoekstra *et al.*, 1984).

We review here the properties of a recently introduced (Cranney *et al.*, 1983; Parente and Lentz, 1985) probe that meets most, if not all, of these requirements, namely 1-palmitoyl-2[[2-[4-(6-phenyl-*trans*-1,3,5-hexatrienyl)phenyl]ethyl]carboxyl]-3-*sn*-phosphatidylcholine (DPHpPC; see Figure 2). We review also our use of this probe not only to follow lipid mixing due to the fusion process but also to detect local domains or lipid redistributions that may be crucial to the fusion process (Parente and Lentz, 1986a; 1986b). In addition, we summarize some of the data leading to a model of the photophysics of DPHpPC that explains the observed concentration-dependence of the fluorescence lifetime in terms of fluorophore dimerization. It is this property that makes DPHpPC useful for reporting membrane fusion.

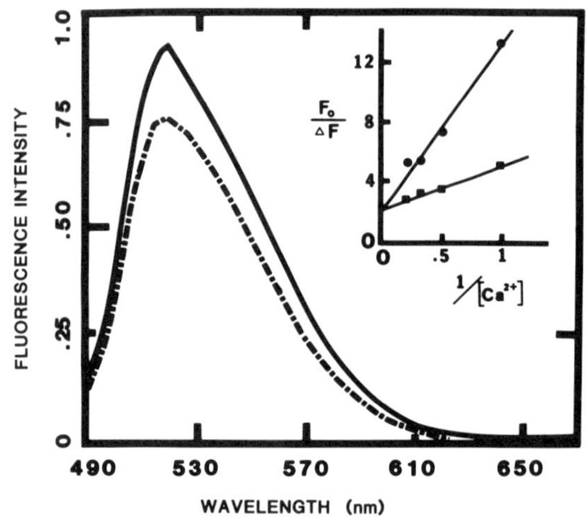

Fig. 1. Effect of Ca^{2+} addition on the fluorescence intensity of $DC_{15}PC$ REV containing 5 mol % NBD-PE. The solid line represents the emission spectrum recorded in the absence of Ca^{2+}. The dashed line is the spectrum resulting from vesicles incubated with 5 mM Ca^{2+} for 10 min. Excitation wavelength = 435 nm. Spectra were recorded at room temperature from 490-700 nm. The inset shows the dependence of Rh-PE (squares) and NBD-PE (circles) relative fluorescence intensity on the concentration of added Ca^{2+} plotted according to a modified Stern-Volmer equation (Lehrer, 1971). F_o is the fluorescence intensity in the absence of Ca^{2+} and ΔF is the difference between F_o and the fluorescence observed in the present of indicated Ca^{2+} concentrations. The y-intercept indicates that the outer leaflet is quenched by Ca^{2+}. Adapted from Parente and Lentz (1986a).

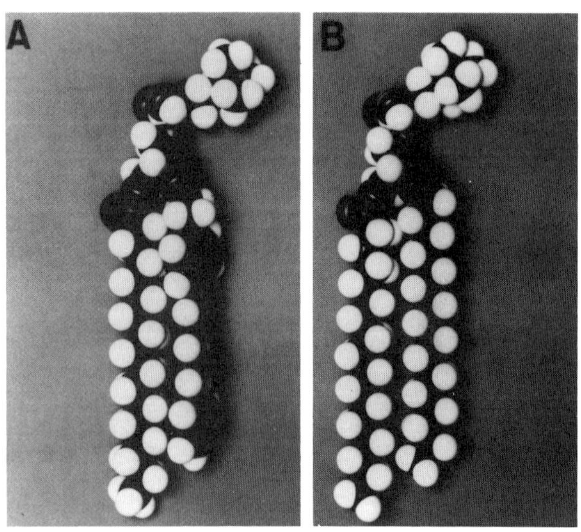

Fig. 2. Space-filling models of DPHpPC(A) and DPPC(B) molecules as they would be oriented in the bilayer. Fatty acid chains are in the *trans* configuration. Reproduced from Parente and Lentz (1985).

MATERIALS AND METHODS

Materials

Phospholipids (1,2-dipalmitoyl-3-*sn*-phosphatidylcholine [DPPC]; 1,2-dipentadecanoyl-3-*sn*-phosphatidylcholine [$DC_{15}PC$]; bovine phosphatidylserine [PS] were obtained from Avanti Polar Lipids, Inc. (Birmingham, AL) and treated as described earlier (Parente and Lentz, 1985; 1986a, 1986b).

Methods

Vesicle preparation. Lipid and probe stocks were mixed in various molar ratios and dried under argon onto the wall of an amberized round-bottom flask (Reliance Glass Works, Inc., Bensenville, IL). The lipid film was re-dissolved in a small amount of benzene and lyophilized under high vacuum (0.5 mmHg) for 8-12 h to yield a white powder. Large, multilamellar vesicles (LMV) were prepared in 50 mM KCl as described previously (Lentz *et al.*, 1976). Large, unilamellar vesicles (LUV) were prepared by reverse evaporation (Szoka *et al.*, 1980). LUV and LMV were stored dark and under argon, and LMV were swirled at 50°C in a rotary incubator for at least 2 hr prior to use.

Fluorescence measurements. All fluorescence spectral and fusion measurements were made on an SLM 4800 spectrofluorometer (SLM Instruments, Urbana, IL). Excitation was with the 366-nm mercury line of a 200-W mercury-xenon lamp (Canrad-Hanovia, Newark, NJ) for emission spectra and fusion measurements or with a 150-W high pressure xenon lamp (Osram, Munich Germany) for excitation spectra.

The detailed fluorescence lifetime measurements reported in Table I were performed with the multifrequency phase and modulation fluorometer described elsewhere (Gratton & Limkeman, 1983). The light source was an argon ion laser whose intensity was sinusoidally modulated using a Pockels cell. The modulation frequency could be varied continuously from 1 to 250 MHz. For the present study, 11 to 15 modulation frequencies are used in the range from 10 to 150 MHz. For each frequency, the phase and the modulation of the fluorescence were measured with respect to a reference fluorophore of known lifetime. Data were collected and analyzed using an IBM PC computer as previously described (Parasassi *et al.*, 1984).

RESULTS AND DISCUSSION

DPHpPC as Probe of Lipid Mixing

Our lipid mixing and phase separation assays are based on the observation that the fluorescence lifetime of DPHpPC is sensitive to its concentration in the bilayer (Parente and Lentz, 1986a; Figure 3). We have found a similar behavior for 1,6-diphenyl-1,3,5-hexatriene (DPH), the parent fluorophore of DPHpPC (Barrow and Lentz, 1985). As applied to the Ca^{2+}-induced fusion of bovine PS vesicles, the design of the assay is illustrated in Figure 4. One vesicle population containing 2 mol % probe (fluorescence lifetime roughly 6.5 nsec) was combined with a ten-fold excess of a second, non-fluorescent vesicle population. Bilayer mixing induced by the addition of Ca^{2+} (left hand arrow) resulted in a reduction of the probe concentration in the fused vesicles and, thus, an increase in lifetime to roughly 7.5 nsec. Errors in our lifetime measurements were typically in the range of 0.05-0.1 nsec, which means that this technique is sensitive to small changes in the local probe concentration. The fluorescence lifetime is an intensive quantity

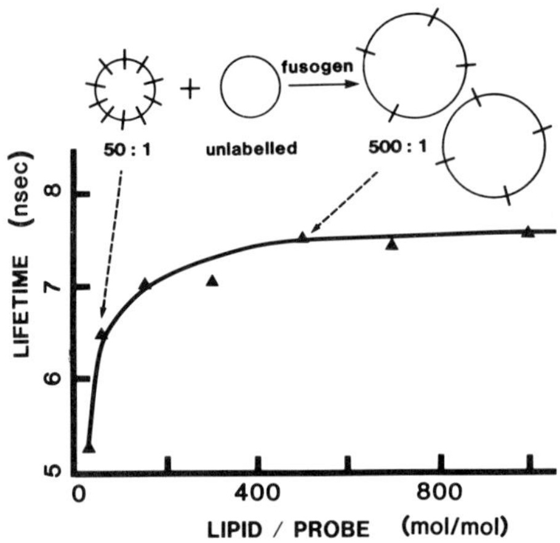

Fig. 3. Dependence of DPHpPC fluorescence lifetime on lipid-to-probe ratio in DPPC LMV. The lipid concentration was 0.15 mM and the vesicles contained varying amounts of DPHpPC. Phase-derived lifetimes recorded at a 30 MHz modulation frequency are plotted. Above this plot is a drawing illustrating how lifetime changes at varying lipid-to-probe ratios can be used as an assay of intervesicular lipid mixing during fusion. Arrows point to the expected lifetime value of vesicles at the beginning and end of the fusion process. Adapted from Parente and Lentz (1986a).

and is not dependent on the concentration of vesicles present in any given experiment. This greatly facilitates comparison of data from different experiments and allows us to distinguish between fusion reactions involving the mixing of bilayer components from several vesicles and those involving only two vesicles.

Figure 4 also illustrates the ability of the concentration-dependence of DPHpPC fluorescence lifetime to detect a phase separation. Thus, following the Ca^{2+}-induced fusion of PS vesicles, a dramatic drop in fluorescence lifetime was observed. According to Figure 3, this implies an increased local concentration of DPHpPC molecules within some portion of the bilayer. The observed lifetime decline is consistent with the previously reported phase separation induced by Ca^{2+}-sequestering of PS molecules (Papahadjopoules, 1978).

Addition of Na_2EDTA (11 mM) after 20 min to chelate Ca^{2+} increased the DPHpPC lifetime to that expected for complete mixing of the lipids in the labelled and unlabelled vesicle populations, as observed at the 5 min time point (see Figure 4). This final lifetime value remained unchanged over the next 20 min. The events presumed to occur in conjunction with the observed lifetime changes are diagrammed in the upper portion of Figure 4, with arrows to indicate the addition of Ca^{2+} and Na_2EDTA.

Aside from its ability to detect lipid mixing associated with Ca^{2+}-induced fusion of PS vesicles, we have shown that DPHpPC is capable also of reporting vesicle fusion induced by PEG (Parente and Lentz,

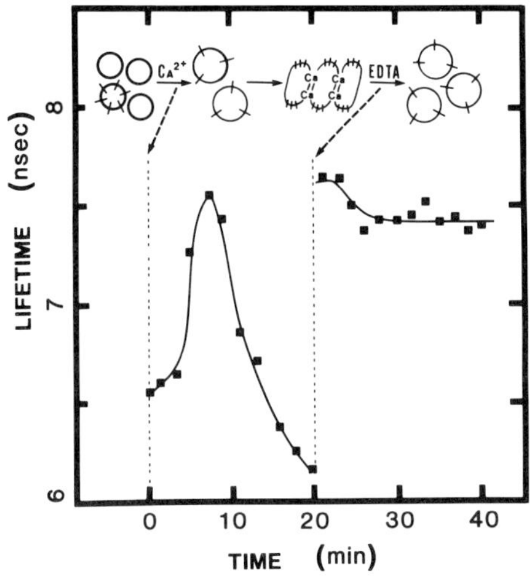

Fig. 4. DPHpPC lifetime vs. time after Ca^{2+} addition to induce membrane fusion of bovine phosphatidylserine large unilamellar vesicles. Vesicles containing 2 mol % DPHpPC (0.01 mM total lipid) were mixed with a 10-fold excess of non-fluorescent vesicles (0.1 mM). Dashed lines indicate times of Ca^{2+} and Na_2EDTA addition, respectively. The accompanying diagram summarizes our interpretation of the Ca^{2+}-induced fusion process in bovine PS REV. The decline in lifetime from 10-20 min is attributed to a phase separation of PS by Ca^{2+} which was reversible by Na_2EDTA addition. Adapted from Parente and Lentz (1986a).

1986b; Figure 5). Most significantly, the extent and rate of lipid mixing reported by DPHpPC agreed well with the mixing of vesicle contents as reported by the ANTS/DPX assay (Ellens *et al.*, 1985; see Figure 5). The agreement of our lipid-mixing assay with this contents-mixing assay implies that fusion-independent exchange of DPHpPC between membranes is minimal. We have demonstrated this directly in studies with aggregating but non-fusing PS-containing vesicles (Parente and Lentz, 1986a).

These results make it clear that measurements of DPHpPC fluorescence lifetime can provide a quantitatively accurate measure of lipid mixing associated with vesicle fusion. However, the use of this probe is not without some limitations (Parente and Lentz, 1985). First, DPHpPC does not partition equally between membrane domains, but favors fluid phase relative to solid phase regions with a partition coefficient of $K_{f/s} \cong 3$. Second, studies of lipid phase behavior indicate that DPHpPC perturbs slightly its local environment in the bilayer. Finally, DPHpPC appears to form small clusters or domains in a solid-phase bilayer. However, these are minor limitations; it is clear from the sum total of our results that DPHpPC is a good probe of lipid mixing associated with fusion.

DPHpPC Dimers: A Model for Photophysical Behavior

Our model to explain the concentration dependence of DPHpPC fluorescence lifetime is shown in Figure 6. The essential element of this model is the existence of dimers in which the normal symmetries of

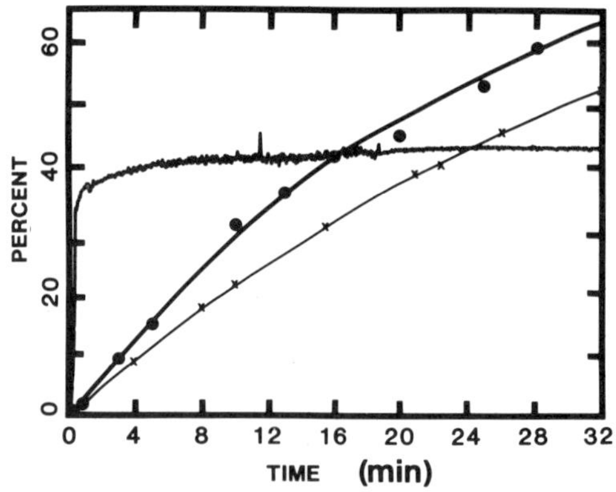

Fig. 5. Percent of leakage (solid line), contents mixing (circles), and lipid mixing (x's) as a function of time after addition of vesicles to 3.8% (w/w) PEG. Maximum leakage (100%) was determined by addition of detergent to ANTS/DPX-containing vesicles. 100% mixing for both lipids and contents was taken to be the value (fluorescence lifetime or intensity) expected for one complete cycle of fusion. Adapted from Parente and Lentz (1986b).

ground and excited state molecular wavefunctions are assumed to be modified. This allows the otherwise symmetry-disallowed transitions between A symmetry states, as shown in Figure 6. The other features of this model, such as the relationship between the Au, Bu* and Ag* states, derive from the literature on polyene photophysics, as recently reviewed by one of us (Lentz, 1987).

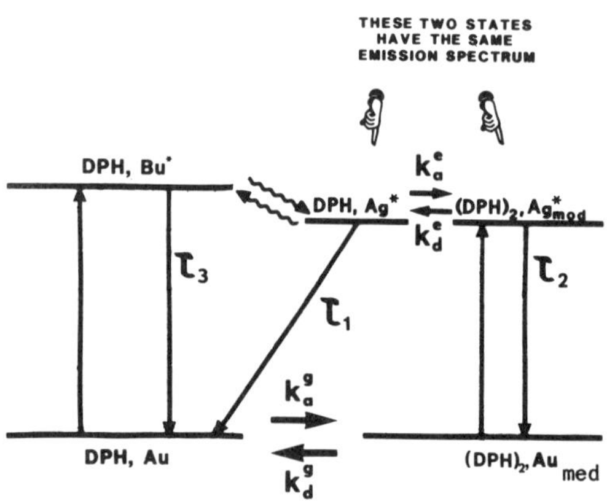

Fig. 6. Model to account for the photophysical properties of DPHpPC in a bilayer. Au, Ag, and Bu are symmetry representations for ground and excited state molecular orbitals. In the dimer, the molecular orbital symmetries are assumed to be modified, allowing what would otherwise be symmetry-disallowed transitions.

The model shown in Figure 6 can account for the decrease in fluorescence lifetime of DPHpPC observed at high probe concentrations, as shown in Figure 3. This explanation requires the lifetime of the dimer (τ_2) be less than that of the normal decay pathway (τ_1). The model also predicts that a second, short lifetime component should appear in membrane samples rich in DPHpPC. Our analysis of multifrequency phase and modulation fluorescence data (Table I) confirmed this prediction. By contrast, only a single lifetime component was necessary to fit the phase and modulation data at low probe concentration (Table I). This effect was most dramatic below the gel-to-fluid phase transition (20° data in Table I), in agreement with our previous phase behavior studies suggestive of probe aggregation in the gel phase (Parente and Lentz, 1985).

The appearance of a second fluorescence lifetime at high probe concentration could be due to excited state quenching of some fraction of DPHpPC probes, as opposed to the existence of a distinctly different excited state species, as suggested by the model shown in Figure 6. The evidence against simple excited state quenching is two-fold. First, the observed fluorescence intensity didn't vary with probe concentration in the same way that lifetime did, as indicated in Figure 7. Second, the long lifetime component resolved in our membrane samples didn't remain constant with varying probe concentration (Table I). Instead, the τ_1 component dropped in magnitude with increasing probe concentration, consistent with an excited state reaction, such as shown in Figure 6. Indeed, efforts to date have shown the model in Figure 6 to be quantitatively consistent with all of our lifetime observation (unpublished calculations).

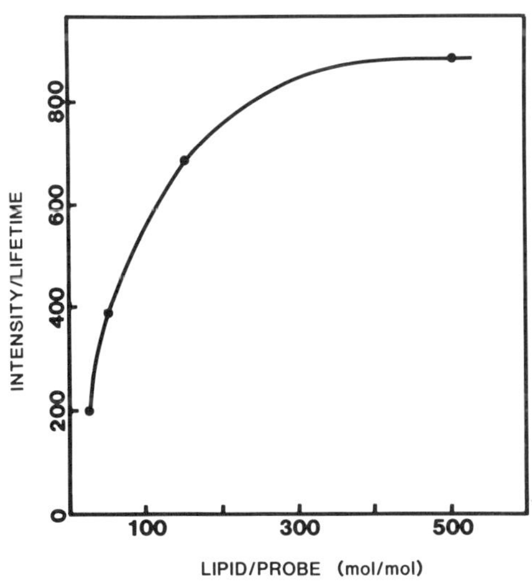

Fig. 7. Fluorescence intensity (arbitrary units)/lifetime ratio as a function of lipid/probe molar ratio for DPHpPC in DPPC LMV. Vesicles containing different mole fractions of DPHpPC (0.25 mM DPPC) were suspended in 50 mM KC1. DPHpPC fluorescence was excited at 366 nm and the fluorescence intensities were recorded at the maximum of the emission spectrum (435 nm). Lifetimes were the average lifetime for the sample obtained using an isochronal standard of DPH in heptane (τ = 6.77 ns; 2 x 10^{-7} M) as a reference for lifetime measurement (Barrow and Lentz, 1983).

The final observation in support of the model shown in Figure 3 is the excitation spectrum shown in Figure 8. The shift in the excitation spectrum seen at high probe concentration (Curve A of Figure 8) provides evidence for the ground state association suggested in our model. It is also worth noting that no other observation that we have made (excitation spectrum, excitation and emission polarization spectra, dynamic polarization measurements) has proven inconsistent with this model (unpublished results).

CONCLUSIONS

1. Ground and excited state self-association of DPHpPC molecules can explain the increased contribution of short lifetime components at high probe concentration.

2. Since dimer formation is clearly concentration dependent, this justifies the empirical use of DPHpPC lifetime to monitor lipid mixing during fusion.

3. If the model is correct, calibration on known systems should yield the equilibrium constants for dimer formation. These could be used in a simple calculation to obtain local probe concentrations as a function of observed lifetimes in any membrane.

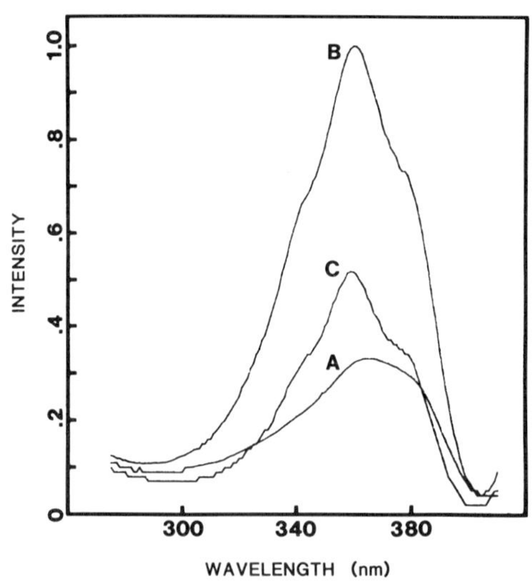

Fig. 8. Fluorescence excitation spectra of DPHpPC in DPPC LMV at 45°C in 50 mM KCl. Spectra are presented for samples at lipid/probe molar ratios of (A) 25:1, (B) 150:1, and (C) 500:1 (0.25 mM DPPC). The excitation spectra were recorded in ratio mode using a rhodamine quantum counter so as to correct for instrument response. Fluorescence intensity is in arbitrary units. Fluorescence emission was detected using a Schott KV418 cut-off filter in the emission path.

Table I: Analysis of Fluorescence Lifetime Components for DPHpPC in DPPC Multilamellar Vesicles

Sample DPPC/DPHpPC Ratio	Temperature	Analysis	Blank τ	Blank f	Components τ_1	Components f_1	Components τ_2	Components f_2	χ^2
500/1	45°C	Blank + 1 component	3.7	.005	6.8	.995	---	---	33.1
	45°C	Blank + 2 components	3.7	.005	8.7	.72	4.4	.28	4.4
150/1	48°C	Blank + 1 component	3.7	.005	6.3	.995	---	---	9.7
	48°C	Blank + 2 components	3.7	.005	6.5	.97	2.6	.03	8.2
25/1	45°C	Blank + 1 component	3.7	.005	2.3	.995	---	---	105
	45°C	Blank + 2 components	3.7	.005	3.2	.68	1.3	.32	2.2
20/1	48°C	Blank + 1 component	3.7	.005	1.9	.995	---	---	153
	48°C	Blank + 2 components	3.7	.005	2.8	.73	.81	.27	5.6
500/1	20°C	Blank + 1 component	3.7	.005	7.1	.995	---	---	8.1
	20°C	Blank + 2 components	3.7	.005	7.5	.91	5.0	.09	5.8
25/1	20°C	Blank + 1 component	3.7	.005	1.9	.995	---	---	637
	20°C	Blank + 2 components	3.7	.005	4.2	.58	1.0	.42	12.5

Heterogeneous lifetime analysis was performed on data sets composed of phase shift and modulation ratio measurements obtained at 11-15 frequencies between 10 and 150 MHz. The average blank lifetime was determined using a one component fit and was held fixed for all analysis. The blank fractional intensity was held fixed at 0.5% as measured directly. Student's T-tests were performed on the analysis results, using the reduced summed-squared-residuals shown (χ^2). In all lipid-rich samples a single-component fit was adequate (even for the 500/1 sample at 45°C), while two lifetime components were necessary to describe the probe-rich samples.

REFERENCES

Barrow, D.A., Lentz, B.R., 1983, The use of isochromal reference standards in phase and modulation fluorescence lifetime measurements, J. Biochem. Biophys. Methods, 7:217.

Barrow, D.A., and Lentz, B.R., 1985, Membrane structural domains. Resolution limits using diphenylhexatriene fluorescence decay, Biophys. J., 48:221.

Cranney, M., Cundall, R.B., Jones, G.R., Richards, J.T., Thomas, E.W., 1983, Fluorescence lifetime and quenching studies on some interesting diphenylhexatriene membrane probes, Biochim. Biophys. Acta, 735:418.

Ellens, H., Bentz, J., Szoka, F.C., 1985, H^+ and Ca^{2+}-induced fusion and destabilization of liposomes, Biochemistry, 24:3099.

Gratton, E., Limkeman, M., 1983, A continuously variable frequency cross-correlation phase fluorometer with picosecond resolution, Biophys. J., 44:315.

Hoekstra, D., de Boer, T., Klappe, K., Wilschut, J., 1984, Fluorescence method for measuring the kinetics of fusion between biological membranes, Biochemistry, 23:5675.

Lehrer, S.S., 1971, Solute perturbation of protein fluorescence. The quenching of the tryptophyl fluorescence of model compounds and of lysozyme by iodine ion, Biochemistry, 10:3254.

Lentz, B.R., Barenholz, Y., Thompson, T.E., 1976, Fluorescence depolarization studies of phase transitions and fluidity in phospholipid bilayers. 1. Single component phosphatidylcholine liposomes, Biochemistry, 20:4521.

Lentz, B.R., 1987, Membrane "fluidity" from fluorescence anisotropy measurements, in "Spectroscopic Membrane Probes," L.M. Lowe, ed., CRC Press, Boca Raton, FL, in press.

Morris, S.J., Gibson, C.C., Smith, P.D., Greif, P.C., Stirk, C.W., Bradley, D., Haynes, D.H., Blumenthal, R., 1985, Rapid kinetics of Ca^{2+}-induced fusion of phosphatidylserine/phosphatidylethanolamine vesicles. The effect of bilayer curvature on leakage, J. Biol. Chem., 260:4122.

Papahadjopoulos, D., 1978, Calcium-induced phase changes and fusion in natural and model membranes, in: "Membrane Fusion", G. Poste and G.L. Nicholson, eds, Elsevier/North-Holland Biomedical Press, Amsterdam, pp. 765-790.

Parasassi, T., Conti, F., Glaser, M., Gratton, E., 1984, Detection of phospholipid phase separation. A multifrequency phase fluorimetry study of 1,6-diphenyl-1,3,5-hexatriene, J. Biol. Chem., 259:14011.

Parente, R.A., Lentz, B.R., 1985, Advantages and limitations of 1-palmitoyl-1,2-[2-[4-(6-phenyl-*trans*-1,3,5-hexatrienyl)phenyl]ethyl]carbonyl]-3-*sn*-phosphatidylcholine as a fluorescence membrane probe, Biochemistry 24:6178.

Parente, R.A., Lentz, B.R., 1986a, Fusion and phase separation monitored by lifetime changes of a fluorescence phospholipid probe, Biochemistry, 25:1021.

Parente, R.A., Lentz, B.R., 1986b, Rate and extent of poly(ethylene glycol)-induced large vesicle fusion monitored by bilayer and internal contents mixing, Biochemistry, 25:6678.

Struck, D.K., Hoekstra, D., Pagano, R.E., 1981, Use of resonance energy transfer to monitor membrane fusion, Biochemistry, 20:4093.

Szoka, F., Olson, F., Heath, T., Vail, W., Mayhew, E., Papahadjopoulos, D., 1980, Preparation of unilamellar liposomes of intermediate size (0.1-0.2 microns) by a combination of reverse phase evaporation and extrusion through polycarbonate membranes, Biochim. Biophys. Acta 601:559.

CONTRIBUTORS

Q. F. Ahkong
Department of Biochemistry and Chemistry
Royal Free Hospital School of Medicine
University of London
Rowland Hill Street
London, NW3 2PF, UK

T. M. Allen
Department of Pharmacology
University of Alberta
Edmonton, Alberta, Canada
T6G 2H7

W. Almers
Department of Physiology and Biophysics
University of Washington Medical School
Seattle, WA 98195

K. Arnold
Institute of Biophysics
Department of Physics of the Karl Marx University
Leipzig
GDR

J. Bentz
Department of Pharmaceutical Chemistry
University of California
San Francisco, CA 94143

R. Blumenthal
Section of Membrane Structure and Function
LMMB
National Cancer Institute
NIH
Bethesda, MD 20892

L. J. Breckenridge
Department of Physiology and Biophysics
University of Washington Medical School
Seattle, WA 98105

P. M. Brown
Department of Biochemistry
McGill University
Montreal, Quebec, Canada H3G 1Y6

S.W. Burgess
Department of Biochemistry
231H
University of North Carolina at Chapel Hill
Chapel Hill, NC 27514

A.L. Burns
Laboratory of Cell Biology and Genetics
National Institutes of Diabetes, and Digestive Diseases
National Institutes of Health
Bethesda, MD 20205

M.K. Chaudhury
Dow Corning Corporation
Midland, MI 49686

V. Citovsky
Department of Biological Chemistry
Institute of Life Sciences
The Hebrew University of Jerusalem
91904 Jerusalem, Israel

D. Collins
Department of Biochemistry
University of Tennessee
Knoxville, TN 37996-0840

C.E. Creutz
Department of Pharmacology and the Programs in Biophysics, Neuroscience and Cell and Molecular Biology
University of Virginia
Charlottesville, VA 22908

P.R. Cullis
Biochemistry Department
University of British Columbia
Vancouver, British Columbia
Canada V6T 1W5

P. P. da Silva
Membrane Biology Section
Laboratory of Mathematical Biology
Bldg. 538, Room 104
National Cancer Institute
Frederick Cancer Research Facility
Frederick, MD 21701

R.W. Doms
Department of Cell Biology
Yale University School of Medicine
333 Cedar Street
New Haven, CT 06510

D.J. Doyle
Department of Biological Sciences
State University of New York at Buffalo
Buffalo, NY 14260

N. Düzgünes
Cancer Research Institute and
Department of Pharmaceutical
Chemistry
University of California
San Francisco, CA 94143

O. Eidelman
Section on Membrane Structure and
Function
LMMB
National Cancer Institute
NIH
Bethesda, MD 20892

H. Ellens
Department of Pharmacology
University of California
San Francisco, CA 94143

J.D. Ernst
The Rosalind Russell Arthritis
Research Laboratory and Department
of Medicine
University of California
San Francisco, CA 94143

E.A. Evans
Pathology and Physics
University of British Columbia
Vancouver, British Columbia
Canada V6T 1N5

J. Fedor
Cancer Research Institute
University of California
San Francisco, CA 94143

K. Gawrisch
Institute of Biophysics
Department of Physics of the Karl
Marx University
Leipzig, GDR

R.J. Good
Department of Chemical
Engineering
State University of New York
at Buffalo
Buffalo, NY 14260

I.M. Goldstein
The Rosalind Russell Arthritis
Research Laboratory and
Department of Medicine
University of California
San Francisco, CA 94143

U. Gras
Faculty of Biology
University of Konstanz
P.O. Box 5560
D-7750 Konstanz
Federal Republic of Germany

E. Gratton
Department of Physics
University of Illinois
Urbana, IL 61801

M. Gratzl
Abteilung Anatomie und
Zellbiologie
Universitat Ulm
Bundesrepublik Deutschland

A.M. Haywood
Departments of Pediatrics,
Medicine, and Microbiology
Box 689
University of Rochester Medical
Center
Rochester, NY 14642

A. Helenius
Department of Cell Biology
Yale University School of
Medicine
333 Cedar Street
New Haven, CT 06510

A. Herrmann
Institute of Biophysics
Department of Biology of the
Humboldt University
Berlin, GDR

D. Hoekstra
Laboratory of Physiological
Chemistry
University of Groningen
Bloemsingel 10
9712 KZ Groningen
The Netherlands

B. Hohne
Faculty of Biology
University of Konstanz
P.O. Box 5560
D-7750 Konstanz
Federal Republic of Germany

K. Hong
Cancer Research Institute
University of California
San Francisco, CA 94143

M.J. Hope
Biochemistry Department
University of British Columbia
Vancouver, British Columbia
Canada V6T 1W5

L. Huang
Department of Biochemistry
University of Tennessee
Knoxville, TN 37996

S.K. Huang
Biophysics Department
Roswell Park Memorial Institute
Buffalo, NY 14263

S.W. Hui
Biophysics Department
Roswell Park Memorial Institute
Buffalo, NY 14263

V. Kapoor
Biomedical Research and Development
American Red Cross
Rockville, MD 20855

H. Kersken
Faculty of Biology
University of Konstanz
P.O. Box 5560
D-7750 Konstanz
Federal Republic of Germany

K. Klappe
Laboratory of Physiological Chemistry
University of Groningen
Bioemsingel 10
9712 KZ Groningen
The Netherlands

B.R. Lentz
Department of Biochemistry
231 H
University of North Carolina at Chapel Hill
Chapel Hill NC 27514

R. Leventis
Department of Biochemistry
McGill University
Montreal, Quebec
Canada H3G 1Y6

D. Lichtenberg
Department of Physiology and Pharmacology
Tel Aviv University
Tel Aviv, 69978, Israel

M.M.S. Lo
National Institute on Drug Abuse
Addiction Research Center
4940 Eastern Avenue
Baltimore, MD 21224

A. Loyter
Department of Biological Chemistry
Institute of Life Sciences
The Hebrew University of Jerusalem
91904 Jerusalem, Israel

J.A. Lucy
Department of Biochemistry and Chemistry
Royal Free Hospital School of Medicine
University of London
Rowland Hill Street
London, NW3 2PF, UK

C.J. Lumpert
Faculty of Biology
University of Konstanz
P.O. Box 5560
D-7750 Konstanz
Federal Republic of Germany

R.C. MacDonald
Department of Biochemistry, Molecular Biology and Cell Biology
Northwestern University
Evanston, IL 60201

E. Mayhew
Departments of Experimental Pathology and Experimental Therapeutics
Roswell Park Memorial Institute
Buffalo, NY 14263

P.R. Meers
Cancer Research Institute
University of California
San Francisco, CA 94143-0128

M. Momayezi
Faculty of Biology
University of Konstanz
P.O. Box 5560
D-7750 Konstanz
Federal Republic of Germany

M.P. Murari
Department of Chemistry
Yale University
New Haven, CT 06511

M. Murata
Department of Biophysics
Faculty of Science
Kyoto University
Sakyo-ku, Kyoto 606, Japan

T.J. Murtaugh
Department of Anatomy and Structural Biology
Albert Einstein School of Medicine
Bronx, NY 10461

D. Needham
Mechanical Engineering and Material Science
Duke University
Durham, NC 27706

S. Nir
Seagram Center for Soil and Water Sciences
Faculty of Agriculture
The Hebrew University of Jerusalem
Rehovot, 76100, Israel

O. Nussbaum
Department of Biological Chemistry
Institute of Life Sciences
The Hebrew University of Jerusalem
91904 Jerusalem, Israel

M.P. O'Brien
Department of Chemistry
Yale University
New Haven, CT 06511

S. Ohki
Department of Biophysical Sciences
State University of New York at Buffalo
Buffalo, NY 14214

S-I Ohnishi
Department of Biophysics
Faculty of Science
Kyoto University
Sakyo-ku, Kyoto 606, Japan

D. Papahadjopoulos
Cancer Research Institute and Department of Pharmacology
University of California
San Francisco, CA 94143-0128

R. Pape
Faculty of Biology
University of Konstanz
P.O. Box 5560
D-7750 Konstanz
Federal Republic of Germany

R.A. Parente
Department of Pharmacy and Pharmaceutical Chemistry
School of Pharmacy
University California
San Francisco, CA 94143-0128

C. Parra
Department of Physiology and Biophysics
Faculty of Medicine
University of Chile
Santiago, Chile

V.A. Parsegian
DCRT
National Institutes of Health
Bethesda, MD 20892

J. Petell
Department of Biological Sciences
State University of New York at Buffalo
Buffalo, NY 14214

H. Plattner
Faculty of Biology
University of Konstanz
P. O. Box 5560
D-7750 Konstanz
Federal Republic of Germany

H.B. Pollard
Laboratory of Cell Biology and Genetics
National Institutes of Diabetes and Digestive Diseases
National Institutes of Health
Bethesda, MD 20205

L. Pratsch
Institute of Biophysics
Department of Biology of the Humboldt University
Berlin, GDR

J.H. Prestegard
Department of Chemistry
Yale University
New Haven, CT 06511

A. Puri
Section on Membrane Structure and Function
LMMB
National Cancer Institute
NIH
Bethesda, MD 20892

R.P. Rand
Biological Sciences
Brock University
St. Catherines, Ontario, Canada

E. Rojas
Laboratory of Cell Biology and Genetics
National Institutes of Diabetes, and Digestive Diseases
National Institutes of Health
Bethesda, MD 20205

D.S. Roos
Department of Biological Sciences
Stanford University
Stanford, CA 94305

B.H. Satir
Department of Anatomy and Structural Biology
Albert Einstein School of Medicine
Bronx, NY 10461

J. Sawyer
Department of Biological Sciences
State University of New York at Buffalo
Buffalo, NY 14214

R.A. Schlegel
Department of Molecular and Cell Biology
The Pennsylvania State University
University Park,PA 16802

D.P. Siegel
Procter & Gamble Co.
P.O. Box 39175
Cincinnati, OH 45247

J.R. Silvius
Department of Biochemistry
McGill University
Montreal, Quebec, Canada H3G 1Y6

A.E. Sowers
Biomecical Research and Development
American Red Cross
Rockville, MD 20855

B. Stecher
Faculty of Biology
University of Konstanz
P.O. Box 5560
D-7750 Konstanz
Federal Republic of Germany

T. Stegmann
Laboratory of Physiological Chemistry
University of Groningen
Bloemsingel 10
9712 KZ Groningen
The Netherlands

D.A. Stenger
Biophysics Department
Roswell Park Memorial Institute
Buffalo, NY 14263

N.K. Subbarao
Department of Pharmacy and Pharmaceutical Chemistry
School of Pharmacy
University of California
San Francisco, CA 94143

F.C. Szoka
Department of Pharmacy and Pharmaceutical Chemistry
School of Pharmacy
University of California
San Francisco, CA 94143

T.E. Thompson
Department of Biochemistry
University of Virginia
Charlottesville, VA 22908

M. Tomita
Department of Biological Chemistry
The Johns Hopkins University
School of Medicine
Baltimore, MD 21205

T.Y. Tsong
Department of Biological Chemistry
The Johns Hopkins University
School of Medicine
Baltimore, MD 21205

C.J. Van Oss
Department of Microbiology and Chemical Engineering
State University of New York at Buffalo
Buffalo, NY 14214

J. Vilmart-Seuwen
Faculty of Biology
University of Konstanz
P.O. Box 5560
D-7750 Konstanz
Federal Republic of Germany

A. Walter
Section on Membrane Structure and Function
LMMB
National Cancer Institute
NIH
Bethesda, MD 20892

P. Williamson
Department of Biology
Amherst College
Amherst, MA 01002

J. Wilschut
Laboratory of Physiological Chemistry
University of Groningen
Bloemsingel 10
9712 KZ Groningen
The Netherlands

W.J. Zaks
Department of Pharmacology and the Programs in Biophysics, Neuroscience and Cell and Molecular Biology
University of Virginia
Charlottesville, VA 22908

J. Zimmerberg
Physical Sciences Laboratory
Division of Computer Research and Technology
National Institutes of Health
Bethesda, MD 20892

U. Zimmermann
Lehrstuhl fur Biotechnologie
Universitat Wurzburg
FRG

ATTENDEES

Q.F. Ahkong
Royal Free Hospital, London, UK

Z. Ahmed
SUNY at Buffalo

A.D. Albert
SUNY at Buffalo

T.M. Allen
University of Alberta, Canada

J.J. Alletto
SUNY at Buffalo

W. Almers
University of Washington

M. Anthony
Roswell Park Memorial Institute

K. Arnold
Karl-Marx-Universitat, GDR

T. Arvinte
Cornell University

B. Asgharian
SUNY at Buffalo

W. Asche
Fruss GmbH, FRG

A.L. Auerback
SUNY at Buffalo

O.P. Bahl
SUNY at Buffalo

R.F. Baker
University of Southern California

M. Bender
Farleigh Dickinson University

R. Blumenthal
National Institutes of Health

K. Boesze
SUNY at Buffalo

L. Boni
The Liposome Company, Inc., NJ

W.F. Boss
North Carolina State University

B.A. Brown
SUNY at Buffalo

P.M. Brown
McGill University, Canada

M. Bryszewska
McMaster University, Canada

T. Bukowski
SUNY at Buffalo

K.N.J. Burger
State Univ. of Utrecht,The Netherlands

J. Byczkowski
Roswell Park Memorial Institute

D.A. Cadenhead
SUNY at Buffalo

L. Chamberlin
SUNY at Buffalo

D. Chang
Baylor College of Medicine

N.R. Chen
Cornell University

V. Chen
SUNY at Buffalo

B. Cheng
Case Western Reserve University

K. Cheng
Roswell Park Memorial Institute

L.A. Chung
SUNY at Stony Brook

J. Coorssen
Brock University, Canada

P. Costanzo
SUNY at Buffalo

C.E. Creutz
University of Virginia

A. Cudd
Cornell University

P.R. Cullis
University of British Columbia, Canada

L. Davenport
Brooklyn College

R.B. Bankert
Roswell Park Memorial Institute

M.R. Deziel
VA Medical Center, Buffalo

D.J. Doyle
SUNY at Buffalo

J.R. Drake
Roswell Park Memorial Institute

N. Duzgunes
University of California, SF

O. Eidelman
National Institutes of Health

G. Enhorning
SUNY at Buffalo

E.A. Evans
University of British Columbia Canada

D.S. Faber
SUNY at Buffalo

G. Feigenson
Cornell University

F. Ferro
Roswell Park Memorial Institute

T.D. Flanagan
SUNY at Buffalo

K. Florine
Cornell University

C.L. Forest
Brooklyn College of CUNY

N.L. Fuller
Brock University, Canada

A.O. Fuller
University of Chicago

H. Gigiotti
California State University

R. Goldstein
Cornell University

R.J. Good
SUNY at Buffalo

M. Gratzl
Universitat Ulm, FRG

C.C. Gregorio
Roswell Park Memorial Institute

Y. Guo
Roswell Park Memorial Institute

M.L. Hammarskjold
SUNY at Buffalo

A.M. Haywood
University of Rochester

W. Heacock
Perkin Elmer Company, Rochester, NY

A. Helenius
Yale University

T.M. Hennessey
SUNY at Buffalo

D. Hoekstra
Univ. of Groningen, The Netherlands

G. Hofmann
Biotechnologies & Exper.Res., Inc., CA

S.K. Hong
SUNY at Buffalo

W. Hong
SUNY at Buffalo

K.S.K. Huang
Roswell Park Memorial Institute

L. Huang
University of Tennessee

H. Hui
SUNY at Buffalo

S.W. Hui
Roswell Park Memorial Institute

T.V. Isac
Roswell Park Memorial Institute

L. Izu
SUNY at Buffalo

D.M. Jacobs
SUNY at Buffalo

B. Jhun
SUNY at Buffalo

I. Jo
SUNY at Buffalo

C.Y. Jung
SUNY at Buffalo

R.J. Kelleher
SUNY at Buffalo

U. Kim
Roswell Park Memorial Institute

L. Kirkpatrick
SUNY at Buffalo

R. Kubiniec
Roswell Park Memorial Institute

N.R. Kumar
SUNY at Buffalo

Y.W. Kwon
SUNY at Buffalo

M. Ladjadj
Wayne State University

P.C. Lee
Children's Hospital, Buffalo

S. Leibler
Cornell University

P.I. Lelkes
National Institutes of Health

B.R. Lentz
University of North Carolina

C. Lewis
SUNY at Buffalo

M.J. Litmus
SUNY at Buffalo

M. Liu
Roswell Park Memorial Institute

S. Liu
Instit.of Zoology, Beijing,China

D. Lloyd
SUNY at Stony Brook

P. Lorge
Univ. Libre de Bruxelles,Belgium

A. Loyter
Hebrew Univ. of Jerusalem, Israel

J.A. Lucy
Royal Free Hospital, London, UK

D.V. Lynch
Cornell University

R.C. MacDonald
Northwestern University

R. MacDonald
Northwestern University

J. Marrone
SUNY at Buffalo

E.J. Massaro
US Environ. Protection Agency, NC

G.L. Mayers
Roswell Park Memorial Institute

E.G. Mayhew
Roswell Park Memorial Institute

L. McEvoy
Pennsylvania State University

D.J.L. McIver
Univ. of Western Ontario, Canada

P. Meers
University of California, SF

J.M. Merrick
SUNY at Buffalo

K. Metz
Pennsylvania State University

M.Da Graca Miguel
National Institutes of Health

D. Miles
Duke University

V. Milisauskas
SUNY at Buffalo

L.A.F. Minasi
SUNY at Buffalo

S.J. Morris
National Science Foundation

T.J. Murtaugh
Albert Einstein School of Medicine

D. Needham
Duke University

B.J. Nicholson
SUNY at Buffalo

S. Nir
Hebrew Univ. of Jerusalem, Israel

E.K. Novak
Roswell Park Memorial Institute

M. Nowak
SUNY at Buffalo

S. Ohki
SUNY at Buffalo

S.I. Ohnishi
Kyoto University, Japan

C.J. van Oss
SUNY at Buffalo

C.V. Paganelli
SUNY at Buffalo

J. Paiement
Universite de Montreal, Canada

D. Paphadjopoulos
University of California, SF

C.K. Park
SUNY at Buffalo

M.T. Paternostre
National Institutes of Health

H. Plattner
Universitat Konstanz, FRG

H. Pollard
National Institutes of Health

V. Polonis
SUNY at Buffalo

D. Pradhan
Pennsylvania State University

J.H. Prestegard
Yale University

R.M. Price
SUNY at Buffalo

A. Puri
National Institutes of Health

P. Rand
Brock University, Canada

P. Reddy
University of Minnesota

D. Rekosh
SUNY at Buffalo

E. Repasky
Roswell Park Memorial Institute

K. Rittle
Roswell Park Memorial Institute

D.S. Roos
Stanford University

R.T. Root
E.I. Dupont and Company, DE

D. Sarkar
National Institutes of Health

J.J. Sawyer
SUNY at Buffalo

R.K. Scheule
University of Texas

R.A. Schlegel
Pennsylvania State University

A. Schneider
Albany Medical College

H. Schuel
SUNY at Buffalo

R. Schuel
SUNY at Buffalo

S. Schullery
Eastern Michigan University

D. Scott
Veterans Administration, Buffalo

A. Scotto
Cornell University

A. Sen
Roswell Park Memorial Institute

D.P. Siegel
Proctor and Gamble Company, OH

S.M. Simasko
SUNY at Buffalo

J.R. Silvius
McGill University, Canada

G. Smutzer
Buffalo, NY

A.E. Sowers
American Red Cross, MD

R.A. Spangler
SUNY at Buffalo

D. Stenger
Roswell Park Memorial Institute

P.L. Steponkus
Cornell University

M.W. Stinson
SUNY at Buffalo

I.P. Sugar
University of Virginia

D.K. Sukumaran
SUNY at Buffalo

L.A. Surgalla
SUNY at Buffalo

Y. Suzuki
Shizuoka College of Pharmacy,Japan

R.T. Swank
Roswell Park Memorial Institute

F. Szoka
University of California, SF

A.I. Tauber
Boston University

T.E. Thompson
Max-Planck-Institut, FRG

T.Y. Tsong
Johns Hopkins University

M. Uemura
Cornell University

V. Vaidhyanathan
SUNY at Buffalo

A. Walter
National Institutes of Health

Z.X. Wang
Zhejiang Medical University, China

N. Waxman
University of Texas

J. Wheeler
North Carolina State University

G. Williams
Pennsylvania State University

G. Willsky
SUNY at Buffalo

J. Wilschut
Univ. of Groningen, The Netherlands

A. Wishnia
SUNY at Stony Brook

X.C. Yang
SUNY at Buffalo

M. Yatvin
University of Wisconsin

M. Yeager
Cornell University

P.L. Yeagle
SUNY at Buffalo

H.Y. Yeh
SUNY at Buffalo

J. Zimmerberg
National Institutes of Health

U. Zimmermann
Universitat Wurzburg, FRG

INDEX